A HANDBOOK ON

BIOTELEMETRY AND RADIO TRACKING

A HANDBOOK ON
BIOTELEMETRY AND RADIO TRACKING

Proceedings of an International Conference on
Telemetry and Radio Tracking in Biology and Medicine
Oxford, 20-22 March 1979

Editors:
CHARLES J. AMLANER, Jr.
and
DAVID W. MACDONALD
Animal Behaviour Research Group, University of Oxford

PERGAMON PRESS
OXFORD · NEW YORK · TORONTO · SYDNEY · PARIS · FRANKFURT

U.K.	Pergamon Press Ltd., Headington Hill Hall, Oxford OX3 0BW, England
U.S.A.	Pergamon Press Inc., Maxwell House, Fairview Park, Elmsford, New York 10523, U.S.A.
CANADA	Pergamon of Canada, Suite 104, 150 Consumers Road, Willowdale, Ontario M2J 1P9, Canada
AUSTRALIA	Pergamon Press (Aust.) Pty. Ltd., P.O. Box 544, Potts Point, N.S.W. 2011, Australia
FRANCE	Pergamon Press SARL, 24 rue des Ecoles, 75240 Paris, Cedex 05, France
FEDERAL REPUBLIC OF GERMANY	Pergamon Press GmbH, 6242 Kronberg-Taunus, Pferdstrasse 1, Federal Republic of Germany

First edition 1980

British Library Cataloguing in Publication Data

International Conference on Telemetry and Radio Tracking in Biology and Medicine, *University of Oxford, 1979*
A handbook on biotelemetry and radio tracking
1. Telemeter (Physiological apparatus) - Congresses
2. Electronics in biology - Congresses
3. Tracking and trailing - Congresses
I. Title II. Amlaner, Charles J
III. Macdonald, David W
591.1'028 QP55 79-41234

ISBN 0 08 024928 0

In order to make this volume available as economically and as rapidly as possible the authors' typescripts have been reproduced in their original forms. This method has its typographical limitations but it is hoped that they in no way distract the reader.

Printed and bound in Great Britain by A. Wheaton & Co. Ltd., Exeter

Contents

Preface

The development of ever more obscure jargon is divisive among scientists, inhibiting communication between people who might otherwise solve at least some of each other's problems. Nowhere is this unnatural rift more obvious than between biologists and engineers and yet with a little patience it can be bridged. It was to bring together people of diverse training, but with similar interests, that we organized the International Conference on Telemetry and Radio Tracking in Biology and Medicine held in The University of Oxford, Oxford, U.K. The papers presented there gave rise to this Handbook. We have tried to include papers which on the one hand illustrate the most recent advances connected with every aspect of biotelemetry and radio tracking while, on the other hand, trying to ensure that the novice could find guidance on how to solve even elementary problems. In this sense we hope that these papers truly constitute a Handbook.

The organization of the conference and the subsequent editing of this book have relied on the enthusiasm and good will of many people and organizations, to all of whom we are grateful. The conference was supported by the Biological Engineering Society, the International Society on Biotelemetry and the Association for the Study of Animal Behaviour. Through its Honorary Secretary Keith Copeland, the Biological Engineering Society courageously acted as financial guarantors during the gestation period preceding the arrival of a gratifyingly large number of delegates. The Department of Zoology, Oxford, afforded a fine venue. The staff of Pergamon Press have worked hard to speed the production of the book, as have our readers, Eileen Morrell and Eileen McDowell. We are also grateful to the authors who responded with good humor to our nearly impossible deadlines. Above all, we are indebted to Beverly Amlaner for her diligence and tenacity throughout every stage of the conference and the production of this Handbook.

When we began to explore our respective interests in telemetry there was little published guidance of a practical nature. We hope this book will be the sort of Handbook which we wish had been available then.

Charles J. Amlaner, Jr.
David W. Macdonald
Department of Zoology
Oxford
4 July 1979

Conference Participants

K. M. Adams, Animal Ecology Research Group, Department of Zoology, South Parks Road, Oxford OX1 3PS, U.K.
C. T. Alabaster, Developmental Biology Department, Pfizer Central Research, Sandwich, Kent CT13 9NJ, U.K.
B. R. Amlaner, Animal Behaviour Research Group, Department of Zoology, South Parks Road, Oxford OX1 3PS, U.K.
C. J. Amlaner, Jr., Animal Behaviour Research Group, Department of Zoology, South Parks Road, Oxford OX1 3PS, U.K.
F. W. Anderka, Canadian Wildlife Service, National Wildlife Research Centre, Ottawa, Ontario K1A OE7, Canada
S. C. Anderson, Sea Mammal Research Unit, British Antarctic Survey, Madingley Road, Cambridge CB3 OET, U.K.
C. M. Andrew, Institute of Psychiatry, De Crespigny Park, Denmark Hill, London SE5 8AF, U.K.
M. Artois, Centre National d'Etudes mu la Rage, B.P. 9, 54220 Malzeville, France
K. R. Ashby, Zoology Department, University of Durham, Durham DM1 3LE, U.K.
M. Baettig, Le Bochet, CH-1349 Daillens, Switzerland
R. V. Baker, 1 Welwyn Road, Hertford, Herts, U.K.
F. G. Ball, Room 231, 25 Wellington Square, Oxford, U.K.
N. J. Ball, Animal Behaviour Research Group, Department of Zoology, South Parks Road, Oxford OX1 3PS, U.K.
J. Bannister, Medical Physics Unit, St. James's Hospital, Beckett Street, Leeds LS9 7TF, U.K.
T. Barker, Department of Zoology, South Parks Road, Oxford OX1 3PS, U.K.
J. W. Barlow, 14 Cromwell Park, Modbury, Devon, U.K.
R. F. W. Barnes, Sub-department of Animal Behaviour, Madingley, Cambridge CB3 8AA, U.K.
E. B. M. Barrett, Department of Physical Anthropology, Downing Street, Cambridge CB2 3DZ, U.K.
R. C. Barrett, Tregarthen, Tenbury Road, Cleehill, Ludlow, Salop, U.K.
M. R. Baxter, Scottish Farm Buildings Investigation Unit, Craibstone, Bucksburn, Aberdeen AB2 9TR, U.K.
M. J. Beach, Sonicaid Ltd., Terminus Road, Chichester, U.K.
C. Bearder, The Crossways, Hertford Heath, Hertford, U.K.
S. K. Bearder, The Zoological Society of London, Wellcome Laboratories, Regent's Park, London NW1 4RY, U.K.
P. C. W. Beatty, Fisheries Laboratory, Pakefield Road, Lowestoft, U.K.
D. J. Bell, Department of Zoology, Park Place, University College, Cardiff CF1 1XL, U.K.

P. M. G. Bell, University Department of Pharmacology, South Parks Road, Oxford OX1 3QT, U.K.
B. C. R. Bertram, Research Centre, King's College, Cambridge CB2 1ST, U.K.
J. D. S. Birks, Department of Biological Sciences, Prince of Wales Road, Exeter, Devon, U.K.
L. Boitani, Institute of Zoology, University of Roma, Viale dell'Università 32, Roma 00100, Italy
P. G. Bowen, Pfizer Central Research, Sandwich, Kent CT13 9NJ, U.K.
J. Boyce, Institute of Terrestrial Ecology, Hill of Brathens, Banchory AB3 4BY, U.K.
S. Broekhuizen, Research Institute for Nature Management, Kemperbergerweg 67, Arnhem, The Netherlands
D. A. Burchard, Engineering Office Burchard, Tiruler Weg 2, D-7801 Ebringen, F.R.G.
D. G. Burnside, P.O. Box 434, Carnarvon 6701, Western Australia 6701
P. J. Butler, Department of Zoology, University of Birmingham, Birmingham B15 2TT, U.K.
D. C. Catt, Institute of Terrestrial Ecology, Hill of Brathens, Banchory AB3 4BY, U.K.
L. A. Causby, 104 Ross, Northeastern State University, Science and Math Division, Tahlequah, Oklahoma 74464, U.S.A.
D. J. Chapple, Pharmacology Department, The Wellcome Research Laboratories, Beckenham, Kent BR3 3BS, U.K.
C. L. Cheeseman, St. Annes, Brownshill, Stroud GL6 8AP, U.K.
K. E. Church, Rural Route 1, Plymouth, Wisconsin 53073, U.S.A.
K. L. Church, Rural Route 1, Plymouth, Wisconsin 53073, U.S.A.
K. A. Collis, Institute for Research on Animal Diseases, Compton, Newbury, Berks RG16 0QY, U.K.
R. H. Colson, Department of Medical Electronics, St. Bartholomew's Hospital, West Smithfield, London EC1A 7BE, U.K.
J. Compton, Silicon Microsystems Ltd., Market Cross, Malmesbury SN16 9AS, U.K.
K. Copeland, Biophysics Department, University College, London, Gower Street, London WC1E 6BT, U.K.
L. K. Corbett, Institute of Terrestrial Ecology, Hill of Brathens, Banchory AB3 4BY, U.K.
M. D. Craggs, MRC Neurological Prostheses Unit, Institute of Psychiatry, De Crespigny Park, London SE5 8AF, U.K.
D. E. Creighton, British-American Tobacco Company Ltd., Regent's Park Road, Southampton SO9 1PE, U.K.
J. P. Croxall, British Antarctic Survey, Madingley Road, Cambridge CB3 0ET, U.K.
M. G. Curry, Sea Mammal Research Unit, British Antarctic Survey, Madingley Road, Cambridge CB3 0ET, U.K.
S. Daan, Zoological Laboratory, University of Groningen, Haren, The Netherlands
K. L. Dawes, 2010 Wolfe Lane, Fort Smith, Arkansas 72901, U.S.A.
A. Deat, 7 Rue Giannesini, 79270 Frontenay R. R., France
M. Dolan, Department of Zoology, South Parks Road, Oxford OX1 3PS, U.K.
B. A. C. Don, 85 Edgeway Road, New Marston, Oxford, U.K.
N. de N. Donaldson, MRC Neurological Prostheses Unit, 1 Windsor Walk, London SE5 8BB, U.K.
D. D. Dow, Department of Zoology, University of Queensland, St. Lucia, Australia 4067
T. S. G. Dreyfert, University of Stockholm, Box 6807, S-113 86 Stockholm, Sweden
I. J. H. Duncan, Poultry Research Centre, West Mains Road, Edinburgh EH9 3JS, U.K.
N. Dunstone, Department of Zoology, South Road, Durham DH1 3JB, U.K.
P. J. Edwards, 26 Walton Drive, High Wycombe, Bucks, U.K.
H. Ellenberg, Lehrstuhl für Biogeographie, Universität des Saarlandes, D-6600 Saarbrücken, F.R.G.
M. A. El Messiery, Biomedical Engineering Unit, Medical Institute, Hartshill, Stoke-on-Trent, U.K.
M. J. English, School of Engineering, University of Sussex, Falmer, Brighton BN1 9QT, U.K.
S. G. E. Erlinge, Ecology Building, S-223 62 Lund, Sweden
J. Esser, Zoologische Staabsaminlung, Maria-Wardstrabe 16, D-8 Munchen 19, F.R.G.

P. G. H. Evans, Edward Grey Institute, Department of Zoology, South Parks Road, Oxford OX1 3PS, U.K.
C. O. R. Everard, Leptospirosis Reference Laboratory, Rear of Colindale Hospital, Colindale Avenue, London NW9 4DX, U.K.
N. Fairall, Mammal Research Institute, University of Pretoria, Pretoria, South Africa
P. N. Ferns, Zoology Department, University College, Cathays Park, Cardiff CF1 1XL, U.K.
A. Fiechter, St. Martin du Mont, 01160 Pont d'Ain, France
R. J. Fifield, New Scientist, King's Reach Tower, Stanford Street, London SE1 9LS, U.K.
J. H. Filshie, Poultry Research Centre, West Mains Road, Edinburgh EH9 3JS, U.K.
M. J. Follenfant, The Wellcome Research Laboratories, Langlay Court, Beckenham, Kent BR3 3BS, U.K.
G. R. Francis, Beecham Pharmaceuticals, Coldharbour Road, Harlow, Essex CM19 5AD, U.K.
R. L. Francis, Psychology Department, Queen's University, Belfast, N. Ireland
T. B. Fryer, 14029 Saratoga Hills Road, Saratoga, California 95070, U.S.A.
R. W. Furness, Zoology Department, Glasgow University, Glasgow G12 8QQ, U.K.
G. W. Gabrielsen, Wildlife Research Station, Holtveien, 9000 Tromsø, Norway
J. P. Gautier, Biological Station of Paimpont, 35380 Plélan-le-Grance, France
G. K. Göransson, Ecology Building, Helgonav 5, S-223 62 Lund, Sweden
L. M. Gosling, Coypu Research Laboratories, Ministry of Agriculture, Fisheries and Food, Jupiter Road, Norwich, Norfolk, U.K.
T. G. Grassie, Strathcylde University, Bioengineering Unit, Wolfson Centre, Glasgow G4 0NW, U.K.
J. M. Green, Department of Biology, Memorial University, St. John's, Newfoundland, Canada
D. F. Hackett, Department of Biology, Laurentian University, Sudbury, Ontario P3E 2C6, Canada
P. Hague, Institute of Animal Physiology, Babraham, Cambridge, U.K.
T. R. Hardy, Rodent Pests Department, Ministry of Agriculture, Fisheries and Food, Hook Rise South, Tolworth, Surbiton KT6 7NF, U.K.
S. Harris, Department of Zoology, University of Bristol, Woodland Road, Bristol BS9 1UE, U.K.
R. E. Hawkins, Rural Route 3, Carbondale, Illinois 62901, U.S.A.
G. F. Hayward, Animal Ecology Research Group, Department of Zoology, South Parks Road, Oxford OX1 3PS, U.K.
H. I. Helmersson, Grimsö Forskningstation, S-77031 Riddarhyttan, Sweden
C. A. van't Hoff, Rosandepolder 39, Arnhem, The Netherlands
N. G. Hough, Animal Behaviour Research Group, Department of Zoology, South Parks Road, Oxford OX1 3PS, U.K.
C. J. Houghton, Department of Zoology, University College, P.O. Box 78, Cardiff CF1 1XL, U.K.
L. C. Ireland, 2221 E. Court Street, Flint, Michigan 48503, U.S.A.
G. L. Janeau, Ministère de l'Agriculture, INRA Laboratoire de la Faune Sauvage, 31320 Castanet Tolosan, France
M. B. Jansen, Technical and Physical Engineering Research Service, Mansholtlaan 12, Wageningen, The Netherlands
D. J. Jefferies, Nature Conservancy Council, Godwin House, George Street, Huntingdon, U.K.
J. G. Jennings, Southwest Fisheries Center, National Marine Fisheries Service, P.O. Box 271, La Jolla, California 92038, U.S.A.
P. B. Johnsen, Monell Chemical Senses Center, University of Pennsylvania, 3500 Market Street, Philadelphia, Pennsylvania 19104, U.S.A.
C. E. Johnson, Royal Holloway College, University of London, Egham, Surrey, U.K.
E. Johnson, Zoology Department, The University, Whiteknights, Reading RG6 2AJ, U.K.
A. Kacelnik, Animal Behaviour Research Group, Department of Zoology, South Parks Road, Oxford OX1 3PS, U.K.
J. W. Kanwisher, Woods Hole Oceanographic Institution, Woods Hole, Massachusetts 02543, U.S.A.
V. Keil, Judenbuhlweg 31, D-8700 Würzburg, F.R.G.

R. E. Kenward, Institute of Terrestrial Ecology, Monks Wood Experimental Station, Abbots Ripton, Huntingdon PE17 2LS, U.K.
Barbara C. Kermeen, 3101 West Clark Road, Champaign, Illinois 61820, U.S.A.
M. Kilvington, Hull Royal Infirmary, Salthouse Road, Hull, U.K.
H. P. Kimmich, Department of Physiology, University of Nijmegen, Nijmegen, The Netherlands
P. C. King, Hatfield Polytechnic, Hatfield, Herts AL10 9AB, U.K.
J. K. Kirkwood, Department of Animal Husbandry, University of Bristol, Langford House, Langford, Bristol BS18 7DU, U.K.
W. H. Ko, Engineering Design Center, Case Western Reserve University, Cleveland, Ohio 44106, U.S.A.
J. C. Kroll, Box 6109 SFA, Nacogdoches, Texas 75962, U.S.A.
F. J. Kuhn, C. H. Boehringer Sohn, Department of Pharmacology, D-6507 Ingelheim am Rhein, F.R.G.
F. Labhardt, Gemeindeholzweg 18, CH-4103 Bottmingen, Switzerland
A. N. Lance, Nature Conservancy, 12 Hope Terrace, Edinburgh, EH9 2AS, U.K.
R. A. Lancia, Forestry Department, Box 5488, North Carolina State University, Raleigh, North Carolina 27650, U.S.A.
T. E. Langford, C.E.R.L. Marine Biology Laboratory, Fawley Power Station, Fawley, Hants, U.K.
V. I. Lebedeb, Gorky Street, Kislovodsk, U.S.S.R.
S. J. Legg, Army Personnel Research Establishment, Royal Aircraft Establishment, Farnborough, Hants, U.K.
P. A. Lemnell, Grimsö Wildlife Research Station, S-77031 Riddarhyttan, Sweden
C. K. Leuze, Culterty Field Station, Newburgh, Ellon, Aberdeenshire AB4 OAA, U.K.
R. F. Levy, Strathclyde University, Bioengineering Unit, Wolfson Centre, Glasgow G4 ONW, U.K.
O. Lind, Alfa Laval AB, P.O. Box 39, S-14700 Tumba, Sweden
I. Lindsay, Animal Behaviour Research Group, Department of Zoology, South Parks Road, Oxford OX1 3PS, U.K.
D. D. Lock, 43 Grove Road, Stevenage, Herts SG1 3NT, U.K.
J. Lodal, Statens Skadedyrlaboratorium, 14 Skovbrynet, DK 2800 Lyngby, Denmark
T. W. I. Lovel, Garvery, Hurstbourne Tarrant, Andover, Hants, U.K.
D. W. Macdonald, Animal Behaviour Research Group, Department of Zoology, South Parks Road, Oxford OX1 3PS, U.K.
J. MacGregor, Strathclyde University, Bioengineering Unit, Wolfson Centre, Glasgow G4 ONW, U.K.
P. J. Mallinson, Easter Park Cottage, Nympsfield, Stonehouse, Gloucester, U.K.
R. C. Malpas, Department of Applied Biology, Pembroke Street, Cambridge CB2 3DX, U.K.
R. Mani, Department of Zoology, University of Aberdeen, Aberdeen AB9 2TN, U.K.
C. Mansard, CNRS-CEBAS, Villiers en Bois, 79360 Beauvoir, France
V. Marcstrom, Institute of Zoophysiology, Box 560, S-751 22 Uppsala, Sweden
G. Markus, Zollgässli 4, CH-3177 Laupen, Switzerland
R. Mauget, CEBAS, Villiers en Bois, 79360 Beauvoir, France
D. Maurel, CNRS-CEBAS, Villiers en Bois, 79360 Beauvoir, France
R. H. McCleery, Edward Grey Institute, Department of Zoology, South Parks Road, Oxford OX1 3PS, U.K.
D. J. McFarland, Animal Behaviour Research Group, Department of Zoology, South Parks Road, Oxford OX1 3PS, U.K.
M. J. McKee, Department of Zoology, Ulster Museum, Belfast BT5, N. Ireland
C. N. Mcleod, Nuffield Institute for Medical Research, Headley Way, Headington, Oxford, U.K.
L. D. Mech, U.S. Fish and Wildlife Service, North Central Forest Experiment Station, 1992 Folwell Avenue, St. Paul, Minnesota 55108, U.S.A.
W. F. Miller, Water Research Centre, Elder Way, Stevenage, Herts, U.K.
F. Morris, North Staffordshire Polytechnic, College Road, Stoke-on-Trent, U.K.
J. A. Morris, Institute of Terrestrial Ecology, Hill of Brathens, Banchory AB3 4BY, U.K.
W. M. Morris, Silicon Microsystems Ltd., Market Cross, Malmesbury SN16 9AS, U.K.

M. Newdick, Animal Behaviour Research Group, Department of Zoology, South Parks Road, Oxford OX1 3PS, U.K.
V. Niebuhr, Animal Behaviour Research Group, Department of Zoology, South Parks Road, Oxford OX1 3PS, U.K.
F. J. J. Niewold, Research Institute for Nature Management, Kemperbergerweg 67, Arnhem, The Netherlands
N. Nowaira, Bio-Medical Engineering Unit, c/o Medical Institute, Stoke-on-Trent ST4 7NY, U.K.
D. W. Osgood, Department of Zoology, Butler University, Indianpolis, Indiana 46208, U.S.A.
F. Packman, Pfizer Central Research, Sandwich, Kent CT13 9NJ, U.K.
T. Parish, Institute of Terrestrial Ecology, Hill of Brathens, Banchory AB3 4BY, U.K.
J. B. Payne, Sub-department of Veterinary Anatomy, Tennis Court Road, Cambridge CB2 1QS, U.K.
B. Pells, Roslyn Road, London N15 5JB, U.K.
P. D. Penning, Grassland Research Institute, Hurley, Maidenhead, Berks, U.K.
T. A. Perkins, MRC Neurological Prostheses Unit, 1 Windsor Walk, London SE5 8BB, U.K.
B. Petch, Developmental Biology Department, Pfizer Central Research, Sandwich, Kent CT13 9NJ, U.K.
D. G. Pincock, Electrical Engineering Department, University of New Brunswick, P.O. Box 4400, Fredericton E3B 5A3, Canada
J. W. Piper, Medelec Ltd., Manor Way, Old Woking, Surrey, U.K.
N. V. C. Polunin, Department of Zoology, Downing Street, Cambridge CB2 3EJ, U.K.
I. G. Priede, Department of Zoology, University of Aberdeen, Aberdeen AB9 2TN, U.K.
P. A. Prince, British Antarctic Survey, Madingley Road, Cambridge CB3 OET, U.K.
N. J. Reeve, Zoology Department, Royal Holloway College, Bakeham Lane, Englefield Green TW20 9TY, U.K.
A. Saint Joseph, Wildfowl Trust, Slimbridge, Gloucester GL2 7BT, U.K.
S. Salmons, Department of Anatomy, The Medical School, University of Birmingham, Birmingham B15 2TJ, U.K.
W. M. Sansen, Katholik University of Leuven, Elektrotechniek, 94 K. Mercierlaan, 3030 Heverlee, Belgium
A. B. Sargeant, 334 15th Avenue N.E., Jamestown, North Dakota 58401, U.S.A.
W. M. Schleidt, 13022 Seventh Street, Bowie, Maryland 20715, U.S.A.
F. Schober, Forschungsinstitut für Wildtierkunde der Veterinärmedizinischen Universität, Linke Bahngasse 11, A-1050 Vienna, Austria
A Sempéré, CEBAS, Villiers en Bois, 79360 Beauvoir, France
D. M. Sequeira, 14 Latimer Road, Wimbledon, London SW19 1EP, U.K.
J. P. Shaffery, Animal Behaviour Research Group, Department of Zoology, South Parks Road, Oxford OX1 3PS, U.K.
W. M. Shearer, DAFS, 6 California Street, Montrose, Angus, U.K.
M. D. Shelley, Pharmacology Department, The Wellcome Research Laboratories, Beckenham, Kent BR3 3BS, U.K.
G. E. Shelvey, Pfizer Central Research, Sandwich, Kent CT13 9NJ, U.K.
N. G. Sheret, 14 Torver Way, Orpington, Kent BR6 8DX, U.K.
L. J. Shields, 3751 Multiview Drive, Hollywood, California 90068, U.S.A.
R. M. Sibly, Department of Zoology, University of Glasgow, Glasgow G12 8QQ, U.K.
A. J. Smale, Ormed Engineering Ltd., 32 Hydeway, Welwyn Garden City, Hertfordshire AL7 3AW, U.K.
P. Smale, Ormed Engineering Ltd., 32 Hydeway, Welwyn Garden City, Hertfordshire AL7 3AW, U.K.
E. N. Smith, Division of Natural Science and Mathematics, Northeastern State University, Tahlequah, Oklahoma 74464, U.S.A.
G. W. Smith, DAFS Marine Laboratory, Victoria Road, Aberdeen AB1 5RB, U.K.
H. R. Smith, 151 Sanford Street, Northeast Forest Experiment Station, Hamden, Connecticut 06514, U.S.A.
J. C. Smith, Pest Infestation Control Laboratory, Ministry of Agriculture, Fisheries and Food, Hook Rise South, Tolworth, Surbiton KT6 7NF, U.K.

L. J. Smith, Department of Psychology, Prudhoe Hospital, Prudhoe, Northumberland, U.K.
R. H. Smith, Department of Zoology, University of Reading, Whiteknights, Reading RG6 2AJ, U.K.
A. M. Solomons, Department of Electrical Engineering, Dover Street, The University, Manchester M13 9PL, U.K.
J. F. Stout, Biology Department, Andrews University, Berrien Springs, Michigan 49104, U.S.A.
D. A. Stubbs, 15 Eagle Close, Alton, Hants, U.K.
Y. V. Stupin, Lermoniov Street, Kislovodsk, U.S.S.R.
J. P. Swaddling, Ham House, Church Street, Marcham, Oxon, U.K.
K. D. Taylor, Warren House, Albury Heath, Guildford GU5 9DB, U.K.
N. Teall, The Bourne, Cranes Road, Sherborne St. John, Basingstoke, Hants RG24 9HY, U.K.
H. Theib-Krämer, Muhlrain 8, D-6305 Buseck-Trohe, F.R.G.
J. N. Toner, Institute of Animal Physiology, Babraham, Cambridge CB2 4AT, U.K.
M. J. Tonkin, Merlewood Research Station, Grange-over-Sands, Cumbria LA11 6JU, U.K.
B. Trevor-Deutsch, Biology Department, Laurentian University, Sudbury, Ontario P3E 2C6, Canada
B. J. Trowbridge, Culterty Field Station, Newburgh, Ellon, Aberdeenshire AB4 0AA, U.K.
A. N. Tuddenham, University of Bristol, Department of Animal Husbandry, Langford House, Langford, Bristol BS18 7DU, U.K.
C. H. Tuite, Wildfowl Trust, Slimbridge, Gloucester GL2 7BT, U.K.
G. G. Urquhart, DAFS Marine Laboratory, Victoria Road, Aberdeen AB9 8DB, U.K.
D. R. Voigt, Wildlife Research, Ontario Ministry Natural Resources, P.O. Box 50, Maple, Ontario, Canada
M. P. Waller, c/o P. J. Mallinson, Easter Park Cottage, Nympsfield, Stonehouse, Gloucester, U.K.
P. M. Waring, Park Farm House, Banbury Road, Kidlington, Oxford OX5 1AH, U.K.
P. R. Watt, Brockham Park, Betchworth, Surrey RM3 7AF, U.K.
B. Watts, Department of Biology, University of Stirling, Stirling, U.K.
W. Webber, Lower Frys, St. Mary's Street, Axbridge, Somerset BS26 2BN, U.K.
H. Westerberg, Oceanografiska Institutionen, Box 4038, S-400 40 Göteborg, Sweden
J. E. Whitten, 357 King's College, Cambridge CB2 1ST, U.K.
T. J. Whitten, 357 King's College, Cambridge CB2 1ST, U.K.
H. Wijnandts, Zoological Laboratory, Groningen University, Groningen, The Netherlands
R. J. Wilkie, Poultry Research Centre, West Mains Road, Edinburgh EH9 3JS, U.K.
H. Witt, Institut für Haustierkunde, Ohlshausenstr. 40-60, D-23 Kiel, F.R.G.
A. J. Woakes, Department of Zoology, University of Birmingham, Birmingham B15 2TT, U.K.
T. G. Wolcott, Department of Marine Science and Engineering, North Carolina State University, Box 5577, Raleigh, North Carolina 27607, U.S.A.
B. Woodward, Electronic and Electrical Engineering Department, Loughborough University, Loughborough, U.K.
D. J. Worth, 212 Ross Hall, Northeastern State University, Tahlequah, Oklahoma 74464, U.S.A.
R. M. Wright, 280 Beacon Road, Loughborough, U.K.
B. J. Yates, 22 Enstone, Tawhouse, Skelmersdale, U.K.
M. J. Yerbury, Environmental Physics Department, School of Physics, University of Sydney, Sydney, N.S.W., Australia 2006
I. Y. Yourgenc, 42 Lenid's Avenue, Moscow, U.S.S.R.
R. Zerzawy, Medizinische Poliklinik, Ostl. Stadtmauerstr. 29, D-8520 Erlangen, F.R.G.
F. Ziesemer, Winterbeker Weg 26, D-2300 Kiel, F.R.G.

Commercial Exhibitions

AVM Instrument Company
3101 West Clark Road
Champaign, Illinois 61820
U.S.A.

Biological Engineering Society
Biophysics Department
Faculty of Medical Sciences
University College London
Gower Street
London WC1E 6BT
U.K.

Dynamic Electronics Limited
Roslyn Road
Tottenham N15 5JB
U.K.

Fenlow Electronics Limited
Ham House
Church Street
Marcham, Oxon
U.K.

Pergamon Press Limited
Headington Hill Hall
Oxford OX3 OBW
U.K.

Telemetry Systems, Incorporated
5830 North Shore Drive
Milwaukee, Wisconsin 53217
U.S.A.

Wildlife Materials, Incorporated
Rural Route 2
Reed Station-Dillinger Ros.
Carbondale, Illinois 62901
U.S.A.

I. Introductory Reviews

Artifact Free Measurement of Biological Parameters: Biotelemetry, a Historical Review and Layout of Modern Developments

H. P. KIMMICH

Department of Physiology, University of Nijmegen, Geert Grooteplein Noord 21, Nijmegen, The Netherlands

Abstract – The beginning of the development of biotelemetry is difficult to assess. It depends on the meaning that is attributed to the word 'biotelemetry'. The first successful transmission of biological information from a living animal by Marey, 110 years ago, was certainly not recognized as telemetry in those days. However, when translating 'biotelemetry' literally to mean measurement of biological information from afar, then that experiment was certainly the first successful telemetry transmission. The initial recording of an ECG by Einthoven in 1903 was again a telemetric experiment with a transmission distance of more than 1.5 km.

Based on the definition above, almost every biological measurement could be qualified as biotelemetry. Since the word biotelemetry should define a measurement from an unrestrained animal, subject or patient, biotelemetry was often required to be radiotelemetry, a method initiated by Winters, 1921, and with animal and patient borne equipment by Fuller and Gordon, 1948 and by Holter and Gengerelli, 1949, respectively. Today it is obvious that this definition is too narrow for two reasons. Other transmission techniques are also very important in biotelemetry, e.g. ultrasonics for application in seawater, as well as light which may be a major carrier of biomedical information in the future. In addition it is often not the transmission system that restrains the animal, subject or patient, but the interface, i.e. the sensors and transducers.

Today biotelemetry may be described as a method that allows assessment or control of biological parameters from animals, subjects and patients with relatively little disturbance and restraint, resulting in disturbance free measurement of physiological parameters. This review will give a historical development of the different fields of biotelemetry and their classification; it will point out some achievements and will finally suggest ideas to overcome some of today's problems, as well as general problems in medicine that could probably be solved by incorporating biotelemetry techniques.

INTRODUCTION

The task of this introductory paper is to trace the beginnings of biotelemetry and to review the history of continuous monitoring and/or controlling of physiological parameters with a minimum of impediment and interference.

DEFINITION AND PRINCIPLE OF BIOTELEMETRY

The meaning attributed to the word biotelemetry has changed from time to time. The original meaning obtained by translating biotelemetry literally, namely 'measurement of biological parameters from afar' is not really a definition, since far is a relative term. By this so called 'definition' most modern measurements of biological parameters could be classified as biotelemetry, while a later restriction to 'radio transmission' would exclude a large number of modern applications of wireless transmission. The meaning of biotelemetry in its modern usage is not shaped by the manner of transmission such as radio, sound, or light (even temporary storage of the biological information on miniature tape recorders or solid state memories is possible — storage telemetry), nor by the distance of transmission. Transcutaneous transmission of biological parameters in the form of electrical signals over a distance of only a few millimeters is biotelemetry in its proper form, while data transmission over thousands of kilometers, e.g. from a space craft may not necessarily be telemetry at all. In its modern context, biotelemetry means assessment or control of biological parameters from animals, subjects and patients with relatively little disturbance and restraint of the animal/subject, resulting in undisturbed and noise-free measurement of physiological parameters.

Assessment of physiological parameters by telemetry includes measurement of the information source (biological parameters) with the aid of electrodes or transducers, signal processing and amplification, multiplexing, and modulation of a carrier, transmission or temporary storage, reception, signal conditioning and demultiplexing, amplification, and data processing, and finally display (Fig. 1). Furthermore, it is *not* sufficient to assess and transmit correctly the biological information, it must also be displayed in such a way that it is unambiguous (correctly interpreted by the observer). Paper recordings may not necessarily be superior to screen display. The huge amount of data collected tempts the observer to scan through it quickly instead of exactly. Important changes may, however, be missed in screen display. A combination of both systems seems desirable, but much more important is a method for subsequent data reduction. But before getting into details of modern biotelemetry, let us concentrate on the early activities and historical development of medical telecommunications.

EARLY DEVELOPMENTS

The roots of biotelemetry are difficult to find and mainly depend on the meaning attributed to the word biotelemetry. The first successful transmission of biological information from a living animal by Marey, 110 years ago, was certainly not recognized as biotelemetry in those days, but according to the 'definition' when the term was first used (presumably around 1920), Marey's was indeed the first successful telemetry experiment. The next important step in the development of biotelemetry was the recording and transmission of an ECG over a distance of 1.5 km by Einthoven in 1903, using the wires of the public telephone system. In 1910, Barker connected several wards of a hospital to an ECG monitoring station with a specially designed wire system. The major problem in this application was similar to Einthoven's, that currents were induced along the lines, since signal transmission was direct without the use of subcarrier frequencies.

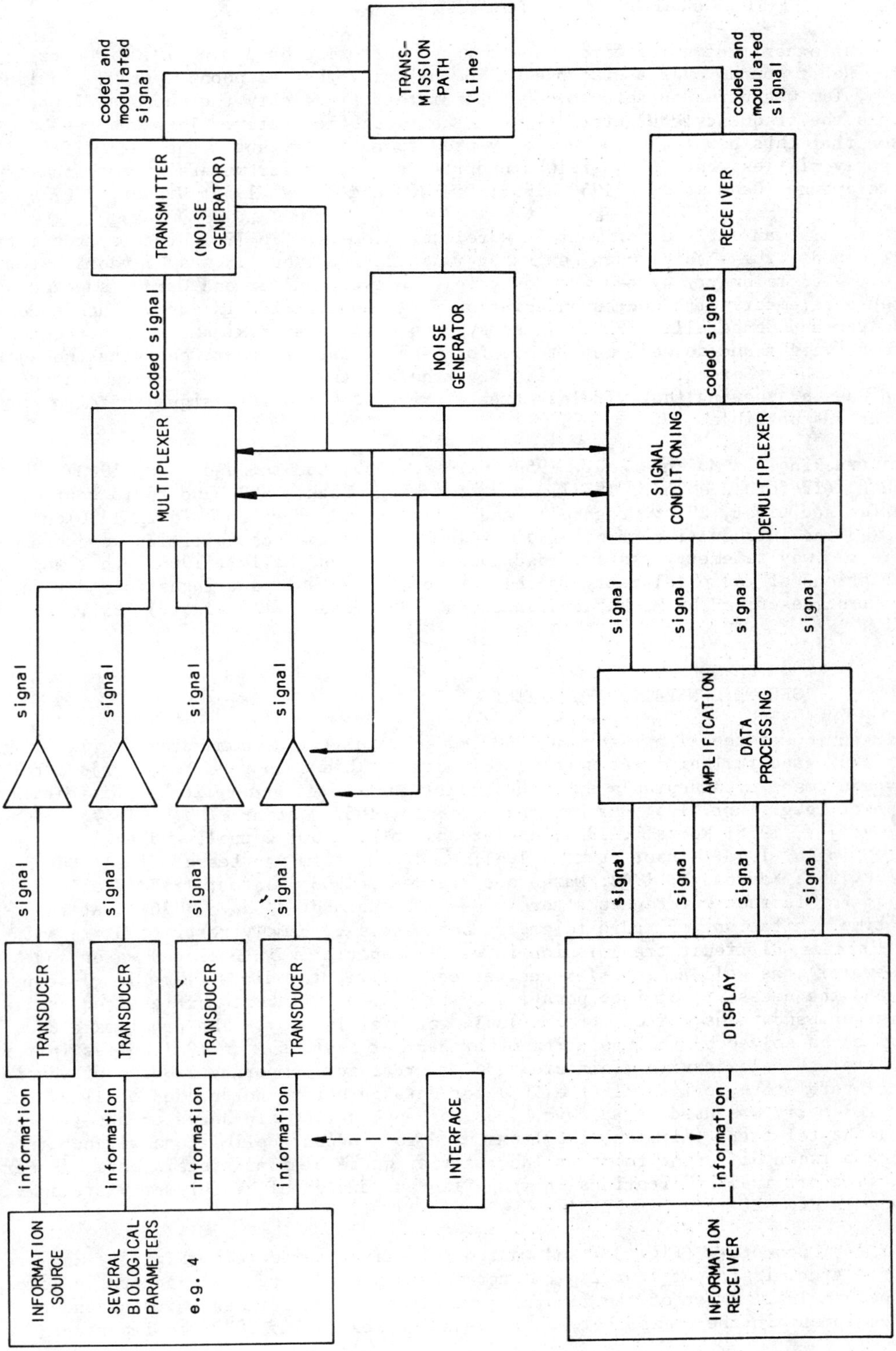

Fig. 1. Principle layout of a modern biotelemetry system. Note that at certain stages of data processing noise is picked up from surrounding noise generators.

FIRST GENERATION OF BIOTELEMETRY

The first experiment using wire telemetry was performed by Brown, 1910, who transmitted heart sounds over a distance of 150 km using the telephone system. Obviously this system was not adaptable to other parameters since only the phonocardiogram matches the frequency characteristics of the telephone system. This was the main reason that this new technique was only used sporadically until the late fifties and early sixties when ECG's (telecardiogram) and other parameters were transmitted via telephone (Rahm *et al.*, 1952, 1953; Diamond and Berry, 1953; Winsor, 1961).

There is also an early experiment in wireless telemetry; in 1921 Winters transmitted heart sounds from a ship, using existing radio devices, but it was 25 years before a 'portable' telemetry system was designed. In 1948, Fuller and Gordon succeeded in radio telemetry with non-human animals, while application in man was achieved by Holter and Gengerelli, 1949. These systems weighed approximately 50 percent of the body weight and so were unsuitable for measurements at rest; carrying the equipment meant heavy exercise in itself. Nevertheless they introduced biotelemetry techniques by showing that wireless transmission of biological signals from a moving subject was possible.

Radio tracking (Le Muyen *et al.*, 1959; Harper, 1959; Cochran and Lord, 1963), sonic tracking (Trefethen *et al.*, 1957), radar tracking (Lack, 1959) and radio control (Chaffee and Light, 1934; Gengerelli and Kallejan, 1950; Mauro *et al.*, 1950) closely followed the introduction of bio-radio telemetry and the possibilities were exhausted by the two way telemetry system, described by Young and Naylor, 1964. Challenging applications of radio telemetry in the form of implantable and ingestible devices were first described by Mackay and Jacobson, 1957, and Farrar *et al.*, 1957, respectively.

SECOND GENERATION EQUIPMENT

The invention of the transistor in 1948 and especially its commercial availability after 1952 gave further impetus to biotelemetry. This soon resulted in radio telemetry equipment of acceptable size and weight the second generation of biotelemetry equipment (e.g. Barr, 1954; Mackay and Jacobson, 1957; Farrar *et al.*, 1957; Bassan and Lovdijev, 1958; Morrell, 1959; Essler and Folk, 1960; Connell and Rowlands, 1960; Shipton, 1960; Winsor *et al.*, 1961; Kamp and Storm van Leewen, 1961; Davis *et al.*, 1961; Kavanaugh, 1963; Marko and McLeuman, 1963; Slater, 1963; Vreeland, 1963). For further references after 1963 see Ysenbrandt *et al.* (1976). At the same time limitations of radio telemetry became clear, namely restricted transmission distance, difficult transmission in water (especially in polluted water and in seawater), as well as interference between different radio systems in close proximity and the necessity of incorporating systems with different carrier frequencies and larger bandwidths, often not available for biotelemetry. The problems could sometimes be solved by incorporating other carrier systems. In 1961 Holter introduced storage telemetry (a miniature tape recorder for temporary storage of physiological parameters, originally the ECG) for data gathering not needed on-line. Ultrasonic telemetry was used by Baldwin, 1963 for applications in seawater and (infra red) light telemetry (Kimmich, 1969) was used for indoor applications without interference between different rooms or laboratories where similar systems were in operation. The original publications in the different fields of biotelemetry are summarized in Fig. 2.

Except for locomotive (Fig. 3c) and gastro-intestinal parameters (Fig. 3e) measurement and transmission of 1 or 2 parameters was achieved with first generation equipment before 1952. Most of the remaining parameters were assessed with second generation equipment in the period between 1952 and approximately 1972, an important

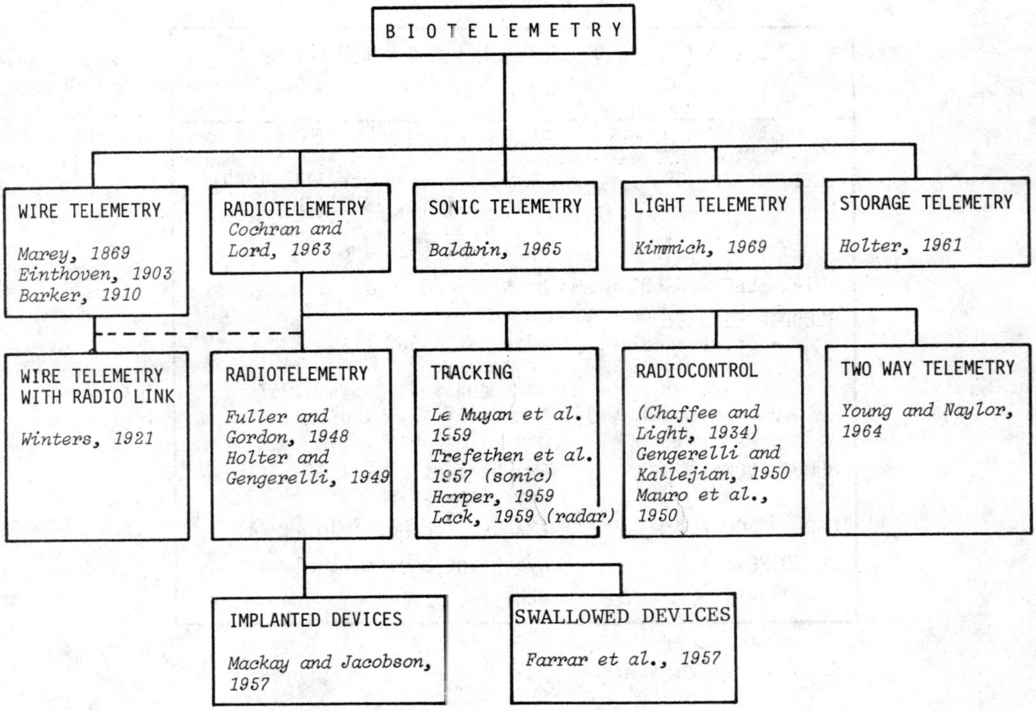

Fig. 2. Early developments of biotelemetry. The first developments in each field of biotelemetry have been traced as far as possible, and the corresponding publications stated.

exception being the non-invasive continuous telemetry of blood pressure. Third generation equipment, incorporating CMOS integrated circuits and modern coding techniques such as PCM (Smith *et al.*, 1970; Fryer and Westbroek, 1974; Hull and Mote, 1974; Kimmich and Ysenbrandt, 1974; Kucushkin, 1976) mainly show improved reliability, often together with reduced package size and system costs. Improvement of transmission equipment is still possible in many fields but the major problem has been shifted to the interface, i.e. methods and types of transducers and display.

CLASSIFICATION OF BIOTELEMETRY EQUIPMENT

Classification of biotelemetry has been tried (Kimmich, 1975; Sandler, 1976) but it is difficult and a generally acceptable classification may even be unnecessary. Three major divisions are generally accepted: (a) fields of application, (b) applied transmission links, and (c) engineering and design criteria. A matrix (Fig. 4) with a possible subdivision shows applicability of different methods in the major fields of biotelemetry. An actual classification has, however, deliberately been avoided.

CARDIOVASCULAR PARAMETERS

The telemetry of cardiovascular parameters has a prominent role in wireless acquisition of physiological parameters. However, a large number of publications and applications are restricted to the assessment of the ECG or even heart rate, which includes

CARDIOVASCULAR PARAMETERS	
PHONOCARDIOGRAM:	Brown, 1910
ELECTROCARDIOGRAM:	(Holter and Gengerelli, 1949) (Kanatsoulis, 1951 (AM)) Winsor et al., 1961 Bellet et al., 1961
FETAL ELECTROCARDIOGRAM:	Hess, 1962
BLOOD PRESSURE:	
- intravascular:	Eliassen, 1960 In man: Bachmann and Thebis, 1967
- on vascular wall:	Mackay et al., 1967
- non invasive	---
BLOOD FLOW:	Franklin et al., 1964 Moskalenko, 1964
CARDIAC OUTPUT:	(Mackay and Hechtman, 1974)
BLOOD PO_2:	Rybak and Dedien, 1967
BLOOD pH:	Uchiyama, 1964

Fig. 3a.

RESPIRATORY PARAMETERS	
RESPIRATORY FLOW:	Fuller and Gordon, 1948
TIDAL VOLUME:	Morrow and Vosten, 1953
RESPIRATORY PO_2:	Kimmich and Kreuzer, 1967
RESPIRATORY TEMPERATURE:	Kimmich and Kreuzer, 1972
OXYGEN UPTAKE:	(Kissen, 1967) Kimmich, 1969
OXYGEN UPTAKE PER BREATH:	Kimmich and Kreuzer, 1972

Fig. 3b.

LOCOMOTIVE PARAMETERS	
ELECTROMYOGRAM:	Brannick, 1963
FORCE:	Vos and Kimmich, 1971 Ishiko, 1971
DISPLACEMENT:	Rawson and Hartline, 1974
VELOCITY:	Kimmich, 1969
ACCELERATION:	(Fujie, 1965) Moon et al., 1971

Fig. 3c.

NEUROPHYSIOLOGICAL PARAMETERS	
NERVE ACTION POTENTIALS:	Morrell, 1959 (AM)
ELECTROENCEPHALOGRAM:	Breaksell and Parker, 1949

Fig. 3d.

GASTROINTESTINAL PARAMETERS	
PRESSURE:	Farrar et al., 1957 Mackay and Jacobson, 1957
pH	Jacobson and Mackay, 1957
TEMPERATURE:	Mackay and Jacobson, 1957

Fig. 3e. Original publications in different fields of biotelemetry, namely (a) cardiovascular parameters, (b) respiratory parameters, (c) locomotive parameters, (d) neurophysiological parameters and (e) gastro-intestinal parameters. Publications that do not fulfil the requirements of modern biotelemetry, but are important steps in its development are noted in brackets, e.g. equipment that is heavy (Holter and Gengerelli), unsuitable transmission technique (Kanatsoulis), on-line principle without telemetry (Kissen) etc.

those ECG components used for rate detection only. This may arise for several reasons, including the special medical importance of the ECG, as well as the long history of availability of ECG telemetry equipment (both phonocardiogram and ECG telemetry were developed as first generation equipment). In situations of reduced muscular activity (especially of the muscle groups connected with arm movement), ECG telemetry is a routine operation with no theoretical, and few practical problems.

The continuous telemetry of direct blood pressure was first described by Eliassen, 1960, who took measurements in the aorta and in the pulmonary artery of ducks and gulls. Since then a large number of experiments have been performed (e.g. Brannick, 1963; Van Citters, 1966). Today blood pressure can be transmitted from all major arteries in various animals. At the beginning the main problem was the necessity of transmitting d.c. measurements. Mackay *et al.* (1967) have measured blood pressure on the vascular wall, a method useful for longterm experiments, but certainly no less invasive. A review of these experiments has been published by Hörnicke *et al.* (1970), who has transmitted blood pressure of exercising horses.

The problems connected with direct continuous blood pressure measurement in man during physical exercise were first solved by Bachmann and Thebis, 1967. These problems were only partly of a technical nature, and include such factors as *in vivo* compatibility of the measuring catheter. Fibrine deposition, for example, largely depends on the blood flow which is substantially influenced by pathophysiological as well as exercise conditions. Since 1966, Bachmann *et al.* have performed approximately 400 telemetric recordings each year with individual experiments of between 2 or 3 and 28 hours duration. The majority of these measurements have been made under routine

Transmission Link / Applied Method — Field of Application	Wire Telemetry	Combined Telemetry (wire telemetry and wireless telemetry)	Radiotelemetry	(Ultra)sonic Telemetry	Light Telemetry	Storgae Telemetry	Implants	Ingestible Devices	Portable Devices	Fixed Devices
ANIMAL APPLICATIONS										
Remote measurement	(*)	-	*	*	*	-	*	*	*	-
Tracking	-	-	*	*	(*)	-	(*)	(*)	*	-
Control	-	-	*	*	(*)	*	*	-	*	-
HUMAN APPLICATIONS										
Clinical Surveillance (Patient Monitoring	*	*	*	-	*	-	(*)	-	*	(*)
Function tests	-	-	*	*	*	-	*	*	*	-
Rehabilitation	-	-	*	-	*	*	-	-	*	-
Remote Diagnosis	*	*	*	-	-	-	-	-	*	-
Remote Medication	(*)	(*)	*	*	-	*	(*)	-	(*)	*
Mobile Clinical Emergency Systems	-	-	*	-	-	-	-	-	(*)	*
Biological Research	-	(*)	*	*	*	*	(*)	*	*	-
Sport and Work	-	(*)	*	*	*	*	(*)	*	*	-

Fig. 4. Classification of biotelemetry equipment in major groups (field of application, transmission link and method applied) with subclassifications, incorporation of the different methods and transmission links in the different fields of application (* = often used, (*) = rarely used, - = practically never used).

clinical circumstances, during daily activity of cardiac patients (including running, swimming, skiing, driving a car, and flying). Bevan *et al.* (1969) and Krönig *et al.* (1974) have described methods for prolonged telemetry of arterial blood pressure.

Non-invasive continuous blood pressure measurement and its telemetry has unfortunately not yet been accomplished. Intermittent measurements, e.g. from an aircraft, using conventional techniques are unsatisfactory (Pircher, 1967).

The remaining parameters of the cardiovascular system where telemetry has been accomplished have not yet achieved clinical importance. Telemetry of the cardiac output (Mackay and Hechtmann, 1974) uses telemetric methods for data assessment but measurement can only be performed at reduced physical activity, maybe even only at rest. In addition system accuracy is not yet entirely established. They measure cardiac output by integrating aortic flow, measured according to the Doppler principle, and by estimating aortic diameter. The same principle is applied for the telemetry of blood flow, either extravascularly or intravascularly with a transducer mounted on a catheter tip (Franklin *et al.*, 1964; Moskalenko, 1964). Continuous blood pH (Uchiyama, 1964) and arterial PO_2 (Rybeck and Dedieu, 1967) measurements are certainly sophisticated techniques but their telemetry do not present additional difficulties. The different cardiovascular parameters that can be transmitted are summarized in Fig. 3a together with the pioneering publications.

RESPIRATORY PARAMETERS

While telemetric assessment of cardiovascular parameters generally uses conventional methods of data acquisition, the situation is entirely different for respiratory parameters. The use of 'miniaturized' spirometers and Douglas bags has been tried but contradict the philosophy of biotelemetry. In special circumstances incorporation of these means may supply acceptable results, but the applications are rather limited. Douglas bags have been used, for example, for the collection of the expiratory gas of exercising horses. Personal oxygen uptake equipment using a spirometer has been commercialized and may be applied in certain circumstances, but is certainly not suitable for application in sport. Telemetric methods measure respiratory flow as a primary parameter. It is easy to obtain respiratory volume by integrating the flow signal, but telemetry of flow involves the solution of many difficult problems.

Respiratory flow was the first parameter to be transmitted (Fig. 3b). In 1948 Fuller and Gordon presented a pneumo-inductive pick-up unit with which they transmitted the human pneumogram, together with other physiological parameters such as carotid pulse. The performance of this so-called 'radio inductograph' was, however, never determined completely. To transmit respiratory flow Morrow and Vosten (1953) modified the basic principle of Fleisch (1925) by incorporating a fine mesh nickel screen with 2/3 of free area. The power necessary for heating the mesh, in order to prevent water deposition, is generally not available in telemetric applications. At the same time data processing was introduced, in the form of analog integration of the flow signal, which lead to the first telemetric assessment of tidal volume.

Early investigations have also been performed by Bassan and Lovdjiev, who achieved telemetry of respiratory rate in 1955 and of respiratory flow in 1958. However, they measured only the expiratory flow by incorporating a valve, a principle they did not pursue further. It was soon evident that respiratory flow was the most difficult parameter to transmit for several reasons. One of the difficulties was the large dynamic range of about 30 between rest and maximum exercise, another was the deposition of water on the transducer during expiration. Many new ideas have been published concerning telemetry respiratory flow (e.g. Lewillie and Sneppe, 1968; Krobath and Reid, 1964; Shiraishi *et al.*, 1967; Kimmich and Kreuzer, 1972) but all of them have more or less severe restrictions (generally of flow range due to sensitivity to acceleration), and application has generally been restricted to the specific case of the publication. An exception was the principle of Kimmich and Kreuzer which has been adapted for different purposes, e.g. for telemetry of respiratory function in exercising horses (Hörnicke *et al.*, 1973). These flow sensors have been reviewed by Kimmich, 1972. Our group has put great effort into improving respiratory flow measurements which led to a lightweight system with good characteristics: zero drift = 20 parts per thousand per ^{o}C, sensitivity to acceleration = 0.6 parts per thousand per g, and response time to a step function = 10 ms (Kimmich, 1976).

To follow the variations of the blood gas concentrations of respiratory gases by telemetry, the transducers must not only be fast responding but also small. Only polarographic cells fulfil both conditions. CO_2 electrodes are not yet fast enough for telemetric application. Continuous telemetry of respiratory PO_2 was first described by Kimmich and Kreuzer, 1967. They circumvented the temperature sensitivity of the polarographic cell, by incorporating a sampling system and a miniature solid state heating element, a method yielding accurate results without problems. In addition Küchler *et al.* (1974) designed a system where the temperature sensitivity of the polarographic cell was compensated electronically.

The parameter of major interest in respiratory telemetry is oxygen uptake or oxygen consumption. Kissen (1967) has described the first continuous on-line assessment of $\dot{V}O_2$. He mixed the expiratory air in a mixing chamber with continuous in- and out-let of the gas and calculated the $\dot{V}O_2$ from the flow and PO_2 in the mixing chamber. Supposing there exists nitrogen equality between inspiration and expiration (which may not necessarily be the case in all circumstances, see Cissik, 1972) then exact values could only be obtained by assessing CO_2 in the mixing chamber for correction of differences between inspiratory and expiratory volumes. In addition, transient changes of $\dot{V}O_2$ can only be measured correctly if the volume of the mixing chamber is increased, but then the system would hardly be suitable for telemetric application. The first principle of continuous telemetric measurement of oxygen exchange was described by Kimmich, 1969 and its practical realization by Kimmich and Kreuzer (1972). PO_2, flow and temperature are continuously measured in an open system (so-called half mask) and the oxygen uptake calculated, breath by breath, and per minute according to the following formula:

$$\dot{V}O_2 = \frac{1}{P_0(t_n - t_m)} \int_{t_m}^{t_n} \dot{V} \cdot PO_2 \frac{PB - PH_2O(F)}{PB - PH_2O(E)} \cdot \frac{T_0}{T_0 + T_F} \, dt \; .$$

$\dot{V}O_2$ = oxygen consumption in ml O_2 s^{-1}

PO_2 = partial pressure of oxygen in mm Hg

PB = barometric pressure in mm Hg

$PH_2O(E)$ = partial pressure of water vapor in mm Hg at the site of the oxygen electrode

t = time in s

t_m, t_n = times at the beginning of inspirations m and n

$\dot{V}$ = ventilatory flow in ml s^{-1} at ambient temperature, pressure, and water vapor

T_0 = 273^oK

T_F = temperature of the flowmeter in ^oC

$PH_2O(F)$ = partial pressure of water vapor in mm Hg at the site of the flowmeter

P_0 = 760 mm Hg

The accuracy of the system (Fig. 5) depends mainly on the response time of the oxygen measuring system and the accuracy of the flow detection. To achieve reasonable accuracies the original system could only be used in a small flow range, which had to be chosen before the experiment. Incorporation of the new flowmeter (Kimmich, 1976) has considerably expanded this range, allowing telemetry from moderate up to maximal exercise without altering the measuring range. An example, of $\dot{V}O_2$ telemetry, using the original system is given in Fig. 6.

LOCOMOTIVE PARAMETERS

Radio location (as a function of time) has not been included in Fig. 3c, since this is generally known as tracking and will be discussed by Broekhuizen *et al.* (1980, this volume) Macdonald and Amlaner (1980, this volume) and Mech (1980, this volume).

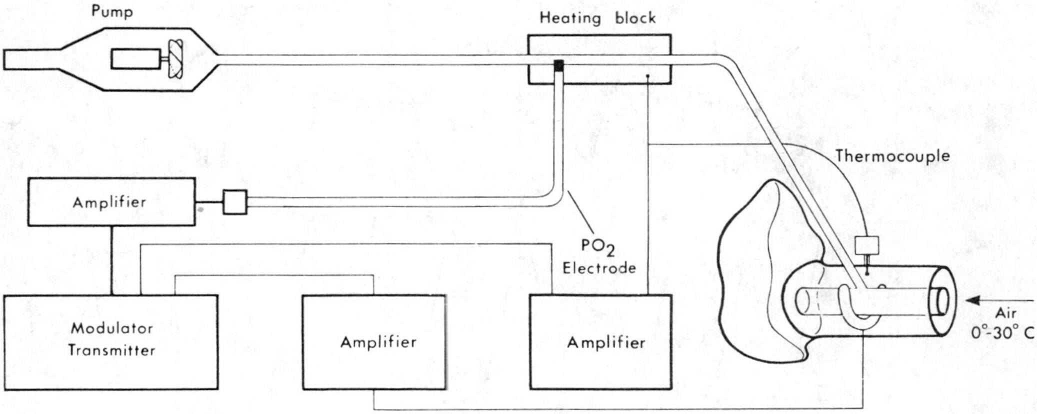

Fig. 5. Schematic diagram of the basic principle of $\dot{V}O_2$ measurement breath by breath, according to the dynamic principle. The flow is measured in a respiratory tube from which a small fraction of the respiratory gas is continuously sucked by a miniature pump through a heating block, where the PO_2 is measured. The oxygen uptake is calculated at the receiving side from flow, PO_2 and temperature (see text).

Telemetry of EMG and acceleration was relatively simple and were achieved in the case of EMG in 1963 by Brannick and of acceleration in 1965 by Fuije. Fuije has transmitted acceleration from an aircraft rather than from the biological subject itself, but the method was established and applied in many different fields, first during competition football by Moon *et al.*, 1971 and later in other sports such as running, skiing and gymnastics (e.g. Unold *et al.*, 1974; Voroshkin, 1974; Haberl and Prokop, 1974). EMG was used for gait analysis (Baumann and Baumgartner, 1974), in skiing (Grimm *et al.*, 1976) and in many other situations.

The telemetry of displacement, initiated by Rawson and Hartline, 1964 and of velocity (Kimmich, 1969) seemed relatively simple. No standard technique was used but many – often very original – ideas were incorporated. Herron and Ramsden mounted a switch into the heel of a shoe, which monitored contact between the foot and the ground. The 'rubberband goniometry' of Neukomm (1974) used 1 to 3 stretched rubber bands to connect a subject to the axle arm of 1 to 3 potentiometers, such that the axle arms were always pointing towards the subject. In this way 3 dimensional positions, displacements, etc. could be calculated from the resistance(s) of the potentiometer(s). The continuous recording of vertical jaw movements by Dibdin and Griffiths (1975) was another challenging system. More sophisticated systems for 1-3 dimensional detection of position, movements, velocities and accelerations have been described by Lindholm and Oeberg (1974) and by Woltring (1974, 1975). They incorporate infra red cameras and LED's fixed to the subject. On-line processing is possible.

Telemetry of biological forces is extremely difficult and has so far only been accomplished in special laboratory situations. There are two approaches, namely the measurement of pressure, often lacking correlation to the causal force (e.g. Carlson *et al.*, 1974; Couvillon *et al.*, 1976; Bretz, 1978) or the detection of forces on secondary equipment, enervated by the biological system. A beginning was made during rowing where force was measured on the footplate (Vos and Kimmich, 1971) or on the oar blade or paddle (Ishiko, 1971; Vos *et al.*, 1974; Schneider *et al.*, 1978), followed by other applications in skiing where the force is measured on the boot or ski

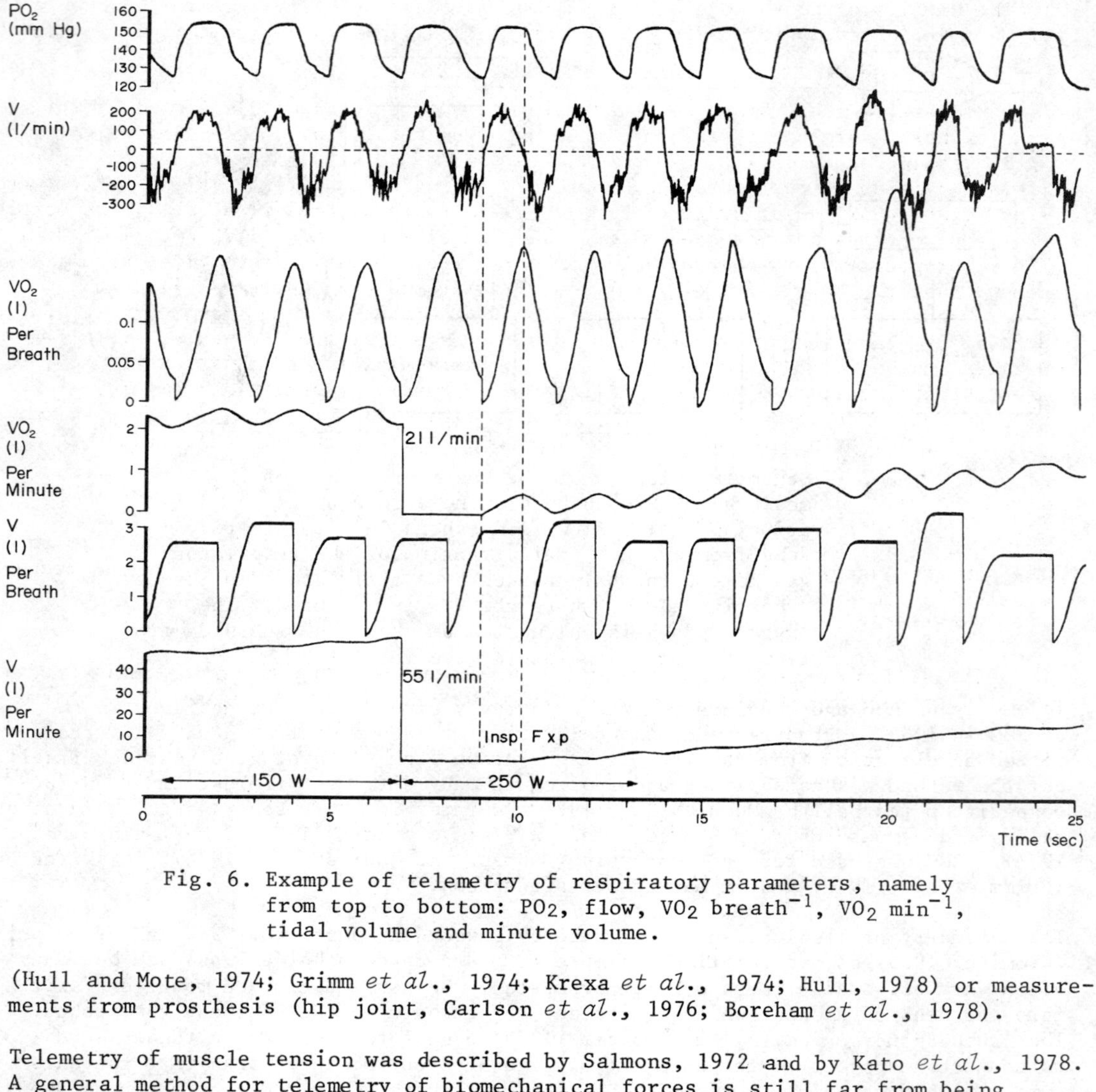

Fig. 6. Example of telemetry of respiratory parameters, namely from top to bottom: PO_2, flow, VO_2 $breath^{-1}$, VO_2 min^{-1}, tidal volume and minute volume.

(Hull and Mote, 1974; Grimm *et al.*, 1974; Krexa *et al.*, 1974; Hull, 1978) or measurements from prosthesis (hip joint, Carlson *et al.*, 1976; Boreham *et al.*, 1978).

Telemetry of muscle tension was described by Salmons, 1972 and by Kato *et al.*, 1978. A general method for telemetry of biomechanical forces is still far from being realized.

NEUROPHYSIOLOGICAL AND GASTRO-INTESTINAL PARAMETERS

Some of the important neurophysiological and gastro-intestinal parameters and the corresponding original articles on their telemetric assessment are listed in Fig. 3d and e. From a technical point of view, design and construction of gastro-intestinal telemetry devices (radiopill) are similar to implantable equipment and the physiological parameters measured are also often identical (temperature, pressure, pH, etc.).

EEG telemetry is used in two different fields, namely in pharmacological research (often with implanted, or semi-implanted devices) and in long term monitoring of psychiatric patients (e.g. Stalberg, 1969; Dymond *et al.*, 1976). Data acquisition and transmission have to be done carefully to avoid artifacts and noise pickup due to the small amplitude EEG signal (μV), but the major problem lies in the huge amount of data collected, which calls for extensive data reduction: a vast field for further activities.

Light telemetry shall stand as an example of modern biotelemetry. In the clinical sphere, both biotelemetry and patient monitoring still play only a limited role. While simple data acquisition systems with one or two wirelinks are generally used for supervision of post infarction patients during their first physical activities, incorporation of larger systems (Kimmich, 1974) with several wireless links, possibly in two directions has not been accomplished on a large scale, for several reasons. The major problems are the availability of sufficient radio frequencies with adequate bandwidths, the complexity of government regulations and, extremely important, the interference with other users. A solution to the bandwidth and frequency allocation problem is being sought by several scientific organizations. However, even if an adequate number of VHF channels would be allocated internationally for clinical wireless patient monitoring, there still remains a severe problem of interference. We have therefore decided to investigate the possibilities of light telemetry (Kimmich and Klijn, 1979), which theoretically has none of the problems mentioned.

Infra red light has the advantage of not requiring governmental licensing for operation thus allowing transmission of signals with large bandwidths. Light can be shielded easily which results in no interference between two separate systems on the same frequency located in different rooms. The disadvantage of optical transmission is the directed propagation of the carrier signal, however reflection is such that light transmission is quite possible within rooms.

We used Gallium Arsenide infra red emitting diodes as a transmitter. Small individual lenses ensured a hemispherical radiation of the infra red signal, and a semi fish-eye lens created similar characteristics at the receiver. Satisfactory performance, with little attenuation, is achieved up to approximately 200 KHz. The most important test measures light radiation as a function of the angle between the optical axis of the transmitter and the interconnecting line transmitter/receiver (Fig. 7). Besides direct transmission, there is a considerable amount of reflection on the walls. Except for glass the reflection material has little influence on the total amount of reflection. The large amount of reflection renders this system ideal for

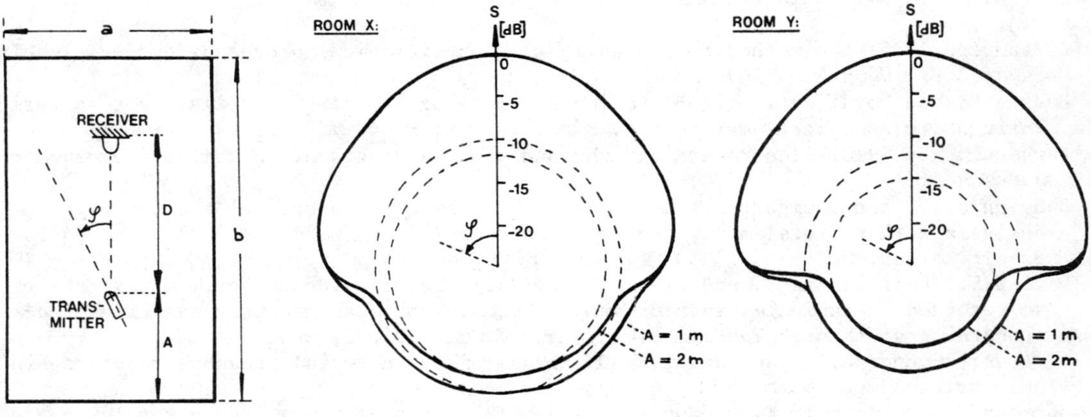

Fig. 7. Infra red telemetry: receiving signal strength (S) as a function of the angle ζ between the optical axis of the transmitter and the interconnecting line transmitter/receiver, measured in two different rooms: room X = 2.7 × 4.5 m and room Y = 6.1 × 6.5 m

wireless transmission of biological data. Reception is possible independent of the transmission angle, even with an obstacle (man) between transmitter and receiver (Fig. 8).

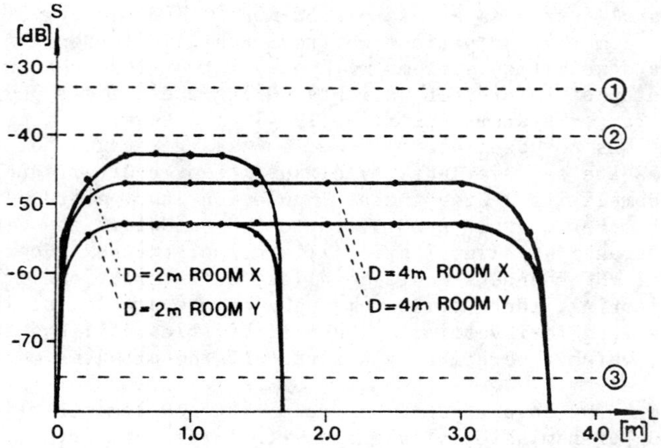

Fig. 8. Infra red signal strength with an obstacle (man) between transmitter and receiver as function of the distance L (transmitter/obstacle). (1) and (2) are the corresponding levels without obstacle and (3) is the noise level. For explanation of D see Fig. 7.

REFERENCES

Bachmann K. and Thebis J. (1967) Die drahtlose Uebertragung kontinuierlicher direkter Blutdruckmessungen. *Zt. Kreislaufforschung* 56, 188-191.

Baldwin H. A. (1963) Development of telemetry devices for dental research. Technical Documentary Report No. SAM-TDR-63-36. USAF School of Aerospace Medicine, Brooks Air Force Base, Texas.

Barr N. L. (1954) Radio transmission of physiological information. *Military Med.* 114, 79.

Bassan L. (1955) New method for investigating physiological processes during exercise. *Physiol. J. USSR* 41, 95-96.

Bassan L. and Lovdijev I. (1958) Radio-transmitting set for investigating respiration during work and sport performance. *Fiziol. Z. SSSR* 44, 773.

Barker L. F. (1910) Electrocardiography and phonocardiography, a collective review. *Johns Hopkins Hosp. Bull.* 20, 358.

Baumann J. U. and Baumgartner R. (1974) Synchronized photo-optical and EMG gait analysis with radiotelemetry. In P. A. Neukomm (ed.), *Biotelemetry II*, pp. 94-96. Basel: Karger.

Bellet S., Deliyiannis S. and Eliakin M. (1961) The electrocardiogram during exercise as recorded by radioelectrocardiography: comparison with the post-exercise electrocardiogram (Master Two-step Test). *Am. J. Cardiol.* 8, 385.

Bevan A., Honour A. J. and Stott F. D. (1969) Direct arterial pressure recording in unrestricted men. *Clin. Sci.* 36, 329.

Boreham J. F., Markolf K. L. and Willis J. (1978) Biotelemetry from a total hip replacement prosthesis. In H.-J. Klewe and H. P. Kimmich (eds.), *Biotelemetry IV*, pp. 195-198. Braunschweig: Klewe/Kimmich.

Brannick C. J. (1963) Recording physiological functions from unrestrained dogs. In L. E. Slater (ed.), *Bio-telemetry*, pp. 303-309. New York: Pergamon Press.

Breaksell C. C. and Parker C. S. (1949) Radio transmission of the human electroencephalogram and other electrophysiological data. *Electroencephalog. Clin, Neurophysiol.* 2, 243.

Bretz K. (1978) Telemetry of limb forces. In H.-J. Klewe and H. P. Kimmich (eds.), *Biotelemetry IV,* pp. 207-210. Braunschweig: Klewe/Kimmich.

Broekhuizen S., Van't Hoff C. A., Jansen M. B. and Niewold F. J. J.(1980) Application of radio tracking in wildlife research in the Netherlands. In C. J. Amlaner, Jr. and D. W. Macdonald (eds.), *A Handbook on Biotelemetry and Radio Tracking,* pp. 65-84. Oxford: Pergamon Press.

Brown S. (1910) A telephonic stethoscope. *Sci. Am.* 102, 496.

Carlson C. E., Mann R. W. and Harris W. H. (1974) A radio telemetry device for monitoring cartilage surface pressures in the human hip. *IEEE Trans. Biomed. Engng. BME* 21, 257.

Carlson C. E., Mann R. W. and Harris W. H. (1976) Design features of a pressure telemetering hip prosthesis. In T. B. Fryer, H. Z. Miller and H. Sandler (eds.), *Biotelemetry III,* pp. 127-130. New York: Academic Press.

Chaffee E. L. and Light R. V. (1934) A method for the remote control of electrical stimulation of the nervous system. I. The history of electrical excitation. *Yale J. Biol. Med.* 7, 83-128.

Cissik J. H. and Johnson R. E. (1972) Myth of nitrogen equality in respiration: its history and implications. *Aerospace Med.* 755-758.

Cochran W. W. and Lord R. D., Jr. (1963) A radio-tracking system for wild animals. *J. Wildl. Mngmt.* 27, 9-24.

Connell A. M. and Rowlands E. M. (1960) Wireless telemetering from the digestive tract. *Gut* 1, 266.

Couvillon L. A., Jr., Baker C. D., Gover T. P. and Konigsberg E. (1976) Telemetry monitoring of intraocular pressure. In T. B. Fryer, H. A. Miller and H. Sandler (eds.), *Biotelemetry III,* pp. 123-126. New York: Academic Press.

Davis D. A., Thornton W., Grosskreuz D. C., Sugioka K. and McKnight R. (1961) Radio telemetry in patient monitoring. *Anesthesiology* 22, 1010.

Dibdin G. H. and Griffiths M. J. (1975) An intra-oral telemetry system for the continuous recording of vertical jaw movements. *Phys. Med. Biol.* 20, 355-365.

Diamond E. G. and Berry F. M. (1953) Transmission of electrocardiographic signals over telephone circuits. *Am. Heart J.* 46, 906.

Dymond A. M., Coger R. W., Martin J. P. and Searfetinides E. A. (1976) Multichannel biotelemetry of electrophysiological parameters from unrestraint psychiatric patients. In T. B. Fryer, H. A. Miller and H. Sandler (eds.), *Biotelemetry III,* pp. 157-160. New York: Academic Press.

Einthoven W. (1903) Galvanometrische Registrierung des Menschlichen Elektrokardiograms. *Archs. ges. Physiol.* 99, 472.

Eliassen E. (1960) A method for measuring the heart rate and stroke/pulse pressures in birds by normal flight. *Arbook Univ. Bergen, Mat.-Nat.-Sci.* 12, 1-22.

Essler W. O. and Folk G. E., Jr. (1960) Monitoring heart rates and body temperatures of unrestrained mammals. by radio transmitter. *Anat. Rec.* 137, 353.

Farrar J. T., Zworkin V. K. and Baum J. (1957) Pressure-sensitive telemetering capsule for study of gastrointestinal motility. *Science* 126, 975.

Fleisch A. (1925) Der Pneumotachograph: Ein Apparat zur Geschwindigkeitsregistrierung der Atemluft. *Pflügers Arch.* 209, 713-722.

Franklin D. L., Watson N. W. and van Citters R. L. (1964) Blood velocity telemetered from untethered animals. *Nature* 203, 528-530.

Fryer T. B. and Westbroek R. M. (1974) A multichannel biotelemetry transmitter utilizing a PCM subcarrier. In P. A. Neukomm (ed.), *Biotelemetry II,* pp. 202-204. Basel: Karger.

Fujie Z. (1965) Physiological monitoring in flight. *Digest. 6th Int. Conf. Med. Biol. Engng.* 11, 195.

Fuller S. L. and Gordon T. M., Jr. (1948) The radio inductograph, a device for recording physiological activity in unrestrained animals. *Science* 108, 287.

Gengerelli J. A. and Kallejian V. (1950) Remote stimulation of the brain in the intact animal. *J. Psychol.* 29, 263-269.

Grimm C.-M., Krexas H. and Asang E. (1974) Applied biotelemetry in skiing. In P. A. Neukomm (ed.), *Biotelemetry II*, pp. 121-123. Basel: Karger.
Haberl R. and Prokop L. (1974) Physiological aspects of synthetic field tracks. Foot-motions during supportphase of various running types on different track materials. In P. A. Neukomm (ed.), *Biotelemetry II*, pp. 109-111. Basel: Karger.
Harper W. G. (1959) Roosting movements of birds and migration departures from roosts as seen by radar. *Ibis* 101, 201-208
Herron R. E. and Ramsden R. W. (1967) A telepedometer for the remote measurement of human locomotor activity. *Psychophysiology* 4, 112-115.
Hess O. W. (1962) Radiotelemetry of fetal heart energy. *Obstet. Gynecol.* 20, 516.
Holter N. J. (1961) New method for heart studies. *Science* 134, 1214.
Holter N. J. and Gengerelli J. A. (1949) Remote recording of physiologic data by radio. *Rocky Mountain Med. J.* 46, 747.
Hörnicke H., v. Engelhardt W., Ehrlein H.-J. and Tolkmitt G. (1970) Blutdrucktelemetrie in der Tierphysiologie. In L. Demling and K. Bachmann (eds.), *Biotelemetrie*, pp. 241-252. Stuttgart: Thieme.
Hörnicke H., Ehrlein H.-J., Tolkmitt G., Nagel M., Epple E., Decker D., Kimmich H. P. and Kreuzer F. (1973) Method for continuous oxygen consumption measurement in exercising horses by telemetry and electronic data processing. *Proc. VI. Symp. on Energy Metab. of Farm Animals*, Stuttgart.
Hull M. L. and Mote C. D. (1974) Precision PCM telemetry in ski injury research. In P. A. Neukomm (ed.), *Biotelemetry II*, pp. 115-117. Basel: Karger.
Hull M. L. and Mote C. D. (1974) Pulse code modulation telemetry in ski injury research. I. Instrumentation. *Biotelemetry* 1, 182-192.
Hull M. L. and Mote C. D. (1978) Spectral analysis of leg loading in snow skiing. In H.-J. Klewe and H. P. Kimmich (eds.), *Biotelemetry IV*, pp. 195-198. Braunschweig: Klewe/Kimmich.
Ishiko T. (1971) Biomechanics in rowing. In *Biomechanics II*, pp. 249-252. Basel: Karger.
Jacobson B. and Mackay R. S. (1957) A pH-endoradiosonde. *Lancet* 272, 1224.
Kamp A. and Storm van Leewen W. (1961) A two channel EEG radio telemetering system. *Electroencephal. Clin. Neurophysiol.* 13, 458.
Kanatsoulis A. (1951) Le Télé-électrocardiographe. *Presse Méd.* 59, 458.
Kato M., Murakami S., Ohta S., Ando Y. and Matsumoto G. (1978) Stiffness of a human muscle during concentration. In H.-J. Klewe and H. P. Kimmich (eds.), *Biotelemetry IV*, pp. 199-202. Braunschweig: Klewe/Kimmich.
Kavanaugh D. A. (1963) Development of a miniature telemetry system. Technical Documentary Report No. SAM-TDR-63-80. USAF School of Aerospace Med. Brooks Air Force Base, Texas.
Kimmich H. P. (1969) Telemetry of Respiratory Oxygen Pressure in Man. Ph.D. thesis, Nijmegen, 1-79.
Kimmich H. P. (1972) Radiotelemetry of respiratory parameters. *Bull. Physiopath. Resp.* 8, 807-857.
Kimmich H. P. (1975) Multichannel biotelemetry. *Biotelemetry* 2, 206-254.
Kimmich H. P. (1976) Telemetry of respiratory flow with a new magnetic type of flowmeter. In T. B. Fryer, H. A. Miller and H. Sandler (eds.), *Biotelemetry III*, pp. 25-28. New York: Pergamon Press.
Kimmich H. P. and Klijn J. A. J. (1979) Infrared propagation and reflection properties for application in patient monitoring (in press).
Kimmich H. P. and Kreuzer F. (1967) Telemetry of respiratory oxygen pressure in man during exercise. *Digest. 7th Int. Conf. Med. Biol. Engng.*, Stockholm, 90.
Kimmich H. P. and Kreuzer F. (1972) Telemetric determination of oxygen uptake by man during exercise. *Int. J. Biomed. Engng.* 1, 27-36.
Kimmich H. P. and Ysenbrandt H. J. B. (1974) Personal PDM/PCM biotelemetry system. In P. A. Neukomm (ed.), *Biotelemetry II*, pp. 2-4. Basel: Karger.
Kissen A. T. (1967) General description and evaluation of an on-line oxygen uptake computer. AMRL-TR-67-17: 1-45.
Krexa H., Grimm C.-M. and Asang E. (1974) Experimental biotelemetry in alpine skiing. In P. A. Neukomm (ed.), *Biotelemetry II*, pp. 124-126. Basel: Karger.

Krobath H. and Reid C. (1964) A new method for the continuous recording of the volume of inspiration and expiration under widely varying conditions. *Am. J. Med. Electron.* 3, 105-109.

Krönig B., Dufey K., Reinhardt P., Jahnecke J. and Wolff H. P. (1974) Long-term telemetry of direct arterial blood pressure in unrestricted hypertensives. *Biotelemetry* 1, 117-131.

Küchler G., Wagner W., Schneiderreit R. and Wolburg I. (1974) Studies to compensate temperature effects in measurements of respiratory oxygen. In P. A. Neukomm (ed.), *Biotelemetry II*, pp. 22-24. Basel: Karger.

Kucushkin I. A. (1976) On frequency selection of signal quantization of arterial pulse in biotelemetry systems with pulse-code modulation. In Rosenblat and Fraden (eds.), *Bioradiotelemetry* (in Russian), pp. 113-116. Sverdlovsk: USSR Academy of Sci.

Lack D. (1959) Migration across the North Sea studies by radar. Part I. Survey through the year. *Ibis* 101, 209-234.

Le Muyan C. D., White W., Nyberg E. and Christian J. J. (1959) Design of a miniature radio transmitter for use in animal studies. *J. Wildl. Mngmt.* 23, 107-110.

Lewillie L. and Sneppe R. (1968) Telemetric measurement of the respiratory function. *Ergonomics* 11, 77-81.

Lindholm L. E. and Oeberg K. E. T. (1974) An optoelectronic instrument for remote on-line movement monitoring. In P. A. Neukomm (ed.), *Biotelemetry II*, pp. 100-102. Basel: Karger.

Macdonald D. W. and Amlaner C. J., Jr. (1980) A Practical guide to radio tracking. In C. J. Amlaner, Jr. and D. W. Macdonald (eds.), *A Handbook on Biotelemetry and Radio Tracking*, pp. 143-159. Oxford: Pergamon Press.

Mackay R. S. (1967) Telemetering studies from aquatic animals. *Digest 7th Int. Conf. Med. Biol. Engng.*, Stockholm, 85.

Mackay R. S. and Hechtmann H. B. (1974) Continuous monitoring of cardiac output. *Biotelemetry* 1, 21-30.

Mackay R. S. and Jacobson B. (1957) Endoradiosonde. *Nature* 179, 1239.

Marey M. (1869) Phenomenon of Flight in the Animal Kingdom. Smithsonian Ann. Rept., pp. 226-285. Washington, D.C.: U.S. Gov. Print. Office.

Marko A. R. and McLeuman M. A. (1963) A seven channel personal telemetry system using pulse duration modulation. *Proc. 16th Ann. Conf. Eng. Med. Biol.*, Baltimore, 154.

Mauro A., Davey W. L. M. and Scher A. M. (1950) Central nervous stimulation by implanted high frequency receiver. *Fedn. Proc.* 9, 86.

Mech L. D. (1980) Making the most of radio-tracking — a summary of wolf studies in Northeastern Minnesota. In C. J. Amlaner, Jr. and D. W. Macdonald, *A Handbook on Biotelemetry and Radio Tracking*, pp. 85-95. Oxford: Pergamon Press.

Moon D. W., Beedle C. W. and Kovacic C. R. (1971) Peak head acceleration of athletes during competition — football. *Med. Sci. Sports* 3, 44-50.

Morrell R. (1959) Amplitude modulation radio telemetry of nerve action potentials. *Nature* 184, 1129.

Morrow P. E. and Vosten R. E. (1953) Pneumotachographic studies in man and dog incorporating a portable wireless transducer. *J. appl. Physiol.* 5, 348-360.

Moskalenko Y. E. (1964) Telemetrical equipment for studying blood circulation of the brain. *Proc. Physiol. Soc.* 20/21, 3-4.

Neukomm P. A. (1974) The rubber band goniometry. *Biotelemetry* 1, 12-20.

Pircher L. (1969) Telemetry of the blood pressure in the aircraft. *Wehrmed. Monatsschr.* 13, 75-77.

Rahm W. E., Barmore J. L. and Dunn F. L. (1952) Electrocardiographic transmission over standard telephone lines. *Neb. St. med. J.* 37, 222.

Rahm W. E., Barmore J. L., Ellestad I. M. and Dunn F. L. (1953) Local and long distance transmission and storage of electrocardiograms and other low frequency signals. *Circulation Res.* 1, 518.

Rawson K. S. and Hartline P. H. (1964) Telemetry of homing behavior by the deermouse, *Peromyscus*. *Science* 146, 1596-1598.

Rybeck B. and Dedieu J. M. (1967) Radio-transmission de la pression partielle d'oxygène. *Life Sci.* 6, 1417-1421.

Salmons S. (1972) A telemetric technique for measuring muscle tension in conscious unrestraint animals. In H. P. Kimmich and J. A. Vos (eds.), *Biotelemetry*, pp. 335-336. Leiden: Meander NV.

Sandler H. (1976) Biotelemetry: its first 50 years. In T. B. Fryer, H. A. Miller and H. Sandler (eds.), *Biotelemetry III*, pp. 1-11. New York: Academic Press.

Schneider E., Morell F. and Sidler N. (1978) Long-term investigation in rowing by multi parameter telemetry. In H.-J. Klewe and H. P. Kimmich (eds.), *Biotelemetry IV*, pp. 211-214. Braunschweig: Klewe/Kimmich.

Shipton H. W. (1960) A simple telemetering system for electrophysiological data. *Electroencephal. Clin. Neurophysiol.* 12, 922.

Shiraishi N., Shirahata A., Nakamura K. and Shiraishi Y. (1967) Telemetering system for measuring minute volume, tidal volume and respiratory frequency. *Digest 7th Int. Conf. Med. Biol. Engng.*, Stockholm, 3/11, 93.

Slater L. E. (ed.) (1963) *Bio-telemetry*. Oxford: Pergamon Press.

Smith R. E., Wekstein D. R. and Sullivan R. H. (1970) A multi-channel pulse-coded multiplexing and demodulating telemetry system. *IEEE Trans. Biomed. Engng.* 17, 151-156.

Stalberg E. (1969) Telemetric long-term EEG recording. *Electroencephal. clin. Neurophysiol.* 26, 341.

Trefethen P. S., Dudley J. W. and Smith M. R. (1957) Ultrasonic tracer follows tagged fish. *Electronics* 30, 156-160.

Uchiyama A. (1964) Continuous recording of blood pH and its radiotelemetry. *Jap. J. Physiol.* 14, 630-637.

Unold E. M., Neukomm P. A. and Nigg B. M. (1974) Continuous acceleration measurements in humans during locomotions on different surfaces. In P. A. Neukomm (ed.), *Biotelemetry II*, pp. 112-114. Basel: Karger.

Van Citters R. L. and Franklin D. L. (1966) Telemetry of blood pressure in free ranging animals via an intravascular gauge. *J appl. Physiol.* 21, 1633-1636.

Voroshkin G. (1974) Radio-telemetric method of evaluating force-dynamic values in alpine skiing. In P. A. Neukomm (ed.), *Biotelemetry II*, pp. 118-120. Basel: Karger.

Vos J. A. and Kimmich H. P. (1971) Telemetry of biomechanical forces during exercise. *Digest. Int. Symp. Biotelemetry*, Nijmegen. 59-60.

Vos J. A. and Kimmich H. P. (1972) Telemetry of biomechanical forces during exercise. In H. P. Kimmich and J. A. Vos (eds.), *Biotelemetry*, pp. 279-288. Leiden: Meander NV.

Vos J. A., Kimmich H. P., Mäkinen J., Ysenbrandt H. J. and Vrijens J. (1974) Telemetry of dynamic forces in endurance sports. In P. A. Neukomm (ed.), *Biotelemetry II*, pp. 106-108. Basel: Karger.

Vreeland R., Collins C., Williams A. A., Yeoger C., Gianascol A. and Henderson J. (1963) A subminiature radio EEG telemeter for studies of disturbed children. *Electroencephal. Clin. Neurophysiol.* 15, 327.

Winsor T., Sibley E. A. and Fisher E. K. (1961) Electrocardiograms by telemetry. *Calif. Med.* 94, 284.

Winters S. R. (1921) Diagnosis by wireless. *Sci. Am.* 124, 465.

Woltring H. J. (1974) New possibilities for human motion studies by real-time light spot position measurements. *Biotelemetry* 1, 132-146.

Woltring H. J. (1975) Calibration and measurement in 3-dimensional monitoring of human motion by optoelectronic means. *Biotelemetry* 2, 169-196.

Young J. I. and Naylor W. S. (1964) Implanted two way telemetry in laboratory animals. *Am. J. Med. Electron.* 3, 28.

Ysenbrandt H. J. B., Selten Th. A. J., Verschuren J. J. M., Kock T. and Kimmich H. P. (1976) Biotelemetry, literature survey of the past decade. *Biotelemetry* 3, 145-250.

The Advantages of Short Range Telemetry through the Intact Skin for Physiological Measurements in Both Animals and Man

T. B. FRYER

NASA, Ames Research Center, Mail Stop 213-3, Moffett Field, California, U.S.A.

Abstract — The development of the transistor, slightly more than 20 years ago, has made the ability to surgically implant electronic devices in the body of man and animals a reality, for both prosthetic and diagnostic purposes. The electronic pacemaker is probably the most widely known such device. It is usually preprogramed and not a telemetry device, although some of the earlier systems did provide external rate control, a form of command telemetry, through the intact skin. The surgical implantation of electronic devices within the body presents an exciting new frontier in the advancement of medical science.

Although radio telemetry is commonly conceived of as a system for telemetry or communications over very great distances, it can also have very useful short range applications. Specifically an implant system is just as remote and inaccessible after surgery as a space satellite is and ideally requires data communications (down link) and command telemetry (up link) as in a satellite system. Wireless telemetry through the intact skin provides both a data and command link for such prosthesis as nerve stimulators, bladder control, pain suppression, artificial vision, artificial hearing, artificial pancreas, to name only a few that are currently in various stages of development. Diagnostic systems have received less attention but could be extremely useful in determining the efficacy of a particular drug treatment. Telemetry for gathering physiological data from small unrestrained animals is now widely used. When a long range is required, as might be the case in wild animals in the natural habitat, the short range telemetry can be combined advantageously with an external relay system to achieve virtually an unlimited operating distance.

INTRODUCTION

The advent of the transistor just over 20 years ago, made the design and development of microminiature battery operated telemetry systems a reality. The usefulness of telemetry for physiological monitoring, free of the restraint of wires and cables, was recognized by various investigators, Mackay (1970) being one of the foremost.

In the ensuing years a wide variety of telemetry systems have been developed and reported. The spectrum of biomedical telemetry systems covers a wide gamut from microminiature systems with an operating range of less than one foot to larger systems that are usually worn or strapped to man or animals and have a capability of transmitting hundreds of miles or more. The purpose of this review is to explore the advantages of the short range telemetry and describe some of the systems that are currently being developed.

The definition of short range and long range is relative and could be confusing since systems with operating distances of less than one foot to thousands of miles are in use. For the purposes of this review, systems that operate anywhere from 50 feet to many miles will be considered long range. The primary reason for this definition is that almost all of these systems operate with radio transmitters that are licensed or controlled by government regulations which also vary from country to country. Consequently the transmitter power, frequency control, and bandwidth, principal factors affecting the operating distance, are all tightly regulated variables beyond the control of the designer and placing very severe constraints on the design. Also, in most countries biomedical applications seem to receive a low priority in the allocation of the available RF spectrum. Transcutaneous short range systems in contrast are usually outside such regulations giving the designer more flexibility and greater opportunity to achieve an optimum design limited only by the current state of the art. The higher powered transmitter also is usually of standard design for use in many applications thereby not requiring extensive specialized discussion for biomedical usage.

Investigators new to the field of biomedical telemetry might question the need for telemetry where the operating range is only a few feet. Two major applications exist. One is the totally implanted system that can be used on man and animals and eliminates any transcutaneous wires with consequent healing and infection problems. A transmission distance of a fraction of an inch, just sufficient to go through the intact skin, to a foot or so is all that is needed for such applications. The other situation is the small caged animal, such as a rat, where an operating range of a few feet is more than adequate. Either surgically implanted systems or externally mounted ones can be used with small caged animals. An implanted system, after the initial surgery, is just as remote as an orbiting satellite, barring additional undesirable surgical trauma. They share the same needs for data communications and command control and differ only in the power and type of transmitter used in the RF carrier to span the differences in operating distance.

Total implant systems have some very unique design requirements that cannot be solved with standard telemetry engineering as is the case with transmitters used in long range systems. Particularly in the case of an implant design there are many interface, interdisciplinary physiological, and engineering design problems. For example, body fluids seem to have an uncanny way of working into almost any device unless a true hermetic seal is used. Because of the surgical trauma involved in placing a device in the body, a design for long operating life is essential to make the system reasonably useful. This in turn means the circuits must be designed to operate at very low power, preferably in microwatts, since the self-contained battery usually incorporated into the system must be small in size. Besides the problems of power consumption and hermetic sealing, there are the additional mechanical problems of lead breakage and tissue compatibility. A very important aspect of an implant system is that once surgically implanted the unit is completely inaccessible for any knob adjustments or similar alterations. All these considerations then make the design of total implant systems quite unique and worthy of the special discussion that will be attempted in this review article.

CARRIER FREQUENCY

The primary factors affecting the range of a telemetry system are the transmitter power, frequency, bandwidth, and antenna gain. Electromagnetic energy is propagated by two means, one is induction and the other RF radiation. The field strength of inductive coupling diminishes as the cube of the distance so the operating range is very limited, while RF radiation is inversely proportional to the distance (first power). Consequently, inductive coupling is effective only for very short range applications of a few feet, while RF radiation is essential for distances greater than 5 to 10 feet. To achieve RF radiation it is necessary to have an antenna comparable to $\frac{1}{4}\lambda$ of the carrier frequency. For man and larger animals it is possible ot utilize antennas extracorporeally in the 100-200 MHz region with effective radiation, since $\frac{1}{4}\lambda$ equals 15-30 inches. But for an implant such a length is not usually feasible. As the antenna length is significantly reduced the effective radiated RF power drops rapidly. For an implant the effective antenna length is also affected by being placed within conductive tissue as well as energy losses in the tissue. When these factors are considered along with the need to operate the system from a selfcontained battery at very low power, it can be seen that long range implant systems are not very practical. The problems are further compounded by government regulations that would limit the designer's choice of frequency and bandwidth. If considerable range is necessary then the logical choice is an extracorporeal system. The frequency choice is then not so restrictive; larger batteries that can be changed periodically and more or less standard authorized types of transmitter can be used. When required, external systems can also be utilized in conjunction with implant systems to relay the short range implant signal over a greater distance. Each link can then operate in its most optimum mode.

For short range systems, where inductive coupling is effective, a lower carrier frequency typically in the 1-10 MHz range, can be used. This has a number of advantages. First a simple easily fabricated, low power, single transistor oscillator can be used in the transmitter. Although simple single transistor oscillators in the higher frequencies, 50-200 MHz, are also common, they usually require more power, are marginal without crystal frequency control, and often do not meet regulatory requirements. The second advantage of the lower frequencies is that a simple LC oscillator will usually have more than adequate frequency stability at low frequency without the need for crystal control. All these assumptions are based on the premise that inductive coupled systems are in the same class as other standard electronic-electrical equipment which normally do not require government telecommunications approval. Thirdly when using inductive coupling the lower frequencies are equally as effective as the higher ones and require less complicated receivers. The smallest possible radiating coil or antenna that is feasible is essential in small animals as well as being desirable in man and larger animals. The use of a small coil in conjunction with frequencies of 10 MHz or less ensures that there is little or no RF radiation and consequent regulatory control. Small coils of $\frac{1}{2}$ to $\frac{3}{4}$ in diameter have been used by Fryer, Deboo and Winget (1966) and Mackay (1970) for transmitters in the 100 MHz region and even though this size is a very inefficient RF antenna at 100 MHz it can provide operating ranges up to 50 feet with a milliwatt transmitter. Devices of this nature fall into the gray area of government regulation while the same size coil operating at 10 MHz will have essentially no RF radiation and not be regulated. An operating distance of one to five feet is possible at 10 MHz with a very low powered oscillator.

An additional advantage of inductive coupling is that because of the cube law reduction in field strength, many systems can be operated in close proximity to each other. In contrast, radio transmitters because of their potentially great operating range, must be government regulated to make the radio spectrum useful. A RF transmitter with a short tightly prescribed operating range is a concept that is being employed in some cases to utilize more completely and effectively the limited RF spectrum.

Once it is determined that a system does not fall under the rules governing radio transmission the design is limited only by the current state of the art. The selection of transmitter power, frequency, and bandwidth can be made on the basis of optimum performance. Any of the standard carrier modulation schemes such as FM, AM, pulse width (PWM) and pulse interval (PIM) modulation, etc., can be used in conjunction with normal data handling techniques of multiplexing and encoding.

ELECTRONIC SIGNAL CONDITIONING

A short range or long range system can utilize the same electronic signal conditioning techniques but with the difference that in a short range system the designers latitute of choice is unlimited, while standard RF transmission is government restricted. There are many applications where very short range is more than adequate and when not, it can often be advantageously combined with external tape recorders, solid state memories, or more powerful extracorporeal RF transmitters. A combination system also allows the use of more complex data conditioning extracorporeally such as microprocessors to digest data and reduce the data bandwidth. This then makes it easier to use RF transmitters that fit within government regulations. Reduced bandwidth as a result of preliminary data reduction is also advantageous for maximum utilization of tape recorders and solid state memories.

The inaccessibility of an implant system for repair, knob adjustment, and circuit modification as well as the requirements for low power makes it desirable to minimize the complexity of the implant. As much as possible of an overall system should be located externally. Integrated circuits, microprocessors, and the like can be utilized most effectively in the extracorporeal location where they are accessible for repair, circuit adjustment, and programing changes. Short range systems provide a relatively high signal to noise ratio, which coupled with an unrestricted bandwidth and frequency choice, allows the easy transfer of wide bandwidth unprocessed data. The implant segment of a total system then ideally consists of the transducers whether they be chemical sensors, pressure cells, electrodes or the like to interface with the physiological system and the necessary coding to provide accurate data transfer. Coding with PWM, PCM, PIM, and FM modulation of a subcarrier as well as multiplexing is normally utilized in the same manner as with long range telemetry transmission. Once the data is transferred to the extracorporeal receiver-demodulator the signal can be observed, recorded, further processed, or retransmitted as required. Signals can also be transmitted to the implant for command and control purposes. The overall system is then a highly versatile two way data link similar in many ways to the up and down link in satellite telemetry systems.

Figure 1 shows a multichannel telemetry system (Fryer *et al.*, 1975) that has been used in larger animals such as chimpanzees and dogs, and illustrates many of the aspects of implant telemetry systems that have been discussed so far. The system provides for two pressure measurements, one electromagnetic flow and an EKG for use in cardiavascular research. The heavy power requirements of the electromagnetic flow meter is a major factor in the size of this unit. Two 500 MAH rechargeable NICAD batteries are included in the implant to operate the flowmeter. If an ultrasonic flowmeter had been selected, a significant reduction in power would have been possible, but the electromagnetic flowmeter was selected in this case because of its demonstrated accuracy for blood flow measurements, particularly in the case of the ascending aorta. The rechargeable batteries allow the system to operate continuously for up to two hours. Recharging is done through a close coupled induction loop operating in the 250 KHz region. Although it is normally desirable to minimize the power consumption of the implant this example illustrates that significant amounts of power, up to a few watts or more, can be transferred by a close coupled transcutaneous induction system when required. The choice of flow transducer in this case dictated the heavy power requirements. Considerable signal conditioning, including amplification, data encoding, and multiplexing can be accomplished at very low milliwatt power levels and the overall power needs are usually determined by the power requirements of the transducers that are available or specifically selected for various reasons.

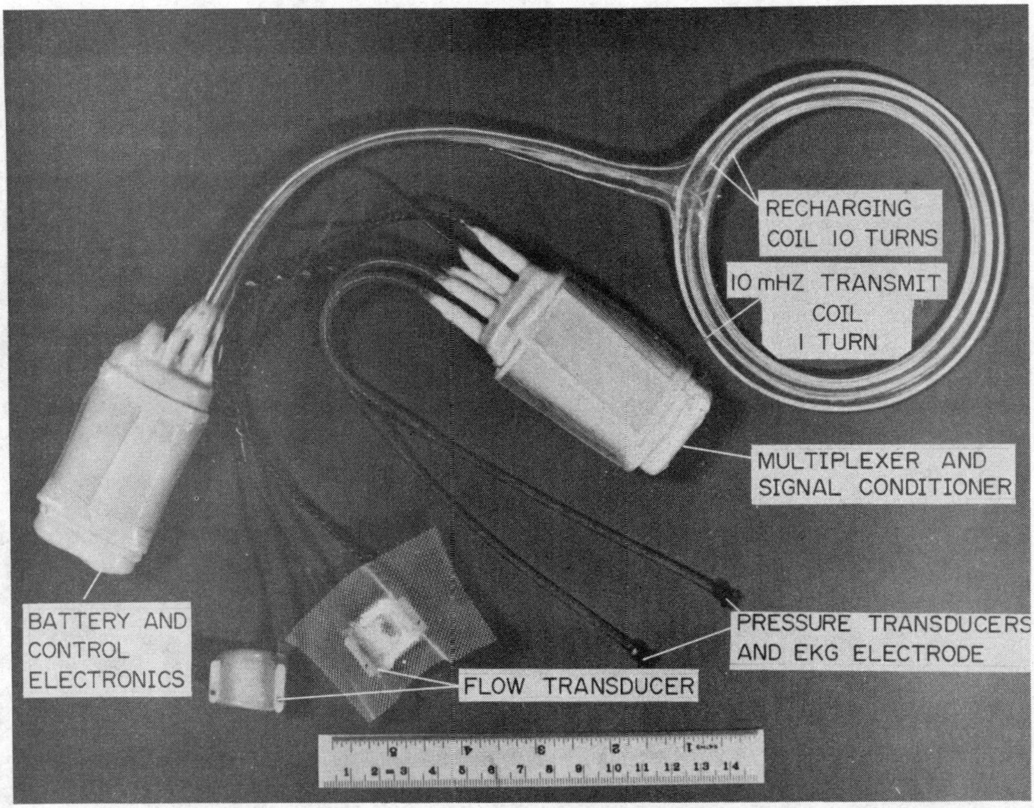

Fig. 1. Multichannel telemetry system for flow, pressure and EKG measurements.

The electromagnetic flowmeter uses a 400 Hz wave carrier and this signal is demodulated internally to an analog signal proportional to flow. The signal is time multiplexed, along with the other pressure and EKG signals, into a single channel that is then PWM encoded for transmission. Both a 10 MHz carrier suitable only for transcutaneous use and a higher frequency 100 MHz transmitter suitable for modest range up to 50 feet have been incorporated into the system. Regulations governing the use of short range RF transmitters in the 100-200 MHz region vary widely from country to country. This system besides providing for re-energizing the battery and transmission of physiological data also has an elaborate control scheme. By momentary interruption of the 250 KHz energizing power an electronic stepping switch is advanced. Four switch positions are provided for different flowmeter operation modes. Specifically a flow 'on' and 'off' mode is provided as well as a 'calibration' and 'zero' condition. The flow 'off' mode allows the operation of the system with pressure and EKG data for over 24 hours without battery recharging. A command system, mounted in a vest on a dog, has been used successfully to obtain one minute of data every hour including advancing the flowmeter through its zero and calibration modes. Such an operation allows for many days of use without recharging. A similar vest mounted extracorporeal package can be used to charge the battery when required. The use of the command control system to activate internal switches for the flowmeter operation is illustrative and the switching circuits can be used for any generalized application that may be required. This example shows the high degree of flexibility possible with short range transcutaneous telemetry systems. Figure 2 illustrates a similar but smaller system. The flowmeter has been deleted as well as the recharageable batteries in order to allow the use of the system in small rhesus monkeys. Two pressures, EKG, and temperature can be monitored with this system and it utilizes a similar

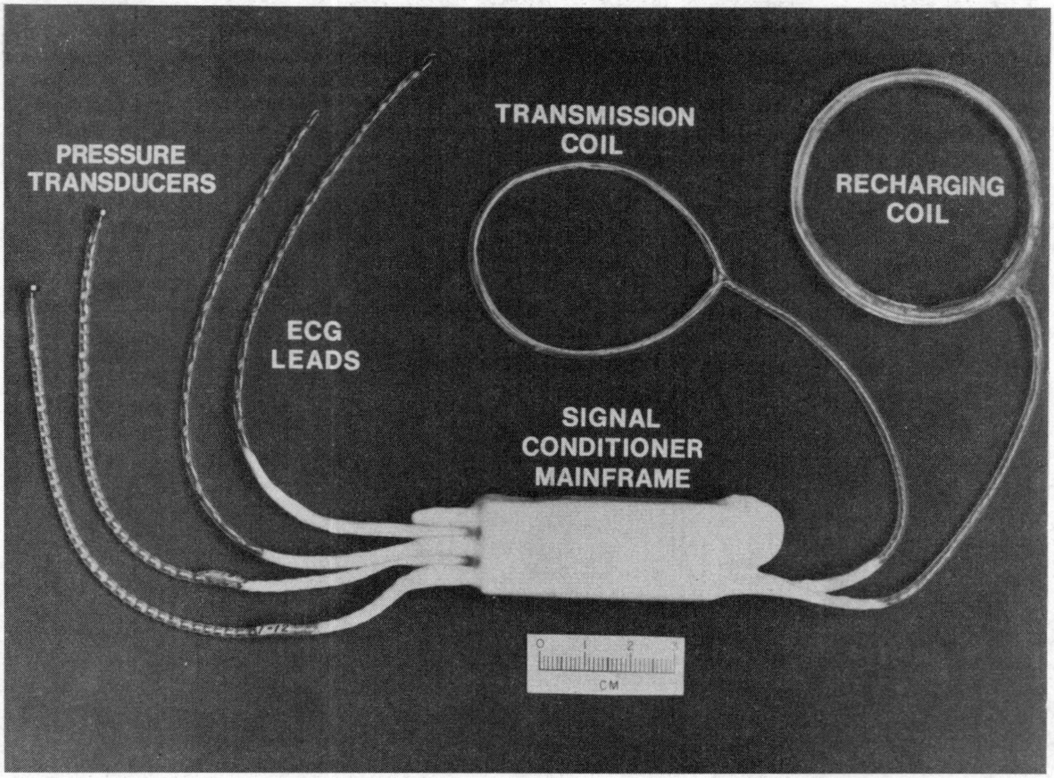

Fig. 2. Batteryless multichannel telemetry system for pressure and EKG measurements.

time multiplexed PWM encoded data transmission. Elimination of the rechargeable batteries means that the system is only activated when an external coil mounted in a vest is energized. No command control has been provided in this system except for the cessation of operation when the external coil is removed or de-energized. A control system similar to that of Fig. 1 could be incorporated if desired. By eliminating the batteries and using hybrid circuit construction a significantly smaller unit was possible. The carrier frequency for the data in this example is in the 100 MHz region.

Recently a transmitter operating at 8.5 MHz was designed for use in rats (Fryer, Lund and Williams, 1978). Since these animals are housed in very small cages, normally only a fraction of a cubic foot, no difficulty is observed in obtaining strong signals. The overall system is shown pictorially in Fig. 3. The transmitting coil because of the size of the animal must be made as small as possible. Either a $\frac{3}{4}$ in diameter air core coil or a $\frac{1}{4}$ in diameter × $\frac{1}{4}$ in long ferrite core has been used in the transmitter. The frequency and coil dimensions provide essentially only inductive radiation. Since the radiation field is polarized various orientations of the animal will give a null or zero output to the receiver. To overcome this a two receiver diversity arrangement is used, but since at this frequency a very simple receiver is possible, the use of two receivers is quite simple compared to the needs at higher frequencies. The receiver, because of the frequency selected, can be fabricated essentially from a single IC device and a dual unit is very simple. A superhet design is not used or needed in this case. A pulse interval type of data

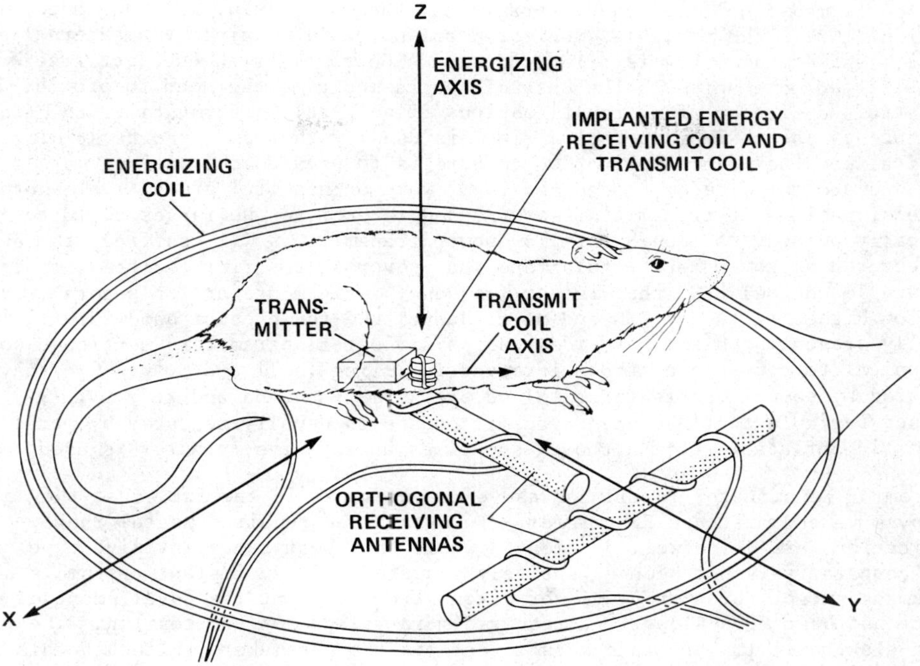

Fig. 3. An inductive powered telemetry system for use with small caged animals such as rats.

transmission is used and an average power consumption of 100 µW works satisfactorily. The implant is designed to measure temperature and EKG but activity can also be derived indirectly by measuring the variations in field strength.

The circuit is designed to operate from a 3 V battery (1 lithium or 2 silver oxide) or by inductively induced power. The low power consumption of 100 µW and the small cage size, 25 cm diameter, made it possible to surround the cage with a coil energized at 250 Khz and use a pick-up coil wound on the $\frac{1}{4}$ in diameter × $\frac{1}{4}$ in ferrite transmitter core as the energy source. Although the efficiency is low, a few watts of oscillator power at 250 KHz was sufficient to provide a 0.12 gauss field at the cage center, more than adequate to provide the needed 100 µW of power. The plane of the pick-up coil and the energizing coil are horizontal so as the animal turns around there is no change. If the animal stands then the coil is inclined but there is sufficient over powering in the favorable mode that the system will still operate with a 70-80° inclination, consequently the loss of signal time is low. Most inductive powering systems are designed for use where the energizing coil is fixed to the skin of the animal (Figs. 1 and 2) or human. This is a novel example where the animal is completely free and unrestrained within the cage, and a batteryless telemetry system is then able to transmit various physiological data. Rats are used in large quantities for laboratory drug studies and by eliminating the need to replace the battery the procedure in taking the transmitter out of one animal and putting it in another is greatly simplified. The system can easily be modified to measure other physiological parameters.

HUMAN APPLICATIONS

The use of telemetry with humans to date has primarily involved the measurement of such variables as EKG (Larsen *et al.*, 1972), EMG (Quanbury *et al.*, 1976; Winter and

Quanbury, 1975) and EEG (Dymond *et al.*, 1976) in both the clinical situation and for sports medicine (Fleischer, Zerzawy and Bachmann, 1974; Hull and Mote, 1974; Kimmich, 1972). The variables monitored are normally obtained via external electrodes and the use of telemetry is a logical extension to allow the subject free and unrestricted movement. In the case of sports medicine the need to provide unimpeded, unrestrained freedom of motion is obvious if physical performance is to be accurately monitored. The maximum operating range is usually that of a sports stadium. For clinical use the major application to date is to provide the patient with more comfort and greater freedom around the room, ward or hospital grounds. In both of these situations RF transmitters are needed to achieve the ranges of 10 to 500 m typically needed. Although very low power transmitters will suffice, these devices are covered by government regulations that severely restrict the freedom of design. For single channel EKG, the present frequency assignments are in general workable, but for higher frequency EMG and multichannel EEG the present bandwidth limits usually present problems. Besides the physiological parameters mentioned so far, mechanical forces such as those involved in skiing (Hull and Mote, 1974) have been measured by some investigators. Blood pressure (Bachmann and Zerzawy, 1972) is another variable that has been transmitted and is usually measured by means of an automatic inflatable cuff although sometimes an invasive catheter is used.

The common denominator in all the above measurements is the fact that they are usually noninvasive and must use RF telemetry to obtain the required useful range. The above applications are quite well documented while the short range invasive type of systems have comparatively not been extensively reported. As an implant system is not used on humans except for therapeutic or diagnostic treatment the development of this aspect has been much slower and required more validation and testing. The most widely known electronic implant in humans to date is the pacemaker. Although this is not a telemetry device but an electronic oscillator, it has many features in common with telemetric implants. The development of other implant devices that do require transcutaneous telemetry for optimum performance are in progress in many research laboratories at present.

Prosthesis such as nerve stimulators (Hallgren and Potter, 1972; James, 1968), bladder control (Bradley *et al.*, 1962; Timm and Bradley, 1973), pain suppression, (Burton and Maurer, 1974), artificial vision (Donaldson and Sayer, 1973), artificial hearing (Merzenich, Schindler and Sooy, 1974) and artificial pancreas (Bessman and Schultz, 1972) to name a few, are devices currently in various stages of development that will probably utilize telemetry links. Most of these devices primarily require a transmit 'in' control link. But often the addition of a feedback loop, with physiological data provides the most effective operation. This then would be a two-way telemetric link with features much like the animal systems of Figs. 1 and 2. These systems are only in the experimental stages at present because of their complexity and the need for safety. If the problems were electronic or of instrumentation complexity they no doubt could be rather easily solved with much of the sophisticated instrumentation including microprocessors that are currently available, but that is not the primary problem. It is more the complexity of the living organism that one is trying to replace and the many compatibility problems that occur when various materials are placed in contact with body tissue. For example external electrodes can be replaced when they dry out or otherwise become ineffective, but surgically implanted systems must be designed to function for many years. Although they do not dry out, the body normally walls off any foreign body with a membrane, or scar tissue, often impairing its intended function. Difficulties with miniature low power transducers are often another major impediment to progress. For instance in the case of an artificial pancreas there is the need for a transducer to measure the blood glucose level. Although such chemical type sensors have been reported (Bessman and Schultz, 1972, 1973) ones that are not either walled off by the body or saturated after a period of time are still not available. The problems of storing the insulin to be dispensed is a further problem. Although the proposed devices are usually preprogramed self-contained implants, the use of a transcutaneous telemetry feedback loop might be advantageous.

This is especially true of this example and many others where continuous operation is not necessary as in the case of a pacemaker. Even in the pacemaker application the pioneer devices (Cammilli, Pozzi and Pizzichi, 1964) used a transcutaneous induction coupling from an external oscillator whose rate could be controlled. Artificial vision and hearing are hindered by the difficulty of even approximating the complexity of the body and interfacing to the body's nervous system. Pain control and suppression is one area in which commercial units are currently being marketed. The complexity is not extreme since only a stimulating signal much like a pacemaker, to the nervous system is needed. Systems vary from preprogrammed total implants to inductively coupled transcutaneous ones as well as totally extracorporeal ones.

The development of various electronic implant prosthesis has been more extensively pursued to date than diagnostic devices. It is believed that implant diagnostic devices could also serve a very useful medical function in improving the treatment of various afflictions. This would allow the efficacy of various drugs or other therapeutic treatment to be more accurately measured and controlled, especially when external measurements are inadequate or 24 h measurements under different environmental and stress conditions are required. For instance a safe reliable blood pressure monitor that could be implanted could prove very useful in the improved treatment of some hypertension patients. A glucose sensor that could telemeter the data through the intact skin could be useful to diabetics in achieving a better and more uniform blood glucose level. Such would be similar to an artificial pancreas but less complex and would require a similar sensor but not the insulin dispenser. The use of myo signals transmitted via implant devices (Hirsch, Kaiser and Petersen, 1966) to operate artificial arm or leg prostheses have been experimentally pursued. The use of short range telemetry to monitor the engineering performance of an artificial hip prosthesis (Carlson, Mann and Harris, 1974) is another slightly different example. The possibilities in each specialty of medicine are numerous and varied.

Figure 4 shows an implant telemetry device (Fryer *et al.*, 1978) for monitoring intracranial pressure (ICP) in the case of patients with head trauma, cerebral edema, or other situations likely to produce abnormal intracranial pressure. The measurement of ICP is important so that the proper therapeutic treatment can be instituted for its control. The implant device contains a pressure transducer, electronic signal conditioner transmitter and energy pick-up coil. The pressure cell is a capacity

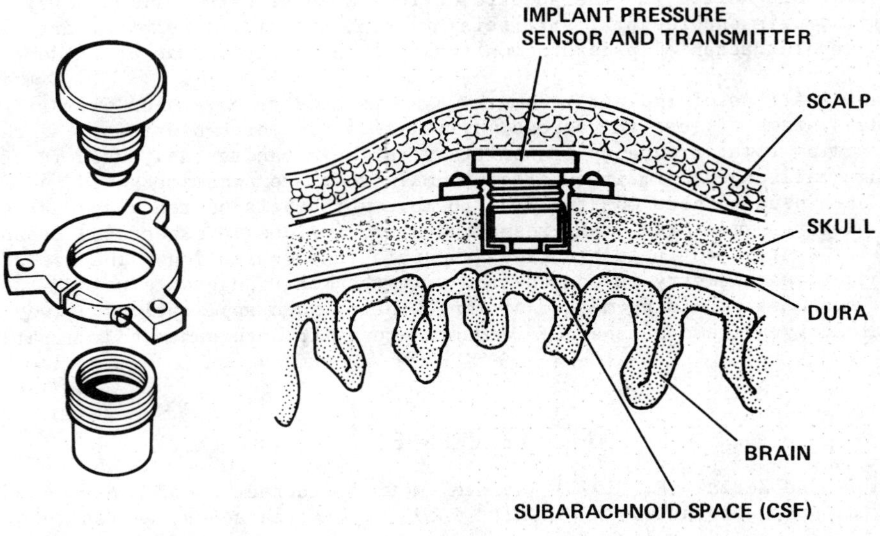

Fig. 4. Batteryless pressure telemetry system for intracranial measurements in clinical applications.

type of about 7 mm diameter and 2 mm thickness. A thin diaphragm spaced approximately 0.025 mm from a fixed plate results in a capacity of approximately 10 pf and a ΔC of 1 pf for 50 mm Hg pressure change. The pressure cell forms the capacitive element of the LC tank circuit of an oscillator operating in the region of 10 MHz. Because of the minute capacitance involved it is necessary to provide a complete metal shield around the transducer and circuitry and to provide a buffer amplifier between the oscillator and the transmitter radiating coil. The oscillator frequency changes as a function of pressure with a change of about 500 KHz for 50 mm Hg. Two small coils of about $\frac{3}{4}$ in diameter are used, one to pick up energy from a 250 KHz induction field and the other to radiate the 10 MHz region variable oscillator frequency. As seen in Fig. 4 the device is surgically placed in a 16 mm burr hole and the required transmission and energy pick-up distance is only skin depth. An operating distance of 2 cm can easily be provided at low power to allow for normal depth variations and horizontal alignment errors. The implant device is a completely inert, devoid of batteries, hermetically sealed unit that is activated by an induction field in the 250 Khz region and will then radiate out a variable frequency that is proportional to the surrounding pressure. Because there are no batteries or other components with a limited shelf life, the device can be left in the body for years and activated at any time on command. Devices of a similar nature to measure other parameters can certainly be devised.

CONCLUSION

Very short range, transcutaneous, induction coupled telemetry systems for use in both human and animal applications have been described. These systems contrast with RF coupled systems that will operate over a much greater distance but which have many limitations because of government regulations and controls. Total implant systems for physiological data acquisition and internal command control are most effectively operated with short range inductive coupling. In human applications such systems will broaden the role of electronics implanted within the body from the well known pacemaker to a vast range of electronic devices to supplement and supplant body functions that have weakened or failed. Telemetry transmission both 'in' and 'out', in much the same manner as 'up' and 'down' telemetry links are used in space satellites, can be used effectively with implant devices to expand their usefulness and scope. Both diagnostic and therapeutic devices such as nerve stimulators, bladder controllers, pain suppressors, artificial vision, artificial hearing, artificial pancreas and intracranial pressure monitors are currently in use or in development.

Animal applications of implant telemetry systems to date have been primarily data gathering systems for use in various physiological research programs. As described in this review lower frequency transmitters can be advantageously used in studies involving small caged animals. Larger animals that are sometimes wild and free ranging have usually been monitored as to movement habits or tracking, using external modest power transmitters. Many researchers are anxious to expand the capability of tracking systems to include various physiological monitoring. The use of short range inductive telemetry in conjunction with the long range external RF transmitters provides the most promising method of implementing such applications. Two stage relay telemetry systems can be used advantageously in both human and animal applications.

REFERENCES

Bachmann K. and Zerzawy R. (1972) Radiotelemetry of direct blood pressure measurements in aorta, pulmonary artery and heart. In H. P. Kimmich and J. A. Vos (eds.), *Biotelemetry I*, pp. 183-195. Leiden: Meander N.V.

Bessman S. P. and Schultz R. D. (1973) Prototype glucose-oxygen sensor for the artificial pancreas. *Trans. Am. Soc. Artif. Int. Organs* XLX, 361-364.

Bessman S. P. and Schultz R. D. (1972) Sugar electrode sensor for the 'artificial pancreas.' *Horm. Metab. Res.* 4, 413-417.

Bradley W. E., Wittmers L. E., Chou S. N. and French L. A. (1962) Use of radio transmitter receiver unit for treatment of neurogenic bladder. *J. Neurosurg.* 19, 782.

Burton C. and Maurer D. D. (1974) Pain suppression by transcutaneous electronic stimulation. *IEEE Trans. Biomed. Engng. BME* 21, 81-88.

Cammilli L., Pozzi R. and Pizzichi G. (1964) Radio frequency pacemaker with receiver coil implanted on the heart. *Ann. N.Y. Acad. Sci.* 111, 1007-1029.

Carlson C. E., Mann R. W. and Harris W. H. (1974) A radio telemetry device for monitoring cartilage surface pressures in the human hip. *IEEE Trans. Biomed. Engng. BME* 21, 257-264.

Donaldson P. E. K. and Sayer S. (1973) Toward a high definition visual prothesis. Conf. on Hybrid Microelectronics. *IERE* No. 27, Canterbury, pp. 261-262.

Dymond A. M., Coger R. W., Martin J. F. and Serafetinides E. A. (1976) Multichannel biotelemetry of electrophysiological parameters from unrestrained psychiatric patients. In T. B. Fryer, H. A. Miller and H. Sandler (eds.), *Biotelemetry III*, pp. 157-160. New York: Academic Press.

Fleischer H., Zerzawy R. and Bachmann K. (1974) Combined telemetry of cardiovascular parameters in sports. In P. A. Neukcmm (ed.), *Biotelemetry II*, pp. 115-117. Basel: Karger.

Fryer T. B., Deboo G. J. and Winget C. M. (1966) Miniature long-life temperature telemetry system. *J. Appl. Physiol.* 21, 295-298.

Fryer T. B., Lund G. F. and Williams B. A. (1978) An inductively powered telemetry system for temperature, EKG, and activity monitoring. *Biotelemetry and Patient Monitoring* 5-2, 53-76.

Fryer T. B., Sandler H., Freund W. R., McCutcheon E. P. and Carlson E. L. (1975) A multichannel implantable telemetry system for flow, pressure and ECG measurements. *J. Appl. Physiol.* 39, 318-326.

Fryer T. B., Corbin S. D., Silverberg G. D., Schmidt E. V. and Ream A. K. (1978) Telemetry of intracranial pressure. *Biotelemetry and Patient Monitoring* 5-2, 88-112.

Hallgren R. C. and Potter A. G. (1972) Contactless nerve stimulating transducer. *IEEE Trans. Biomed Engng. BME* 19, 316-317.

Hirsch C., Kaiser E. and Petersen I. (1966) Telemetry of myo-potentials. A preliminary report of telemetering of myo-potentials from implanted microcircuits for servo control of powered prostheses. *Acta orthop. scand.* 37, 156-165.

Hull M. L. and Mote C. D. (1974) Pulse code modulation telemetry in ski injury research. *Biotelemetry* 1, 182-192.

James E. D. (1968) Equipment and methcd involved in the treatment of urinary incontinence by electrical stimulation. *Med. Biol. Engng.* 6, 595-602.

Kimmich H. P. (1972) Oxygen consumption measurements during exercise by means of radiotelemetry. In H. P. Kimmich and J. A. Vos (eds.), *Biotelemetry I*, pp. 130-136. Leiden: Meander N. V.

Larsen J. L., Dillman R. F., Nardizzi A. M. and Tverdoch R. N. (1972) An effective ECG telemetry system. *Hewlett-Packard J.* 23, 2-9.

Mackay R. S. (1970) *Biomedical Telemetry*. New York: Wiley.

Merzenich M. M., Schindler R. A. and Sooy F. A. (eds.) (1974) Proc. 1st Int. Conf. on Electrical Stimulation of the Accustic Nerve as a Treatment for Profound Sensorneural Deafness in Man. University cf California, San Francisco 1973, VeloBind.

Quanbury A. O, Foley C. D., Winter D. A., Letts R. M. and Steinke T. (1976) Clinical telemetry of EMG and temporal information during gait. *Biotelemetry* 3, 129-137.

Timm G. W. and Bradley W. E. (1973) Technologic and biologic considerations in neurourologic prosthesis development. *IEEE Trans. Biomed. Engng. BME* 20, 208-212.

Winter D. A. and Quanbury A. O. (1975) Multichannel biotelemetry systems for use in EMG studies, particularly in locomotion. *Am. J. Phys. Med.* 54, 142-147.

Physiology of Large Mammals by Implanted Radio Capsules

G. E. FOLK, JR. and M. A. FOLK

Department of Physiology and Biophysics, The University of Iowa, Iowa City, Iowa 52242, U.S.A.

Abstract — Our experience with implant telemetry of a few simple physiological measurements (body temperature, heart rate, and ECG) was aimed at not only recording interesting physiological data but also understanding the daily behavior of the animal. About 30 species of mammals were radio tagged, some weighing 400 kg. First it was necessary to realize how valuable simple heart rates can be, if recorded every 30 min; it is not necessary to know the stroke volume of the heart at the same time. Body temperature and heart rate data have to be used appropriately, however, with full knowledge of the day-night physiological setting. Examples will be given of this setting as seen in new data from wolverines and from American opossums. Our technical changes and progress over many years will also be reported.

INTRODUCTION

Our reviews of implant radio telemetry published in 1964 (Folk) and 1972 (Folk and Copping) described typical results and listed plausible and important experiments for the future. We have not found many papers on these key experiments in recent years. In addition to describing necessary experiments of the future, our review emphasized: (1) when to use physiological radio telemetry, (2) the arguments for intraperitoneal implantation and its history, (3) the design of a successful radio capsule, (4) what functions are the most profitable to measure, (5) what is the effect upon the animal while carrying the implanted capsule, and (6) some typical results. This information was obtained from Iowa radio capsules implanted in about 30 species of mammals, including three species of bears weighing up to 400 kg (Folk, Folk and Minor, 1972; Folk, Larson and Folk, 1976). This radio capsule is a long-life, short range type which has provided physiological data from mammals for periods up to 1.5 years (Shook and Folk, 1965); it is especially intended for animals weighing one kg or more.

In the present paper we will review what can be learned from daily heart rate patterns if recorded every 30 min, describe the need for understanding the day-night physiological setting when interpreting telemetry data, and discuss modifications of our implant technique. Future critical experiments will be described.

HEART RATE AND DAILY BEHAVIOR RECORDING

The capsules are designed to transmit heart rates, ECG, and abdominal body temperature. First will be provided a list of what can be determined from the heart rate capsule alone. In our earlier review (1972) illustrations of each of the following types of experiments are provided. One may: (1) determine whether the mammal has a nocturnal or day-active behavior pattern, (2) record basal resting heart rates of unrestrained and unexcited mammals, (3) measure the 24 h (or nearly 24 h) daily rhythm of activity and physiological function, (4) measure the accuracy of 'physiological clocks', (5) use the pattern of cardiac activity to represent energy expenditure over the 24 h period, (6) study cardiac physiology: which species regularly doubles or triples its resting heart rate due to spontaneous activity like feeding? What is the relation between heart rate and size of an animal? How do the intervals of the ECG spikes compare between groups of hibernating and non-hibernating mammals? What is the pattern of the work of the heart each day; which 'hearts' frequently return to baseline both day and night while other 'hearts' show two distinct baseline settings at two times of the day? What is the cardiac activity of animals in hibernation in deep burrows, or in extreme environments such as deep snow in the forest, or in darkness?

Some of the experiments described above can be done by measuring body temperature. It is clear, however, that changes in behavior of the animal are quickly evident in heart rate data while there is more lag in changes in the body temperature of the animal. Often it is desirable to measure both of these functions but if only one must be selected, more can be learned from heart rates. Furthermore, there has been a tendency to attempt a few elaborate experiments by telemetry with the result that many basic experiments are missing. An example of valuable contributions from heart rate measurements alone are found in the papers on growth and bradycardia of deer by Jacobsen (1979a, 1979b). A classic hibernation study has been done in the field over the entire winter season by the telemetry of body temperature, showing that the torpor season provides an energy saving of 87.8 percent (Wang, 1978). More detailed information could have been collected with a heart rate capsule in that study. There has been some reluctance to measure heart rate alone because of the impression that such figures must be combined with simultaneous measurements of stroke volume and possibly blood pressure. It now seems reasonable to interpret patterns of animal activity based upon heart rate alone because of the work of Rushmer (1970). He showed that normally there is little change in stroke volume even in spite of exercise in man and in those other mammals which have been studied. This has been confirmed in the experiments of Franklin and Van Citters (1963) with free ranging baboons in Africa: "The cardiac response was as in dogs; startle reactions, postural changes and onset of exercise were accompanied by increased heart rate and reduced duration of ejection. Stroke volume remained essentially constant in spite of wide variations in cardiac output and heart rate." Even under reversed circumstances (reduced heart rate) such as in diving bradycardia (Elsner *et al.*, 1964) stroke volume does not appear to change. Therefore it seems reasonable to emphasize in this paper the value of heart rate capsules in elucidating many biological phenomena.

INTERPRETATION OF PHYSIOLOGICAL TELEMETRY DATA

When daily patterns of heart rate or body temperature are recorded, they are much more easily interpreted if one visualizes the remarkable changing physiological condition of the person or the animal over the 24 h period. This is not a phenomenon related to sleep. It probably does not occur with all species of mammals but a definitive answer to this question must be determined in the future. Let us suppose that we are asked to select the lowest basal values for heart rate or body temperature. We might make the mistake of assuming that just because the animal is sleeping, this will represent the basal resting state. We found with human subjects that there is a 27 percent difference in quiet resting heart rates between 10 a.m. and 10 p.m., after food intake, body position, activity, and emotion had been controlled. This physiological day-night setting is associated with the individual physiological circadian rhythm, not with sleep (Folk, 1974, p. 76). We could not study isolated organs of human subjects but we could study organs from rodents, namely, cultured adrenal glands (six days) and perfused hearts (three days). For days every 12 h these organs would change their physiological basal activity by as much as 60 percent above or below the mean (Folk, 1974, p. 66). The same observation was made with resting cultured heart cells. Thus, when you examine the pattern of heart rate or body temperature collected from most mammals, you can easily find two classes of resting or sleeping heart rates or body temperatures, each at two different times of the day, one representing a high physiological setting and one representing a low one. This principle is now illustrated from recent work on two species of mammals.

DAY-NIGHT PHYSIOLOGICAL SETTINGS RECORDED FROM WOLVERINES AND OPPOSSUMS

Recordings of body temperatures of wolverines by radio capsule showed a sinusoidal curve over 24 h with a valley of approximately 38°C and a peak of nearly 40°C; yet the animal might sleep at a high setting of 39°C or at a low setting of 38°C (Folk, Folk and Craighead, 1977). These results emphasize that behavioral observations in such experiments are necessary to correlate with the physiological data. It should be emphasized that activity is merely superimposed upon these settings. A setting can only be revealed when feeding, activity, excitement, and a changing light cycle has been removed as an influence. Similar results were obtained by recording the body temperatures of American opossums; these animals could be sleeping at a body temperature of 33°C or at 34°C (Treagust *et al.*, 1979). In other words, it is necessary to define in terms of the day-night setting *which* resting body temperature is being reported. It is possible, however, that some mammals do not have this day-night setting; heart rates of the arctic fox appear to return to a base line approximately every 30 min both day and night, so that its basal resting condition does not vary with the time of day (Folk, 1964b).

THE IOWA IMPLANT PROCEDURE

In our earlier review in 1972, we described in detail the advantages of using implant telemetry. Before 1959 we had used harnesses outside the animal to carry the radio capsule, but in 1959 learned that the peritoneal cavity was a superior place to carry a small electronic instrument. For many years after that we sutured the transmitters (which weighed from 10 to 40 g) at the midline of the interior of the peritoneal cavity. We did this to have a standardized position for recording the ECG. The midline incision was always made because cutting through the linear alba results usually in a bloodless field. More recently we have been placing the capsule to one side of the midline for the following reason: although we have done several hundred operations, placing the capsule in the midline incision position without any harmful effects, on three occasions the capsule did work its way, over a two-month period, from the abdominal cavity through the abdominal wall and out through the skin. This happened with three different species. It is of interest that there was no evidence of a patent fistula at a time of finding the transmitters; the tissues had closed and healed from inside the animal to the outside as the object progressed. After the third occasion, we decided to place the capsule (for consistency's sake) always on the right-hand side of the midline by about two centimeters. Except for the three cases described, there have been no difficulties with the peritoneal implantation; about 50 percent of the capsules have had a tough sac of connective tissue form around them, but surprisingly this has never interfered with obtaining a good ECG record.

THE MIDLINE CLOSE-POSITIONED ELECTRODES

There are some advantages in using ECG electrodes which are one to two centimeters apart with the implanted radio capsule between them; such electrodes touch the ventral peritoneal cavity near the midline. We have not sutured the electrode itself to the ventral wall, just the radio capsule. Some investigators wish to have a long insulated electrode which passes ventrally from the capsule out of the peritoneal cavity near to the base of the tail under the skin, with the other electrode passing under the skin to the pectoral girdle region. Another method is to pass an electrode to each end of the sternum. Our objections to this are that much more surgical operating is required, and such electrodes must be an irritant to the animal. Especially with animals that curl in a ball when they sleep. It is a faster and simpler operation to have the two electrodes as an organic part of the radio capsule. We have found little attention in the literature given to this question of how much the radio device irritates the animals carrying the capsules. It is appropriate to give close attention to this question now.

As we stated in an earlier paper, under natural conditions the ventral abdominal wall, especially in the female, must be of sturdy design since in some cases an animal weighing 300 g will carry a litter of young weighing 67 g. Today such an animal is asked to carry a transmitter in the peritoneal cavity which weighs only approximately 10 g, at the most. We have been impressed with the apparent disregard by the animal of the radio capsule after surgery. We have seen a beagle dog, the day after the implanting of a glass and stainless steel capsule weighing 20 g, leap up repeatedly against the attendant, spin around on the leash, and gaily walk or trot one mile outdoors without evident fatigue or discomfort. A large rodent carrying an unnecessarily heavy transmitter will climb up to a high shelf, leap from the high shelf to the floor, and will perform its usual running on an exercise wheel. We have never observed an operated animal to scratch, worry, or give any attention to the incision on the ventral abdominal wall. A series of animals were bred while carrying transmitters weighing 80 g, and several litters were nursed to weaning by these instrumented mothers. The above remarks apply to close-positioned electrodes on the implanted radio capsule.

Perhaps more attention should be given to the possible irritating effect of transmitters and electrodes upon instrumented animals. For example, the whale 'Gigi' had a transmitter sutured to the flesh above the backbone for tracking, and walruses for several days were tracked by a 'limpet' device with fixed hooks attaching a package to the skin behind the skull above the neck (Sebester and Lund, 1979). We do not know whether the transmitters brought about longer migrations due to the driving effect of an irritating stimulus; perhaps when possible an abdominal implant would be preferable. It has already been determined that tracking capsules can be designed for transport in the peritoneal cavity (Shirer and Downhower, 1968).

Although we feel that there are advantages in not running wires around the animal to serve as ECG electrodes, it now needs to be established as to the type of record obtainable with electrodes very close together. Such records are surprisingly similar to records obtained from electrodes on the limbs. Most of our records look like those from the human Lead II, or a record from the exploring electrode. In other words, the ECG electrodes from an implanted capsule do not have to be positioned one on either side of the heart. Since this is not generally known, it occurred to us to study this further by using skin electrodes on human subjects. Three illustrations are provided of three human subjects, (1) a normal male (Fig. 1), (2) a normal female (Fig. 2), and the third a patient demonstrating atrial flutter (Fig. 3). In each case, a control record is shown with electrodes on the right wrist and the left ankle. Except in one case, the gain of the instrument was not altered and the electrodes were placed on the abdominal skin to simulate where the electrodes are found in the Iowa implant procedure. In many cases the craniad electrode was placed on the xiphoid process which is situated well caudad to the heart.

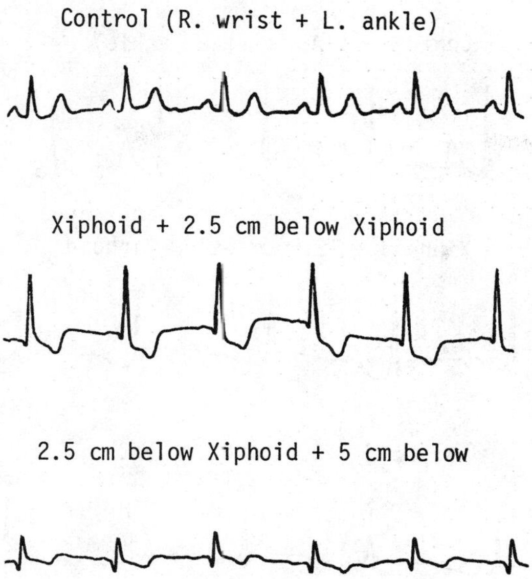

Fig. 1. ECG record from a human subject using midline electrodes on abdominal skin to show that electrodes need not be on two sides of the heart and can be in close proximity. The first record is a normal control. The last was recorded from opposite sides of the navel. The gain used for the control record was not changed.

In two cases, both electrodes were well below the xiphoid process on the abdominal skin and only 2.5 cm apart (Figs. 1 and 3). These abdominal recordings are similar to the control readings and in most cases, intervals between the ECG spikes could be measured. In one case, the gain was turned up (Fig. 2) to show that a record even more like the control can be obtained. In summary, our experience has shown that the electrodes do not need to be distant from the radio capsule in the peritoneal cavity, and the electrodes do not need to be positioned on both sides of the heart, at least for obtaining certain types of ECG recordings.

EXAMPLES OF MIDLINE ECG RECORDING

One of the more interesting advantages of ECG recording from wild animals by radio capsules is the study of the relative positions of the three spikes recorded with each heart beat (P, R and T waves). We have demonstrated, as have earlier individuals (Dawe and Morrison, 1955) that the second and third spikes are very close together in hibernator mammals when the animals are active and not hibernating (Fig. 4). The second spike (R wave) is easily recognized as being the one with greatest amplitude (Folk, Hunt and Folk, 1979). The principle of this observation is that the third spike (T wave) represents the relaxation of the heart; therefore hibernators have a rapid relaxation time. There has been considerable debate as to whether the winter dormancy of bears can be called hibernation. By using the ECG records we found evidence that, from the standpoint of cardiac physiology, the bears are closer to hibernators than they are to non-hibernators. This is demonstrated by the relatively long interval between the second and third spikes

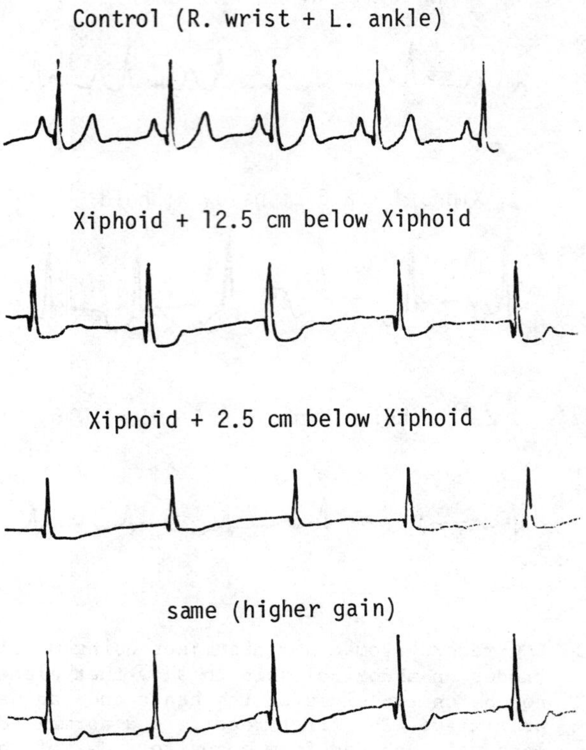

Fig. 2. ECG records from a human subject using midline electrodes on abdominal skin. The last record was made with the gain used for the control record increased slightly.

(R-T interval) of each heart beat in the macaque monkey and the dog, compared to the relatively short interval found in the three species of bears (Figs. 5 and 6).

FUTURE APPLICATIONS OF TELEMETRY

In order to review the field of physiological telemetry, one must try to diplomatically rank the importance of the types of comparative physiology being done. Professor Ladd Prosser once stated that Comparative Physiology must not become a process of simply applying old techniques to a new or different species. Radio telemetry should not be used in this fashion, because there is far more important work to be done. Furthermore, research is not necessarily important just because it is difficult.

There appear to be several classes of work being done; first, there is an abundance of fundamental theroetical work which is not comparative. This is sometimes done because it is easier to do difficult, modern physiology on 40 white rats (carefully standardized) than it is to do simple experiments on bears and eagles. Secondly, there is an abundant class of challenging, interesting, publishable work which is not aimed directly at a principle of biology. Such work describes new physiological adaptations. For example, there are many papers which record simultaneous measurements on a new species of animal, with recordings of body temperature, oxygen

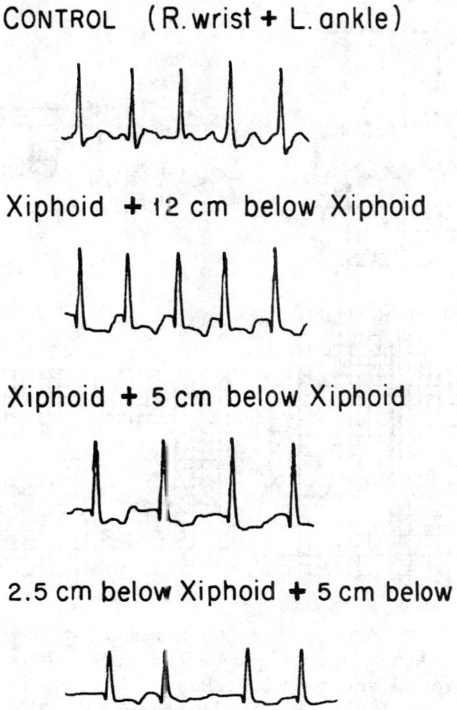

Fig. 3. ECG records from a patient with atrial flutter, using midline electrodes placed on abdomen. The last one was recorded from opposite sides of the navel. The gain used for the control record was not changed.

consumption, evaporative water loss, water balance, and heart rate. Some of this work is being done by radio telemetry. To many biologists, these papers can be enjoyed like reading a novel, even at midnight, but the question is whether they are quite as important as our third class. We find a dearth of the third class of publication which is aimed directly at contributing to an important biological principle; in many cases comparative, whole-animal mammalian physiology is the key to this type of contribution. Two illustrations will suffice: (1) The estimation of the energy consumption of animals in the field by recording their heart rate by radio telemetry makes an important contribution to not only biological science but to the solving of some of the problems of humanity today, because of the food shortages around the world. Unfortunately, illustrations of this kind of work are rare: Two of the first contributions were by Morhardt and Morhardt (1971), and by Lund and Folk (1976). Since then, the work of Moen (1978) and of Wooley and Owen (1978) stand out; they calculated by use of telemetry the energy costs for free ranging animals varying in size from ducks up to white-tailed deer. It would seem that we need more funds and more teams of biolcgists and electrical engineers focusing on such problems.

My second example is the lack of information concerned with the hibernation of bears. Only one worker has recorded the low heart rate of these animals in hibernation. This is a very conspicuous and important biological event; the sleeping heart rate of many bears in summer is a regular 50 bpm (Folk, Berberich and Sanders, 1973).

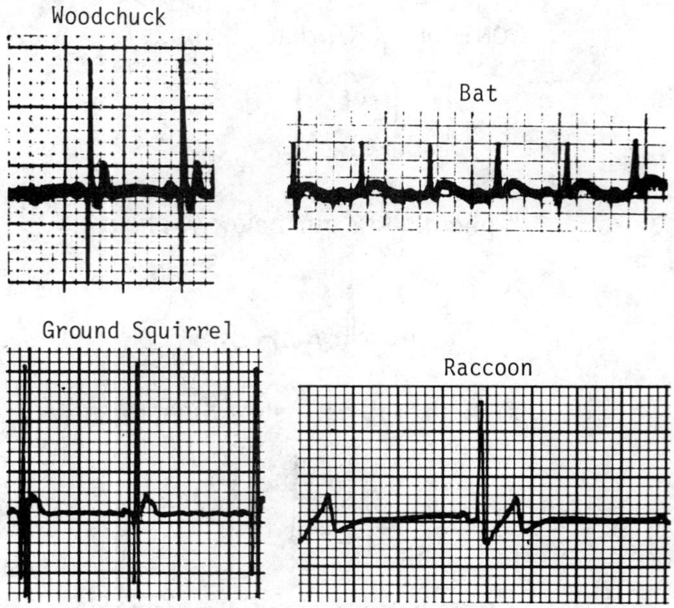

Fig. 4. ECG records of active hibernator mammals by implanted midline radio capsule (except the bat) to show that the second and third spikes of each heart beat (R wave and T wave) are in close proximity. The species were, from left to right: *Marmota monax*, *Myotis lucifugus*, *Citellus tridecemlineatus*, and *Procyon lotor*.

During hibernation the heart rate becomes slower and slower until it is as low as 8 bpm. This is so important that confirmatory studies should be made on all species of bears. Why are bears particularly important? This appears to be the only type of animal that can remain in hibernation for as long as six or seven months without feeding, drinking, urinating, or defecating. Again, only a few workers have shown categorically with a few individual bears, that this seems to be the rule. There are many medical implications to this. Furthermore, these animals appear to remain quiescent during this period of time. Such a possibility relates, of course, to understanding bed rest in human subjects and the problems of the astronauts who appear to have metabolic difficulties inspite of attempting to take exercise in their capsules. We have only the hearsay evidence from a few observers who noted bears during part of their hibernation period; they state that the bears appear to remain immobile most of the winter. Even the possibility of calcium leaching out inappropriately from the bones into the circulation of these animals during hibernation is of fascination to medical scientists (Nelson *et al.*, 1973). There should be teams of engineers and biologists using motion sensitive transmitters on the heads of these animals during the long hibernation period to determine definitively the length of this spectacular period of bed rest. The interest shown by medical scientists in this hibernation of bears implies that a principle of bio-medical science is at stake, not simply a listing of new adaptations. Another important point about this form of hibernation is that these animals at will can double their weight by adding adipose tissue in the fall, and under their own physiological control can remove this excess weight at the end of the hibernating season (Nelson *et al.*, 1973).

The question now arises why some important work using telemetry has not been applied

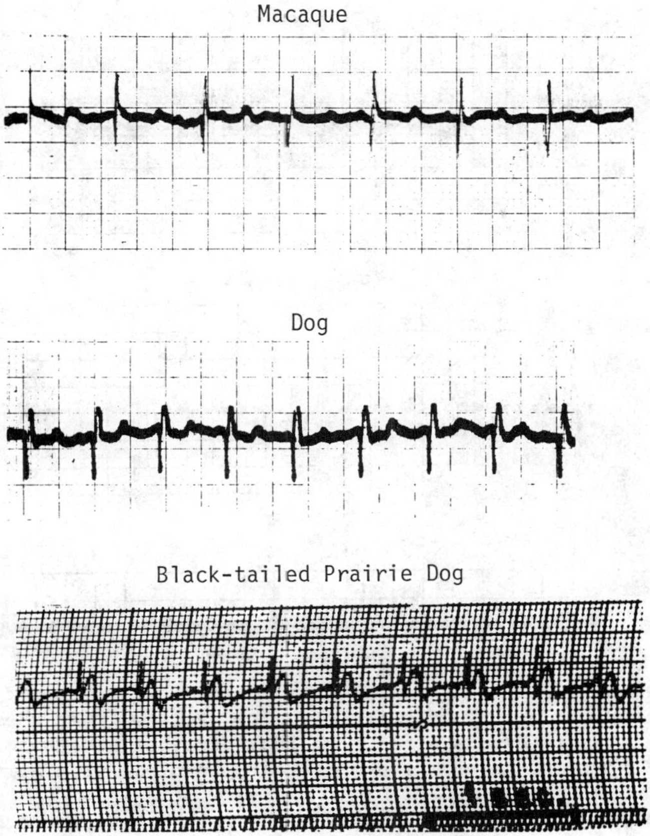

Fig. 5. ECG records made by implanted midline radio capsules from non-hibernators (macaque monkey and dog) to show the long interval between the second and third spikes of each heart beat (R wave and T wave). The third record shows the close proximity of the second and third spikes in an active hibernator, the prairie dog (*Cynomys ludovicianus*).

to these principles and to other similar needs. Many physiologists and administrators are misled by their belief that information obtained from bears and wild-caught animals is only of interest to the naturalist. Actually much of the basic work in understanding the physiology of hibernating bears has been accomplished with a colony of black bears at the Mayo Clinic. From these results, there already have been applications on how to manage patients whose kidneys have failed. Perhaps there is also some conflict between the opinions on the value of working with caged animals, compared with free animals. Tester (1978) and Amlaner (1978) have correctly made the point that physiological data from animals in enclosures are often very different from those obtained from the same species when free ranging. Yet, some important and successful comparative physiology can be measured on caged animals. For example, the depression in heart rate and the depression in metabolism (Hock, 1960) recorded from caged bears in hibernation are an important first step, and probably when these measurements are made in free living bears, the depression will be more marked rather than less marked.

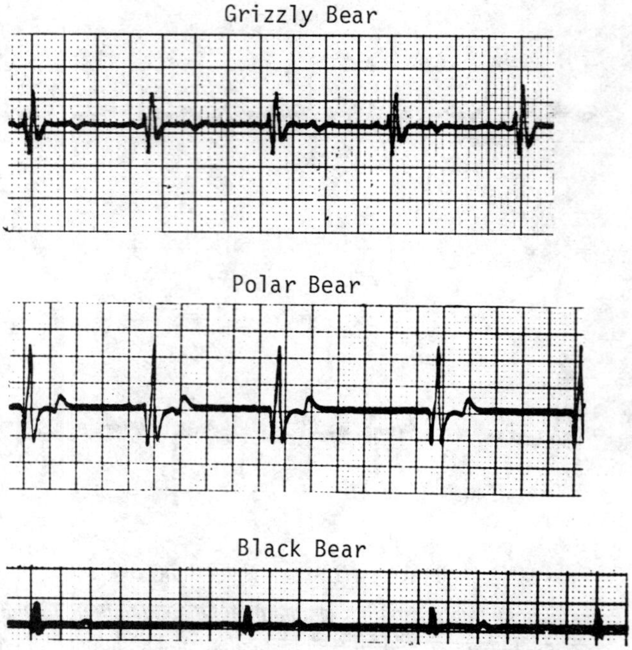

Fig. 6. ECG records made by implanted midline radio capsules from three species of bears, to show that their ECG pattern is similar to that of other hibernators. The second and third spikes are in close proximity.

The trained biologists are there; they are accomplishing interesting and complicated physiology. Either they have not: (a) learned about, or have not been taught, those special areas which are crying for attention, or (b) they have been discouraged by the radio telemetry equipment which has been designed but not developed, or (c) they have been unable to afford the cost of radio telemetry equipment. If we look out over the panorama of the future physiological research which is to be assisted by radio telemetry, let us hope we will see many examples of team research directed toward those biological principles which we are just beginning to understand and use.

Acknowledgements — The original work reported here was supported by a contract with The Office of Naval Research.

REFERENCES

Amlaner C. J. (1978) Biotelemetry from free ranging animals. In B. Stonehouse (ed.), *Animal Marking: Recognition Marking of Animals in Research*, pp. 205-228. London: Macmillan.

Dawe A. R. and Morrison P. R. (1955) Characteristics of the hibernating heart. *Am. Heart J.* 49, 367-384.

Elsner R. W., Franklin D., Van Citters R. and Watson N. (1964) Circulatory adaptations in diving mammals. 15th Alaskan Sci. Conf. Alaska Div. AAAS. Abstract No. 30.

Folk G. E., Jr. (1964a) The problem of electrodes for use with electrocardiogram radio capsules. In Murry and Salisbury (eds.), *Biomedical Sciences Instrumentation* 2, pp. 235-265. New York: Plenum Press.

Folk G. E., Jr. (1964b) Daily physiological rhythms of carnivores exposed to extreme changes in Arctic daylight. *Fedn. Proc.* 23, 1221-1228.

Folk G. E., Jr. (1974) *Textbook of Environmental Physiology*, 2nd ed., Philadelphia: Lea and Febiger.

Folk G. E., Jr., Berberich J. J. and Sanders E. E. (1973) Bradycardia in the polar bear. *Arctic* 26, 78-79.

Folk G. E., Jr. and Copping J. R. (1972) Telemetry in animal biometeorology. *Int. J. Biometeorol.* 16, Suppl. 5: Part II, 153-170.

Folk G. E., Jr., Folk M. A. and Craighead D. (1977) A comparison of body temperatures of least weasels and wolverines. *Comp. Biochem. Physiol.* 58A, 229-234.

Folk G. E., Jr., Folk M. A. and Minor J. G. (1972) Physiological condition of three species of bears in winter dens, pp. 107-125. S. Herrero (ed.), Pub. series of IUCN, Morges, No. 23.

Folk G. E., Jr., Hunt J. M. and Folk M. A. (1979) Further evidence for hibernation of bears. C. J. Martinka (ed.), Pub. series of IUCN, Morges, No. 42.

Folk G. E., Jr., Larson A. and Folk M. A. (1976) Physiology of hibernating bears, pp. 373-381. M. Pelton *et al.* (eds.), Pub. series of IUCN, Morges, No. 40.

Franklin D. L. and Van Citters R. L. (1963) Stroke volume changes in baboons during spontaneous activity. *Am. J. Cardiol.* 12, 816-821.

Hock R. J. (1960) Seasonal variations in physiologic functions of Arctic ground squirrels and black bears. *Bull. Mus. Comp. Zool.* 124, 155-173.

Jacobsen N. K. (1979a) Changes in heart rate with growth and activity of white-tailed deer fawns (*Odocoileus virginianus*). *Comp. Biochem. Physiol.* (in press).

Jacobsen N. K. (1979b) Alarm bradycardia in white-tailed deer fawns (*Odocoileus virginianus*). *J. Mammal.* (in press).

Lund G. F. and Folk G. E., Jr. (1976) Simultaneous measurements of heart rates and oxygen consumption in black-tailed prairie dogs (*Cynomys ludovicianus*). *Comp. Biochem. Physiol.* 55A, 201-206.

Moen A. N. (1978) Seasonal changes in heart rates, activity, metabolism, and forage intake of white-tailed deer. *J. Wildl. Mngmt.* 42, 715-738.

Morhardt J. E. and Morhardt S. S. (1971) Correlations between heart rate and oxygen consumption in rodents. *Am. J. Physiol.* 221, 1580-1586.

Nelson R. A., Wahner H. W., Jones J. D., Ellefson R. D. and Zollman P. E. (1973) Metabolism of bears before, during, and after winter sleep. *Am. J. Physiol.* 224, 491-496.

Rushmer R. F. (1970) Cardiovascular telemetry implants. *Telemetry J.* 6, 15-20.

Sebester P. D. and Lund G. F. (1979) Overview of NASA wildlife sensing projects. Pecora IV Symp. on Applications of Remote Sensing Data to Wildlife Management. *The Wildl. Soc.* (in press).

Shirer H. W. and Downhower J. P. (1968) Radio tracking of yellow bellied marmots. *Trans. Kansas Acad. Sci.* 71, 463-480.

Shook G. L. and Folk G. E., Jr. (1965) Body moisture and the operating life of implantable heart rate transmitters. *Trans. Biomed. Engng.* 12, 44-46.

Tester J. R. (1978) Analysis of circadian rhythms of free-ranging mammals. *Proc. 4th Int. Symp. Biotelemetry*, p. 11, Garmisch Partenkirchen.

Treagust D. F., Folk G. E., Jr., Randall W. and Folk M. A. (1979) The circadian rhythm of body temperature of unrestrained opossums (*Didelphis virginiana*). *Thermal Biol.* (in press).

Wang L. C. H. (1978) Energetic and field aspects of mammalian torpor: The Richardson's ground squirrel. In L. C. Wang and J. W. Hudson (eds.), *Strategies in Cold.* New York: Academic Press. 715 p.

Wooley J. B. and Owen R. B. (1978) Energy costs of activity and daily energy expenditure in the black duck. *J. Wildl. Mngmt.* 42, 739-745.

Physiological Radio Telemetry of Vertebrates

E. N. SMITH

Northeastern State University, Division of Natural Science and Mathematics, Tahlequah, Oklahoma 74464, U.S.A.

Abstract — Recent advances in the technology of multichannel radio telemetry have made possible the simultaneous measurement of several physiological parameters from free ranging or captive animals. Data obtained under natural conditions are essential to our understanding of the behavior, ecology and environmental physiology of any species. A subcarrier radio transmitter is described in detail. It operates on the 88-108 MHz broadcast band or in the 150 MHz VHF range. Subcarrier frequency is 1 KHz, facilitating the use of inexpensive portable receivers and cassette tape recorders. A magnetic switch and CMOS integrated circuitry are used to provide on/off switching of the surgically implanted radio transmitter. Problems requiring attention in the future include transmitter costs, improvement of 'battery free' transmitters, radio telemetry frequency band allocations and data analysis.

INTRODUCTION

The solution to a biologist's problem is often lying, dust covered, on an engineer's shelf. One aim of this conference is to show what is available, what our various needs are, and how they may be met. The purpose of this paper is to demonstrate the diverse applications of physiological radio telemetry. Several thorough reviews are available (Slater, 1963; Caceres, 1965; Mackay, 1970, 1974; Folk and Copping, 1973; Fryer and Sandler, 1974; Amlaner, 1978).

ADVANTAGES OF RADIO TELEMETRY

Investigators using radio tags have gained insight into the movement, activity patterns, and home ranges of secretive animals. Measurement of physiological parameters from undisturbed free ranging animals in the wild requires radio telemetry. Also biotelemetry may offer a better way to make laboratory measurements, eliminating wire connections between the animal and other electronic equipment. This frees the animal and reduces line noise interference. Wireless operation also reduces the hazard of shock. Restraint, even of highly inbred laboratory animals, often introduces stress and alteration of physiological and psychological parameters (Keller and Unbreit, 1956; Folk and Copping, 1974) which may otherwise invalidate experiments.

Radio telemetry is not a panacea: transmitters fail, range in the field seldom lives up to expectations, recording systems malfunction and telemetry is expensive and often difficult to use. Simplicity is best. If the required data can be obtained by more direct methods the use of radio telemetry should be discouraged, but it often becomes the method of choice by default when no other technique is adequate.

PHYSIOLOGICAL RADIO TELEMETRY

Virtually any physiological parameter that can be converted to an electrical potential and measured in the field may be transmitted, e.g. temperature, biopotentials such as ECG, EEG and myopotential, blood or glandular flow rates, blood pressure, respiration rate and volume. Various biochemical parameters such as pH (gastric or blood) Pco_2, Po_2 may also be transmitted.

Individuals interested in home range or habitat utilization can profit from measurement of physiological parameters. Conversely, it is often impossible for the laboratory physiologist working with highly inbred laboratory animals, to interpret his measurements in a way that is ecologically meaningful. We need an integrated approach to studying the interrelation of natural history, ecology, physiology and behavior. Measurement of heart rate can help to establish social hierarchies (Ruff, 1971) and measurement of body temperature can aid understanding of habitat utilization.

RADIO TELEMETRY IN THE LABORATORY

Requirements for laboratory radio telemetry are generally less demanding than field requirements (transmission over meters instead of thousands of meters). Nevertheless, there are problems: most laboratory environments are electrically noisy, often necessitating increased transmitter power, sophisticated antennas or Faraday shielding. Laboratories are often in major metropolitan areas with high utilization of the radio frequency bands used in radio telemetry. Commercial transmission often interferes with radio telemetry and conversely the possibility of accidental interference by the telemetry transmitter becomes more likely.

PHYSIOLOGICAL RADIO TELEMETRY IN THE FIELD

Captive animals often behave differently from free ranging animals. For example, measurements of body temperature and heart rate under natural conditions differ from similar measurements in the laboratory (Folk and Copping, 1974; Smith, Peterson and Thigpen, 1975). Qualitative differences also exist. An example from my own research is a case in point.

Many vertebrates that feed underwater show a marked reduction in heart rate upon submergence. Diving bradycardia has been observed in many species of reptiles, birds and mammals (Andersen, 1966). The heart rate response of alligators was measured in the laboratory using tethers and ambiguous results were obtained. Forced and voluntary dives produced bradycardia (Wilber, 1960; Andersen, 1961). Gaunt and Gans (1969) viewing dives with closed circuit TV observed voluntary dives with no bradycardia while fear-evoked dives always resulted in marked bradycardia in *Caiman crocodilus*. Radio telemetry studies of a free ranging alligator clearly showed the bradycardia to be a response to fear or disturbance and not to submergence (Smith, Allison and Crowder, 1974). The previously reported 'diving bradycardia' for alligators was an artifact of the experimental conditions. Free ranging alligators do not exhibit diving bradycardia upon submergence unless frightened.

Telemetry provides the only way of obtaining physiological data from many animals. Some species cannot be maintained in captivity for extended periods of time and many animals are dangerous to handle or impossible to tether (Folk, 1980, this volume). Other animals, such as many cetacians, are simply too large to maintain under laboratory conditions. Telemetry is especially useful for studying hibernation where disturbance might awaken the animal (Folk, 1967; Folk, Folk and Minor, 1972; Folk, Larson and Folk, 1976).

PROBLEMS OF FIELD RADIO TELEMETRY

Transmitted range under field conditions seldom approaches theoretical calculations. Due to capacitive loading effects of the body a transmitter's range may drop from one km to tens of meters when placed inside or attached to an animal.

The RF oscillator is usually the part of the transmitter circuit that consumes the most power and transmits the signal to the receiver. There are two approaches to RF oscillator design. The use of crystal controlled oscillators fixes the operating frequency of the RF oscillator. This simplifies receiver operation and facilitates automatic data collection. Radio frequency oscillators contain tuned circuits which must correspond to the crystal frequency for efficient operation and maximum transmitted range. Unfortunately the frequency of the tuned circuit is affected by the proximity of conductors including the tagged animal. If the crystal controlled oscillator and tuned circuit frequency do not match, efficiency and range fall off sharply. This is precisely what happens when a radio transmitter is placed on (or inside) an animal. The circuit may, however, be detuned in the laboratory in anticipation of the effects of the animal. Optimum efficiency and maximum range are difficult to achieve with most crystal controlled oscillators.

The other approach is to use a free running RF oscillator where the tuned circuit itself determines the oscillator frequency. Once again, attachment or implantation alters the tuning of the circuit. This usually results in a downward shift in the operating frequency, which makes reception and automatic recording difficult because the receiver must be tuned to find and follow the transmitted signal. Efficiency and transmitted range however, remain high.

The kind and amount of vegetation, soil type, soil moisture and terrain each affects transmitted range. In working with western diamond back rattlesnakes in western Oklahoma, I found that ranges varied from several meters to over 2 km depending on vegetation and terrain (Smith, 1970). There is no ready solution to this problem.

DESIGN PROBLEMS

Unfortunately, the simultaneous solution to the ideal design criteria cannot be achieved with current technology. For example, minimum size and maximum range are almost mutually exclusive. Accuracy and complexity must be compromised for cost, especially in field studies with the risk of losing the transmitter. Nevertheless, biotelemetry systems are available for field studies of vertebrates as small as a few hundred grams or smaller.

ENVIRONMENTAL EFFECTS ON EQUIPMENT

Ambient temperature and humidity alters performance of receivers, recorders and demodulators. Such equipment must be designed to operate accurately in the wide range of environmental extremes to which it is exposed or it must be protectively housed in a less severe microhabitat (such as a thermostatically controlled ice chest). Telemetry equipment design often needs to approach military specifications but unfortunately this increases equipment cost.

USING BIOTELEMETRY

ATTACHMENT OF THE TRANSMITTER

External Attachment

Various methods of direct attachment have been developed for radio tag studies and some of these are adaptable to biotelemetry. If, for example, skin surface temperature or movement data are being transmitted then collars or harnesses are acceptable. If transducers or electrodes are to be surgically implanted these must be connected to the transmitter by wires. The point of exit of the leads may produce irritation and infection. Movement of the transmitter (such as rotation of the collar) can result in broken wires and transmitter failure. For large animals, where long range is necessary, externally mounted transmitters have been used successfully. An externally mounted transmitter must be temperature compensated because it is often exposed to a harsh thermal environment.

Advantages of external attachment are: (a) The transmitter provides immediate visual identification of the animal (easily facilitated by the use of brightly colored tape or paint). (b) The transmitter is accessible for battery replacement or repair. (c) A more efficient antenna can be used thus increasing range (since tuned antennas represent a tuned circuit, proximity of the animal or movement detunes it resulting in loss of range). (d) External radio packages also permit the use of solar power (Patton, 1973; Williams and Burke, 1973; Church, 1980, this volume).

Tethered Transmitters

Attachment of transmitters to fish may result in altered swimming performance and problems with infection while internal implantation may be difficult. Thus, some investigators resort to a tethering arrangement by which the fish pulls a neutrally buoyant transmitter package. Obviously this is beset with problems concerning the behavior of the animal.

Surgical Implantation

Folk perfected an implantation operation and demonstrated that the abdominal cavity is an acceptable site for radio transmitters (Essler and Folk, 1961). Surgical procedure is simple, requiring from less than 30 min to several hours depending on animal size. Side effects are few. The implanted transmitter must be sealed from body moisture (Shook and Folk, 1965) and implantable electrodes must be used (Folk, 1964). The transmitter is often encapsulated within connective tissue and becomes attached to the abdominal wall. Body temperature may be telemetered directly and other physiological parameters may be measured by the use of appropriate transducers and leads (Fryer, 1970). I often place transmitters in the abdominal cavity and run two ECG leads out of the cavity, then subdermally to each end of the sternum to ECG electrodes. The transmitted signal is attenuated only slightly in passing through the abdominal wall and the transmitter is well protected. Subcutaneous implants near the neck are possible on larger animals but again, leads must be placed subdermally to various transducers.

Disadvantages of surgical implantation are: (a) Battery replacement or transmitter repair involves surgery and recovery. (b) The transmitting antenna is restricted to the magnetic dipole of the internal tank circuit and ranges in excess of 100 m become difficult.

Repeaters

One can obtain the best of both methods by surgically implanting a small short range transmitter and externally attaching a transponder or repeater. The internal transmitter sends a weak signal to a receiver carried externally on a collar or harness. The signal is received and modulates a more powerful long range radio transmitter.

Such systems have been used to transmit data to satellites. Information was temporarily stored in the satellite and recovered when the satellite passed over the appropriate tracking station (Craighead *et al.*, 1971).

REQUIREMENTS FOR PHYSIOLOGICAL RADIO TELEMETRY

The simplest physiological telemetry transmitter might consist of an RF oscillator directly modulated by a biopotential or transducer. Such transmitters have been used with varying success (Mackay, 1970; Folk and Copping, 1973; Smith and Crowder, 1974). Two problems prevail: (a) The RF oscillator operates continuously and so consumes power and reduces battery life. (b) Since many physiological events have subaudio components, special receivers, amplifiers and recorders are needed to process the data.

An improved and widely used physiological telemetry system is shown in Fig. 1 (Roy and Hart, 1966; Fryer, 1970; Smith and Salb, 1975). In addition to the RF oscillator and transducer it contains a subcarrier oscillator which reduces the duty cycle. Since the RF oscillator often consumes more than 10 times the power of the remainder of the circuit (even in sophisticated multi-channel systems), a reduction of RF duty cycle reduces battery drain and extends operating life in spite of increased complexity. Duty cycles can be made very short (1 to 10 percent) with little degradation of transmitted range. The upper frequency component of the measured physiological parameters dictates the subcarrier frequency. Theoretically, the subcarrier frequency should be at least 10 times the upper frequency of the biological signal (Strong, 1970). In practice good fidelity demands that the subcarrier frequency is 20 to 50 times greater than the highest frequency of the physiological signal.

Commercial telemetry systems often employ subcarrier frequencies from several KHz to 100 KHz or higher. Special receivers and wide band recorders must be employed which add to the expense of the telemetry system. As many physiological parameters of interest are low frequency events, a biological signal passband of DC — to 50 or

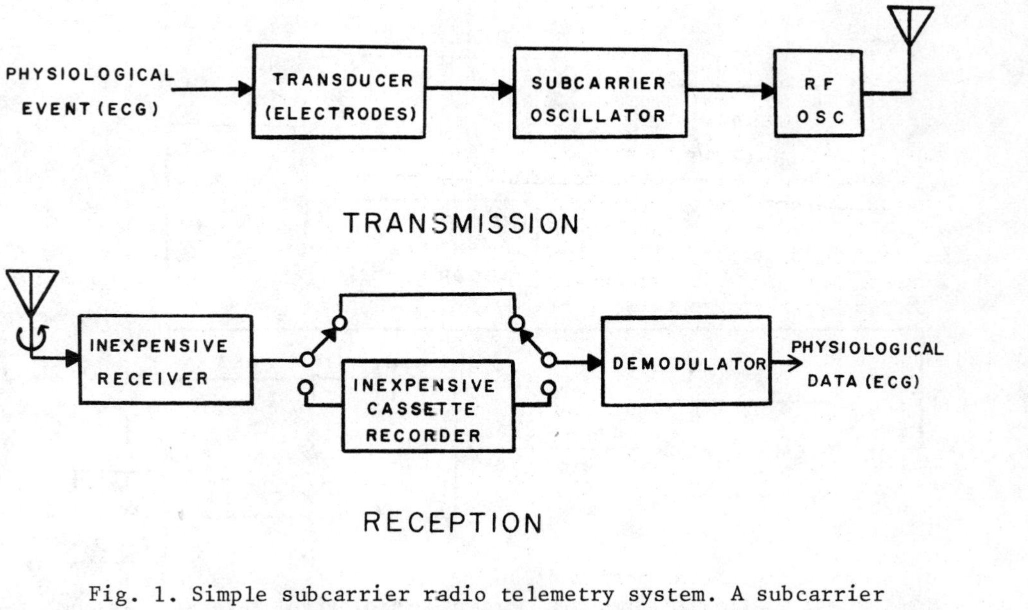

Fig. 1. Simple subcarrier radio telemetry system. A subcarrier oscillator reduces RF duty cycle and simplifies information storage.

100 Hz is an adequate compromise. A subcarrier system operating at a few KHz (Smith and Salb, 1975) provides several advantages. The transmitted signal may be received by readily available inexpensive receivers (US $50.00). The biological signal (such as ECG or respiration) causes the audible tone to vary in pitch. One can often count breathing or heart rate directly from the receiver, thus simplifying demodulation. The audio tone containing the biological data may be recorded on an inexpensive cassette recorder. Data may be demodulated (by audio frequency detection) directly or information stored on magnetic tape for later demodulation. Thus, the subcarrier system not only results in a considerable increase of battery life in the transmitter but also simplifies reception and data recording.

Multichannel Radio Telemetry

As various physiological events often differ in frequency content, it is possible to modify the simple subcarrier system described above to carry more than one channel of biological information. For example, body temperature changes very slowly and can be easily contained within a bandwidth of DC to 0.05 Hz. An ECG signal is, biologically speaking, a rapid event and can be contained in a passband of 1.0 to 100 Hz. A reasonable quality ECG occupies a bandwidth of 10-50 Hz. Since the two physiological events, temperature and ECG, occur at different frequencies they may both be transmitted through the simple subcarrier system described above.

Figure 2 shows the modified subcarrier system. Both the slow changing temperature and rapid changing ECG signal modulate the frequency of the subcarrier oscillator and both appear at the demodulator output. Simple RC low and high pass filters are then used to separate the biological signals and each appears unaffected by the other (Venables and Smith, 1972).

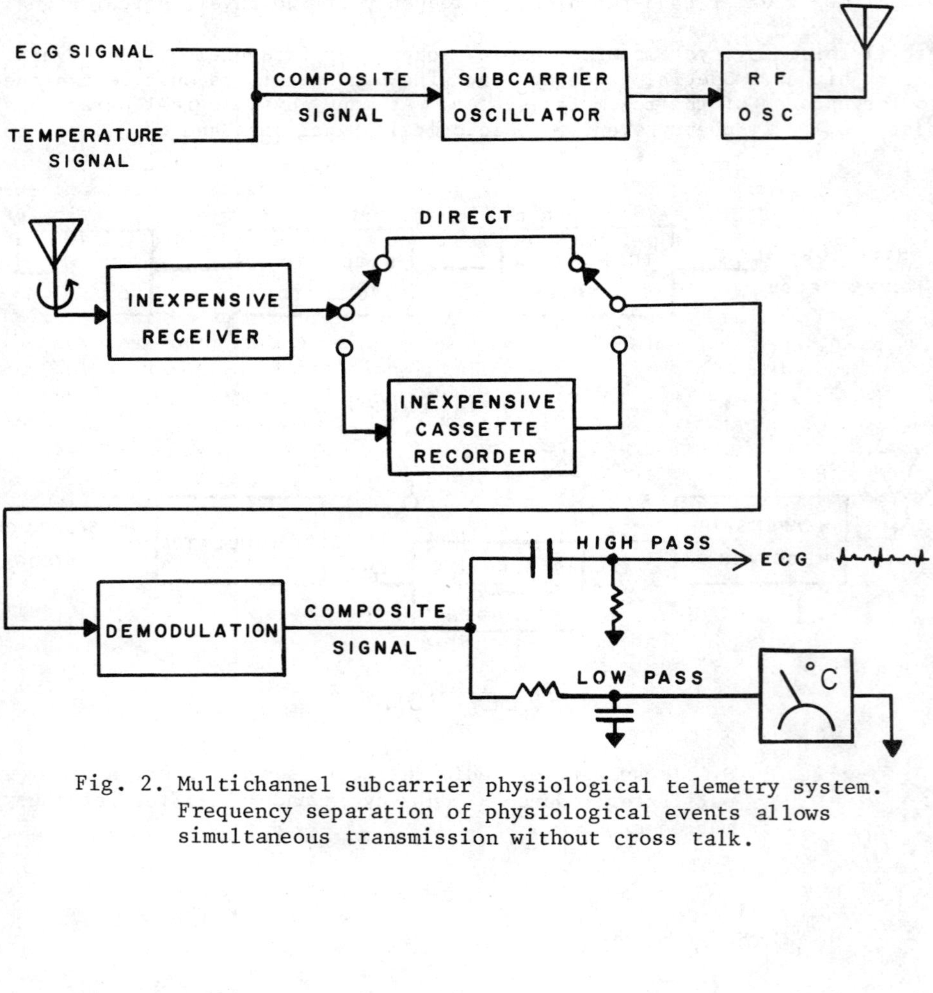

Fig. 2. Multichannel subcarrier physiological telemetry system. Frequency separation of physiological events allows simultaneous transmission without cross talk.

This kind of multichannel approach is ideal because it adds little complexity or power drain to the transmitter, and may be extended to many channels of information. Such a system has been described for telemetering body temperature, ECG, and respiration rate for birds (Smith and Salb, 1975), and results obtained for a domestic duck (Smith *et al.*, 1975), free ranging woodchucks (Smith, 1978; Smith and Causby, 1980, this volume), and cottontail rabbits (Smith and Worth, 1980, this volume). The system is supplied by Biotelemetry Systems, Inc. (P.O. Box 10, Rush, New York 74464, U.S.A.).

Reduction of Duty Cycle

Rapid events such as ECG, EEG, myopotentials and neural potentials require sampling of hundreds to thousands of times each second. Slow events such as temperature may be sampled once each second or less. Locating and tracking rapidly moving animals requires at least a signal every second or so. This is especially true if highly directional antennas are used because the antenna must be pointing toward the animal when the transmitter is transmitting. Slow pulsing transmitters are difficult to locate.

With rapid events such as ECG only a portion of the information may be needed. For example, ECG is often used to determine heart rate. Since the R-wave is the fastest rise time wave and often the largest amplitude component of an electrocardiogram it may be used directly to trigger a transmitter. Occasionally the T-wave approaches the amplitude of the R-wave but its frequency is lower and its amplitude may be easily reduced by an appropriate high pass filter. The R-wave then could be used to trigger the RF oscillator with each heart beat. This provides a simple method of recording heart rate with an extremely low duty cycle. Problems exist, however. Failure of the ECG signal to trigger the circuit from reduced QRS amplitude (e.g. low body temperature) or broken leads results in failure of the transmitter to operate. Alternately, muscular activity may introduce spurious triggering of the transmitter. Nevertheless, this method has been successfully used on small vertebrates.

I used a low duty cycle multichannel telemetry system for measuring three different temperatures and heart rate in alligators (Smith, 1974). The transmitter transmitted 4 short (0.5 ms) clicks and the time between the various pulses indicated the three temperatures. Heart rate was obtained from the time interval of the first click to the next first click. A voltage was developed from the R-wave interval that was proportional to heart rate.

Processing the alligator ECG in the transmitter posed an interesting problem. ECG signal to myopotential noise ratio was seldom greater than 5:1. This meant amplifier gain was critical. The height of the QRS signal (and to a lesser extent amplifier gain) diminished by a factor greater than 10 with low temperature. Several audio automatic gain control circuits were tried. Most were unstable and all used more than a few microamps of current. Finally, a thermistor was used directly as a temperature sensitive gain control to reduce preamplifier supply voltage at high temperatures. Amplifier gain was then increased at low temperatures to compensate for reduced QRS height.

Automatic Switching

The development of CMOS technology enables timing circuits to operate with nearly negligible power requirements. Transmitters may be designed for example, to turn on one min out of every ten min to extend battery life by approximately a factor of ten.

Magnetic Switching

Miniature magnetic reed switches have been used to switch physiological ECG transmitters (Fryer, 1970; Smith and Crowder, 1974). However, it was necessary to keep the magnet near the animal or use a magnetic latching switch to keep the transmitter on. Magnetic reed switches may be used in conjunction with CMOS latches to electronically switch the circuit on or off. Magnetic switching may be used to control transmitters in free ranging animals by hiding a magnet near the burrow or den.

Temperature Switching

Body temperature can be used to switch a transmitter on or off during onset of fever, torpor, or hibernation. Again, this would be of value for studies inside dens or burrows and results in greatly extended battery life.

Other Switching Techniques

Light or darkness could be used to control transmitters as could bursts of RF energy or even ultrasound. Activity or lack of activity could be used as well. In each case the appropriate transducer (photo cell, antenna or microphone) would trigger a CMOS latching circuit and change the status of the transmitter until another burst of energy arrives.

AREAS REQUIRING ATTENTION

Biological Batteries

Conventional batteries limit size, range and operational life of telemetry transmitters. Recent advances in biological batteries that use body fluid as the electrolyte, look promising and have already been used experimentally (Roy, 1971; Wan and Tseung, 1974). Radioactive batteries with half lives of years also appear promising (Greatbatch and Bustard, 1973; Ko and Hymecek, 1974). RF powered sources are now used to change implanted NiCd batteries for long term operation (Ko, 1979, this volume).

Cost Reduction

Field studies are often limited by the cost of transmitters. The cost of mass produced, complex function integrated circuits such as those used in electronic watches and calculators is low, often less than a U.S. dollar. Unfortunately the cost of a comparably complex custom designed integrated circuit for use in radio telemetry might exceed US $10,000 in small quantities. Instead of 50 different laboratories and a dozen manufacturing companies each building their own circuits we need a united effort. We could then go to an integrated circuit manufacturer with a large volume order and obtain a sophisticated multichannel physiological telemetry transmitter for less than US $10.00. We simply need a versatile circuit that most of us could use either in complete form or use portions of it. For example, one group might use the chip for an ultra small radio tag. Someone else might use it for temperature and ECG, while others could use it for something else.

Frequency Allocations

Present frequency allocations are useable for narrow band crystal controlled RF oscillators but more telemetry bands are needed for the utilization of smaller, more efficient, wide band LC oscillators.

Data Processing

Biotelemetry provides thousands of data points in a short time. Data reduction necessitates the use of computer systems. The advent of microprocessors enables early reduction of data and could be incorporated into telemetry demodulation equipment. Microprocessors could also be used to demodulate data, make preliminary calculations or corrections and prepare the data directly for more sophisticated computer processing.

Third Generation Radio Telemetry

If we consider radio tags as comprising a first generation of telemetry and biotelemetry forming a second generation, we might consider radio telemetry to be approaching a third generation. We now have the technology for communicating in both directions between animal and investigator. The animal might carry both a transmitter and receiver while the investigator measured physiological responses to various physiological, pharmacological, or neural stimuli. For example, the investigator could, by remote control, inject drugs, electrically stimulate specific areas of the brain, or alter the temperature of certain well defined areas in the animal (such as blood temperature to the hypothalamus). Once again, this would provide new meaning to the kinds of experiments that have been done only under laboratory conditions.

SUMMARY

The capabilities of modern physiological radio telemetry are exciting. Radio tracking studies have demonstrated that wild animals can be radio monitored and surveillance maintained. Radio telemetry enables measurement of a variety of important physiological variables under natural conditions with minimal disturbance. Radio telemetry technology, already developed, can help provide solutions to many difficult areas of ecology, behavior, and physiology. Two areas requiring immediate improvement are a reduction in cost, especially of the transmitter, and improvement of communications between engineers and biologists.

Acknowledgements — This research was made possible by support from National Institutes of Health Minority Biomedical Support Program (MBS) Grant No. 5 S06 RR08123 and Minority Access to Research Careers (MARC) Grant No. 5 T32 GM07694. Travel funds were supplied by the College of Arts and Sciences and a faculty grant from Northeastern State University. Debbie Boatman and Nell Knight typed the manuscript and Jackie Cashion prepared the artwork. Edgar Folk, Jr., University of Iowa Medical School reviewed an early draft and made many helpful suggestions. I am indebted to Charles Amlaner, Jr. and David Macdonald, Oxford University, for organizing this conference and for the opportunity to participate.

REFERENCES

Amlaner C. J., Jr. (1978) Biotelemetry from free-ranging animals. In B. Stonehouse, *Animal Marking: Recognition Marking of Animals in Research,* pp. 205-228. London: Macmillan.

Andersen H. T. (1961) Physiological adjustments to prolong diving in the American alligator, *Alligator mississippiensis*. *Acta. physiol. scand.* 53, 23-45.

Andersen H. T. (1966) Physiological adaptations in diving vertebrates. *Physiol. Rev.* 46, 212-243.

Caceres C. A. (ed.) (1965) *Biomedical Telemetry,* New York: Academic Press.

Church K. E. (1980) Expanded radio telemetry potential in wildlife investigations with the use of solar transmitters. In C. J. Amlaner, Jr. and D. W. Macdonald (eds.), *A Handbook on Biotelemetry and Radio Tracking,* pp. 247-250. Oxford: Pergamon Press.

Craighead J. J., Craighead F. C., Jr., Varney J. R. and Cote C. P. (1971) Satellite monitoring of black bear. *BioScience* 21, 1206-1213.

Essler W. O. and Folk E. G., Jr. (1961) Determination of physiological rhythms of unrestrained animals by radio telemetry. *Nature* 190, 90-91.

Folk G. E., Jr. (1964) The problem of electrodes for use with electrocardiogram radio capsules. In Murray and Salisbury (eds.), *Biomedical Sciences Instrumentation No. 2,* pp. 235-265. New York: Plenum Press.

Folk G. E., Jr. (1967) Physiological observations on subarctic bears under winter den conditions. In K. Fisher and F. South (eds.), *Mammalian Hibernation*. Edinburgh: Oliver and Boyd. 535 p.

Folk G. E., Jr. and Copping J. R. (1973) Telemetry in animal biometerology. *Int. J. Biometerology* 16, Supplement 5: Part II, 153-170.

Folk G. E., Jr. and Copping J. R. (1974) Implanted radio-capsules to avoid restraint of mammals. In H. Magalhaes (ed.), *Environmental Variables in Animal Experimentation*, pp. 58-74. Lewisburg: Bucknell University Press. 152 p.

Folk G. E., Jr. and Folk M. A. (1980) Physiology of large mammals by implanted radio-capsules. In C. J. Amlaner, Jr. and D. W. Macdonald (eds.), *A Handbook on Biotelemetry and Radio Tracking*, pp. 33-43. Oxford: Pergamon Press.

Folk G. E., Jr., Folk M. A. and Minor J. G. (1972) Physiological condition of three species of bears in winter dens. In S. Herrero (ed.), *Bears — Their Biology and Management*, pp. 107-125. Switzerland: Morges IUCN, New Series 23.

Folk G. E., Jr., Larson A. and Folk M. A. (1976) Physiology of hibernating bears. In M. R. Pelton, J. W. Lentfer and G. E. folk (eds.), *Bears — Their Biology and Management*, pp. 373-380. Switzerland: Morges IUCN, New Series 40.

Fryer T. B. (1970) Implantable biotelemetry systems — A report. NASA SP-5094. U.S. Government Printing Office, Washington, D.C. 113 p.

Fryer T. B. and Sandler H. (1974) A review of implant telemetry systems. *Biotelemetry* 1, 351-374.

Gaunt A. S. and Gans C. (1969) Diving bradycardia and withdrawal bradycardia in *Caiman crocodilus*. *Nature* 223, 207-208.

Greatbatch W. and Bustard T. S. (1973) A $^{238}Pu\ O_2$ nuclear power source for implantable cardiac pacemakers. *IEEE Trans. Biomed. Engng. BME* 20, 336-345.

Keller D. L. and Unbreit W. W. (1956) "Permanent" alteration of behavior in mice by chemical and physiological means. *Science* 124, 723.

Ko W. H. (1980) Power supply for implant telemetry devices. In C. J. Amlaner, Jr. and D. W. Macdonald (eds.), *A Handbook on Biotelemetry and Radio Tracking*, pp. 225-245. Oxford: Pergamon Press.

Ko W. H. and Hymecek J. (1974) Implant evaluation of a nuclear power source, Betacel battery. *IEEE Trans. Biomed. Engng. BME* 21, 238-241.

Mackay R. S. (1970) *Bio-medical Telemetry*. 2nd Ed. New York: Wiley.

Mackay R. S. (1974) Field studies on animals. *Biotelemetry* 1, 286-312.

Patton D. R. (1973) Solar panels: an energy source for radio transmitters on wildlife. *J. Wildl. Mngmt.* 37, 236-238.

Roy O. Z. and Hart J. S. (1966) A multichannel transmitter for the physiological study of birds in flight. *Med. Engng.* 4, 457-466.

Roy O. Z. (1971) Biological energy sources. A review. *Biomed. Engng.* 6, 250-256.

Ruff R. L. (1971) Telemetered heart rates of free-living uinta ground squirrels in response to social interactions. Ph.D. dissertation. Utah State University, Logan, Utah.

Shook G. L. and Folk G. E., Jr. (1965) Body moisture and the operating life of implantable heart rate transmitters. *Trans. Biomed. Engng. BME* 12, 44-46.

Slater L. E. (ed.) (1963) *Bio-telemetry: The use of Telemetry in Animal Behavior and Physiology in Relation to Ecological Problems*. New York: Macmillan.

Smith E. N. (1970) A telemetry system for tracking rattlesnakes. *BioScience* 41, 21-23.

Smith E. N. (1974) Multichannel temperature and heart rate radio-telemetry transmitter. *J. Appl. Physiol.* 36, 252-255.

Smith E. N. (1978) Fear-evoked bradycardia of free ranging woodchuck. *The Physiologist* 21 (4), abstract.

Smith E. N., Allison R. D. and Crowder W. E. (1974) Bradycardia in a free ranging American alligator. *Copeia* 1974, 770-772.

Smith E. N. and Causby L. A. (1980) Respiratory sinus arrhythmia in free ranging woodchucks, *Marmota monax*. In C. J. Amlaner, Jr. and D. W. Macdonald (eds.), *A Handbook on Biotelemetry and Radio Tracking*, pp. 557-562. Oxford: Pergamon Press.

Smith E. N. and Crowder W. E. (1974) Implantable ECG transmitter employing a magnetic switch. *J. Appl. Physiol.* 36, 634-635.

Smith E. N., Peterson C. and Thigpen K. (1975) Body temperature, heart rate and respiration of an unrestrained domestic mallard duck, *Anas platyrhynchos domesticus*. *Comp. Biochem. Physiol.* 54A, 19-20.
Smith E. N. and Salb T. J., Jr. (1975) Multichannel subcarrier ECG, respiration, and temperature biotelemetry system. *J. Appl. Physiol.* 39, 331-334.
Smith E. N. and Worth D. J. (1980) Atropine effects on fear bradycardia of the eastern cottontail rabbit, *Sylvilagus floridanus*. In C. J. Amlaner, Jr. and D. W. Macdonald (eds.), *A Handbook on Biotelemetry and Radio Tracking*, pp. 549-555. Oxford: Pergamon Press.
Strong P. (1970) *Biophysical Measurements*, pp. 426-430. Beaverton, Oregon: Tektronix.
Venables B. and Smith E. N. (1972) A simple method for recording opercular and heart rate of unrestrained fish. *Prog. Fish Culturist* 34, 233-234.
Wan B. Y. and Tseung A. C. C. (1974) Some studies related to electricity generation from biological fuel cells and galvanic cells *in vitro* and *in vivo*. *Med. Biol. Engng.* 12, 14-27.
Wilber C. G. (1960) Cardiac responses of *Alligator mississippiensis* to diving. *Comp. Biochem. Physiol.* 1, 164-166.
Williams T. C. and Burke E. H. (1973) Solar power for wildlife telemetry transmitters. *Am. Birds* 27, 719-720.

Approaches, Field Considerations and Problems Associated with Radio Tracking Carnivores

A. B. SARGEANT

Northern Prairie Wildlife Research Center, Jamestown, North Dakota 58401, U.S.A.

Abstract — The adaptation of radio tracking to ecological studies was a major technological advance affecting field investigations of animal movements and behavior. Carnivores have been the recipients of much attention with this new technology and study approaches have varied from simple to complex. Equipment performance has much improved over the years, but users still face many difficulties. The beginning of all radio tracking studies should be a precise definition of objectives. Study objectives dictate type of gear required and field procedures. Field conditions affect equipment performance and investigator ability to gather data. Radio tracking carnivores is demanding and generally requires greater time than anticipated. Problems should be expected and planned for in study design. Radio tracking can be an asset in carnivore studies but caution is needed in its application.

INTRODUCTION

The adaptation of radio tracking to ecological studies in the early 1960s was a major technological advance affecting field investigations of animal movements and behavior that is now widely used to study many species (Slater, 1963; Macdonald, 1978). Carnivores have received much attention with this new technology; some of the earliest radio tracking studies were on the grizzly bear (*Ursus horribilis*), striped skunk (*Mephitis mephitis*), red fox (*Vulpes vulpes*), and badger (*Taxidea taxus*) (Craighead and Craighead, 1965; Storm, 1965; Verts, 1967; Sargeant and Warner, 1972).

There have been spectacular advances in radio tracking such as the early development of the Cedar Creek Automatic Radio Tracking System (Cochran *et al.*, 1965) and the recent satellite tracking of polar bears (*Thalarctos maritimus*) (Kolz, Lentfer and Fallek, 1978). Such advances illustrate the sophistication of telemetry technology. Of greater importance to most biologists have been the many subtle improvements in transmitter design, power sources, encapsulation and attachment, receiver capabilities, and tracking procedures. These advances have improved reliability to the extent that investigators with little electronic background can conduct independent field studies. However, in spite of the nearly two decades of radio tracking and numerous advances, novice and experienced users still face many difficulties in undertaking radio tracking work. The purpose of this paper is to discuss some practical approaches to using conventional methods to radio track carnivores and to discuss some factors that influence the outcome of a study.

APPROACHES TO RADIO TRACKING STUDIES

Useful data are not easily obtained by radio tracking. Successful studies are the results of careful designs, numerous compromises and realistic assessments of circumstances and capabilities. In most studies, investigators must be present to collect data and must adapt to the target species' time schedule. The return is nearly always less than one animal-hour data collected for each investigator-hour invested. The wise investigator allocates time carefully and avoids temptations to overextend capabilities. It is easy to overlook the effort needed to collect the supplementary data necessary to interpret the results of radio tracking.

Study objectives dictate type of gear, field procedures, and data needed. Hence, all radio tracking studies should begin with precise delineation of study objectives. This fundamental first step is all too often slighted because some investigators apparently still believe that use of radio tracking ensures collection of useful data. It is likely that more money and effort have been wasted on ill-conceived telemetry studies than on the use of any other field technique. Investigators should be sure of their reasons for using telemetry and identify the exact data sought. Examples of a few well planned radio tracking studies involving different approaches are Mech *et al.* (1971) concerning timber wolf (*Canis lupus*) movements, Seidensticker *et al.* (1973) on mountain lion (*Felis concolor*) social organization, Fritzell (1978a) regarding raccoon (*Procyon lotor*) habitat use, and Houseknect and Tester (1978) on striped skunk denning habits. In these studies radio tracking was the principal method for data collection and its use dominated field activities; this is generally true. In some studies radio tracking is used to provide supplementary data. For example, Storm *et al.* (1976) effectively used radio tracking to study red fox dispersal to aid in interpretation of tag recovery data.

At Northern Prairie Wildlife Research Center (NPWRC), U.S.A., we employ radio equipment to track red fox, raccoon, and striped skunks as well as ducks and some other migratory birds. Nearly all of our carnivore tracking is from vehicles with roof mounted antennas supplemented by foot tracking with hand held antennas. In other studies we rely on aircraft tracking and occasionally use boats or fixed field antennas.

Our use of radio tracking on carnivores is diverse. One approach involves enhancing visual observations. A small multiple receiver system mounted in an observation booth in a 4.4 ha pen is used to direct visual attentions. Signals from up to four radio equipped animals are monitored audibly from strategically located speakers. An animal activity is known continuously and its direction from the observer is easily determined with a conveniently located directional antenna. The system reduces observer fatigue and extends observation ability further into twilight periods. Preston (1975) used this system to study red fox home range defense. In a similar manner we recently used radios to locate free ranging striped skunks to observe foraging behavior and in a slightly different manner to locate skunk bedding sites to recover individually identifiable fecal material. These approaches result in gathering data which does not come directly from the tracking gear and have been referred to as predictive tracking (Macdonald, 1978).

Another approach was to study red fox prey handling during pup rearing by radio tracking prey rather than foxes. Ducks equipped with durable transmitters were made available to free ranging foxes. The disposition of ducks killed by foxes was tracked and data necessary for interpretation of duck remains found at fox dens was obtained (Johnson and Sargeant, 1977). The potential of this approach is also apparent in studies of radio tagged pheasants (*Phasianus colchicus*) and white-tailed deer (*Odocoileus virginianus*) by Dumke and Pils (1973) and Cook *et al.* (1971), respectively.

An approach that we have not used extensively but which has much potential is gathering activity rhythm and time budget data only from interpretations of signal modulations or signal type. Mech, Heezen and Siniff (1966) used this approach to study activity of cottontail rabbits (*Sylvilagus floridanus*) and snowshoe hares (*Lepus americanus*). Because knowledge of animal location is not required, data collection is simpler than in tracking studies. Transducers and recording devices can be employed to aid in signal interpretations and to reduce observer time (Knowlton, Martin and Haug, 1968; Swanson, Kuechle and Sargeant, 1976). A similar approach can be used to determine presence or absence of radio tagged animals at specific sites (Gilmer, Kuechle and Ball, 1971).

We have used telemetry in more conventional ways to track raccoons (Fritzell, 1978a, b), and in two recently completed studies to track red fox and striped skunks. The latter studies had some unique aspects. We integrated our fox study with continuing study of coyote (*Canis latrans*) ecology being conducted by another agency. We were interested in spatial relations between foxes living among or near the radio tagged coyotes. Field activities and data recording were coordinated to save time, money and manpower.

In the striped skunk study an entire population living on a 21-32 km^2 study area was radio tagged during each of three springs. Up to 38 skunks were systematically tracked from two vehicles during each of about 90, 12-24 h tracking periods by use of a newly developed programmable memory receiver (Winter *et al.*, 1978) and by carefully planned tracking schedules. Skunk bedding sites were located daily with hand held tracking gear. The fulltime efforts of four investigators were required to maintain daily contact with the study animals; these efforts yielded over 20,000 locations on 113 skunks.

FIELD CONSIDERATIONS

Many factors interact to determine the quality and quantity of data obtained from radio tracking studies. Equipment limits the type of data that can be obtained, but field conditions and the investigator determine how effectively equipment can be and is used. That combination of particulars is sought that will best achieve study objectives. Each study is, of course, unique, but the recent red fox-coyote and striped skunk studies at NPWRC provide good examples of field considerations.

In these studies, objectives and species' characteristics greatly affected choice of gear. The fox-coyote study was concerned almost exclusively with general spatial relationships whereas the skunk study was concerned with precise spatial relationships, resource utilization and foraging behavior. From previous experience we knew that continuous signals provided more qualitative activity data and more accurate triangulation locations than pulsing signals, but that more battery power was required for continuous transmission. We were also aware that closed loop transmitter antennas would provide activity data because of signal nulls not experienced with whips. Frequency choice of 30 and 164 MHz was important too. The 30 MHz frequency provided better obstacle penetration than 164 MHz but 30 MHz restricted vehicle and foot tracking to use of lower gain loop receiving antennas because Yagi antennas were too large for practical use. The 164 MHz frequency provided maximum although more variable range and allowed use of small, easily transported Yagi antennas.

For the skunk study we chose the low frequency band because of better transmission from burrows and through dense vegetation where skunks spend much time. We also chose continuous signals from closed loop antennas because we needed high quality activity data and maximum location accuracy. However, continuous transmissions required the recapture of all study animals midway through each field season for the replacement of transmitter batteries; the shorter range closed loop antennas required additional effort to locate animals. For the fox-coyote study we chose the high frequency band and used pulsing signals and protruding whip antennas to maximize transmitter life and range. With this choice we sacrificed signal quality and location accuracy. The compromise was acceptable and was necessary to avoid recapture of study animals which is very difficult, and to facilitate maintaining contact with the foxes on a 360 km^2 study area with limited road network.

We were aware of study area influences on telemetry gear performance and on our ability to track animals. Our day and night tracking required good roads and access to private lands. We were especially concerned with triangulation accuracy on the skunk study and recognized the effects of inherent tracking errors (Slade, Cebula and Robel, 1965; Heezen and Tester, 1967). The error effects were minimized by choosing a study area with large or isolated habitat blocks and a good road network where most locations could be determined from distances of 0.8 km or less. Of special concern was suitability of roads during wet weather and factors affecting signal reception. Roads and interference from power lines were field checked and study area residents were contacted regarding attitudes toward trespass and night-long traffic. Small factors such as barking farm dogs could prevent night use of strategic tracking sites. Potential study areas with heavy or fast traffic would increase chances of accident.

Fewer constraints were imposed on selection of the fox-coyote study area because of the suitability of long range tracking. Choice, however, was limited to a few areas because of species distributions. Long range transmitters permitted avoidance of difficult tracking sites and permitted use of an area with limited road network.

PROBLEMS

Difficulties should be anticipated in radio tracking carnivores. Equipment performance should be analyzed in the study design phase so that steps can be taken to minimize problems. Investigators should consult both suppliers and users of telemetry gear for information about suitability of equipment for particular need, performance and manner of use. Be aware that equipment is seldom as good or bad as is often portrayed. Consultants need to know actual study constraints (i.e. minimum life, weight, and range of transmitters needed) and the intended use of equipment to provide useful recommendations. If possible, test equipment at field sites to determine actual performance.

Be sure to make firm arrangements with suppliers regarding service. A further precaution is to have backup, spare pieces of equipment readily available. In our fox-coyote and skunk studies, standby receivers, batteries, cables, earphones, antennas, and a variety of components and repair materials were immediately available at a field station. Considering the overall expense of radio tracking studies, procuring backup equipment is a small investment to ensure data collection.

Equipment problems will be minimized by careful handling. Radio tracking equipment is delicate with many components sensitive to temperature, moisture, and shock. Carrying straps, padding, mounting brackets, weather proofing and other precautions can be utilized to reduce stresses. Problems can easily arise during final stages of transmitter preparation especially when conducted at field sites. Short circuiting batteries, excessive heat, dropping, mistuning, and careless potting are common errors that can reduce performance. Field tests of each transmitter prior to use should be part of the routine. This procedure identifies poor performance tags and allows matching transmitters to particular animals. Before releasing animals always check operation of each newly attached transmitter and mark exact tuning location on the receiver. A simple checklist will ensure that these and other final steps are taken.

Transmitter encapsulation should be artfully conducted. The combination of time, environment, and animal abuses all act to reduce performance. Our general experience is that most transmitter failures occur because of difficulties associated with encapsulation and attachment. Moisture, corrosive agents used in construction, gasses from batteries, and antenna breakage from flexing can cause considerable difficulty.

Environmental factors are also a source of difficulty. Atmospheric noise (i.e. electrical storms) and interference from power lines and electrical devices mask signals thereby reducing range. High winds and vibrations break antennas and bad weather may make tracking impossible. Metal objects such as culverts and junk piles detune transmitters or block signals. On our skunk study we experienced range reduction of about 50 percent from April to July due to vegetation growth. Transmitter pulse rates and duration were also of importance in the fox-coyote study. Rates less than 90 pulses min^{-1} or durations less than 20 ms were unsuitable for rapid audio identification and triangulation. A new problem with using the 30 MHz frequency band in our area is sideband interference from very popular Citizen Band radios.

The study animals themselves pose problems. Some species or individuals are more abusive to radio tags than others. Furthermore, investigators must consider the effects of radio tagging on the animal; few tests have been conducted in this regard (see Amlaner, 1978, for review). Physical problems can develop from collar or harness fit. If attachments are too loose radio tags may slip off animals or entangle them. If they are too tight, irritation can develop and lead to serious difficulties. Growing animals and species subject to seasonal weight changes pose the most serious problems. In some cases expandable collars can be used (Follmann and Buitt, 1978) but these are not suitable for all species and are not easily adapted to closed loop antennas. Wrap-around collars are generally used for whip antenna transmitters because they are adjustable and easily attached. Collars with closed loop antennas must be literally built around an animal's neck or body (Mech *et al.*, 1965) or slipped over the head or body and held in place by straps (Verts, 1963) or inner spacers. We found preformed slip-over collars with inner spacers especially useful on skunks but four sizes were needed and contacting surfaces had to be polished to avoid skin irritation. This method was unsuitable for foxes because of extreme head-neck size differences. Periodic handling of animals subject to growth or substantial weight changes is advisable to check attachment fit. Removal of radio tags upon study completion should be attempted if possible.

CONCLUSIONS

Radio tracking is a technique that is difficult to use but which is sought by field biologists because of the unique advantages it offers in understanding animal ecology. Many factors interact to affect the outcome of radio tracking studies, only some of which were identified in this paper. Good equipment, although important, does not ensure study success. Careful planning, ingenuity, and strong commitment are integral parts of successful radio tracking studies. Radio tracking can be an asset in carnivore studies but caution is needed in its application.

Acknowledgements — I appreciate the assistance of V. B. Kuechle and staff at the Cedar Creek Bioelectronics Laboratory, University of Minnesota, for providing radio tracking gear and advice and to V. B. Kuechle, R. G. Greenwood and D. S. Gilmer for manuscript review.

REFERENCES

Amlaner, C. J., Jr. (1978) Biotelemetry from free ranging animals. In B. Stonehouse (ed.), *Animal Marking: Recognition Marking of Animals in Research*, pp. 205-228. London: Macmillan.

Cochran W. W., Warner D. W., Tester J. R. and Kuechle V. B. (1965) Automatic radio-tracking system for monitoring animal movements. *BioScience* 15, 98-104.

Cook R. S., White M., Trainer D. O. and Glazener W. C. (1971) Mortality of young white-tailed fawns in south Texas. *J. Wildl. Mngmt.* 35, 47-56.

Craighead F. C., Jr. and Craighead J. J. (1965) Tracking grizzly bears. *BioScience* 15, 88-92.

Dumke R. T. and Pils C. M. (1973) Mortality of radio-tagged pheasants on the Waterloo Wildlife Area. *Wis. Dep. Nat. Resour. Tech. Bull.* 72.

Follmann E. H. and Buitt J. M. (1978) An adjustable radio collar for foxes. *J. Wildl. Mngmt.* 42, 949-951.

Fritzell E. K. (1978a) Habitat use by prairie raccoons during the waterfowl breeding season. *J. Wildl. Mngmt.* 42, 118-127.

Fritzell E. K. (1978b) Aspects of raccoon (*Procyon lotor*) social organization. *Can. J. Zool.* 56, 260-271.

Gilmer D. S., Kuechle V. B. and Ball I. J., Jr. (1971) A device for monitoring radio-marked animals. *J. Wildl. Mngmt.* 35, 829-832.

Heezen K. L. and Tester J. R. (1967) Evaluation of radio-tracking by triangulation with special reference to deer movements. *J. Wildl. Mngmt.* 31, 124-141.

Houseknecht C. R. and Tester J. R. (1978) Denning habits of striped skunks (*Mephitis mephitis*). *Am. Midl. Nat.* 100, 424-430.

Johnson D. H. and Sargeant A. B. (1977) Impact of red fox predation on the sex ratio of prairie mallards. *U.S. Fish Wildl. Serv. Res. Rep.* 6.

Knowlton F. F., Martin P. E. and Haug J. C. (1968) A telemetric monitor for determining animal activity. *J. Wildl. Mngmt.* 32, 943-948.

Kolz A. L., Lentfer J. W. and Fallek H. G. (1978) Polar bear tracking via satellite. In 15th Rocky Mountain Bioengineering Symposium and 15th International ISA Bio-Medical Sciences Instrumentation Symposium, pp. 137-144. Pittsburgh: Instrument Society of America.

Macdonald D. W. (1978) Radio-tracking: some applications and limitations. In B. Stonehouse (ed.), *Animal Marking: Recognition Marking of Animals in Research*, Chapter 20. Baltimore: University Park Press.

Mech D. L., Frenzel L. D., Jr., Ream R. R. and Winship J. W. (1971) Movements, behavior, and ecology of timber wolves in northeastern Minnesota. In L. D. Mech and L. D. Frenzel, Jr. (eds.), *Ecological Studies of the Timber Wolf in North-eastern Minnesota*, pp. 1-35. U.S. For. Serv. Res. Pap. NC-52, St. Paul.

Mech L. D., Heezen K. L. and Siniff D. B. (1966) Onset and cessation of activity in cottontail rabbits and snowshoe hares in relation to sunset and sunrise. *Anim. Behav.* 14, 410-413.

Mech L. D., Kuechle V. B., Warner D. W. and Tester J. R. (1965) A collar for attaching radio transmitters to rabbits, hares, and raccoons. *J. Wildl. Mngmt.* 29, 898-902.

Preston E. M. (1975) Home range defense in the red fox (*Vulpes vulpes l.*), *J. Mammal.* 56, 645-652.

Sargeant A. B. and Warner D. W. (1972) Movements and denning habits of a badger. *J. Mammal.* 53, 207-210.

Seidensticker J. C., IV, Hornocker M. G., Wiles W. V. and Messick J. P. (1973) Mountain lion social organization in the Idaho primitive area. *Wildl. Monogr.* 35.

Slade N. A., Cebula J. J. and Robel R. J. (1965) Accuracy and reliability of bio-telemetric instruments used in animal movement studies in prairie grasslands of Kansas. *Trans. Kans. Acad. Sci.* 68, 173-179.

Slater L. E. (1963) *Bio-telemetry: the Use of Telemetry in Animal Behavior and Physiology in Relation to Ecological Problems*. Oxford: Pergamon Press.

Storm G. L. (1965) Movements and activities of foxes as determined by radio-tracking. *J. Wildl. Mngmt.* 29, 1-13.

Storm G. L., Andrews R. D., Phillips R. L., Bishop R. A., Siniff D. B. and Tester J. R. (1976) Morphology, reproduction, dispersal and mortality of midwestern red fox populations. *Wildl. Monogr.* 49.

Swanson G. A., Kuechle V. B. and Sargeant A. B. (1976) A telemetry technique for monitoring diel waterfowl activity. *J. Wildl. Mngmt.* 40, 187-190.

Verts B. J. (1963) Equipment and techniques for radio-tracking striped skunks. *J. Wildl. Mngmt.* 27, 325-339.

Verts B. J. (1967) *The Biology of the Striped Skunk.* Urbana: University of Illinois Press.

Winter J. D., Kuechle V. B., Siniff D. B. and Tester J. R. (1978) Equipment and methods for radio tracking freshwater fish. Univ. Minn. Agric. Exp. Stn. Misc. Rep. 152-1978.

Application of Radio Tracking in Wildlife Research in the Netherlands

S. BROEKHUIZEN,* C. A. VAN'T HOFF,*
M. B. JANSEN** and F. J. J. NIEWOLD**

*Research Institute for Nature Management, Arnhem, The Netherlands
**Technical and Physical Engineering Research Service, Wageningen, The Netherlands

Abstract — Radio tracking is practised in ecological research on red fox (*Vulpes vulpes*), free roaming farm cat (*Felis catus*), beech marten (*Martes foina*), polecat (*Putorius putorius*), European hare (*Lepus europaeus*), harbor seal (*Phoca vitulina*), black grouse (*Lyrurus tetrix*) and little owl (*Athene noctua*). Transmission at 27 MHz was used for land mammals and birds because of low attenuation of the signal by obstacles, and at 154 MHz for the harbor seal because of longer range signal capabilities. Schematics and attachment of the transmitters are described. Data concerning the batteries and the antennas are given. For tracking wide ranging animals, a receiver was developed, equipped with digital stabilization of the VFO including a signal strength- and tuning-indicator with memory. Applications, properties of the equipment, methods applied, reliability of the transmitters and the influence of capturing and radio tracking on the animals' behavior are dealt with.

INTRODUCTION

Wildlife research, carried out at the Research Institute for Nature Management mainly concerns investigations into population dynamics, movements, activity patterns, dispersal, interrelations, size and utilization of home ranges of species which are subject to nature management. Radio tracking is indispensible for the study of species which are elusive owing to their life style, protective coloring, nocturnal behavior or shyness. Problems occurring with the use of this method, can be subdivided into four interrelated categories:

1. problems concerning the animals such as capturing and recapturing, handling, anesthetizing, size and attachment of the transmitter;

2. problems concerning the construction of the transmitter, operative period, distance covered, batteries, antennas and the application of a practical and adjustable receiving unit;

3. problems concerning legal and social restrictions such as maximum allowable transmitting frequencies and restricted accessibility of the study areas;

4. problems concerning the accuracy of the data, such as precise localization in relation to the distance and topography.

Not only is collaboration with electronic engineers important for the development of the required equipment, but also a technical service is required which supports the biologists when the equipment fails or needs modification.

To date radio tracking has been applied to (a) fox, social structure and dispersal, home range and activity radius; (b) hare, utilization of home range and position of doe and young in the nursing period; (c) polecat and beech marten, utilization of home range and habitat requirements relative to availability of food and resting-places; (d) free roaming farm cat, and the human impact on population size, social structure and dispersal; (e) harbour seal, wintering quarters and influences of disturbance; (f) black grouse, factors determining the reproduction rate; and (g) little owl, territory size relative to habitat requirements. Radio tracking is mainly used to localize a radio tagged animal in order to visually observe it later. The activities of common seal, polecat, beech marten and black grouse are now being recorded automatically as well.

GENERAL PRINCIPLES AND REQUIREMENTS

For our radio tracking program only the 27 and 154 MHz frequencies were available. The signal could not be attenuated and reflected too much by obstacles such as trees, buildings and small hills. Therefore, the 27 MHz transmitting frequency was chosen for land mammals and birds. For the seals, however, a range of at least 15 km was required. As the problems of signal attenuation are minor in a flat coastal area, the 154 MHz frequency was used for those animals.

A pulse frequency of 40-60 min^{-1} proved to be satisfactory for the localization of an active animal. Smaller animals, however, were given a frequency of 20-30 min^{-1} in order to extend the transmitter lifetime, and also allow a good determination of signal availability on an activity penrecorder. The seal's transmitter produces, during the first 5 s after emersion, five pulses (80 ms long) s^{-1} and afterwards only 1 s^{-1}. When the seal dives, the transmitter is switched off in order to save energy.

27 MHz transmitters are attached around the body with a band which serves as a loop antenna. It consists of a metallic core and a plastic cover. The band is attached around the neck for mammals and for birds it is wrapped around the body behind the wings. The loop antenna has the advantage that its position affects the strength of the received signal which makes animal's activity more noticeable. A disadvantage is that signal direction is more difficult to identify when the signal strength fluctuates.

A special netting harness was designed which kept the transmitter on the back of the seal. As the signal can be received only when the seal has surfaced, a rod antenna placed on the foremost part of the back switches on the transmitter as soon as the antenna tip comes out of the water.

The weight of a transmitter is adjusted depending on the species involved and sometimes on the sex, status, and behavior of the individual. As batteries account for a major part of the package weight, often one has to experimentally determine the maximum permissible weight for smaller species (Table 1).

If radio tracking is done on foot or by bicycle, a portable receiver and batteries are required. For tracking animals which cannot be followed continuously, a receiver provided with VFO stabilization, a signal strength meter and tuning indicator with memory is recommended.

Table 1.
Total weights of transmitter packages used in radio tracking, compared with the body weights of the animals

Species	Body weight (kg)	Total package weight (g)	Package-body weight (%)
Harbor seal	80-100	3000	3-3.8
Fox adult	4-7	200	2.9-5
Fox juvenile	1.5-3.5	65	1.9-4.3
Fox cub	0.6-1.5	25	1.7-4.2
Domestic cat	2.5-5	65	1.3-2.6
Beech marten, ♀ + subadults	1.5	40	2.7
Polecat ♂♂	1.2	40	3.3
Polecat ♀♀	0.75	25	3.3
Hare	3.5	65	1.9
Black grouse	1.0	29	2.9
Little owl	0.19	15	7.9

Mobile receivers provided with a tuned loop antenna are appropriate for determining signal direction. In permanent applications where the activity pattern is to be assessed, a whip antenna, sensitive in all directions, is preferable. Should the animal's position also be required, a combination of two or more rotating antennas might be used.

EQUIPMENT

27 MHz TRANSMITTER

Circuit Diagram

A standard crystal controlled transmitter circuit is used (Fig. 1). The pulse 'on' time and interpulse period can be varied by changing the values of R1, R2 and C1. The transmitting frequency can be chosen within a 50 KHz band between 27.5 and 28.0 MHz (according to Dutch regulations). Since the diameter of the loop is dictated by

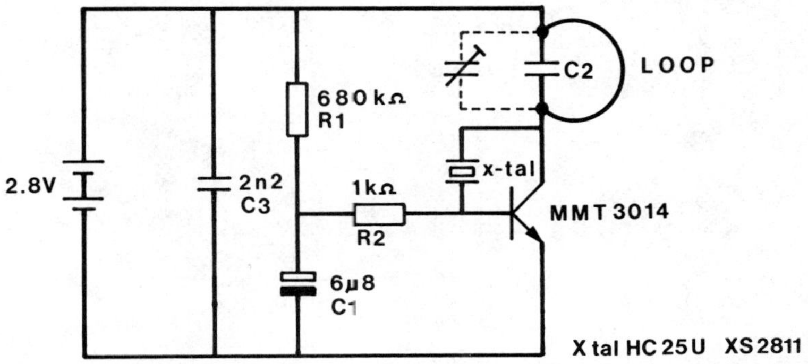

Fig. 1. Circuit diagram of the 27 MHz transmitters. Component values give an on time of 80 ms and a interpulse interval of 1.5 s.

the size of the animal's neck or body, it is necessary to tune the loop with the trimming capacitor C2. Used on tiny animals with a neck circumference smaller than 25 cm, the tuning of the loop becomes very critical. To avoid this problem, a loop consisting of two windings is used. To achieve as compact as possible a transmitter, the electronic components are soldered directly together. For very small transmitters even the trimming capacitor is left out. Obviously this means that the size of the loop antenna cannot be varied widely. Each transmitter is individually adjusted (tuned) for its own size loop antenna.

Mechanical Design

After testing several possible modes, a design was chosen, constructed of polyvinyl chloride (PVC) plastic, which was machined to obtain the desired outer shape. Holes for the transmitter and the batteries (Fig. 2) were drilled. PVC is easy to machine, is not hygroscopic, can easily be glued, has good electrical insulating properties, is strong and waterproof and is comparatively light. The holders for the batteries are sealed with a screw cap so that the batteries are easily replaceable. The transmitter compartment can be filled with a two part epoxy adhesive which guarantees good protection of the transmitter against moisture and mechanical vibrations.
The loop antenna on mammals consists of a copper or stainless steel strip, which is easily attachable to the transmitter by means of threaded bolts. When a transmitter is attached to a bird, the harness consists of a chest and body loop of flexible 4 mm copper wire with insulation. The two loops are connected on the bird's back by a canvas strip, and the transmitter is situated on the chest. When attaching the transmitter, the canvas strip must be cut to fit, and the body loop is soldered together. All the transmitters and collars used on mammals are provided with Scotchlite 3M reflecting tape, so that the animals are easy to identify with a spotlight or infra red binoculars.

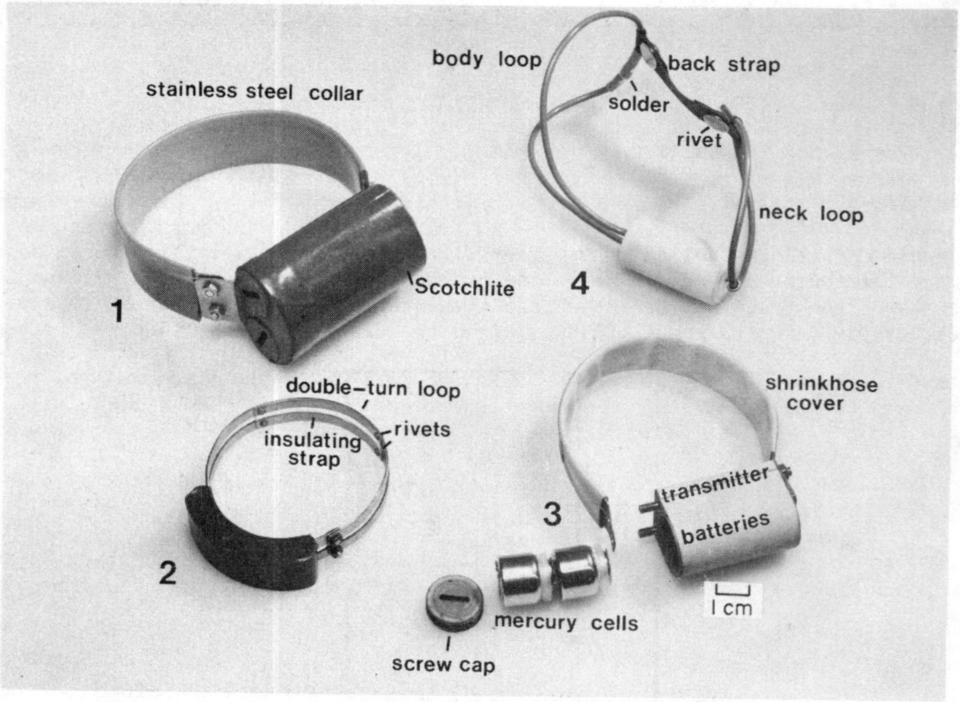

Fig. 2. Housings of 27 MHz transmitters. 1. adult fox; 2. fox cub, polecat; 3. juvenile fox, house cat, hare; 4. black grouse.

154 MHz TRANSMITTERS

Block Diagram

Figure 3 shows the block diagram of the total seal transmitter.

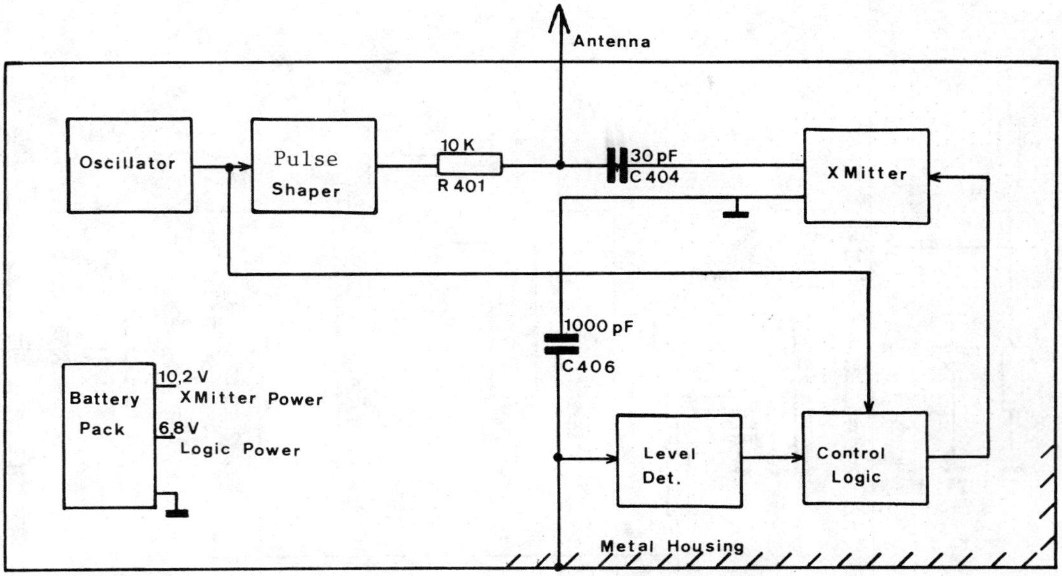

Fig. 3. Block diagram of the 154 MHz transmitter.

The two stage transmitter is capable of delivering 50 mW of RF energy at a battery current consumption of approximately 12 mA and can be switched by a logic level supplied from the digital control logic. The digital logic of the transmitter is powered by a 6.8 V tap of the battery pack in order to reduce the current consumption. The oscillator runs at a speed of 10 pulses s^{-1} and the output is fed to a pulse shaper, that produces narrow, negative going control pulses of 3 ms width. The output of the pulse shaper is connected to the antenna via a 10 K resistor, so besides the RF signal of the transmitter there is also a pulse signal present at the antenna.

The transmitter and logic ground are connected via C406 to the metal case of the transmitter. For RF signals this capacitor almost represents a short circuit, but it effectively blocks the control pulses that appear on the transmitter case when, due to the conducting seawater, there is electrical conduction between the antenna tip and the case. The received control pulses then continuously reset the control logic so the transmitter receives no keying signal, and remains switched off.

When the seal emerges, first the antenna tip comes out of the water, the conductive path through the seawater disappears, and the control logic no longer receives reset pulses.

During the first five seconds, the control logic generates a pulse signal at a rate of 5 pulses s^{-1}, and after this period a pulse signal of 1 pulse s^{-1}. The extra energy required for the increased pulse rate is compensated for by the time the seal is under water.

Circuit Description of the Transmitter Section (Fig. 4)

Special attention has been paid to reducing feedback effects of the final amplifier when the antenna tuning is disturbed, for instance when the antenna is only partially above sea-level.

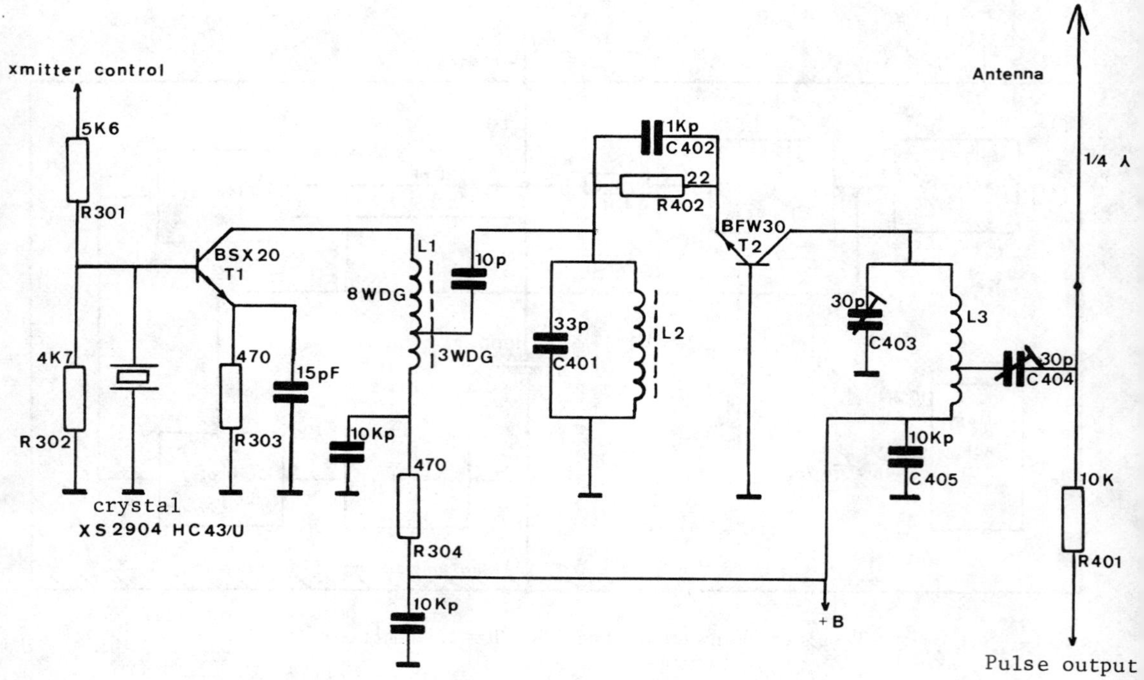

Fig. 4. Circuit of the 154 MHz transmitter.

Transistor T1 operates with L1 and a crystal in parallel resonance at half the antenna frequency, T2 serves as a doubler/power amplifier stage in a grounded base configuration. This configuration guarantees the necessary isolation between antenna and oscillator. The oscillator is switched on and off by the pulse signal from the control logic by providing the correct base current. The antenna is a quarter wave whip antenna and the RF antenna input is approximately 50 mW at a battery voltage of 10 V.

With the oscillator emitter resistor RE the RF output can be adjusted to the desired value. The circuit around L1 resonates at 77 MHz while L2/C401 and L3/C403 resonate at 154 MHz. C404 facilitates the antenna tuning. This must be done with the transmitter installed in its final housing, and with the original antenna.

Circuit Description of the Control Logic (Fig. 5)

The control logic is built up around four CMOS integrated circuits. Gates 4093 A and B operate as an astable oscillator with a period time of 0.1 s 4093 C is used with C103 and R103 as a pulse shaper, that provides negative going pulses of 3 ms. These pulses are fed to the antenna through R401, which has a negligible effect on the RF output of the transmitter.

Gate 4093 D serves as a Schmitt trigger for the control pulses, that appear on the transmitter housing when there is a conductive path between the antenna and the housing.

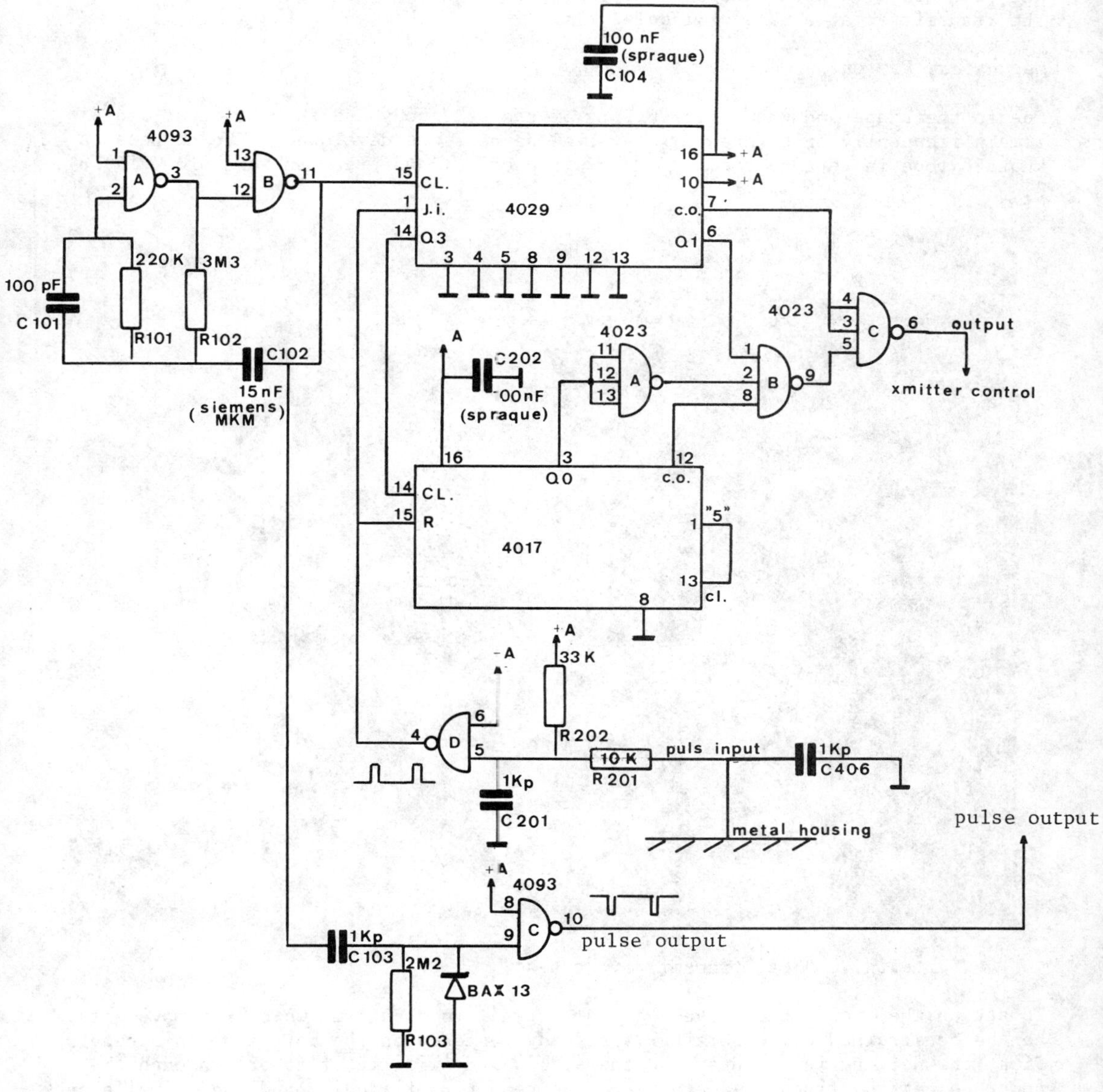

Fig. 5. Circuit of the control logic.

The output of this Schmitt trigger resets the counters 4029 and 4017 every 0.1 s so that these counters remain in the '0' state. Since Q0 of the 4017 is high, the output of 4023A is low, thus keeping the output of 4023B high. C.O. of the 4029 is high, so the output of 4023C remains low and thus the transmitter remains off.

When the conducting path between antenna and transmitter housing disappears, the output of 4093D remains low, so the 4029 and 4017 counters receive a reset signal no longer, and are allowed to count. Q1 of the 4029 provides control pulses of 100 ms, at a rate of 5 pulses s^{-1}, and during this time the 4017 counter needs to reach the '5' state, these pulses are allowed to switch the transmitter. After the 4017 has reached the '5' state, the C.O. of the 4017 goes low and blocks gate 4023B, the output of which goes high, and now only the pulses of the C.O. of the 4029 can switch the transmitter at a rate of 1 pulse s^{-1}.

Mechanical Design

Due to the shape and great flexibility of the seal body and the fact that seals remain frequently on sandbanks, a special harness was developed. The harness is knotted rope in netting form from one piece of 4 mm neoprene rubber cord (Fig. 6).

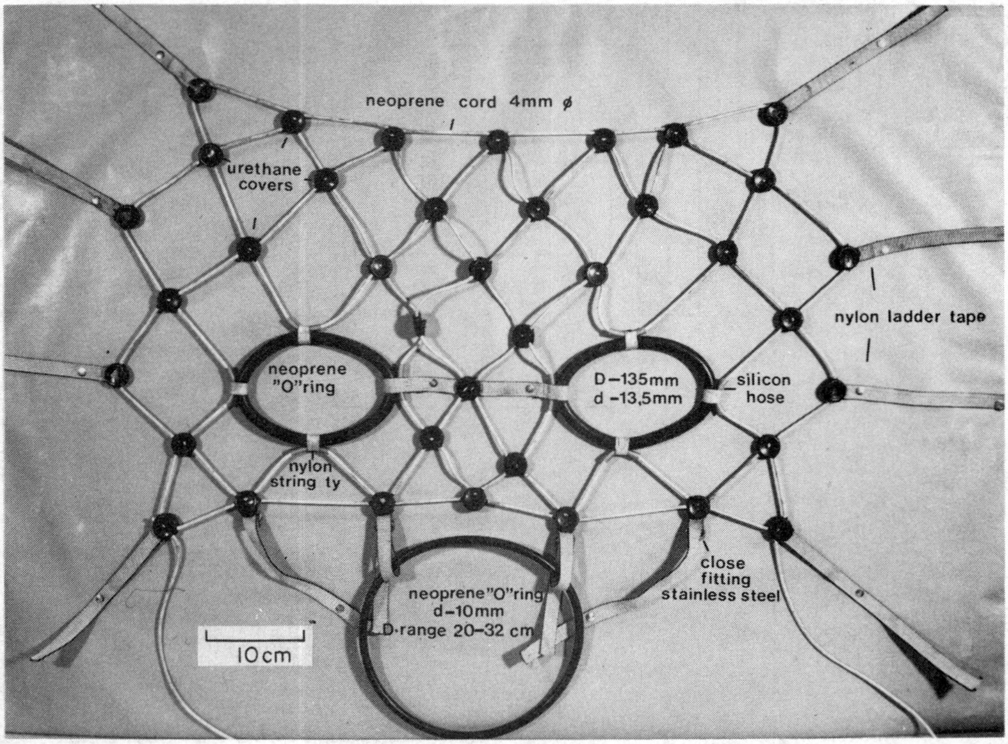

Fig. 6. Seal transmitter harness.

To attach the transmitter, the seal is forced into a cage in which its movements can be constrained. The netting is already fastened on the bottom of this cage. First the netting is fixed around the seal's body with its flippers through the neoprene netting rings. A detachable neoprene ring fitted around the animal's neck is pulled over the head and fastened to the netting by leather straps. After that the transmitter is placed on the seal's back, and also fastened to the netting with nylon straps. During this procedure the netting is closely pulled around the seal's body. The transmitter housing is made of two part urethane rubber (Fig. 7). Two stainless steel cylinders with screw caps contain the transmitter and the three batteries (Fig. 7). The transmitter is built in a cylindrical form, and consists of four printed circuit boards, containing the control logic and the RF-transmitter. The separate boards are first tested and pre-adjusted and then assembled into a compact unit.

The antenna must be composed of a stainless steel spring with very flexible cable in it and connected to a small rod. The entire package including the rod aerial

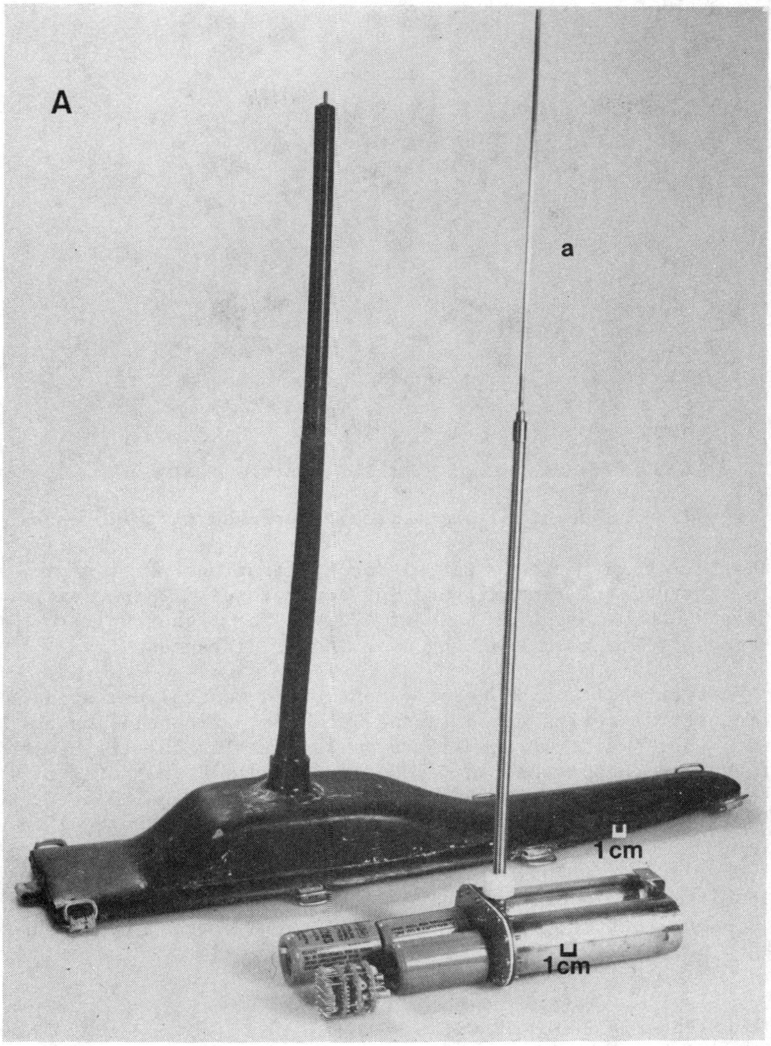

Fig. 7a. Housing of the 154 MHz transmitter for seals.

can be moulded in the urethane compound. A piece of stainless steel plate soldered to the metal case and the non-covered top-end of the antenna act as contacts for switching the transmitter on and off when the seal emerges.

BATTERIES

After the first experiments, it became obvious that transmitters, equipped with commercial Mallory-mercury batteries, did not reach the calculated operating time. We changed to pacemaker mercury cells with a capacity of 1000 mAh (type MR85 made by Leclanché, Switzerland).

Since bigger cells of the same quality were not available, Leclanché advised us to

Fig. 7b. The separate pre-assembled printed circuit board.

connect the batteries in series/parallel for applications where a greater capacity is needed, for instance in transmitters for adult foxes. This configuration results in a capacity of 2000 mAh at a voltage of 4.2 V. The higher voltage causes a greater RF output, necessary for good tracking over longer distances.

After testing the expensive and strictly controlled medical cells, we switched to the cheaper industrial series of Leclanché, without noticeable loss of transmitter operating time. For smaller transmitters we use smaller capacity types according to the manufacturers recommendations. Due to the pulsed character of the battery discharge, and a slightly longer interpulse interval than calculated, remarkably longer battery life was achieved (e.g. fox transmitter calculated life — 138 days, realized life — 320 days).

The seal transmitter is powered by three TADIRAN lithium cells type HE-D (3.4 V, 12 Ah), which have a better capacity-volume ratio than other types, and a very low self-discharge.

RECEIVER FOR RADIO TRACKING

In addition to the normally used receivers, we take advantage of a receiver which has no channel selector but a variable frequency oscillator (VFO) and which is frequency-stabilized by a digital control circuit. The antenna frequency is read out with a digital frequency meter. The signal strength meter has an analog memory. To provide correct tuning before the narrow crystal filter is switched on, the receiver is also equipped with a tuning indicator. The continuous frequency coverage enables us to use a very large number of transmitters, each a few KHz apart in frequency.

Block Diagram

The receiver is a double superheterodyne with a first I.F. of 9 MHz and a second IF of 455 KHz (Fig. 8). The RF amplifier and first mixer are equipped with dual gate mosfets which have good cross modulation properties. In the first I.F. stage two crystal filters can be selected to provide a 500 Hz or 3 KHz IF passband. A RF volume potentiometer controls the amplification of the RF and IF stages, so

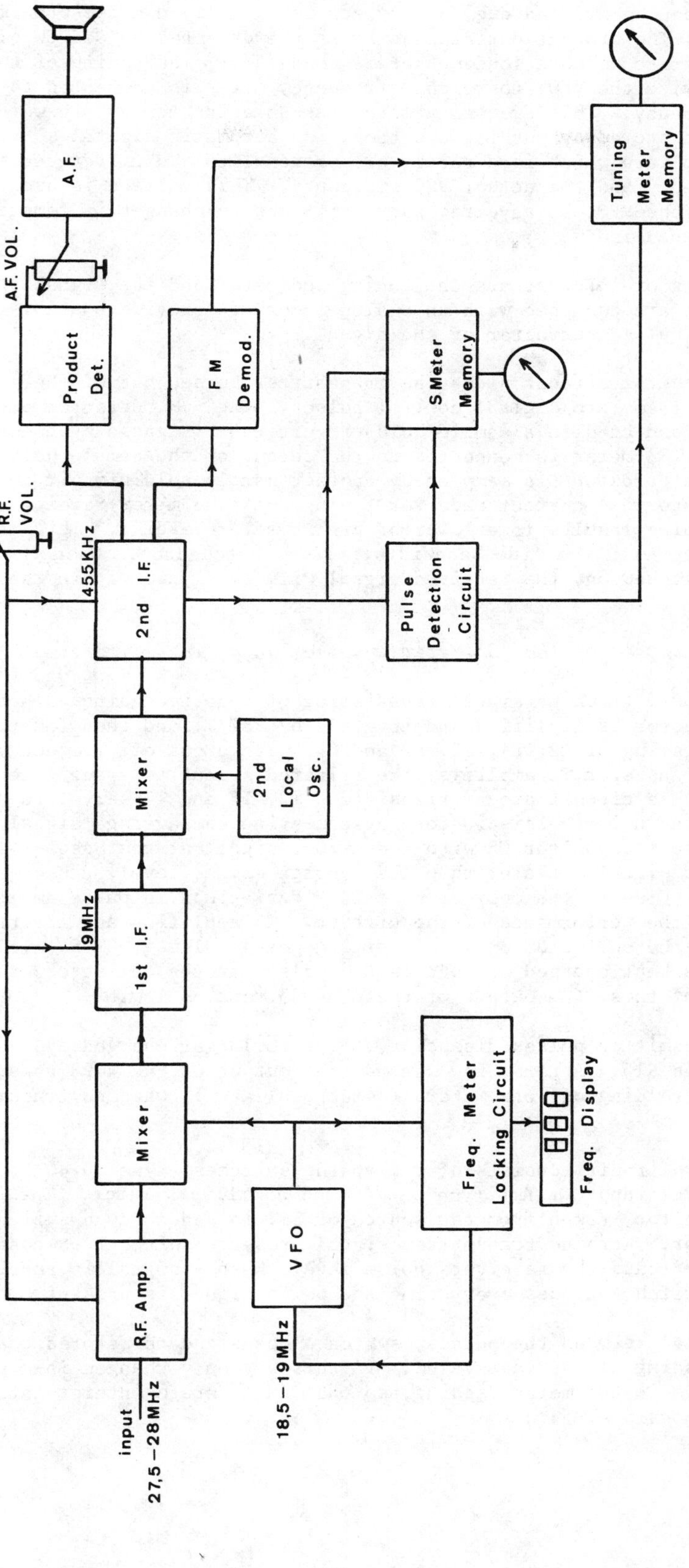

Fig. 8. Block diagram of the advanced receiver for radio tracking.

overloading can be avoided due to very strong signals in close proximity to the transmitter. The product detector delivers the detected pulsed CW signal to the A.F. amplifier to which a loudspeaker is connected. The tuning of the receiver is carried out with the VFO, to which a frequency meter is connected to read out the antenna frequency. This frequency meter has an extra memory, in which the VFO frequency can be stored by pushing a button. An automatic digital subtraction circuit produces a voltage which is directly proportional to the difference between the stored frequency and the actual VFO frequency. This voltage is used to correct the frequency of the VFO, so any frequency drift due to changes in temperature or supply voltage is cancelled.

Other sections of interest are the tuning indicator and the signal strength meter. Both circuits are equipped with an analog memory to provide a stable reading in spite of the pulsed character of the signal.

The pulse detector circuit picks the transmitter pulses out of the receiver noise and converts them into digital control pulses. The A.M. detector output of the receiver is connected to a sample/hold circuit that is gated by the digital control pulses. The 'S' meter is connected to the output of the sample/hold circuit. The F.M. demodulator output is sampled by another sample/hold circuit that drives the tuning indicator. A correct receiver tuning puts the meter reading midscale while incorrect tuning results in a lower or higher meter reading. This provision enables correct tuning with the wide bandwidth filter switched on. When the narrow bandwidth filter is switched on, the received signal will surely be within the passband of this filter.

Circuit Description of the 'S' Meter and the Tuning Indicator

A received radio tracking signal, consisting of negative going pulses, derived from the A.M. detector is amplified and inverted by FET T1 and then fed to a low pass filter, consisting of C2, L1, C3, L2 and C4, with a cut off frequency of 150 Hz (Fig. 9). Transistor T2 amplifies the filtered signal to a sufficient level and inverts it. The circuit around transistors T3, T4 and T5 serves to separate the pulses from the noise. This is done by comparing the average signal level that is developed across capacitor C5 with the direct signal at the base of T3. That part of the direct signal, greater than the average signal level, causes a clean negative going output pulse at the collector of T4. Variations in pulse amplitude have no influence on the performance of the circuit. T6 amplifies and inverts the output of T4, so at the collector of T6 a positive going pulse at CMOS level is available. Via D3, C6 is kept charged as long as the pulses at the collector of T6 are present and because of this, the output of the CD 4093 remains low.

When the transmitter pulses disappear, C6 is no longer charged and begins to discharge through R15. After a few seconds the output of the 4093 goes high, to permit a continuous reading of both meters (described later), when no transmitter signal is present.

The 'S' meter circuit consists of two analog switches, type CD 4016, the hold capacitor C8, a FET input Op Amp type μA 740 and a milliamp meter. The signal from the A.M. detector, taken from the source of T1, is fed through switch A and B to the hold capacitor. With no transmitter signal present, switch B is continuously closed and the meter reads the receivers noise level. When a signal is received, switch B opens, and switch A closes every time the pulse signal is present.

In the interval between the pulses, switch A opens and the stored charge in C8 keeps the meter reading at the last value. In this way only changes in signal strength cause a change in the meter reading, so that an accurate interpretation of the signal strength is possible.

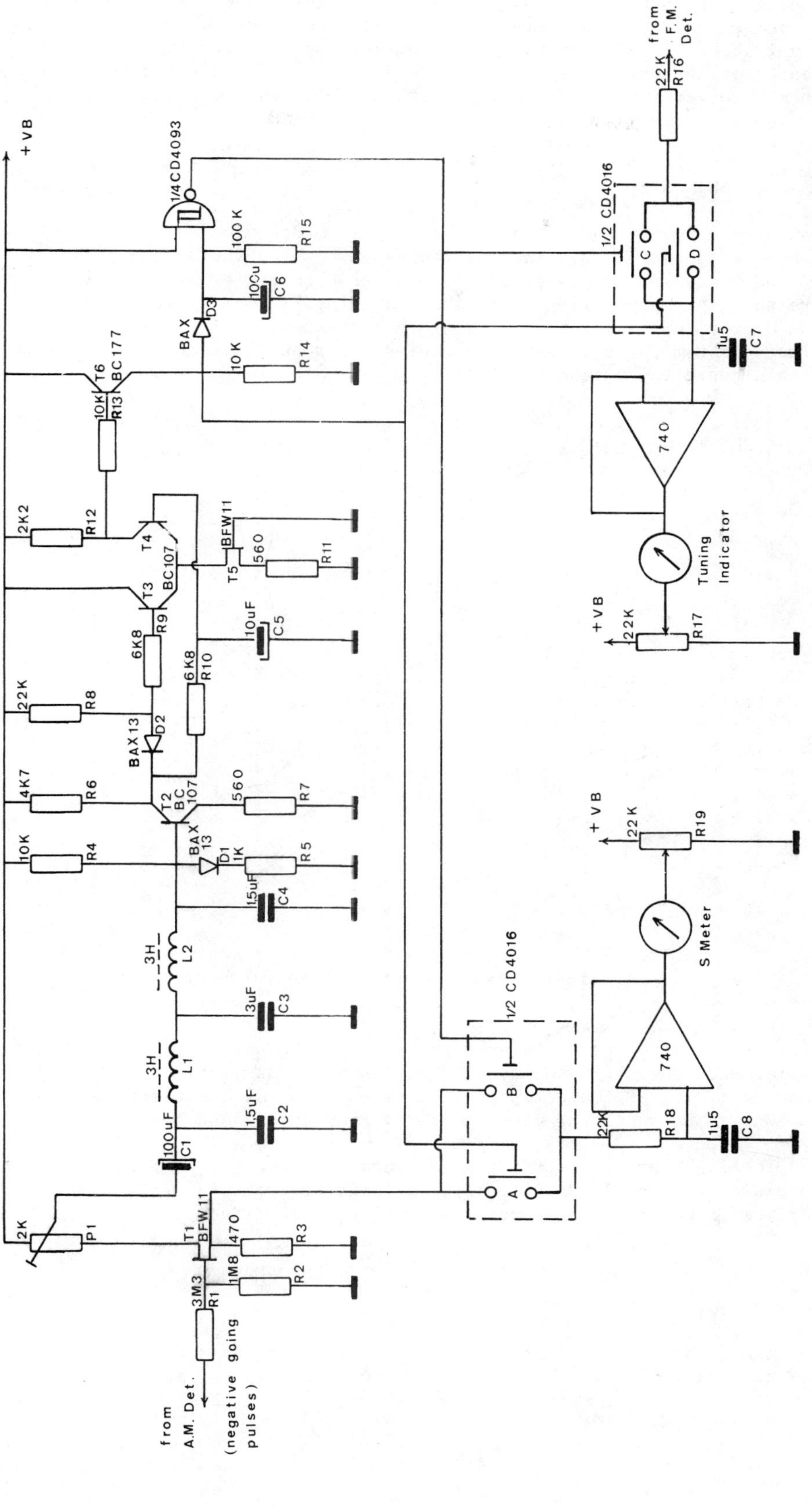

Fig. 9. Pulse detection circuit and memory circuit of the receiver.

The tuning indicator circuit operates in the same way as the 'S' meter circuit does. The signal here is derived from a ratio detector, which indicates a zero output level when the receiver tuning is correct and the I.F. signal frequency is at the center of the crystal filters passband. This causes a midscale meter reading. When the receiver is tuned too high or too low, a lower or higher meter reading results.

ANTENNAS

For radio tracking 27 MHz transmitters, use is made of adjusted loop antennas with a diameter of about 50 cm. These antennas have the advantage to be handy and light, and besides they are extremely sensitive to differences in direction which renders them very appropriate for making bearings. The adjustment of the antenna is done on the top by means of a 25 pf tuning capacitor (Fig. 10). The rectangular loop inside the main loop serves to achieve a proper impedance matching with the coax cable connected to the receiver.

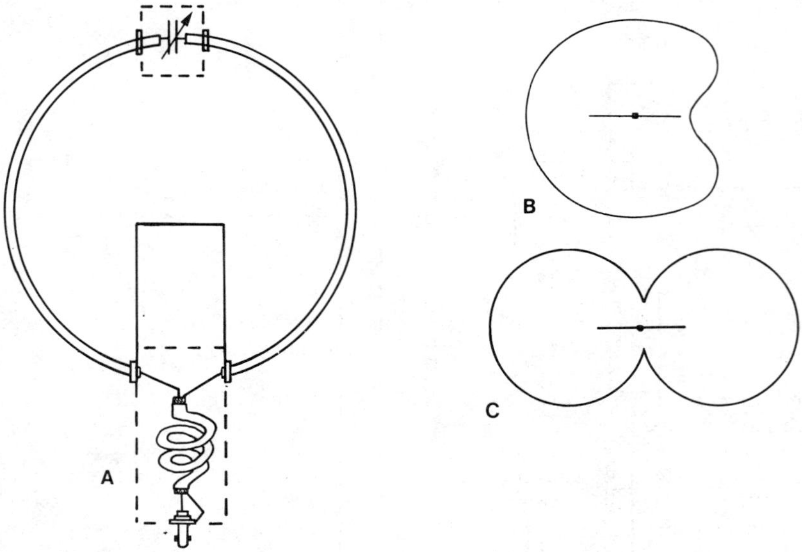

Fig. 10. 27 MHz antenna. (A) Construction scheme and radiation pattern with (B) and without (C) coax coil.

Normally the antenna has a radiation pattern as shown in Fig. 10B. Due to asymmetry the null is not sharp. However, by providing the coax coil with about 10 windings, the coax shielding of the antennas is disconnected which makes the antenna work as a true loop. The radiation pattern is now symmetrical and has a sharp null.

For the localization of 154 MHz transmitters a vertically polarized yagi antenna with 8 elements is applied (Fig. 11). In order to register animal activity automatically, a fixed station is used with a 5/8 λ omnidirectional ground-plane antenna.

APPLICATION OF 27 MHz TRANSMITTERS

ANIMALS

In 1968 the first red foxes raised in captivity, were fitted with prototype transmitters and released into the field. The question was put forward of whether control

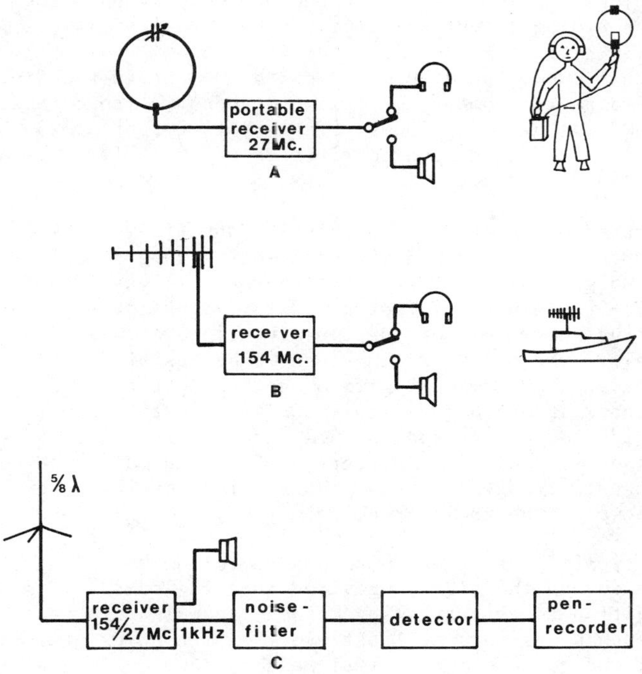

Fig. 11. Receiving systems. A: mobile tracking 27 MHz frequency (loop antenna); B: mobile tracking 154 MHz frequency (yagi antenna); C: fixed station activity registration (ground plane antenna).

of foxes in an area induced a migration from a higher density area, and consequently a dispersion of rabies. This led among other things to investigations concerning migration and social relations in foxes, during which time the use of radio tracking became necessary. After some early problems, due to failures of badly adapted transmitter models, poor battery operation and problems in catching foxes, the research was well under way by 1972. During 1970 to 1977, 86 cubs and young foxes up to 5 months of age (to September) were tracked, 72 foxes in their first year (September-April) and 78 (32 males, 46 females) adults.

Furthermore 22 hares were fitted with transmitters and studied from 1973 to 1977. During 1976 four little owls were radio tracked, during 1977 nine free roaming domestic farm cats and since 1978 two beech martens, four polecats and seven black grouse hens.

PROPERTIES OF THE EQUIPMENT AND METHODS

As the radio tracking methods and the equipment used during our studies with foxes served as a model for later investigations with other species, the following account describes mainly the experiments with foxes.

In flat and open country the biggest transmitters, used on adult foxes, had a range of 2 km while the smallest ones, used on fox cubs and little owls reached 1 km maximum. Due to the small diameter of the polecat collar, this type only reached about 500 m. Hills, heavy trees, metal wire fences and buildings often caused unexpected reflections which could only be recognized with experience. However, signal strength was attenuated only slightly. At a distance of 500 m in open and flat country the big transmitters could be triangulated with an accuracy of 25 m.

The position of the tracked animals was plotted on maps in the field. The scale of these maps depended on the home range size of the animal concerned. For instance a scale of 1:12,500 was used for foxes with territories of 50-250 ha, while a scale of 1:25,000 was used for foxes with territories of about 1,000 ha and a scale of 1:10,000 for hares and polecats.

Bearings were carried out by a mobile receiving unit from conveniently situated spots along roads and paths, marked with reflecting tape so as to find them more easily during the night. The direction of the animals was determined either by turning the antenna in such a way that a maximal signal strength was obtained (peak signal) or by taking the middle of the two directions at which no signal could be heard (null signal). The null signal proved to be easier within distances of about 500 m. In most cases a cross-bearing from another spot could be achieved within 2-3 min, while signal strength gave additional information about the position. For fast moving individuals the interval between two bearings appeared to be too long. In most cases, however, the moving direction of the animal could be determined from differences in signal strength. Visual observations by infra red binoculars provided regular checks of the radio tracking method.

With a high density of foxes it was possible to follow 3-5 simultaneously, locating them every 30 min during the night, provided that the foxes did not show abnormal behavior. In addition to this tracking method, animals were followed more or less continuously in such a way that at least every 15 min an exact location was plotted, while in between, the route of the animal was mapped using signal strength differences and visual observations. To determine the home range during a given period, the plots of that period were combined on a map. In the area with small territories a radio fix interval was made each 30 min and each 15 min for the area with large territories. Although very labor intensive, this method also proved to be satisfactory regarding the study of social structure in fox populations. Automatic devices for bearing and plotting the locations of the animals can save a lot of labor. However, direct observations in the field are of essential importance to reach the necessary basic knowledge of the habits of animals.

During the inactive period, the individuals were in general tracked only once a day. Underground animals, such as foxes and polecats could be located within one meter by using the coax cable of the receiver as an antenna near the den. This accuracy is important to restrict damage to the den by digging when the animals have to be dug out. Foxes without a transmitter could often be located exactly in their den by releasing a dachshund with a transmitter whose position underground was then pinpointed.

Air force airplanes or helicopters were used to track lost animals. The loop antenna was fastened under the plane perpendicular to its longitudinal axis and the receiver could be tended by the crew. Flying at a height of 300-400 m the range of the big fox transmitters appeared to be about 4 km. In spite of this range the results of these trials were unsatisfactory: only once were a lost fox and hare recovered.

It is worth mentioning that, one could assess when foxes went into water, for instance to cross a canal. The wet coat inside the collar caused a higher pulse frequency.

RELIABILITY OF THE TRANSMITTERS

The calculated transmitter life-time, based on battery capacity, could be achieved in the field. A broad variation in life-time could be expected since the loop antennas of the transmitters had to be individually adjusted for the animals. Some transmitters, however, failed before the batteries were discharged. The distribution and the causes of failure in fox transmitters are given in Table 2.

The construction of the collar is of great importance, and must be sufficiently sturdy. Deformation of collars by other foxes occurred seven times, in three of which an animal was injured. In young foxes a rather loose collar was chosen to prevent injuries to animals that could not be recaptured at the proper time to replace the transmitter with an adult type. Initially the growth of foxes sometimes caused transmitter failure because a thicker neck detuned the loop antenna. Most of these difficulties could be avoided by the use of a collar loop antenna with a double winding. A loose collar proved a possible risk to young foxes which hooked a front paw through the loop antenna. Since this too changed the pulse frequency, an individual could be identified and recaught. In one case we were unsuccessful and the fox died. In two more cases the cubs could not be recaptured; one of which was shot after six months with its collar deeply embedded into its neck.

Of the juvenile foxes, 23 could not be recaptured on time, 12 of these were shot or caught afterwards. In two of these 12, both large males, the transmitter caused light skin abrasions. As the recovery rate of ear-tagged juveniles has proved to be about 50 percent, it is unlikely that the collar caused a remarkable mortality. The reliability of the transmitters, a continual field control and a large labor input in recapture efforts resulted in what we judged to be an acceptable risk for the cubs and juveniles with the small transmitters.

The transmitters used on cats, beech martens, polecats and hares were highly reliable. In the few cases in which the transmitter failed early it was due to an inferior connection between transmitter and collar. Although little owls were radio tracked for only a week, the observations were successful and the feeding territories of the males during May and June could be determined satisfactorily.

The first four black grouse hens lost their transmitters because of weak soldering joints between body loop and transmitter. After improvements were made on this connection one transmitter stopped after 120 days. A second hen was swept down by a hawk after which the carrion could be traced by the still working transmitter, and a third hen disappeared while she was incubating a clutch of eggs.

INFLUENCE OF RADIO TRACKING ON THE ANIMAL'S BEHAVIOR

Immediately after release, animals with a radio collar scratched their necks frequently. This behavior disappeared in one or two days after which, we believe, they did not react further to the transmitter. This impression is also based on observations with infra red binoculars and a remote controlled TV camera with video recorder situated in the field, by which the behavior of animals with and without transmitters, could be compared.

Although the transmitter of the little owl was relatively heavy (Table 1), observations showed no abnormal behavior. One bird still bearing the transmitter was caught in the same nest cavity after one year. Transmitters on red foxes showed some clear punctures of teeth and a few weak collars were deformed. Direct influence of a transmitter on the behavior of cats and hares took the form of difficulty in jumping. After a few jumps, however, the animals adapted to the new addition to their weight.

Due to their capture, foxes were often rather inactive during the first 24 hours. Sometimes they went out of their normal home range, a phenomenon also observed in hares, polecats and black grouse. However it did not lead to a permanent home range shift. Possibly this may happen in young animals that are not yet established, although this could not be shown.

Table 2.
Distribution and causes of failure in fox transmitters

CUB type: longest operation time 34 days

Days	0-10	10-20	20-30	30-40	total
Recapture		8	15	13	36
Transm. stop	4	6	8		18
					54 (13 litters)

stop after 21 days 4

Causes of failure: exterior damage 3
front paw within collar 1
bad contacts 10
18

JUVENILE type: longest operation time 155 days

Days	0-25	25-50	50-75	75-100	100-125	125-150	150-175	total
Recapture	9	20	7	5	6			47
Transm. loss	4	1	1					6
Transm. stop	6	2	1	4	3	4	3	23
								76

stop after 3 months 12

Causes of failure: exterior damage 1
migration 1
critical tuning and bad contacts etc. 9
23

ADULT type: longest operation time 320 days

Days	0-50	50-100	100-150	150-200	200-250	250-300	300-350	total
Recapture	14	13	27	6	4	5	3	72
Migration, obs. stop	1	1	3	2	2	1		10
Transm. stop	25	11	7	16	11	8	1	79
								161

stop after ½ year 29

Causes of failure: exterior damage 9
bad prints, contacts, crystals, batteries 13
production faults (1974-1975) 14
unknown 14
79

One instance of shock occurred in foxes. This happened more often in hares; nearly 5 percent of the animals died of shock shortly after being captured. Cats and mustelids were anesthetized in their traps with Rompun and Ketamine respectively to facilitate the attachment of a transmitter and no complications were observed. The black grouse all suffered a certain degree of shock which often resulted in restrained movements after release. Therefore they were put in the dark for some hours before release. Restricted movement even after a few days may imply that their shock subsided slowly.

Remarkable differences in shyness among several individuals of the same species occurred. Some foxes for instance were seen only occasionally or watched the tracking car continuously from considerable distances while others behaved normally even in the beam of the headlights. Similar individual differences could be observed in hares and polecats. Cats were rarely observed in the field. The behavior of all animals seemed less restricted if the observer was on the observation spot before the animal appeared, so mobile radio tracking may influence an animal's behavior.

AUTOMATIC ACTIVITY RECORDING

Simultaneous visual observations and radio tracking showed that some components of the animal's behavior could be recognized by variations in pattern and strength of the signal. To make use of this in an efficient way a device was made to register the signal automatically on a penrecorder (Fig. 11). The structure is rather simple and our preliminary experience with beech marten, polecat and black grouse is encouraging. Much labor can be saved in this way. Nest visitation and foraging times could be followed in a black grouse hen during the whole breeding period.

Atmospheric activity (storms, lightning) form a problem, severely disturbing the recording pattern locally. Therefore, a noise filter was developed and is to be tested in the field this year.

APPLICATION OF 154 MHz TRANSMITTERS

In 1978 the first two adult seals were provided with a transmitter, attached to their backs with a harness made of neoprene. An 'O'-ring around the neck, also made of neoprene, prevents the net from slipping backwards (Fig. 6).

In the first released seal a breakable piece of sisal was attached within the neoprene 'O'-ring. Based on calculations this sisal should decay after about 9 months in seawater. After 3 weeks, however, transmitter and harness were found on a sandbank. The sisal connection was broken.

In August a second seal (adult female) was captured on a sandbank and provided with a harness without sisal in the 'O'-ring. The first day the animal moved some distance away and could not be observed on a sandbank. Although the animal returned to its normal range on the second day it did not rest on the bank as usual. From the third day onwards no abnormal behavior was seen. By visual observations on the bank the animal was watched rolling over on the sand without disturbance to the position of the transmitter.

The transmitter range depended on the position of the transmitting antenna and the height of the receiving antenna. Signal strength is maximal when the antenna is completely above sea-level. From an airplane, flying at a height of 2 km the maximal range was about 100 km. At a height of 100 m this figure amounted to 25 km and from a boat or dike about 8 m above sea-level, 10 km.

Automatic recording was achieved using the equipment in Fig. 12c, and results are shown in Fig 12. A whip antenna was positioned on a high chimney (106 m) on the nearby coast. Unfortunately the transmitter stopped working after one month, too short a time to develop a good and reliable radio tracking method to use at high tide when the seals are in the water. Triangulation was hampered furthermore by the short duration of the seals being above sea-level and transmitting. Apart from following seals at close distances and visual observations from a boat, radio tracking simultaneously from two stations or from two boats seems to be a possibility.

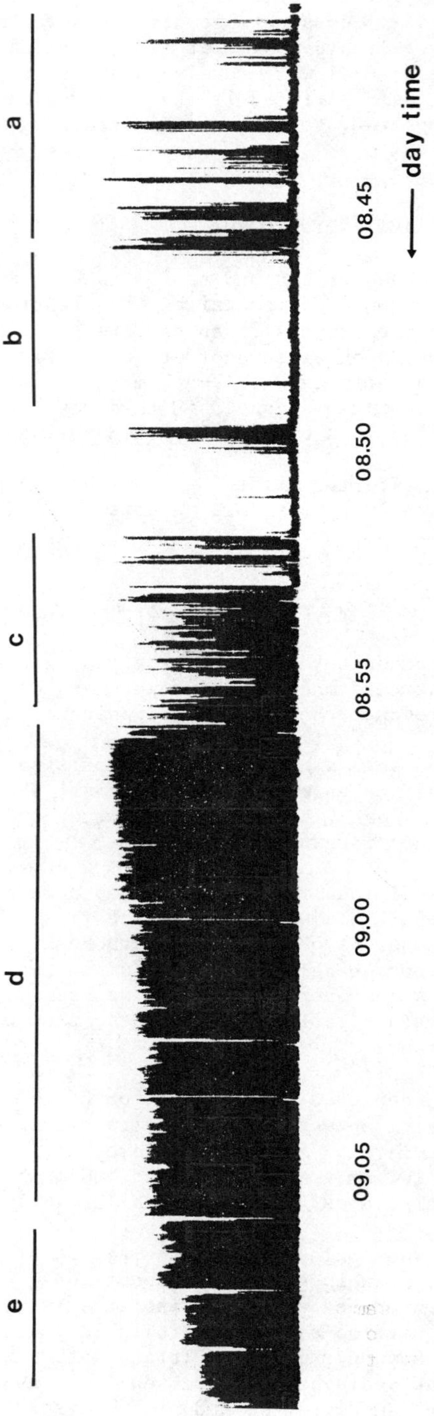

Fig. 12. Activity registration of a seal. a: alternately under water and on the surface, b: under water, c: on the sandbank in about 30 cm water, d: on the dry sandbank, e: as d, crawling to the border (signal screened by the bank).

Making the Most of Radio Tracking— A Summary of Wolf Studies in Northeastern Minnesota

L. D. MECH*

Patuxent Wildlife Research Center, U.S. Fish and Wildlife Service, Laurel, Maryland 20811, U.S.A.

Abstract — In a continuing study of wolves (*Canis lupus*) and their major prey, white-tailed deer (*Odocoileus virginianus*), in northeastern Minnesota (U.S.A.), radio tracking has been the primary technique used since 1968. Some 182 wolves and 106 deer have been captured in one study area, and most of them radio tracked. All individuals captured for radio tagging were also blood-sampled so that their nutritional and physiological state could be determined. Basic information has been gathered about the socioecology of both species and about interactions of the two. Seasonal movements, home range and territory sizes, migrations, and dispersal of individual wolves and deer have been studied, and histories have been accumulated for as long as 4 years for a deer, 7 years for an individual wolf, and 11 years for a pack. Composite data over the study period has been accumulated for over 16 contiguous wolf packs.

Studying large numbers of both predators and prey over a long enough period, however, has also made it possible to gather data on population parameters of these species. Thus productivity, density, population trend, survivability, mortality, and predation rates have also been measured, and details of the interaction between wolves and deer have been described from the standpoint of both species.

With the identities and locations of individuals and social units known, a wealth of supplementary information could also be gathered. Individual wolf packs were tracked and their scent-marking described. Packs of known compositions were induced to howl under various circumstances, and their responses analyzed. Carcases of study animals were given thorough autopsies, so that causes of death could be ascertained. The spatial distribution of wolf-killed deer were plotted in relation to deer-concentration areas and to wolf-pack territories, and a relationship was discovered between deer and wolf socioecology.

*Mailing address: North Central Forest Experiment Station, 1992 Folwell Avenue, St. Paul, Minnesota 55108, U.S.A.

This attempt at combining radio tracking of both predator and prey, of blood-sampling study animals, and of supplementing the radio tracking studies as much as possible, produced a matrix of information about the entire predator-prey system that far exceeds the value of merely recording a series of radio locations for its constituent animals.

INTRODUCTION

Radio tracking has been in common use for wildlife research since about 1963 when Cochran and Lord (1963) devised a practical radio tracking system. Since then, this technique has revolutionized wildlife ecology studies. Earlier I traced the history of radio tracking (Mech, 1968) and discovered that several schools of investigation had been developing parallel techniques at about the same time. Now, some 15 years after perfection of this important method, a great variety of species are being radio tracked routinely in most parts of the world. One of the leading commercial suppliers of radio tracking equipment had sold over 17,500 transmitters as of March, 1979.

Because radio tracking was such a revolutionary technique, its application even to common and well studied wildlife has provided new insights into many aspects of the lives of these species. Furthermore, its application to elusive animals, wilderness species, large carnivores, and other animals difficult to study, has been especially productive. Even fish are commonly studied through the use of radio tracking, and marine mammals under the Antarctic shelf ice are routinely investigated via this technique.

Several of the applications of radio tracking are immediately obvious to any researcher. However, there are many other highly productive uses that may only occur to an investigator when he or she is deeply involved in an intensive radio tracking study. In this paper I will discuss the obvious uses of radio tracking as well as some less obvious ones. I will take as an example a series of my own studies on wolves and their prey in northeastern Minnesota, U.S.A., supplemented with other examples that seem to be particularly good examples of the creative use of radio tracking.

The most obvious use of radio tracking is in following the movements of the subject animals. With enough persistence, ingenuity, funding, and/or hard work, it is now theoretically possible to keep constant track of the movements of most individual animals. One highly sophisticated variation of this technique involves automatic monitoring of the movements of radio tagged animals (Cochran *et al.*, 1965). Those not fortunate enough to have such a space-age system at our disposal must be content with obtaining only a location per day or sometimes even less if our subjects are wider ranging or more elusive. Some people have even solved this problem with large, elusive species such as the polar bear (*Ursus maritimus*) by placing their receiving apparatus in satellites (Kolz, Lentfer and Fallek, 1978, 1980).

Regardless of whether the locations are obtained via an automatic tracking system, a satellite, an airplane, a vehicle or by an observer on foot, certain types of information can be extracted from the data. A series of continual radio fixes, regardless of the sampling interval, provides at least a rough approximation of an animal's travel route. A composite of such points over a long enough period gives an idea about the extent of the individual's travel. After recognizing the possible biases involved in the sampling intervals or times of day or season, one can also obtain an approximation of an animal's intensity of use of its range from such a composite of radio locations. An idea of the size, shape and geographic location of an animal's home range or territory can also be formed. Mosaics of such data from several individuals in the same population provide considerable insight into the spacing and relationships among social units of the population.

Dispersal of individuals from local populations are especially well studied through the use of radio tracking. We have followed dispersing wolves for straight-line distances of up to 208 km. Much wider ranging movements of, for example, migrating birds have been studied with extremely interesting results. The longest path of travel that I know of which has been described through the use of radio tracking is the southward migration of a peregrine falcon followed by Cochran (1975) from Wisconsin into Mexico during a 16 day journey, although possibly some of the recent satellite studies have broken this record.

If enough members of a local population are studied over a long enough period, the composite results of all of the individual parameters discussed above will amount to more than the mere sum of those parameters. The net result will be a description of the socioecology of the species. To achieve this the study must be continued long enough to allow the following of individuals born to radio tagged parents, the radio tagging of those offspring, and the following of the offspring through dispersal and into the next generation, either within the population or in newly colonized areas. A few such studies have already been conducted, e.g. those on black bears (Rogers, 1977), wolves (Mech, 1972, 1973, 1974, 1977a; Peters and Mech, 1975; Rothman and Mech, 1979), and white-tailed deer (Nelson and Mech, submitted).

In another situation, we combined radio tracking studies of the socioecology of the wolf and its major prey the white-tailed deer. This combined research is leading to a new understanding of the co-evolution of the social lives of these species (Hoskinson and Mech, 1976; Mech, 1977a, 1977c; Nelson and Mech, submitted).

A new dimension can be added to radio tracking studies by collecting as much data as possible from the animal to which the radio will be attached, for example, sex, approximate age, and weight of the subject animal. However much more can be done. With many species, a tooth can be extracted for precise aging by means of dental annulations. This has proven extremely useful for studies of black bears (Rogers, 1977), deer (Nelson and Mech, submitted), and many other species. At the very least, care can be taken to retrieve carcases of radio tagged animals that die during the study, and teeth then can be removed for aging, along with other obvious tissue specimens such as reproductive tracts, stomach contents, etc.

One of the most valuable pieces of information about the subject animal can be obtained through collecting and analyzing a blood sample. Although so far blood analyses and interpretation has required collaboration with a blood physiologist, an impressive literature is now growing, giving baseline data and examples of blood values of many species, along with suggestions on interpreting the values. The advantage of knowing whether your study animal is in good, moderate, or poor nutritional condition, or is infected with a certain disease, is immeasurable.

Several other pieces of information or specimens can be collected from the subject animal. A small handful of hair can give an idea of its nutritional state and possible mineral deficiences (Franzmann, Flynn and Arneson, 1975). Saliva, urine, milk and pieces of hoof, claw, and various tissues can be collected and analyzed for nutrients. Hormone stimulation tests, which we are now performing routinely on our wolves, can give an idea about the state of maturity and reproductive status of the subject animal.

Another category of information that can be obtained through radio tracking involves using the radio to locate a subject animal so that further observations can be made. A behaviorist could use the radio to locate and identify a subject animal and then watch it for long periods. In our study we note the activity and the number of individuals in each radio tracked wolf pack and coupled with information on the territory size of each pack, these data allow population estimates to be made. If this is done for a long enough period a population trend can be seen and attempts then made to relate it to other conditions.

Kills, natal dens, nests, winter dens, calving grounds and other areas of special importance in an animal's life can be found through the use of radio tracking. In the case of bears, location of winter dens is of extreme importance both for determining the number of cubs produced and for allowing easy replacement of the radio collar each year (Rogers, 1977). With wolves, our ability to find dens and rendezvous sites (nests above ground where wolf pups live until autumn after leaving the den) allows us to observe from an airplane the number of pups produced and surviving for each wolf pack. Early winter observations of the same pack when the cubs are traveling with it, yields an estimate of the summer and fall pup mortality (Mech, 1977b).

Just as studying productivity is greatly aided by the use of radio tracking, so too is the investigation of mortality. By knowing how many individuals have survived for certain periods and also knowing what percentage of radio tagged animals died, one can determine mortality rates and the seasonal distribution of mortality (Mech, Barnes and Tester, 1968; Trent and Rongstad, 1974; Mech 1977c). The most important aspect of mortality studies, i.e. determining the cause of death, is greatly aided by radio tracking by watching closely for a lack of change in an animal's location. Usually a fresh carcase can be found which can be examined for cause of death. Unless the mortality cause is obvious, the carcase should always be submitted to a veterinary pathologist for autopsy (Mech *et al.*, 1968; Mech, 1977c).

A special case of mortality, e.g. predation, can also be studied through radio tracking (Mech, 1967). In fact, anyone radio tracking prey animals will automatically study predation, for sooner or later some radio tagged prey will be killed by predators. By keeping very close track of each radio tagged prey animal, a researcher can quickly determine when an animal is dead and can then make a direct check to gather information about the instance of predation.

To facilitate the study of predation through radio tracking, however, one can use a special activity transmitter that indicates when an animal is active (Knowlton, Martin and Haug, 1968). Then by monitoring the signal, the investigator can determine whether or not the animal is active. If it is inactive for a long enough period, mortality is suspected, and a direct check is made. A more sophisticated type of transmitter is made especially to indicate when an animal has died. Such a transmitter is set to produce a certain kind of pulsing signal as long as the animal continues to move intermittently. However, if the animal stops moving for a specified period, for example 1 or 2 hours, the pulse rate changes, indicating that the animal has probably died (Kolz, 1975). Extensive use of this type of collar has been made in studies of predation on young ungulates (Schlegel, 1976; Ballard and Taylor, 1978).

Obviously the more background information obtained about the radio tagged prey animal, the more useful is the predation study. For example we recently learned that in spring white-tailed deer fawns having a significantly lower amount of non-esterified fatty acids had a greater probability of being killed by wolves in the Superior National Forest, U.S.A. (Seal *et al.*, 1978). The following is a summary of predator-prey research on wolves and white-tailed deer fawns.

WOLF STUDIES

The wolf research has focused on ecology, behavior, movements, spatial organization, and prey relationships of several contiguous wolf packs in an area of about 2,600 km^2 in the east central part of the Superior National Forest of northeastern Minnesota, U.S.A. The Superior National Forest (Fig. 1) lies at 92° west longitude and 48° north latitude and has been well described by Stenlund (1955) and Mech and Frenzel (1971).

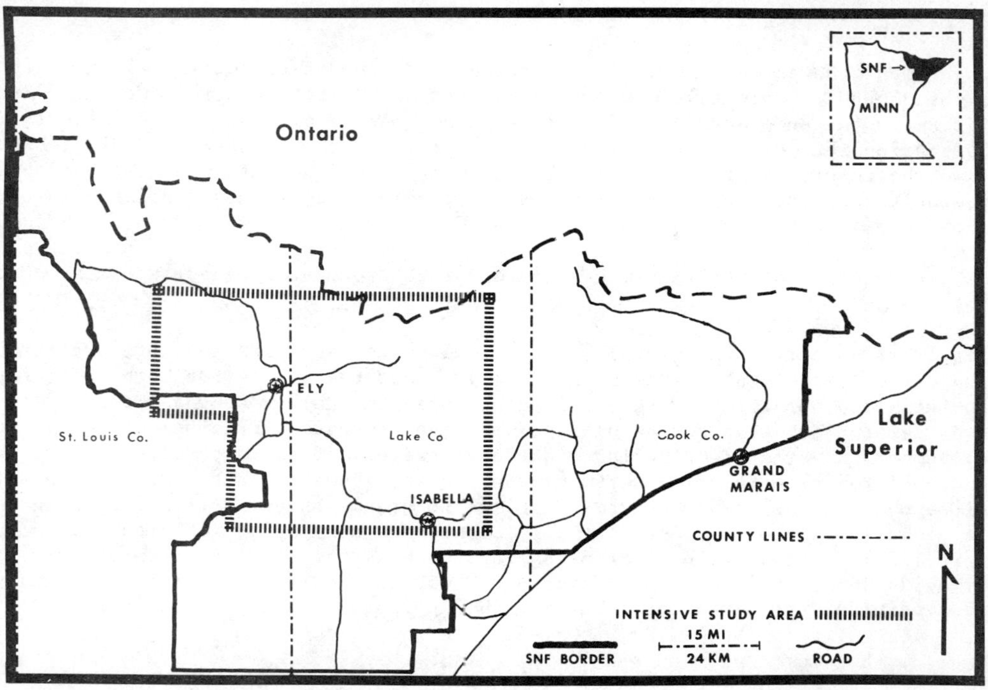

Fig. 1. The study area.

The primary prey of the wolf in the study area is white-tailed deer, with moose (*Alces alces*) and beaver (*Castor canadensis*) supplementing this diet. All three prey species are harvested by humans, although much of the intensive study area is inaccessible.

METHODS

Wolves were captured in steel foot traps (Kolenosky and Johnston, 1967; Mech and Frenzel, 1971; Mech, 1974). They were anesthetized via a jabstick with a syringe of sernylan and promazine (Seal, Erickson and Mayo, 1970; Mech, 1974), and were then blood-sampled, weighed, and measured. A radio collar (Cochran and Lord, 1963) was attached to each wolf, after which the animal was released. Most of the radio tracking was done from an aircraft; usually one flight per week throughout the year and daily flights during winter, weather permitting. During winter the rate of observation of wolves and their packs was about 75 percent, whereas in summer it was about 10 percent (Mech, 1974). Records were kept on the number of wolves seen at each location, their behavior, and whether or not they were at a kill. When possible, we later landed and examined the remains of kills made by our radio tagged wolves and by any other wolves in the area.

Additional information was obtained on the wolves' primary prey in the area, the white-tailed deer. Records were kept on the location and distribution of the deer and their numbers. An aerial census was also conducted in part of the study area during later years (Floyd, Mech and Nelson, 1979). From 1974 on, my colleagues and I also live-trapped deer via dart gun, clover traps and rocket nets, and radio collared them (Hoskinson and Mech, 1976; Nelson and Mech, submitted). We also blood-sampled, weighed, and measured the deer and radio tracked them several times per week, depending on season. The location error for this type of tracking was measured at 7 to 40 m (Hoskinson, 1976).

To supplement the aerial studies, we carried out a number of additional investigations from the ground. Radio tagged wolves and their pack mates were followed for a total of 373 km in the snow, and records were kept of the number of times they scent-marked during that distance (Peters and Mech, 1975; Rothman and Mech, 1979). We also studied the functions of wolf howling. Our main technique was to approach the radio tagged wolves on the ground and howl towards them and record any replies obtained. Replies were then analyzed by sonagraphing (Harrington and Mech, 1978a, 1979). Another part of this study involved placing microphones within 0.5 km of active wolf homesites and recording any howling that occurred around them (Harrington and Mech, 1978b).

An important aspect of the overall study involved the analysis and interpretation of the blood samples taken from both the wolves and the deer. For both species the following blood parameters were examined: hemoglobin, number of red blood cells, percent hematocrit, mean corpuscular volume, mean corpuscular hemoglobin concentration, number of white blood cells, and concentrations of cholesterol, calcium, phosphorus, bilirubin, uric acid, serum urea nitrogen, glucose, lactic dehydrogenase, alkaline phosphatase, serum glutamic oxalacetic transaminase, total protein, albumin, gamma globulin, thyroxine, and cortisol (Seal, Mech and Van Ballenberghe, 1975; Seal *et al.*, 1978). In addition other parameters such as thyroxine and non-esterified fatty acids were examined when necessary. Thus, the health and nutritional status of the individual study animals that were being radio tracked was monitored.

RESULTS

Through early 1979, a total of 182 wolves and 106 deer had been captured, and most of these were radio tracked. Over 50 recaptures of radio collared wolves have been made; one wolf (No. 2407) was radio tagged six times over a 7 year period. One deer (No. 98) was studied over a 4 year period. Data are now available on tens of thousands of locations of wolves and deer, and on several hundred wolf-killed deer. The following results, however, pertain only to the data analyzed for the specified periods.

PREDATION ON DEER

During the study the deer population declined markedly, and the following description of that decline is from Mech and Karns (1977). The deer decline resulted from a combination of three factors: (1) gradual maturation of the forest resulting in a deterioration of deer habitat, (2) a series of seven severe winters from 1966 through 1972, and (3) a high wolf density approximating one wolf 26 km^{-2} (Mech, 1973).

The deer decline became most apparent after the severe winter of 1968-69, and by 1974 deer were essentially gone from an area of about 3,000 km^2 of the poorest deer habitat. On the edge of this area deer densities in 1975-76, 76-77, and 77-78 were 0.4 to 0.7 deer km^{-2} (Floyd *et al.*, 1979).

Throughout the study, wolves had been killing primarily fawns, deer aged 5 years old or older, and individuals with various abnormalities (Mech and Frenzel, 1971; Mech and Karns, 1977). In general, radio tagged fawns killed by wolves in winter had significantly lower non-esterified fatty acids in their blood than those who survived (Seal *et al.*, 1978). Under usual conditions the prime members of the deer herd were still left to produce enough offspring to enable the herd to continue to survive in a moderate density. However, probably because of both habitat degeneration and the series of severe winters, the productivity of the deer herd dropped markedly and the high density of wolves brought extra pressure on the remaining older deer and on the fawns that were produced (Mech and Karns, 1977).

The surviving deer had several interesting characteristics. First, their blood values indicated that at least in late winter and early spring the animals were nutritionally deficient in several ways (Seal *et al.*, 1978). Secondly, the survivors were unusually old, averaging about 4.6 years of age (Hoskinson and Mech, 1976; Nelson and Mech, submitted). Furthermore, during the study these deer continued to survive at an unusually high rate; for example, the study animals survived on the average a minimum of 2.2 additional years.

The third characteristic of the surviving deer that may well have caused their longevity was the fact that most of them lived, both in winter and summer, in the buffer zones of wolf pack territories (Hoskinson and Mech, 1976; Mech, 1977b; Nelson and Mech, submitted). Buffer zones are strips about 2 km wide along the edges of wolf pack territories where packs from either side can venture but where neither spend very much time (Peters and Mech, 1975). Apparently wolves do not kill many deer in their buffer zones until they have depleted the vulnerable prey within their pack territories and then become desperate (Mech, 1977a).

Before deer declined drastically in the study area, the kill rate by wolves averaged about 18 deer per wolf per year, or 2.5 kg per wolf per day, based on winter kill estimates, excluding beavers and moose in the calculations (Mech and Frenzel, 1971). However, when consumption of beaver and moose are considered, the estimated kill rate decreased to about 15 deer per wolf per year (Mech, 1971). With wolves at a density of about one 26 km^{-2}, which was the approximate density in 1971 (Mech, 1973), it would take a deer density of about one km^{-2} to produce the number of deer consumed by wolves per year. There probably has not been an average deer density this high in the study area since about 1970. After the deer decline, the average daily consumption rate dropped as low as 1.5 kg per wolf per day (Mech, 1977a).

WOLF SPATIAL ORGANIZATION

Approximately 16 contiguous wolf packs have been studied in the central Superior National Forest, U.S.A. Based on an analysis of data through 1973, the pack territories ranged in size from 125 to 310 km^2 each (Mech, 1974), and winter pack sizes averaged 4.1 to 7.3 members from 1970 through 1975 (Mech, 1977c). Each pack territory is surrounded by the territories of several other packs (Mech, 1973). The territories are maintained by two methods. First, scent marking by raised-leg urination provides a network of olfactory locations, indicating which wolves occupy the particular area. These marks occur at an average frequency of about one fresh mark 450 m^{-1}, and the rate along the edges of the territory (that is in the buffer zone) is twice as high as that in the territory center (Peters and Mech, 1975).

The second method that wolves use to help maintain their land tenure is howling. Little is still known about the rate of spontaneous howling by packs during all times of the year. However, two packs studied during summer had a spontaneous howling rate around their homesites of 0.75 to 1.0 howling sessions 24 h^{-1} period (Harrington and Mech, 1978b). Furthermore, wolf packs tended to reply at high rates to our simulated wolf howling when they had kills, dens or rendezvous sites to defend (Harrington and Mech, 1979). Apparently howling is used to advertise the immediate location of the pack when it has something it is prepared to defend.

All wolves are born into a pack, and pups can mature sexually as early as 9 or 10 months of age (Medjo and Mech, 1976). However more often wolves do not mature until they are 2 or 3 years old (Mech, 1970). Some of the maturing wolves leave the pack and become lone wolves. Lone wolves may be nomadic around the wolf population in which they were born, covering areas up to 5000 km^2 and concentrating their movements around the edges of wolf pack territories (Mech and Frenzel, 1971; Mech, 1973). Or they can disperse directionally from their population for distances of 200 km or more (Mech and Frenzel, 1971). In either case if they find a vacant territory and a wolf of the opposite sex, they may settle in the territory with that individual.

Probably because lone wolves will be chased and killed if they are caught by resident packs, they keep a low profile by responding at a very low rate to simulated howling (Harrington and Mech, 1979), and only rarely scent marking by raised-leg urination (Rothman and Mech, 1979). However, within at most a few days of encountering a member of the opposite sex, the lone wolves begin scent-marking, and they mark at an unusually high rate during the first few weeks of bonding, so apparently scent-marking is part of the pair-bonding process (Rothman and Mech, 1979). Meanwhile they have heavily marked their new area and claimed it for their territory. They are then ready to breed.

WOLF REPRODUCTION

As breeding season approaches, wolves scent-mark more often (Peters and Mech, 1975; Rothman and Mech, 1979), and pair members sleep increasingly closer to each other (Mech and Knick, 1978). In northeastern Minnesota, wolves breed as early as January 28 or as late as March 4, and the pups are born in April and early May. During the present study individual litter sizes ranged up to seven pups per litter, but mean litter sizes gradually declined (Mech, 1977c). Pup growth rates and sizes varied considerably; some members weighed almost twice as much as some of their litter mates in autumn (Van Ballenberghe and Mech, 1975). An increasing number of wolf pups were underweight, as the deer herd declined, and many of their blood values were deviant from the norm (Seal *et al.*, 1975). Underweight pups usually perished by November of their first year (Van Ballenberghe and Mech, 1975). Immature and subordinate pack members tended to have more deviant blood values than the 'alpha' or breeding pair (U.S. Seal, in Mech, 1977a). As litter sizes decreased, the proportion of male pups that were born and survived increased, and in 1973 males comprised 83 percent of the pups captured in the study area (Mech, 1975).

MORTALITY

Three major types of mortality affected the wolves in this study. During the early part of the investigation, human-caused mortality was one of the most important factors. However, as legal protection took effect, human-caused mortality diminished, and natural mortality became much more significant (Mech, 1977c). Of equal importance were two natural mortality factors: (1) pup malnutrition (Seal *et al.*, 1975) and (2) intraspecific strife, that is, killing of wolves by strange packs, usually during trespasses (Mech, 1977,a,c). Annual over-winter mortality of pups and adults varied from 5 to 51 percent; the highest mortality rates occurring within a few years after the deer declined.

ACTIVITY

On the basis of winter observations of five of our radio tagged wolves during winter 1968-69, Mech and Frenzel (1971, p. 11) concluded the following: "the wolves were resting 62 percent of the time, traveling 28 percent and feeding 10 percent. They tended to travel more before 11:00 a.m. and after 3:00 p.m., although resting still composed at least 45 percent of the activity during every hour." In this analysis, 'resting' and 'sleeping' were pooled and presented as 'resting'.

An intensive analysis of the winter daytime activity of one of the study packs over

6 year period showed interesting changes. As the deer population declined throughout the period, this pack traveled less and slept more. They slept only 17 percent of the time in 1968-69, but in 1974-75, they slept 70 percent of the time (Mech, 1977a).

POPULATION TREND

Within 2 years after the beginning of the deer decline, the wolf population also began to drop (Mech, 1977c). As indicated above, mortality of pups due to malnutrition was one of the first factors recognizable in this decline. However, that was soon followed by a decrease in average litter sizes and an increase in mortality. The result was a decline in population of about 55 percent from 1969 to 1974. However, this severe decline in wolf numbers evidently amounted to an overadjustment, for the next year the population increased (Mech, 1977c), and during the next 2 years it remained approximately at the same level.

Although it is possible that the increase in wolf numbers was preceded by an increase in deer numbers, our best indications are that the deer population was continuing to decline, or at least had declined and remained stable, during those 3 years (Floyd *et al.*, 1979). Perhaps a switching by the wolves to a greater reliance on beavers during April through November, and on moose (Mech, 1977a) throughout the year, helped their numbers to increase modestly, but this remains as speculation.

POPULATION REGULATION

Regulation of the Superior National Forest wolf population is a complex process, involving a number of factors (Mech, 1972, 1977c; Packard and Mech, in press). Social factors that helped regulate the population, are territoriality, intraspecific strife, dispersal, reproductive failure of subordinate females, delayed maturity, and disparate sex ratios, but these factors were influenced to varying degrees by nutrition. Furthermore, nutrition acting directly through starvation and disease, or indirectly through influencing litter size and pup survival certainly also played an important role in the regulation process. Thus social and nutritional factors combined in a complex way in affecting the regulation of this wolf population, and the continuing investigations should help further to distinguish the precise roles of each.

CONCLUSION

During the decade of this study, the wolf and deer populations have been declining, and the research has produced much information about the details and mechanics of the population dynamics. One of the most significant environmental causes of the decline was a series of severe winters. Conceivably the study area could eventually experience a period of mild winters, and the deer herd population may increase. If so, an increase in the wolf population would also be expected, so future research would be directed toward describing the details and mechanisms of that change.

Acknowledgements — This continuing study has been supported by the USDI Fish and Wildlife Service, the USDA North Central Forest Experiment Station, the Ober Charitable Foundation (now the Mardag Foundation), and the World Wildlife Fund. Numerous subprojects referred to in this article were also financed by several other agencies, institutions, and individuals which were acknowleged in the publications cited. Logistical help, cooperation, and encouragement for this study were also granted by the USDA North Central Forest Experiment Station, the Superior National Forest, and the Minnesota Department of Natural Resources, with L. F. Ohmann, A. K. Kemp, J. H. Ohmann, L. T. Magnus, K. Siderits, G. B. Joselyn, and P. D. Karns being particularly helpful. Numerous assistants and students helped with the field work and were acknowledged in previous publications. Biological Technician J. J. Renneberg assisted with the research throughout most of the study. Within the USDI Fish and Wildlife Service, strong administrative support came from R. C. Erickson, L. F. Stickel, K. M. Schreiner, and L. A. Greenwalt. I thank all of the above.

REFERENCES

Ballard W. B. and Taylor K. P. (1978) Moose calf mortality study. Alaska Dept. of Fish and Game Prog. Rept. Vol. I, Projects W-17-9 and W-17-10, Job 1.23R (multilith).

Cochran W. W. (1975) Following a migrating peregrine from Wisconsin to Mexico. *Hawk Chalk* 14, 28-37.

Cochran W. W. and Lord R. D., Jr. (1963) A radio-tracking system for wild animals. *J. Wildl. Mngmt.* 27, 9-24.

Cochran W. W., Warner D. W., Tester J. R. and Kuechle V. B. (1965) Automatic radio-tracking system for monitoring animal movements. *Bioscience* 15, 98-100.

Floyd T. J., Mech L. D. and Nelson M. E. (1979) An improved method of censusing deer in deciduous-coniferous forests. *J. Wildl. Mngmt.* 43, 258-261.

Franzmann A. W., Flynn A. and Arneson P. D. (1975) Levels of some mineral elements in Alaskan moose hair. *J. Wildl. Mngmt.* 39, 374-378.

Harrington F. H. and Mech L. D. (1978a) Wolf vocalization. In R. L. Hall and H. S. Sharp (eds.), *Wolf and Man: Evolution in Parallel*, pp. 109-132. New York: Academic Press.

Harrington F. H. and Mech L. D. (1978b) Spontaneous howling in wolves. *Can. J. Zool.* 56, 2024-2028.

Harrington F. H. and Mech L. D. (1979) Wolf howling and its role in territory maintenance. *Behaviour*, in press.

Hoskinson R. L. (1976) The effect of different pilots on aerial telemetry error. *J. Wildl. Mngmt.* 40, 137-139.

Hoskinson R. L. and Mech L. D. (1976) White-tailed deer migration and its role in wolf predation. *J. Wildl. Mngmt.* 40, 429-441.

Knowlton F. F., Martin P. E. and Haug J. C. (1968) A telemetric monitor for determining animal activity. *J. Wildl. Mngmt.* 32, 943-948.

Kolenosky G. B. and Johnston D. H. (1967) Radio-tracking timber wolves in Ontario. *Am. Zool.* 7, 289-303.

Kolz A. L. (1975) Mortality-censusing wildlife transmitters. 12th International ISA Biomedical Sciences Instrumentation Symposium, April 28-30, 1975. Denver, Colorado: 57-60.

Kolz A. L., Lentfer J. W. and Fallek H. G. (1978) Polar bear tracking via satellite. Proc. 15th International ISA Biomedical Sciences Instrumentation Symposium, April 17-18, 1978. Ames, Iowa: 137-144.

Kolz A. L., Lentfer J. W. and Fallek H. G. (1980) Satellite radio tracking of polar bears instrumented in Alaska. In C. J. Amlaner, Jr. and D. W. Macdonald (eds.), *A Handbook on Biotelemetry and Radio Tracking*, pp. 743-752. Oxford: Pergamon Press.

Mech L. D. (1967) Telemetry as a technique in the study of predation. *J. Wildl. Mngmt.* 31, 492-496.

Mech L. D. (1968) Radio-tracking. *Nat. History* 77, 20-24.

Mech L. D. (1970) *The Wolf: The Ecology and Behavior of an Endangered Species.* New York: Natural History Press, Doubleday Publishing Co. 389 pp.

Mech L. D. (1971) Wolves, dogs, coyotes. In M. Nelson (ed.), *Proc. Symp. on White-Tailed Deer in Minnesota*, pp. 19-22. Minnesota Chapter, The Wildlife Society. Minn. Dept. Nat. Resources, St. Paul, Minnesota.

Mech L. D. (1972) Spacing and possible mechanisms of population regulation in wolves. *Am. Zool.* 12, 642 (abstract).

Mech L. D. (1973) Wolf numbers in the Superior National Forest of Minnesota. USDA For. Serv. Res. Paper NC-97. North Central Forest Exp. Sta., St. Paul, Minnesota. 10 pp.

Mech L. D. (1974) Current techniques in the study of elusive wilderness carnivores. *Int. Congr. Game Biol.* 11, 315-322.

Mech L. D. (1975) Disproportionate sex ratios of wolf pups. *J. Wildl. Mngmt.* 39, 737-740.

Mech L. D. (1977a) Population trend and winter deer consumption in a Minnesota wolf pack. In R. L. Phillips and C. Jonkel (eds.), *Proc. 1975 Predator Symp.*, pp. 55-83. Montana For. and Cons. Expt. Sta., Missoula.

Mech L. D. (1977b) Wolf pack buffer zones as prey reservoirs. *Science* 198, 320-321.

Mech L. D. (1977c) Productivity, mortality and population trend in wolves from northeastern Minnesota. *J. Mammal.* 58, 559-574.

Mech L. D. and Frenzel L. D., Jr. (1971) Ecological studies of the timber wolf in northeastern Minnesota. USDA For. Ser. Res. Paper NC-52. North Central For. Exp. Stn., St. Paul, Minnesota. 62 pp.

Mech L. D. and Karns P. D. (1977) Role of the wolf in a deer decline in the Superior National Forest. USDA For. Serv. Res. Paper NC-148. North Central For. Exp. Stn., St. Paul, Minnesota. 23 pp.

Mech L. D. and Knick S. T. (1978) Sleeping distance in wolf pairs in relation to breeding season. *Behav. Biol.* 23, 521-525.

Mech L. D., Barnes D. M. and Tester J. R. (1968) Seasonal weight changes, mortality, and population structure of raccoons in Minnesota. *J. Mammal.* 49, 63-73.

Medjo D. and Mech L. D. (1976) Reproductive activity in nine and ten-month-old-wolves. *J. Mammal.* 57, 406-408.

Nelson M. E. and Mech L. D. Deer social organization and wolf predation in northeastern Minnesota. *Wildl. Monogr.* (Submitted).

Packard J. and Mech L. D. Population regulation in wolves. Proc. Symp. on Natural Regulation of Animal Numbers. (Vancouver, B.C.) and *Proc. Symp. on Biosocial Mechanisms of Population Regulation*. New York: Plattsburgh, in press.

Peters R. P. and Mech L. D. (1975) Scent-marking in wolves: a field study. *Am. Scient.* 63, 628-637.

Rogers L. L. (1977) Social relationships, movements, and population dynamics of black bears in northeastern Minnesota. Ph.D. Dissert. U. Minn. 203 pp.

Rothman R. J. and Mech L. D. (1979) Scent-marking in lone wolves and newly formed pairs. *Anim. Behav.* In press.

Schlegel M. (1976) Factors affecting calf elk survival in north central Idaho. A progress rept. *Proc. 56th Ann. Conf. W. Assoc. State Game Fish Comm.* pp. 342-355.

Seal U. S., Erickson A. E. and Mayo J. G. (1970) Drug immobilization of the carnivora. *Int. Zoo Yb.* 10, 157-170.

Seal U. S., Mech L. D. and Van Ballenberghe V. (1975) Blood analyses of wolf pups and their ecological and metabolic interpretation. *J. Mammal.* 56, 64-75.

Seal U. S., Nelson M. E., Mech L. D. and Hoskinson R. L. (1978) Metabolic indicators of habitat differences in local deer populations. *J. Wildl. Mngmt.* 42, 746-754.

Stenlund M. H. (1955) A field study of the timber wolf (*Canis lupus*) on the Superior National Forest, Minnesota. Minn. Dept. Cons. Tech. Bul. 4. 55 p.

Trent T. T. and Rongstad O. J. (1974) Home range and survival of cottontail rabbits in southwestern Wisconsin. *J. Wildl. Mngmt.* 38, 459-472.

Van Ballenberghe V. and Mech L. D. (1975) Weights, growth, and survival of timber wolf pups in Minnesota. *J. Mammal.* 56, 44-63.

Radio Monitoring Birds of Prey

R. E. KENWARD

Institute of Terrestrial Ecology, Monks Wood Experimental Station, Abbots Ripton, Huntingdon PE17 2LS, U.K.

Abstract — This paper is a consideration of field techniques rather than of equipment design. The first of three parts reviews applications of radio monitoring in bird of prey studies, with suggestions concerning terminology. There has been some radio telemetry of activity and incubation parameters, but most monitoring of raptors has involved radio tracking. This has included location of birds for investigation of movements, ranges, habitat utilization and roost sites, but could also be used for surveillance of what birds were doing.

The second section describes techniques used in radio surveillance of a particularly secretive raptor, the goshawk, and presents original data on prey selection which were collected during studies of predation on woodpigeons and pheasants. Finally, there is an outline of how selection data were combined with predation rate measurements and hawk density estimates, obtained using radio tags as a Lincoln Index marker during hawk sightings, to determine the impact of predation on these economically important prey.

INTRODUCTORY REVIEW

Birds of prey, or raptors, are excellent subjects for radio monitoring, a term which may be used to include radio telemetry (the transmission of measurements of activity, temperature, etc.) and radio tracking of an unmodulated transmitter. Many raptors are big, which enables large batteries and transmitter antennas to be carried. Bald eagles (*Haliaetus leucocephalus*) were among the first birds to be radio tracked (Southern, 1964), and although the smallest raptors can now be tagged with miniature transmitters, only large birds can support packages suitable for tracking by satellites (Craighead and Dunstan, 1976). Tail feathers on birds of prey are usually long and do not detach easily, so that transmitters may be attached by tail mounting, which has several advantages over other techniques (Dunstan, 1973; Fuller and Tester, 1973). Most birds of prey spend the majority of their time perched well above the ground, or in flight, which ensures better signal transmission than for birds which live mainly on or near the ground.

Moreover, there are many fruitful applications for radio monitoring among birds of prey. Many live solitary lives, such that radio transmission provides an unambiguous means of long range identification and, for the more secretive species such as accipiters, provides the first comprehensive study technique away from the nest. Thus Platt (1973) recorded habitat utilization of a breeding pair of sharp-shinned hawks (*Accipiter striatus*), Bendock (1975) found roost sites of an Alaskan goshawk (*A. gentilis*), and an extensive study of sparrowhawks (*A. nisus*) before and throughout the breeding season is providing valuable information on range size variation with habitat quality, sex and stage of breeding cycle (Newton and Marquiss, in preparation). The nocturnal habits of owls make them a difficult group to study, but activity and locations have been recorded automatically for owls and diurnal raptors by fixed stations at Cedar Creek Natural History Area in Minnesota (Nicholls and Warner, 1968, 1972; Fuller and Tester, 1973; Forbes and Warner, 1974). In Europe, Hardy (1977) and Nilsson (1978) have investigated range used and activity of radio tagged tawny owls (*Strix aluco*).

Radio monitoring can be useful in the conservation of endangered birds of prey, for instance by providing information on habitat and food requirements (Fuller, Nicholls and Tester, 1974). Thus T. Dunstan (personal communication) showed that raptors nesting in the Snake River reserve in Idaho hunt far out in the adjoining country, and he points out that protection of the nesting sites is useless if the hunting areas are not also taken into account. Elsewhere Dunstan has radio located nests themselves with transmitters in food taken by adults.

Knowledge of migration may be important for conservation. Cochran (1972a, 1975) pioneered work in this area, and followed one radio tagged peregrine (*Falco peregrinus*) by plane from the Great Lakes to Mexico. Another conservation application is in release projects, both for injured wild birds after veterinary treatment (Serveen and English, 1976) and for re-stocking programs with domestic progeny (e.g. Meng and Kaufmann, 1975; Sherrod and Cade, 1978). Telemetering of eggs has assisted in the development of domestic breeding itself (Ellis and Varney, 1973; Schwartz *et al.*, 1977).

Application in falconry has enhanced demand for radio tracking equipment and stimulated its development; some useful instructions on tracking techniques have appeared in falconry journals (Grier, 1970; Cochran, 1972b) and equipment is custom built for falconers in both North America and Great Britain. Radio tracking has also been used to investigate the survival of lost falconers' birds, which have helped inadvertently to restock Britain with goshawks (Kenward, Marquiss and Newton, in preparation).

Most radio tracking studies of raptors have concentrated on the radio location of individuals. Less use has been made of the opportunity for radio surveillance of what they were doing. Radio surveillance can be used for accurate assessment of predation, which causes controversy about some birds of prey. The goshawk, for instance, is accused of sufficient predation on game to justify the annual destruction of thousands in Scandinavia, whereas in Britain it has been suggested that the species might be a useful controller of pests such as woodpigeons (*Columba palumbus*) and grey squirrels (*Sciurus carolinensis*). The second part of this paper describes radio surveillance techniques for recovery of fresh goshawk kills, which could be analyzed for an investigation of selection effects. Radio surveillance was also used for measuring predation rates, and the final section outlines how these were combined with the selection data to assess the impact of goshawk predation on pigeons in Britain and pheasants (*Phasianus colchicus*) in Sweden.

SELECTIVE PREDATION BY GOSHAWKS

MATERIALS AND METHODS

Since goshawks had not yet established themselves in a study area in the British lowlands, four imported hawks were released. Two had been trapped in late autumn in Finland, one in Germany, and one had received falconry training after being taken from a Finnish nest. Hawks radio tagged in Sweden were all wild birds trapped in the study areas.

The birds were equipped with tail mounted radio packages weighing 1-2.5 percent of body weight, the main antenna being a 1 mm Teflon (PTFE) coated wire fastened along the feather shaft. This technique protects a lightweight antenna, keeping it straight and off the ground for optimal signal transmission with minimal risk of entanglement, and ensures loss of the package at the next moult. Harness mountings (Nicholls and Warner, 1968; Dunstan, 1972) can entangle the feet of birds of prey and on other birds can cause abrasions, weight loss or behavioral abnormalities (Boag, 1972; Ramakka, 1972; Greenwood and Sargeant, 1973; Lance and Watson, 1977). However, goshawks with tail mounted transmitters did not differ in weight changes or tendency to leave an area from hawks marked only with rings (Kenward, 1978a). Initially, four transmitters were attached to single moulted feathers which could then be 'shaft-imped' into a bird (Kenward, 1976). This type of mounting, in which the transmitter feather was glued and tied into the proximal 3 cm of hollow shaft left after the end had been cut off an existing feather, could be completed in five minutes. The transmitters remained attached for up to four months, but a supply of moulted feathers was required and, since three such feathers eventually broke off, the method was less secure than that used for 74 subsequent attachments in which two feathers were used to support the package. Attachment threads were then sewn through one rectrix and tied round another so that the feathers could moult separately (Kenward, 1978a). Seven of these packages were recovered and all had detached satisfactorily, although four of the 74 were shed after a few days, possibly because the area where a feather entered the body had been bruised during radio mounting.

The first four hawks were monitored with a 3-channel RB4 'Falconer' receiver from Custom Electronics, Urbana, Illinois. For use with many hawks in Sweden, a 12-channel LA12 receiver was purchased from AVM Instrument Company, Champaign, Illinois. Both receivers were operated with a 3-element hand-held yagi antenna.

The four hawks released in Britain were followed continuously, to record as many of their flights and kills as possible. Although this provided comprehensive behavior data, it restricted the number of birds observed. In Sweden emphasis was placed on obtaining material from a large number of hawks. Only experience can tell how many birds can be kept under surveillance at a time in a given area; a beginner should start with one or two, preferably with stronger transmitters than a more experienced operator might need. Good access to terrain is essential, preferably with few areas more than 1 km from vehicle access points and no barriers necessitating long detours. In the most favorable circumstances up to six hawks could be monitored at a time. To standardize range-use data, their positions were recorded four times a day: mid-morning, mid-day, mid-afternoon and after dark. They were checked more frequently for kills, with the aim of recovering all large prey (weighing more than about 500 g), which took well over an hour and often several visits to pluck and consume. Hawk behavior could be inferred from signal variations. Flight was typified by large signal amplitude fluctuations, which varied rhythmically for soaring and were usually best received with a horizontal polarization of the yagi antenna. Perched hawks gave a fairly steady signal, usually with a more vertical polarization of the receiving antenna. Bathing wetted the transmitter antenna and caused slight frequency fluctuations.

Cues for approaching located kills were cessation of the typical regular hunting flights, low signal strength at short range indicating that a hawk was on the ground, and, with some transmitters, a fluctuation of signal frequency as the antenna touched the ground when the hawk braced its tail to pull at food. The decision to stalk was crucial, because approaching on foot was time consuming and could result in failure to complete a day's range-use data for another bird (a transmitter giving an unambiguous feeding signal would be invaluable). The aim of a stalk was to minimize the area to be searched for a carcase or other remains when a hawk flushed. Signal strength (an index of distance) and direction were monitored every few paces, using vertical and horizontal receiving antenna polarizations. Horizontal yagi polarization often gave greatest accuracy, especially in conifer plantations where the vertical signal component might contain many reflections (see Lemnell, in preparation). With thick vegetation or good topographic cover it was sometimes safe to move to the side of a hawk to triangulate it at short range without causing it to flush. Otherwise, the feeding bird was approached directly until the signal or a sighting indicated that it had flown. In thick cover a hawk might fly up 10 m away, but at other times a segment 100 m long needed to be searched. Snow cover made kills easier to spot.

Hawks usually carried away small kills when flushed, in which case pluckings were used for identification. The four hawks radio tagged in Britain were males and each had difficulty carrying pigeons, which were usually recovered after little had been eaten and replaced with a similar amount of other food. The Swedish hawks rarely moved pheasant kills, and returned to feed up to four more times after being disturbed so that carcases could be measured. If possible a *pectoralis minor* (*supracoracoideus*) muscle was removed as a predictor of body weight and an index of nutritive condition (Kenward, 1977, 1978b). The excised muscle was about 3 percent of the kill weight, so its removal did not greatly reduce the food available to a hawk.

RESULTS AND DISCUSSION

Bodyweight estimates for pigeons captured by the goshawks or shot in Oxfordshire between November, 1974 and March, 1975 are shown in Table 1. Shot pigeon samples, obtained in November, December, January and February, did not differ significantly in bodyweight, probably because shed grain was unusually available on unploughed stubbles as a result of a wet harvest and autumn. *Pectoralis minor* dry weights of captured pigeons were significantly lower than those of shot birds, both among adults (Mann-Whitney U test; $z = 3.50$, $P < 0.001$) and among juveniles ($z = 2.50$, $P < 0.01$). Although there was selection for poor condition, there was no evidence of any between particular age or sex classes.

Table 1.
The sex and estimated bodyweight of woodpigeons captured by goshawks or shot in Oxfordshire between November 1974 and March 1975

		Sex subtotals			Adults			Juveniles		
						Bodyweight (g)			Bodyweight (g)	
	Total	Female	Male	?	No.	Mean	Range	No.	Mean	Range
Captured	21	8	7	6	14	471	396-579	5	443	342-517
Shot	38	19	17	2	22	540	415-589	16	528	487-587

The falconry trained hawk did not differ from the others in his hunting behavior or kills, but remained on prey when approached and could, with care, be enticed onto the fist. Other released trained birds have behaved similarly at kills and, by weighing individuals and the food consumed, this tameness might be exploited for energetics studies.

Table 2 gives bodyweight estimates for pheasants captured by ten goshawks or shot at an estate in Sweden where about 4,000 poults were released annually. The mean weight of shot pheasants differed significantly between months, but to simplify

Table 2.
The estimated bodyweight of juvenile pheasants captured by goshawks or shot at Frötuna estate between October 1976 and January 1977

	Females, shot or captured at same sites between October & December			Males, shot or captured at same sites in October			Males, shot or captured at different sites in January		
	No.	Mean (g)	Range (g)	No.	Mean (g)	Range (g)	No.	Mean (g)	Range (g)
Captured	22	957	758-1110	3	906	650-1106	4	1294	1226-1388
Shot	298	967	701-1200	6	1109	1031-1166	43	1288	1047-1500

comparisons the mean of the largest shot sample (in December) was used as a standard; for each month all pheasant weights were adjusted by the difference between the shot sample mean in that month and the shot sample mean in December. The weight of captured females was not significantly lower than that of shot birds ($z = 0.97$, $P = 0.17$). Too few captured males were available for an adequate analysis in the period when shot samples were obtained. Although there was some evidence of selection among males in October ($P = 0.05$), there was no sign of particularly poor birds being captured in January, when lack of marked inter-site bodyweight differences made comparison with samples shot at other sites more acceptable.

The sex ratio among 399 released poults, and 222 shot at random in November, did not differ significantly from 1:1. However, as the released pheasants grew to full size the proportion of females increased markedly among hawk kills: in August 33 percent of kills were female ($n = 15$), in October 67 percent ($n = 97$), and in November 93 percent ($n = 15$).

Why were there these differences in selection between pigeons and pheasants? Selection for condition probably depended on whether the prey's escape reaction provided opportunity for a chase by the goshawk. Pigeons attempt to outfly an attacking hawk and, in arranged attacks (Kenward, 1978b), the further away a goshawk was when pigeons flushed the greater was the tendency to select poor condition stragglers. There was little selection when the pigeons were taken by surprise. Among the four released hawks' kills, two of the heaviest pigeons were taken at roost trees just after first light, and three of the adults taken considerably further from cover (> 50 m) than any others had significantly lower *pectoralis minor* dry weights ($z = 1.79$, $P = 0.04$). It is unlikely that birds feeding far from cover tended to be below average in condition, because woodpigeons usually feed in open fields in winter (Murton, 1965).

Pheasants, on the other hand, respond to a goshawk attack by seeking cover. They rarely fed far from shelter, which they could rapidly enter with little chance of a chase if a hawk did not take them by surprise. Only 18 percent of 79 pheasant kills were more than 5 m from moderately dense cover. It is unlikely that the absence of marked selection for poor condition was a result of released pheasants being particularly vulnerable, because captured wild pheasants too showed no more selection than other population samples (Kenward, Marcström and Karlbom, in preparation). However, the abundance of released pheasants may have enabled hawks to choose to attack mainly females, the smaller sex. Goshawks flown in falconry often eventually refuse to seize hold of cock pheasants. There was no significant selection for female wild pheasants, but they were present more frequently among kills when snow was on the ground than at other times.

It is clearly unwise to assume that predators always take poor quality prey, but what is their effect on a prey population when they do? If a relationship can be found between prey survival and one or more parameters which can be measured at kills, it is possible to separate the 'net' effect of predation from the 'gross' number of prey killed. From a reanalysis of marking data collected by Murton, Isaacson and Westwood (1971), R. Sibly (personal communication) has produced linear relationships between a survival index (y) and mid-winter bodyweight of woodpigeons (x), as follows:

$$y = 0.0019\,x - 0.60 \text{ for adults, and} \qquad \text{(i)}$$

$$y = 0.0012\,x - 0.28 \text{ for juveniles.} \qquad \text{(ii)}$$

Using these equations, the mean survival prospects of captured pigeons were 72 percent those of shot birds. For example, if a gross predation of 100 pigeons occurred, the net removal from the population would only have been 72, because 28 would have died anyway. In a more extreme case, net predation may be negligible because almost all prey taken would have died anyway (Jenkins, Watson and Miller, 1964). It is therefore important for predation assessments to include an investigation of selection effects.

RADIO TAGGING FOR PREDATION ASSESSMENT

As well as providing fresh kills for selection analysis, radio monitoring of several hawks enables estimation of the average number of a particular prey taken per hawk-day of radio tracking. This predation rate, modified if necessary by the selection effect, can be used with data on hawk and prey populations to estimate the proportion of the prey population removed. An example is the woodpigeon study of Kenward (1976, 1979).

Woodpigeons were the most frequent goshawk diet item (40 percent), with one killed per 0.27 hawk-days. Using pigeon density data from Murton (1965, 1974), goshawk densities ranging from 1 to 4 per 1000 ha could cause a 7 percent to 28 percent net reduction in the pigeon breeding population, but would have no effect on the size of the mid-winter population which damages brassica crops. The highest goshawk density recorded is about 1 pair per 1000 ha (Bednarek, 1975).

The impact of predation on British pigeons could only be predicted, whereas wild goshawk predation on Swedish pheasants could be accurately assessed, using radio tagging not only to determine predation rates and investigate selection but also to estimate goshawk numbers. Except for hawks being tracked at the time, each one seen could be recorded as radio tagged or not even if it was glimpsed only momentarily. Knowing the number of tagged birds hunting the area, the total number could be estimated from their proportion in the sightings. Combining this total with the predation rate indicated that close to 40 pheasants per week were being taken (about 1 percent per week). During the same period the gamekeepers noted all the kills they found, without being told where monitored hawks had killed. The proportion of monitored hawk kills in these finds was the same as the proportion of tagged hawks in the sightings, which gave confidence in the radio tag estimate of hawk numbers (Kenward, 1977). The same technique was therefore used to estimate hawk numbers in a wild pheasant study area, where the prey was censused using men in line abreast with dogs to search the undergrowth in autumn and spring. The goshawks there reduced the overwinter pheasant population by a net 25 percent, being responsible for 37 percent of the overwinter mortality (Kenward *et al.*, in preparation).

These examples show how predation can be assessed by radio tagging a small number of predators. Other approaches are to radio monitor a large number of prey to determine what proportion is predated (Stoddart, 1970; Dumke and Pils, 1973; Brand,

Vowles and Keith, 1975), or simply to search an area for kills. In such prey-based investigations there may be difficulty ascertaining the cause of death, the extent of selection, and whether cessation of a radio signal indicates loss from the area, battery failure or destruction of a radio at the time of death. Predator tagging overcomes these problems and provides additional data on predator behavior, but requires separate collection of prey population data and only assesses predation by the species tagged. Deciding whether to use radio tagging on prey, predator, or another technique entirely, will depend on the circumstances and information required in a particular predation study.

REFERENCES

Bednarek W. (1975) Vergleichende Untersuchungen zur Populationsökologie des Habichts (*Accipiter gentilis*): Habitatbesetzung und Bestandsregulation. *Dt. Falkenorden jb.* 47-53.

Bendock T. (1975) Winter roosting behavior of a northern goshawk — a radio-telemetry study. Appendix III. In J. D. McGowan (ed.), *Distribution, Density and Productivity of Goshawks in Interior Alaska.* Alaska Department of Fish and Game Report.

Boag D. A. (1972) Effect of radio packages on behaviour of captive red grouse. *J. Wildl. Mngmt.* 36, 511-518.

Brand C. J., Vowles R. H. and Keith L. B. (1975) Snowshoe hare mortality monitored by telemetry. *J. Wildl. Mngmt.* 39, 741-747.

Cochran W. W. (1972a) A few days of the fall migration of a sharp-shinned hawk. *Hawk Chalk* 11, 39-44.

Cochran W. W. (1972b) More on falconry telemetry equipment and technique. *Hawk Chalk* 11, 26-31.

Cochran W. W. (1975) Following a migrating peregrine from Wisconsin to Mexico. *Hawk Chalk* 14, 28-37.

Craighead F. C. and Dunstan T. C. (1976) Progress towards tracking migrating raptors by satellite. *Raptor Res.* 10, 112-120.

Dumke R. T. and Pils C. M. (1973) Mortality of radio-tagged pheasants on the Waterloo wildlife area. Department of Natural Resources, Wisconsin, Technical Bulletin, 72, 1-52.

Dunstan T. C. (1972) A harness for radio-tagging raptorial birds. *Inland Bird Banding News* 44, 4-8.

Dunstan T. C. (1973) A tail feather package for radio-tagging raptorial birds. *Inland Bird Banding News* 45, 3-6.

Ellis D. H. and Varney J. R. (1973) A fully automated egg for telemetering adult attentiveness and incubation temperatures. *Raptor Res.* 7, 73-77.

Forbes J. E. and Warner D. W. (1974) Behaviour of a radio-tagged saw-whet owl. *Auk* 91, 783-795.

Fuller M. R. and Tester J. R. (1973) An automated radio tracking system for biotelemetry. *Raptor Res.* 7, 105-106.

Fuller M. R., Nicholls T. H. and Tester J. R. (1974) Raptor conservation and management applications of bio-telemetry studies from Cedar Creek Natural History Area. In F. N. Hamerstrom, B. E. Harrell and R. R. Olendorff (eds.), *Management of Raptors*, Raptor Research Foundation.

Greenwood R. J. and Sargeant A. B. (1973) Influence of radio packs on captive mallards and blue-winged teal. *J. Wildl. Mngmt.* 37, 3-9.

Grier J. W. (1970) Radio telemetry for locating lost hawk. *Hawk Chalk* 9, 17-27.

Hardy A. R. (1977) Hunting ranges and feeding ecology of owls in farmland. Ph.D. thesis, University of Aberdeen.

Jenkins D., Watson A. and Miller G. R. (1964) Predation and red grouse populations. *J. appl. Ecol.* 1, 183-195.

Kenward R. E. (1976) The effect of predation by goshawks, *Accipiter gentilis*, on woodpigeon, *Columba palumbus*, populations. D. Phil. thesis, Oxford.

Kenward R. E. (1977) Predation on released pheasants (*Phasianus colchicus*) by goshawks (*Accipiter gentilis*) in central Sweden. *Viltrevy* 10, 79-112.

Kenward R. E. (1978a) Radio transmitters tail-mounted on hawks. *Ornis scand.* 9, 220-223.

Kenward R. E. (1978b) Hawks and doves: factors affecting success and selection in goshawk attacks on woodpigeons. *J. Anim. Ecol.* 47, 449-460.

Kenward R. E. (1979) Winter predation by goshawks in lowland Britain. *Br. Birds* 72, 64-73.

Lance A. N. and Watson A. (1977) Further tests of radio-marking on red grouse. *J. Wildl. Mngmt.* 41, 579-592.

Meng H. and Kaufmann J. (1975) *Falcons Return.* New York: William Morrow & Co.

Murton R. K. (1965) *The Woodpigeon.* London: Collins New Naturalist.

Murton R. K. (1974) The impact of agriculture on birds. *Ann. appl. Biol.* 76, 358-365.

Murton R. K., Isaacson A. J. and Westwood N. J. (1971) The significance of gregarious feeding behaviour and adrenal stress in a population of woodpigeons *Columba palumbus*. *J. Zool.* 165, 53-84.

Nicholls T.H. and Warner D. W. (1968) A harness for attaching radio transmitters to large owls. *Bird Banding* 39, 209-214.

Nicholls T. H. and Warner D. W. (1972) Barred owl habitat use as determined by radiotelemetry. *J. Wildl. Mngmt.* 36, 213-224.

Nilsson J. N. (1978) Hunting in flight by tawny owls *Strix aluco*. *Ibis* 120, 528-531.

Platt J. B. (1973) Habitat and time utilisation of a pair of nesting sharp-shinned hawks (*Accipiter striatus velox*) — a telemetry study. M.Sc. thesis, Brigham Young University.

Ramakka J. M. (1972) Effects of radio-tagging on breeding behaviour of male woodcock. *J. Wildl. Mngmt.* 36, 1309-1312.

Schwartz A., Weaver J. D., Scott N. R. and Cade T. J. (1977) Measuring the temperature of eggs during incubation under captive falcons. *J. Wildl. Mngmt.* 41, 12-17.

Serveen C. and English W. (1976) Bald eagle rehabilitation techniques in western Washington. *Raptor Res.* 10, 84-87.

Sherrod S. and Cade T. (1978) Release of peregrine falcons by hacking. In T. A. Geer (ed.), *Bird of Prey Management Techniques*, pp. 121-136. British Falconers' Club.

Southern W.E. (1964) Additional observations on winter bald eagle populations: including remarks on biotelemetry techniques and immature plumages. *Wilson Bull.* 76, 222-237.

Stoddart L. C. (1970) A telemetric method for detecting jackrabbit mortality. *J. Wildl. Mngmt.* 34, 501-507.

An Analysis of Objectives in Telemetry Studies of Fish in the Natural Environment

I. G. PRIEDE

Department of Zoology, University of Aberdeen, Tillydrone Avenue, Aberdeen AB9 2TN, U.K.

Abstract — Most fish weigh 0.1-5 kg and require small biotelemetry transmitters. Acoustic underwater transmission is used most often since radio is attenuated by conductive natural waters. Radio is however preferable in shallow freshwater. Satellite tracking can be used for large animals near the surface. Stress spectra are discussed and it is shown that fish spend very little time at extremes of their scope of activity. The small amount of time spent near the lethal limits is important but difficult to measure; this is only possible by means of biotelemetry. Both in behavior and physiology, biotelemetry increases the resolution of measurements on the time scale.

INTRODUCTION

Fish are generally not readily visible in their natural environment. Knowledge of their biology and movements has been assembled from many isolated discontinuous observations. Only since the advent of ultrasonic tracking and biotelemetry techniques has continuous observation of movements and behavior of individuals become possible. In studies on terrestrial animals radio tracking is often used as a means of finding an individual which can then be subsequently visually observed for detailed measurements. In fish tracking the biologist is generally entirely dependent on his instruments as the only means of observing the animal. In addition to simple tracking therefore much effort is directed tc telemetry of physiological and behavioral data from the animal.

Despite the large variety of fish (from minnows to large sharks) commercially significant species weigh mainly between 100 g and 5 kg. The main developments of telemetry have been in devices weighing 5 tc 50 g for this sort of size range of fish. Atlantic salmon (*Salmo salar* L.) at the smolt stage weigh about 50 g when they migrate from the rivers into the sea. These have provided the stimulus for development of some of the smallest tracking transmitters.

Acoustic underwater transmission is the most commonly used technique in studies on fish but radio is coming into its own fcr fresh water work. Two major review papers in the last two years (Stasko and Pincock, 1977; Ireland and Kanwisher, 1978) give thorough accounts of the technology used. This paper will therefore briefly review the scope of available techniques and then go on to consider the philosophy behind application of the more sophisticated telemetry techniques.

ACOUSTIC TRANSMISSION

Underwater acoustic signals in this context are almost universally generated by means of lead zirconate titanate (PZT) cylindrical transducers. The cylinder is hollow and is manufactured as a piezoelectric ceramic which responds to an oscillating voltage applied between the inner and outer wall by expanding and contracting in diameter (Fig. 1). This generates an approximately omnidirectional field of sound in the surrounding water (Mitson and Young, 1975). Absorption of sound by water increases with frequency and 300 KHz is the highest practical frequency. The upper hearing threshold of fish varies between 1 and 7 KHz (Hawkins, 1973) which is below the frequency used in telemetry devices. Some marine mammals use frequencies up to 100 KHz for communication; this is a potential source of noise and other problems.

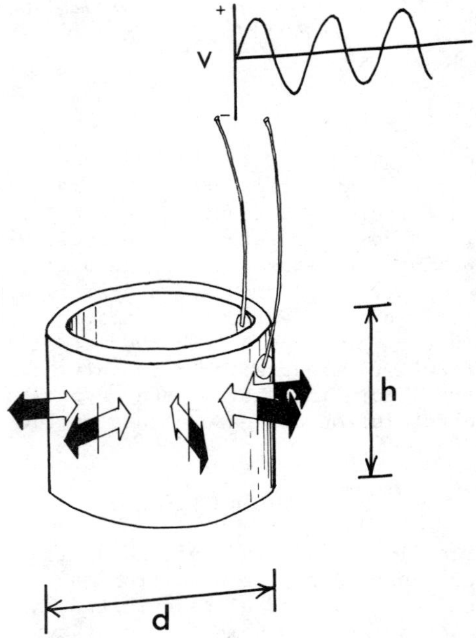

Fig. 1. Generation of sound by a cylindrical PZT ceramic transducer. The inner and outer walls are coated with a conductive metallic coating to which the leads are attached. An applied oscillating voltage generates radial mechanical oscillations (indicated by the arrows). The reciprocal piezoelectric effect is used in hydrophones. h/d is generally less than 1.

The greatest efficiency of sound radiation is achieved if the transducer cylinder is driven at its resonant frequency. A 20 KHz cylinder is about 40 mm diameter and is too large for most fish to carry. At higher frequencies the cylinder can be smaller and a 2 mm diameter cylinder at 300 KHz is the smallest in use (Mitson and Storeton-West, 1971). Choice of frequency is therefore a trade-off between increased range at low frequencies and compactness of high frequency transducers. An 80 KHz transducer 12.5 mm diameter is convenient for most uses with 200 to 300 KHz being used for the smallest transmitters.

Signals are detected by hydrophones placed in the water and usually made from piezoceramic elements which generate an electrical signal equivalent to the transmitted sound signals. The PZT receiver transducer can be identical to the cylinder used in the transmitter with appropraite parabolic reflectors to provide directional sensitivity if necessary. High efficiency of the hydrophone is essential and the best signal to noise ratios are obtained if the transducer is resonant at the frequency of the transmitter. The output from the hydrophone can be very small, so a tuned preamplifier close to the transducer element is useful. Broad band hydrophones are generally not satisfactory; background noise being a problem even if sufficient sensitivity can be achieved.

Sound propagates in straight lines underwater but is subject to refraction, and reflection as well as absorption. Underwater obstacles block the signal; a direct 'line of sight' being required between transmitter and hydrophone. Obstacles and shore line can reflect signals and both reflection or refraction can occur at interfaces between waters of different densities (i.e., different salinities or temperatures). Air bubbles in fast flowing streams or surf zones of beaches can very severely attenuate sound signals. Certain species of algae contain air sacs (e.g. *Anaboena* spp.); when in bloom these produce a green appearance in the water and can severely attenuate sound signals. Sea noise in rough weather and the noises of animals such as snapper shrimp (*Alpheus* spp.) can be a problem in certain areas.

For location and tracking purposes a number of different hydrophones systems are possible (Fig. 2).

SIMPLE DIRECTIONAL HYDROPHONE

Most commonly used is a small portable hydrophone mounted on a vertical rod which is dipped in the water and the direction of maximum signal is determined audibly using a loudspeaker or headphones on a portable receiving unit. This can be used either on foot from the shore or from a boat. Permanent mountings can be used for extended work on larger boats. Movement of the hydrophone through the water can generate noise and a streamlined housing is required for tracking 'on the move'. Noise from the engine can also be a problem.

Signal strength can provide a measure of range (Holliday *et al.*, 1973) but two or more hydrophones are required for properly triangulating fixes on the fish. In areas where fish are intensively studied in their home range permanent hydrophone stations set on pylons with remote control and signal monitoring from a shore based laboratory have been used (Young *et al.*, 1972).

FISH PASS INDICATOR

A single fixed hydrophone scanning across a linear system such as a river can detect passage of fish equipped with sonic tags. This is particularly effective if the hydrophone is placed immediately above a weir. Upstream movement is then characterized by sudden appearance of the signal as the fish comes over the weir. Downstream movement is characterized by gradual increase in the signal which then suddenly disappears (T. Langford, unpublished). A number of such listening stations in a river system can be readily automated to record individual fish.

OMNIDIRECTIONAL HYDROPHONE

With 3 or more omnidirectional hydrophones the arrival of a sonic pulse from a fish can be timed at each hydrophone. From this the position of the fish can be calculated (Hawkins *et al.*, 1974). With an on-line computing facility, fixes can be obtained at intervals of less than a second.

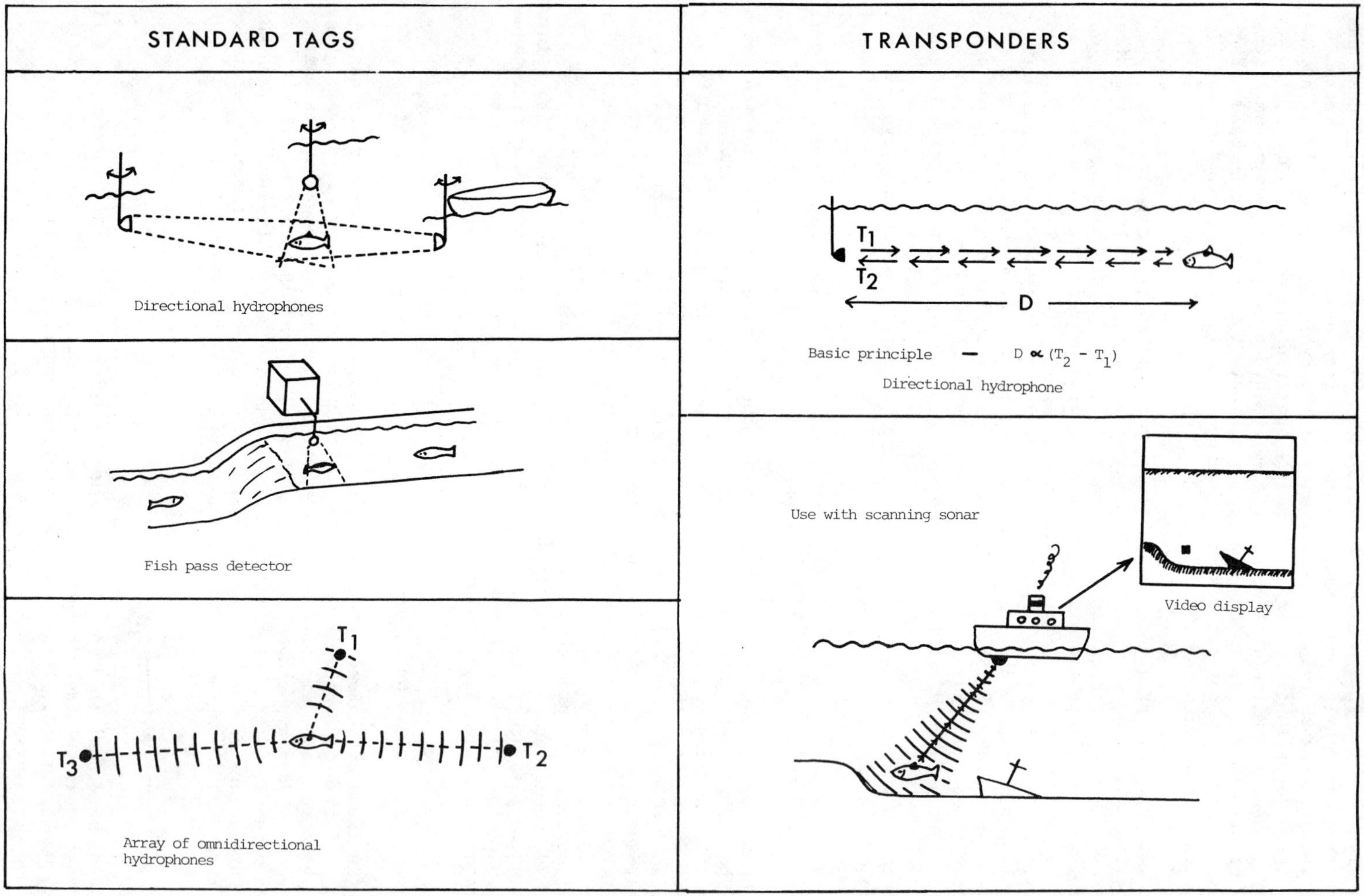

Fig. 2. Hydrophone systems used for location of fish. Standard tags are simple transmitters on the fish which emit a regular pulsed signal. Transponders only emit a pulse if interrogated by an incoming sonic pulse.

TRANSPONDERS

This system employs a transponding tag on the fish. The hydrophone emits an interrogation pulse which elicits a response from the fish tag. The time interval between interrogation and response is then a measure of range. If direction can be detected this gives an absolute fix using only one hydrophone station.

In the fish tag it is convenient to use one PZT transducer for both receiving and sending signals so the two signals are generally on the same frequency. Tags of this type tend to be more bulky than the simple pinger tags but can have very low power consumptions when not interrogated.

The most elegant application of transponders is in conjunction with sector scanning sonar gear. The fish tag is triggered by the normal sonar pulses and its signal therefore is received by the scanner as an ultrareflective target which appears on a display screen as a bright spot against the features of the sea bed (Fig. 2) (Mitson and Storeton-West, 1971; Greer-Walker *et al.*, 1971).

DEPTH DETERMINATION

Depth can be determined by means of vertical triangulation with any of the above tracking systems. Given a hydrophone at some distance from the fish, vertical angular displacement is small and difficult to resolve, accurately. Gardella and Stasko (1974) describe a simple portable linear hydrophone array which can provide an inexpensive means of depth measurement. Otherwise depth must be determined by means of pressure telemetry.

SHIP POSITION

Location of the fish relative to the hydrophones can be as accurate as 1 m. For tracking in open water the ship or boat can rarely be so accurately located. Manual navigational position fixing at frequent intervals poses problems. Most European merchant ships and fishing vessels carry the DECCA navigator system. This works on a hyperbolic principle; a shipboard receiver measures the phase difference between synchronized radio signals from fixed shore stations. The receiver continually automatically displays coordinates of the ship's position. The coordinates can be directly fed onto an x-y plotter. Repeatability of fixes is about 50 m although quoted typical navigational accuracy is 0.25 nautical miles (DECCA Navigator Co. Ltd., 9 Albert Embankment, London, U.K.). This is a very convenient system for fish tracking applications available in many parts of the world but not North America. A more accurate version (Decca Hi-fix) has portable transmitters which can be deployed by the user in any area of interest. For tracking from small boats without sophisticated aids approximate fixing of the fish relative to the boat with a simple directional hydrophone is sufficient, as the limiting factor is inability to locate the boat to an accuracy of better than 100 m.

RADIO TELEMETRY

Early success with sonic tags (Trefethen, 1956) directed attention away from radio frequencies. Radio waves are severely attenuated by seawater but in freshwater of low electrical conductivity a useful signal emerges from the water above a submerged transmitter (Fig. 3). Rays with an angle of incidence in excess of 6° from the normal are reflected back from the surface. The rest of the rays are refracted and take the form of an omnidirectional field emanating from a circle above the fish. The receiving antenna should be as high as possible in the air and McCleave *et al.* (1978) have successfully tracked migrating salmon from an aircraft.

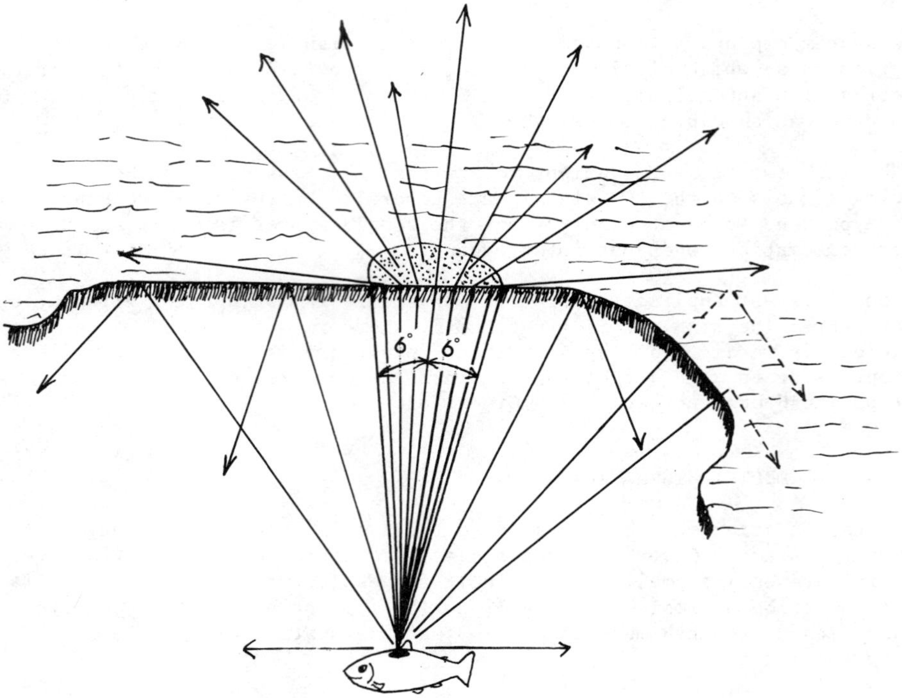

Fig. 3. Radio transmission from underwater. Note how rays are reflected and refracted at the surface of the water. Only approximately perpendicular rays to the surface pass through and can be received by an elevated antenna.

Radio signals are not affected by air bubbles, water turbidity or obstructions in the way that sonic signals are. Therefore radio is preferable to acoustic tracking particularly in fast flowing shallow rivers. It is also reported that radio signals can be detected through ice. Radio location is not as accurate as is acoustic tracking but data transmission rates may be much faster for radio telemetry.

The technology for radio tracking is the same as for other radio systems described elsewhere in this symposium so no special descriptions are required. Design criteria for efficient coupling of the underwater transmitting antenna into the aquatic medium have not been rigorously described. A shorter antenna than the equivalent one in air is required (Stasko and Pincock, 1977).

SATELLITES

Tracking of animals by use of satellites has been demonstrated in polar bears and marine mammals (Lentfer *et al.*, 1977; Jennings *et al.*, 1979). Development work so far has been carried out using the NIMBUS-6 satellite random access measurement system (RAMS) but from 1979 Service ARGOS is the main operational system for this kind of work. A simple earth based radio transmitter (carried by an animal) (Fig. 4) is located relative to the orbit of the satellite by measuring the Doppler shift of the signal due to the motion of the satellite.

Frequency stability of the transmitter oscillator is therefore important to ensure

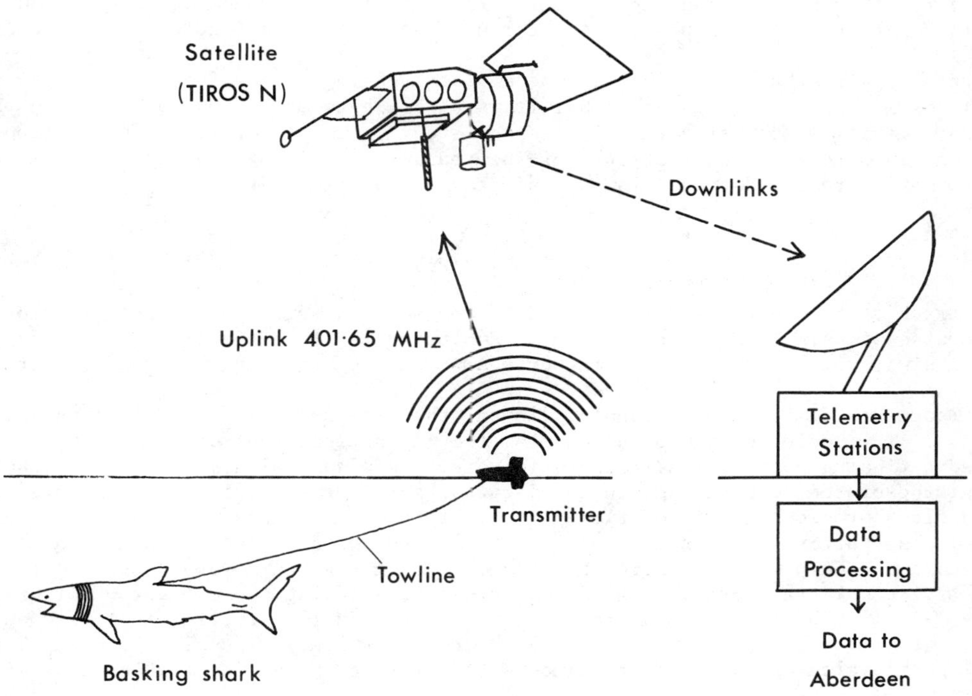

Fig. 4. Elements of a satellite tracking system as used on basking sharks (*Cetorhinus maximus*) (Priede, 1979).

location accuracy which can be as good as a few hundred meters. At present such stability is difficult to achieve for animal transmitters and errors up to several kilometers are likely. Good antenna design and signal strength are also important and details are available in literature issued by Service ARGOS, CNES, Toulouse, France.

For work at sea the radio transmitter must obviously be on the surface of the water and present designs weigh at least 2 to 3 kg. Applications for fish work are limited but trials have been carried out on basking sharks using a towed transmitter enclosed in a buoyant capsule (Priede, 1979).

The ARGOS system can also be used for ship location and data transmission. This may be useful in an ancillary role in tracking studies. Use of satellite based earth surface imagery may be a good way of obtaining environmental data to correlate with animal behavior. Development of satellite technology is likely to make a contribution to various aspects of field studies of animals even if tracking *per se* is restricted mainly to large terrestrial animals.

ATTACHMENT OF TRANSMITTERS TO THE FISH

The techniques used are the same as for attachment of conventional fish tags (Laird and Stott, 1978; Laird, 1978). Attachment can be external with wire or flexible monofilament bridles. The epidermis of fish is delicate and harnesses which cause excessive abrasion should be avoided; a firm subdermal attachment or anchorage passing through the musculature is preferred. Transmitters can also be placed internally in the stomach or body cavity (Hart and Summerfelt, 1975). The body tissues are essentially transparent to accoustic transmitters. Radio tags can be arranged to have a trailing external antenna passing out through the gill slits (McCleave *et al.*, 1978).

External tags increase drag on fish whereas stomach tags may affect feeding behavior. Equilibrium may also be disturbed but it is possible to make tags neutrally buoyant so as to minimize such effects. Before any study proceeds careful consideration must be given to the likely effects of the transmitter package and considerable effort may be required to develop satisfactory attachment methods.

APPLICATIONS

Fish tracking is now a well established technique used in many fisheries studies in the course of normal management policy. Equipment is commercially available from a number of suppliers particularly in North America for both sonic and radio systems.

The importance of tracking technology in an otherwise unobservable animal is self evident. It should however be noted that the major migrations of fish have been worked out using conventional techniques; e.g. the classical discovery of the spawning ground of the European eel in the Sargasso sea (Bertin, 1956). Electronic tracking devices provide details of timing and pattern of movements on a shorter time scale. The pattern of migration of plaice (*Pleuronectes platessa*) in the Southern North Sea has been well known for some time. Sonic tagging however revealed the selective tidal transport system used by these fish during the migration (Greer-Walker *et al.*, 1978). The fish rests on the bottom during adverse tidal flow and moves into mid water on favorable tides. Tracking similarly reveals the diurnal and tidal periodicity of movement of salmon smolts through estuaries (McCleave, 1978).

TELEMETRY DATA

Telemetry of data from the fish in addition to simple position fixing is the major area in which advances are likely to take place in the next decade. Satchell (1971) looked to telemetry as a means of extending physiological studies to fish too large or difficult to handle in the laboratory. It has since become apparent that telemetry may be of more fundamental importance in our understanding of the biology of fishes.

Telemetry of data can fall into two categories, external and internal.

External or Environmental Data

Telemetry of pressure has been used for determination of depth of fish (Standora *et al.*, 1972; Stasko and Rommel, 1974). Temperature transmitters are available and are particularly useful in studies of behavior of fish in relation to thermal discharges (Kelso, 1974, 1976). Telemetry of light level was used by Gayduk and Malinin (1971) and developments are likely in telemetry of salinity, pH, oxygen and a number of other environmental factors.

Internal Data

Both radio (Nomura *et al.*, 1972) and sonic (Priede and Young, 1977) heart beat biotelemetry devices have been demonstrated. Whilst not strictly internal, compass orientation of the fish can be detected as well as swimming speed (Standora *et al.*, 1972). Tail beat frequency as a measure of swimming activity has also been biotelemetered by Young *et al.* (1972).

Breathing and feeding movements of the jaws have been biotelemetered by detection of the electromyogram of the *adductor mandibulae* of brown trout (Oswald, 1978) and biotelemetry of the EMG of other muscles is possible. EEG biotelemetry from the brain may be possible and biotelemetry of a variety of internal physiological variables is probable in the future in both the single and multichannel mode. Standora *et al.* (1972) demonstrated multichannel biotelemetry for a shark but current developments suggest that more compact transmitters will become available for use with the smaller sizes of fish.

A RATIONALE FOR FUTURE WORK

External environmental factors can be regarded as *stress* applied to the animal. By mechanical analogy the physiological or behavioral response can be regarded as *strain*. In stretching a wire the relationship between stress and strain is well known and measurement of either variable can be used to define the state of the system. Similarly if the response of a fish to temperature is well understood one can choose to monitor either the environmental temperature or the response. For any animal a number of external and internal variables or dimensions can be considered as important with respect to survival of the individual.

Suppose we monitor the level of such a variable, D, relevant to the animal's survival. In course of time D will fluctuate as in Fig. 5. Limits can be placed on the value of D which can be considered as the upper and lower lethal limits. If D is temperature then these are the upper and lower lethal temperatures. It is convenient to normalize data so that the lower limit is 0 and the upper limit 1. These limits can be expressed as the distribution of probability of mortality with respect to D as in the graph on the left in Fig. 5. As the animal approaches the extremes of its range of tolerance, probability of mortality increases eventually to 1. The mortality probability may be unimodal: e.g. levels of environmental toxins where there will be an upper lethal limit but no lower limit.

Over a period of time the distribution of time T_D spent at different levels of D can be determined. This, in terms of reliability theory, is the stress spectrum (Bazovsky, 1961). If M_D is the mortality as a function of D then the total mortality attributable to D is:

$$M = \sum_{D=0}^{D=1} M_D \cdot T_D \ .$$

Mortality is equal to the time spent at each D level multiplied by the rate of mortality at that D level. If a number of independent D factors are considered then the mortalities attributable to each can be simply added to give a total mortality figure M_{total}.

$$M_{total} = M_B + M_1 + M_2 + M_3 \ \ldots\ldots\ldots \ M_n \ .$$

Where M_B is a background mortality not attributable to any identifiable factor. (In practice such simple arithmetic addition would rarely be possible.)

It can generally be assumed that animals will behave so as to minimize M_{total} so far as aims of growth and ultimate reproduction allow. If a shift in stress spectra is observed such as to change M_{total} this should be an early indicator of population changes before changes in mortality rate are noticed.

D can also be a spatial measure such as radius from the center of activity where the probability of mortality increases when the animal leaves its home range. If D is a factor such as light level which varies in an obviously cyclical manner then it may be more revealing to plot the activity of the animal against time of day or any other cycle of interest (Fig. 5). This distribution can then be substituted

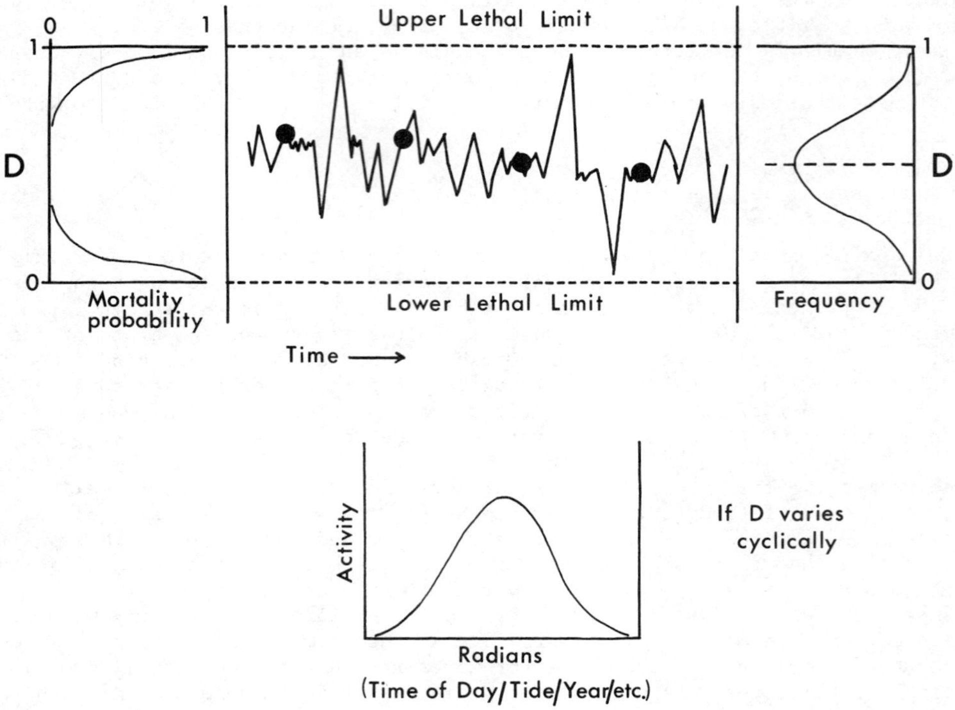

Fig. 5. *D* is any variable relevant to the animal's survival. *D* fluctuates continuously and this can only be traced by biotelemetry but intermittent sampling at regular intervals may be possible (dots on the center graph) using conventional techniques. The graph on the left shows the increase in mortality probability as the upper and lower lethal limits are approached. The graph on the right shows a typical frequency distribution of *D* or stress spectrum.

for T_D in the analysis. Diurnal fish for example may well suffer predation at dawn and dusk and it may be possible to express mortality as a function of time. In analysis of cyclical data circular statistics should be used (Mardia, 1972; Priede, 1978).

Niche

The fish can be considered as living in a multidimensional space with dimensions (*D*), the limits of which it is hazardous to approach. This parallels the concept of a niche as theoretically developed by a number of authors (e.g. May, 1974; Pianka, 1976). These authors have discussed the methods of dealing with the multidimensional problem quantitatively. Telemetry is probably more useful for investigating the abiotic aspects of the niche rather than problems such as interspecific competition. Telemetry methodology also allows analysis of the short term time scale relevant to day by day survival rather than probability of ultimate reproduction.

HEART RATE BIOTELEMETRY

Priede and Young (1977) describe the techniques in a study of heart rate of brown trout (*Salmo trutta* L.) in the wild in a Scottish loch. Heart rate was used as a measure of metabolic rate R (Priede and Tytler, 1977). If R_a is the active or maximum metabolic rate and R_s is the resting or standard metabolic rate then the data can be normalized:

$$\frac{R - R_s}{R_a - R_s} = S \ .$$

S then represents the metabolic scope (Fry, 1947) with 0 for resting and 1 for maximum metabolism. The distribution of time spent at different rates is plotted in cumulative form in Fig. 6. The heart rate to metabolism conversion has unidirectional error characteristics (Priede and Tytler, 1977) so the true metabolism values should lie to the left of the lines on the graph. Heart rates and metabolic rates are much higher in summer than in winter but it can be seen that after normalization the data from 3 fish at different seasons of the year all look very similar. Both extreme high and low rates are rare, most time being spent at intermediate levels of S.

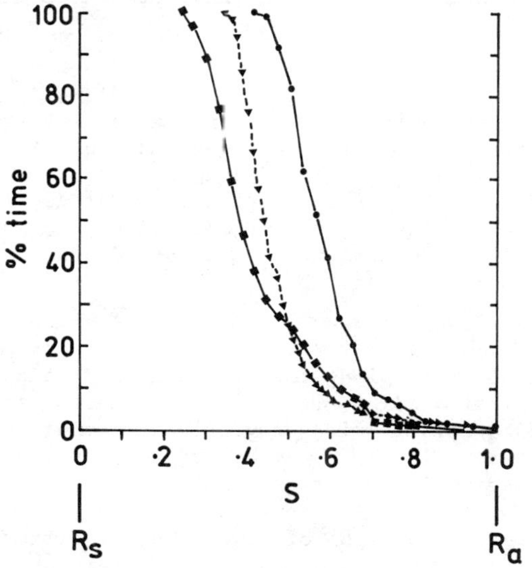

Fig. 6. Cumulative frequency distributions of S. S is a relative measure of metabolic rate based on heart rate telemetry. D is shown for 3 different brown trout. Triangles — fish in the summer at 15°C (Priede and Young, 1977). Squares and circles are two different fish at 5.5°C in the winter (Priede, 1978). R_s — standard metabolic rate. R_a — active metabolic rate.

A generalized log-normal distribution can be fitted to this data and this shown in Fig. 7 with a plausible mortality distribution superimposed. The mortality curve is constructed so as to give a realistic value for natural mortality rate of 0.1 percent per day (Priede, 1977). In this model 90 percent of the mortality is accounted for by the time spent at S values below 0.1 and above 0.9. The mortality function is a notional distribution which has been sketched in, but despite possible errors, the essential feature is that most mortality occurs at extremes of the stress or strain spectrum.

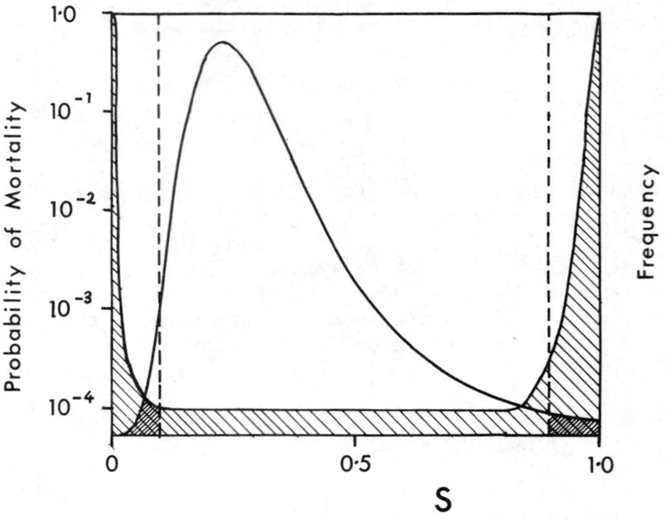

Fig. 7. Brown trout in a loch as in Fig. 6. The frequency distribution (based on log-normal) of time at different levels of *S*. The U shaped curve is the corresponding mortality distribution (Priede, 1977). The vertical lines represent *S* = 0.1 and *S* = 0.9, in this model most mortality occurs outside those limits.

SAMPLING OF STRESS SPECTRA

Using ordinary observation or sampling techniques it may be possible to measure *D* at regular intervals as indicated in Fig. 5. For example fish position could be determined by sonar and correlated with conventional temperature profiling techniques. A good estimate of the mean value of *D* may be obtained but nothing would be known about the tails of the stress distribution function.

The tails of the distribution function of time spent at different values of *D* determine the rate of mortality. The rate of mortality determines the fate of the population and is ultimately the basis of natural selection. It seems that biotelemetry is one of the few techniques with precise enough resolution on the time scale to measure accurately the extremes of stress or strain spectra. Biotelemetry therefore has a central role to play in future advances in the study of fish biology.

CONCLUSIONS

Tracking and biotelemetry techniques are already contributing to our understanding of fish biology. The broad scope of life cycles, behavior and physiology can be studied by conventional means. The main benefit of tracking and biotelemetry lies in the ability to resolve changes on a time scale of minutes rather than weeks or months. At this level many fundamental mechanisms are operative which cannot be investigated by any other means.

REFERENCES

Bazovsky I. (1961) *Reliability Theory and Practice*. New Jersey: Prentice-Hall.
Bertin L. (1956) *Eels, a Biological Study*. London: Cleaver-Hume.

Fry F. E. J. (1947) Effects of the environment on animal activity. *Publ. Ont. Fish. Res. Lab.* 55, 62 pp.

Gardella E. S. and Stasko A. B. (1974) A linear array hydrophone for determining swimming depth of fish fitted with ultrasonic transmitters. *Trans. Am. Fish. Soc.* 103, 635-637.

Gayduk V. V. and Malinin L. K. (1971) An informational utrasonic transmitter for biotelemetric investigations. *Biol. Vnutr. Vod. Inform. Byull.* 22, 74-78 (Fish Res. Bd Can. Transl. Ser. No. 2278).

Greer-Walker M., Harden-Jones F. R. and Arnold G. P. (1978) The movements of plaice (*Pleuronectes platessa* L.) tracked in the open sea. *J. Cons. int. Explor. Mer.* 38, 58-86.

Greer-Walker M., Miston R. B. and Storeton-West T. (1971) Trials with a transponding acoustic fish tag tracked with an electronic scanning sonar. *Nature, Lond.* 229, 196-198.

Hart L. G. and Summerfelt R. C. (1975) Surgical procedures for implanting ultrasonic transmitters in flathead catfish (*Pylodictus olivaris*). *Trans. Am. Fish. Soc.* 104, 56-59.

Hawkins A. D. (1973) The sensitivity of fish to sounds. In H. Barnes (ed.), *Annual Review of Oceanography and Marine Biology*, pp. 291-340. London: Allen and Unwin.

Hawkins A. D., MacLennan D. N., Urquhart G. G. and Robb C. (1974) Tracking of cod *Gadus morhua* L. in a Scottish sea loch. *J. Fish Biol.* 6, 225-236.

Holliday F. G. T., Tytler P. and Young A. H. (1973) Activity levels of trout (*Salmo trutta*) in Airthrey Loch, Stirling, and Loch Leven, Kinross. *Proc. R. Soc. Edin.* 74B, 315-331.

Ireland L. G. and Kanwisher J. W. (1978) Underwater acoustic biotelemetry: Procedures for obtaining information on the behaviour and physiology of free-swimming aquatic animals in their natural environments. In *Behaviour of Fish and Other Aquatic Animals*. New York: Academic Press.

Jennings J. G., Gandy W. F. and Vanselous T. M. (1979) Development of a satellite linked marine mammal transmitter system. In *Data Collection and Location by Satellite*. ARGOS Utilization Conference, Toulouse, France, March 1979.

Kelso J. R. M. (1974) Influence of a thermal effluent on movement of brown bullhead (*Ietalurua nebulosus*) as determined by ultrasonic tracking. *J. Fish. Res. Bd Can.* 31, 1507-1513.

Kelso J. R. M. (1976) Movement of Yellow Perch (*Perca flavescens*) and White Sucker (*Catostomus commersoni*) in a near shore Great Lakes habitat subject to a thermal discharge. *J. Fish. Res. Bd Can.* 33, 42-53.

Laird L. M. (1978) Marking fish. In B. Stonehouse (ed.), *Animal Marking: Recognition Marking of Animals in Research*, pp. 95-101. London: Macmillan.

Laird L. M. and Stott B. (1978) Marking and tagging. In T. Bagenal (ed.), *Methods for Assessment of Fish Production in Fresh Waters*, pp. 84-100. Oxford: Blackwell.

Lentfer J. W., Fallek H. G. and Kolz A. L. (1977) Satellite tracking of Polar Bears. *2nd ARGOS Users Meeting*, Paris, November 1977.

Mardia K. V. (1972) *Statistics of Directional Data*. New York: Academic Press.

May R. M. (1974) Ecosystem patterns in randomly fluctuating environments. *Prog. Theoret. Biol.* 3, 1-50.

McCleave J. D., Power J. H. and Rommel S. A. (1978) Use of radio telemetry for studying upriver migration of adult Atlantic salmon (*Salmo salar*). *J. Fish. Biol.* 12, 549-558.

Mitson R. B. and Storeton-West T. (1971) A transponding acoustic fish tag. *Radio. Electron. Engng.* 41, 483-489.

Mitson R. B. and Young A. H. (1975) A survey of the engineering problems of developing small acoustic fish tags. *Proc. IERE Conf. Instrum. Oceangr.* Bangor, Wales, pp. 163-174.

Nomura S., Ibaraki T., Hirose H. and Shirahata S. (1972) Application of back-pack cardiotelemeter for fishes. I. Heart rate and cardiac reflex in fishes during unrestrained swimming. *Bull. Jap. Soc. Scient. Fish.* 38, 1105-1117.

Oswald R. L. (1978) The use of telemetry in study light synchronization with feeding and gill ventilation rates in *Salmo trutta*. *J. Fish. Biol.* 13, 729-739.

Pianka E. R. (1976) Competition and niche theory. In R. M. May (ed.), *Theoretical Ecology*. Oxford: Blackwell.
Priede I. G. (1977) Natural selection for energetic efficiency and the relationship between activity levels and mortality. *Nature Lond.* 267, 610-611.
Priede I. G. and Tytler P. (1977) Heart rate as a measure of metabolic rate in teleost fishes: *Salmo gairdneri* Richardson, *Salmo trutta* L. and *Gadus morhua* L. *J. Fish. Biol.* 10, 231-242.
Priede I. G. and Young A. H. (1977) The ultrasonic telemetry of cardiac rhythms of wild brown trout (*Salmo trutta* L.) as an indicator of bioenergetics and behaviour. *J. Fish. Biol.* 10, 299-318.
Priede I. G. (1978) Behavioural and physiological rhythms of fish in their natural environment as indicated by ultrasonic telemetry of heart rate. In J. Thorpe (ed.), *Rhythmic Activity of Fishes*. New York: Academic Press.
Priede I. G. (1979) Satellite tracking of Basking Sharks (*Cetorhinus maximus*). In *Data Collection and Location by Satellite*. ARGOS Utilization Conference, Toulouse, France, March 1979.
Satchell G. H. (1971) *Circulation in Fishes*. Cambridge University Press.
Standora E. A., Sciarrotta T. C., Ferrel D. W., Carter H. C. and Nelson D. R. (1972) Development of a multichannel, ultrasonic telemetry system for the study of shark behaviour at sea. Calif. State Univ. Long Beach Found. Tech. Rept. 5, 69 pp.
Stasko A. B. and Pincock D. G. (1977) Review of underwater biotelemetry, with emphasis on ultrasonic techniqes. *J. Fish. Res. Bd Can.* 34, 1261-1285.
Stasko A. B. and Rommel S. A. (1974) Swimming depth of adult American eels (*Anguilla rostrata*) in a salt water bay as determined by ultrasonic tracking. *J. Fish. Res. Bd Can.* 31, 1148-1150.
Trefethen P. S. (1956) Sonic equipment for tracking individual fish. *U.S. Fish. Wild. Serv. Spec. Sci. Rept. Fish.* 179, 11 pp.
Young A. H., Tytler P., Holliday F. G. T. and MacFarlane A. (1972) A small sonic tag for measurement of locomotor behaviour in fish. *J. Fish. Biol.* 4, 57-65.

Biotelemetry with Radionuclide Tracers

J. BOJSEN

The Finsen Laboratory, The Finsen Institute, Copenhagen, Denmark

Abstract — A survey of the field of biotelemetric measurements of radionuclides is presented. Clinical and biological examinations using radionuclides and measured by stationary equipment such as the gamma-camera or whole body counters are generally limited to time periods of less than one hour. In contrast, biotelemetric methods do not require that the patient or animal be restrained and thus such techniques can be used for longterm physiological investigations as an extension of the capabilities of conventional methods. Ten years ago, only nuclear detectors with very low gamma-counting efficiencies were available, for example, the GM detector, the $CaSO_4$: Dy thermoluminescent dosimeter and the SI(Li)-semiconductor detector. Recently a significant improvement of the gamma-counting efficiency has been achieved by small, reliable, single crystal detectors, made from materials like Cadmium Telluride (CdTe) and Mercuric Iodide (HgI_2). These transducers promote the field of biotelemetry with radionuclides from an experimental tool to include clinical routine and applied physiology. A variety of useful external and implantable biotelemetric equipment has been developed by several groups, changing the actual problems from the users point of view, to analysis concerning tracer stability tests, counting geometry adjustments and interpretation of the results.

INTRODUCTION

Tracer analysis has become an extremely valuable tool in both basic research and clinical medicine. With the aid of a stationary γ-scintillation camera in combination with radiopharmaceutical agents, a great variety of characteristics in tracer-distribution images and tracer-disappearance patterns have been described which are useful for the clinical routine.

Until recently the stationary γ-camera has been the most significant imaging device for the distribution of radionuclides in diagnosis and patient monitoring. However, dramatically good results have been achieved in clinical work by the use of computerized tomography (CT-scan) and ultrasound, both with a considerable improvement in spatial resolution. As a logical consequence of this competition the future γ-camera, although still stationary and imaging, will probably be further improved to show its superiority in dynamic, physiological function studies combined with appropriate pharmacokinetics.

These are the conditions with which biotelemetric measurements using radionuclides as tracers have to compete. Although it seems at first very difficult to establish an alternative to such sophisticated apparatus as described above, it has been proven possible, when supported by the fundamental ideas of biotelemetry. The basic idea is to keep the patient in his normal circumstances during the investigation, and this requirement cannot be fulfilled by huge machines.

Telemetry equipment which has been developed in the past should have been refined not only on the basis of technology and circuitry, but also by experience from applications in controlled biological investigations. Such experiments were sometimes lacking, although they are necessary for developing a reliable construction.

This report is a survey of biotelemetry in relation to the measurement of radionuclide tracers as a tool in physiological studies. The methods are described, evaluated and compared to conventional techniques commonly used in clinical physiology, nuclear medicine or radiography.

CONVENTIONAL MEASURING TECHNIQUES

As a general characteristic the conventional γ-camera or whole body counter requires immobility of the patient during the investigation, and consequently precludes studies lasting more than one hour, or during active physiological conditions like work, exercise and sleep. The fundamental need for keeping the patient immobilized has to be seen in connection with a general design philosophy based on big and heavy lead shielded NaI(Tl) detectors with photo-multipliers, held stationary with respect to the patient, a short distance from the body. If the patient, who is the radioactive source, moves in relation to the detectors, serious artifacts will be introduced, which cannot easily be distinguished from the disappearance of the tracer

BIOTELEMETRY TECHNIQUES

In the use of biotelemetry, patients or experimental animals remain unrestricted in their normal surroundings during the physiological investigation, which then can last for days or weeks. The condition is, however, that the biotelemetry equipment be portable and miniaturized to fulfil the general requirements of small weight, small size and longterm operation. In human applications the equipment is generally kept external (non-invasive) contrary to the implantable measuring devices used in experiments with animals.

MATERIALS AND METHODS

RADIONUCLIDE TRACERS

When the γ-camera is used for external counting, the tracer has to be labeled with a suitable γ-energy-emitter. The most commonly used radionuclide today is ^{99m}Tc with a γ-energy photo-peak at 140 keV, and a physical half lifetime of 6 h. Where biotelemetry is used, it is possible to move the detector closer to the radioactive source and apart from the γ-emitters, measurements with β-emitters are possible, especially when equipment is placed directly in contact with the organ under study.

RADIOACTIVITY DETECTORS

Biotelemetry detectors can be attached to the skin as is most common in human applications or fixed in tissue after implantation in experiments with animals in order to gain a constant counting geometry independent of motor activities etc. Few nuclear detectors, which could work at body temperature with reasonable β-counting efficiency were available for biotelemetry applications a decade ago. Only the GM-detector (Mackay, 1970; Gugnin *et al.*, 1972; Bojsen and Wallevik, 1972), the $CaSO_4$:DY thermoluminescent dosimeter (Bojsen *et al.*, 1974, 1977) and the Si(Li)-semiconductor detector (Kobayashi *et al.*, 1974). These detectors have a very low γ-counting efficiency. Recently a significant improvement of the γ-counting efficiency has been achieved by small, reliable, single crystal detectors, made from semiconductor materials with high atomic numbers like cadmium telluride (CdTe) and mercuric iodide (HgI_2).

CdTe- and HgI_2-detectors

Semiconductor detector characteristics in relation to biotelemetry are given in this paper, but details about crystal growth and detector fabrication can be read in publications by Siffert, Ponpon and Cornet (1975); Siffert (1978) and Schieber (1977). Three detector types have recently been examined in our laboratory and found to be useful in biological experiments:

1. The cadmium telluride (CdTe), made by Centre de Recherches Nucleaires, Strasbourg, France, $3^{\phi} \times 0.6$ mm, P-type, 10^6-10^4 Ω cm resistivity, without external chemical compensation. γ-energies > 35 keV and 100 keV are measured with an energy resolution of about 10 percent at 60 keV, but energies up to 200 keV can be detected. A ^{99m}Tc point source will give 2 cps/μCi and a volume source: 200 counts per second (cps)/μCi/ml. Bias voltage and current are 100 V and 10 nA, respectively.

2. The cadmium telluride detector (CeTe(Cl)), made by Radiation Monitoring Devices Inc., Watertown, Mass., U.S.A., $7^{\phi} \times 1.5$ mm, P-type, 10^8-10^5 Ω cm resistivity with chlorine compensation. γ-energies > 50 keV and < 600 keV are detected with nearly no resolution, but due to the detector volume the counting efficiency is high. A ^{99m}Tc point source gives 500 cps/μCi. The volume source gives rise to 11.000 cps/μCi/ml. The bias voltage and current are 30 V and 0.1 μA respectively.

3. The mercuric iodide detector (HgI_2), made by Hebrew University of Jerusalem, Israel (Schieber, 1977), 44 $mm^2 \times 0.75$ mm. γ-energies > 10 keV and < 700 keV are detected with energy resolution up to 200 keV, depending on the bias voltage, which should be about 2 kV to get the best resolution (6 percent at 60 keV) and charge collection. However, when bias voltage is decreased to 200 V, which is more reasonable in relation to a patient's safety, the counting efficiency of ^{99m}Tc decreased to 92 percent, but even at 200 V the X-rays in the 20 keV range were detected with resolution in energy. When the lower discriminator level is adjusted to 10 keV, the counting efficiency is comparable to that of the CdTe(Cl)-detector. Bias current was measured to be about 50 pA.

No polarization effects were found in any of the detectors. Polarization in the detector crystal means a decrease in charge collection efficiency with time and a change in energy resolution. It should be noted that the linear range of the count rate detected from calibrated ^{99m}Tc sources, has been increased and was found to be maximum 12,000 cps. The detectors mentioned are only examples of a great variety of sizes and characteristics available today (Bojsen *et al.*, 1976; Bojsen *et al.*, 1977).

Any biotelemetric device consists of: the transducer, the radio transmitter or the storage device, and at a remote location: the radio receiver or the storage replay unit, with different read-out devices. Only the transducer, the radio transmitter and storage device will be discussed.

'RADIOPILLS'

In longterm experiments the biotelemetry equipment has to be implanted subcutaneously or intraperitoneally if appropriate count rates cannot be obtained percutaneously. More than 10 years ago, a radio telemetry unit was developed to circumvent this problem for detection of radionuclide tracers in the gastrointestinal tract in humans. The unit was constructed as a pill and was introduced into the intestinal tract by swallowing (Mackay, 1970). The pill consisted of a GM-detector, DC/DC converter for the high voltage supply, short range HF radio transmitter and batteries. Gugnin *et al.* (1972) described a similar pill equipped with a piezoelectric transformer to generate the 300 V high voltage for the GM-detector. The pill measured 29 × 12 mm in diameter, and the battery lifetime was about 30 h. Recently a radiation pill 27 × 10 mm with a small HgI_2-detector, high input impedance preamplifier and 10 m range VHF radio transmitter (106 MHz) has been constructed (Hassan, Pearce and Edwards, 1978). The HgI_2-detector volume was 2 × 2 × 0.4 mm with a volume source sensitivity of ^{99m}Tc equal to 70 cps/μCi/ml. The lifetime of the battery (1.35 V) was about two days of continuous operation.

EXTERNAL RADIO TELEMETRIC DEVICES

The use of a portable external two-channel radio telemetric GM-detector unit with a weight of 260 g for transcutaneous measurements of radionuclide tracers in humans has been published by Bojsen and Vadstrup (1974). It consisted of two 26^{ϕ} mm end window GM-detectors, a high voltage supply for generation of the 600 V operation voltage from a 1.35 V (1 Ah) battery, two subcarrier oscillators one for each channel of 5 KHz and 20 KHz for FM/FM modulation, and a VHF transmitter at 106 MHz with a range of 10-15 m. A separate battery supply of 1.35 V (1 Ah) for the modulator and transmitter circuits was used. The maximum count rate at linear operation was found to be 100 cps. With continuous transmission the operation time was about 5 months. With a battery supply with no reference to ground potential and a high output impedance of the high voltage supply, the requirements for the safety of the patient were fulfilled.

EXTERNAL STORAGE TELEMETRIC DEVICES

When ECG or blood pressure in the ambulant patient is monitored continuously the simultaneous registration by radio telemetry can be of vital importance. However, in biotelemetry with radionuclide tracers the need of simultaneous registration of data is generally lacking and the alternative storage telemetry technique may be used with advantage, as different technical problems concerning RF noise and interference can be avoided.

Bojsen and Vadstrup (1976) published a storage telemetric system with two GM-detectors as a substitute for the external radio telemetry unit mentioned above. The recording medium was a commercially available pocket-sized cassette tape recorder with 24 h continuous registration. The maximum count rate was still limited to 100 cps and the total weight 600 g.

Due to electrical noise from the DC-motor and in particular the very limited counting capacity of 100 cps, we decided to develop a semiconductor memory system based on CMOS-technology as a realistic alternative to other recording media (Bojsen, 1976). A portable semiconductor memory system carried in a belt with two 1024 bit static random access memories (RAM) organized in two 64 sub-channels, each sub-channel having a storage capacity of 5 decades, was made as a first version ('LOGIK', Engineering Co. Ltd., Denmark). Dwell-time periods of 0.3 s, 5 min, 10 min or 20 min could be preselected. The battery capacity was sufficient for continuous 48 h measurements. The size of the memory unit was about that of a cigarette package and the weight 250 g. By this recording method the maximum count rate of registration has recently been extended to 12000 cps, only limited by the dwell-time selected and the proper semiconductor detector (CdTe or HgI_2).

When CdTe- or HgI_2-detectors are used, charge sensitive preamplifiers are needed and they should be built adjoining the detector itself, in order to reduce the influence from external noise sources. However, if the preamplifiers are built from discrete components, it has been necessary to change the design philosophy to keep the size of the attached skin surface detector small. In principle Bojsen *et al.* (1976) and Hassan *et al.* (1978) have used similar preamplifier circuits with a FET source follower and an AC-amplifier, which seems to fulfil the demands, at least when measurements in the counting mode are performed.

IMPLANTED RADIO TELEMETRIC DEVICES

More than 10 years ago a circadian variation in ^{32}P concentration or uptake was recognized in human breast cancer by several investigators (Bullen *et al.*, 1963; Calcutt *et al.*, 1967; Stoll and Burch, 1968; Taylor *et al.*, 1968; Wolley-Hart *et al.*, 1968), and that the inhibition of the cancer cells by cytotoxic drugs or irradiation should be scheduled in relation to these fluctuations in the ^{32}P-content. However, in an attempt to produce evidence in support of these theories it became quite obvious that systematic research had to be performed on an experimental tumor model.

In animal experiments, the usual radionuclide tracer techniques are generally limited by the following procedures: (1) Blood sampling: this procedure is stress provoking and furthermore repeated sampling will change the erythrocyte and protein concentrations in the blood of the animal. (2) Urine and feces sampling involve great uncertainty of measurement due to loss. (3) Whole body counting is stress provoking, like blood sampling, as it is generally performed during restraint. However, the circadian variation in ^{32}P uptake in hormone dependent mammary tumors is easily disturbed and it would then be of advantage to keep the animals unrestrained. Therefore, we had to develop a biotelemetric measuring technique, which could be used repeatedly on an unrestrained mouse or rat in longterm studies with radionuclides. Such biotelemetry methods would, in addition, permit each animal to be its own control, and thereby diminish the uncertainty due to individual variation.

The first equipment developed was an implantable radio telemetry GM-detector which could be applied for continuous registration of the disappearance of radionuclide-tracers in both short and longterm experiments (Bojsen, Wallevik and Møller, 1972; Bojsen and Wallevik, 1972). It was, however, so bulky that only rabbits could be used as experimental animals. The high voltage supply (600 V) and the GM-detector are nearly identical to these previously described in the external two channel device. The end window detector (26$^\phi$ mm) pointed toward the center of the rabbit in which a fraction of the whole body was monitored with a maximum count rate of 100 cps. The VHF transmitter (106 MHz) had a range of about 10 m, and with 1.35 V mercury batteries of 'pacemaker' quality continuous measurements could be performed within a period up to five months. When the device was equipped with a magnetic switch, the measurements could intermittently be extended to about one year. The weight of the implantable unit, coated in Araldite, was 135 g. An important point is the unavoidable requirement, that the encapsulation should be as waterproof as possible, as a high voltage (600 V) is generated inside the implant. The lifespan of one year is due to a slow diffusion of water into the electronic circuitry.

Semiconductor detectors were introduced into biotelemetry several years ago, when a partially implantable radio telemetry system was published by Kobayashi *et al.* (1974). It was developed for experiments with dogs. A small windowless silicon radiation detector (Si(Li)), 5 mm in diameter, and implantable (e.g. in the liver), was used for β-detection in connection with an external transmitter unit 2 × 6 × 9.5 cm in size and a weight of 170 g. The power consumption was 100 mW and continuous operation could be performed for 20 h. Like the CdTe- and HgI_2-detectors, this type of semiconductor also requires a low noise, wide-band preamplifier adjoining the detector itself.

IMPLANTED STORAGE TELEMETRIC DEVICE

The intention was to measure the ^{32}P uptake in unrestrained rats. Radio telemetry would have been an attractive technique, but it was not possible to reduce the volume and weight of the implantable GM-detector unit (Bojsen and Wallevik, 1972), without drastically reducing the counting efficiency as well. Therefore an alternative storage telemetry method was developed utilizing thermoluminescent (TL)-dosimeters ($CaSO_4$:Dy) for longterm measurements of incorporated radioactive substances (Bojsen *et al.*, 1977). By sequential replacement of double TL-dosimeters (60 mg) through a permanently implanted silicon tube, the concentration of different radionuclides (e.g. ^{32}P), could be determined simultaneously in tumors and normal tissue over several 24 h periods in each unrestrained animal. Replacement of dosimeter catheters may stress the animal and influence the results of the ^{32}P uptake measurements. Miles (1962) found that the persistence of the circadian body temperature rhythm or the body temperature itself over short terms are indicators, which indicate when the animals are affected by the experimental conditions (reviewed by Amlaner, 1978). Therefore the TL-dosimeters have been used in combination with an implanted temperature transmitter for continuous registration (max. 2 months) of the circadian body temperature rhythm (Bojsen, Møller and Faber, 1971). The weight of the system in total was about 6 g or 3 percent of the body weight of the rat.

APPLICATIONS

'RADIOPILLS'

The literature on measurements of gastrointestinal radioactive tracers by 'radiopills' is very limited. Mackay (1970), Gugnin *et al.* (1974) and Hassan *et al.* (1978) have published different versions of a radiation sensitive radio transmitter as a pill, primarily for location of the site of bleeding along the gastrointestinal tract. Mackay (1970) and Gugnin *et al.* (1974) claimed that sites of hemorrhage can be detected by the release of ^{32}P labeled erythrocytes into the gut. On the other hand could gastrointestinal tumors be recognized by an increased ^{32}P uptake in the tissue itself? The high energy of the ^{32}P β-particles permitted recognition of the nuclide. Hassan *et al.* (1978) used a HgI_2-detector instead of GM-tube and suggest the use of ^{99m}Tc, which is a γ-emitter, although the detector is also sensitive to ^{32}P (Hassan, 1978).

Apart from laboratory tests only Mackay (1970) has made reference to some preliminary biological experiments with dogs in which a reliable sensing of labeled cells was impossible at moderate doses due to absorption in the coating and the window thickness of the GM-detector.

EXTERNAL GM-DETECTORS

A number of longterm measurements on humans have been made possible by the following techniques. Radio telemetry equipment with two GM-detector channels (Bojsen and Vadstrup, 1974) was primarily developed to measure the ^{32}P uptake in breast cancer over several 24 h periods in ambulatory patients, in order to extend investigations by Stoll and Burch (1968), who used stationary equipment. It could, however, also be used for counting low-energy γ-emitters in local clearance studies with ^{133}Xe or in the continuous measurements of ^{125}I uptake in the lobes of the thyroid gland.

When the storage telemetry equipment became available (Bojsen and Vadstrup, 1976), the continuous detection of ^{125}I-labeled fibrinogen was suggested during its disappearance from the legs of patients with thrombophlebitis. The percutaneous detection of ^{125}I-labeled insulin disappearance from depots in diabetics has proven to be of great interest. Very recently Kølendorf, Bojsen and Jørgensen (1978) published a clinical investigation on the subcutaneous absorption of ^{125}I-NPH-insulin from both legs in ten newly diagnosed, juvenile ketotic diabetics. The measurements were performed with 24 h continuous registration to be compared with plasma insulin measured by sequential blood sampling after subcutaneous injection of unlabeled NPH-insulin and were found to agree. Comparable studies with a stationary NaI-detector equipment have been performed by Binder (1969). The greatest advantage of using a biotelemetry technique is continuous detection and completely undisturbed conditions of the patient during the investigation, as the release of insulin from subcutaneous tissue is a function of different physiological conditions (Kølendorf, Bojsen and Nielsen, 1979).

EXTERNAL CdTe-DETECTORS

The appearance of the CdTe-semiconductor detectors in combination with the CMOS-semiconductor memory have extended the practicability of biotelemetry with radionuclide tracers. In relation to the tape recorder, the CMOS-memory has a number of advantages. It has a higher maximum count rate, no acoustic noise and tape costs, and a lower weight, power consumption, and read-out time due to the calculation of count rates before storing.

A number of local clearance studies has been performed with ^{133}Xe in man during rest and exercise in adipose tissue and muscles. After injection in the anterior tibial muscle the disappearance of Xenon was simultaneously measured at rest with the CdTe-detector attached to the skin surface and a 3"-NaI-detector fixed 10-20 cm apart from the injection site (Bojsen *et al.*, 1977). Due to the small tissue volume monitored by the CdTe-detector, it measures more correctly the local disappearance rate of the Xenon depot, even without the need of lead shielding. Contrary to this, the NaI-detector monitors a rather big tissue volume even when restricted by lead shielding, so the radioactive Xenon is detected with high efficiency also after it has been moved away from the depot by the blood. In a similar experiment the blood flow was measured in the anterior femoral muscles during rest and ergonometry it was not possible to monitor the disappearance rate of Xenon with the NaI-detectors during the exercise, but the small detector was not disturbed by motor activity.

There is a need for determination of cardiac output during exercise. Preliminary investigations were performed in patients (Bojsen *et al.*, 1977) with the CdTe-detector attached to the skin surface over the heart, it was possible to measure the wash-out from the heart of a ^{99m}Tc radioactive dose injected i.v. as a bolus. The results were compared with simultaneous measurements with a computerized γ-camera with digitized regions of interest. The count rate curves were found alike.

Recently a method for determination of glomerular filtration rate (GFR) with ^{99m}Tc-DTPA(Sn) and skin surface semiconductor detectors was introduced for routine measurements in the clinic (Bojsen and Rossing, 1978; Rossing, Bojsen and Frederiksen, 1978). By single injection and external counting in a 3 h period with CdTe-detectors attached to the chest just below the clavicles only one blood sample is necessary for an exact determination of GFR compared to the conventional ^{51}Cr-Ethylenediamine-tetra acetic acid (EDTA) technique based on 4 blood samples in a 5 h period. By using the conventional blood sampling technique with ^{51}Cr-EDTA the determination of GFR will take about 9 h from the time of injection of the tracer, due to preparation and counting of plasma samples. In comparison to this, the GFR value can be calculated about 3 h after the injection, when a continuous measurement by telemetry is used. The correlation coefficient between the biotelemetry and reference method is $r = 0.98$ (29 patients, range 30 to 140 ml min^{-1}) (Bojsen *et al.*, 1978). McLeod (1977) and McLeod, Sampson and Houston (1977) confirm our findings of the external GFR measurements using a stationary external arm counting intermittently after single injection i.v. of ^{99m}Tc-(Sn)DTPA. The final part of the external measured curve is monoexponential (final slope) after 45 min. We found that in external measurements on the chest with CdTe-detectors the final slope was already attained 30 min after injection. When plasma sampling techniques are used the final slope of the plasma concentration curve is reached after 3 h.

IMPLANTED GM-DETECTORS

The purpose of using implantable radionuclide detectors especially in animals has three primary objects. One is to extend the field of monitoring physiological phenomena while the animal is completely undisturbed. The second is to extend the measurements over long time periods. The third is to obtain continuous measurements during this time. The possibilities of the methods have, however, not yet been fully elucidated.

Implanted radio telemetry GM-detector units have been used in studies of circadian rhythms in renal function in rabbits (Vadstrup and Bojsen, 1976a). The detector measured the body disappearance of ^{125}I-iodide (1 cps per 1 nCi/ml) following an injection i.v., and since the food contained carrier iodide, the excretions of iodide were solely renal and followed first order kinetics. Such studies show that undisturbed rabbits have a circadian rhythm in the renal iodide clearance synchronized to the light-dark cycle. Iodide excretion is determined both by glomerular and tubular factors, but was chosen because of a suitably low total disappearance rate, which allows continuous measurements during more than 24 h. As the circadian rhythm is labile and will disappear if the animal is disturbed, a radio telemetry technique would be especially suited for this type of study. The method could also be used in a similar way to determine the renal clearance of ^{125}I-iothalamate (GFR) and ^{125}I-hippuran (renal plasma flow = RPF) in rabbits following single injections i.v. of the tracers (Vadstrup and Bojsen, 1976c). When normal patterns of excretion have been determined, the effects of different pharmaceuticals and hormones can be investigated by i.v. or s.c. administration during the excretion phase of the tracer and the excretion curve can be analyzed for changes. This has been the basis for studies of the renal effect of pituitary hormones on the renal iodide excretion in rabbits (Vadstrup and Bojsen, 1976b; Vadstrup, 1977).

IMPLANTED Si(Li)-DETECTORS

Partially implantable radio telemetry equipment with a Si(Li)-detector published by Kobayashi *et al.* (1974) was developed for detection of β-ray emitting radioisotopes in dogs. It was suggested to use the equipment in the study of liver metabolism by i.v. injected ^{14}C-labeled alcohol, but no biological results were presented. Although ^{14}C-spectra were measured in the laboratory by the Si(Li)-probe, it is questionable that measurements *in vivo* would succeed because of tissue reaction, which will always appear around the implants. It was, however, admitted that improvements of the system still remain.

IMPLANTED THERMOLUMINESCENT DOSIMETERS (TLD)

The combination of TL-dosimeters, which could be replaced, for instance every second hour through an implanted silicon catheter, and an implanted temperature transmitter (Bojsen *et al.*, 1977) have been used in 48 h experiments with normal and DMBA-tumor bearing rats. Both ^{32}P (E_{max} = 1.7 MeV) and ^{42}K (E_{max} = 3.6 MeV), which are β-emitters, were used in doses of 0.02 μCi g^{-1} body wt, and the coefficient of variation of each 2 h integrated dose was found to be ± 2 percent (SD). Due to the short range of β rays in tissue the results could be related to specific tissues around the dosimeters. In normal subcutaneous tissue the ^{32}P uptake was monoexponential with a half-life time of about 6 days, similar to that found in external GM-detector measurements in restrained rats (Møller and Bojsen, 1976). In experiments with ^{42}K in subcutaneous tissue and muscles systematic 6 to 8 h changes in the ^{42}K concentration appeared. These results need further investigations. During the initial phase of some of the experiments the circadian temperature rhythm disappeared 6 to 7 h as an expression of the reaction of the animals to the experimental procedures.

DISCUSSION

In biotelemetry, as in any other technique, the transducer of the measuring equipment is of paramount importance for the capability of the measuring method. The use of small, reliable, single crystal semiconductor detectors with high atomic numbers like CdTe and HgI_2 are now possible in biotelemetry with radionuclide tracers, where a small detector volume is a prerequisite. Due to their high stopping power to γ-rays, small power consumption, and uniquely small size, they seem to expand the field of biotelemetry with radionuclides from an experimental tool to include areas such as routine clinical work and physiology.

In studies with radionuclide tracers, *in vivo*, the measuring geometry has to be kept constant in order to diminish the number of unknowns, and thereby improve the possibilities of interpretation of the biological results. Before biological investigations are started it is advisable to perform an examination of the integrated semiconductor detector probes and simulate typical geometry situations as they may appear *in vivo*. Isocount curves should be obtained with radioactive point sources, in tissue equivalent Perspex, and they can be of importance in the valuation of, e.g. local clearance studies with ^{133}Xe or absorption of ^{125}I-NPH-insulin. Count response curves should be obtained with a stratified radioactive source in Perspex, and can be used in the valuation of results from different depths in tissue of a homogeneous well mixed tracer as in kidney function studies with ^{99m}Tc-(Sn)-DTPA.

Because of the small size of the semiconductor detector it has a spatial resolution in itself, without any lead collimation. Detectors with a sufficiently high counting efficiency (^{99m}Tc) could additionally be geometrically adjusted using a lead collimator (2 mm thick, about 7 g) in order to achieve a reduced influence from the skin just outside the detector window or the tissue reaction surrounding the detector, when it is implanted.

Several small, indigestible telemetry radiopills have been developed in the past for measurements of temperature, pressure or pH in the gastrointestinal tract after swallowing in both animals and man. Based on the same philosophy Mackay (1970); Gugnin *et al.* (1972) and Hassan *et al.* (1978) described analogous radiopills for radioactivity detection of bleeding sites in the gastrointestinal tract. However, the value of such a radionuclide tracer technique is questionable as a clinical or experimental tool, primarily due to changes which can occur in counting geometry when the detector passes through the intestinal tract. Even when ^{32}P labeled erythrocytes are used with a relatively short range of maximum 8 mm in tissue, an influence from big blood vessels of the intestine cannot be be discriminated in the results. When γ-emitting tracers are used — as proposed by Hassan *et al.* (1978) — the long range (e.g. of ^{99m}Tc) makes the interpretation of the biological results even worse. Therefore more reliable conventional techniques like radiography with i.v. injected contrasts, ultrasound scanning or CT-scan seem at present to be superior in the clinical routine, when diagnosis of bleeding sites or tumor recognition in the gastrointestinal tract are going to be performed.

The small radiopill described by Hassan *et al.* (1978), seems nevertheless, to have several potential applications when implanted permanently in small laboratory animals like rats or even mice. The storage telemetry thermoluminescent dosimeter method (Bojsen *et al.*, 1977) is time consuming in all steps of the experimental procedure. Therefore, using the radiopill, which combines a HgI_2-detector and a radio transmitter, has the advantage of continuous measurements of γ- or β-emitting tracers with relatively high efficiency (Hassan, 1978). Because of the power consumption of the preamplifier, inductive powering of the implanted radiotelemetry unit could be preferential in longterm investigations (Fryer, Lund and Williams, 1978).

In renal function studies with rabbits (Vadstrup and Bojsen, 1976c) some of the main problems are the limited possibility of checking to what extent non-steady state conditions occur, and to what extent they will affect the disappearance rate determinations. An ideal condition to fulfil is, that direct comparisons between the telemetry obtained values and values obtained by conventional clearance technique, including urine collections should be performed for a wide range of disappearance rates. This is technically very difficult and it is hampered by the fact that such direct clearance comparisons affect the physiological conditions, and only can be done in anesthetized or restrained rabbits. However, a supplementary use of HgI_2 radio transmitters to measure disappearance rates in specific rabbit tissue could improve the further elucidation of circadian rhythms in renal function.

A successful biological investigation depends on both capable measuring equipment and the behavior of the radioactive labeled substance used as the tracer. Several requirements of the reliability of the tracer have to be fulfilled, as it has to retain the characteristics of the parent substance throughout the course of the investigation. Using biotelemetry methods, the investigation periods have been considerably extended compared to conventional short term techniques (e.g. the γ-camera). This means that further controlled examinations of tracer stability, *in vivo*, by control substances and simultaneous multiple blood sampling and external detection over longterm periods are needed before conclusive biological results of, for instance, short term variation in glomerular filtration rates over 24 h periods with ^{99m}Tc-(Sn)-DTPA can be stated (Bojsen *et al.*, 1978).

One of the potentials of the HgI_2-detector is the low internal noise level, which means that γ- and X-ray energies > 10 keV can be detected. Contrary to this the CdTe-detector noise limits the detectable energy level to > 30-40 keV excluding investigations with ^{125}I, as in absorption studies of ^{125}I-NPH-insulin in diabetics. The question of replacing GM- or CdTe-detectors in the biotelemetry equipment of the future will depend on results from longterm reproducibility tests with HgI_2-detectors.

The prospects of biotelemetry with radionuclides cover a considerable improvement in the design of the measuring equipment. By thick-film or alternative technology a great number of single crystal detectors each connected through preamplifiers to its own CMOS-solid state memory, can be utilized for clinical routine or experimental biotelemetric investigations. Each detector could be fixed to the skin surface at locations of interest and an enormous number of simultaneous and continuous data (e.g. from kidney and bladder in dynamic renal function studies) could be achieved stimulating the interests of pharmacokinetic studies by such techniques and further compartmental analysis.

Biotelemetry with radionuclide tracers has established its own field as a useful clinical and experimental tool. Equipment is relatively simple in circuitry design, as it has to be portable to fulfil the requirements for longterm investigations at normal physiological conditions. Although many of the conventional radionuclide tracer methods in the clinic today are based on the use of stationary apparatus, like the γ-camera, biotelemetry techniques appear to extend the capability of clinical and experimental investigations based on radionuclide tracers.

Acknowledgements — I wish to thank Prof. M. Faber, M.D., Head of the Finsen Laboratory, for valuable help in the preparation of the manuscript. Also valuable comments and suggestions given by K. Wallevik, M.D., K. Kølendorf, M.D., and S. Vadstrup, M.D., are gratefully acknowledged, and also T. Nørager for skillful typing of several drafts and the manuscripts. I thankfully acknowledge Dr. P. Siffert and his group at CRN, Strasbourg, France, for supplying the CdTe-detectors. The other CdTe- and HgI_2-detector types were supplied by a grant from the Danish Council for Scientific and Industrial Research.

REFERENCES

Amlaner C. J., Jr. (1978) Biotelemetry from free-ranging animals. In B. Stonehouse (ed.), *Animal Marking: Recognition Marking of Animals in Research*, pp. 205-228. London: Macmillan.

Binder C. (1969) Absorption of injected insulin. (Dissertation) Copenhagen, Munksgaard.

Bojsen J. (1976) A semiconductor memory system for data storage in biotelemetry of radionuclide tracers. In T. B. Fryer, H. A. Miller and H. Sandler (eds), *Biotelemetry III*, pp. 283-286. New York: Academic Press.

Bojsen J., Møller U., Christensen P. and Lippert J. (1974) Telemetry of radionuclide tracers by implantable thermoluminescent dosimeters on rats. In P. A. Neukomm (ed.), *Biotelemetry II*, pp. 43-45. Basel: Karger.

Bojsen J., Møller U., Christensen P. and Lippert J. (1977) Storage telemetry of radionuclide tracers by implantable thermoluminescent dosimeters. *Am. J. Physiol.* 233, E479-E482.

Bojsen J., Møller U. and Faber M. (1971) Radiotelemetrical equipment for continuous subcutaneous measurements of the circadian temperature rhythm in rats. *Pflügers Arch.* 328, 176-184.

Bojsen J., Nielsen P. E., Rasmussen S. and Parving H.-H. (1978) Storage telemetry of the disappearance rate of DTPA-(Sn)-^{99m}Tc and glomerular filtration rate by external skin surface CdTe-detectors. In H. J. Klewe and H. P. Kimmich (eds.), *Biotelemetry IV*, pp. 251-254. Braunschweig: Verlag.

Bojsen J. and Rossing N. (1978) Biotelemetry of the glomerular filtration rate (GFR) in man by external detection of DTPA-(Sn)-^{99m}Tc with CdTe detectors. *Nucl. Instrum. Meth.* 150, 49-53.

Bojsen J., Rossing N., Soeberg O. and Vadstrup S. (1977) Biotelemetry based on CdTe-detectors. *Rev. Phys. Appl.* 12, 361-363.

Bojsen J. and Vadstrup S. (1974) A portable external two-channel radiotelemetrical GM-detector unit, for measurements of radio-nuclide tracers *in vivo*. *Int. J. Appl. Radiat. Isotop.* 25, 161-166.

Bojsen J. and Vadstrup S. (1976) Storage telemetric system for detection of radionuclide tracers in humans. *J. Appl. Physiol.* 41, 416-418.

Bojsen J., Vadstrup S., Siffert P., Regal R. and Cornet A. (1976) The cadmium-telluride semiconductor detectors as a transducer for biotelemetry of radionuclide tracers. In T. B. Fryer, H. A. Miller and H. Sandler (eds.), *Biotelemetry III*, pp. 247-250. New York: Academic Press.

Bojsen J. and Wallevik K. (1972) A radiotelemetrical measuring device, implantable on animals, for long term measurements of radionuclide tracers. *Int. J. Appl. Radiat. Isotop.* 23, 501-511.

Bojsen J., Wallevik K. and Møller U. (1972) An implantable radiotelemetrical measuring device for simultaneous long term measurements of body temperature and turnover of rabbit serum albumin-^{125}I in unrestrained rabbits. In H. P. Kimmich and U. Møller (eds.), *Biotelemetry*, pp. 337-344. Leiden: Meander N.V.

Bullen M. A., Freundlich H. F., Hale B. T., Marshall D. H. and Tudway R. C. (1963) The activity of malignant tumours and response to therapeutic agents, studied by continuous records of radioactive phosphorus uptake. *Postgrad. Med. J.* 39, 265-277.

Calcutt G. M., Bullen M. A., Marshall D. H. and Godden I. J. (1967) The continuous counting of phosphorus-32 from transplanted rat tumours and the effects of radiosensitisers and radioprotective agents. *Br. J. Cancer.* 21, 438-447.

Fryer T. B., Lund G. F. and Williams B. A. (1978) An inductively powered telemetry system for temperature, EKG, and activity monitoring. *Biotelemetry Patient Monitg.* 5, 53-76.

Gugnin Yu. Ya., Kalantorov K. D., Pluzhnikov V. M., Zanskov V. P. and Semenov V. S. (1972) Radiocapsule for recording the ionizing radiation in the gastrointestinal tract. *Biomed. Engng.* 6, 19-22.

Hassan M. A. (1978) A semiconductor radiation pill. In H.-J. Klewe and H. P. Kimmich (eds.), *Biotelemetry IV*, pp. 61-64. Braunschweig: Verlag.

Hassan M. A., Pearce G. and Edwards J. P. N. (1978) A radio-telemetry pill for the measurement of ionising radiation using a mercuric iodide detector. *Phys. Med. Biol.* 23, 302-308.

Kobayashi T., Sugita T., Sasaki T., Iwase T., Sassa R. and Iio M. (1974) A windowless Si radiation detector and biomedical telemetry system for studying the function of organs of free-ranging animals. *IEEE Trans. Nucl. Sci.* 21, 735-740.

Kølendorf K., Bojsen J. and Jørgensen K. (1978) Storage telemetry of subcutaneous absorption of ^{125}I-NPH-insulin. In H.-J. Klewe and H. P. Kimmich (eds.), *Biotelemetry IV*, pp. 239-242. Braunschweig: Verlag.

Kølendorf K., Bojsen J. and Nielsen S. L. (1979) Adipose tissue blood flow and insulin disappearance from subcutaneous tissue. *J. Clin. Pharmacol. Therap.*, in press.

Mackay R. S. (1970) Radioactivity and bleeding site. *Bio-Medical Telemetry*, 2nd ed. pp. 212-215. New York: John Wiley.

McLeod M. A. (1977) Assessment of renal function using ^{99m}Tc-DTPA-(Sn). *J. R. Naval Med. Service.* 63, 97-100.

McLeod M. A., Sampson W. F. and Houston A. S. (1977) Urinary clearance of ^{113m}In-DTPA and ^{99m}Tc-(Sn) DTPA measured by external arm counting. *Urol. Res.* 5, 71-74.

Miles G. M. (1962) Telemetry techniques for periodicity studies. *Ann. N.Y. Acad. Sci.* 98, 858-865.

Møller U. and Bojsen J. (1976) Circadian variations in ^{32}P uptake of a DMBA-induced mammary tumour and Walker carcinosarcoma in rats. *Br. J. Cancer.* 33, 646-654.

Rossing N., Bojsen J. and Frederiksen P. L. (1978) The glomerular filtration rate determined with ^{99m}Tc-DTPA and a portable cadmium telleride detector. *Scand. J. Clin. Lab. Invest.* 38, 23-28.

Schieber M. (1977) Fabrication of HgI_2 nuclear detectors. *Nucl. Instrum. Meth.* 144, 469-477.

Siffert P. (1978) Current possibilities and limitations of cadmium telluride detectors. *Nucl. Instrum. Meth.* 150, 1-12.

Siffert P., Ponpon J. P. and Cornet A. (1975) Détecteurs de rayonnements nucléaires à semiconducteurs. *L'onde électrique* 55, 281-297.

Stoll B. A. and Burch W. M. (1968) Surface detection of circadian rhythm in ^{32}P content of cancer of the breast. *Cancer* 21, 193-196.

Taylor D. M., Parker R. P., Field E. O. and Greatorex C. A. (1968) An interpretation of the results of measurements of the uptake of ^{32}P in human tumors. *Br. J. Radiol.* 41, 432-439.

Vadstrup S. (1977) A comparison of the effect of TSH, GH and prolactin on the renal iodide excretion in unrestrained rabbits. *Acta Endocr.* (*Kbh.*) 85, 488-496.

Vadstrup S. and Bojsen J. (1976a) The diurnal variation in renal iodide excretion rate in rabbits. *Acta Endocr.* (*Kbh.*) 81, 716-722.

Vadstrup S. and Bojsen J. (1976b) The immediate effect of TSH on renal iodide excretion rate in rabbits. *Acta Endocr.* (*Kbh.*) 81, 723-728.

Vadstrup S. and Bojsen J. (1976c) Biotelemetrical detection of radionuclide tracers in the study of biological rhythms in renal function. In T. B. Fryer, H. A. Miller and H. Sandler (eds.), *Biotelemetry III*, pp. 79-82. New York: Academic Press.

Wolley-Hart A., Twentyman P., Corfield J., Joslin C., Morrison R. and Fowler J. F. (1968) Changes in ^{32}P counting-rate in human and animal tumours. *Br. J. Radiol.* 41, 440-447.

Radar as an Aid to the Study of Insect Flight

J. R. RILEY

Centre for Overseas Pest Research, Radar Entomology Unit, Royal Signals and Radar Establishment, Malvern, Worcestershire, U.K.

Abstract — This paper describes the events which led to the first direct application of radar to the study of insect flight and traces the subsequent development of radar entomology. Results obtained from a variety of field trials are reviewed and the current capabilities and limitations of this new subject are discussed. It is concluded that a good qualitative understanding of the behavior of aerial insect populations is readily obtained by the use of radar but that experiments must be planned with great care if insuperable problems of radar identification are to be avoided.

INTRODUCTION

The first application of radar to biotelemetry was made by ornithologists whose experience of wartime radar operations had alerted them to the potential of radar as a means of studying the flight of birds (Lack and Varley, 1945). The subsequent development of radar ornithology (Eastwood, 1967) vastly increased the range and accuracy of observations of bird migration and led to new and detailed descriptions of their migratory patterns.

The exciting possibility of making similar measurements on targets as small as individual insects was demonstrated by Crawford in 1949 and almost coincidentally, serious proposals were made in 1950 to use radar as a means of observing insect flight (Rainey, 1955). Rainey was unaware of Crawford's work, but deduced from consideration of the water content of locusts that it should be possible for meteorological radars to detect swarms of these insects at least as easily as they detected heavy precipitation. Accordingly he secured the cooperation of the U.K. Meteorological Office in arranging that the G.L.III radars, then in operation as wind finders at a number of meteorological stations in the Middle East, would be used to attempt radar observations of any passing swarms. In the event, the first locust swarm detection was made by a 10 cm naval radar in the Persian Gulf (Rainey, 1955).

The interest of the Anti-Locust Research Centre* (ALRC) of London and the Desert Locust Control Organisation for East Africa was stimulated by these developments,

*Now the Centre for Overseas Pest Research.

and they together investigated the possibility of using airborne radars for locust swarm detection. As a result, the first measurements of the effectiveness of locusts as radar reflectors (i.e. their radar scattering cross sections) were made for ALRC by W. G. Harper at the Royal Radar Establishment in 1962 (communication to the Director ALRC), and it became clear that a commercial, X band, airborne weather radar could be expected to detect locust swarms as far away as 100 km. It was however, anticipated that difficulties would be experienced in differentiating between the radar echoes returned by locusts and those returned by precipitation and by the ground. Harper suggested that the wing motion of locusts in flight might produce characteristically different radar returns, but that firm conclusions could not be reached without an airborne trial. Partly because of the expense of carrying out such a trial, and partly because the desert locust entered a period of recession, plans for airborne operations lapsed.

Meanwhile, reports had begun to appear which strongly suggested that individual insects were responsible for some radar echoes (Bonham and Blake, 1956; Plank, 1958; Tolbert, Straiton and Britt, 1958) and evidence had continued to accumulate that insects often contributed to unidentified radar echoes (Deam and Lagrone, 1965; Browning and Atlas, 1966). New measurements of insect cross section (Hajovsky, Deam and Lagrone, 1966) left no doubt that several species of insects would individually produce detectable radar echoes and it had even been demonstrated that single insects in flight could be followed by automatic tracking radars (Glover *et al.*, 1966). One result of these tracking experiments was the observation that a dragonfly impressed a small but perceptible and regular modulation on its radar returns. It was suggested that this modulation in the 9-7 Hz range might have been generated by wingbeat action which had been slowed down by the low ambient temperature (~ 8°C) from the value of 41-46 Hz expected at room temperature. This interpretation has been justifiably criticized (Schaefer, 1976, p. 164) on the grounds that the frequency was too low and that wing beating was unlikely at 8°C. Although the tracking data obtained by Glover showed that the dragonfly was apparently ascending at ~ 4 ms^{-1} during part of the modulation record, this observation may have been due to the effects of an updraught rather than vigorous wing flapping, so Glover's association of the modulation with wing beating must remain in doubt. Nevertheless his appreciation that "the uniqueness of cross section fluctuation spectra for a given species" would be useful in entomology, was to prove correct (Schaefer, 1976; Riley and Reynolds, 1979).

None of the work reported above had actually produced entomologically useful information about insect flight, but observations by a meteorological radar of a locust swarm in India (Ramana Murty *et al.*, 1964; Mazumdar, Bhaskara Rao and Gupta, 1965) demonstrated that this could be done and had provided new information about locust swarm density and vertical extent.

RADAR ENTOMOLOGY FIELD TRIALS

GROUND RADAR

Interest in the use of radar had in the meantime been maintained at ALRC and it was finally decided to attempt to use a mobile, ground based radar to study the flight behavior of solitary desert locusts (*Schistocerca gregaria*) in Africa. The results of the radar field trial in Niger subsequently sponsored by ALRC in 1968 were not written up in detail, but it appeared from the summaries that the experiment was a success (Roffey, 1969; Schaefer, 1969). The modified, 3.2 cm marine radar displayed parts of the flight trajectories of individual insects out to ranges of more than 2 km by day and night and provided a measure of aerial density, height of flight (up to 1200 m) and ground speed (up to 54 km h^{-1}), as well as showing dramatic increases in aerial density at dusk and at nocturnal wind shift lines. One result of great interest was that the insects detected by the radar at night often showed a tendency to adopt a downwind orientation. Radar returns from insects flying through the radar beam when it was stationary (or being manipulated to follow individuals), were often found to contain spectral components corresponding to the wingbeat frequencies of the desert locusts and butterflies found in the area. It was claimed that locusts could be identified on the basis of these components, and even males could be distinguished from females. Prior to the advent of radar, the maximum reported range of nocturnal observation of insects had been the 100 m achieved by searchlights (Roffey, 1963). The range of the observations made in Niger greatly exceeded this limit, and the results were correspondingly new to entomologists.

Not surprisingly, attempts were soon made to exploit the entomological potential of radar which the Niger field trial had demonstrated. Unfortunately, the first recorded attempt in 1971 (in Saudi Arabia) was frustrated by a lack of insects (Riley, 1974) and the second (made on behalf of the Entomology Division of the Commonwealth Scientific and Industrial Research Organisation, CSIRO) in Australia, by an excess of unwanted species (Schaefer, 1971). The Australian work did however allow observations to be made of the effect of atmospheric convergence processes on insect aerial density (Schaefer, 1976; Roffey, 1972).

The success of the Niger experiments also encouraged speculation that chains of strategically placed, locust detection radar stations would make possible cheaper and more effective locust control operations (Schaefer, 1970). Nevertheless, radar continued to be used in research, rather than as a direct aid to control operations, and new results about the flight behavior of *Aiolopus simulatrix*, other grasshoppers and moths in the Sudan were announced (Schaefer, 1975, 1976). Amongst these results were estimates of average rate of climb, time of flight and migration ceiling as well as descriptions of the insect collecting effect of storm outflows and of the Intertropical Convergence Zone.

A three year program of radar studies started in 1973 in the Middle Delta of the river Niger in Mali resulted in a description of several aspects of the Autumn migratory flight of grasshoppers in the area (Riley and Reynolds, 1979) and in a detailed observation of nocturnal, up-wind orientation by high flying insects presumed to be grasshoppers (Riley, 1975). The Niger delta studies demonstrated that intra-species spread and inter-species overlap of wingbeat rates limited the successful application of wingbeat identification schemes to situations in which the 'wanted' species were known to numerically dominate any other species present of similar size and morphology. Incidental observations made during this study showed that insects often accumulated at altitudes of up to 200 m over the river Niger (Reynolds and Riley, 1979).

Meanwhile in America, Frost and Downing (1972) had used a 1.9 cm radar to detect clouds of mosquitos and other small insects and had monitored their movement at very low altitude. They found that radar returns varied in a manner which corresponded to the number and composition of catches made by vehicle and boat mounted nets which were operated in the radar surveillance area. Richter *et al.* (1973) used a high resolution FM-CW radar to reveal the simultaneous occurrence of undulating atmospheric and insect layers, as well as post-sunset and post-sunrise increases in aerial density.

Further work was also undertaken by the Entomology Division of CSIRO in Australia, which resulted in successful observations in 1974 of *Chortoicetes terminifera* flight (Reid, Wardaugh and Roffey, 1979), and two insect immigration investigations in 1977 and 1979 (J. R. Riley, unpublished report; V. A. Drake, personal communication). Radar derived information about *Chortoicetes* flight behavior was subsequently used by Farrow (1975) in an analysis of the effects of off-shore migration.

AIRBORNE RADAR

Although the effective range of radar greatly exceeds that of all other forms of nocturnal observational techniques used in entomology, it is still very limited when compared to the apparent range of flight of many insects. As part of a strategem to overcome this problem during a migration study of the spruce budworm moth (*Choristoneura fumiferana*) in Canada, Schaefer successfully resurrected the idea of using an airborne radar (Schaefer, in press). His system projected a beam vertically downward and was able to determine at a series of altitudes, the average alignment direction and density of airborne insect concentrations over which the instrumented aircraft flew. The results achieved with this equipment combined with ground radar measurements, have yet to be published, but apparently did much to elucidate the migratory behavior of the budworm moth (Greenbank, Schaefer and Rainey, in press). Observations of dense, migrating moth concentrations are reported to have led to suggestions that radar-directed air to air spraying would lead to an improved and more ecologically sound means of control (Anon, 1974).

CURRENT CAPABILITIES AND LIMITATIONS

It is apparent from the work briefly described in this review that valuable new observations of insect flight behavior have been made by radar. On the other hand, in the author's opinion, the assumptions made in the interpretation of radar data have not always been made clear, and in consequence the degree of confidence which can be placed on some of the results is open to question. Accordingly, I describe below some of the interpretative methods used in radar entomology and comment on the way in which they affect the validity of the measurements attempted.

IDENTIFICATON

Two stages of target identification are required in radar entomology. Firstly it is necessary to discriminate between the echoes returned by insects and those returned by birds, bats and by precipitation, and secondly, it is usually necessary to identify the species of insect detected. Current target identification schemes include examination of the spatial extent of the received echoes, their amplitude, temporal variation and their response to polarization changes, as well as estimation of the target air speed.

Insects and Precipitation

In the case of low aerial densities of insects ($< 10^{-4} m^{-3}$) observed at close range (< 3 km) the spatially discrete returns detectable from individual large and medium sized insects are very clearly different from the semicontinuous echoes generated by precipitation and no confusion is therefore likely. Returns from insect concentrations may on the other hand be very similar to rain echoes and reliable discrimination between the two is difficult, especially at long range when supplementary evidence (i.e. rain at the ground) may not be available. It has been suggested that the use of circular polarization, which substantially reduces radar sensitivity to rain echoes might alleviate this difficulty (Schaefer, 1976) but no supporting measurements have been reported to date.

Insects and Birds

Although the radar cross-sections of birds (Edwards and Houghton, 1959) and presumably of bats, are in general much larger than those of insects (Riley, 1973; Schaefer, 1976) the difference between the average amplitudes of echoes received from the larger species of insect and those from the smallest birds may be very small (or even nonexistent) and are certainly not large enough to produce obvious differences on the display of a scanning radar system. On the other hand, reported air speeds of birds (Meinhertzagen, 1955) and of bats (Pye, 1978) are higher than those of most insects (Johnson, 1969), so that differences in flight trajectories might be reasonably expected to provide a means of discriminating between small birds and insects (Riley, 1974; Schaefer, 1976). Recent radar measurements by Larkin *et al.* (1979) suggest however that migrating birds sometimes displace through the air at speeds well below 5-6 ms^{-1} and so it cannot always be safely assumed that all slow fliers are necessarily insects. It is appropriate here to emphasize the fact that accurate estimates of target flying speed require simultaneous measurements of wind velocity at the same altitude as the target, and preferably close to it.

The modulation of radar echo amplitude caused by wing beating provides a valuable means of discrimination between birds and insects. Schaefer (1976) points out that although there is a considerable overlap of insect and bird wingbeat frequency in the range 8-30 Hz, only those birds which maintain continuous, rather than intermittent or irregular wing beats, are likely to be confused with insects. The upper wingbeat frequency for these birds is ~14 Hz (Schaefer, 1976) so the range of overlap is effectively reduced to 8-14 Hz.

In conclusion aerial targets found to have flying speeds below 6-7 ms^{-1} and generating continuous wingbeat modulation above 14 Hz will almost certainly be insects.

Resolving Insect Species

Some degree of 'automatic' discrimination will always occur between species differing substantially in size, larger insects being more readily detected than smaller ones (Riley, 1979). In fact radar thresholds of detection may be adjusted to ensure that small insects are not (individually) detected at all at a selected range of interest. Apart from crude categorizations of this sort, little use may be made of the average amplitude of insect radar echoes because echo size depends on the insects' aspect (Riley, 1973) and their position in the radar beam, both of which are usually unknown. Even if position and aspect were resolved by the use of high resolution tracking radar, much overlap in radar echoing area between species could be expected (Schaefer, 1976).

Insects with longer wings tend in general to have lower wingbeat frequencies than those with shorter wings (Greenwalt, 1962). In the case of some Acrididae (grasshoppers) this tendency has been found to take the form of a well defined inverse power relation between wing length and frequency (Schaefer, 1976). Spectral analysis of radar returns from an airborne population of Acridids thus allows one to make estimates of the distribution of their winglengths and hence, in the special case of a population containing only a few differently sized Acridid species, to deduce the species present and even to distinguish the sexes (Schaefer, 1976; Riley and Reynolds, 1979). Application of this procedure will be frustrated in any but the simplest of entomological environments and great care should be exercised in planning radar observations to ensure that measurements are made in an area in which only a few species are present, and preferably in which the 'wanted' species is believed to be numerically dominant (Riley and Reynolds, 1979). In the case of measurements made on overflying populations which have origins several hundred km from the experimental site, these criteria may be very hard to realize.

Less information is available about possible relations between wing length and frequency in other insect Orders, and it seems in any case very probable that the abundance of species of very similar size, for example in the Noctuid moths, will make identification from wing beat frequencies tentative at best. It is thus even more important to ensure 'wanted species dominance' when planning radar experiments on these insects than is the case with Acridids.

A technique intended to improve identification procedures by simultaneously determining wingbeat frequency and a 'body shape factor' has been described by Riley (1979). Preliminary results from field trials of this technique suggest that clear distinction between insects of the same wingbeat frequency but different shape (e.g. moths and grasshoppers) is readily achieved (Riley, unpublished data). These results also showed that a small proportion of the received signatures did not contain any perceptible wingbeat modulation, although they were very similar in amplitude and 'shape factors' to the majority which were modulated. It thus seems likely that at least some insects glide for parts of their nocturnal flight.

Results from both conventional wingbeat frequency and 'body shape' techniques are subject to biases which favor both large and fast moving targets at the expense of small and slowly displacing ones. Corrections may be made for the displacement bias but only indirectly for size bias, so that assessments of species composition of airborne populations are usually only approximate.

It must be emphasized that all attempts at radar identification rely heavily on the accumulation of supporting information, particularly that obtained from conventional trapping and survey techniques. Trapping from aircraft offers a direct means of identification, but the low sampling rate normally achieved (~2 m^3s^{-1}) limits its use to circumstances when very high aerial densities are expected. It is relevant to point out that to permit aerial trapping at a satisfactory rate (say at least 1 insect caught every 10 min), densities greater than 1 insect every 1000 m^3 are required — and this density is at least an order of magnitude greater than the maximum from which satisfactory radar signatures can be obtained with equipment currently in use.

AERIAL DENSITY MEASUREMENTS

Individually Resolved Targets

Provided that there are no variations in radar performance, changes in 'dot' density registered on a radar screen give a good qualitative measure of changes in insect aerial density. Quantitative measurements are more difficult, because the aerial volume effectively scanned by a radar is a function of the radar cross section presented by the targets it is intended to sample (Riley, 1979). Accurate measurements of aerial density thus require an accurate knowledge of the radar cross section distribution exhibited by the aerial population in the area of interest. In many cases this population will contain a mixture of species, with different radar scattering properties and different heading distributions. The best that can be done in these circumstances is to estimate a representative cross section, based in part on the species presumed to be present and on their heading distributions, and in part on the amplitude distribution of the echoes recorded by the radar. Errors in assessment of scanned volume then depend on the accuracy of this estimate. I have found in my work that errors of up to ± 50 percent in estimates of scanned volume can usually be expected. In fairness it may be said that radar measurements often show changes of aerial density of 2 or 3 orders of magnitude so that errors of up to 50 percent are not insupportable.

Dense Concentrations

If insect concentrations are so dense that many (> 10) targets simultaneously occupy the radar pulse volume, measurements of volume reflectivity may be used to estimate aerial density. As in the case of similar measurements of precipitation (Battan, 1973), interpretation of reflectivity requires some estimate of the cross sections presented by the individual targets. In radar entomology this means that information about the identity and orientation of the insects is required, and the accuracy of the measurements achieved will be limited by the degree to which this information is available. Unfortunately signatures from representative individual insects are not usually possible in the presence of high aerial densities, so identities must be established by alternative means (e.g. trapping).

For aerial densities intermediate between those permitting the resolution of individual targets and the case considered above, display measurements are difficult to interpret, and 'time above threshold' methods must be used (Marshall and Hitschfeld, 1953; V. A. Drake, in preparation). The requirement for information about individual presented cross sections however remains.

MEASUREMENTS OF HEADING DISTRIBUTION

The movement of the dots displayed on a radar plan position indicator screen (PPI) provides a good, semiquantitative indication of the velocity of the targets detected by a scanning radar. If the trajectories of wind finding balloons are also registered on the PPI, vector subtraction of balloon from target velocity may be used to give target air speed and heading. When insect collective orientation occurs, an indication of average body axis alignment is also readily discernable from the resulting 'polarization' of the dot distribution on the screen (Riley, 1975). In many experiments these observations will be of adequate precision to describe the aspects of flight behavior under investigation. However, there are some instances, for example, in the study of the collective orientation effect, where accurate measurements of the heading distribution of an overflying population may be needed and in these cases PPI derived data will usually be unsatisfactory. This is because the trajectories displayed on radar PPI system are subject to serious bias effects which cannot be eliminated without a detailed knowledge of the target's angular variation of radar cross section (Riley, 1979). Equally, the degree of 'polarization' on the PPI is the result of the combination of spread both in target radar scattering properties and in heading. In the absence of comprehensive scattering data, accurate heading distributions cannot therefore be determined from the 'polarization' effect.

It is appropriate to point out here that the published measurements of the angular variation of cross sections (Riley, 1973; Schaefer, 1976; Greneker and Corbin, 1978) were taken at zero elevation (i.e. two dimensional), and are applicable only as approximations to the finite elevation results always obtained in field work. This is the case even for cylindrically symmetric targets, because their scattering properties are dependent on the angle between body axis and the propagation direction, as well as that between body axis and electric vector. Complete specification of target scattering properties thus requires measurements in three dimensions.

These difficulties may be overcome to a large extent by the supplementary use of a vertically looking radar, in which the plane of polarization is rotated. The advantage of this system is that the mechanism by which it determines target alignment is insensitive to angular variations in scattering properties, and is subject only to biases which are simply related to displacement speed and maximum radar cross section. The method is capable of precise measurements of the alignment of individuals and thus of resolving subtle aspects of common orientation phenomenon — for example the simultaneous adoption of two mean headings by an aerial population (Riley and Reynolds, 1979).

DURATION AND RANGE OF FLIGHT

Although the radars currently used in entomological work are capable of detecting insects of moderate size out to ranges of 2-3 km, individual insects very rarely produce tracks for more than a few hundred meters of their flight paths. This is because targets in horizontal flight remain in the narrow elevation range covered by the scanning 'pencil beams' used in these radars for only short sections of their trajectories. In consequence, information about the extent of insect flights which may last many hours and cover hundreds of kilometers, must usually be derived from indirect measurements of the type described below.

The spectacular increases of aerial density at dusk noted by most radar entomology workers provide good evidence that many insects take off for their night's flight within half an hour of sunset. The decline in aerial population usually observed after the take-off peak has been interpreted in terms of subsequent mean flight duration (Schaefer, 1976). However, application of this interpretation to measurements made by a single ground radar, presupposes that the insects overflying the radar during the course of an evening have come from an aerial population which was uniformly distributed over at least the length of the longest flight paths. Otherwise changes in overflying populations reflect the insects' original geographical distribution, rather than their flight times. In many situations this requirement for an homogeneous population distribution will not be met.

Non-uniform geographical distributions may sometimes be used to advantage if they lead to the production of discrete, overflying populations. The differences between times of population overflight and of their (presumed) take-off give a measure of minimum flight times; back-trajectories may be calculated from measured displacement velocities and used to identify probable take-off areas (Riley and Reynolds, 1979). Some recent observations showing apparent sequential grasshopper over-flight of two radar stations separated by 120 km have supported the validity of this approach (Riley, unpublished data). It has also become clear however that in the case of moths, substantial take-off may occur several hours after dusk, so the 'overflight method' must be used with caution.

CONCLUSIONS

The limitations described in the preceding paragraphs demonstrate that radar is not a universally applicable means of studying the flight behavior of all types of insect. In particular, the restrictions imposed by difficulties of identification are substantial, and largely dictate the situations in which the technique can be successfully used to study individual species. Radar nevertheless remains by far the most powerful method of observing many types of insects in natural flight and when used with care in an appropriate environment can be expected to yield accurate descriptions of this otherwise inaccessible aspect of their behavior.

Acknowledgements — I gratefully acknowledge the helpful comments and suggestions made by Dr. D. R. Reynolds and Dr. R. C. Rainey during the preparation of this manuscript.

REFERENCES

Anon (1974) Taking advantage of locusts and moths. *New Scientist* 653.

Battan L. J. (1973) *Radar Observations of the Atmosphere*. Chicago: University of Chicago Press.

Bonham L. L. and Blake L. V. (1956) Radar echoes from birds and insects. *Scient. Monthly* 82, 204-209.

Browning K. A. and Atlas D. (1966) Velocity characteristics of some clear air dot angels. *J. atmos. Sci.* 23, 592-604.

Crawford A. B. (1949) Radar reflections in the lower atmosphere. *Proc. IRE* 37, 404-405.

Deam A. P. and Lagrone A. H. (1965) Quantitative observations of dot angels at two frequencies. *Radio Sci.* 1, 537-543.

Eastwood E. (1967) *Radar Ornithology*. London: Methuen.

Edwards J. and Houghton E. W. (1959) Radar echoing area polar diagrams of birds. *Nature, Lond.* 184, 1059.

Farrow R. A. (1975) Off-shore migration and the collapse of outbreaks of the Australian plague locusts (*Chortoicetes terminifera* Walk.) in south eastern Australia. *Aust. J. Zool.* 23, 569-595.

Frost E. L. and Downing J. D. (1972) Recent radar observations of diurnal insect behavior. *Proc. 59th Annual Meeting of the New Jersey Mosquito Extermination Ass.* 114-132.

Glover K. M., Hardy K. R., Konrad T. G., Sullivan W. N. and Michaels A. S. (1966) Radar observations of insects in free flight. *Science* 154, 967-972.

Greenbank D., Schaefer G. W. and Rainey R. C. Spruce budworm moth flight and dispersal new understanding from canopy observations, radar and aircraft. *Mem. Can. ent. Soc.*, in press.

Greenwalt, C. H. (1962) Dimensional relationships for flying animals. *Smithsonian Misc. Collect.* 144.

Greneker E. F. and Corbin M. A. (1978) Radar reflectivity of airborne insects: a literature survey. Prepared for Western Cotton Research Lab. USDA, Phoenix, Arizona.

Hajovski R. G., Deam A. P. and Lagrone A. H. (1966) Radar reflections from insects in the lower atmosphere. *IEEE Trans.* AP-14, 224-227.

Johnson C. G. (1969) *Migration and Dispersal of Insects by Flight.* London: Methuen.

Lack D. and Varley G. C. (1945) Detection of birds by radar. *Nature, Lond.* 156, 446.

Larkin R. P., Griffin D. R., Torre-Bueno J. R. and Teal J. (1979) Radar observations of bird migration over the western north Atlantic. *Behav. Ecol. Sociobiol.* 4, 225-264.

Marshall J. S. and Hitschfeld W. (1953) Interpretation of fluctuating echoes from randomly distributed scatterers. Part 1. *Can. J. Phys.* 31, 962-994.

Mazumdar S., Bhaskara Rao N. S. and Gupta G. R. (1965) Radar and synoptic study of locust swarms over Delhi. *WMO Tech. Note* 69, 162-188.

Meinertzhagen R. (1955) The speed and altitude of bird flight. *Ibis* 97, 81-117.

Plank V. G. (1958) Atmospheric angels mimic radar echoes. *Electronics* 31, 140-144.

Pye J. D. (1978) Some preliminary observations of flexible echolocation systems. *Proc. 4th Int. Bat. Res. Conf., 1975 Kenya Nat. Acad. for Arts and Sciences.*

Rainey R. C. (1955) Observation of desert locust swarms by radar. *Nature, Lond.* 175, 177.

Ramana Murty BH. V., Roy A. K., Biswas K. R. and Khemani L. T. (1964) Observations of flying locusts by radar. *J. Sci. ind. Res. New Delhi* 23, 289-296.

Reid D. G., Wardaugh K. G. and Roffey J. (1979) Report on radar studies of insect flight, at Benalla, Victoria in February 1974. CSIRO Aust. Div. Entomol. Tech. Paper No. 16.

Reynolds D. R. and Riley J. R. (1979) Radar observations of concentrations of insects above a river in Mali, West Africa. *Ecol. Entomol.* 4, 161-174.

Ricther J. H., Jensen D. R., Noonkester V. R., Kreasky J. B., Stimmann M. W. and Wolf W. W. (1973) Remote radar sensing: atmospheric structure and insects. *Science* 180, 1176-1178.

Riley J. R. (1973) Angular and temporal variations in the cross section of insects. *Proc. IEE* 120, 1229-1232.

Riley J. R. (1974) Radar observations of individual desert locusts (*Schistocerca gregaria*) (Forsk). *Bull. ent. Res.* 64, 19-32.

Riley J. R. (1975) Collective orientation in night flying insects. *Nature, Lond.* 253, 113-114.

Riley J. R. (1979) Quantitative analysis of radar returns from insects. In C. R. Vaughn, W. Wolf and W. Klassen (eds.), *Proc. Workshop on Radar, Insect Population and Pest Management* held 2-4 May 1978, Wallops Flight Center, Wallops Island, Virginia.

Riley J. R. and Reynolds D. R. (1979) Radar-based studies of the migratory flight of grasshoppers in the middle Niger area of Mali, West Africa. *Proc. R. Soc.* B204, 67-82.

Roffey J. (1963) Observations on night flight in desert locust. *Anti Locust Bull.* 39, 1-32 ALRC London.

Roffey J. (1969) Radar studies of the desert locust. Niger, Sept-Oct 1968. ALRC Occasional Report No. 17.

Roffey J. (1972) Radar studies of insects. *PANS* 18, 303-309.

Schaefer G. W. (1969) Radar studies of locust, moth and butterfly migration in the Sahara. *Proc. R. ent. Soc. (Lond.)* 34, 39-40.

Schaefer G. W. (1970) Radar detection of individual locusts and swarms. Proc. Int. Study Conf. on current and future problems of Acridology.

Schaefer G. W. (1971) Preliminary report on radar field expedition March 4-16, 1971. CSIRO Division of Entomology Canberra.

Schaefer G. W. (1975) Radar studies of the flight activity and the effects of wind-fields on the dispersal of Sudan Gezira insects. From Ciba-Geigy Seminar on the Strategy for cotton pest control in the sudan gezira. Wad Medani, Sudan.

Schaefer G. W. (1976) Radar observations of insect flight. In R. C. Rainey (ed.), *Insect Flight Symp. R. ent. Soc.* (*Lond.*), 157-197.

Schaefer G. W. An airborne radar technique for the investigation and control of migratory pest insects. R. C. Rainey and D. L. Gunn (eds.), Proc. R. Soc. Discussion Meeting 29-30 July 1977, London.

Tolbert C. W., Straiton A. W. and Britt C. O. (1958) Phantom targets at MM wavelengths *IRE Trans*. AP-6, 380-384.

II. Design, Construction and Performance

A Practical Guide to Radio Tracking

D. W. MACDONALD and C. J. AMLANER, Jr.

Animal Behaviour Research Group, Department of Zoology, University of Oxford, South Parks Road, Oxford OX1 3PS, U.K.

Abstract — The art of practical radio tracking has more to it than the would-be tracker might anticipate. We try to provide some practical guidance in the choice and construction of equipment, in its use and its shortcomings.

INTRODUCTION

Radio tracking should not be undertaken lightly; it is expensive, time-consuming and often, frustrating; but it can lead to answers to biological questions which are otherwise unassailable. We describe here practical aspects of selecting, building and using radio tracking equipment, in the hope of helping the beginner to anticipate and to avoid some problems. For example, few manufacturers' catalogs tell you that cold weather is likely to change the transmitting frequency of the transmitter, and/or the receiver's performance and hence the channel on which you hear a signal, yet it is worth knowing about these possibilities when your subject apparently disappears in the small hours of a winter's night.

TRANSMITTERS

The history of radio tracking effectively began in the early Sixties when Cochran and Lord (1963) published the design of a miniature crystal controlled transmitter. The circuit was simple, robust, adaptable to both loop and whip antennas and could be packaged to fit many animals. A similar circuit (Fig. 1) was developed to capitalize on the more useful features of Cochran and Lord's circuit. By placing a variable capacitor (5-25 pf) in the 'tank circuit' (an oscillator consisting of a capacitor and inductor), it can be tuned optimally for different antenna loads (10-50 cm) thus ensuring the highest power output. Furthermore, if a resistive device is incorporated in the circuit at R_T (see Fig. 1) with an approximate value of 150-400 K, physiological and environmental parameters can be measured by pulse interval modulation (PIM). For instance, a 200 K thermistor (at 25°C) may be substituted to measure linear temperatures between 31 and 45°C, and nonlinear extremes from 15 to 45°C, accurate to 0.1°C. This temperature transmitter has been used to measure 24 h internal body temperatures of foxes (*Vulpes vulpes*), drey and ambient temperatures from free ranging squirrels (*Sciurus carolinensis*), and egg and body temperatures of incubating seagulls (*Larus argentatus* and *L. fuscus*). Other environmental parameters such as light level (R_T = photoresistor), presence of moisture, and pressure (R_T = strain gauge) can also be readily measured. As this circuit is not temperature compensated, it is advisable not to locate R_T remotely from the main transmitter, to avoid inaccuracies.

If the circuit of Fig. 1 is built with subminiature components, it can weigh as little as 1.0 g and operate for 7-10 days using a 0.3 g E212E Eveready mercury hearing aid battery. These transmitters can have R_T values of 150-400 K which produce pulse rates of 50 to 150 s^{-1}. The approximate fixed value of the capacitor in the tank circuit is 15 pf for a 10 cm antenna. This transmitter has been field tested on birds as small as great tits (*Parus major*) weighing approximately 20 g and effective tracking ranges are 100 m in thick forest and 500 m in line of sight.

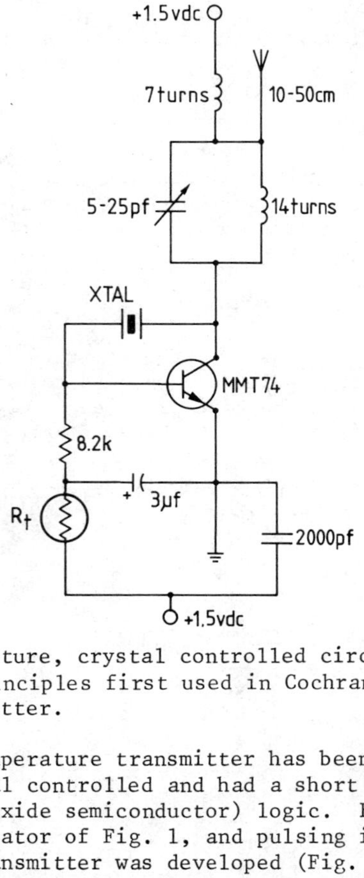

Fig. 1. A miniature, crystal controlled circuit, incorporating the principles first used in Cochran and Lord's (1963) transmitter.

A more stable design of temperature transmitter has been reported by Skutt *et al.* (1973). This was non-crystal controlled and had a short range RF stage triggered by CMOS (complementary metal oxide semiconductor) logic. By adapting the long range single stage crystal oscillator of Fig. 1, and pulsing it by the same logic of Skutt *et al.* (1973), a stable transmitter was developed (Fig. 2), which transmitted at least one km with a 0.25 λ whip antenna. With this design, the thermistor can be positioned remotely at distances of 10-20 cm from the transmitter. Again, the thermistor can be replaced by other suitable resistive elements within the 1.5 MΩ range to measure the other parameters previously mentioned. The transmitter's output pulses are very narrow (0.1 ms) and sound like sharp clicks when heard on a commercial non BFO (beat frequency oscillator) type receiver. On specialist built tracking receivers (e.g. AVM, Telemetry Systems Inc., Wildlife Materials, etc.), pulses are more easily counted because of the receivers' BFO capabilities. Pulse width is determined by the 510 K resistor and 1000 pf capacitor. Because narrow pulses are difficult to discriminate against background noise, the capacitor value may be increased to lengthen pulses, with a consequent increase in current demands on the battery.

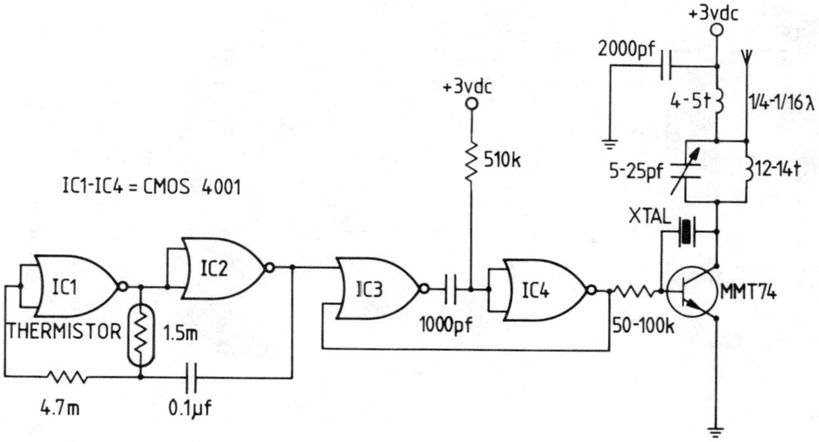

Fig. 2. A transmitter circuit which can be used to telemeter body temperature and other physiological variables.

For temperature telemetry the transmitters in Figs. 1 and 2 can be calibrated by placing the unit in a temperature controlled bath. At each set temperature, allowing time for equilibration, 'bleep' rate is counted and a calibration curve plotted. Each unit must be individually calibrated. The calibration can also be made on the basis of electronically measuring inter pulse interval, giving accurate measures of temperature after sampling only 3 or 4 pulse intervals. An inter pulse interval timer is commercially available from AVM (Champaign, Illinois, U.S.A.).

Powerful multistage transmitters are required for long range tracking (3-10 km). Sophisticated CMOS controlled transmitters are described by Anderka (1980, this volume) and Lotimer (1980, this volume). Stable pulse intervals and pulse widths are achieved through the use of low power logic circuits which maximize the lifetime and give the possibility for pulse code modulation (PCM). By pulsing the transmitter 2 or 3 times every few seconds individual animals can be recognized, much as the alphabet and numbers are recognized, through Morse Code transmission. PIM also allows activity, temperature, and mortality sensing as described above for the CMOS temperature transmitter (Fig. 2).

A two stage activity sensitive transmitter, which may be built easily, is shown in Fig. 3. The values given result in a pulse rate of 60 min^{-1} with a 30 ms pulse length (3 percent duty cycle) when the mercury switch is shorted. Opening the contacts of the mercury switch results in a 120 min^{-1} rate and 15 ms pulse length (also a 3 percent duty cycle). Pulse repetition rates are also adjustable by changing the value of the 1 μF capacitor and 4.7 K resistor. Changes in pulse rate caused by opening and closing the mercury switch allows movement to be detected. The sensitivity of the device to particular movements usually depends on where the transmitter is fitted to the animal and hence, how it moves relative to the plane of animal movement. Unambiguous distinction can be made between inactive resting, sporadic and continuous activity and can be recorded automatically on a magnetic tape recorder by the clock switching circuit described below.

This transmitter will emit pulses of radio waves on carrier frequencies between 30 to 250 MHz. To determine the proper tuning of the oscillator stage, the inductor (0.3 to 1 mH) and trimmer capacitor (5 to 25 pf) values must be calculated to oscillate at approximately the desired fundamental crystal frequency:

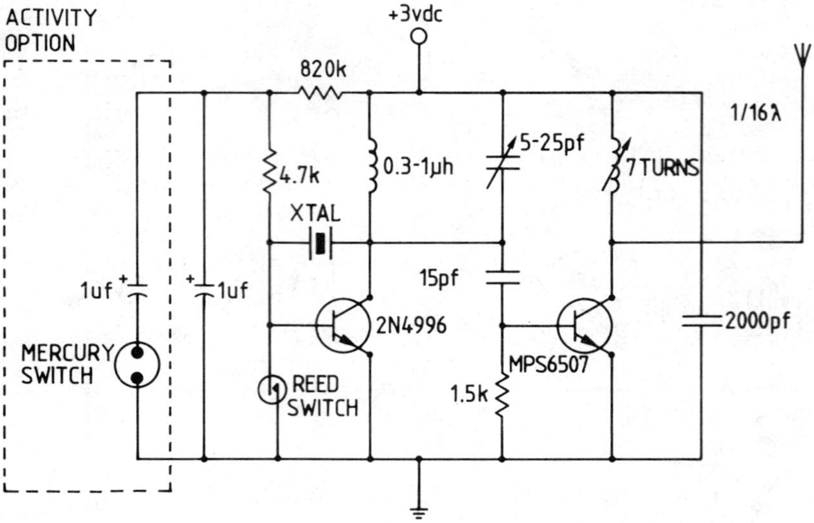

Fig. 3. A two stage transmitter for use in longer range radio tracking studies and including a mercury switch whose movement enables the animal's activity to be monitored through changes in 'bleep' rate.

$$F_{(MHz)} = \frac{10^3}{2\pi\sqrt{LC}} \tag{1}$$

where L = mH and C = pf. Rearranging Equation 1 and solving for inductance yields:

$$L = \frac{10^6}{C(2\pi f)^2} \ . \tag{2}$$

Only approximate values (± 150 μH) of L are required as final tuning is effected by the trimmer capacitor after encapsulation. Taylor and Lloyd (1978) provide details of another 'do it yourself' radio tracking circuit, while Leuze (1977, 1980, this volume) describes a simple non-crystal controlled circuit (see also Morris, 1980, this volume).

BATTERIES

A compromise between battery weight, increased transmitter life and range, limits the use of many transmitters. Relatively expensive lithium cells have roughly twice the milliamphour (mAh) capacity per unit weight of mercury cells (Ko, 1980, this volume). Battery failure either through shelf deterioration or penetration by moisture have been the commonest causes of transmitter failure in our experience (followed by poor quality crystals). Battery manufacturers will guarantee the age of batteries, but at additional expense. The extent of shelf deterioration of a mercury battery can be determined by X-ray photography (with a medical X-ray machine; Harding, Chute and Doell, 1976). The potassium hydroxide electrolyte region gradually diminishes as the electrolyte is expended and ultimately the zinc anode (seen as a central dark spot surrounded by a white ring on the X-ray) collapses.

We have tried this same technique on lithium cells but due to different construction practices, the results are not as clear (i.e. amount of electrolyte is not readily detected by X-rays in lithium batteries). Recently, advances have been made using rechargeable solar cells (e.g. Aucouturier *et al.*, 1977; Church, 1980, this volume), where benefits were obviously only to studies of diurnal animals unless rechargeable batteries were incorporated into the solar transmitter.

It is often convenient to have a completely built transmitter, encapsulated and ready for attachment on an animal. In the past, biologists have soldered a final connection to the batteries in the field. This required a portable soldering iron and skill, especially as heat reduces battery lifetime and it is difficult to carry out delicate soldering in the field. We have gold plated tabs spotwelded to the battery surface to facilitate soldering and to minimize the risk of heat damage. If a normally open (NO) Reed switch is connected between the base of transistor 2N4996 and ground (Fig. 3), then when switch contacts are closed by a small bar magnet, transmission stops. A minimal amount (approximately 4 μA) of current then flows through the 820 K and 4.7 K resistors. This technique is used on commercially available transmitters from Telemetry Systems Inc. (Milwaukee, Wisconsin, U.S.A.).

TRANSMITTER LIFETIME CALCULATIONS

It is usually important to the field biologist to know with some degree of certainty, how long a particular transmitter and battery combination will be expected to work. The calculated lifetime is dependent on accurate measurements of pulse length, peak current during pulsing and pulse rate. We shall use the two stage transmitter of Fig. 3 above as an example to illustrate each one of these measurements. An oscilloscope is required to measure pulse length. A high impedance probe is placed at either the collector of transmitter 2N4996, or at the positive battery terminal. Short leads are required to reduce detuning effects on the transmitter. Oscilloscope triggering from the battery is a consequence of sudden current demands during a pulse. Peak pulse current can be measured by shorting the base of transistor MPS6507 to ground. This turns on the RF stage to maximum output and by connecting a milliampmeter in series with the positive battery terminal and transmitter, a current of approximately 7 mA will be measured.

The proportion of on time (T_{ON} in seconds) between the onset of one pulse to the next is given by:

$$\frac{T_{ON}}{T_{ON} + T_{OFF}} \quad . \tag{3}$$

To determine the milliamp 'rating' of a transmitter over 24 h based on the current during a single pulse, the following equation is employed:

$$\Delta I_{ON} = \left(\frac{T_{ON}}{T_{ON} + T_{OFF}}\right) \times I_{ON} \times 24 \text{ h} \quad . \tag{4}$$

Pulse rate is simply measured by counting the number of pulses in a given time unit. Pulse rate (R) is usually expressed as the number of pulses min^{-1} and can be expressed as a relationship between pulse durations. T_{ON} and T_{OFF}:

$$R = \frac{60}{T_{ON} + T_{OFF}} \quad . \tag{5}$$

Because T_{ON} is relatively easy to measure with an oscilloscope, Equation 5 can be solved for T_{OFF}:

$$T_{OFF} = \frac{60}{R} - T_{ON}$$

substituting this into Equation 4 and simplifying yields:

$$\Delta I_{ON} = \left[\frac{T_{ON}}{T_{ON} + (\frac{60}{R} - T_{ON})} \right] \times I_{ON} \times 24\ \text{h}$$

$$= \frac{T_{ON}\ R}{60} \times I_{ON} \times 24\ \text{h}$$

$$\Delta I_{ON} = 0.40\ T_{ON} \times R \times I_{ON}\ . \qquad (6)$$

Equation 6 is an expression for average current consumption during a 24 h period. It can be easily proven that a similar equation describes average current of the transmitter between pulses (T_{OFF}):

$$\Delta I_{OFF} = 0.40\ T_{OFF} \times R \times I_{OFF}\ . \qquad (7)$$

Therefore the total average current during 24 h is the combination of Equations 6 and 7:

$$\Delta I = \Delta I_{ON} + \Delta I_{OFF}$$

$$\Delta I = 0.40\ R\ (T_{ON} \times I_{ON} + T_{OFF} \times I_{OFF})\ . \qquad (8)$$

To calculate the critical transmitter lifetime, the battery capacity must also be determined. Ko (1980, this volume) has reviewed the problems associated with this variable and we concluded that the manufacturer's specifications are normally the most accurate assessment. Longterm storage and heat lowers battery capacity below that specified by the manufacturers. Battery capacity is expressed as milliamp hours (mAh). To calculate transmitter lifetime, this variable (BmAh) is incorporated into Equation 8:

$$T_{days} = \frac{\text{BmAh}}{0.40\ R\ T_{ON} \times I_{ON} + T_{OFF} \times I_{OFF})} \qquad (9)$$

It is usually the case that $I_{OFF} \ll I_{ON}$. By this we mean that I_{OFF} is at most 0.1 percent of I_{ON} and then, $T_{OFF} \times I_{OFF} \approx 0$ therefore, Equation 9 simplifies to:

$$T_{days} = \frac{\text{BmAh}}{0.40\ R\ (T_{ON} + I_{ON})}\ . \qquad (10)$$

This is the expression for transmitter lifetime in days (*T* days). Representative values of the two stage transmitter (Fig. 3) discussed above result in a theoretically derived lifetime of approximately two years:

$$R = 60\ \text{s}^{-1}$$

$$T_{ON} = 0.32\ \text{s}$$

$$I_{ON} = 7\ \text{mA}$$

$$\text{BmAh} = 4000$$

$$T_{days} = \frac{4000}{0.40 \times 60 \times 0.32 \times 7} = 744\ \text{days.}$$

This value is 3 percent greater than Equation 9 would give to T_{OFF} of 0.968 and

I_{OFF} of 7 μA. If I_{OFF} > 0.1 percent I_{ON}, then Equation 9 must be used to get a realistic value of T days.

There are many reasons why a particular transmitter will not fulfil this expectancy. For instance, temperature fluctuations, moisture invading the encapsulation causing longer pulses and higher current demands, and use of batteries of different rating to that specified by the manufacturer. A conservative correction would reduce the theoretical value by 25 to 50 percent depending on research constraints (e.g. a specified number of weeks for which the transmitters must function).

ENCAPSULATION AND ATTACHMENT

Transmitter components should be encapsulated in light weight, durable and waterproof materials such as dental acrylic, epoxy resin or liquid plastic (Donaldson, 1980, this volume). Components should be coated in beeswax which increases waterproofing and facilitates the removal and replacement of batteries. The potting material usually sets exothermally, hence, precautions must be taken to prevent heat from damaging components. As a general rule, the total package (transmitter, battery, antenna, harness or collar, and encapsulation), should weigh less than 3-5 percent of the animal's body weight (see Welfare).

Mammal studies most frequently use collar designs (e.g. Erlinge, 1980, this volume; Sargeant, 1980, this volume). If the species' anatomy precludes the use of a collar as for badgers, *Meles meles*, a backpack harness may be used (Kruuk, 1977; see also Cheeseman and Mallinson, 1980, this volume). Whip antennas (Amlaner, 1980, this volume) may protrude above the collar, or be built into it. Collars made of inflexible acrylic must have a pliable hinge which is secured with acrylic after fitting. The more pliable nylon collars may be rivetted in place once fitted. Working at night in remote areas and in severe weather conditions puts a premium on simplicity, as does the need to minimize disturbance to the animal. These requirements led to the development for fox work of a collar loop antenna design where the transmitter is switched on simply by fastening the collar. Copper braiding serves as both antenna and collar. The braided wire is very pliable in every dimension and is durable (normally lasting more than 6 months). We cover it with several layers of brightly colored insulating tape (color coded for recognition) to prevent strands of wire being broken during grooming, and causing abrasion.

Kruuk (1978) used a loop antenna attached around the perimeter of the circuit board of a badger backpack and then completely encapsulated it which obviated the risk of water leaking through the antenna connection. Radio tags used for tracking rhino have been embedded in their horns. Hitchins (1971) studied black rhinoceros (*Diceros bicornis*) in Zululand where its preference for dense vegetation precluded detailed study without radio tracking. A 6-7 cm hole was drilled into the horn and the transmitter fitted within. A loop antenna was put in a groove cut horizontally around the horn. Before finally sealing the cavity, the transmitter trimmer capacitor had to be adjusted to the dielectric constant of the horn.

Radio harnesses for birds include (a) chest packs attached by a loop antenna that circles the bird's body (Nicholls and Warner, 1968; Forbes and Warner, 1974), (b) backpacks attached by harness loops under the wings with a whip antenna trailing down the bird's back (Amlaner, Sibly and McCleery, 1978; Hardy, 1977), and (c) tail mounted either on imped or natural feathers (Dunston, 1977; Kenward, 1978).

The back mounted harness used on herring gulls (*Larus argentatus*) (Amlaner, Sibly and McCleery, 1978) was made from non-abrasive nylon coated elastic, measuring 3.2 mm wide (obtainable from sewing shops). The harness was glued to the transmitter and then wrapped around the bird in a figure of eight pattern, crossing the breast area, with the radio positioned at the base of the neck. The average harness weighed 0.5 g.

Feather imping involves securing transmitters onto a molted central rectrice using epoxy resin and thread. The whip antenna is also attached with thread and epoxy along the length of the feather. An equivalent feather is then removed from the bird and the transmitter feather shaft-imped into place and held with more epoxy resin and sutures (Kenward, 1978). This technique requires familiarity with feather imping and is probably not as generally useful as mounting transmitters on the proximal end of natural tail feathers (Dunstan, 1977; Kenward, 1978). Furthermore, rough handling at the base of the feather shaft may cause total transmitter loss in a few days after imping. Swanson and Keuchle (1976) mounted a transmitter above a duck's bill using a nasal pin. Their transmitter included a switch so positioned that it changed the pulse rate depending on whether the duck's head was up or down (as when feeding).

WELFARE

It is imperative for both humanitarian and scientific reasons that the transmitter does not hamper or damage the animal in any way. Boag (1972) found that grouse (*Lagopus lagopus*), wearing transmitters fed less than those without. Sargeant, Swanson and Doty (1973) suggested that fitting a radio pack to a teal (*Anas discors*), contributed to its being predated by mink (*Mustela vison*) (see Siegfried *et al.*, 1977). Signs of collar abrasion (Beal, 1967) or weight loss indicate unsatisfactory design. Comparisons can be made between observations on the same animal before and after radio tagging, but Amlaner, Sibly and McCleery (1978) show that some effects may be very subtle, e.g. on longterm reproductive potential. Hamilton (1976) fitted leopards (*Panthera pardus*), with transmitters camouflaged with black and yellow spots to avoid attracting attention to the predator and Macdonald and Apps (1978) similarly painted cat (*Felis catus*), transmitters to match coat color. Kenward (1977) found no difference in weight loss or dispersion tendency of goshawks (*Accipiter gentilis*), wearing transmitters from those wearing leg rings and noted that hourly rate of bringing prey to the nest was the same for a sparrow hawk (*Accipiter nisus*), before and after being fitted with a transmitter (3.5 percent of body weight). Amlaner, McCleery and Sibly (1978) have conducted a detailed study of the survival and hatching success of herring gulls carrying dummy transmitters weighing between 10-50 g and found that more severe treatments (i.e. wearing heavier transmitters) decreased survival of the offsping.

BASIC RADIO TRACKING

The basic operation of the radio tracking system involves a battery powered transmitter which emits low powered pulsed signals via a transmitting antenna. These signals are received by another, directionally sensitive antenna which connects to the receiver. Ideally, the directional properties (Amlaner, 1980, this volume; Cedurlund and Lemnell, 1980, this volume; Parish, 1980, this volume) of the receiving antenna allow bearings to be taken on the animal's position from two places and the point at which these bearings intersect marks the animal's location. However, as Tester (1971) points out there are three categories of error all of which can act together to frustrate this ideal: (1) System error, inherent in the receiving equipment and its operation. (2) Movement error, resulting from the animal moving while bearings are being taken. (3) Topographical error, resulting from anomalies such as reflection and refraction of the signal creating false bearings.

SYSTEM ERROR

(a) Inaccuracies in the directionality of the antenna: assuming an antenna has been perfectly made, it is unlikely to remain so in the field. We have suffered inaccuracies induced by wind twisting the elements of a Yagi antenna, and overhanging tree limbs buckling the elements of an Adcock and Yagi mounted on a vehicle. The frequent bending of antenna leads during direction finding, or pinching leads in vehicle doors may lead to intermittent loss of contact that deceptively mimic signal nulls. Suspected transmitter failure should always prompt a thorough check of the antenna connections and leads (which can be tested with a known transmitter).

(b) Inaccuracies imposed by the geometry of triangulation: Table 1 shows the magnitude of system error on the accuracy of bearings taken with different antennas from different locations and distances on the position of a sleeping fox lying on the edge of woodland and open habitats. The error does not increase significantly with distance from the fox and is 3-5°. The error is significantly greater for bearings taken from within the woodland to those taken from farmland when the fox slept at the woodland's edge. The dipole's performance was marginally worse than the Adcock's. A similar test under urban conditions was less accurate, as buildings both reflect and diminish the signal (see Harris, 1980, this volume; Hough, 1980, this volume). While the system error does not increase with distance from the fox, the triangulation error does (Heezen and Tester, 1967). The area encompassed by the 'beam' of inaccuracy increases with distance from the animal. With a 3° error this constitutes a decrease in the accuracy of the supposed position of 52 m km^{-1} calculated by:

$$\tan \theta \times \text{distance} = \text{displacement} .$$

Furthermore, 'error polygons' increase in area for a given distance between transmitter and receiver depending on the extent to which the intersecting bearings depart from right angles (and hence the intersecting 'beams' of inaccuracy become less of a square and more of a parallelogram). Thus, bearings should be taken at approximately right angles to each other and as close to the animal as possible. Deat *et al.* (1980, this volume) describe an automated solution to this problem.

Table 1.
A comparison of the mean angular error of sets of bearings taken from different positions with different antennas under ideal conditions (near = 100 m and far = 500 m approximately)

		Woodland	Field	Town
Adcock	Near	4.2	3.1	20.9
	Far	5.1	3.3	20.1
Dipole	Near	7.4	3.8	12.0
	Far	12.6	3.9	25.1

(c) Additional inaccuracies can result from misreading the bearing, for instance, because of holding a magnetic compass too near to a vehicle or the antenna. The antenna should be mounted at least one wavelength above a vehicle roof.

MOVEMENT ERROR

(a) If there is a delay between taking radio bearings used to pinpoint the animal's position, the position of their intersection will bear little relationship to the animal's actual position. The magnitude of this error will depend on the time lag involved, the animal's speed and the positions from which the bearings are taken. Using automated equipment (e.g. Cochran *et al.*, 1965; Deat *et al.*, 1980, this volume; Lemnell, 1970) or two sets of observers with walkie talkies (Reeve, 1980, this volume; Smith and Trevor-Deutsch, 1980, this volume) minimizes this risk. However, people working alone can reduce error by being alert to changes in the radio signal (PIM) which signify the animal's movement.

TOPOGRAPHICAL ERROR

(a) Very high frequencies (200-500 MHz) are seriously attenuated by vegetation and are more inclined to reflect off features of the landscape, but may also give better nulls when not deflected. Hence, VHF transmitters are more suitable in open country or for ground to air transmission whereas high frequencies (30-50 MHz) are better for ground work in rough terrain. Using an intermediate frequency (102-105 MHz) we have found topographical errors to be unquestionably the greatest source of inaccuracy.

(b) Inaccuracy stemming from local topography obviously varies between areas and can only be learned by experience. Valuable practice can be gained in advance by the biologist tracking a surrogate radio tagged animal, such as his wife. The following features pose risks:

(i) Valleys, hillsides, cliffs, and buildings may reflect the signal.
(ii) Metal fences, electrical and telephone wires sometimes pick up and retransmit the signal along their length.
(iii) Burrows, such as fox earths and badger setts, may attenuate the signal while others may broadcast it in narrow beams from their entrances. Such beams are both easy to miss and prone to reflection, e.g. a normal range of up to 5 km can be reduced to 100 m from a deep burrow with one entrance.
(iv) Thick or wet vegetation attenuates the signal, as does the ground. This effect varies seasonally with leaf cover in deciduous woodland. Transmitters designed for smaller mammals should be tested from positions on or close to the ground and not while hanging from the branch of a tree or similar high vantage point. Using a dipole antenna a strong signal can sometimes be received when it is held vertically but not horizontally and sometimes vice versa; while partly attributable to the animal's position, vegetation structure is also important; for instance, large uniform conifer stands appear to polarize the signal (Amlaner, 1980, this volume).

The extent of this problem is illustrated by the case of a radio tagged fox that spent its days in an earth at the head of a deep valley. From there the signal emerged at an acute angle to the opposite side of the valley and ricochetted down to the valley mouth 2 km away. Tracking this fox involved taking a series of zig-zagging bearings of which only the last pair pointed at its real position. Verts (1963) and Anderson and Hitchins (1971) have noted the same problem while tracking skunks (*Mephitis mephitis*) and black rhinoceros (*Diceros bicorius*) respectively. As Hamilton (1976) states "... the receiving system could only tell from which direction the signal arrived and not from which direction it left the transmitter...." Although not commonly stressed in the literature the radio tracker's subjective 'feel' for the radio landscape of his study area is widely acknowledged as an important facet of accurate radio tracking. The following suggestions help to minimize the risks of both movement and topographical errors:

(1) Take consecutive bearings with as little delay as possible. For direction finding, nulls are preferable to peaks as it is easier to detect the difference between signal and silence than between signal and louder signal. Amlaner (1980, this volume) and Parish (1980, this volume) discuss techniques for deciding whether a transmitter is in front of or behind the tracker.

(2) Know the individual features of the landscape well, and how they affect radio signals.

(3) If errors are suspected take several alternative bearings; at least three are necessary to detect movement errors. When radio tracking from a car we have found it helpful to continue listening to the signal while travelling between reception sites so that fluctuations in signal strength along the route can be noted in the context of the landscape.

(4) Rank bearings and the derived radio locations in terms of their accuracy depending on both the width of the null points and the variation of successive bearings that do not intersect at the same spot. Depending on the questions being posed and the size of the animal's home range a given inaccuracy may be tolerable. An inaccurate bearing in the middle of an animal's home range might be a useful datum but if one is interested in tracing the outline of a territory, a similarly inaccurate bearing would be unacceptable. Hamilton (1976) proposed the use of accuracy indicators, i.e. circles of varying diameter which include in the smallest possible area, all the various intersections drawn on the map between several bearings taken on one animal.

Keeping alert for nonsensical locations is difficult in the field at night with maps flapping in the wind. Thus we find it useful to carry a work map fixed to a clipboard and protected from rain by a clear plastic sheet. After each radio fix is taken, the details are marked directly on the map. These maps also provide a uniform record from which data can subsequently be decoded (see 'Data Transcription').

AUTOMATIC DATA RECORDING

Automatic sampling may be achieved without sophisticated data logging equipment. Accurate clock timing is possible with simple, inexpensive integrated CMOS circuits which can be used to build switching devices with variable 'on' and 'off' times. These miniature circuits can operate on low voltages and have many applications both in the field and the laboratory. For example, in the field the presence of radio tagged animals can be systematically sampled using an omni directional antenna (Amlaner, 1980, this volume), receiver, tape recorder and clock switch. One or five minute samples of the radio signal can be recorded every 15, 30 or 60 min. Similarly, sampling physiological variables from biotelemeters can reduce the amount of redundant data collected while preserving the continuity of longterm samples.

Figure 4 presents a circuit using two precision timers (ZN1034E, Ferranti Electronics, U.K.). The frequency of an internal oscillator is determined by an externally connected capacitor and resistor(s). Pulses from the oscillator are fed through a 12-stage binary divider which triggers control logic. The chip also has an internal 5 vdc regulator. By using the internal calibration resistor in this device a timing accuracy of ± 0.01 percent can be achieved, but for long time intervals, the capacitor connected to pin 13 must have a low leakage current.

The circuit operates as follows: timer B determines the sampling frequency (i.e. 15 min setting gives four samples h^{-1}). When timer B triggers timer A the relay contacts close and so turn on any equipment wired to them. Timer A stays 'on' for 1 or 5 min, depending on the pulse length selected. Any common type of relay (12 vdc, 185 Ω) may be used, and by using multiple contacts, several pieces of apparatus can be automatically controlled by one timing device. For instance, a receiver and tape recorder motor can be controlled simultaneously using a double pole relay. Switch S1 turns on the relay circuitry when closed.

DATA TRANSCRIPTION — AN INTERACTIVE X, Y COORDINATE PLOTTER

The conversion of large numbers of radio locations into x, y coordinates for mapping and analysis is a laborious procedure by hand. The process can be speeded-up by the

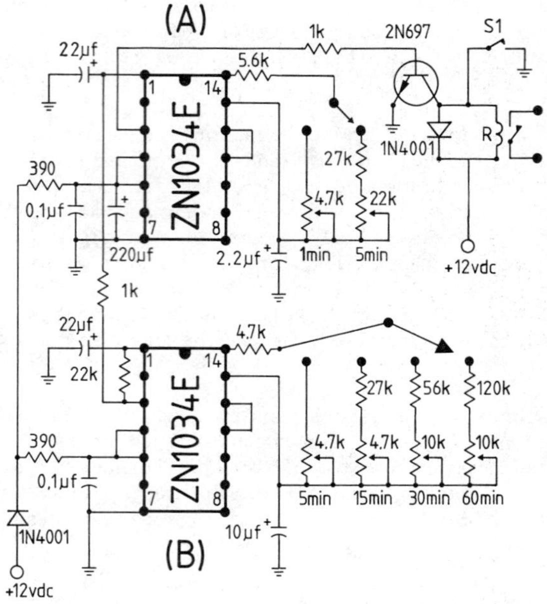

Fig. 4. Circuitry for a precision timer which may be used for efficient automatic data recording.

use of a coordinate plotter, such as that described by Partridge and Cullen (1977) and Partridge, Dawkins and Amlaner (1978). A modified version of the machine and Algol program described by these authors proved ideal for converting radio locations into x, y coordinates scaled in meters preparatory for home range analysis (see Macdonald, Ball and Hough, 1980, this volume). The equipment is illustrated in Fig. 5A, where it can be seen to consist of a table on which a map is placed. It is fitted with two 50 K linear ten turn potentiometers (P1 and P2) mounted a known distance (d) apart (wider than the map being used). They are connected by inelastic string to the potentiometer which rotates as the pointing cursor is moved, causing their output voltages to vary the output of a voltage-to-frequency converter (VCO). The variations in frequency are fed into the computer (PDP8/E) and using software (Partridge *et al.*, 1978), they are converted into x, y coordinates, by simple geometry (Fig. 5B):

$$x_1^2 + y_1^2 = L_1^2 \quad \text{and} \quad x_2^2 + y_2^2 = L_2^2$$

$$x = (L_2^2 - L_2^2 + D^2)\,2D \quad \text{and} \quad y = (L_1^2 - x^2)^{\frac{1}{2}} .$$

The coordinate plotter gives an accuracy better than 0.1 mm on a 25 × 35 cm table. The plotting table is fitted with a keyboard of push buttons each of which specifies a 'procedure' on the plotting program. In our case, a procedure SCALES calibrates the VCO frequency relative to the distance moved by cursor and converts this calibration into the appropriate scale for the map being used. POINT determines the position of the cursor and then calculates the coordinates of radio fixes in units determined by SCALES. Another procedure, TRIANG, was used for triangulating an animal's position from several compass bearings: in this case the consecutive positions of the observer were plotted with the cursor and then the compass bearings of the fox from each position typed into the console. A large amount of other information can be typed into the computer memory (date, time, habitat, etc.) with each map coordinate. A procedure is also included to compute the animal's speed of

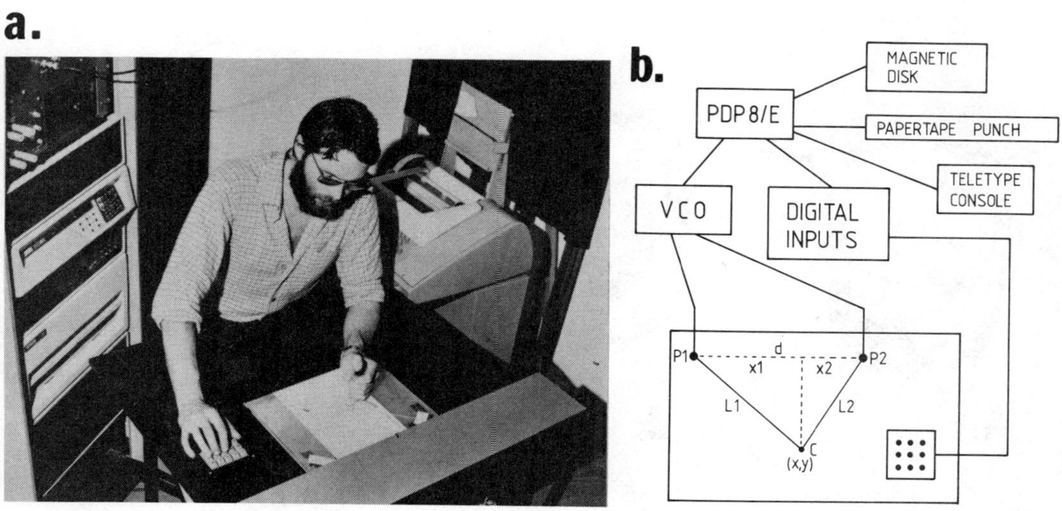

Fig. 5. An interactive *x, y* coordinate plotter, useful for the fast and efficient transcription of large numbers of radio fixes into scaled map coordinates ready for analysis. (a) Shows the plotting table in use while (b) shows the geometry underlying the plotting procedure.

movement between successive radio fixes. This interactive coordinate plotter facilitates very quick and accurate plotting of radio tracking data and as our machine is 'home-made' it is an inexpensive solution to a task that sometimes presents an insurmountable chore to fieldworkers.

CONCLUSION

Radio tracking is an invaluable addition to the biologist's skills and fieldcraft, and has contributed to diverse studies (reviewed by Macdonald, 1978; Mech, 1980, this volume; Sargeant, 1980, this volume). Just as this technique has revolutionized studies of elusive species during the past decade or so, now people look to the possibility of transmitting and receiving additional information, through biotelemetry, as the next methodological advance (reviewed by Amlaner, 1978; Kimmich, 1980, this volume; Smith, 1980, this volume). Some simple but ingenious modifications of tracking transmitters have already been used, of which Charles-Dominique's study of urine marking by bushbabies, *Galago alleni*, is a good example. The Cochran and Lord (1963) transmitter (similar to Fig. 1) was modified so that two bare wires protruded from its housing, which hung on a strap below the animal's waist (Fig. 6). Each time the bushbaby urinated the drops of moisture temporarily shorted these two wires and hence closed a low resistance loop in the circuit, which caused increased 'bleep' rate. So, by detecting these short increases Charles-Dominique (1977) could plot the distribution of scent marking and so discovered that this behavior increased in the neighborhood of territorial borders.

By contrast, Charles-Dominique's study highlights a pitfall into which radio tracking has lured many victims. Namely, there has been a tendency to use it as an excuse to abandon other field skills, rather than to complement them. It is important for many studies that every effort be made to watch animals, as well as to track them. Radio tracking greatly enhances the chances of making observations by

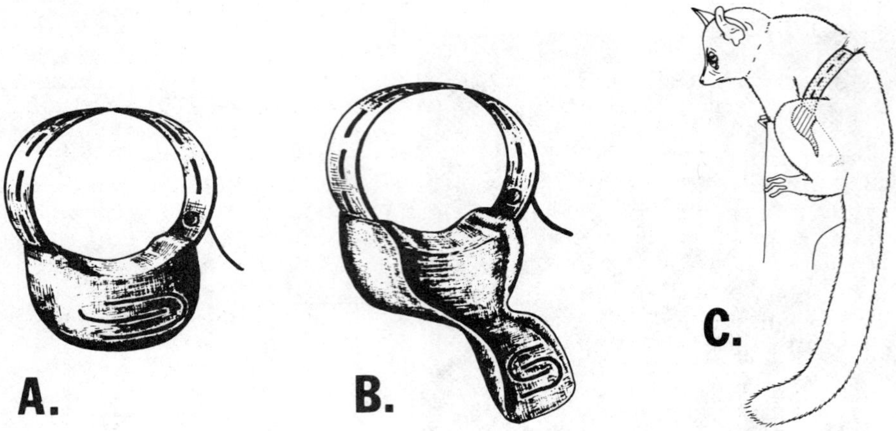

Fig. 6. The transmitter design (A. males and B. females) used by Charles-Dominique (1977) to study the scent marking behavior of (C) *Galago alleni*.

allowing the biologist to predict where the animal can most readily be found and watched without disturbance (termed 'predictive tracking' by Macdonald, 1978; and 'surveillance tracking' by Kenward, 1980, this volume).

Acknowledgements — We are grateful to M. Dolan and T. Barker for help with building transmitters and to M. Cullen and B. Partridge for introducing us to the 'Mad Monster' *x, y* coordinate plotter. N. Hough kindly commented on the manuscript.

REFERENCES

Aucoutourier J. L., Chaillou A., Nicholas G., Canivene R., Marques M., Govaerts R., Mertens R., Lauwers F. and van Overstraeten R. J. (1977) Biotelemetry and radio tracking of wild birds — portable device using solar cells power supply. Report of the *IHSM European Hybrid Microelectronic Conference*. Bad Hamburg, 1976, 7 p.

Amlaner C. J., Jr. (1978) Biotelemetry from free ranging animals. In B. Stonehouse (ed.), *Animal Marking: Recognition Marking of Animals in Research*, pp. 205-228. London: Macmillan Press.

Amlaner C. J., Jr. (1980) Telemetry antenna design. In C. J. Amlaner, Jr. and D. W. Macdonald (eds.), *A Handbook on Biotelemetry and Radio Tracking*, pp. 251-261. Oxford: Pergamon Press.

Amlaner C. J., Jr., Sibly R. and McCleery R. (1978) Effects of transmitter weight on breeding success in herring gulls. *Biot. Pat. Montn.* 5, 154-163.

Anderka F. W. (1980) Miniature radio telemetry transmitter. In C. J. Amlaner, Jr. and D. W. Macdonald (eds.), *A Handbook on Biotelemetry and Radio Tracking*, pp. 181-184. Oxford: Pergamon Press.

Anderson F. and Hitchins P. (1971) A radio-tracking system for the black rhinoceros. *S. Afr. J. Wildl. Res.* 1, 26-35.

Beal R. O. (1967) Radio transmitter collars for squirrels. *J. Wildl. Mngmt.* 31, 373-374.

Boag D. A. (1972) Effect of radio-packages on behaviour of captive red grouse. *J. Wildl. Mngmt.* 36, 511-518.

Cederlund G. and Lemnell P. A. (1980) A simplified technique for mobile radio tracking. In C. J. Amlaner, Jr. and D. W. Macdonald (eds.), *A Handbook on Biotelemetry and Radio Tracking*, pp. 319-322. Oxford: Pergamon Press.

Charles-Dominique P. (1977) Urine marking and territoriality in *Galago alleni* (Waterhouse 1837-Lorisoidea Primates) — a field study by radio telemetry. *Z. Tierpsychol.* 43, 113-138.

Cheeseman C. and Mallinson P. (1980) Radio tracking in the study of bovine tuberculosis in badgers. In C. J. Amlaner, Jr. and D. W. Macdonald (eds.), *A Handbook on Biotelemetry and Radio Tracking*, pp. 649-656. Oxford: Pergamon Press.

Cochran W. W. and Lord R. D., Jr. (1963) A radio-tracking system for wild animals. *J. Wildl. Mngmt.* 27, 9-24.

Cochran W. W., Warner D. W., Tester J. R. and Kuechle V. B. (1965) Automatic radio-tracking system for monitoring animal movements. *BioScience* 15, 98-100.

Church K. E. (1980) Expanded radio telemetry potential in wildlife investigations with the use of solar transmitters. In C. J. Amlaner, Jr. and D. W. Macdonald (eds.), *A Handbook on Biotelemetry and Radio Tracking*, pp. 247-250. Oxford: Pergamon Press.

Deat A., Mauget C., Mauget R., Maurel D. and Sempéré A. (1980) The automatic continuous and fixed radio tracking system of the Chizé Forest: theoretical and practical analysis. In C. J. Amlaner, Jr. and D. W. Macdonald (eds.), *A Handbook on Biotelemetry and Radio Tracking*, pp. 439-451. Oxford: Pergamon Press.

Donaldson N. de N. (1980) Encapsulation techniques. In C. J. Amlaner, Jr. and D. W. Macdonald (eds.), *A Handbook on Biotelemetry and Radio Tracking*, pp. 217-224. Oxford: Pergamon Press.

Dunston T. C. (1977) Types and uses of radio packages for North American falconiform and strigiform birds. In F. M. Long (ed.), *Proc. of First Int. Conf. on Wildlife Biot.*, Laramie, Wyoming.

Erlinge S. (1980) Movements and activity pattern of radio tracked stoats *Mustela erminea* L. In C. J. Amlaner, Jr. and D. W. Macdonald (eds.), *A Handbook on Biotelemetry and Radio Tracking*, pp. 703-710. Oxford: Pergamon Press.

Forbes J. E. and Warner D. W. (1974) Behaviour of a radio-tagged saw-whet owl. *Auk* 91, 783-795.

Hamilton P. H. (1976) Movements of leopards in Travo National Park, Kenya, as determined by radio-tracking. M.Sc. thesis, University of Nairobi.

Harding P. J. R., Chute F. S. and Doell A. C. (1976) Increasing battery reliability for radio transmitters. *J. Wildl. Mngmt.* 40, 357-358.

Hardy A. R. (1977) Hunting ranges and feeding ecology of owls in farmland. Ph.D. thesis, University of Aberdeen.

Harris S. (1980) Home ranges and patterns of distribution of foxes (*Vulpes vulpes*) in an urban area, as revealed by radio tracking. In C. J. Amlaner, Jr. and D. W. Macdonald (eds.), *A Handbook on Biotelemetry and Radio Tracking*, pp. 685-690. Oxford: Pergamon Press.

Heegen K. L. and Tester J. R. (1967) Evaluation of radio tracking by triangulation with special reference to deer movements. *J. Wildl. Mngmt.* 31, 124-141.

Hough N. G. (1980) Movements of a female fox. In C. J. Amlaner, Jr. and D. W. Macdonald (eds.), *A Handbook on Biotelemetry and Radio Tracking*, pp. 691-696. Oxford: Pergamon Press.

Hitchins P. M. (1971) Preliminary findings in a radio telemetric study on the black rhinoceros in Hluhluwe Game Reserve, Zululand. *Proc. Symp. on Biotelemetry*, Pretoria, 1971, Council for Scientific and Educational Research, Pretoria, 15 pp.

Kenward R. E. (1977) Predation on released pheasants (*Phasianus colchicus*) by goshawks (*Accipiter gentilis*) in central Sweden. *Viltrevy* 10, 79-112.

Kenward R. E. (1978) Radio transmitters tail-mounted on hawks. *Ornis Scand.* 9, 220-223.

Kimmich H. P. (1980) Artifact free measurement of biological parameters: biotelemetry — a historical review and layout of modern developments. In C. J. Amlaner, Jr. and D. W. Macdonald (eds.), *A Handbook on Biotelemetry and Radio Tracking*, pp. 3-20. Oxford: Pergamon Press.

Ko W. H. (1980) Power sources for implant telemetry and stimulation systems. In C. J. Amlaner, Jr. and D. W. Macdonald (eds.), *A Handbook on Biotelemetry and Radio Tracking*, pp. 225–245. Oxford: Pergamon Press.

Kruuk H. (1978) Spatial organisation and territorial behaviour of the European badger, *Meles meles*. *J. Zool., Lond.* 184, 1-19.

Lemnell P. A. (1970) Telemetry as a method for studying the home range: examples from a study of the red squirrel, *Sciurus vulgaris*. *Zool. Rev.* 32, 51-56.

Leuze C. (1977) A manual for the construction of the water vole transmitter collar unit. (Obtainable on request from Culterty Field Station, Newburgh, Aberdeenshire, Scotland).

Leuze C. (1980) The application of radio tracking to the study of behavioral ecology of the water vole, *Arvicola terrestris* (L.). In C. J. Amlaner, Jr. and D. W. Macdonald (eds.), *A Handbook on Biotelemetry and Radio Tracking*, pp. 361-366. Oxford: Pergamon Press.

Lotimer J. S. (1980) A versatile coded wildlife transmitter. In C. J. Amlaner, Jr. and D. W. Macdonald (eds.), *A Handbook on Biotelemetry and Radio Tracking*, pp. 185-191. Oxford: Pergamon Press.

Macdonald D. W. (1978) Radio-tracking: some applications and limitations. In B. Stonehouse (ed.), *Animal Marking: Recognition Marking of Animals in Research*, pp. 192-204. London: Macmillan Press.

Macdonald D. W. and Apps P. J. (1978) The social behaviour of a group of semi-dependent farm cats, *Felis catus*. *Carnivore Genetics Newsletter* 3, 256-268.

Macdonald D. W., Ball F. G. and Hough N. G. (1980) The evaluation of home range size and configuration using radio tracking data. In C. J. Amlaner, Jr. and D. W. Macdonald (eds.), *A Handbook on Biotelemetry and Radio Tracking*, pp. 405-424. Oxford: Pergamon Press.

Mech L. D. (1980) Making the most of radio tracking. In C. J. Amlaner, Jr. and D. W. Macdonald (eds.), *A Handbook on Biotelemetry and Radio Tracking*, pp. 85-95. Oxford: Pergamon Press.

Morris P. (1980) An elementary guide to practical aspects of radio tracking mammals. In C. J. Amlaner, Jr. and D. W. Macdonald (eds.), *A Handbook on Biotelemetry and Radio Tracking*, pp. 161-168. Oxford: Pergamon Press.

Nicholls T. H. and Warner D. W. (1968) A harness for attaching radio transmitters to large owls. *Bird Banding* 39, 209-214.

Parish T. (1980) A collapsible dipole antenna for radio location on 102 MHz. In C. J. Amlaner, Jr. and D. W. Macdonald (eds.), *A Handbook on Biotelemetry and Radio Tracking*, pp. 263-268. Oxford: Pergamon Press.

Partridge B. L. and Cullen J. M. (1977) A low-cost interactive coordinate plotter. *Behav. Res. Methods Instrum.* 9, 473-479.

Partridge B. L., Dawkins R. and Amlaner, C. J., Jr. (1978) Voltage-controlled oscillators: inexpensive alternative to analog-to-digital converters. *Behav. Res. Methods Instru.* 10, 712-714.

Reeve N. J. (1980) A simple and cheap radio tracking system for use on hedgehogs. In C. J. Amlaner, Jr. and D. W. Macdonald (eds.), *A Handbook on Biotelemetry and Radio Tracking*, pp. 169-173. Oxford: Pergamon Press.

Sargeant A. B. (1980) Approaches, field considerations and problems associated with radio tracking carnivores. In C. J. Amlaner, Jr. and D. W. Macdonald (eds.), *A Handbook on Biotelemetry and Radio Tracking*, pp. 57-63. Oxford: Pergamon Press.

Sargeant A. B., Swanson G. A. and Doty H. A. (1973) Selective predation by mink, *Mustela vison*, on waterfowl. *Am. Midl. Nat.* 89, 208-214.

Siegfried W. R., Frost P. G. H., Ball, I. J. and McKinney D. F. (1977) Effects of radio-packages on African black ducks. *S. AFr. J. Wildl. Res.* 7, 37-40.

Skutt H. R., Bock F. M., Haugstad P., Holter J. B., Hayes H. H. and Silver H. (1973) Low-power implantable transmitters for telemetry of heart rate and temperature from white-tailed deer. *J. Wildl. Mngmt.* 37, 413-417.

Smith E. N. (1980) Physiological radio telemetry of vertebrates. In C. J. Amlaner, Jr. and D. W. Macdonald (eds.), *A Handbook on Biotelemetry and Radio Tracking*, pp. 45-55. Oxford: Pergamon Press.

Smith R. M. and Trevor-Deutsch B. (1980) A practical, remotely-controlled, portable radio telemetry receiving apparatus. In C. J. Amlaner, Jr. and D. W. Macdonald (eds.), *A Handbook on Biotelemetry and Radio Tracking*, pp.269-273. Oxford: Pergamon Press.

Swanson G. A and Keuchle V. B. (1976) A telemetry technique for monitoring water fowl activity. *J. Wildl. Mngmt.* 40, 187-189.

Taylor K. D. and Lloyd H. G. (1978) The design, construction and use of a radio-tracking system for some British mammals. *Mammal Rev.* 8, 117-141.

Tester J. R. (1971) Interpretation of ecological and behavioral data on wild animals obtained by telemetry with special reference to errors and uncertainties. *Int. Proc. Symp. Biotelemetry S.* 57 CSER, Pretoria, S. Africa, 385-408.

Verts B. J. (1963) Equipment and techniques for radio-tracking striped skunks. *J. Wildlf. Mngmt.* 27, 325-339.

An Elementary Guide to Practical Aspects of Radio Tracking Mammals

P. MORRIS

Department of Zoology, Royal Holloway College, University of London, London, U.K.

Abstract — Though ignorant of the technicalities of radios, my early use of radio tracking has meant that I have received many enquiries from graduate students wishing to make use of this technique; published papers make the methods sound easy, yet to a biologist the technicalities are unfamiliar. This brief introduction to purely practical aspects is based upon the information a naïve graduate needs to have digested before embarking on a radio tracking study.

INTRODUCTION

Today, radio tracking is a familiar technique, widely used in the study of mammal ecology. Despite its early adoption in North America (especially during the 1960s), practical and financial constraints hindered its use in Britain and elsewhere. Consequently relatively little experience was available outside the U.S.A. until recently.

In 1966, with little money and no knowledge of electronics, I set out to develop cheap and simple radio tracking equipment for use on the hedgehog (*Erinaceus europaeus*). The result was a crude and cumbersome system, but moderately successful within the limitations set by time and money. Since then, I have been approached by many postgraduate students and others seeking advice on the use of this apparently simple technique.

The basic problem seems to be that published papers are often so glib in describing their methodology that radio tracking seems to offer the beguiling prospects of considerably extending a study with little practical difficulty, once a few minor technicalities are grasped. Misapprehension is compounded where biologists are ignorant of radio technology and are unable to appreciate elementary technicalities. The present paper is intended as an introduction to the purely practical aspects of setting up a radio tracking study.

The following brief list of questions, roughly in order of their priority, covers the topics the beginner needs to consider and should help to reduce a complex of options to just one or two feasible lines of action. Relevant practical details are discussed later.

1. What radio frequency is permissible in the proposed study area? This is crucial to the whole project as the frequency to be allocated by the Regulating Authority dictates what equipment can or cannot be used.

2. Is suitable ready-made equipment available for use on this frequency? If so, can you afford it? If not, can you build it or pay to have it built?

3. Will your moving animals be located by triangulation? If so, extra equipment and field assistance will be necessary. If not, will disturbance of the animal invalidate any aspect of what is being studied?

4. What type of transmitter will be built (e.g. loop, whip, or integral antenna; depends partly on 1 above)?

5. Precisely what kinds of information will be collected, how often and for how long? Will this body of data then be amenable to analysis providing answers about the species under study?

TERMINOLOGY

'Telemetry' literally means 'measuring at a distance'. The term therefore properly refers to studies of heart beat, respiration rate, activity and other biological parameters, transmitted from sensors on an animal to a near or distant receiver. Biotelemetry is most frequently used for physiological studies, often in confinement, and has been reviewed recently by Fryer and Sandler (1974) and Amlaner (1978).

'Radio tracking' is a means of locating a radio tagged animal using a radio receiver and direction finding antenna. The technique is normally used to study home range, territory and spatial relationships of animals in the field and is reviewed with a useful list of references by Macdonald (1978). With suitable equipment it is sometimes possible to combine biotelemetry and radio tracking to obtain position data and biological measurements simultaneously.

FREQUENCY

The frequency of a radio signal is measured in cycles per second (Hz), in effect the number of 'vibrations' per second performed by the radio wave. Thus a frequency of 104 MHz characterizes a signal with 104×10^6 cycles s^{-1}. In this context, 'frequency' does not refer to the number of 'bleeps' emitted per minute by the transmitter; a feature better described as the 'pulse rate'.

Published accounts of radio tracking show that the commonly used radio frequencies fall into two groups:

1. High frequencies (HF) — Many workers have used transmitters operating in the range of 25 to 35 MHz. Signals on these frequencies may have a considerable range and are not subject to serious attenuation or deflection. However, relatively large transmitter antennas are usually necessary.

2. Very high frequencies (VHF) — More recent research has utilized frequencies of 100 MHz or more. VHF transmitters do not need a cumbersome antenna and the signal propagates fairly well. However, such transmitters are more difficult to build and the signal is liable to be deflected near the ground by solid objects and may be screened by wet vegetation or soil. This is why ultra high frequencies (UHF), are suitable for use on birds which live out in the open or up in the air, but less useful for mammal tracking despite the considerable size advantages of a UHF system (Lawson *et al.*, 1976, suggest 432 MHz).

Perhaps the greatest single advantage of any transmission frequency below 30 MHz is that an ordinary ready made transistor radio receiver can be used for detection. However, with some loss of sensitivity, it is possible to adapt an ordinary receiver so that it can be used to pick up VHF signals above the range for which these cheap receivers were originally designed (Reeve, 1980, this volume).

REGULATIONS

It is often not appreciated that in many countries all radio transmitters must be licensed, and this may curtail the options listed above. In the U.S.A., the allocation of certain bands (e.g. 30, 150 and 164 MHz) considerably eases the problem by making several frequencies freely available for use if transmitter power is less than 100 mW. In Britain, the Regulatory Authority is the Home Office who allocate a band of 104.6 to 105 MHz for radio tracking. Other countries are likely to have their own regulations administered by the Ministry of Posts and Telecommunications. Although officials no longer treat requests for permission to use transmitters on animals as bizarre and frivolous, formalities still take time. To license my hedgehog transmitters took six months. There is a temptation to ignore such formalities but to do so may incur penalities when the work is published which rebound on institutions and on other workers using radio tracking. Also, in populated areas, transmitters on illegal frequencies may interfere with legitimate users or even perhaps safety equipment like navigation aids.

THE TRANSMITTER

In order to keep the transmitter working on the correct frequency, its circuit needs to include a crystal, often the most bulky component. Crystals need to be specially made to match a particular frequency and suppliers often take six weeks or more to produce them. Other components are easily purchased from mail order suppliers.

Transmitters can be built without crystals and are then small and, likely to transmit each on a different frequency, and hence, can be used to give individually recognizable signals. This is most useful, but also illegal in Britain. Transmitters are relatively easy to build, provided a circuit diagram (Fig. 1) and layout are available. The former shows the pattern in which components are connected, the latter is a map of their physical relationships in the transmitter. A description of the 'Ashwell' transmitters, suitable for use on a wide range of mammals down to the size of rats, has been published by Taylor and Lloyd (1978). They give practical details of construction which are not repeated here; copies of this paper are available from the Mammal Society, Harvest House, Reading RG1, U.K. Similar circuit and layout diagrams are also given in this volume by Anderka (1980), Lotimer (1980), Macdonald and Amlaner (1980) and Reeve (1980).

Most transmitters are designed to be expendable (with soldered-in batteries) and are lost when their batteries are exhausted, probably still attached to an animal. With some weight penalty, battery chambers can be incorporated to facilitate replacement of the power source. However, the design must provide reliable battery connections and be waterproof. Battery replacement is only feasible if the animal can be recaptured easily, preferably well before the batteries run down. Transmitter life may be extended by using solar panels to recharge the batteries (Patton, 1973; Church, 1980, this volume), but this is likely to be more useful for diurnally active mammals. Transmitter performance may also be improved by using X-rays to identify defective batteries before they are used (Harding *et al.*, 1976).

Transmitters need to be encapsulated ('potted') in protective resin or silicone rubber. It is probably best to varnish the assembled transmitter first. When this

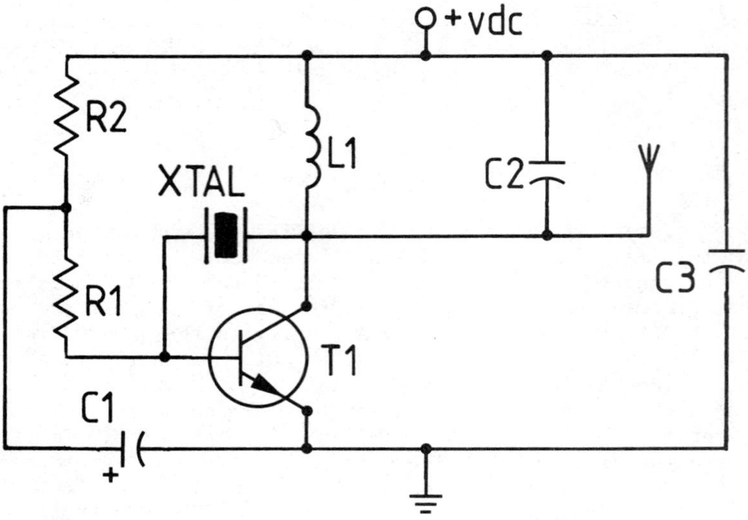

Fig. 1. Circuit diagram for a 27 MHz transmitter used on hedgehogs. Xtal (miniature) 27.12 MHz. Silicon transistor (T1; 2N708). R1, 1K. R2, 100K. C1 electrolytic capacitor, 4-15 μf. C2, 22-68 pf. C3, 1000 pf. Batteries: 4 × Mallory mercury cells (type PX 675) in series to give 6 V. The value of C1 will govern the pulse repetition rate; if it is increased beyond about 15 μf. The pulse rate is so slow that location of the transmitter in the field becomes difficult. The coil (L1) consists of about 12 turns of Litz wire wound on a polystyrene former. A ferrite slug is screwed into the former and to obtain the strongest signal, its position must be adjusted correctly after the transmitter is built.

is dry, the transmitter should be enclosed in a mould (e.g. a flexible one made of 'Plasticene' or modeling clay) and the resin poured over it. The mass of potting compound may alter the capacitance of certain components, causing the transmitter to malfunction or shift frequency. A small tunable capacitor in the circuit offers a solution to the problem, though an access hole needs to be left in the encapsulation to reach it. If the potting compound is transparent, an address tag can be incorporated into it, offering a reward to the casual finder. A beta light or luminous paint may also be embedded into the encapsulation to aid visual detection at night, though with the risk of attracting unwanted attention to the animal.

Early transmitter designs (e.g. Storm, 1965) used a 'tuned loop' collar antenna. This consisted of a rigid ring of metal tube, built into the circuit and tuned with a variable capacitor to obtain a maximal signal. Rigid loops of this kind may impede natural movements; (for example, they would not allow a hedgehog to roll up) A flexible 'whip' antenna seems preferable but may permit moisture to penetrate transmitter potting where the antenna emerges. Integral antennas (dust core) in the Ashwell design overcome these problems (Taylor and Lloyd, 1978).

Ideally a transmitter should have long life, long range and miniscule dimensions. In practice these three factors are interrelated; longer life usually means bulkier batteries. Conversely, smaller size means smaller batteries (batteries are usually at least as large as the rest of the transmitter), reducing range and/or duration (see Ko, 1980, this volume, for a review). The range will also depend upon receiver sensitivity, height of transmitter off the ground and topography of the study area so that it is impossible to predict accurately what range a transmitter will have in the field. 'Line of sight' ranges are likely to be many times greater than if the transmitter is screened by hills or thick vegetation. Thus animals living in open, flat habitats can be tracked at greater ranges than those which live in burrows, forests or among rocks. Large edifices (e.g. buildings) metal objects (such as fences) and electric transmission lines are also likely to distort signals and affect range.

The size of a transmitter (complete with batteries) is also limited by an animal's ability to carry it without impediment. Until recently the smallest transmitters reported in the literature (1.95 g) were those used by Mineau and Madison (1977) for tracking mice but their range was only 100 m (see Thomas, 1979; Macdonald and Amlaner, 1980, this volume; for 1 g transmitter). Even this size may impede the movements of a small animal. Transmitters used on foxes and badgers are normally less than 2 percent of the animal's body weight (and under favorable conditions may have a range of several kilometers); my hedgehog transmitters (30 g) were about 5 percent of body weight. The animals showed no sign of distress, but a similar burden could influence the behavior of a more agile creature. It is wise to plan an experiment using captive animals to assess their reactions to carrying a transmitter. In this way Boag (1972) found that grouse behavior was altered during the first week after transmitter attachment; but Brand *et al.* (1975) showed that wild snowshoe hares were not seriously affected in the longer term, despite signs that some of the animals had attempted to get rid of the radio tag. Amlaner (1978) has reviewed the effects of transmitters on animals, including their breeding success, behavior patterns and physiology.

ATTACHMENT TO A MAMMAL

Most mammals (e.g. fox, squirrel, deer) have relatively thin necks around which a collar may be fitted (made individually adjustable by straps or stitching), to which the transmitter is attached. Some species (e.g. badger, mink) have small heads and thick necks and may require a harness around the fore-quarters. In either case it needs to be tight, but must not chafe; and loose enough to permit growth and movement yet remain on the animal. In some cases an elastic harness may be appropriate. Trials with captive animals should be performed first to gain experience of fitting and show whether chafing is caused; a particular problem with shoulder harnesses which pass across the thin skin of the axilla. Transmitters can be implanted under the skin or into the abdomen, but these must be small and usually have a short range. Surgical implantation may involve trauma and convalescence in captivity, both potentially affecting the validity of subsequent behavior studies. A vivisection license may also be necessary.

THE RECEIVER

While transmitters can be built with minimal knowledge and expertize, receivers cannot. A sensitive receiver is a complex device and unless very sophisticated electronic facilities are available, together with technical assistance, a receiver will have to be purchased ready-made, or commissioned from a specialist supplier. Delivery delays are likely to be considerable, especially with overseas orders, and the cost will be approximately US $750 . Moreover, it would be rash for a study to depend on a single receiver which may be accidentally damaged at any time, especially if it needs to be returned to an overseas manufacturer for repair. The receiver problem is perhaps the *Achilles heel* of any radio tracking study, but with some penalty in sensitivity, there is a partial solution by using a frequency converter to 'translate' the VHF signals used in radio tracking to a lower frequency which a cheap shortwave transistor radio can pick up (Reeve, 1980, this volume). Radio tracking equipment available for 'off the shelf' purchase may operate on frequencies between 30 to 500 MHz. The AVM Instrument Co. (810, Dennison Drive, Champaign, Illinois 61820, U.S.A.) and Telemetry Systems Inc. (5830 N. Shore Drive, Milwaukee, Wisconsin 53217, U.S.A.) manufacture complete radio tracking systems and Amlaner (1978) lists suppliers of biotelemetry equipment.

ANTENNAS

A directional antenna will also be needed but this can be homemade (Taylor and Lloyd, 1978; Amlaner, 1980, this volume). It gives a strong signal when pointed at the transmission source and a weak one when rotated 90° away from the source. Location of a radio tagged animal in the field can be performed in two ways: either by walking towards it until it is visible, or by triangulation. The former method is generally feasible for small animals or those which move slowly or over short distances (e.g. hedgehog, squirrel and perhaps rabbit). It would be less suitable for species (e.g. deer) which are likely to move at least as fast as the investigator and which move away as soon as he is seen or heard. If larger, faster animals are to be located continuously without disturbance, triangulation is necessary. This requires two widely spaced receivers taking simultaneous compass bearings of the tagged animal from a distance at regular intervals. Two receivers require two operators, preferably in contact with each other by walkie-talkie radio in order to achieve proper coordination of fixes.

Radio tracking can be performed automatically by rotary scanners on fixed towers and the data fed into a computer which provides maps of a radio tagged animal's movements (Tester and Siniff, 1965), but such equipment is expensive and very complex. Triangulation is most accurate in locating a transmitter when the bearings from the two receivers intersect at 90°. As the transmitter closely approaches either receiver or the line between receivers, accuracy diminishes. At very long range, signal distortion and limitations inherent in the triangulation method itself also reduce precision; difficulties which are discussed more fully elsehwere (Heezen and Tester, 1967; Taylor and Lloyd, 1978; Macdonald, 1978).

NUMBER OF ANIMALS TO BE TRACKED

Within the small range of frequencies allocated for radio tracking it should be possible to differentiate at *least* 16 separate channels (each 25 KHz wide) and 2 transmitters precisely crystal controlled on each channel. If each channel has a transmitter with a fast and a slow pulse rate, this will permit more animals to be studied simultaneously in the same area. However, if the study requires animals to be tracked continuously, this potential may not be fully utilized, for there are limits to the amount of information an investigator can cope with at one time.

LIMITATIONS OF THE INVESTIGATOR

The greatest single limitation stems from the fact that the majority of mammals are mainly active at night. If radio tracking is intended to reveal what an animal does with its time, the investigator must be prepared to work the same hours regardless of his own normal activity patterns. It may not be necessary to match an animal's time, hour for hour, depending upon the type of information required, but it does underline the necessity of ensuring maximal benefit for the hours expended. Time is well spent in deciding (and practising) the most efficient way of recording data, remembering that it is more difficult in the dark.

CONCLUSIONS

A legal and useful radio tracking system is likely to cost at least US $750 and take upwards of three to six months to acquire. A system purchased 'off the shelf' is likely to be superior (in terms of portability, range, sensitivity etc.) to home-made equipment but it may still be inadequate for a particular species having been perhaps originally designed for use on something else. Where objectives are clearly limited, a cheap, partially home-made system may prove cost-effective. Generally, unless equipment and expertize are already availble, radio tracking studies, especially those requiring work in several seasons, are probably not going to be comfortably accommodated within the normal budget and time span of a Ph.D. project. To be realistic, radio tracking is either a full-time occupation or a waste of resources; it is ill suited to being merely a part of some other investigation. This is not meant to suggest that anything is impossible, especially for a determined person, only that radio tracking is far more expensive in time and resources than is usually evident from the literature. If time, money and manpower are available great advances can be made in the study of mammal ecology using radio tracking. Future technological developments, particularly in further miniaturization of batteries and in the production of cheaper, automatic recording systems, offer the prospect of a major leap forward in the study of mammal ecology.

REFERENCES

Amlaner C. J., Jr. (1978) Biotelemetry from free ranging animals. In B. Stonehouse (ed.), *Animal Marking: Recognition Marking of Animals in Research*, pp. 205-228. London: Macmillan.

Amlaner C. J., Jr. (1980) Telemetry antenna design. In C. J. Amlaner, Jr. and D. W. Macdonald (eds.), *A Handbook on Biotelemetry and Radio Tracking*, pp. 251-261. Oxford: Pergamon Press.

Anderka F. W. (1980) Modulators for miniature tracking transmitters. In C. J. Amlaner, Jr. and D. W. Macdonald (eds.), *A Handbook on Biotelemetry and Radio Tracking*, pp. 181-184. Oxford: Pergamon Press.

Boag D. A. (1972) Effects of radio packages on behaviour of red grouse. *J. Wildl. Mngmt.* 36, 511-518.

Brand C. J., Vowles R. H. and Keith L. B. (1975) Snowshoe hare mortality monitored by telemetry. *J. Wildl. Mngmt.* 39, 741-747.

Church K. E. (1980) Expanded radio tracking potential in wildlife investigations with the use of solar transmitters. In C. J. Amlaner, Jr. and D. W. Macdonald, *A Handbook on Biotelemetry and Radio Tracking*, pp. 247-250. Oxford: Pergamon Press.

Fryer T. B. and Sandler H. (1974) A review of implant telemetry systems. *Biotelemetry* 1, 351-374.

Harding P. J. R., Chute F. S. and Doell A. C. (1976) Increasing battery reliability for radio transmitters. *J. Wildl. Mngmt.* 40, 357-358.

Heezen K. L. and Tester J. R. (1967) Evaluation of radio-tracking by triangulation with special reference to deer movements. *J. Wildl. Mngmt.* 31, 124-141.

Ko W. H. (1980) Power sources for implant telemetry and stimulation systems. In C. J. Amlaner, Jr. and D. W. Macdonald (eds.), *A Handbook on Biotelemetry and Radio Tracking*, pp. 225-245. Oxford: Pergamon Press.

Lawson K., Kanwisher J. and Williams T. C. (1976) A UHF radio-telemetry system for wild animals. *J. Wildl. Mngmt.* 40, 360-362.

Lotimer J. S. (1980) A versatile coded wildlife transmitter. In C. J. Amlaner, Jr. and D. W. Macdonald (eds.), *A Handbook on Biotelemetry and Radio Tracking*, pp. 185-191. Oxford: Pergamon Press.

Macdonald D. W. (1978) Radio-tracking: some applications and limitations. In B. Stonehouse (ed.), *Animal Marking: Recognition Marking of Animals in Research*, pp. 192-204. London: Macmillan.

Macdonald D. W. and Amlaner C. J., Jr. (1980) A Practical guide to radio tracking. In C. J. Amlaner, Jr. and D. W. Macdonald (eds.), *A Handbook on Biotelemetry and Radio Tracking*, pp. 143-159. Oxford: Pergamon Press.

Mineau P. and Madison D. (1977) Radio tracking of *Peromyscus leucopus*. *Can. J. Zool.* 55, 465-468.

Patton D. R. (1973) Solar panels: an energy source for radio transmitters on wildlife. *J. Wildl. Mngmt.* 37, 236-238.

Reeve N. (1980) A simple and cheap radio tracking system for use on hedgehogs. In C. J. Amlaner, Jr. and D. W. Macdonald (eds.), *A Handbook on Biotelemetry and Radio Tracking*, pp. 169-173. Oxford: Pergamon Press.

Storm G. L. (1965) Movements and activities of foxes as determined by radio tracking. *J. Wildl. Mngmt.* 29, 1-13.

Taylor K. D. and Lloyd H. G. (1978) The design, construction and use of a radio-tracking system for some British mammals. *Mammal. Rev.* 8, 117-139.

Tester J. R. and Siniff D. B. (1965) Aspects of animal movement and home range data obtained by telemetry. *Trans. 30th N. Amer. Wildl. Nat. Res. Conf.* 379-392.

Thomas D. (1980) Plans for a lightweight inexpensive radio tracking transmiiter. In C. J. Amlaner, Jr. and D. W. Macdonald (eds.), *A Handbook on Biotelemetry and Radio Tracking*, pp. 175-179. Oxford: Pergamon Press.

A Simple and Cheap Radio Tracking System for Use on Hedgehogs

N. J. REEVE

Zoology Department, Royal Holloway College, Egham, Surrey, U.K.

Abstract — To the field zoologist radio tracking is undoubtedly a valuable technique, however many projects are denied it either because of the high cost of the equipment, or because of a lack of technical knowledge. A receiver was developed which can be built without detailed knowledge of electronics, using commercially available parts. A 'frequency converter' was used to receive the high frequency signal from the transmitter (104.7 MHz), and convert it to a much lower frequency (7.3 MHz) which could then be picked up by a conventional transistor shortwave radio (Sony ICF 5900W). An existing design of transmitter was used (Taylor and Lloyd, 1978) with minor modifications; i.e. the use of a battery chamber to allow easy battery replacement, and a tunable capacitor which allowed tuning for maximum efficiency after encapsulation in resin. The entire system costs less than £150 (US $300) (excluding labor), is reliable, and has proved its worth over the past year in tracking the nocturnal movements and daytime nesting habits of the hedgehog and may well be suitable for other medium sized animals.

INTRODUCTION

To the field zoologist radio tracking is undoubtedly a valuable technique, however many projects are denied it, either because of the high cost of the equipment, or because of a lack of technical knowledge. For a project studying the movements of wild hedgehogs (*Erinaceaus europeaus*) in the field it was necessary to find a cheap and simple radio tracking system which could be built with only a limited knowledge of electronics. The frequency allocated by the British Home Office was 104.7 MHz.

THE RECEIVER

While it is possible to obtain custom built receivers with excellent performance, these are expensive to buy, and costly to repair. With a low budget project, low cost and simplicity were all important and so the high performance and low bulk of a specially made unit had to be compromised.

The receiver consists of two main parts:

1. A frequency converter, commercially available to amateur radio, modified at little extra cost by the manufactuers (Microwave Modules Ltd., U.K.) to be most sensitive to an input of 104.7 MHz and to provide an output of 7.3 MHz, which is very suitable for input into an ordinary shortwave receiver, readily available commercially at low cost. The unit is compact, has a sensitivity of 1 μV, good signal specifity and image rejection and runs from a 12 V DC supply.

2. A Sony FM/AM multi-band receiver (ICF-5900W) was chosen because of its accuracy of tuning, built in beat frequency oscillation (BFO), battery check, headphone socket, signal strength meter, and dial illumination (very useful at night). It was also of low weight (2.2 kg) and good quality. The signal from the converter was fed into the shortwave antenna input.

When trying to locate a detached transmitter, an animal in undergrowth or in a nest, at close range I found it very useful to switch to FM, tune to 104.7 MHz and pick up the signal as breakthrough, while using the signal strength meter to locate the transmitter with great precision. Without this facility it would have been very difficult to find lost transmitters, or the near invisible day nests. An alternative method was to unplug the antenna and to use the much reduced signal on SW. But one had to be very close (1 or 2 m) for the signal to be audible at all.

A test transmitter of each frequency used (see later) was built into the receiver unit for tuning purposes; if a system is to be used without crystal controlled switched channels, these test transmitters are essential.

The advantages of a system such as this are low cost, easy assembly, and because the parts are unmodified in any way they are guaranteed and may be returned to the suppliers for repair.

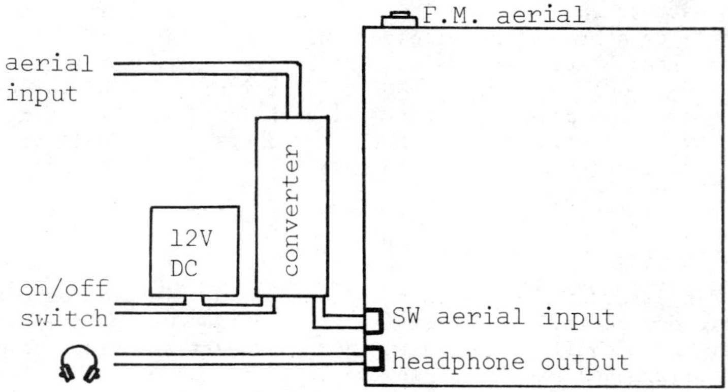

Fig. 1. Simplfied diagram of receiver.

THE TRANSMITTER

The Ashwell transmitter (F75) as described by Taylor and Lloyd (1978) was found to fulfil most requirements, capable of operating at 104.7 MHz, of relatively cheap and simple construction, providing a versatile, compact unit with a good performance which, owing to its integral iron dust core antenna, can be completely encapsulated in resin. This latter feature is of special importance in hedgehog work as the absence of a well defined neck or waist and their habit of rolling up in defense, makes them very unsuitable for harnesses or collars. Hence a loop antenna incorporated into a harness or collar is impractical, neither can a whip antenna be used as hedgehogs need to move freely in undergrowth and retire to a nest during the day.

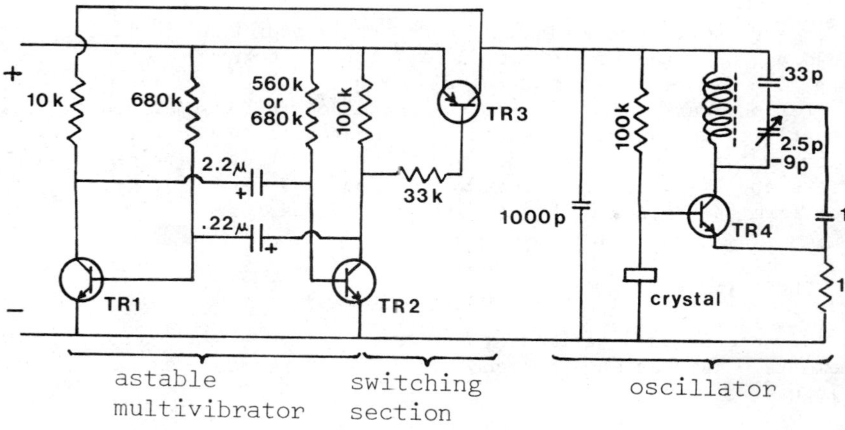

Fig. 2. Circuit diagram of Ashwell transmitter (after Taylor and Lloyd, 1978). N.B. Variable capacitor in oscillator circuit.

One minor alteration was made to the Ashwell design. C_5 (see Fig. 2) was replaced by a tunable capacitor (Verospeed 2.5-9 p*F*); with the adjusting screw projecting from the resin encapsulation, the oscillator circuit can easily be tuned after encapsulation and the screw later covered over with resin. So that batteries could be easily changed in the field a battery chamber was included in the design (see Fig. 3). The positive contact was made by a screw-in brass cap, eliminating the need for external wires. While in the field the cap was sealed over with an air drying silicone rubber solution. Two RM 13H mercury cells (in series) were used to minimize weight and size; however, these cells only supply 85 mAh and a maximum battery life of only around 14 days was achieved. Within the scope of this project this short life does not matter as animals were frequently examined anyway. The entire unit including batteries and resin weighs about 20-25 g and the external dimensions do not exceed 30 × 30 × 15 mm. The overall size could be reduced by the omission of the battery chamber and using a more compact layout, but ease of construction took priority. Restricted to a 50 KHz bandwidth only allowed two frequencies to be used, 104.680 MHz and 104.720 MHz. These two channels were quite distinct on the tuning dial of the receiver and with two pulse rates on each channel four transmitters could be used in the same tracking area at once.

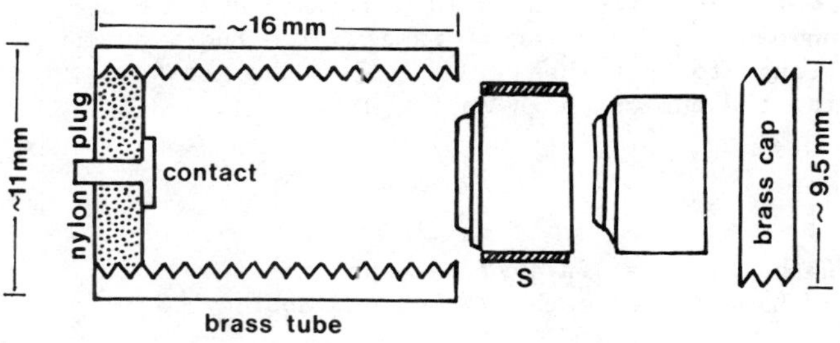

Fig. 3. Diagram of battery chamber showing RM 13H batteries. (N.B. Insulating sleeve (S) on rear battery). The brass tube was threaded using a 3/8" 26 T.P.I. tap (British Standard Brass Thread). A wire soldered to the brass tube provides a +ve contact.

ATTACHMENT

Initially the transmitter was glued to a carefully made elastic harness fitting over the forelegs and with the transmitter dorsal to the thorax. However, some discomfort and chafing did occur and the animals frequently escaped from the harness after a matter of hours. The harness was abandoned in favor of glueing the transmitter directly to the animal's spines. The hedgehogs showed little or no discomfort and were not impeded in rolling up. The transmitters remained attached for several weeks and few were lost; they were removed by carefully prizing off the transmitter from the spines or by clipping the spines away.

RECEIVING ANTENNA

An Adcock 'H' antenna as described by Taylor and Lloyd (1978) was used, with telescopic elements. This antenna has good directional properties and can be collapsed when not in use.

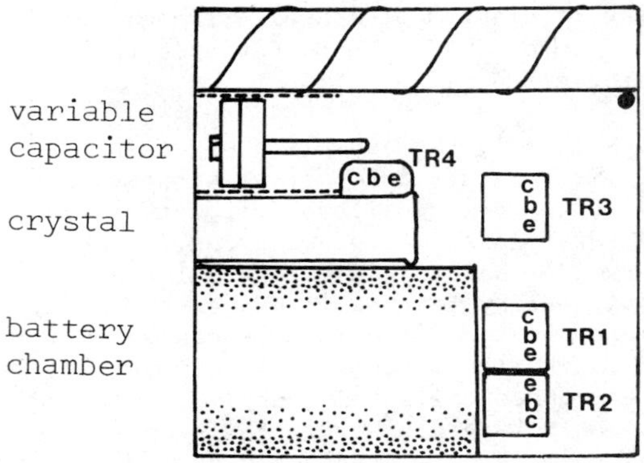

Fig. 4. Modified layout of transmitter. N.B. Variable capacitor was surrounded by a plastic sleeve filled with rubber solution to prevent seizure during potting.

PERFORMANCE

The above combination of equipment gave a maximum working range of 400 m when impaired by obstructions, wet vegetation and buildings. Because a hedgehog is so low to the ground even wet unmown grass can reduce range. Two individuals nested in burrows which reduced therange to less than 100 m. When the grass was wet range could be improved by holding the antenna high in the air.

COST

The total cost of the receiver in 1977 was approximately £120 (US $240). The transmitters cost under £5 (US $10) each, excluding labor.

METHODS IN THE FIELD

The tracking of hedgehogs was carried out on foot with the receiver carried on a shoulder strap and the antenna hand held. The animal was located by walking in the direction indicated by the antenna until the animal was either seen or heard.

Triangulation is impractical without help unless the animal is stationary, and even then is subject to a considerable degree of error. Once contacted the animal was then observed at a distance often a β light was temporarily taped to the transmitter on the animal's back to help to maintain visual contact.

RESULTS

Field work was carried out on a suburban golf course (area approximately 40 ha) during 1977 (without radios) and 1978 (with radios) and continues in 1979. Results so far have revealed that hedgehogs may travel further and use larger areas than previously thought. Using a convex polygon method to estimate home range (for 8 males and 8 females — adults), there appears to be a marked difference in home range size between the sexes. The mean home range size for males is 20.7 ha and for females 7.9 ha. A 't' test shows a significant difference ($0.01 > P > 0.002$).

These figures are derived from all fixes over one year for each animal. Preliminary results suggest that males may expand their ranges when sexually active, but this period may last for most of the active year and is poorly defined. Males radio tracked at this time often follow routes of 2.5 km or more in a night, at an average speed of around 350 m hr^{-1}. These routes are measured without the inclusion of minor meanders made during foraging and courting. Locating an animal more frequently increases the accuracy of distance (and speed) measurements, and an attempt is being made to determine the optimum interval between fixes. Hedgehog home ranges overlap extensively and often completely, irrespective of sex. There is a conspicuous absence of territorial boundaries and associated behaviors.

The use and distribution of day nests shows great variation between individuals. Hedgehogs may use several nests, some only once and others many times (often returned to after many days elsewhere). Day nests are usually used by one individual only but I have found several cases where other hedgehogs used them.

Acknowledgements — This work has been carried out as part of a three year Ph.D. project at Royal Holloway College, Englefield Green, Surrey. I gratefully acknowledge the help of Dr. P. A. Morris and the invaluable assistance of Mr. G. Rentmore in the development of the radio tracking system.

REFERENCE

Taylor K. D. and Lloyd H. G. (1978) The design, construction and use of a radio-tracking system for some British mammals. *Mammal Rev.* 8, 117-141.

Plans for a Lightweight Inexpensive Radio Transmitter

D. W. THOMAS

Department of Zoology, University of Aberdeen, Aberdeen AB9 2TN, U.K.

Abstract — This paper presents details for the construction of a small (1.0 g), inexpensive (US $12), crystal controlled radio transmitter appropriate for use in tracking small non-aquatic vertebrates. The transmitter will function at any selected frequency in the 140 to 180 MHz range, will permit tracking at distances of up to one kilometer, and has a life expectancy of 13 days using a 0.3 g battery.

INTRODUCTION

Despite the increasing use of radio telemetry in ecological studies, few 'do it yourself' plans have been made available to the electronically inexperienced biologist. In this paper I present the necessary details for the construction of an inexpensive (US $12), 1.0 g, crystal controlled radio transmitter. The transmitter is designed to function at any selected frequency in the 140 to 180 MHz range and to permit tracking at distances of up to one kilometer when equipped with a 10 cm antenna, although this depends on the transmitter height, topography, and obstacle density. The life expectancy with a 0.3 g Mallory RM212 battery is 13 days.

COMPONENTS, SUPPLIERS*, AND EQUIPMENT

Capacitors: 2 × 0.047 mfd (radio leads), 1 × 0.1 mfd (axial leads), and 1 × 22 pfd. The 0.047 mfd and 0.1 mfd capacitors are dry tantalum models available from Way-Com Ltd., Bracknell, Britain (part W473 and W104) and Components Inc., Maine, U.S.A. (part MMU-020-473-R40 and MMU-020-104-A40). The 22 pfd value is a standard 6-10 V ceramic capacitor made by a number of companies.

Resistors: 1 each of 10 kΩ, 82 kΩ, 120 kΩ, 820 kΩ, and 18 MΩ values. The resistors are 1/16 W Fuji Sangyo River Ohm MR-004E models available through Hakuto International (U.K.) Ltd., Essex, U.K. At last contact Hakuto did not stock the 18 MΩ value and I substituted a standard 1/8 W carbon resistor.

Transistors: Four transistors of two models (3x BCW 66 and 1x BFS 20) are required and can be purchased from Siemens Ltd., Ottawa, Canada, as parts Q62702-C460 and Q62702-F350, specifying the SOT 23 case. The BCW 66 has surface designation marks *EF* and the BFS 20 has *NA*.

*The addresses of suppliers are appended.

Crystal: The transmitter uses a third overtone crystal to generate the tenth harmonic. To calculate the crystal frequency, divide the desired output frequency by 10 and multiply by 3. When ordering the crystal specify: HC-45/U-2 microminiature can, third overtone mode, series resonant, and no particular temperature characteristics. Suppliers in Britain, the U.S., and Canada are: Interface Quartz Devices Ltd., Crewkerne, U.K.; International Crystal Manufacturing Co., Oklahoma, U.S.A.; and Croven Ltd., Whitby, Ontario, Canada.

Magnet wire: Use No. 38 or 40 *self-fluxing* magnet wire for forming both coils and for connecting the components.

PREPARATION OF COMPONENTS

To save weight and volume the casings of the River Ohm resistors and the 22 pfd capacitor may be trimmed. Properly done, this does not affect their functioning.

With a pair of flat-nosed pliers grasp the end of one of the River Ohm resistors and crush. The tip of the acrylic case should shear off, leaving a flat face at right angles to the leads. Now sand one of the remaining two broad faces flat and call this latter face *A* for convenience. Repeat this procedure on all the River Ohm resistors.

Take the 22 pfd capacitor and, with a pair of heavy forceps, bend each lead 90° to the side, so that it projects straight out from the body. Chip off the ceramic coating from the base of each lead and clip them to about one millimeter length. Gently sand each of the broad faces of the capacitor flat to reduce its thickness, ensuring that you do not expose the metal capacitance plates.

CONSTRUCTION

Assembling the transmitter requires no particular electronic talent, although you should make sure that you can recognize a 'good' solder joint and that you are thoroughly familiar with soldering the self-fluxing magnet wire before you begin.

1. Fix the crystal under a binocular microscope with the frequency designation numbers up and the leads to your right. Make a note of the frequency since this will be covered up and later become irretrievable. A small vice can be made from an aquarium air hose clamp to hold the crystal secure. Following the format and number sequence indicated in Fig. 1, mount the components on the two crystal faces using very small amounts of cyanoacrylate glue (e.g. Krazy Glue or Powa Bond). All the River Ohm resistors should be mounted with face *A* to the crystal and the leads to the right. Transistors 3, 4, and 10 are BCW 66 and transistor 6 is the BFS 20. All the transistors should be mounted with the lettered face down. When mounting transistor 6 and capacitor 7 make sure that there is enough clearance around the capacitor to permit soldering of the transistor leads.

2. When all the components are in position, begin making the connections with resistor 8. Bend the upper lead of the resistor tightly around the body so that it meets the collector of transistor 10 and solder this contact.

3. Bend the closest leads of both resistor 9 and capacitor 11 to meet the base of transistor 10. Cut these to the appropriate length and solder all three leads together. Clip the remaining lead of capacitor 11 to about 1 mm and solder on a 2 cm length of magnet wire. Turn the crystal over and run the magnet wire lead to the base of transistor 4. Loop the wire around the lead, but do not solder it yet.

4. Bend the upper lead of resistor 1 tightly over the body to meet the collector of

transistor 3 and now solder the four leads (base of transistor 4, collector of transistor 3, lead from capacitor 11, and the lead from resistor 1) together.

5. Bend the lower lead of resistor 2 and the closest lead of capacitor 7 to meet the base of transistor 3 and solder this contact.

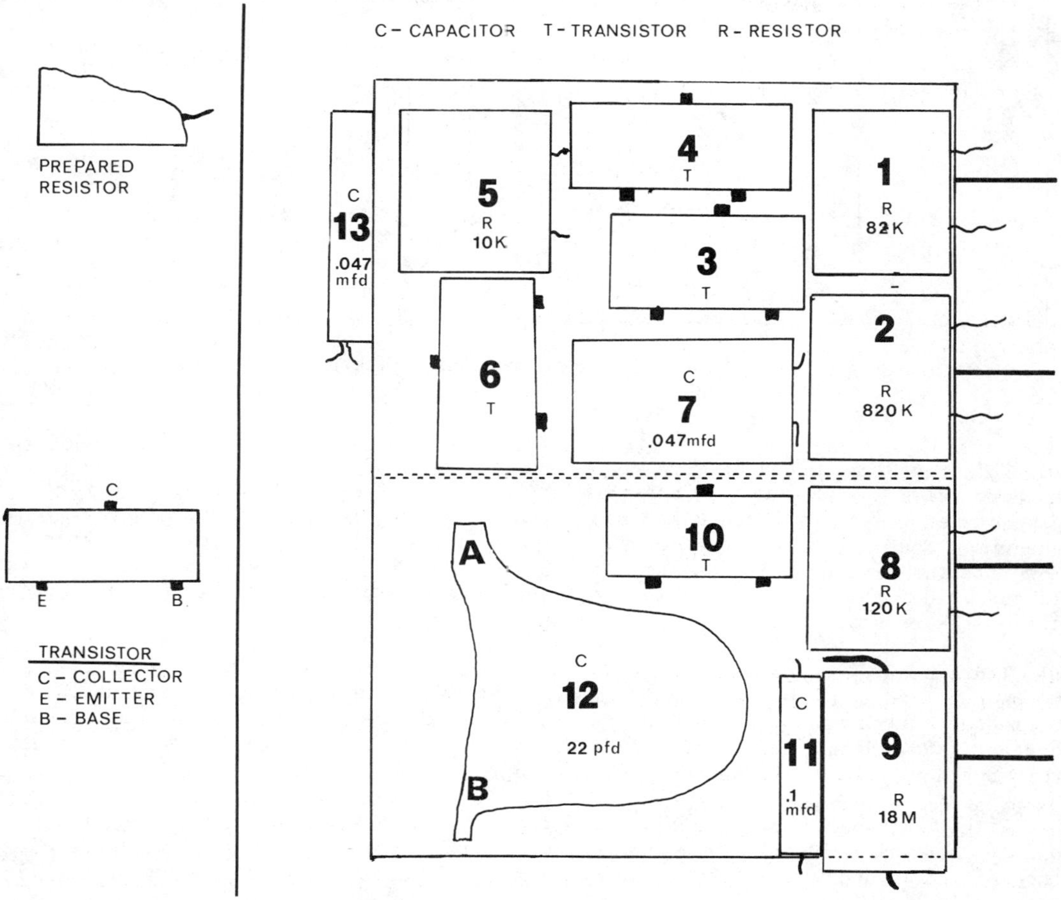

Fig. 1. Diagram showing the placement sequence and positioning of the components on both sides of the crystal.

6. Solder first the upper lead of resistor 5 to the emitter of transistor 4 and then the lower lead of this resistor to the base of transistor 6.

7. Clip both leads of capacitor 13 to 1 mm and solder on two 2 cm lengths of magnet wire. Run the lead marked with the red dot to the emitter of transistor 6, loop it around, and solder. Run the remaining lead from capacitor 13 to the collector of transistor 4 and solder.

8. Clip the remaining lead of capacitor 7 to 1 mm and solder on a 2 cm piece of magnet wire. Run this lead to the opposite side of the crystal and solder it to the collector of transistor 10.

9. Now take a 10 cm piece of wire and solder one end to the emitter of transistor 3. Stretch the wire across to the emitter of transistor 6, loop it around the lead and solder. Run the remaining length straight to the emitter of transistor 10, loop, and solder. The remaining wire will be the *negative* battery lead.

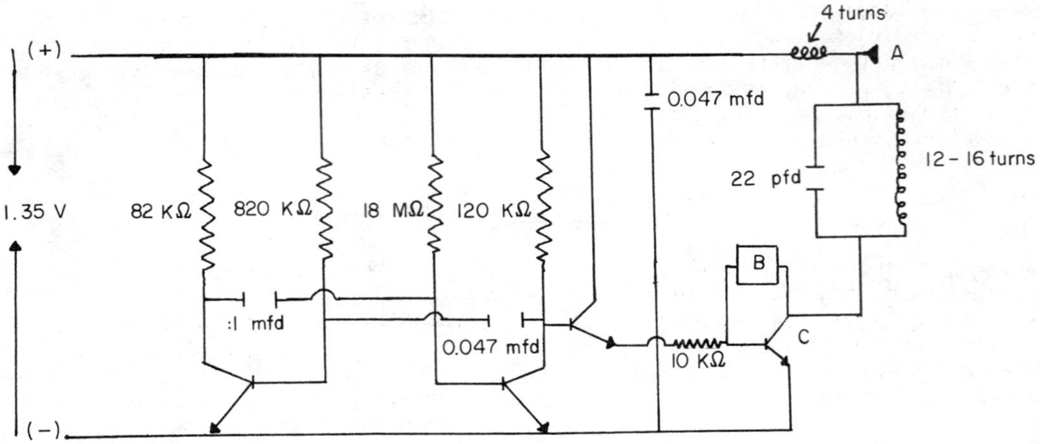

Fig. 2. The circuit plan and component values for the transmitter. A — antenna; B — crystal; C — BFS 20 transistor (all others BCW 66).

10. Solder a 3 cm length of wire to the base of each of the crystal leads and cut the crystal leads as short as possible. Fit the two free wires through the small gap between resistors 8 and 9 and run the wires around lead *A* of capacitor 12. Pull the wires snug and solder one to the base and one to the collector of transistor 6. Now lift the wire coming from the collector of transistor 6 onto lead *A* of capacitor 12 and solder this contact. Make sure that the leads of capacitor 12 do not short out on the crystal case.

11. There should now be only one lead from each of resistors 1, 2, 8, and 9 free. Clip each lead to 1 mm. Take a 10 cm piece of magnet wire and, beginning at the collector of transistor 4, connect the collector and resistor leads, soldering each contact. Now thread the remaining length of the wire through the gap between resistors 8 and 9. This will be the *positive* battery lead.

12. With the magnet wire wind a 14 turn coil on a 3 mm diameter rod. Slide the coil off the rod and hold the turns together with an alligator clip. Cut the leads of the coil short and solder them to leads *A* and *B* of capacitor 12 and remove the alligator clip.

13. Now wind a 4 turn coil on the same diameter rod and glue the turns tightly together with the cyanoacrylate adhesive. Solder one lead of the coil to lead *B* of capacitor 12 and the other to the collector of transistor 4. Press the coil against the side of the crystal and glue in this position.

14. After checking the transmitter to ensure there are no shorts onto the crystal case, connect the battery leads to a 1.2-1.5 V power source. Tune a receiver to the appropriate frequency and gently begin to squeeze the turns of the 14 turn coil together. At some point the transmitter should start to emit 10 ms pulses at the rate of 1 s^{-1}. Ideally the loudest signal should come when the coils are fully compressed. If this is not the strongest signal experiment with 12-16 turn coils. Now glue the turns together and lay the coil on top of capacitor 12.

15. Solder a 10 cm piece of fine guitar string to lead *B* of capacitor 12 as the antenna. Orient the wire so that it runs the length of the transmitter and glue it in this position for rigidity.

16. A Mallory RM212 battery, preferably with factory welded tabs, fits snuggly between the leads of capacitor 12, sitting on capacitor 13. Glue it in this position and solder one lead, leaving the other free until the transmitter is to be used. Pot the transmitter by dripping *small* amounts of cyanoacrylate over the components first and then by spreading a very small amount of beeswax over the components with a soldering iron. If solder, glue and wax has been kept to a minimum the final weight should be 1.0 g.

TROUBLE SHOOTING

Reluctant transmitters can be checked with a digital multi-meter by first placing the meter in series with one of the battery leads. On the μA setting the transmitter should draw 20 μA steadily, with pulses to 30 μA. If it does not draw this or pulse then you have a misplaced lead or a short. If it does draw the proper current, but does not transmit, connect the meter function (Ω × 10) across the two leads of capacitor 12. If the dial shows regular pulses, then the problem lies in the coil.

MODIFICATIONS

The pulse length and the inter-pulse interval are governed by resistors 2 and 9 respectively. Increasing the value of either increases the length of the corresponding parameter. By substituting an appropriate thermistor the pulse repetition rate can readily be made temperature dependent.

SUPPLIER'S ADDRESSES

Components Inc., 313 Elm St, Biddeford, Maine 04005, U.S.A.
Croven Ltd., 500 Beech St., Whitby, Ontario L1N 5J5, Canada
Hakuto International (U.K.) Ltd., Hakuto House, 557-563 Rayleigh Road, Leigh-on-Sea, Essex, U.K.
Interface Quartz Devices Ltd., 29 Market St., Crewkerne, Somerset TA18 7JU, U.K.
International Crystal Manufacturing Co., 10 North Lee St., Oklahoma City, Oklahoma 73102, U.S.A.
Siemens Ltd., Siemens House, Windmill Road, Sunbury-on-Thames, Middlesex TW16 7HS, U.K.
Siemens Canada Ltd., 1673 Carling Avenue, Ottawa, Ontario K2A 1C4, Canada
WayCom Ltd., Wokingham Road, Bracknell, Berkshire RG12 1ND, U.K.

Modulators for Miniature Tracking Transmitters

F. W. ANDERKA

Canadian Wildlife Service, National Wildlife Research Centre, Ottawa, Canada

Abstract — Two types of simple and efficient pulsing circuits suitable for a variety of transmitters are described. One is a fixed rate design with a very stable rate, operating at microwatt levels, unaffected by voltage or load variations. The second design shares many features of the first, except that the pulse rate can be fixed, gated, or varied by a number of different sensors. Assembly is simple, requiring few adjustments or selection of parts. Either circuit will operate from a single mercury cell and can be constructed to occupy less than 0.75 cm^3 without resorting to wirebonding or other exotic construction techniques.

INTRODUCTION

The pulsing circuits described provide a simple yet efficient and versatile means of modulating wildlife tracking transmitters. The fixed rate design derives its stability from a subminiature low frequency quartz crystal and a 16-stage binary divider integrated circuit. The variable rate design employs an astable multivibrator IC that can be reset, gated or varied by different sensors. Over a -40 to +40°C range the change in pulse rate is less than 0.1 percent for the fixed design and approximately 5 percent for the variable design. Either circuit will operate from a single mercury oxide cell with a current consumption of 2 to 4 μA.

FIXED RATE PULSE GENERATOR

The fixed rate pulse generator (Figs. 1 and 3) uses a subminiature 32.7 KHz watch crystal, a flat pack IC (MC14450L) and a maximum of five other components. The IC contains an oscillator, a 16-stage binary divider and two output buffers. Using a 32.7 KHz crystal, positive going, 31 ms pulses at 0.5 Hz appear at pins 4 and 5, 180° out of phase. Diodes D1 and D2 combine the alternate pulses producing a 1.0 Hz signal with a 3.1 percent duty cycle capable of driving a 10 kΩ load. If a 0.5 Hz signal is required, R2 is connected directly to either pin 4 or 5. The pulse rate and width can be modified by choosing the appropriate crystal frequency (Motorola, 1976). For example, a 65.5 KHz crystal can provide a 1.0 or 2.0 Hz signal with a pulse width of 15.6 ms. The value of R2 varies from 10 to 150 kΩ, depending on the peak currents desired (10 to 2 mA).

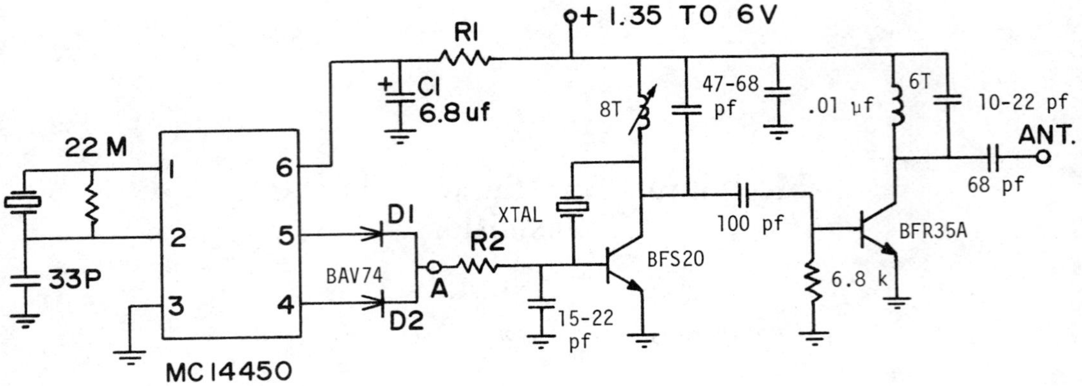

Fig. 1. Diagram of fixed rate pulse generator and transmitter. Variable values will cover frequency range of 145-174 MHz. Crystal (XTAL) is a series resonance type (3rd overtone). Semiconductors are Siemens Manufactured. 8T coil is wound on, and tuned by a Cambion slug No. 515-3218-06-21 (carbonyl SF).

If the transmitter is powered by a single mercury, alkaline or silver oxide cell, R1 and C1 can be deleted. If powered by multiple cells or lithium batteries, R1 must be used to limit the voltage at pin 6 to 1.5 V. The value of R1 is calculated by assuming an average current consumption of 4 µA with the 32.7 KHz crystal and 5 µA with the 65.5 KHz crystal.

Since this circuit is intended to drive the base of the oscillator transistor, the output is not suitable for applications where the entire power supply to the transmitter must be switched. For the latter application the MC14450 (Motorola) can be replaced with an ICM7047 (Intersil). The two ICs are similar in operation except that the output buffers in the Intersil device are normally in the high state, with the signal appearing as negative going pulses 15.6 ms wide (with a 32.7 KHz crystal).

VARIABLE RATE PULSE GENERATOR

The variable rate pulse generator (Figs. 2 and 4) uses a CMOS multivibrator IC (CD4047AK) and five other components. The astable multivibrator function is implemented by connecting the appropriate control input to ground or the supply (RCA, 1975). The pulse rate is determined by R3 and C2, the pulse width by R5 and the pulse direction by D3. As shown in Fig. 2, the orientation of D3 produces a positive going pulse at 'A' suitable for base drive of the oscillator transistor. Reversing D3 will produce a negative going pulse suitable for driving a series pass transistor. The following range of component values will produce pulse rates of 0.5 to 3 Hz and pulse width of 15 to 35 ms: C2, 0.22-0.68 µf; R3, 680 kΩ-2.2 M; R4, 680 kΩ-4.7 M; R5, 18 kΩ-39 kΩ.

The pulse generator can be used to measure temperature with the addition of a thermistor between 'B' and 'C' (Fig. 2). Similarly, light levels can be measured using a photo resistor. Since few studies are conducted after sunset, a photo resistor can be used to reduce the pulse rate to 0.2 Hz or less at night. The maximum and minimum pulse rates are set by R4 and R3 respectively. The average current consumption of the pulse generator is 2 µA at 1.35 V, rising to 10 µA at 6 V.

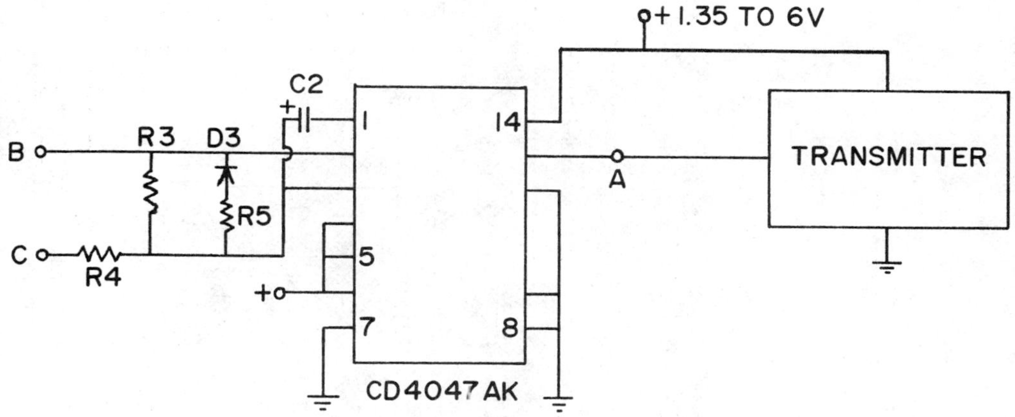

Fig. 2. Diagram of variable rate pulse generator.

CONCLUSION

The timing circuits and transmitters are constructed on double sided epoxy (G-10) board using conventional and chip components. The completed circuit is mounted on a four pin crystal can base and solder sealed in a crystal can (Figs. 3 and 4). Final tuning of the transmitter is performed through a hole in the top of the can that is later solder sealed. Although relatively bulky, this method of encapsulation results in a hermetically sealed package weighing 5 g.

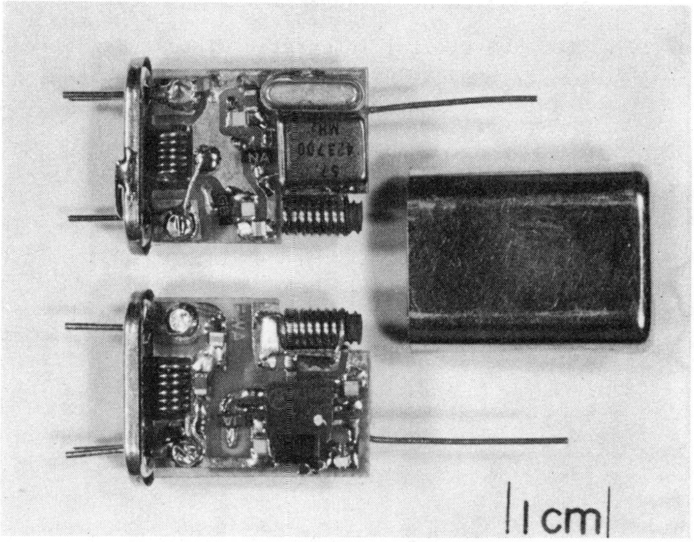

Fig. 3. Top and bottom view of a fixed rate transmitter and case.

The two pulsing systems are being used successfully on various large mammals, i.e. coyote, wolf, elk, caribou, bison and grizzly. Early field results will be available in the spring of 1980 when the projected three year battery supply will be depleted on some of the transmitters now in service.

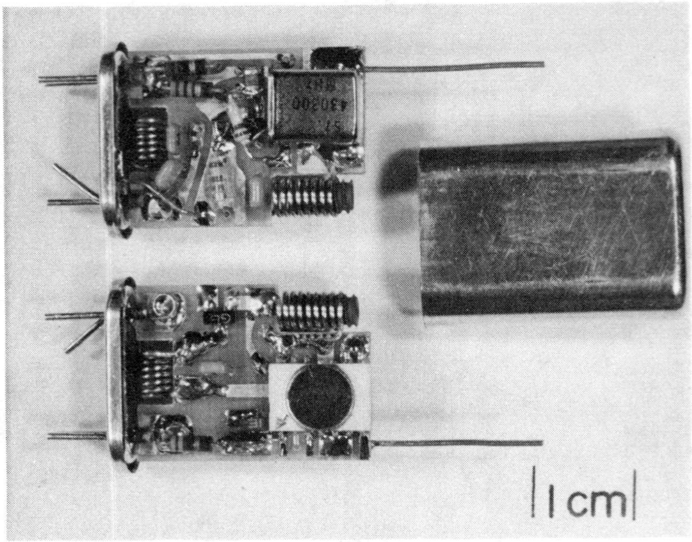

Fig. 4. Top and bottom view of variable rate transmitter and case.

REFERENCES

Motorola Inc. (1976) *Semiconductor Data Library*, CMOS Integrated Circuits, Volume 5/ Series B.
RCA Corporation (1974) COS/MOS Digital Integrated Circuit Manual.
Intersil Inc. (1977) ICM7047 Complementary MOS Quartz Watch Circuit.

A Versatile Coded Wildlife Transmitter

J. S. LOTIMER

Ministry of Natural Resources, Box 50, Maple, Ontario, Canada

Abstract — A wildlife radio tag capable of transmitting physical parameters and digital coding was miniaturized and features increased over earlier circuits. Logic gates required to provide a monostable circuit were halved, and the excess gates available were used to pulse code the transmitter, which in addition to frequency, simplifies tracking a large number of animals. Voltage stabilization was achieved from 2.5 V to greater than 5.0 V. Miniaturization was accomplished through the use of ceramic flatpack integrated circuits and miniature discrete components, which allows ease of construction over hybrid techniques. The circuit allowed the options of activity, temperature and mortality signalling. The completed circuit board weighing 6.5 g was hermetically sealed inside a crystal can 3.0 × 3.3 × 1.0 cm.

INTRODUCTION

To field biologists involved in radio tracking studies, basic transmitter requirements include low power consumption (size and weight), simplicity of identification of animals, and reliability (life and range). Low power integrated circuits (CMOS) have been used for several years to pulse radio frequency transmitters for monitoring wildlife. Present designs include activity, mortality and temperature monitoring (Kolz, Corner and Johnson, 1973; Standora, 1977). This paper describes a transmitter which, in addition to these options, is capable of pulse interval modulation used for identification of individual animals. Reliability and ruggedness are achieved through use of a printed circuit board and hermetic sealing.

CIRCUIT DESIGN

Appropriate placement of a diode resistor network halves the required number of logic gates used in previous designs for pulse generation. Additional gates provided on the integrated chip are then used for digital coding (Fig. 1). This allows multiple transmitters per frequency and we have found it aids greatly in locating and identifying individual and far ranging animals. In addition, the use of a second integrated circuit allows the transmission of activity, mortality, event and temperature data. All options are available on a printed circuit board 2.5 × 2.5 × 0.08 cm which is contained in a hermetic can 3.0 × 3.3 × 1.0 cm. This size was accomplished through the use of miniature components and ceramic integrated circuits. Current consumption of the logic circuits is always less than 6 μA and typically less than 3 μA. This represents less than 3 percent of the transmitter drain and allows operation of the transmitter for at least 2 years with a C size lithium battery. Packaged weight is less than 20 g (without battery).

PULSE GENERATION AND CODING

A monostable circuit using logic gates U_1-A and U_1-B forms the pulse generator (Fig. 2A). Pulse repetition frequency (PRF) is controlled by the charging of C_2 through R_2 while pulse width (PW) is controlled by the discharging of C_2 through the parallel combination of R_1 and R_2. For normal operations the PRF is set between 60 and 90 pulses per minute (PPM) and the PW to 40 ms. The addition of C_1 voltage stabilizes the PRF and PW from 2.5 to greater than 5.0 V. Variations in PRF are one way to identify individual animals operating on the same frequency.

Pulse code modulation (PCM) is achieved through gates U_2-A and U_2-B (Fig. 2B). With only R_{11} and C_7 present, the circuit operates as an astable multivibrator. This circuit is turned on with a positive pulse from U_1-B (RCA, 1972). Voltage stabilization is achieved with the addition of R_9. When the PW from the pulse generator is increased by increasing R_1, the astable circuit is enabled for two or more cycles. Specific values of R_{11} and C_7 will then double- or triple-pulse code the transmitter independent of PRF (Fig. 3). In the example (Fig. 3), the transmitter is activated twice per cycle and the time averaged current drain is doubled. Compensation is achieved by lowering the PRF to less than 50 PPM giving equivalent average drain to a normal transmitter of 100 PPM. It is useful to use a PRF of 20 and have four or more pulses per cycle, easily identifying particular animals and allowing multiple

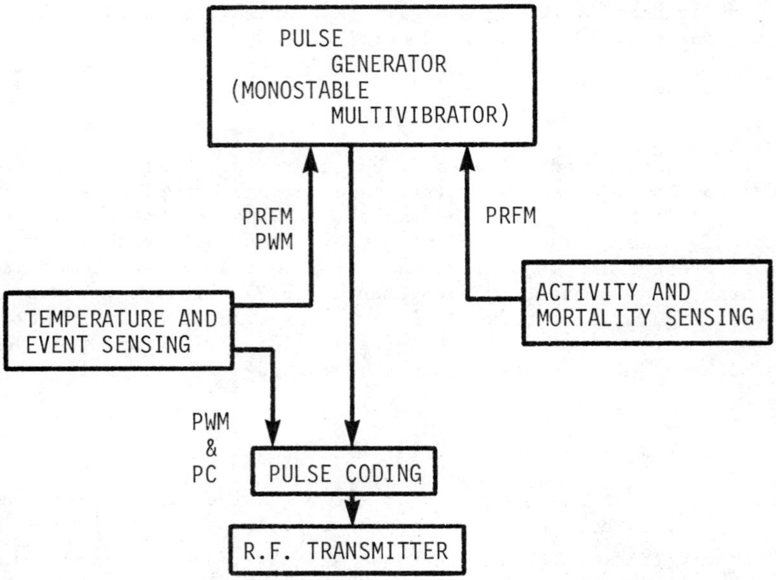

PRFM - pulse repetition frequency modulation

PWM - pulse width modulation

PC - pulse coding

Fig. 1. Block diagram of pulse coded transmitter.

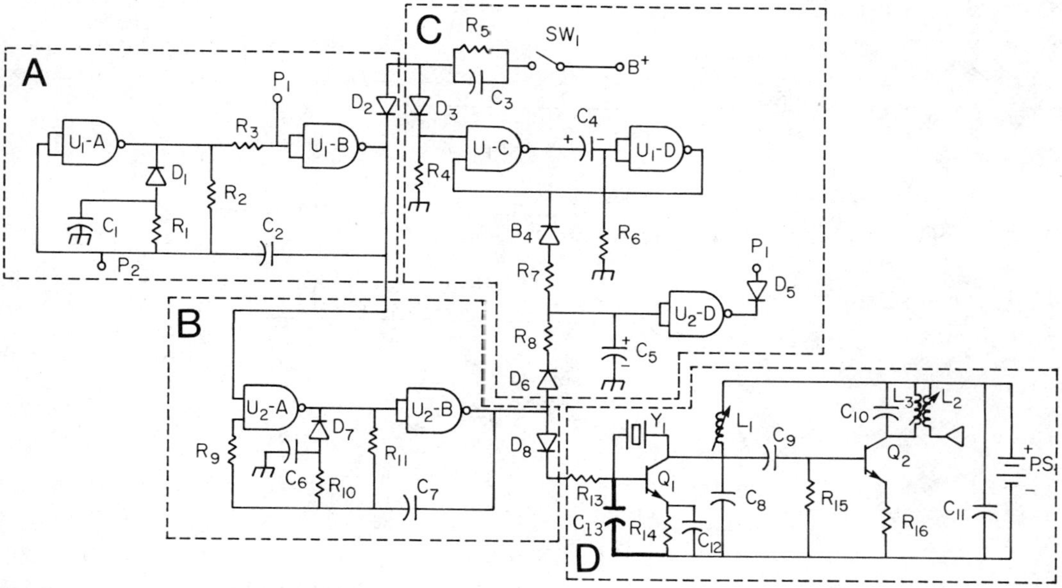

Fig. 2. Circuit schematic of pulse coded transmitter.

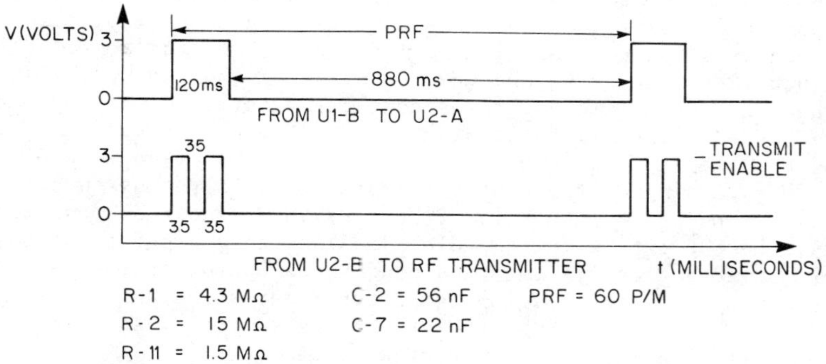

Fig. 3. Example of double pulse coding.

transmitters per frequency. Decoding is done by the operator or in the receiver in the case of more than four pulses per cycle.

Removal of R_9 and inclusion of D_7, R_{10} and C_6 forms a monostable multivibrator with U_2-A and U_2-B (Fig. 2B). This circuit similarly codes the transmitter but in addition, allows control of off-time independent of PW. Pulse interval modulation identifies specific animals by the time period between consecutive pulses as set by R_{11} (Fig. 4). Demodulation can be accomplished in the receiver and read digitally by the operator (Cupal, 1977). Normal single pulse transmission is done with appropriate values of R_{10}, R_{11} and C_7 (Table 1).

ACTIVITY AND MORTALITY MONITORING

Activity and mortality are monitored with the use of three additional logic gates U_1-C, U_1-D and U_2-D (Fig. 2C). Operation is described by Kolz *et al.* (1973) but

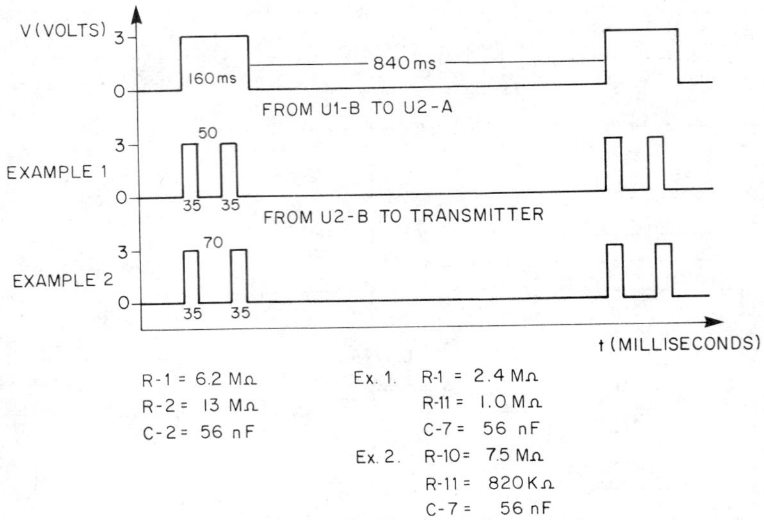

Fig. 4. Example of pulse coding with variable window.

several advantages are realized. Activity is indicated by the motion sensitive mercury switch SW_1 which triggers an extra pulse in the monostable circuit upon movement of the animal. Sensitivity can be set by varying the R_5, C_3 time constant. Mortality is measured by the cessation of movement of SW_1 for a predetermined time (usually 2-4 hours) as set by R_8 and C_5. C_5 is being charged by the square wave pulses from U_2-B and each time SW_1 makes contact it activates the monostable circuit formed by U_1-C and U_1-D, causing the discharge of C_5. The duration of discharge is set by R_6 and C_4. Lack of movement will bias the output of gate U_2-D to ground. This enables free running of the coding circuit (Fig. 2B), thus changing the PRF to that set by C_7 and R_{11}. When this circuit is in a monostable configuration (R_{10} present), the PRF can be for example, doubled and a significant power saving realized over previous circuits in which continuous operation indicated death of the animal. This saving is extremely important for hibernating animals (in our case black bears) and helps sustain the life of the transmitter. As soon as SW_1 is activated the transmitter returns to its original state.

TEMPERATURE MONITORING

Mortality can also be monitored by postmortem temperature drop with the use of a thermistor (Fig. 5). U_2-C is originally biased, input high, output low and as the body temperature drops, thermistor resistance increases, switching the gate output high, providing a second charge path for C_2 through R_{12}. This changes the PRF, as determined by C_2 and the parallel combination of R_2 and R_{12}, and again a power saving is realized through the use of PRF modulation. Comparative temperature monitoring (e.g. ambient outside collar temperature versus collar temperature next to animal) can be achieved by using two thermistors in the biasing network. This may be more advantageous in eliminating false mortality indications due to colder climates and would have to be calibrated for each unit. Other events such as an indication of light intensity can be monitored through similar biasing networks. For example, if a light sensitive resistor were used to replace the thermistor, the transmitter would indicate, through PRF modulation, whether the animal was in the light or dark, or it could be used to save battery energy when the animal was not

Table 1.
Component list for pulse coded radio tracking transmitter

Component	Description	
$R\dagger_{1}, R_{3}, R_{4}, R\dagger_{10}$	Resistor*	1 MΩ
$R_{5}, R_{6}, R\dagger_{11}$	Resistor	4.3 MΩ
$R\dagger_{2}$	Resistor	13 MΩ
R_{7}	Resistor	8.2 KΩ
$R\dagger_{8}$	Resistor	3.0 MΩ
R_{9}	Resistor	7.5 MΩ
R_{12}	Resistor	10 MΩ
R_{13}	Resistor	47 KΩ
R_{14}	Resistor	1.0 KΩ
$R\dagger_{15}$	Resistor	15 KΩ
$R\dagger_{16}$	Resistor	220 Ω
C_{1}, C_{6}	Capacitor, X7R monolithic ceramic	1.0 nF
$C\dagger_{2}, C\dagger_{7}$	Capacitor, X7R monolithic ceramic	53 nF
C_{3}, C_{11}, C_{12}	Capacitor, X7R monolithic ceramic	22 nF
C_{4}	Capacitor, Ultra Miniature Tantalum (Sprague 182 series)**	0.47 µF
C_{5}	Capacitor, mini tantalum (Components Inc. MMJ series)**	100 µF
C_{8}, C_{9}, C_{13}	Capacitor, NPO monolithic ceramic	15 pF
C_{10}	Capacitor, NPO monolithic ceramic	6.8 pF
L_{1}	Tunable Inductor (Piconics ST series)**	0.55-0.68 µH
L_{2}	3 turns AWG 26 magnet wire on 3.5 mm diameter coil	
L_{3}	5 turns AWG 32 magnet wire on 3.5 mm diameter coil	
Tuning core	Ferrite (Micrometals Inc. mix No. 6)**	
Y_{1}	Quartz Crystal third overtone 50 MHz (C.R. Snelgrove Ltd.)**	
D_{1}-D_{8}	Diode MMD70 (Motorola Inc.)**	
Q_{1}	Transistor, MMT74 (Motorola Inc.)**	
Q_{2}	Transistor, MMT72 (Motorola Inc.)**	
SW_{1}	Micro-mini Mercury Switch (Gordos)** nc.)**	
U_{1}, U_{2}	Integrated Circuits CD4011 UBK (RCA)**	
PS	'C' size lithium battery, 3.4 V (Tadiran)**	

*All resistors 1/8 W.
**Use of trade names does not imply endorsement by the Government of Ontario.
†Component values change for specific circuits.

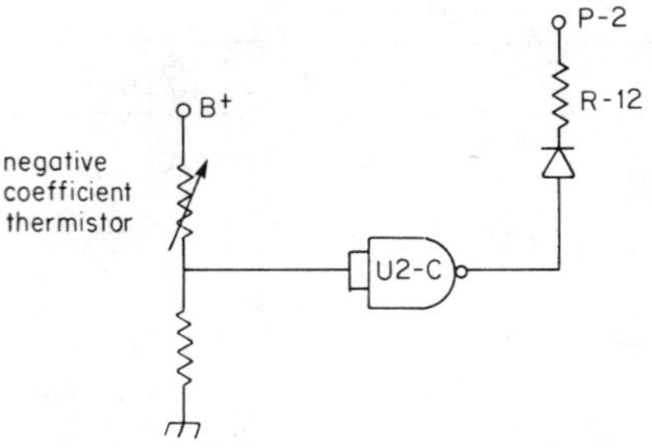

Fig. 5. Circuit schematic of temperature monitoring.

being tracked. Ambient temperatures (or other events) can also be monitored linearly by calibrating either PW modulation or PRF modulation, or both, when sensors are placed in the timing networks (Standora, 1977).

RADIO FREQUENCY TRANSMITTER

The RF transmitter is similar to previous designs (Cochran and Lord, 1963; Cochran, 1967). It consists of a colpitts oscillator at 50 MHz and a tripler amplifier giving the transmission frequency of approximately 150 MHz (Fig. 2D).

Prior to the design of digital pulsing, the transmitter was timed using an R-C circuit in the oscillator base built from discrete components. To keep PRF and PW consistent, the active components and crystals had to be preselected. Temperature changes caused drift in these variables and in frequency. With digital logic pulsing, these problems are significantly reduced, the active components need no longer be selected and the transmitter remains stable from -30°C to +40°C.

We control the RF transmitter on-time to between 30 and 40 ms to ensure good signal quality at the receiver. A longer on-time appears to be a misuse of energy. Current consumption can be controlled with the emitter resistors or with the base resistor R_{15}. We generally keep our peak pulse current drain to less than 4 mA for battery voltages of 3.0 to 3.4 V. This optimizes radiated power versus input power. Slightly more radiated power can be realized from double the input current for specialty purposes. We have used both whip and loop antennas depending upon the species and found radiation from both to be satisfactory. When a 30-40 cm antenna is used, L_3 and L_2 are wound in a bifilar configuration in the ratio of 5:3 and tuned with a ferrite slug (Table 1). Power output is in the range of 2 mW and is dependent upon antenna efficiency. Typical ground-to-ground range with a 40 cm whip is 3-5 km, depending upon terrain.

DISCUSSION

The use of CMOS integrated circuits has greatly improved reliability and stability of our wildlife radio tracking transmitters. It has made construction easier and component selection virtually unnecessary. The miniature ceramic IC's while being considerably more expensive are made to stringent specification and allow great reduction in size. Our collars are constructed such that the batteries can easily be replaced. We expect reuse of the transmitter for many years and this, in addition to size convenience, justifies the additional expense of reliable miniature components. The circuit board is double sided with plated-through holes and electrodeposited with tin lead (Fig. 6). Construction is intricate but size and reliability are well appreciated.

Fig. 6. Printed circuit board layout and final case.

Acknowledgements — I wish to express my thanks to Ching Ang for his assistance on the RF transmitter and Paul Singh for his work on the circuit board designs. This is Ontario Ministry of Natural Resources, Wildlife Br. Contribution No. 79-7.

REFERENCES

Cochran W. W. and Lord R. D. (1963) A radio-tracking system for wild animals. *J. Wildl. Mngmt.* 27, 9-24.

Cochran W. W. (1967) 145-160 MHz beacon (tag) transmitter for small animals. A.I.B.S. BIAC Information Module M15. 12 pp.

Cupal J. J. (1977) A transmitter identifier for use with wildlife biotelemetry. *Proc. First Int. Conf. Wildl. Biotelemetry*, 22-29.

Kolz A. L., Corner G. W. and Johnson R. E. (1973) A multiple-use wildlife transmitter. U.S.D.I., Fish and Wildl. Serv., Spec. Sci. Rep. — Wildl. No. 163. 11 pp.

RCA Corporation (1972) COS/MOS integrated circuits manual. RCA Solid State Div., Somerville, N.J. CMS-271. 222 pp.

Standora E. A. (1977) An eight-channel radio telemetry system to monitor alligator body temperature in a heated reservoir. *Proc. First Int. Conf. Wildl. Biotelemetry*, 70-78.

Hybrid Microcircuits for Implantable Radio Transmitters

J. H. FILSHIE, R. J. WILKIE and I. J. H. DUNCAN

Agricultural Research Council's Poultry Research Centre, King's Buildings, West Mains Road, Edinburgh EH9 3JS, U.K.

Abstract — Hybrid microelectronic techniques provide a means of interconnecting high densities of microminiature discrete components and silicon chips on glass or ceramic substrates. It is thus possible to produce complex circuits on planar arrays of 1 or 2 cm^2 in area and a few mm high. Hybrid microcircuits are of two types depending on the manufacturing method. Thick film circuits have conductor and resistor patterns printed onto ceramic substrates as glazes by screen printing methods whereas thin film circuits have the resistor and conductor materials evaporated onto glass substrates and the patterns subsequently produced by photolithographic etching. The relative merits of these techniques are discussed in the context of small quantity production of implantable radio transmitters.

INTRODUCTION

Hybrid circuits are so called because they consist of a mixed technology where some of the passive components are formed during the initial production of the circuit and the remaining passive and active components are added later. Two separate technologies are used to produce hybrid circuits and although to users the devices are similar, the method of production is different.

Thick films consist of conductor and resistor patterns which are screen printed onto a ceramic substrate and fired to produce a stable circuit configuration. In contrast to the additive method of thick film production, thin film technology is a subtractive process starting with resistive and conductive films deposited onto a finely polished insulating substrate.

Hybrid technology was developed for electronic purposes in the 1950s at which time it was considered as an extension of printed circuit techniques to be used with discrete transistors as a means of miniaturizing circuits.

The development of the planar integrated circuits in the late 1950s produced a decline in the development effort in the hybrid field. Many engineers saw, in the integrated circuit, the panacea to all our electronic problems. Certainly in the digital electronics field their predictions have been shown to be true. The continual development and refinement of integrated circuit techniques has produced in the last few years large scale integrated circuits of such complexity that they are almost entirely self-contained digital systems.

Linear integrated circuits have not been developed to this extent. Nearly all linear circuits require external components and most low power low voltage systems require a mixture of integrated circuits and discrete transistors of different technologies. Hybrid circuit techniques are capable of providing a means of high density interconnection of the components of a complex system.

HYBRID TECHNOLOGY

Whereas it is possible with either technology to produce capacitors and inductors in addition to resistors and conductors, the reactive components are not as good as their discrete component counterparts and usually take up larger substrate areas. The following description of hybrid circuits will therefore be confined to the production of resistors and conductors and the interconnection of discrete components.

THIN FILM CIRCUITS

The production of a thin film circuit starts with the deposition of resistor/conductor metal film pairs on optically polished substrates. Substrate materials range from borosilicate glass through alumina to synthetic sapphire. For low power circuits operating below 800 MHz glass is used almost exclusively.

The deposited metal film pairs must be stable, must not react with each other and must be sufficiently chemically different so that it is possible to selectively dissolve the films. Also, the resistive film which is deposited first must adhere strongly to the substrate preferably via a chemical bond. Common metal pairs used in modern thin film circuits are: (a) nickel-chromium/gold, (b) palladium/gold, and (c) tantalum or tantalum nitride/copper.

The films are deposited in vacuum chambers by either evaporation or sputtering. The metallized substrates are coated with photoresist, exposed to UV light through masks and developed so that the resistor and conductor patterns are protected by the resist. The conductors and resistors are produced by selective chemical removal of the exposed areas.

The circuits are then tested and critical resistors trimmed. In the case of nickel-chromium this is done by spark erosion or laser, whereas tantalum films may be adonized to change resistivity. Adonized tantalum is passivated but nickel-chromium must be protected by an impermeable insulation. These circuits are therefore returned to the vacuum chamber for a coating of SiO_2 which then has holes etched in it to expose bonding areas and mounting pads on the conductors.

Figure 1 shows a typical set of masks for the three stages of photolithography in the production of a thin film temperature radio transmitter, the circuit of which is shown in Fig. 2. The actual size of these masks is 20 × 10 mm.

THICK FILM CIRCUITS

Alumina is the most commonly used substrate material for thick film circuits. Conductor and resistor patterns are printed directly onto the substrates as pastes by screen printing methods. Conductor pastes consist of mixtures of precious metals such as silver, gold, platinum and palladium in an organic binder. Resistor pastes are mixtures of conductor material and glass in a binder. When fired they produce cermet resistors. The resistivity of these films may be varied over several orders

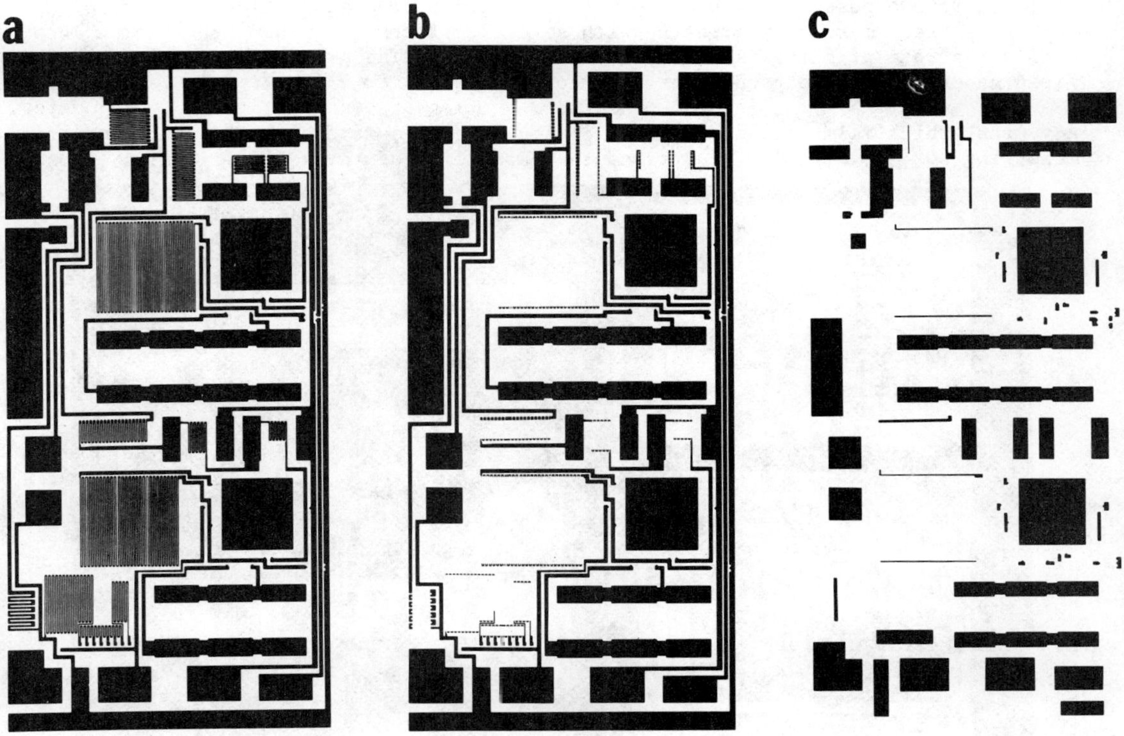

Fig. 1. Masks used in the production of a thin film hybrid microcircuit transmitter. (a) Mask for removal of all unwanted resistor and conductor material; (b) mask for removal of unwanted conductor material overlying the resistors; and (c) mask for exposing bordering areas through the S_iO_2 overlay.

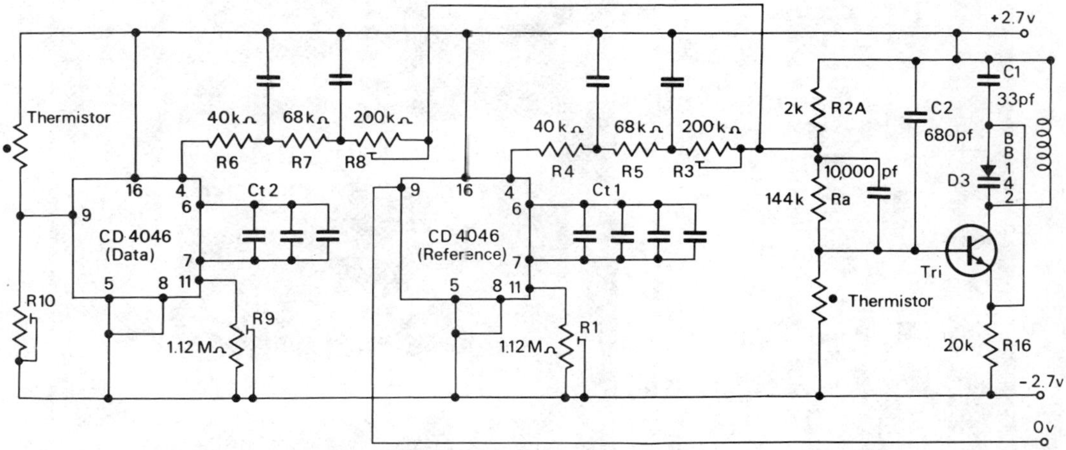

Fig. 2. Circuit diagram of a biotelemetry transmitter for the transmission of temperature. This may be expanded to several channels of information.

of magnitude by varying the type of conductor and its concentration in the paste. It is therefore possible to produce widely different resistor values without occupying large areas merely by printing with several different types of resistors. However a separate printing is required for each type. The printed substrates are then fired under controlled conditions in a zone furnace.

A typical thick film circuit is shown in Fig. 3. This is a high frequency amplifier measuring 30 × 15 mm. It illustrates the capability of thick film circuits to use

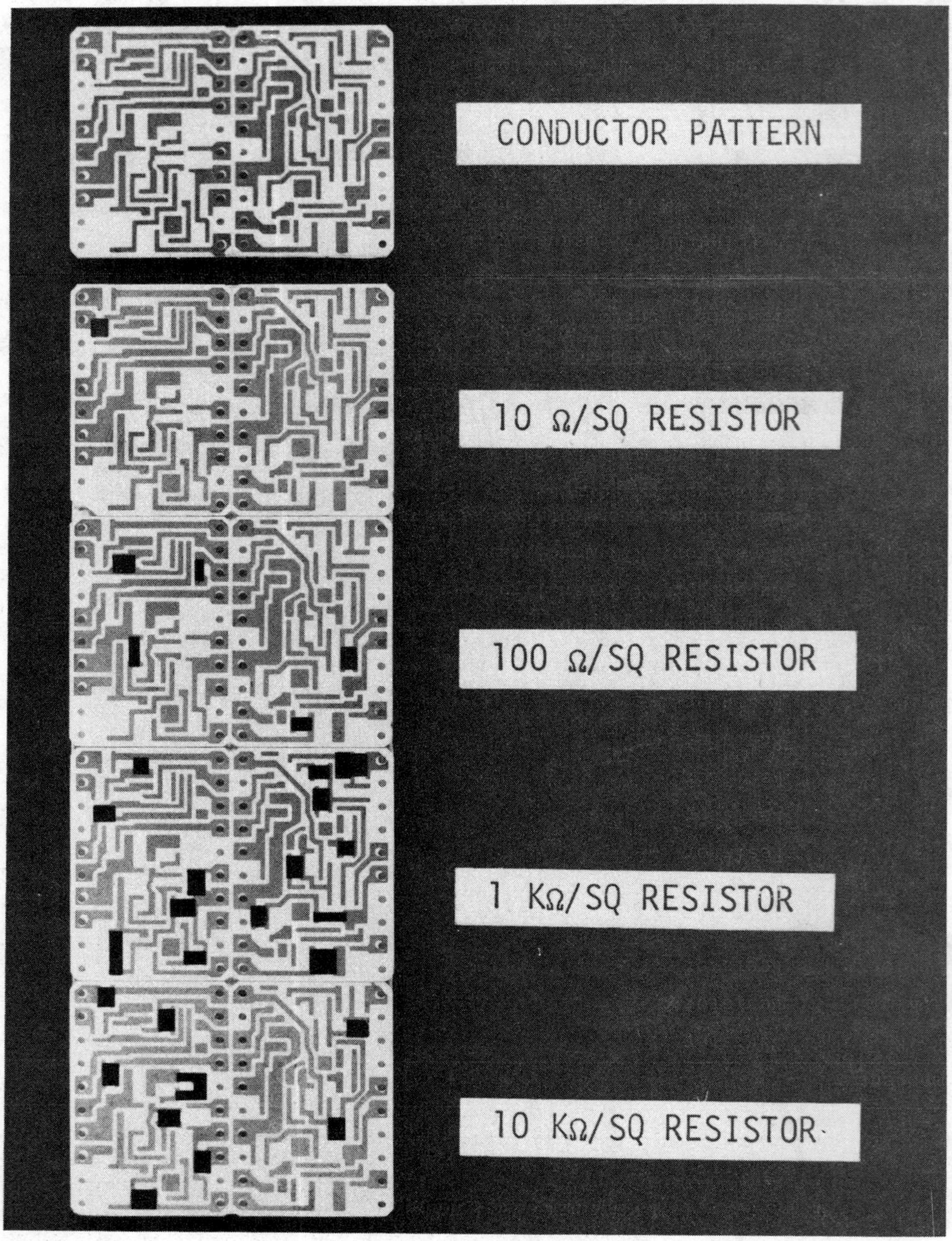

Fig. 3. A series of thick film substrates showing resistors of differing resistivities.

resistor pastes of several resistivities, thus allowing all resistors to be of a similar size. This is to be contrasted with Fig. 1 where resistors occupy areas of greatly differing magnitudes. On the other hand it can be seen that the line resolutions are not as high on thick film circuits.

DISCRETE COMPONENTS

The techniques for attaching components to hybrid circuits are similar for both thin and thick films.

SOLDERING

Capacitors, packaged transistors and integrated circuits and connecting leads are usually soldered onto the hybrid circuits. Reflow soldering methods are used for thick film circuits as they allow attachment of many components simultaneously. In thin film circuits the components are usually attached individually by either hot gas bonding or even by conventional hand soldering.

CONDUCTIVE EPOXY BONDING

Unencapsulated semicondictor components for hybrid circuits are usually gold backed. The best method for mounting these components is by using a silver or gold loaded epoxy resin. This is a reliable method and has almost entirely replaced the earlier method of heating the chips and substrates to the gold/silicon eutectic point, which could cause an alteration in the semiconductor parameters.

WIRE BONDING

Integrated circuits and transistors have connections on their top surfaces. To connect them electrically to the hybrid circuit, jumper wires must be attached. This is achieved by ultrasonic bonding of aluminum wires or thermo-compression bonding of gold wires. The most appropriate method to be used depends on the type of pad on the silicon chip and the conductor material on the thick or thin film circuit. Both methods are very reliable but the wires are fragile, having diameters of less than 25 μm, and require protection before the circuit is encapsulated.

The bonding diagram for the thin film transmitter is shown in Fig. 4, and Fig. 5 is a a photograph of a completed transmitter prior to encapsulation. This is typical of the component densities achievable by this technique. Such circuits may be even further reduced in size by replacing the large resistors by chips.

COMPARISON OF THE HYBRID TECHNOLOGIES

Table 1 compares thick and thin film resistors. It shows that thin film resistors are more accurate and stable, have lower temperature coefficients and are less noisy. The resistivity of thin film materials is low compared with that of thick films. This means that high value thin film resistors are large compared with their thick film counterparts. However, because of the higher line resolution of thin film techniques, low value resistors can be much smaller.

This higher line resolution of thin films allows a much higher packing density of attached components — especially integrated circuits. Thus the circuit area ratio of thin:thick films can be as little as 1:10. However, comparison of the production steps in Table 2 shows that thin film technology is more complex and labor intensive than that of thick film. Thick film techniques are ideally suited to mass production

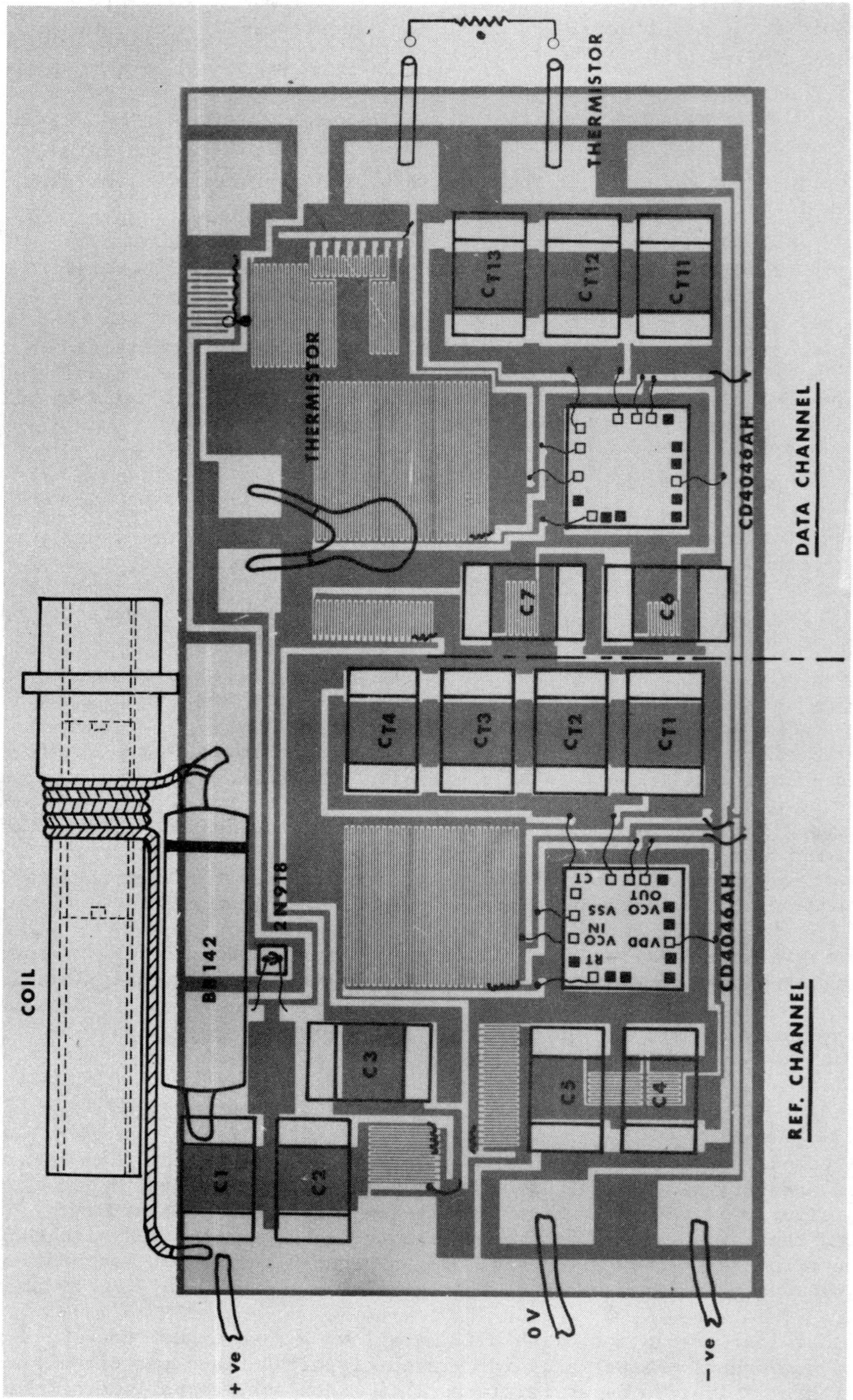

Fig. 4. Bordering diagram of the thin film transmitter of Fig. 2 showing the placement of discrete components.

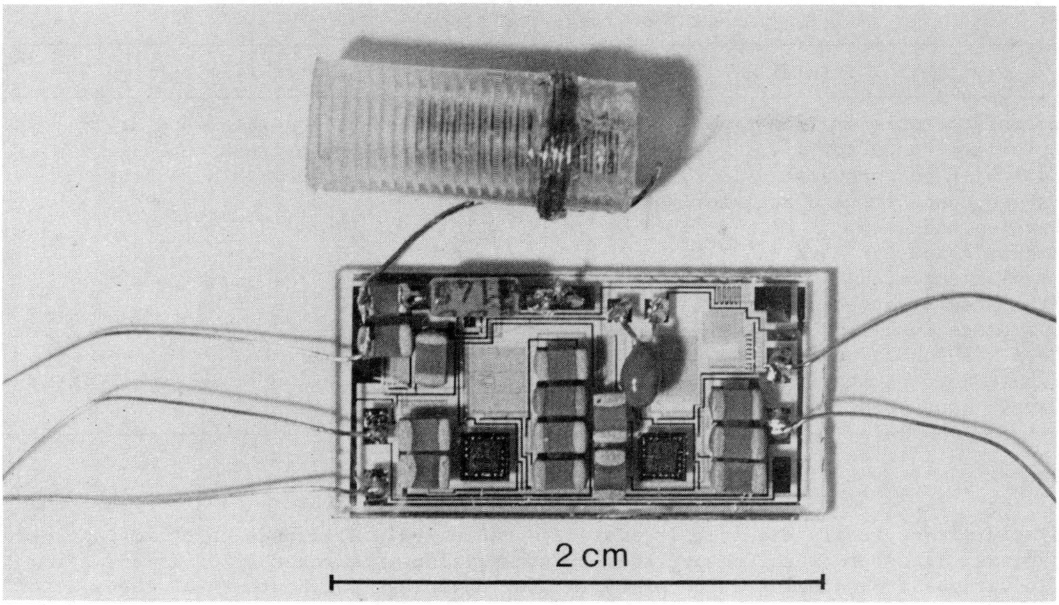

Fig. 5. Photograph of the thin film transmitter prior to encapsulation.

Table 1.
Resistor comparisons

	Thin films	Thick films
Line resolution	1 μm	50 μm
Resistivity	10–400 Ω square^{-1}	10 Ω square^{-1} 10 MΩ square^{-1}
Dimensions	∿ 100 kΩ(mm^2)$^{-1}$	1 mm^2 minimum
Accuracy (untrimmed)	5 percent	20 percent
Accuracy (trimmed)	0.1 percent	1 percent
Temperature coefficient	< 50 ppm $^oC^{-1}$	> 200 ppm $^oC^{-1}$
Noise	low	high for high R values
Voltage sensitivity	negligible	significant for high R values

of thousands of devices whereas thin film methods are better suited to the hundreds. In general it is more realistic to consider thick film technology for miniaturization because commercial production facilities are more widely available. Also the cost of setting up an 'in house' facility is an order of magnitude less for thick film circuits. On the other hand, where size is critical and high stability is required and it is not economically possible to commission a custom silicon integrated circuit, thin film techniques should be considered.

In this latter case it might be possible for limited numbers of circuits to be produced in collaboration with the Authors' Institute if the circuits are relevant to its remit. The Authors would welcome contact from interested parties.

Table 2.
Production steps

Thin film	Thick film
Evaporate resistor film	Print conductor pattern and dry
Evaporate conductor film	Print resistor pattern and dry
Coat with photo-resist	repeat for other value pastes
Expose conductor and resistor mask	Fire
Develop and etch	Print overglaze
Expose conductor mask	Fire
Develop and etch	
Strip resist and clean	
Evaporate insulator	
Coat with photo-resist	
Expose bonding mask	
Develop and etch	

CONCLUSION

Hybrid microcircuits are ideally suited to medium volume production of implantable devices. They are much cheaper than custom silicon integrated circuits and allow the various technologies to be mixed so that advantage can be taken of the special merits of each type of device. Hybrid circuits have much higher packing densities than their component counterparts and can therefore support more complex circuits than otherwise would have been possible. Both thin and thick film circuits are suitable for biological implantation when suitably hermetically sealed and encapsulated in a biocompatible material.

Low Power VHF Transmitter for Multiplexed Telemetry

T. A. PERKINS

MRC Neurological Prostheses Unit, Institute of Psychiatry, London, U.K.

Abstract — Commercial medical band 102.2-102.4 MHz FM transmitters are available, however their high current consumption and limited high frequency response render them unsuitable for some applications. In an implant, long battery life is essential. To avoid unduly cumbersome batteries the current required by this transmitter is kept under 500 μA. A 3 dB 'down point' over 30 KHz enables this transmitter to cope with up to 16 multiplexed channels of biological signals without appreciable cross-talk occurring. A preamplifier enables signals of 30 μV p.p. to 6 mV p.p. to be telemetered while presenting an input impedance of approximately 1 MΩ.

DESIGN CONSIDERATIONS

The range required for the transmission of biological signals from a small animal to a receiver immediately outside its cage or from an implant to the outside world may be quite small — of the order of 1 m. Since, as Carruthers *et al.* (1973) stated, the transition frequency of a transistor operated below 1 mA tends to fall off linearly with collector current, the principal limitation on economy of supply current may then be that due to the frequency required in each stage of the device. The convenience of using readily available commercial FM receivers which cover the medical band is some compensation for the range/power penalty of using a continuous VHF transmission.

The CMOS multiplexing system designed by Craggs and Andrew (1980) requires a 6 V supply rail and an FM transmitter capable of dealing with biological signals of about 100 μV from a source impedance of many kΩ and a large synchronizing pulse of about 3 mV. If, for any particular channel, the residual cross-talk from the previous time slot is to be less than 2 percent at, say, three quarters of the way through the current time slot it can be shown that the 3 dB 'down point' of the transmission system must be greater than 0.83 times the multiplexor clock frequency. For a maximum clock frequency of 40 KHz, as might be the case in a sixteen channel system, the cut-off frequency for the telemeter must be greater than 33 KHz.

CIRCUIT

A complete circuit of preamplifier and transmitter is shown in Fig. 1. A common emitter

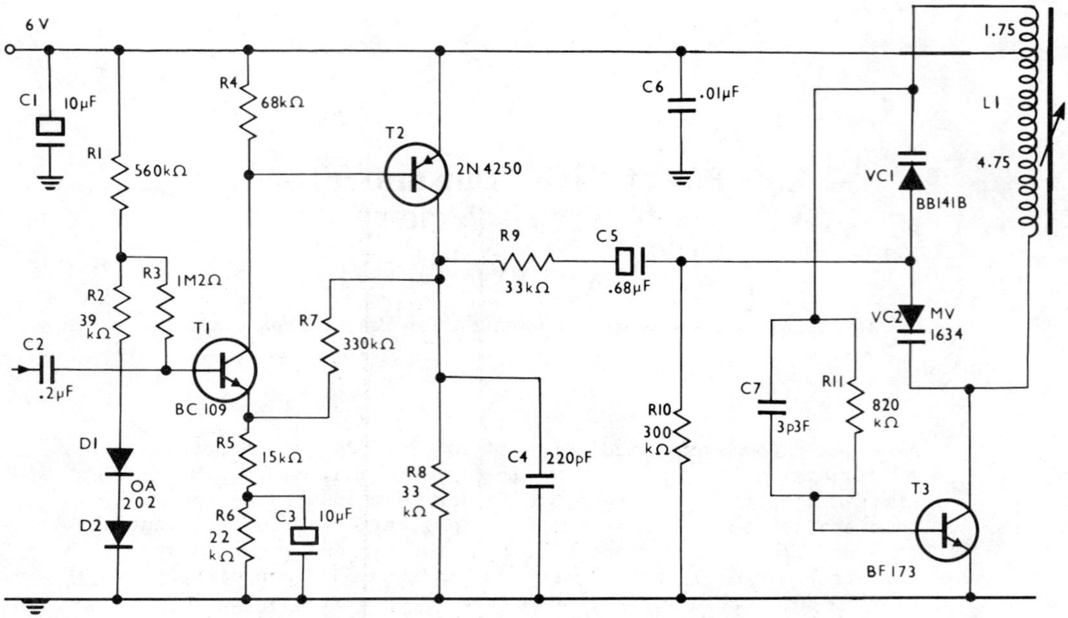

Fig. 1. Circuit of wideband low power VHF FM transmitter.

Hartley oscillator transmitter was chosen, for which a theoretical analysis is given by Meindl (1969). Mackay (1970) pointed out that the tendency for this type of oscillator to squeg is dependent on the time constant of the base feed capacitor and resistor. Since R11 controls the supply current to T3, it is C7 which has to be kept small. The BF173 was chosen as it retains sufficient high frequency response at 300 µA to maintain an oscillation of 100 MHz. The ferrite slug in L1 provides accurate tuning. Frequency modulation is via varactors VC1 and VC2 which have a dc reverse bias of the full supply voltage. The gain of the preamplifier (and hence the sensitivity of the system) is controlled by the ratio of R7 to R5. Temperature compensated bias is provided by R1, R2, D1 and D2 – cf. Horwitz (1974). The input impedance of around 1 MΩ is determined largely by R3. The 10 µA current through T1 determines the preamplifier's high frequency cut-off point of about 43 KHz. Both transmitter and preamplifier operate stably over the supply range of 4 to 8 V. A prototype is shown in Fig. 2.

PERFORMANCE

Performance details using a stereo FM receiver are given in Table 1.

Table 1.

Supply voltage	6 V
Supply current at 6 V	450 µA
Maximum input	6 mV p.p.
Broad band noise level at 2 m range	30 µV p.p.
L.F. 3 dB down point	ca 1.5 Hz
H.F. 3 dB down point	ca 43 KHz
Input impedance	> 1 MΩ

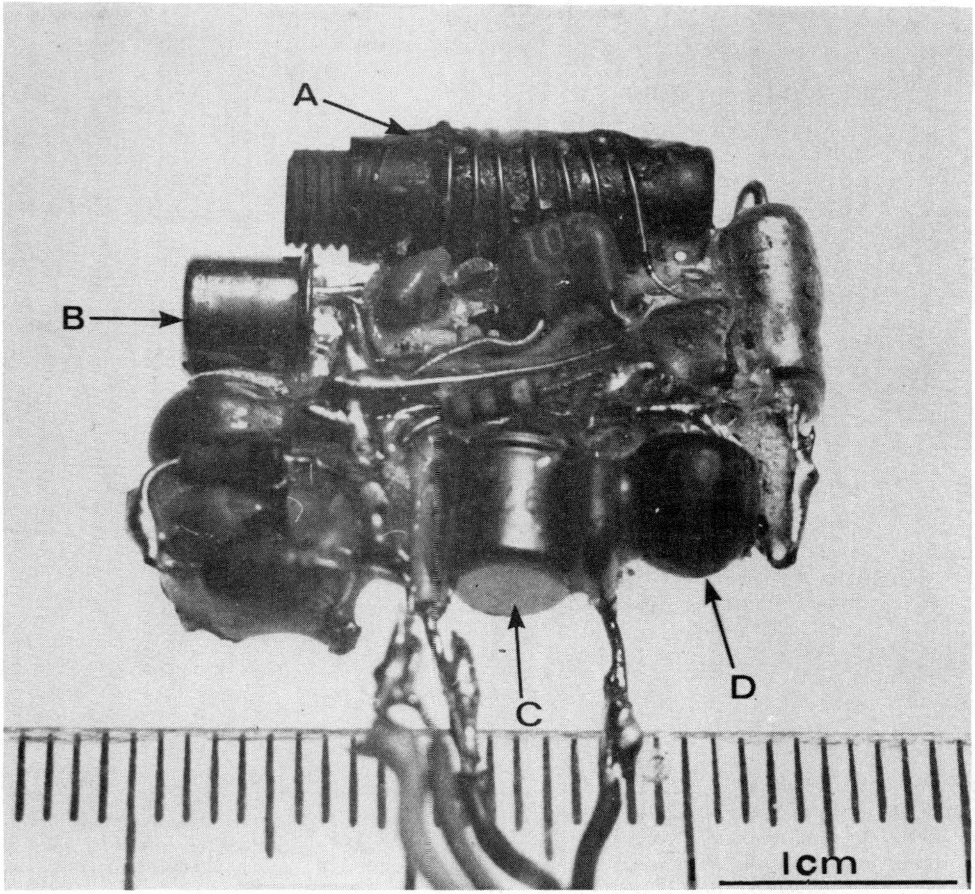

Fig. 2. Prototype of the low power VHF transmitter. A = L1; B = T3; C = T1; D = T2.

Figure 3 shows the output from the receiver for a 20 KHz square wave input to the transmitter.

FUTURE DEVELOPMENT

To make the device implantable a miniaturized thick film version in an hermetic package would be required. As all the semiconductors used in this design are readily available in chip form this should present no difficulty. Applications may occur where longer range is required in which case pulse interval modulation would probably be used.

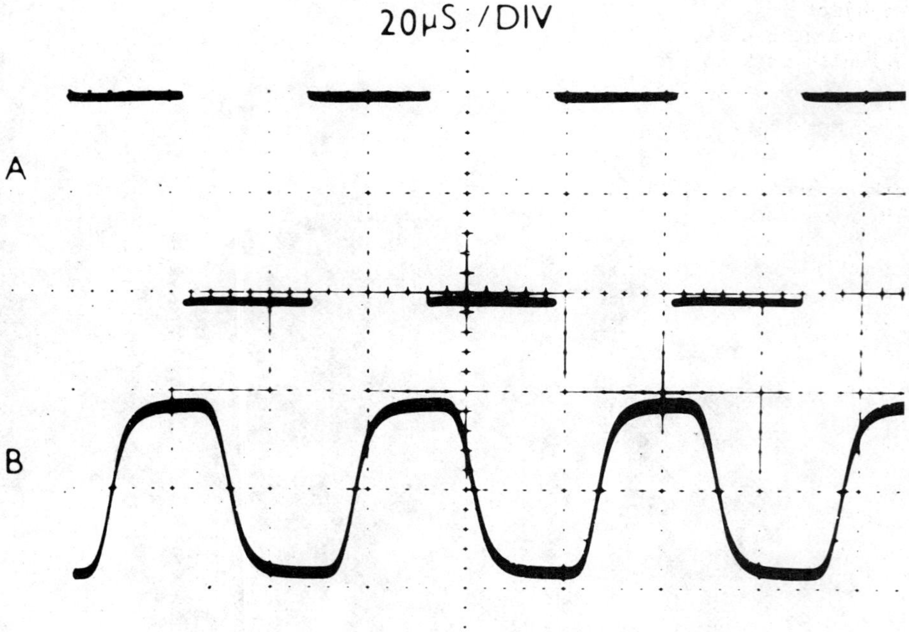

Fig. 3. Signal received from VHF transmission. A — Input to transmitter 0.5 mV/division. B — Receiver output 0.5 V/division.

REFERENCES

Carruthers J., Evans J. H., Kinsler J. and Williams P. (1973) Circards-10: micropower circuits. *Wireless World* 79, 495-496.

Craggs M. D. and Andrew C. M. (1980) Time-sharing for biological signals in radio telemetry, In C. J. Amlaner, Jr. and D. W. Macdonald, *A Handbook on Biotelemetry and Radio Tracking*, pp. 323 - 328. Oxford: Pergamon Press.

Horwitz C. (1974) Micropower low noise amplifier. *IEEE BME* 21, 60-61.

Mackay R. S. (1970) *Bio-Medical Telemetry*, Second Edition, p. 4. New York: John Wiley.

Meindl J. D. (1969) *Micropower Circuits*. Chapter 9. New York: John Wiley.

Assembly of a Small Ultrasonic Transmitter

L. C. IRELAND* and K. D. LAWSON, JR.**

*Bermuda Biological Station for Research, St. George's West, 1-15, Bermuda
**Neil Brown Instruments, Box 498, Cautaumet, Massachusetts 02534, U.S.A.

Abstract — The construction of an acoustic transmitter suitable for use in underwater tracking experiments is described. All the components required are available commercially. Transmitter frequency is 50 KHz. The device projects pulses of sound at a rate of 1.5 s^{-1}; pulse length is 55 ms. Both frequency and pulse rate may be easily changed. Acoustic output is approximately 38 dB re 1 μbar at 1 m. Useful life is approximately 30 days. The transmitter is 1.58 cm in diameter, 4.20 cm long, and weighs 19 g in air.

INTRODUCTION

The first prerequisite for obtaining a comprehensive record of the ethology and ecology of any species is the ability to maintain contact with the animals over long periods of time. Acoustic telemetry enables scientists to track many species of aquatic animals for long periods of time in many environmental situations in which it is impossible to obtain information by means of other techniques (Ireland and Kanwisher, 1978). Unfortunately commercially available acoustic transmitters are expensive, and transmitters are usually not recovered at the end of experiments. These considerations led us to design the simple and inexpensive transmitter described here. Discussion of all the characteristics of acoustic transmitters and the many variables affecting the propagation of sound through water is beyond the scope of this report. For information on these topics see Urick (1975) and Stasko and Pincock (1977).

MATERIALS AND METHODS

The circuit diagram for the transmitter, a one-transistor squegging oscillator design, is shown in Fig. 1. Pulsed transmission is used to conserve battery life. Transmitter frequency (50 KHz), pulse rate (1.5 s^{-1}), pulse width (55 ms), and signal transmission (acoustic output = approximately 38 dB re 1 μbar at 1 m) are controlled by an obviously small number of components, and component values are critical. For best results, therefore, we recommend the use of the exact components specified here.

The transformer is wound on a Ferroxcube 905 F1D bobbin. The primary is 25 turns no. 38 magnet wire, tapped at 9 turns, and the secondary is approximately 205 turns

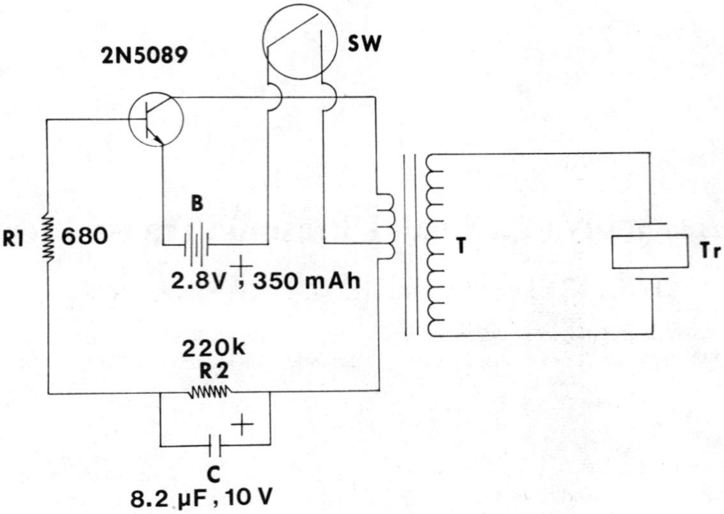

Fig. 1. Circuit diagram for a 50 KHz acoustic transmitter. B is a Mallory TR152 mercury battery. C is a Sprague no. 162D-825X9010CD2 miniature tantalum capacitor. The semiconductor is by Texas Instruments. The switch, SW, is made from two insulated wires left outside the case. R1 and R2 are 1/8 W, ± 5 percent tolerance, Ohmite® resistors. See the text for details on T, the transformer. The transducer, Tr, is a 1.65 cm o.d., 1.49 cm i.d., 0.64 cm long cylinder of lead-zirconate-titanate, TCD-4® material, manufactured by Marine Resources Incorporated.

no. 40 magnet wire. Windings are held in place with a light coating of coil dope. The bobbin is placed inside a Ferroxcube 905C-A100-3B7 adjustable gap pot core, used with the adjustor. One end of the transducer is sealed with a 0.08 cm thick piece of Lexan® (polycarbonate). The piece of Lexan® is the same diameter as the transducer and is held in place with cyanoacrylate adhesive. All the components are now connected together and the circuit is tested. Care should be taken when making connections to the transducer, the transistor, and the battery; excessive heat can damage each of these parts. Some adjustment of the circuit may be necessary. We usually find this to be the case whenever we receive a new order of transducers. Increasing the number of turns on the transformer secondary decreases transmitter frequency; decreasing the turns has the opposite effect. Final tuning is done with the pot core adjustor. Pulse rate is set by C and R2. Increasing the value of R2 decreases the pulse rate and vice versa. When the circuit is satisfactory, the pot core is placed inside the transducer and potted with silicone rubber of the kind used to seal aquariums. The rubber should completely fill the interior of the transducer. Use of this 'pressure-release' material results in greater transducer efficiency. All the components are then potted inside a glass epoxy tube. Tube and transducer must be a close fit if good electric-to-acoustic efficiency is to be achieved. We use 4.20 cm lengths of 1.43 cm inner diameter tubing with a 0.08 cm wall, sealed at one end with a circular piece of glass epoxy, 0.15 cm thick (Stevens Tubing Corp.). Hysol® R9-2039 epoxy resin and H2-3651 hardener (Dexter Corp.) are excellent for potting transmitters. Two insulated leads extending beyond the case serve as a switch and are soldered together to activate the device. A finished transmitter is shown in Fig. 2. It has a weight of 19 g in air and a useful life of approximately 30 days.

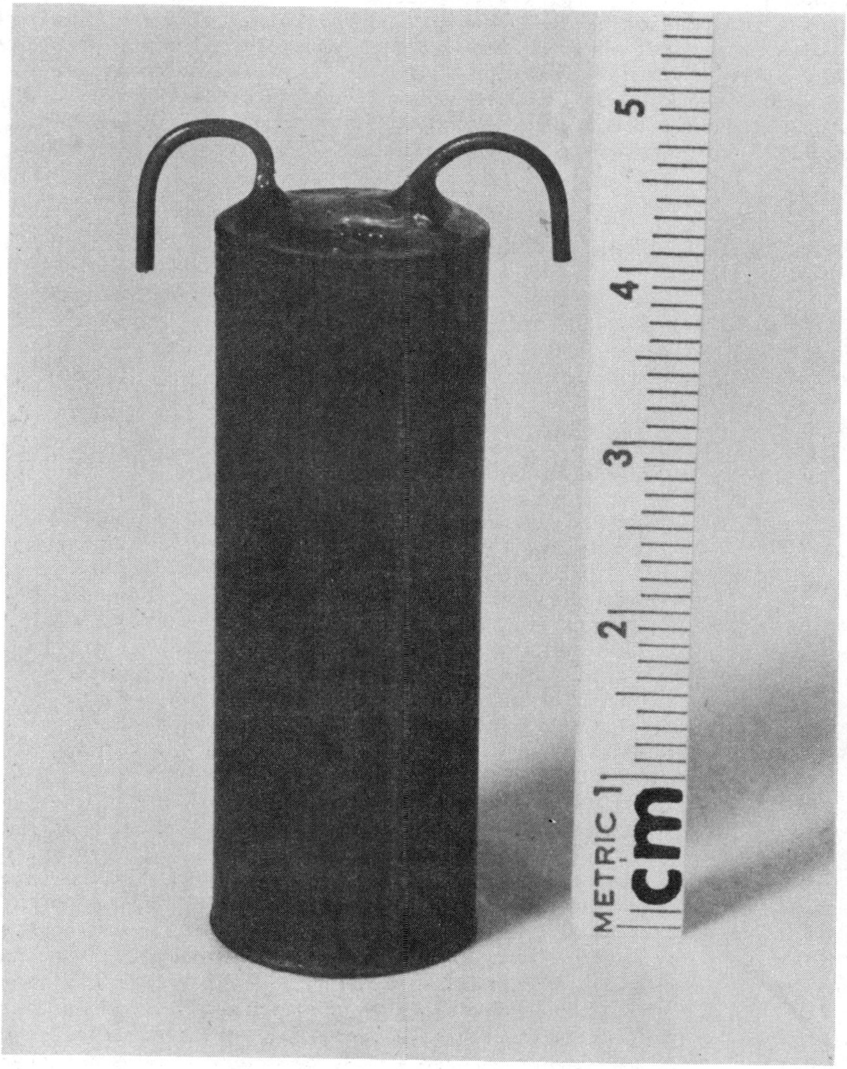

Fig. 2. Finished acoustic transmitter.

Acknowledgements — We thank Dr. John Kanwisher for his many useful suggestions. This material is based upon work partially supported by the National Science Foundation under Grant No. BNS 78-18196. This report is Contribution No. 808 of the Bermuda Biological Station for Research.

REFERENCES

Ireland L. C. and Kanwisher J. W. (1978) Underwater acoustic biotelemetry: procedures for obtaining information on the behavior and physiology of free-swimming aquatic animals in their natural environments. In D. I. Mostofsky (ed.), *The Behavior of Fish and Other Aquatic Animals*, pp. 341-379. New York: Academic Press.

Stasko A. B. and Pincock D. G. (1977) Review of underwater biotelemetry, with emphasis on ultrasonic techniques. *J. Fish. Res. Bd Can.* 34, 1261-1285.

Urick R. J. (1975) *Principles of Underwater Sound*. 2nd ed. New York: McGraw Hill.

Improved Techniques in the Construction of pH-sensitive Radio Pills

R. H. COLSON and B. W. WATSON

Department of Medical Electronics, St. Bartholomew's Hospital, London, U.K.

Abstract — The merits and design difficulties of radio telemetry capsules incorporating glass electrodes for the measurement of pH are briefly reviewed, and details of an improved capsule (7.6 mm diameter, 26.0 mm long) are described. *In vitro* evaluation of the capsule, including accuracy of pH measurement, response to sudden pH changes, and long term stability indicate that the capsule will be suitable for implantation studies for up to three weeks, as well as investigations of pH in the gastrointestinal tract.

INTRODUCTION

The development of small, accurate and stable pH-sensitive radio telemetry capsules has been a continuing project in St. Bartholomew's hospital for some 15 years (Watson and Kay, 1965; Meldrum *et al.*, 1972). Initial interest centered on the investigation of the pH profile of the human gastrointestinal tract, and a pH capsule was developed for this purpose (Meldrum *et al.*, 1972). Difficulties in manufacturing this capsule and the need to increase the operational lifetime in order to undertake short term implantation studies of localized pH changes have led to improvements to the original design, as well as a reduction in overall size.

METHOD

pH-sensitive glass membranes remain the most accurate and stable types of pH transducer for *in vivo* measurement. An antimony electrode has been used in a commercial capsule (the Heidelberg capsule) (Nöller, 1962), but the poor accuracy and short operating life (Connell and Waters, 1964) of this device have been largely attributed to the deterioration of antimony in the presence of oxidizing agents in body fluids. Recently developed pH-sensitive polymers also appear to be unsuitable due to long term drift and limited pH range (LeBlanc *et al.*, 1976).

The principal difficulty encountered with a glass pH transducer is its very high source impedance — typically $10^{9}\Omega$ for a 3.0 mm diameter bulb. Consequently it has been found vitally important that the wire electrode leading into a glass pH bulb be completely sealed from the liquids in contact with the outer surface of the bulb, which can effectively short circuit the potential produced by the transducer.

Previous capsule designs have depended on epoxy resins to form this seal (Meldrum *et al.*, 1972; Kunz, Norby and Rogers, 1971). The slow ingress of moisture into these epoxies resulted in a gradual deterioration in the pH response and a limited lifetime for the capsules. Several other encapsulants have been tried, including paraffin wax and dental acrylic, with no significant improvement. Our present design overcomes the need for a glass/epoxy seal. All the components of the radio transmitter are effectively housed inside the transducer by the use of a glass body for the capsule (Fig. 1). The circuit is manufactured as a single thick film chip (5.5 × 6.5 mm) with solder pads for connection to a 1.6 mH ferrite cored antenna, mercury cell contacts and pH sensing electrode. The circuit assembly is encapsulated in epoxy to form a cylindrical insert, a mixture of silicone oil and paraffin wax providing the necessary moisture barrier between the insert and buffer jel. The reference electrode, comprising a compacted Ag/AgCl disc and salt bridge, is contained in a silicone rubber cap which is held in the capsule by two oversize skirts.

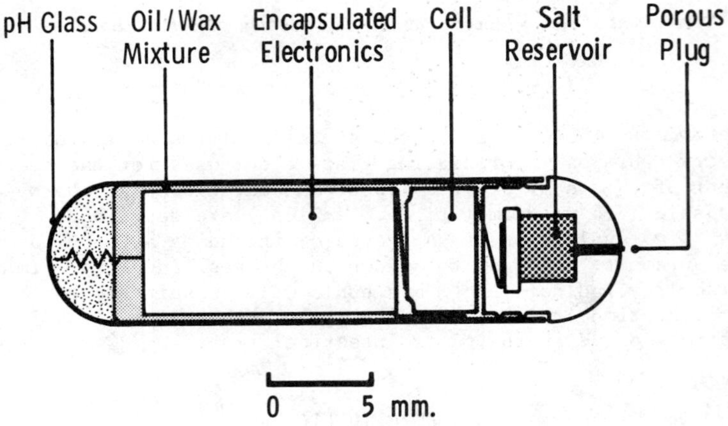

Fig. 1. Circuit of pH capsule.

The circuit (Fig. 2) is based on an earlier design and uses a frequency modulated oscillator with an output of 10 μW in the range 350-550 KHz. Variations in potential difference from the pH transducer cause a change in the capacitance of the back biased diode, with a corresponding change in output frequency. Power for the capsule is provided by a single replaceable mercury cell, type RM212. Stability with small changes in cell voltage has been improved by careful selection of the 70 kΩ oscillator drive resistor; as the cell voltage falls, the anticipated fall in oscillator frequency is counter-balanced by reduced drive and hence faster turn on, storage, and turn off times in the transistor. Inductive coupling between the radiated signal and loop receiving antenna limits the range of the capsule to about 30 cm, effectively precluding any possibility of radio interference to other services. Frequency variations are detected by a simple direct conversion receiver (Rigel Research Ltd., Sutton, Surrey, U.K.), and pH is monitored on a chart recorder.

IN VITRO EVALUATION

The capsule produces an output frequency shift of about 2.5 KHz/pH and is linear (± 0.2 pH) in the range pH 1-12. Response time, as measured by a step change from one buffer to another, is approximately 1 s to reach 95 percent of the final recorded value (Fig. 3). The temperature error is approximately 0.1 pH/°C.

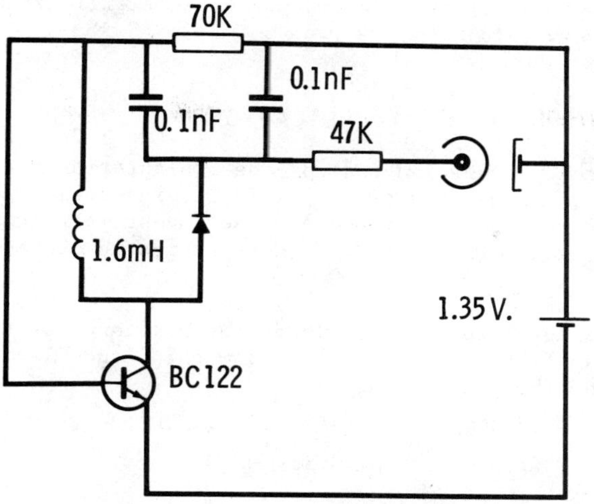

Fig. 2. Circuit of pH capsule.

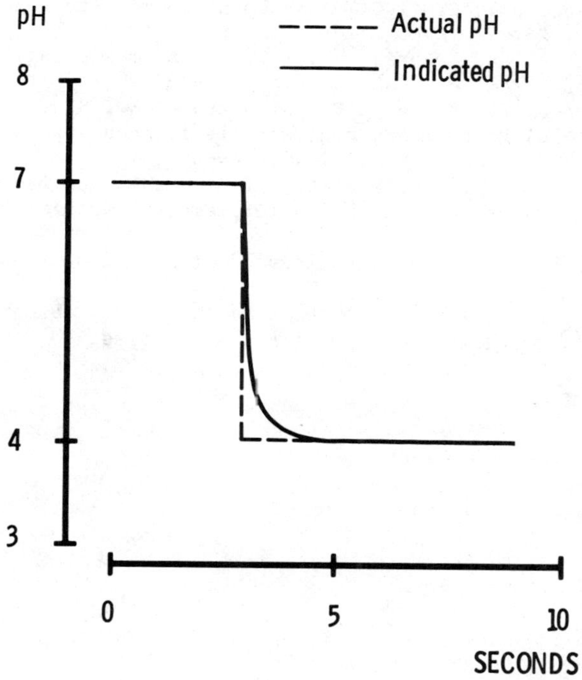

Fig. 3. Step response of pH capsule.

Overall lifetime of the capsules has not yet been established, although two have been operating in buffer solutions with an external power supply and a calomel reference electrode for 3 months without any deterioration in pH response. Preliminary checks on operational stability with a mercury cell and Ag/AgCl reference cap indicate a net drift not exceeding 0.5 pH over a 3 week operating period, with a short term drift of less than 0.1 pH in any 48 h section. It appears that these drifts are due to limitations imposed by the design of the reference cap assembly, and we are currently investigating improvements which will exploit the full 5 week capacity of the mercury cell.

APPLICATION

Clinical studies already scheduled include the implantation of the capsule in rats with standard carcinomas (Bellamy *et al.*, 1978) to investigate the pH of the extracellular environment in growing tumors; and the investigation of duodenal pH in young children with cystic fibrosis and other pancreatic diseases.

Acknowledgements — We gratefully acknowledge the assitance of Pye Dynamics Ltd., in the design and manufacture of the thick film chip, and Electronic Instruments Ltd. who manufactured the glass capsules.

REFERENCES

Bellamy D., Hinsull S., Watson B. and Blache L. A. (1978) Inhibition of the development of Walker 256 Carcinoma with a simple metal-plastic implant. *Europ. J. Cancer* 15, 223-232.

Connell A. M. and Waters T. E. (1964) Assessment of gastric function by pH telemetry capsule. *Lancet* 2, 227-230.

Kunz H. J., Norby T. E. and Rogers C. H. (1971) A pH measuring radio capsule for the alimentary canal. *Am. J. Dig. Dis.* 16, 145-149.

LeBlanc O. H., Brown J. F., Klebe J. F., Niedrach L. W., Slusarczuk G. M. J. and Stoddard W. H. (1976) Polymer membrane sensors for continuous intravascular monitoring of blood pH. *J. Appl. Physiol.* 40, 644-647.

Meldrum S. J., Watson B. W., Riddle H. C., Sladen G. E. and Brown R. L. (1972) pH profile of the gut as measured by radiotelemetry capsule. *Br. Med. J.* 2, 104-106.

Nöller H. G. (1962) Results of examinations of stomach functions with the Heidelberg capsule. *Forsch. Med.* 9, 351.

Watson B. W. and Kay A. W. (1965) *Biomechanics and Related Bioengineering Topics.* R. M. Kennedi (ed.), Chapter 9. Oxford: Pergamon Press.

Construction Techniques Using Subminiature Discrete Components

A. J. WOAKES

Department of Zoology and Comparative Physiology, University of Birmingham, Birmingham B15 2TT, U.K.

Abstract — A variety of transmitter designs have been developed for monitoring heart rate and respiratory frequency in wildfowl. The transmitters were required to have the smallest possible size as they were to be implanted in birds. This was achieved by using subminiature components without printed circuit boards or special substrates. The resulting construction technique enables the non-expert, with the minimum of equipment, to assemble subminiature transmitters that approach the dimensions achieved by the more costly hybrid forms of construction.

Biotelemetric systems are normally designed for a specific application, and the various operating parameters are then tailored to suit the requirements of the experimental situation. Invariably some design compromises have to be made, with the main limitation usually based on the maximum allowable size of the transmitter. This should be small enough, relative to the size of the host animal, so that no interference is caused to the behavioral or physiological responses under investigation. If the animal cannot tolerate the external placement of any equipment, then implantation will be required. The size of the transmitter may be kept to a minimum through the careful design of the circuit. This can be achieved by both reducing the power storage requirements and avoiding the use of bulky components. Attention to the mechanical design of the device can also reduce its bulk and, in addition, improve its physical robustness.

We have developed a number of transmitter designs for monitoring heart rate and respiratory frequency in wildfowl. These variables have been studied during natural diving in tufted ducks (*Aythya fuligula*) and free flight in barnacle geese (*Branta leucopsis*) (Woakes and Butler, 1975; Butler and Woakes, 1979). These transmitters were required to be as small as possible and, through necessity, were constructed with the minimum of special equipment. This was achieved by using subminiature components and by packing them closely together without the use of printed circuit boards or substrates. Through the careful arrangement of components relative to the largest (the mercury cell), wasted space was reduced to an absolute minimum.

The circuit design was first tested using full size components mounted on copperclad stripboard. This allowed most of the initial testing and substitution to be carried out on easily handled and relatively inexpensive components. When the component values had been finalized, the full size components were replaced by their subminiature equivalents. As each was removed from the board, it was replaced by pins mounted at each lead position. The smaller devices were then temporarily soldered to these pins for testing, and then removed as they were required during the construction of the transmitter.

The normal 'cordwood' construction method (Harper, 1977) uses components in matched packages, closely packed together and mounted between printed circuit end-boards. This method was modified by using just one board, for mechanical rigidity only. This was formed from stripboard reduced to 0.3 mm in thickness, but retaining an area of copper strip for applying power and test signals during construction (Fig. 1).

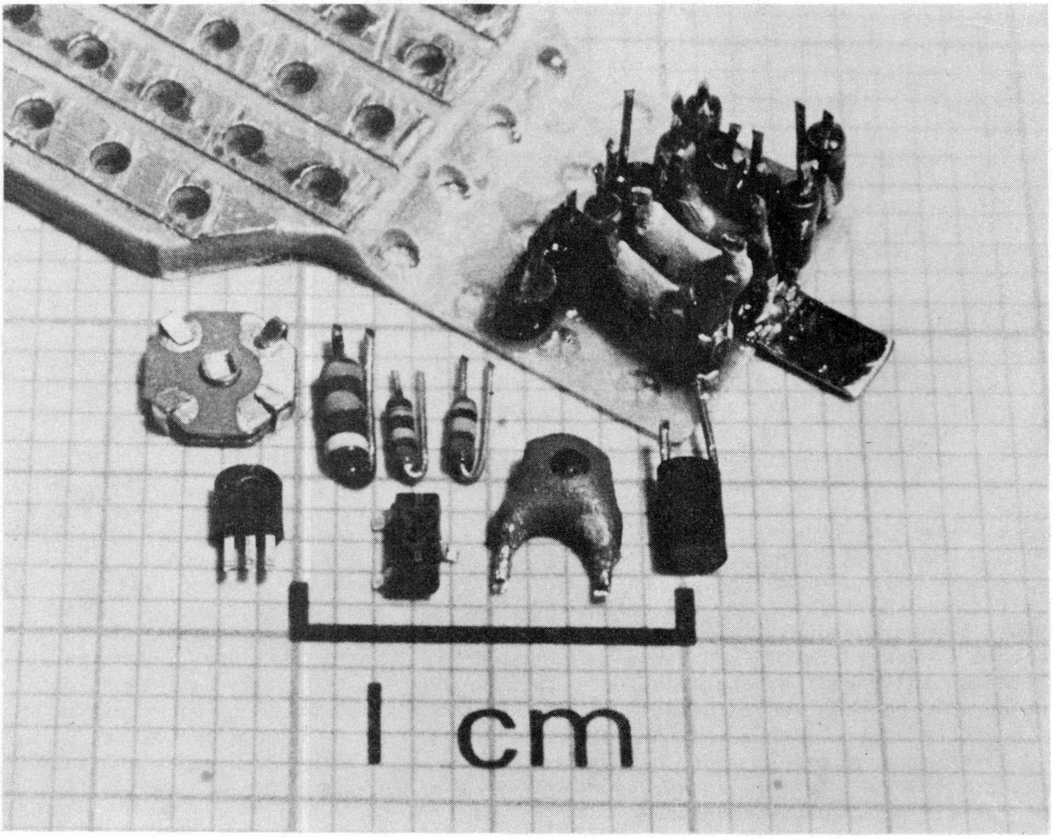

Fig. 1. Partially constructed transmitter showing the method of mounting the components.

Each component was removed from the test jig, and its leads preformed so as to exit from one end of the package. It was then mounted upright on the baseboard with epoxy adhesive. The components were packed as closely together as possible and arranged so that their leads were accessible from the top or side surfaces of the component block. They were then interconnected with fine 0.1 mm diameter uninsulated copper wire, starting from the signal input end of the transmitter and wrapping the wire around each lead. One node was wired at a time and, using fine resin-cored solder or solder cream, the wire was then soldered to the leads. Wires were also connected to the stripboard, so that power and test signals could be applied at each stage of construction. The wiring was then insulated and secured with a light coat of epoxy resin.

The completed and tested electronic package was then mounted in an outer frame carved from acrylic plastic or cast from epoxy resin. This carried the input pins, mercury cell and, in later designs, had the antenna embedded within it. The component block was wired into the frame and then completely embedded in epoxy resin, thus producing a solid and mechanically rigid transmitter package. The mercury cell was soldered to both battery contacts, and the ECG leads and respiration transducer were connected to the input pins. The transmitter was then either encapsulated for subsequent implantation, or mounted in a harness for external use.

This method of construction has proven inexpensive and versatile, and, while requiring care, can be easily mastered by the non-expert. A temperature controlled, fine tipped soldering iron and a limited range of dissecting instruments were the only tools required. However, most of the assembly of the transmitter had to take place under a binocular microscope, for the 35 separate components of a two-channel transmitter were compressed into a block measuring only 12 × 10 × 5 mm. The complete transmitter measured 28 × 15 × 6 mm, weighed 6.5 g and had a range of 30 m. An operational life of 6 days could be extended as required through the use of a magnetically operated switch. A similarly constructed single channel ECG transmitter measured 22 × 12 × 5 mm with a weight of 3.4 g. This technique is ideal where small numbers of relatively simple transmitters are required, while the overall dimensions of the completed transmitter can approach those achieved by the more costly, and less available, hybrid forms of construction.

Acknowledgement — I wish to thank the Science Research Council for their financial support.

REFERENCES

Butler P. J. and Woakes A. J. (1979) Changes in heart rate and respiratory frequency during natural behaviour of ducks, with particular reference to diving. *J. exp. biol.* 79.

Harper C. A. (ed.) (1977) *Handbook of Thick Film Hybrid Microelectronics*, pp. 9-33. New York: McGraw-Hill.

Woakes A. J. and Butler P. J. (1975) An implantable transmitter for monitoring heart rate and respiratory frequency in diving ducks. *Biotelemetry* 2, 153-160.

Encapsulation and Packaging of Implanted Components

N. DE N. DONALDSON

MRC Neurological Prostheses Unit, Institute of Psychiatry, London, U.K.

Abstract — Encapsulation and packaging is a large field which would need an extensive text to be covered comprehensively. For long term implanted electronic devices, it must be the engineers' major consideration. By a brief discussion collected into three topics, I try to show what must be considered when encapsulating circuits, which are not in a large hermetic enclosure. I also describe hermetic packaging, the choice between various types and their performance.

INTRODUCTION

Implanted electrical prosthetic devices usually need to be small and durable in a hostile environment for long periods. They will be subjected to a warm, corrosive aqueous solution with continued movement in the surrounding tissue. Implants are generally more complex than the well tried cardiac pacemaker; the size of components has to decrease and the field strengths correspondingly increase. This has meant that as would-be implant builders, we have had to devote ourselves to the problems of encapsulation and packaging.

The operating conditions for implants are worse than those faced by most other electronic systems with the important exception that the temperature is more or less homeostatic. I am sure that the relevance of the several points discussed below can be assessed with other conditions in mind (e.g. external encapsulation). I shall use 'encapsulation' to mean: the embedding of parts in a soft polymerizing material; a package is a more or less hermetic container in which components are protected.

THE ENCAPSULANT: ADHESION OR IMPERMEABILITY?

All implanted electronic devices require some encapsulation, and for simple circuits using discrete components, working lives of many months can be obtained by good encapsulation of the components alone. The implant shown in Fig. 1 survived inside a patient for twenty months without any additional protection. What is the prime requirement for the encapsulant? Figure 2 shows two electronic parts in different encapsulants, both of which have failed by a shunt current being established between the pairs of adjacent leads. In the first case (Fig. 2a) the material has not

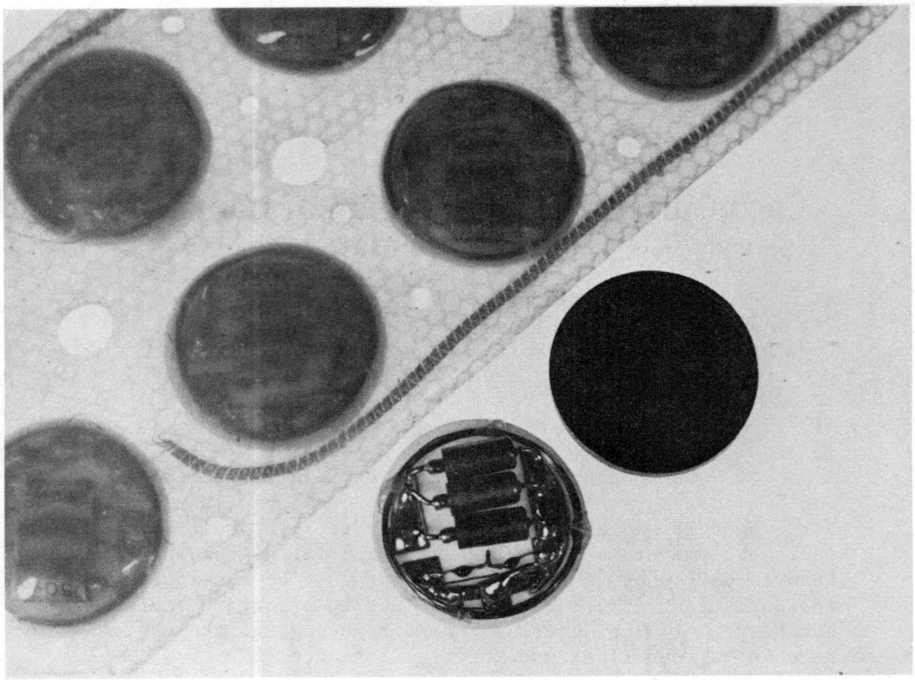

Fig. 1. Part of an implanted array of inductively coupled receivers. These hybrid thick-film circuits are encapsulated in silicone rubber. One unencapsulated receiver and a penny are also shown.

adhered, or has lost its adhesion, to the assembly and saline has penetrated to the region between the conductors. Below (Fig. 2b), the adhesion is good but bubbles have been left in the material and at the interface. Water vapor has diffused through the permeable material, condensed into the voids and again caused failure. Even if the material is a barrier to ions for the saline, we observe, in such cases, that there are enough ions available at the void to enable an electrolytic current to flow. Patently, an encapsulant that does not adhere is useless but one which adheres even though it is permeable to water vapor should suffice if it contains no voids. We have found this to be true; most electrical devices being unaffected by an encapsulant which is saturated with water vapor.

All our encapsulations use industrial adhesive silicone rubbers because they can be formed free of voids and the adhesion to solid surfaces remains unaffected when the rubber is wet. After pouring uncured rubber into the mould, it is subjected to a vacuum cycle which de-aerates the rubber and draws it into crevices within and between the electrical part. This is sometimes aided by simultaneous centrifugation which also flattens the meniscus on the liquid rubber (Donaldson and Sayer, 1975). The mould is then placed in a pressure chamber to discourage bubble formation as the rubber cures and shrinks.

SURFACE PROPERTIES OF AN ENCAPSULATED COMPONENT

The surface condition of the encapsulated part is important both because it partly

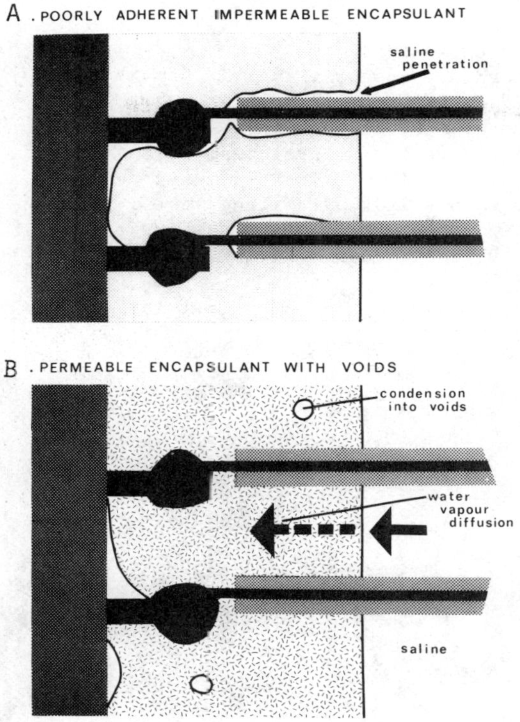

Fig. 2. Types of failure due to moisture in encapsulation.

determines the adhesive bond with the encapsulant and because of its possible tendency to corrode. If the adhesion does fail and water condenses at the interface but in such a place that no electrical failure is caused, then it is desirable that the surface should not corrode under the liquid, for that would enlarge the area of adhesion failure until it reached the electrical leads. On the other hand, it is generally true that for metal surfaces, the least reactive (e.g. Noble metals) make the worst bonds to the encapsulant. During tests, it is difficult to see whether surface corrosion or adhesion failure occurs first. For a surface of a single metal, it is probably best to compromise and have slight reactivity. It is certainly desirable to minimize the number of metals in the surface and thus the electrochemical potential. We found that nominally gold plated Kovar* packages potted in silicone rubber (Dow Corning 3140 which evolves alcohol during cure), soon have rusty patches at the interface after *in vivo* tests. If the package is copper plated, this is prevented presumably because:

(a) The rubber bonds better to the copper than to gold.
(b) It is unlikely that pin-holes in the copper plate coincide with pin-holes in the gold plate and so the exposed metal couple is gold-copper instead of gold-Kovar.

An alternative approach is to use a non-metallic surface such as alumina or glass.

---

*An alloy Fe 54%, Ni 29%, Co 17%.

At the circuit voltages that we have used (Donaldson, 1973), the following components can be potted without additional protection:

(i) Ceramic Chip capacitors.
(ii) Miniature carbon resistors.
(iii) Epoxy coated passivated microdiodes.
(iv) Hermetic tantalum capacitors.
(v) Diodes in hermetic glass envelopes.

We have tried to extend this list by testing sealed-junction transistors (Donaldson, 1977) and standard CMOS Flatpacks. Figure 3 shows a type of implant using Flatpacks which is currently being assessed *in vivo*.

Fig. 3. An implant which uses two CMOS Flatpacks potted in silicone rubber. Around the central (inverted) flatpack, there are four thick-film substrates to which are soldered chip capacitors, hermetic tantalum capacitors, miniature resistors and hermetic glass diodes. The two Morphognostic receiving coils (Donaldson, 1979) and the interconnecting wires are polyurethane-coated copper.

It is important that parts to be potted are thoroughly cleaned of flux residues.

DESIGN CONSIDERATIONS FOR ENCAPSULATED PARTS

Having chosen the encapsulant and the types of components to be used, some thought must be given to choosing the component values, shapes and arrangement which will keep the adhesion high and the electric field strengths low (Donaldson, 1977).

Resistors should be of the lowest possible values and from a wide range of chip capacitor shapes, the longest between the metallized end caps are to be preferred in the required capacitance value.

Since adhesive rubbers contract during their cure (1.5 percent linear for Dow Corning 3140), a tensile stress will be set up which tends to reinforce adhesion on convex objects and oppose it on concave. Thus small spherical components are preferred to large flat ones. A failure occurred in an implant similar to the one shown in Fig. 3 because the adhesion failed between the rubber and the metal lid of the small flat-pack. When the liquid had broken the bond round the side of the package as far as the leads, one input to the circuit was short circuited by an electrolytic path to the lid.

Components on hybrid circuits must have rubber all round them and this normally requires that they are soldered well off the substrate so that rubber can be injected underneath. This also increases the favored convex stress. If a component is too large to ensure that injected rubber covers the downward, invisible side, it is better to have a hole in the substrate there so that rubber may be applied from behind. As, for example, the small flatpack in the implant on Fig. 3.

HERMETIC SEALS FOR LEADS

It probably seems obvious that implanted electronics should be enclosed in a hermetic package which is itself encapsulated rather than the electronic components being encapsulated individually. However, since the encapsulation requirements are no different in kind whether the part be a semiconductor or a wire, we need to know the limiting conditions so that the extent of the hermetic enclosure can be chosen. The incentives for this are the minimization of implant size and the reduction of costs, for commercial hermetic packages are expensive especially if they are custom built. It appears at present that the limiting electric field strength which can be endured by rubber at the surface of a semiconductor is at least 10^6 V m^{-1}.

Cross sectional diagrams of three hermetic lead-out seals are shown in Fig. 4. Figure 4b is a commercially available gold plated Kovar-glass seal. This is the type of package which I mentioned above because, when encapsulated, it lasted longer if the unusual step is taken of copper plating over the gold. Packages are available with leads through the wall (Flatpack) or through the base (plug-in). Although very convenient, these suffer from several disadvantages:

(a) High cost.
(b) The minimal lead pitch is usually 0.05" which is coarse if many leads are required.
(c) The external surface is mostly metal so there must be greater incipient corrosion.
(d) Since the base of the package is metal, a separate substrate bearing the circuit will normally have to be placed inside the package which loses volume.

These disadvantages can be avoided if a hermetic package is made by thick-film techniques on the substrate of the microcircuit. Two ways of doing this are shown in

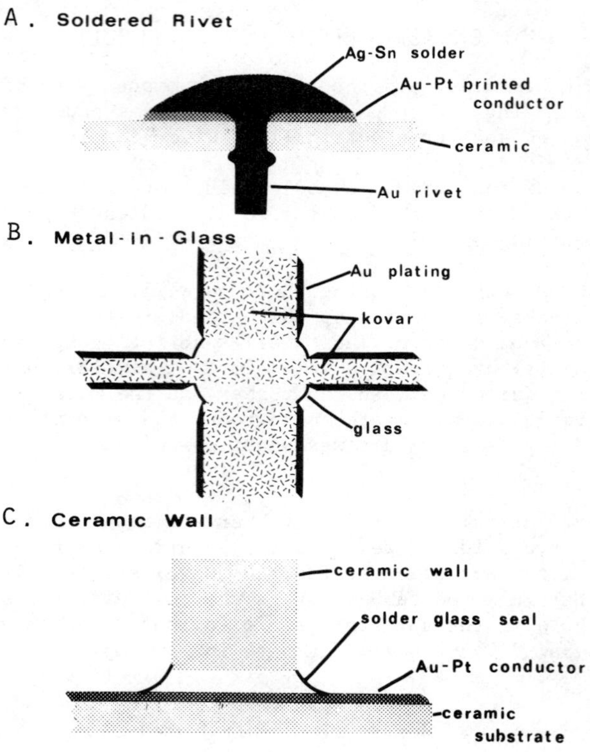

Fig. 4. Three types of hermetic lead out which all rely on a glass-to-metal bond.

Fig. 4a and c. Corkhill (Holmes and Loasby, 1976) developed the latter method in which printed conductors run under the ceramic wall and which is attached by several layers of printed glass. Besides multiple printing, there may be a compromise between lead pitch and lead width versus the substrate's peripheral length. The alternative, the soldered rivet, can have its leads spread over the whole area of the substrate and only requires one printed layer but it does need holes drilled in ceramic (or punched through the substrate before it is first fired) and special rivets of the required size.

LID SEALING MATERIALS

Having connected the electronic components inside the hermetic package, the lid must be attached. The materials for so doing may be divided into metals, glasses and plastics. Metals are approximately four orders less permeable than glasses which are themselves two orders better than the best plastics. The permeability of plastics range over four orders of magnitude.

If a solder glass is employed, it needs to be fired at very high temperature (about 500°C) which severely limits the types of materials inside the package.

All plastic sealants are so permeable that a package sealed with one cannot be considered hermetic for implants (Traeger, 1977). By that I mean the inside of the package will become saturated with water vapor after a time which is short compared to the desired implant life. Special epoxy resins are generally used for plastic lid seals by the electronics industry.

The last category, metal seals, includes soldered and welded joints. The weld is potentially the most satisfactory joint for it requires no flux and the average temperature of the package is low during the operation. However, special welding machines are used which only handle certain types of packages. We have no experience of this technique.

With solders, one has a range of alloys available. These vary from the high melting point eutectics with a noble metal present (e.g. Au/Sn, M Pt 280°C) which can be used fairly easily without flux to the low melting point types which are very difficult to use fluxless (e.g. Sn/Pb M Pt 179°C).

For packages containing hybrid circuits, the solders used on the components inside will limit the melting temperature of the lid-sealing solders. If fluxes are used, they should be as inactive as possible because the activator leaves halide ions in the flux residue. To some extent, vulnerable components in the package can be protected from the flux with epoxies or rubbers.

PERMEABLE PACKAGES

The time taken for water vapor to saturate the air inside a package with a permeable lid seal which is potted in porous silicone rubber is short but it does not follow that the electronics will fail immediately in consequence. Indeed we know that most components survive in an 'atmosphere' of water saturated silicone rubber.

CMOS integrated circuits have been tested continuously in packages which had lids glued on with an epoxy (Epotek H77), the package potted in silicone rubber (Dow Corning 3140), and submerged in baths of 1 percent saline at room temperature. Three out of three such samples have now run successfully for 21 months. Direct measurement of the humidity inside such a package during submersion showed that the inside saturated after about 50 days.

Of another batch of 17 nominally similar chips subjected to the same test, 2 failed in under 20 days and one lasted more than 200 days, the rest fell in between. It is thought that this is due to an unusually high phosphorus fraction in the Topcoat glass of these integrated circuits. This phosphorus is normally added to the silicon dioxide film on the chips because the resulting glass covers the surface more uniformly and for other reasons. However, too much phosphorus will make the layer hygroscopic and water vapor will condense onto it before the atmosphere is saturated. Examination after failure showed that in all fifteen samples which had failed inside the package, the aluminum on the chips had corroded. This indicates the importance of keeping the package contents clean of ionic contaminants which will cause condensation into themselves and thus form electrolytes, particularly in close proximity to the electronic components.

It may be objected that testing packages at room temperature is too favorable even for an implant and that by running the device continuously, the heat dissipated tends to keep the chip dry. To answer this criticism, two samples were placed in a saline bath which was temperature cycled (six hours on, six hours off) between room temperature and 10°C higher. The devices are run for 1/128th of the time in 0.5 s bursts. Neither device has failed after 16 months.

Acknowledgements — Some of the topics discussed above have been dealt with fully by other members of the Unit; most of the ideas result from team effort. I should like to acknowledge the support of the Medical Research Council.

REFERENCES

Donaldson N. de N. (1979) Morphognostic coils: a technique for transmitting several nearfield radio signals through the same space. *Med. Biol. Engng. Comput.* 17, 271-274.

Donaldson P. E. K. (1973) Experimental visual prosthesis. *Proc. IEE.* 120, 281-298.

Donaldson P. E. K. and Sayer E. (1975) A vacuum centrifuge for void-free potting of implantable hybrid microcircuits in silicone. *Med. Biol. Engng.* 13, 595-596.

Donaldson P. E. K. (1976) The encapsulation of microelectronic devices for long-term surgical implantation. *IEEE Trans.* BME-23, No. 4.

Donaldson P. E. K. (1977) Silicone-rubber adhesives as encapsulants for microelectronic implants; effect of high electric fields and of tensile stresses. *Med. Biol. Engng. Comput.* 15, 712-715.

Holmes P. J. and Loasby R. G. (1976) *Handbook of Thick Film Technology.* Electrochemical Publication Ltd.

Traeger R. K. (1977) Nonhermeticity of polymeric lid sealants. *IEE Trans.* PHP-13, No. 2, 147-152.

Power Sources for Implant Telemetry and Stimulation Systems

W. H. KO

Engineering Design Center, Case Institute of Technology, Case Western Reserve University, Cleveland, Ohio 44106, U.S.A.

Abstract — A discussion on various possible power supplies will be given. This includes: (1) mercury, silver oxide, and lithium batteries; (2) nuclear batteries; (3) rechargeable batteries; (4) RF and magnetic induction powering; (5) body fluid galvanic cells; (6) body fluid fuel cells; (7) mechanical convertors; (8) other body energy to electrical power convertors; and (9) combinations of these techniques. The RF powering is the choice for longterm implant telemetry. Either separate frequencies can be used for power in and signal out or a single frequency can be used for power and signal transmission on a time sharing basis. The design of the two frequencies and single frequency RF powering systems will be discussed. The telemetry systems employing each of these schemes will be presented with evaluation data in animals.

INTRODUCTION

For present biomedical implant telemetry or stimulation systems, the power source occupies more than 50 percent of the total space and weight of the implant unit and accounts for more than 50 percent of system failures; according to experience with cardiac pacemakers and biomedical telemetry (Parsonnet, 1977; Ko, 1979). The limited lifetime, lack of dependability and lack of nondestructive test methods to assess the condition of the battery prior to implant operation, are the major problems facing system designers and experimentalists using implant electronic instruments.

Many power sources have been used, reported, and proposed in the literature. These can be grouped as: (1) Primary batteries — including nuclear batteries. (2) Secondary batteries — including various rechargeable batteries. (3) Radio frequency (RF) coupled and magnetic induction power. (4) Body fluid galvanic and fuel cells. (5) Body and environmental energy converters.

This article attempts to survey the sources briefly and to collect useful information on the first three categories for the use of biomedical engineers, and to describe a few examples of RF powered telemetry systems.

PRIMARY BATTERIES

Primary batteries convert chemical or nuclear energy into electrical energy through irreversible processes. In chemical cells, the cathode is chemically reduced inside the cell and becomes the positive terminal for the outside circuit; while the anode is oxidized in the cell and becomes the negative terminal for the outside circuit. The anode is usually a metal that gives up electrons which become (+) charged cations in the electrolyte. The cathode is usually a compound that can accept electrons, forming (-) charged anions in the electrolyte. The primary battery is the most convenient source to use since it is obtained as a component and no design work is involved, except in the selection of a proper type of battery. For implant telemetry and stimulation, the average current drain is usually low, therefore, the battery's internal resistance is not very important, although there may be exceptions. The major considerations are: (1) Safety in the body environment; such as the hermetic seal, out gas possibility, toxicity, radiation level, etc. (2) Size and weight: generally should be less than 2 percent of the body volume and weight. (3) Availability and cost.

Table 1 shows the typical characteristics of primary batteries with parameters that are of interest to electronic designers (Grossman, 1978; Yates, 1978). Figure 1 indicates the typical discharge curves for primary batteries.

The service capacity of a battery is affected by: (1) Discharge rate — current drain. (2) Temperature — higher and lower temperature than 20°C may reduce the capacity significantly. Some dry batteries will have only 50 to 75 percent of 20°C capacity at 40°C. (3) Duty cycle. (4) Allowed voltage range. (5) Shelf time at storage temperature.

Even without considering the shelf time, which varies from unit to unit depending on how long the battery has been stored since it was manufactured, the actual charge capacity of a battery would be smaller than the value specified at 20-25°C by the manufacturer. A safety margin is needed in calculating battery capacity. A factor of 1.5 to 2.0 is generally used (for a 100 mAh requirement, use a *new* 150 to 200 mAh battery) and a derating of the battery due to storage time must be added.

LITHIUM BATTERIES

The lithium battery, which has become available in recent years, is clearly the choice if cost and availability are not considered. The lithium battery has the combined advantages of: (1) High energy density per weight and per volume. (2) High open circuit voltage per cell. (3) Long shelf and service life.

For these batteries, lithium is used as the anode (negative terminal). Iodide and other materials may be used as the cathode. Both solid and liquid electrolytes may be used (Grossman, 1978; Greatbatch, 1971). Table 2 shows typical characteristics of several kinds of lithium batteries used as cardiac pacemaker power sources, with Hg/Zn and Ni/Cd batteries listed for comparisons. Figure 2 shows the discharge curve of Sanyo LF-C (120 mAh) lithium cells as compared to mercury and silver oxide batteries. Lithium cells from 30 mAh up to several A-H are now available.

With the advance of double-wrap techniques for mercury batteries, the lifetime and failure rate has been improved greatly however, small batteries fabricated with this technique are not yet generally available in small quantities. Table 3 lists the dimensions of several primary chemical batteries that are commonly used in biotelemetry systems.

NUCLEAR BATTERIES

Small nuclear batteries were developed for cardiac pacemakers and can be used for

Table 1.
Typical characteristics of primary batteries

Common name	Carbon zinc (Leclanche)	Alkaline-manganese dioxide	Mercuric oxide	Silver oxide	Lithium*
Electrochemical system	Zinc-manganese dioxide (usually called Leclanche or carbon-zinc)	Zinc-alkaline manganese dioxide	Zinc-mercuric oxide	Zinc-silver oxide	Li/various materials
Voltage per cell	1.5	1.5	1.35	1.5	2.2-3.6
Negative electrode (anode)	Zinc	Zinc	Zinc	Zinc	Li
Positive electrode (cathode)	Manganese dioxide	Manganese dioxide	Mercuric oxide	Monovalent silver oxide	Various materials
Electrolyte	Aqueous solution of ammonium chloride and zinc chloride	Aqueous solution of potassium hydroxide	Aqueous solution of potassium hydroxide or sodium hydroxide	Aqueous solution of potassium hydroxide or sodium hydroxide	Various electrolytes
Commercial capacities	60 mAh to 30 Ah	Several hundred mAh to 23 Ah	16 mAh to 28 Ah	35 mAh to 210 mAh	30 mAh to 12 Ah
Energy densities:					
W h/kg		100	110	110	220-330
W h/cm^3	4.5	0.18	0.5	0.5	0.3-0.9
Practical current drain rates:					
Pulse high	Yes	Yes	Yes	Yes	Yes
(more than 50 mA)	100 mA/in^2 of zinc area ('D' cell)	200 mA/in^2 of zinc area ('D' cell)	No	No	Yes
Operating temperature range	20 to 130°F	-20 to 130°F	32 to 130°F	32 to 130°F	-40 to 160°F
Shelf life (20°C)	1-2 yr	1-2 yr	3-4 yr	3-4 yr	10 yr
Cost: initial	Low	Medium	High	High	High

*These are number of basic Lithium battery systems. Values given are typical for Lithium batteries.

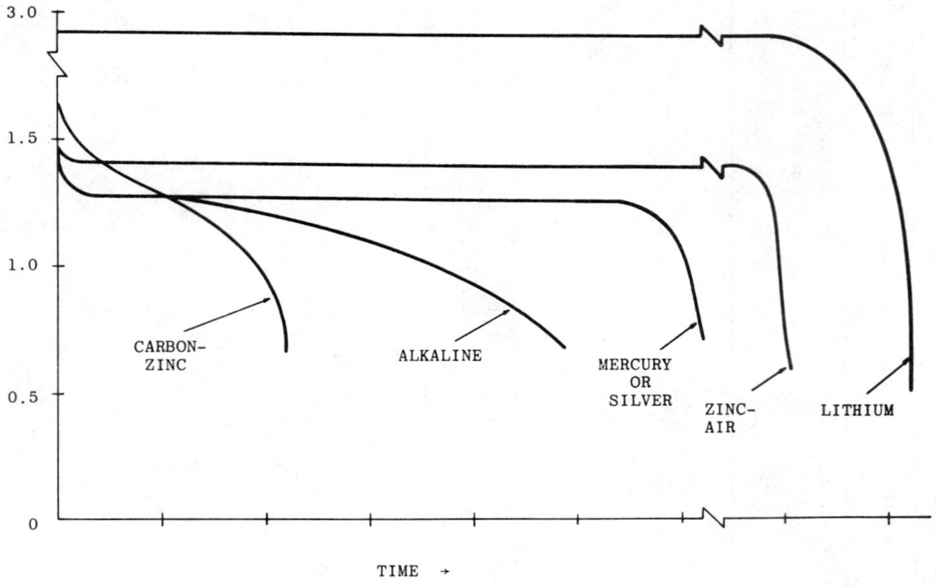

Fig. 1. Representative discharge curves for primary batteries.

other implant instruments. Two types of conversion processes are used: betavoltaic and thermoelectric. The former utilizes the energy of the beta particles emitted by the radio isotope, promethium-147, to interact with silicon p-n junctions to generate electrical energy; the latter utilizes the heat generated by the isotope, plutonium-238, and thermoelectric elements to convert heat into electrical energy. Table 4 summarizes the characteristics of available nuclear batteries (Gasper and Fester, 1975). The maximum radiation level of all devices is below 20 mrem h^{-1} and the average level is below 5 mrem h^{-1}. These values are considered to be acceptable for the safety of the patients and population. Over 2300 nuclear battery-powered pacemakers were implanted between 1970 and the end of 1976, according to Parsonnet (1977), and there has not been a single failure reported since the beginning of the implants.

SECONDARY BATTERIES

Table 5 summarizes the characteristics of secondary batteries that can be recharged. Some of the popular cell dimensions are also given in Table 3. Only the hermetically sealed nickel-cadmium cells have been used in implant electronics, due to cost, availability and cycle lifetime. However, the rechargeable batteries are used extensively in the RF coupled or magnetic induction power sources. A brief summary of the selection and use of these secondary batteries is given below.

The energy density and discharge curves of the secondary cells are given in Fig. 3a and b. Recently, lithium rechargeable cells have been reported and their characteristics are included in Table 5, but at this time they are not generally available.

When using Ni-Cd sealed cells, attention sould be given to charging rate and cell temperature. Most of the Ni-Cd cells are recommended at 10 h charging rate {I = (AH/10)} for 15 h (150 percent recharge or 50 percent overcharge). Many fast charging cells can be charged at the one hour rate but the charging circuit has to have an automatic circuit to reduce charge rate when the battery approaches full charge (Lin *et al.*, 1970). A safety circuit should be included to prevent the cell

Table 2.
Typical characteristics of lithium and other batteries used in cardiac pacemaker

	Wilson Greatbatch		Catalyst	Other lithium						
	702E (+CRC)	755	research 802/23	Saft 210 silver chromate	GTE/ARCO thionyl chloride	Mallory LSA-9006	Dupont cordis	W G Bromine 766	HG/Zn Mallory RMI	NiCAD BF-20
Anode	Li	Li	Li	Li	Li	Li	Li	Li	Zn	Ni
Electrolyte	LiI	LiI	LiI	$LiClO_4$ +propylene	$LIAlCl_4$ $+Al_2O_3$	LiI $+Al_2O_3$	$LiClO_4$	LiB_r	NaOH+Ag	NiOOH
Cathode	I_2+P_2VP	I_2+P_2VP	I_2+P_2VP	Ag_2CrO_4	$SOCl_2C$	PbI_2PbS +Pb	CuS	Br_2P_2VP	HgO	Cd
Voltage-open circuit/cell	2.8	2.8	2.8	3.4	3.66	1.9	2.11	3.45	1.356	1.27
Weight (g)	80	33	30	8.0	16	40	17(51)t	27	13.8	12.0
Volume (cm^3)	31	9.5	11.2	2.88	6.6	10	6.0	9.5	3.0	6.83
Capacity (B.O.L A-H)	3.5	3.0	2.3	0.55	1.9	0.91	1.8	3.5	1.0	0.19
Energy density W-H/kg	110	230	200	230	410	120	0.217	440	100	20
W-H/cm^3	0.28	0.79	0.53	0.65	0.97	0.46	0.61	1.25	0.45	0.04
Self discharge	<10% in 10 yr	<10% in 10 yr	<10% in 10 yr	None measured	<1% in 1 yr	None measured	None measured	<10% in 10 yr	Yes, 23 μA	15%/wk
Hermetic seal	Yes	Yes	Yes	Yes	Yes	Yes	No	Yes	No	Yes

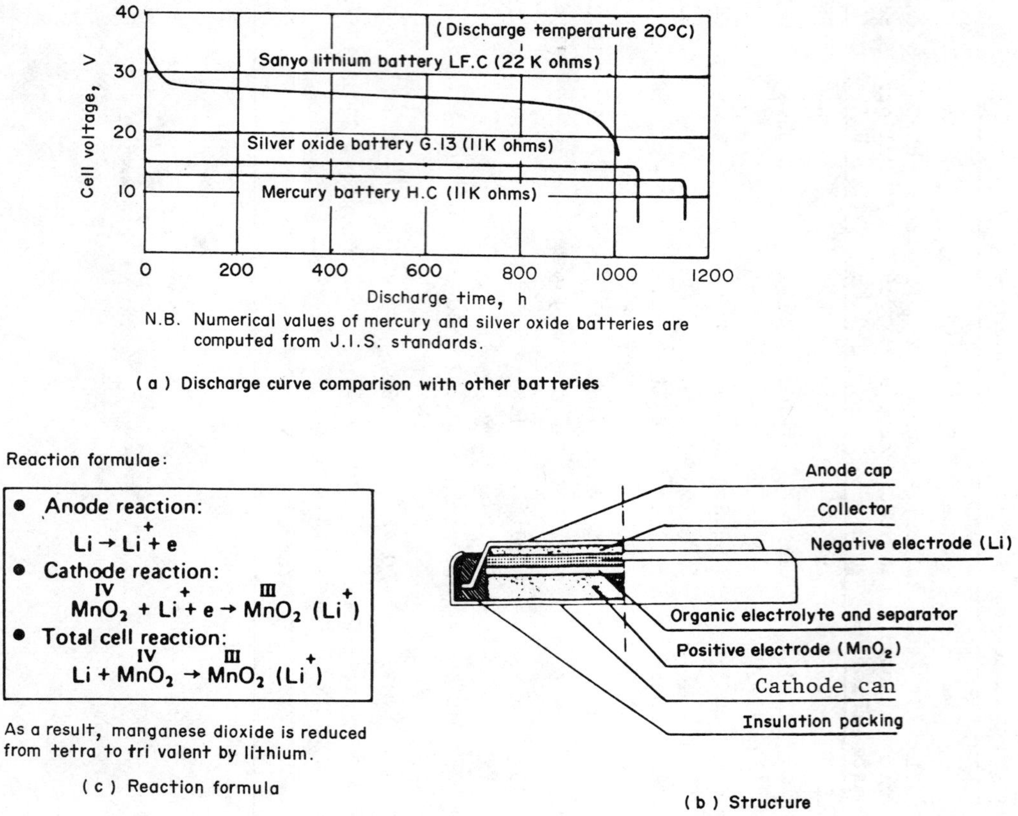

Fig. 2. Lithium (SANYO LF-C) battery discharge characteristic and structure.

temperature from going beyond 45°C to avoid excessive gas pressure build-up in the cell which may cause an explosion.

One of the reversible failure modes of the Ni-Cd cell is referred to as the memory effect which is caused by repetitive discharge to a shallow depth. If a cell is discharged to 20 percent of its capacity and then fully recharged, after many cycles the cell memorizes this and will deliver only 20 percent of its rated capacity before the voltage drops down, when a deep discharge to greater capacity is attempted.

Sometimes there is also the depression of the discharge voltage curve during the latter portion of the discharge period when the cell has been subjected to prolonged overcharge with no use (Yates, 1978).

For implant systems using rechargeable batteries, a remotely coupled charging circuit is needed to charge the implanted battery. Circuits described in the next paragraph can be used for this purpose.

RF COUPLING AND MAGNETIC INDUCTION POWERING

RF powering can be used to power an implant electronic device over an indefinite period inside the body by magnetic coupling. This power may be used to supply the

Table 3.
Dimensions of small batteries used in biotelemetry

Electrochemical system	Type	Capacity mA h	Weight g	Height/ diameter (cm)
Mercury/Zn	RM-212	16	0.3	0.33/0.55
1.4V/cell	RM-312 E-312 Hg-312	36	0.64	0.36/0.79
	RM-13GH	60	0.98	0.533/0.79
	RM-400 E-400 Hg	80	1.15	0.345/1.16
	RM-675 E-695 Hg-675	160	2.22	0.540/1.16
	RM-640	500	7.59	1.11/1.59
Silver oxide/Zn	MS-13H	60	0.98	0.533/0.79
1.5V/cell	E-301 W5-11	100	1.68	0.415/1.16
Lithium/MNO_2	LF-A LF-C	30 120	0.5 1.5	0.36/0.79 0.54/1.16
3.0/cell	LF-1/2 W	200	4.0	0.28/2.45
Lithium	TL-71	80	1.8	0.35/1.15
3.4/cell	TL-73	180	2.45	0.40/1.40
Li/$(CF)_n$	BR-425	20	0.55	2.59/0.42
2.8V/cell	BR-2325	140	3.1	0.25/2.3
Ni/Cd rechargeable	GN-20B	20	1.50	0.55/1.15
1.20/cell	GN-50B	50	2.58	0.63/1.55

RM and MS Mallory, E-Eveready, Hg-Borgess, BR and WS-Panasonic, LF-Sanyo, TL-Tadiran, GN-Gould National.

circuit needs directly or may be used to recharge a battery (Fryer, 1974) or a capacitor (Ko *et al.*, 1979) which may store the energy for some time period. A general block diagram of the RF powering system is shown in Fig. 4a, if $M = k\sqrt{L_1L_2}$ is the mutual inductance between the coils L_1 and L_2, $Q_1 = WL_1/R_1$, $Q_2 = WL_2/R_2$ are the unloaded Qs of the coils.

The power efficiency,

$$y = \frac{P_0}{P_1} = \frac{K^2Q_1Q_2{}^3\ R_0/2}{(\frac{R_0}{2} + Q_2{}^2R_2)\{1 + K^2Q_1Q_2\}(\frac{R_0}{2}\ Q_2R_2)} \quad . \qquad (1)$$

In practical cases, R_0 is determined by the electronic load to be powered; Q_1, Q_2 are nearly constant for a selected wire and configuration; the efficiency (y) will be determined by k which depends on the core size and spacing. There are many ways to optimize the efficiency. If dy/dR_2 is set to zero in the optimization

Table 4.
Characteristics of nuclear batteries

Type and manufacturer	Size (cm)	Weight (g)	Output power (μW)	Conversion elements
Betacel™, McDonnell Douglas	2.4 × 2.3D	98	370 @ 3.6 V BOL 70 μW @ 2.9 V after 5 yr	Silicon pn Junctions
Atomcell™, Nuclear Battery Corp.	4.6 × 1.7D	28	600 @ 0.3 V	Bismuth Telluride Thermopile
General Atomic	3.8 × 1.6D	26	400 @ 0.4 V	Bismuth Telluride Thermopile
ARCO Nu-5	6.5 × 3.2	120	200 @ end of 11 yr	Cupron-Tophel wire tapes
Syncal	3.8 × 1.3D	28	300 @ 5.0 V	Silicon Germanium Thermopile
CIT-Alcatel	4.7 × 2.3D	30	250 @ 0.55 V	Bismuth Telluride Thermopile
Coratomic	6 × 4.70 × 1.92	61	33 @ 4.05 V	Bismuth Telluride Thermopile

$$R_{2op} = \frac{R_0}{2} \cdot \frac{(1 + k^2Q_1Q_2)^{\frac{1}{2}}}{Q_2^2} \tag{2}$$

and

$$y_{op} = \frac{k^2Q_1Q_2}{\{1 + (1 + k^2Q_1Q_2)^{\frac{1}{2}}\}^2} \tag{3}$$

it is seen that for maximum y, $k^2Q_1Q_2$ should be maximized.

Figure 4b shows the typical coil configuration where L_1, L_2 are coaxial coils of diameter d_1 and d_2 with n_1 and n_2 turns respectively. For a given d_2 (receiving coil diameter) and D, in order to maximize k, the primary coil diameter d_1 should be:

$$d_1 = \{d_2{}^2 + (2D)^2\}^{\frac{1}{2}} \tag{4}$$

The procedure for designing the RF coupling circuit can be summarized as follows (Ko, 1977):

1. Derive known conditions:
   (a) $R_o = V_o/L_o$ at the load.
   (b) Determine the mean or maximum spacing D required.
   (c) Determine d_2.
   (d) Determine frequency of RF powering f.
2. Calculate d_1 from Equation (4).
3. Calculate (r_2/r_1); shape factor F_1, F_2, of coils L_1, L_2; and coupling factor N from (Terman, 1947) and calculate $k = 1.27N/\sqrt{F_1F_2}$.
4. Estimate unloaded Q_1, Q_2* and calculate y_{op}.

Table 5.
Typical characteristics of secondary batteries

Common name	Nickel-cadmium (sealed)	Silver-cadmium	Lead-acid	Silver-zinc	Lithium* Titanium Disulfide
Electrochemical system	Nickel-cadmium	Silver-cadmium	Lead acid	Silver zinc	Lithium
Voltage/cell	1.2	1.1	2	1.5-2.1	1.5-1.9
Negative electrode (anode)	Cadmium	Silver oxide	Lead cal-cium/anti-mony	Silver zinc	Lithium
Positive electrode (cathode)	Nickel-hydroxide	Cadmium oxide	Lead dioxide	Zinc	Titanium disulfide
Electrolyte	Aqueous solu-tion of potas-sium hydroxide	Potassium hydroxide	Gelled solution of sul-phuric acid	Potassium hydroxide	—
Cycle life	300 to 2000	150 to 300	200 to 400	25 to 100	—
Typical capacities	20 mAh to 10 Ah	100 mAh to > 50 Ah	30 mAh to 25 Ah	50 mAh to > 50 Ah	70-90 mAh
Energy density					
W h/kg	26-35	48-75	17.5-22	88-110	22-26
W h/cm^3	0.079-0.103	0.091-0.165	0.067	0.153-0.195	0.085
Temperature range:					
Storage	-40-140°F	-85-165°F	-40-100°F	-85-165°F	—
Discharge	-40-140°F	-10-165°F	-76-140°F	-10-165°F	—
Charge	32-113°F	32-115°F	32-113°F	32-115°F	—
Shelf life	Fair	Fair to good	Good	Fair to good	Good
Cost:					
Initial	Medium to high	High	Low	High	—

*Preliminary spec. for Battery Division, Exxon Ent. Somerville, NJ, U.S.A.

5. Calculate $L_2 = Q_2R_2/2$, $C_2 = 1/(2\pi f)^2L_2$.
6. Calculate L_1, C_1, n_1, n_2* and wire size.

Following this procedure, several telemetry and stimulation RF power sources were designed, with satisfactory results.

In designing RF power coupling system, two approaches have been demonstrated, (1) using two different frequencies for powering and signal transmission. This is similar to the frequency division multiplex except that one channel is used to supply power to the telemetry unit. In this case, the two frequencies should be far apart to avoid interference, particularly interference from the power channel to the signal channel. Frequency ratios (f_1/f_2) of 10 to 100 were used. Because the powering channel has much higher power, it should be kept constant by a regulated dc supply which helps maintain constant amplitude. (2) The second approach is similar to the time division multiplex, and involves using the same, or nearly the same, frequency for powering and signal transmission, but time sharing the frequency by turning the power channel on and off and transmitting the signal while the power channel is off (Ko *et al.*, 1979).

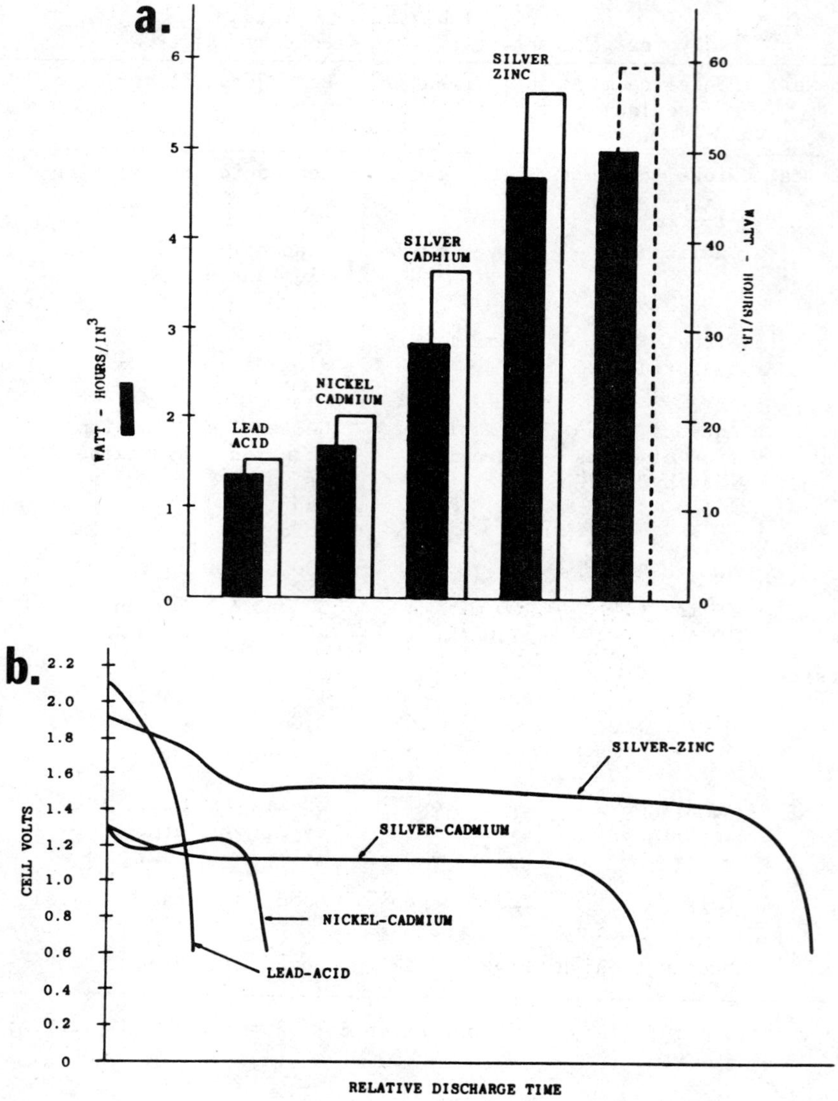

Fig. 3. Characteristics of secondary batteries. (a) Power density, (b) discharge curves.

For telemetry purposes, the load requirement is generally small, in the order of 1 mW. However, for stimulation purposes or for charging secondary batteries, the power required will be several orders of magnitude higher. The principle for the design of higher power coupling circuits and the switched coupling circuit is the same as for low power coupling, but each case will need individual adjustment. The present procedure can be used as a first step for designing all types of RF powering circuits. Three RF powered implant systems covering different cases will be given as examples.

IMPLANT INTRACRANIAL PRESSURE AND TEMPERATURE MONITORING SYSTEM

The implant intracranial pressure (ICP) and temperature monitoring system is aimed at monitoring the ICP of neurosurgical patients for a period from weeks to years. Figure 5

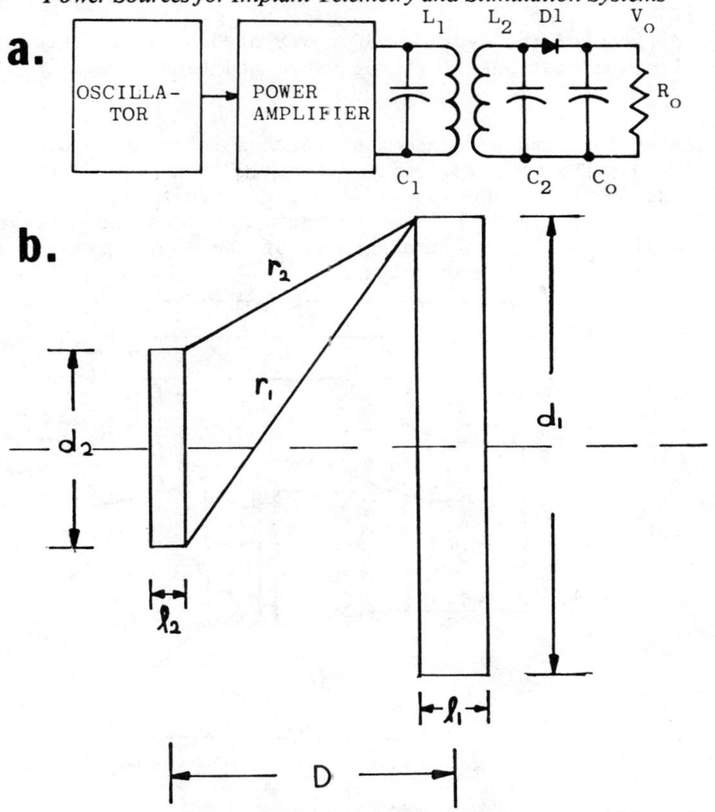

Fig. 4. (a) Basic RF powering block diagram, and (b) coil configuration.

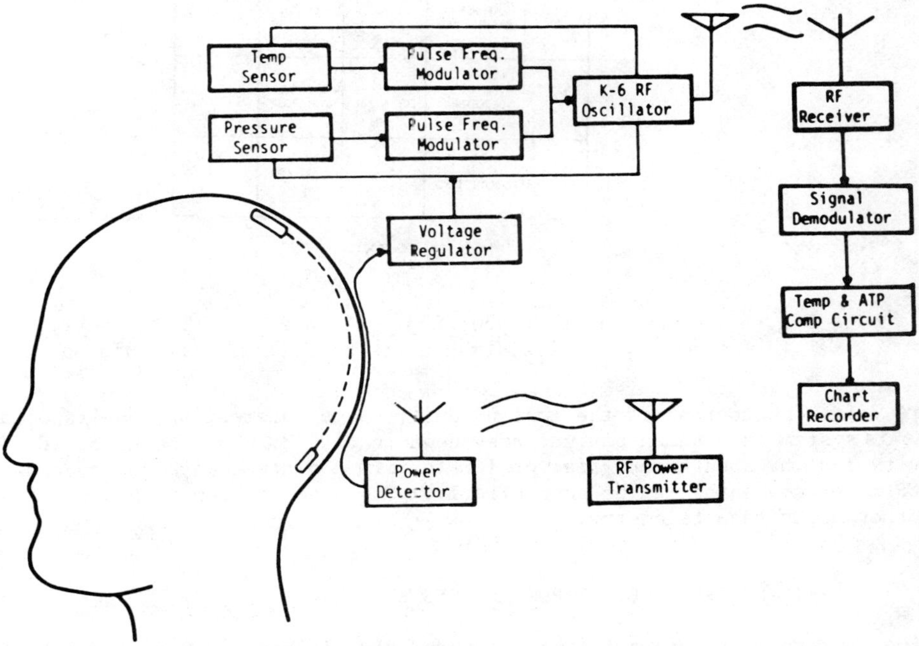

Fig. 5. Block diagram of the ICP telemetry system.

shows the block diagram of the system. RF power at 3.5 MHz is used to power the implant electronics, which transmit the pressure and temperature information at 120 MHz.

Figure 6 illustrates the structure and construction information of: (a) the external unit including the 3.5 MHz powering coil and the 120 MHz receiving coil; and (b) the implant unit including the 3.5 MHz power detector circuit and the 120 MHz transmitting coil. The aim is to allow 0.2 to 1.5 cm separation between the powering coil and detector with a maximum of 1.5 cm misalignment of the axis, giving a 1 mW power and 4 V at the implant power detector terminals.

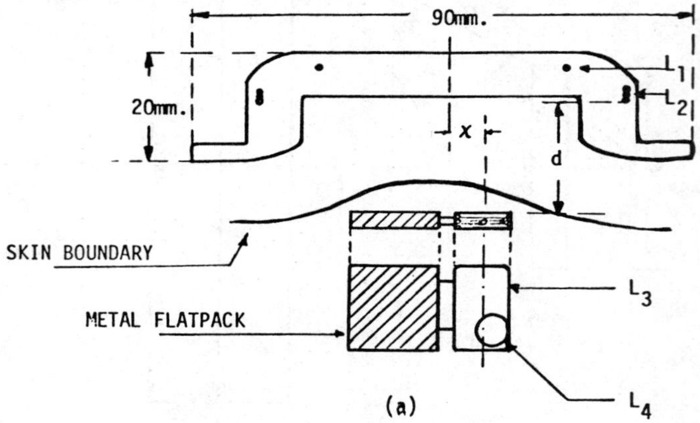

(a)

COIL	FUNCTION	TURNS	DIAMETER	WIRE GAUGE
L_1	SIGNAL RECEIVER	1 T.	43 MM.	20
L_2	POWER TRANSMITTING	12 T.	65 MM.	28
L_3	POWER DETECTOR	2 LAYERS 24T & 23T	RECTANGULAR 16*9.5*3.2 MM.	38
L_4	SIGNAL TRANSMITTING	5 T.	4.5 MM.	26

(b)

Fig. 6. Structure of RF powering coils used in ICP telemetry unit. (a) Coils structure; (b) Coil winding information.

Figure 7 is a photograph of the implant unit. Some interesting results obtained from this system on a human patient are shown in Fig. 8. The change of ICP during activity and the spontaneous rise of ICP to form a plateau wave and apparent decrease of ICP after awakening can be seen clearly. This information is not obtainable by any other means than telemetry.

SINGLE FREQUENCY RF POWERED SYSTEMS

By time sharing a frequency for powering and signal transmission, the interference between the two functions is essentially eliminated; the circuit can be simplified;

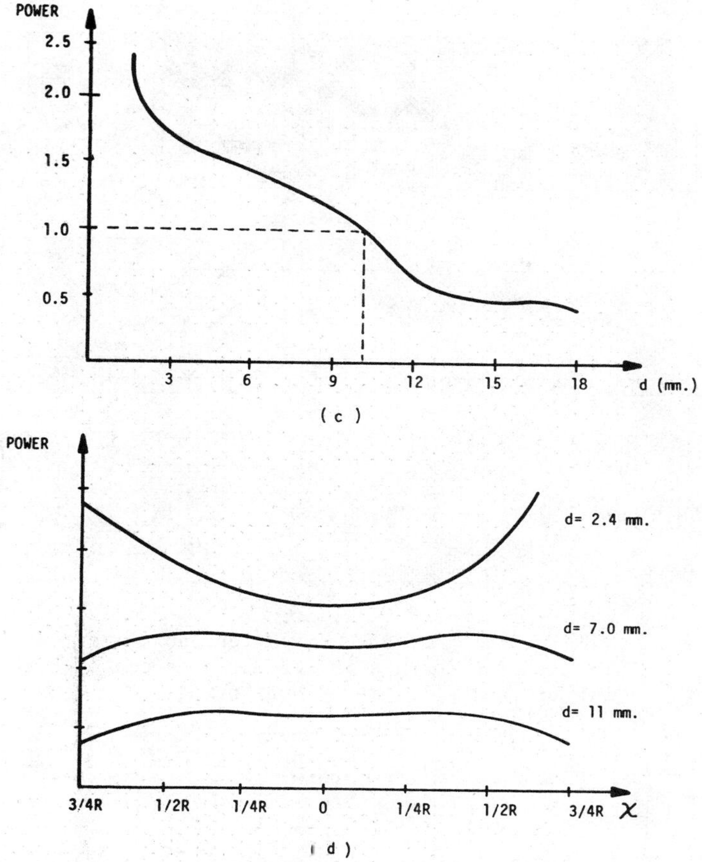

Fig. 6. Performance of the coils. (c) Power versus separation between coils; (d) power versus lateral misalignment.

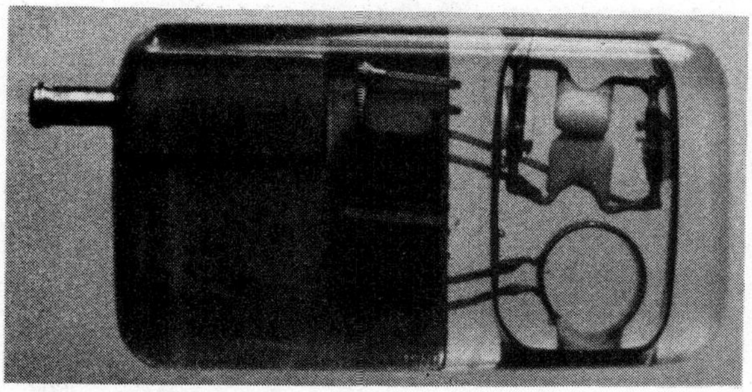

Fig. 7. ICP implant unit.

and since one tank circuit is used for both functions, the volume of the implant unit can be reduced. This RF powering scheme was studied at the Engineering Design Center, Case Western Reserve University (Ko, 1977; Kou, 1976). The block diagram of a three channel telemetry system is shown in Fig. 9 along with the timing diagram. The signal is transmitted as pulse position modulated (PPM) carrier pulses in the time slot when the RF powering oscillator is turned off. Many variations of

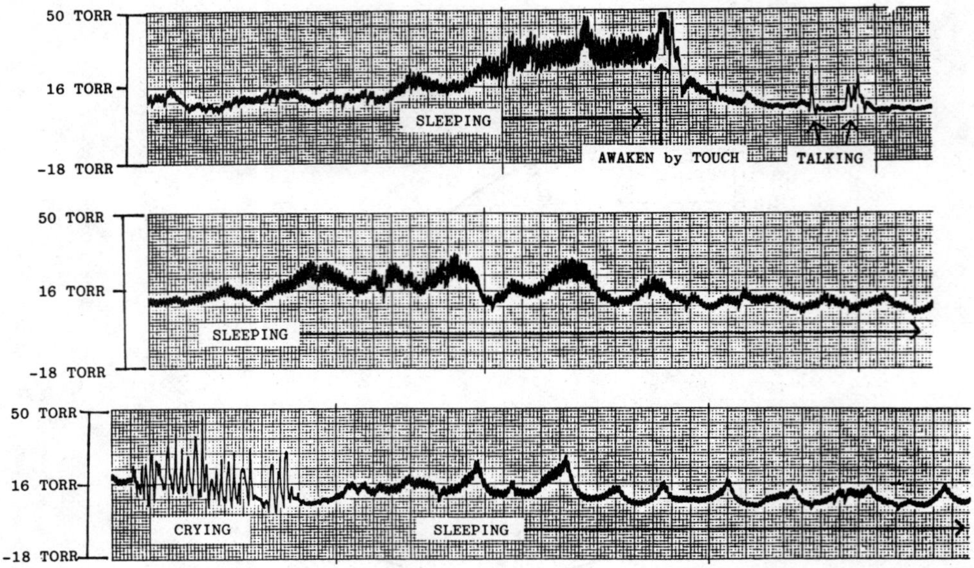

Fig. 8. ICP recording of a human patient during sleep with spontaneous changes in ICP during sleep. (Recorded by Continuous Telemetry Monitoring).

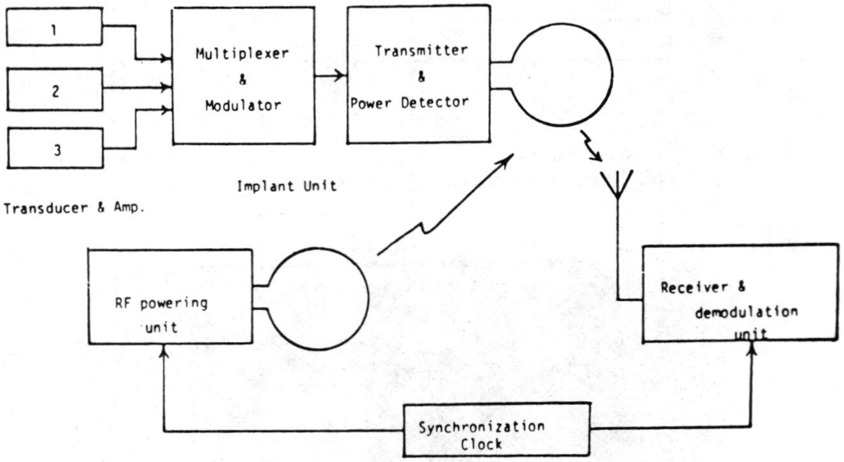

Fig. 9. Single frequency RF powering telemetry system. (a) System block diagram.

design can be pursued. The duty cycle of the RF powering oscillator can be varied and more than one signal channel can be transmitted during one time slot. Both the three channel and single channel unit were constructed and tested in the laboratory. The RF powering coil is a three turn, 12 cm diameter coil and the implant coil is an air core or ferrite core coil of 1 cm diameter.

An interesting design by Hynecek (Ko *et al.*, 1979) is shown in Fig. 10 where the junction FET, Q-1, is used as a diode rectifier during the powering phase, when Q_2

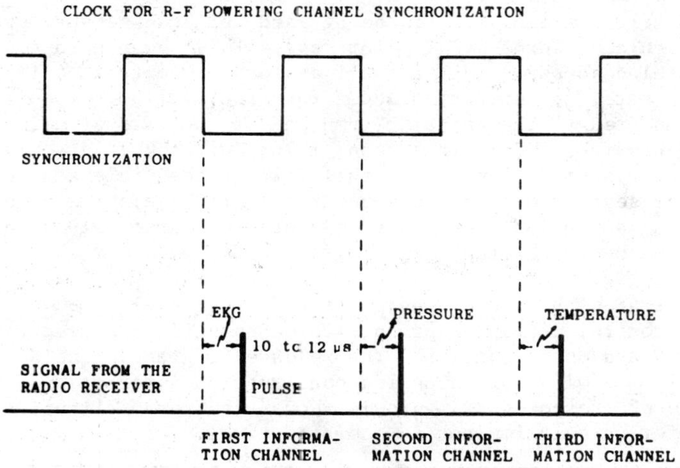

Fig. 9b. Input signal format.

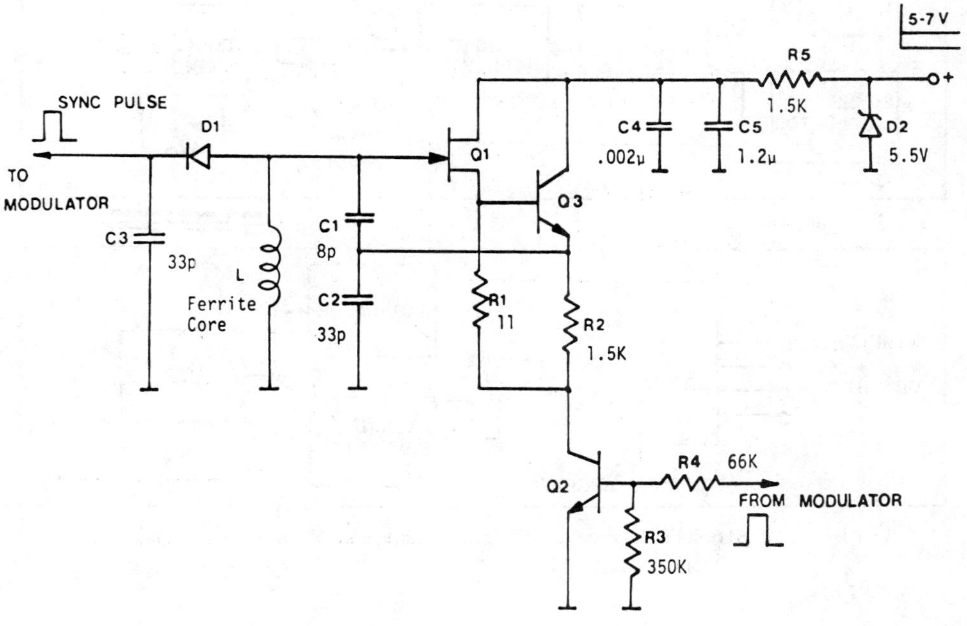

Fig. 10. Oscillator-detector circuit diagram of single frequency RF powering system.

is turned off. The rectified RF energy from the tank circuit L, C_1 and C_2 is stored in C_5, and the voltage is regulated by the Zener diode D_2. D_1 and C_3 are used to generate the synchronizing pulse for the modulator. When Q_2 is turned on by the modulator for a time period τ after the RF powering signal stops, Q_1, Q_3 and L_1, C_1, C_2 become an oscillator, transmitting the position modulated pulse signal to the external receiver through the same tank circuit. This principle was used for single channel ECG telemetry of rhesus monkeys for several months with good results.

RF POWERED NEUROMUSCULAR STIMULATOR

Functional electrical stimulation is being used for both sensory prostheses as well as for neuromuscular disorders. Implant devices have been used for visual and hearing prostheses (Mladejovsky, 1976) as well as for the control of (a) heart rate, (b) respiration rate, (c) urinary bladder function, (d) paralyzed limb muscles, (e) muscles connected to the spinal column and (f) nerve system for pair suppression (Hembrecht and Riswick, 1977; Richardson, 1976; Swiontek *et al.*, 1976). The power level required for stimulation is generally higher than telemetry can provide. Average power of several mW, and peak power at watt levels are needed. RF powering has been used in many chronic implants to avoid the large volume and weight required for primary or secondary batteries.

Figure 11 illustrates the block diagram of a four channel limb muscle stimulator system for the control of finger grasp. Three channels are used for proportional control of grasp and one channel for the release of grasp. The RF powering coils are shown in Fig. 12 with corresponding construction information. Figure 13 shows the performance of the coils. Figure 14 shows the hybrid circuit stimulator before final package for chronic implant in a small animal.

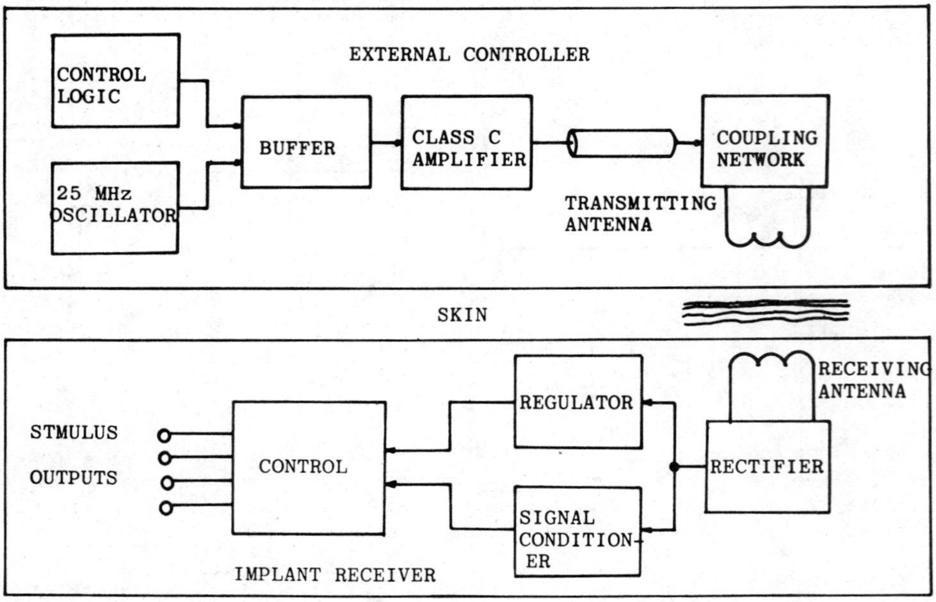

Fig. 11. Block diagram of RF powered multichannel muscle stimulator.

BIOLOGICAL BATTERY, BODY FLUID CELL AND BODY ENERGY CONVERTERS

Besides the power sources discussed above, there are many other means to transfer electrical energy into the body. Audren (1968) has reported on the use of transcutaneous carbon bottoms to provide direct connection through skin boundary, and the use of a skin tunnel transformer. Biological batteries using dissimilar materials to form galvanic cells from blood or body fluid as the electrolyte have been described (Parsonnet, 1964; Racine and Massie, 1966; Reynolds, 1964). Table 6 summarizes the characteristics of some of the body fluid cells. Since the unit functions as a galvanic cell, the anode material will dissolve and there is the problem of corrosion on other electrodes. These metal ions must be expelled by body organs or

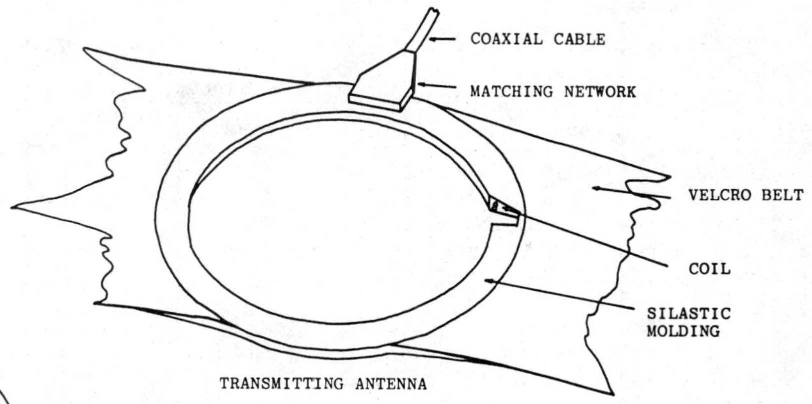

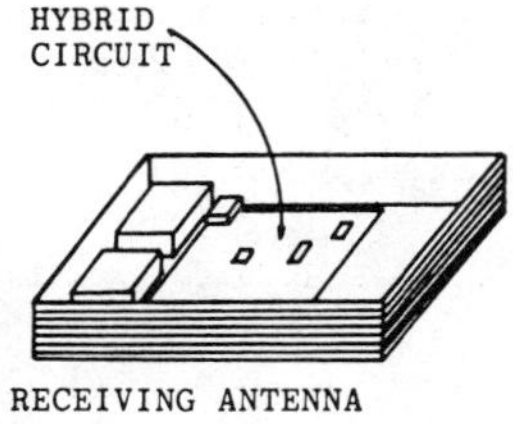

COIL DATA

COIL	SHAPE	SIZE	WIRE AWG	LAYER	TURNS	L, μH
TRANSMIT	ROUND	3" ID 0.08" T	18	1	2	0.9
RECEIVING	RECTANGLE	1.0"x1.4" x0.1" T	28	1	7	5.1

Fig. 12. Transmitting and receiving antenna for muscle stimulator system.

will accumulate and have toxic effects. Therefore, they can not be used where the toxic effect may be damaging. Experiments have been reported on implantable fuel cells which allow oxygen and fuels present in body fluids to react and generate electrical energy (Yao *et al.*, 1969; Dohan *et al.*, 1971). The fuels studied include: glucosamine, other carbohydrates, and polysaccharides. However, these fuel cells are still under development and will not be available for general use for many years.

Many other forms of energy can be transmitted across the body boundary such as light, infra red, vibration and ultrasound. In principle, these forms of energy can be used in a similar way as RF energy to provide a boundary coupling across the body. As yet no practical technique for general application exists in the literature.

Energy converters can be used to transform body energy into electricity. For example, thermal energy contained in a temperature difference between the deep part and the surface of the body can be utilized to generate electric power with thermoelectric elements. The chemical energy associated with pH differences between the stomach and other parts of the body fluid may be explored. The many cyclic motions of the body such as blood flow, heart beat, respiration, and muscle movement may be converted into electrical energy with electromagnetic devices and piezoelectric material. Figure 15 shows a mechanical converter coupled to a heart that can convert the heart movement into the vibration of a piezoelectric beam that generates ac power to be rectified and used to power a pacemaker. Such devices were fabricated and coupled to dogs' hearts that generated 30 μW for several months (Ko, 1966). However, for practical application further research in material and esign must be carried out.

CONCLUSION

In conclusion, lithium cells and other primary cells are simple to use if cost and availability are not serious problems. They are preferred power sources because of

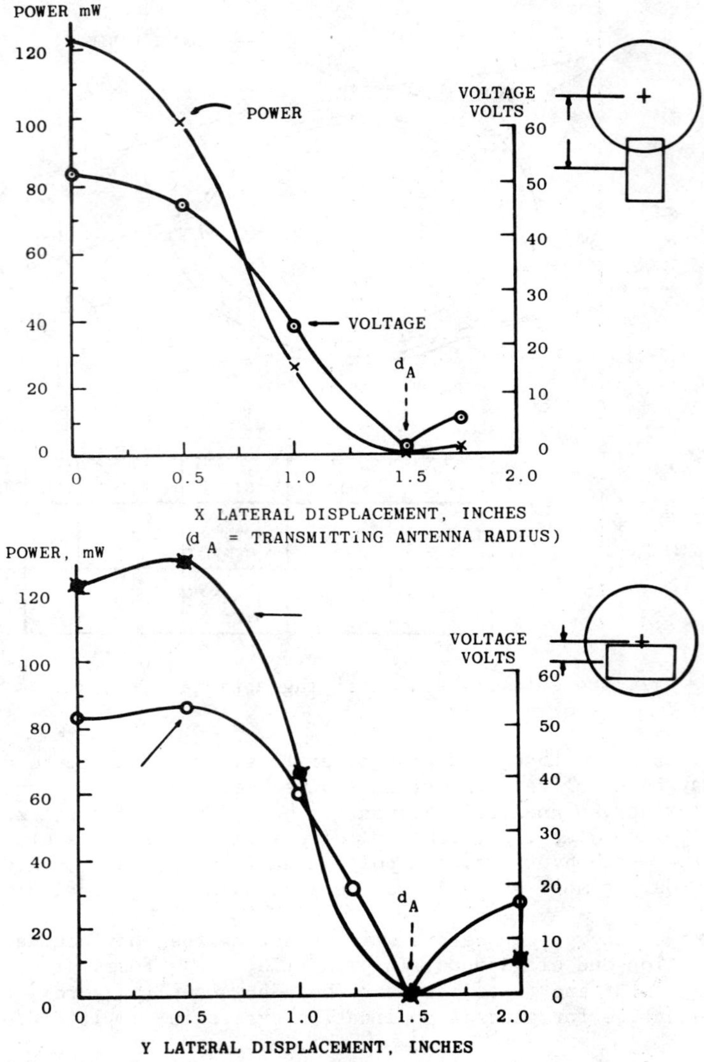

Fig. 13a. Pick-up voltage and power versus lateral displacement (22 kΩ load).

their simplicity and because no special skills are required for their use. Lifetimes from 5 to 10 years can be obtained. For chronic implant operation, or where the implant size or weight is a limiting factor, RF powering, involving either a two frequency or single frequency approach may be used. For future research in power sources, the real fuel cell and mechanical converter seem to offer good possibilities.

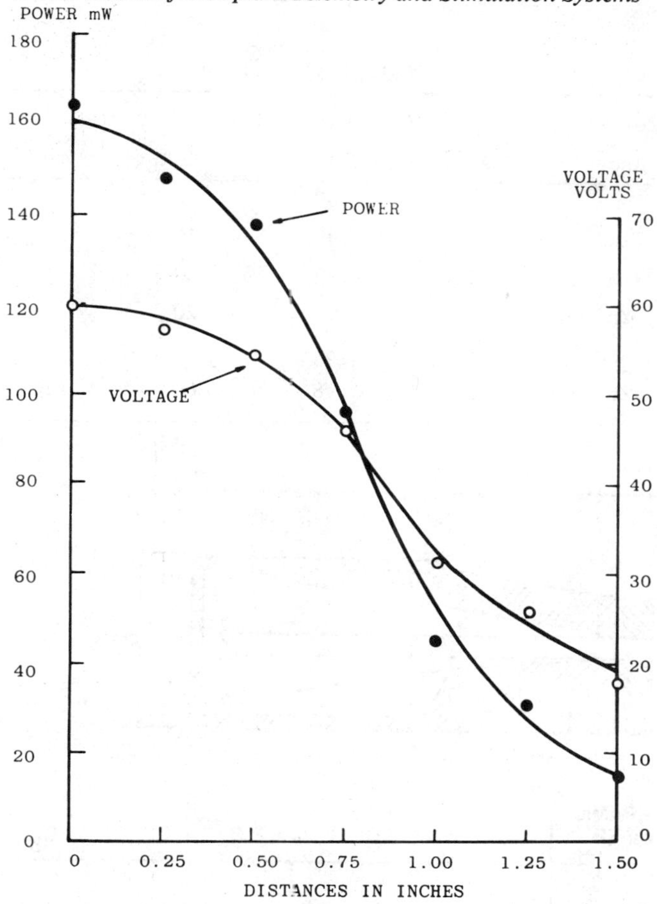

Fig. 13b. Voltage and power versus antenna vertical displacement.

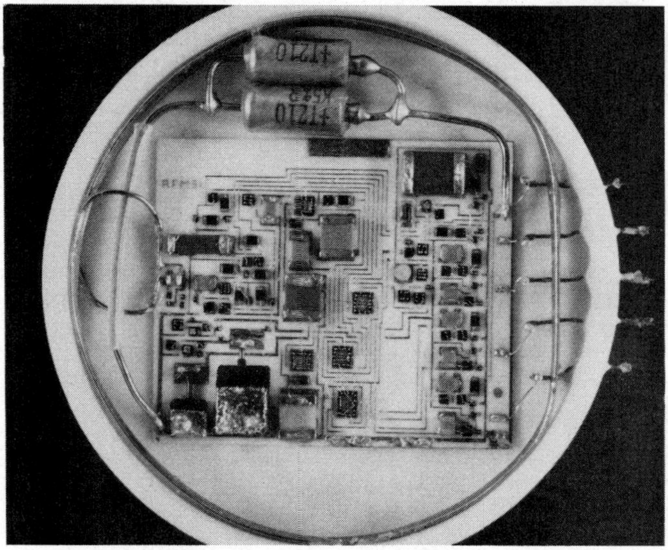

Fig. 14. Hybrid circuit in Macor™ ceramic capsule with antenna and feedthroughs (implant muscle stimulator).

Table 6.
Biological battery

Electrodes		$V_{o.c.}$ (Volts)		P_o (μW)		R_{out}
Cat.	Anode	Meas.	Calc.	$18k\Omega R_L$	$470\Omega R_L$	(kΩ)
Pt-Bk	HSS	0.70	1.67	80		2
Pt	Zn	1.12	1.99	50		1
Pt	Zn	1.0	1.99	30		10
Ag	Zn	0.7	0.91	20		1
AgCl	HSS	0.64	0.66		300	0.6
AgCl	Zn	1.0	0.98	60	1200	0.5

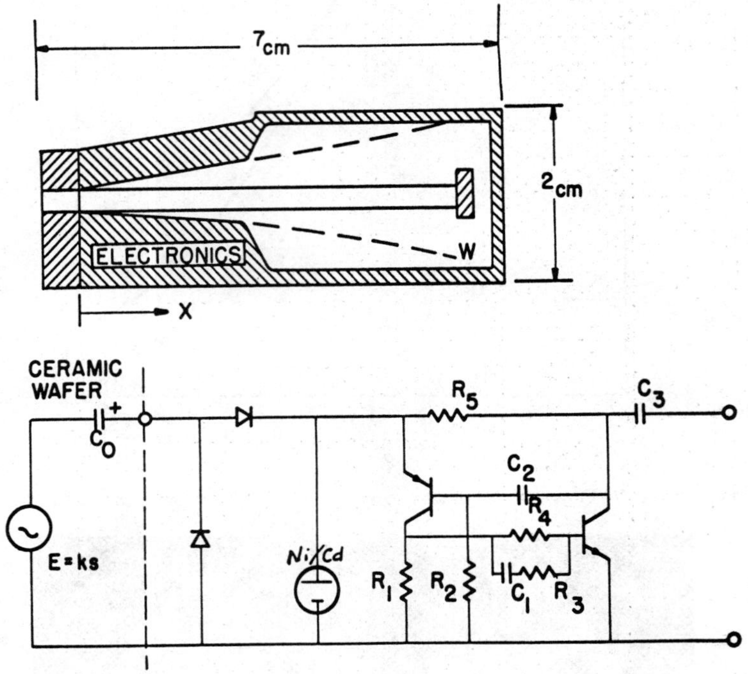

Fig. 15. Mechanical converter for pacemaker.

Acknowledgements — The support of NIH, GM-14267 and RR-00857; and my colleagues at the Engineering Design Center, Case Western Reserve University, whose results are used in this article, are sincerely acknowledged.

REFERENCES

Audren C. T. (1968) The skin tunnel transformer. *IEEE Trans.* BME 15, 4, 278-280.

Birk J. R., Klunder K., Smith J. C. (1979) Super batteries: a progress report. *IEEE Spectrum* 16, 3, 49-55.

Cataldi H. (1975) Performance of a mercuric oxide-zinc battery specifically designed for implantable cardiac pulse generators. *10th AAMI Meeting*, Boston, Mass.

Dohan L. A., Yao S. J. and Wolfson S. K. (1971) An enzymatic converter for the implanted fuel cell. *Trans. Am. Soc. Artif. Int. Organs* XVII, 411-414.

Fester K. and Doty R. (1973) Effects of epoxy encapsulation on performance of pacemaker batteries. *8th AAMI Meeting*, Washington, D.C.

Fester K. E., Doty R. L. and VanderVelden R. (1976) Chemical batteries for cardiac pacemaker. *Proc. 6th Annual Am. Soc. Mech. Engng. Sym. on Energy Alternatives.* Albuquerque, New Mexico, 93-103.

Fryer T. B. (1974) Power sources for implanted telemetry systems. In H. P. Kimmich and J. A. Vos, *Biotelemetry*, pp. 31-40. Leiden: Meander NV.

Gasper K. and Fester K. (1975) Cardiac pacemaker power sources. *Proc. Intersociety Engineering Conversion Conf.*, pp. 1205-1213, Newark, Delaware.

Goldson R. (1974) Pacemaker batteries. *Med. Elec. Data*, May-June, 60-63.

Greatbach W., Lee J. H., Mathias W., Eldridge M., Moser J. R. and Schneider A. A. (1971) The solid state lithium battery: a new improved chemical power source for implantable cardiac pacemakers. *IEEE Trans.* BME 18, 5, 317-323.

Grossman M. (1978) Focus on batteries. *Electron. Design* 12, 88-101.

Hembrecht S. T. and Reswick J. B. (1977) *Functional Electrical Stimulation-Application in Neuroprostheses.* New York: Marcel Dekker.

Jacob R. (1976) Aerial design in inductive loop telemetry. *IEEE Trans.* BME 23, 2, 118-123.

Kadefors R., Kaiser E. and Peterson I. (1969) Energizing implantable transmitters by means of coupled inductance coils. *IEEE Trans.* BME 16, 3, 177-183.

Ko W. H. (1977) Instrumentation for neuro-muscular stimulation. In S. T. Hembrecht and J. B. Reswick (eds.), *Functional Electrical Stimulation-Application in Neuroprostheses.* New York: Marcel Dekker. 515-528.

Ko W. H. (1966) Piezoelectric energy converter for electronic implants, *Proceedings of the 19th Annual Conf. on Engineering in Medicine and Biology*, 67.

Ko W. H. (1977) RF powered cage system for telemetry. *Proc. 30th Annual Conf. on Engineering in Medicine and Biology*, 260.

Ko W. H., Hynecek J. and Hona J. (1979) Single frequency R-F powered ECG telemetry systems. *IEEE Trans.* BME 26, 2, 105-109.

Ko W. H. and Hynecek J. (1974) Implant evaluation of a nuclear power source-betacel battery. *IEEE Trans.* BME 21, 3, 238-241.

Kou A. H. (1976) Single frequency RF powered three channel telemetry system, M.S. thesis, Case Western Reserve University, Cleveland, Ohio, U.S.A.

Lin W. D., Ko W. H. and Agrawal O. (1970) Test design circuits for Ni-Cd batteries used in implant electronic systems. *IEEE Trans.* BME 16, 177-183.

Mladejovsky M. G., Eddington D. K., Evans J. R. and Dobelle W. H. (1976) A computer-based brain stimulation system to investigate sensory prostheses for the blind and deaf. *IEEE Trans.* BME 23, 4, 286-295.

Parsonnet V. (1977) Cardiac pacing and pacemaker VII, power sources for implantable pacemakers, I & II. *Am. Heart J.* 94, 517-528.

Parsonnet V., Meyers G., Zucker R. and Lotman H. (1964) *Ann. N.Y. Acd. Sci.* 3, 922.

Racine P. J. and Massie H. L. (1966) An experimental internally powered cardiac pacemaker. *Med. Res. Engng.* 3, 24.

Reynolds L. W. (1964) Utilization of bioelectricity as power supply for implanted electronic devices. *Aerospace Med.* 35, 115.

Richardson D. E. (1976) Brain stimulation for pain control. *IEEE Trans.* BME 23, 4, 304-306.

Swiontek T. J., Sances A., Larson S. J., Ackmann J. J., Cusick J. F., Meyer G. A. and Millar E. A. (1976) Spinal cord implant studies. *IEEE Trans.* BME 23, 4, 307-312.

Talaat M. E., Kraft J. M., Cowley R. A. and Khazet H. (1967) Biological electrical power extraction from blood to power cardiac pacemakers. *IEEE Trans.* BME 14, 267-269.

Terman I. E. (1947) *Radio Engineers' Handbook*, Section 2, pp. 73-90, Table 12, p. 54, Table 17, p. 69. New York: McGraw Hill.

Yao S. J., Appleby A. J., Geisel A., Cash H. R., Jr. and Wolfson S. K., Jr. (1969) Anodic oxidation of carbohydrates and their derivatives in neutral saline solution. *Nature* 224, 921.

Yates W. (1978) Batteries: cordless power. *Electronic Products* 20, 39-46.

Expanded Radio Tracking Potential in Wildlife Investigations with the Use of Solar Transmitters

K. E. CHURCH

College of Environmental Science, University of Wisconsin—Green Bay, Green Bay, Wisconsin, U.S.A.

Abstract — The longevity of solar transmitters enables the researcher to gather longterm life history information which was previously limited by package weight and battery life associated with conventional primary power sources (i.e. mercury, lithium, and silver oxide batteries). Continuous transmission is possible with relatively low levels of indirect solar radiation, depending upon transmitter design. These range in size and capabilities from daylight only, 6 g, 1.25 V, single stage subminiature transmitter with a power output of 0.25 mW and a transmission range of 2 km (ground to ground); to a rechargeable, 800 g, 2.5 V, two-stage standard transmitter with a power output of 2 mW and a transmission range of 9.5 km (ground to ground). Attention must be given to attachment design to ensure solar panel exposure to solar radiation. Examples of attachment techniques are; fitted collar, slip-on neck collar, dorsal harness, tail-clip, tarsus 'bracelet', and cutaneous adhesion. Gray partridge (*Perdix perdix*) have been monitored with 10.5 g (total package weight) rechargeable solar transmitters in east-central Wisconsin, U.S.A., to gather year-round life history information previously unattainable via conventional primary power batteried transmitters.

INTRODUCTION

Radio telemetry is an established research technique for the study of animal behavior, physiology, and movements (LeMunyan *et al.*, 1959; Marshall, Gullion and Schwab, 1962; Cochran and Lord, 1963). This paper discusses the use of solar cells as an alternative to conventional primary power sources (i.e. mercury, lithium, and silver oxide batteries), for radio transmitters as suggested by Cochran and Lord (1963). Solar transmitters provide a greater energy-to-weight and energy-to-volume ratio in smaller packages than conventional primary battery powered transmitters. The increased longevity of solar transmitter packages enables the researcher to gather longterm life history information on 'weight sensitive' species without having to recapture the radio tagged cohort to replace expired batteries.

METHODS

There are two basic types of solar transmitters, daylight only and rechargeable.

Daylight only transmitters operate from the power emitted by the solar microgenerator panels via a storage capacitor. Transmission occurs only during daylight hours, and may temporarily discontinue when solar radiation is severely restricted (e.g. fog, dense cloud cover, dense vegetation or in den cavities). The field life of daylight only transmitters may be as long as 8 years (D. Royce, personal communication). Major factors limiting longevity are attachment failure, antenna breakage and the eventual failure of the storage capacitor.

Rechargeable solar transmitters operate from power supplied by a nickel-cadmium (NiCd) battery recharged through the solar panels (Fig. 1). Transmission is possible on a 24 h basis and during extended periods of restricted solar radiation. In most instances 2.5 to 5 h of continuous indirect sunlight will sufficiently charge the NiCd battery for 24 to 36 h of transmission, depending upon battery size and drain. The field life of rechargeable solar transmitters is limited by the number of times the NiCd battery can be recharged. The smaller avian size transmitters have a field life of 3 years, and larger mammalian size transmitters have a field life of 5 years (O. Royce, personal communication). The largest battery size that a subject can tolerate is generally recommended to maximize longevity and prevent excessive discharging. Likewise, battery size dictates the solar panel surface area necessary for recharging, and may be a consideration in determining shape and dimensions of the transmitter package.

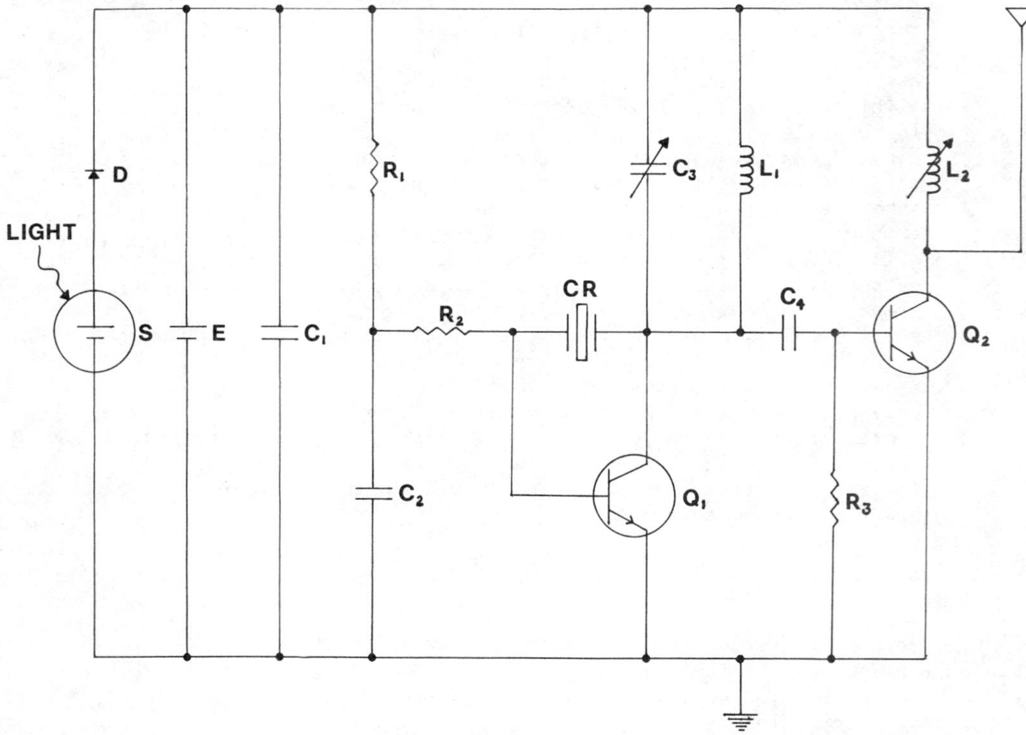

Fig. 1. Schematic for two-stage rechargeable solar transmitter. S: solar panels 3-6 vdc; D: blocking diode, 1N4001; E: 3 vdc rechargeable battery; C_1: capacitor, .001 mfd; C_2: capacitor, 2.2-10 mfd; C_3: capacitor, 6-35 pF; C_4: capacitor, 15-22 pF; R_1: resistor, 270 K-1 MΩ; R_2: resistor, 1.5-8 KΩ; R_3: resistor, 1.5 KΩ; CR: for a crystal frequency of 1-64 MHz = operating frequency, 65-129 MHz = operating frequency-5KHz/2, 130-220 MHz = operating frequency-10 KHz/3; Q_1 and Q_2: transistor, Motorola MMT3014; L_1: inductor, 0.33-1 μH; L_2: inductor, 6 turns, No. 24 AWG on a 5/32" diameter ferrite core.

The shelf-life of completely assembled solar transmitters supplied by Telemetry Systems Inc. (5830 North Shore Dr., Milwaukee, Wisconsin 53217, U.S.A.), is 10 years if kept in total darkness at a temperature range of 4 to 21°C. In addition, solar transmitters are capable of operating between -57 and 65°C, with the efficiency of the solar cells increasing at lower temperatures and decreasing at higher temperatures. Solar transmitters range in size and capabilities from daylight only, 6 g, 1.25 V, single stage subminiature transmitters with a power output of 0.25 mW and a transmission range of 2 km; to a rechargeable, 800 g, 2.5 V, two-stage standard transmitter with a power output of 2 mW and a transmission range of 9.5 km. Ranges are line-of-sight ground-to-ground transmission using a RT-20A receiver (Telemetry Systems Inc.) and a vehicle mounted 8-element Yagi antenna.

The requirement for solar cells to be exposed to solar radiation can limit the applicability of solar transmitters for some species. To ensure ample solar exposure, strict attention must be given to the attachment design. Transmitters attached to gray partridge (*Perdix perdix*), sage grouse (*Centrocercus urophasianus*), and spruce grouse (*Canachites canadensis*) have absorbed sufficient solar radiation despite dorsal feathers being preened over the solar panel surface area. Counter balances on collars attached to whitetail deer (*Odocoileus virginianus*), woodland caribou (*Rangifer tarandus*), and coyote (*Canis latrans*) have successfully maintained solar panels in an upright position. The retrieval of solar transmitters after mortality can be a problem if the solar panels are faced downward on the ground surface, or if the transmitter is located in an area of restricted sunlight. Some examples of attachment techniques which have been used successfully include: a fitted collar on pronghorn antelope (*Antilocapra americana*), slip-on collar on Canada geese (*Branta canadensis*), dorsal harness on blue grouse (*Dendrapagus obscurus*), tail-clip on kestrel (*Falco sparverius*), tarsus 'bracelet' on sandhill crane (*Grus canadensis*), and cutaneous adhesion on armadillo (*Dasypus novemcinctus*).

The cost of solar transmitters is slightly greater than conventional primary battery powered transmitters. Completely assembled a 17 g rechargeable solar transmitter typically used on ring-neck pheasant (*Phasianus colchicus*) cost approximately US $100, a comparable 18 g lithium transmitter would cost approximatley US $85.

DISCUSSION

I have used rechargeable solar transmitters weighing 10.5 g (total package weight) on gray partridge in east-central Wisconsin, U.S.A., for the past two years. Since winter trapping is the only reliable capture technique in North America, it was necessary to monitor gray partridge for approximately 9 months to study reproductive behavior through the brood rearing period. Although it is difficult to assess the effect of transmitter packages on the behavior of free ranging animals, gray partridge in Wisconsin have met the criteria used by Brander (1968) for normal behavior of transmitter-equipped ruffed grouse in Minnesota. Monitored partridge have survived abnormally severe winter conditions, mated, dispersed, successfully reproduced and raised young to adult size.

Other species in North America previously not mentioned on which solar transmitters have been employed are: greater prairie chicken (*Tympanuchus cupido*), ruffed grouse (*Bonasa umbellus*), sharptail grouse (*Pedioeceter phasianellus*), bobwhite quail (*Colinus virginianus*), turkey (*Meleogris gallopavo*), Eastern cottontail (*Sylvilagus floridanus*), and elk (*Cervus elaphus*) (O. Royce, personal communication, 1979).

Acknowledgements — I gratefully acknowledge Owen Royce of Telemetry Systems Inc., Mequon, Wisconsin, U.S.A. for his cooperation and assistance in preparation of the schematic.

REFERENCES

Brander R. B. (1968) A radio-package harness for game birds. *J. Wildl. Mngmt.* 32, 630-632.

Cochran W. W. and Lord R. D., Jr. (1963) A radio-tracking system for wild animals. *J. Wildl. Mngmt.* 27, 9-24.

LeMunyan C. D., White W., Nybert E. and Christian J. J. (1959) Design of a miniature radio transmitter for use in animal studies. *J. Wildl. Mngmt.* 23, 107-110.

Marshall W. H., Gullion G. W. and Schwab R. G. (1962) Early summer activities of porcupine as determined by radio-positioning techniques. *J. Wildl. Mngmt.* 26, 75-79.

The Design of Antennas for Use in Radio Telemetry

C. J. AMLANER, JR.

Animal Behaviour Research Group, Department of Zoology, University of Oxford, South Parks Road, Oxford OX1 3PS, U.K.

Abstract — There are many types of antennas for both receiving and transmitting. This paper briefly describes the design, frequency modification and use of the more important antennas for radio telemetry in the 30-500 MHz band. Much of the material presented here is explained in greater detail in the American Radio Relay League Antenna Handbook (1974).

WAVE THEORY

Wave length (λ) is defined by the relationship

$$\lambda = \frac{\nu}{f} \qquad (1)$$

where ν equals the velocity of the wave and f equals the frequency of current causing the wave. For wavelength in meters and frequency in MHz the equation is conveniently expressed as

$$\lambda_m = \frac{300}{f_{MHz}} \quad . \qquad (2)$$

Equation 2 will be used throughout this chapter when calculating λ for antenna elements and spacings.

TRANSMISSION DISTANCE

The estimated transmission distance to the horizon (D_{km}) in kilometers is given in eq. 3, where H is the antenna elevation in meters above ground.

$$D_{km} = 4.124 \sqrt{H_m} \quad . \qquad (3)$$

This equation takes into account that under normal conditions a slight bending of the radio signal occurs over the earth's sphere. Generally, antennas should be as high as possible above the surrounding countryside and should not be near or adjacent to steep, irregular terrain, power lines, buildings or thick forest. If an antenna is placed on a hill, it should not be located at the top, especially if the transmission originates in front of a hill that has a steep slope. Waves arriving at such hills tend to be defracted above the antenna if it is placed on the summit. This can be alleviated by lifting the antenna higher above the ground surface. Power lines and thick forest tend to 'bend' the radio signal which seriously affects the transmission distance and directional accuracy.

POLARIZATION

Antennas composed of several different elements arranged in parallel, will have the same polarization as any one of their elements. For example, if all elements on the antenna are vertical, then the antenna will be vertically polarized. Vertically polarized signals tend to penetrate thick forests better than horizontally polarized waves which travel further under line-of-sight conditions.

Consider now the association between transmission and receiving antennas. For all practical purposes the properties of a properly tuned (see matching antenna to feed lines) receiving antenna are the same as those used in transmission. The gain, capture pattern and impedance of both receiving and transmitting antennas are the same at any given point of measurement. The only conceptual difference is that a receiving antenna acts as the source of radio power whereas in transmission it is the actual 'load' for power.

TRANSMITTING ANTENNAS

The whip, dustcore, and the simple loop antenna are common transmitting antennas, and are moderately efficient at VHF. The efficiency of a loop antenna is greatly increased by concentrating the electromagnetic flux, and can be achieved by using a ferrite core within the loop itself. This is commonly called a dustcore antenna. Various sizes of ferrite cores are available commercially with approximately a 2-4 mm diameter being standard. A gain factor of approximately 10 can be achieved by using a dustcore instead of a loop of the same size. The electromagnetic radiation pattern from a loop is shaped like a 'figure 8'. Reduced size is the main advantage of using loop and dustcore antennas, which makes them of particular use in implanted transmitters. Larger loop antennas made from wire braiding may be incorporated into the transmitter as a collar. These must be tuned to resonance to optimize power output. This is achieved by placing a tuning capacitor parallel to the loop collar.

Straight wire antennas (whips) are more efficient signal radiators than loop antennas. The choice of a whip antenna is affected by several different factors including: (1) animal suitability, (2) operating frequency, and (3) method of external connection between antenna and transmitter. Whip antennas are particularly suitable for bird radio telemetry, since there are proven methods for attaching these radio packs to birds. (a) A radio pack can be tail mounted, thereby incorporating it into the tail feather. (b) The pack can be mounted on the animal's back using a harness (e.g. made from underwear elastic). In this configuration the antenna can either be horizontally or vertically mounted. Whip antennas may be made from spring steel (piano wire or guitar strings). Where the antenna will be immersed in salt water, stainless steel is the best choice.

The calculation of λ for a whip antenna is given in eq. 2. Most practical applications will necessitate that the antenna length be as short as possible, therefore optimal lengths of 5/8 or 1/2 λ cannot be achieved. If the transmitter radio frequency (RF) stage has the facility for tuning the antenna to resonance (which optimizes power output), then the longest practical length of antenna should be used. This is usually impossible to achieve in miniature transmitters which utilize fixed values of capacitance in the RF antenna section. On these transmitters, a 1/4, 1/8 or 1/16 λ is usually chosen to resonate in the circuit. Antenna efficiency is drastically reduced at shorter lengths. For example, at 1/16 λ, transmission range for a low power RF oscillator may be 100's of meters. Whip antennas may be shortened by the use of a loading coil between the RF section and the base of the antenna. This loading coil is made up of approximately 2-6 turns of enameled wire (20-40 gauge) and may be encapsulated within the transmitter.

Power output is greatly dependent upon the efficient coupling of the RF section to the antenna and so special care should be taken at this step of design. In practice, the best antenna RF coupling can only be achieved through experiment where signal strength is checked on a sensitive power meter. Transmitters used for long range radio tracking or physiological measurements should be checked in field conditions over the required range, as theoretical calculations rarely anticipate suboptimal field conditions.

RECEIVING ANTENNAS

Differing laboratory and field applications require specific receiving antennas. For instance, long range reception may be desirable, but huge antennas cannot be carried through thick undergrowth, therefore a compromise is inevitable. The specific aim and operating frequency, e.g. tracking fish using HF, terrestrial animals using VHF, or birds on VHF and UHF bands, will dictate the antenna's shape and size. The following sections will describe several types of receiving antenna with suggestions on their construction.

DIRECTIONAL ANTENNAS

The antenna types discussed in this section all have a plane in which the signal strength is greatest, and a secondary plane in which the signal strength is minimal. These are most useful in locating the direction of a specific transmitter.

Dipole Antenna

The dipole antenna is simple to build and can produce fairly high gain. The gain of the dipole will be used as a standard to which other antennas will be compared. This antenna is potentially portable and appropriate for field work in thick underbrush and is commercially available. Minimum signal strength is directly in line with the two poles and a possible ambiguity exists as to the direction of the signal source. A dipole will generally yield an accuracy of 5-10°, but may also be used in the laboratory, where direction is unimportant. The dipole resonant length (0.5 λ) can be calculated from eq. 4.

$$0.5\ \lambda_m = \frac{150}{f_{MHz}} \times k \qquad (4)$$

where k is a constant, calculated from the ratio

$$\frac{0.5\ \lambda_m}{d_m}\ . \qquad (5)$$

Here d equals the diameter of material used to make the dipole. A ratio of 10,000 yields a k value of approximately 0.98 from Fig. 1 and in effect it means that the smaller the ratio, the shorter the electrical length of the antenna. In practice, metal tubing or solid rod of similar dimensions have the same constant k.

Loop Antenna

Loop antennas are useful where the subject to antenna distance is short and the

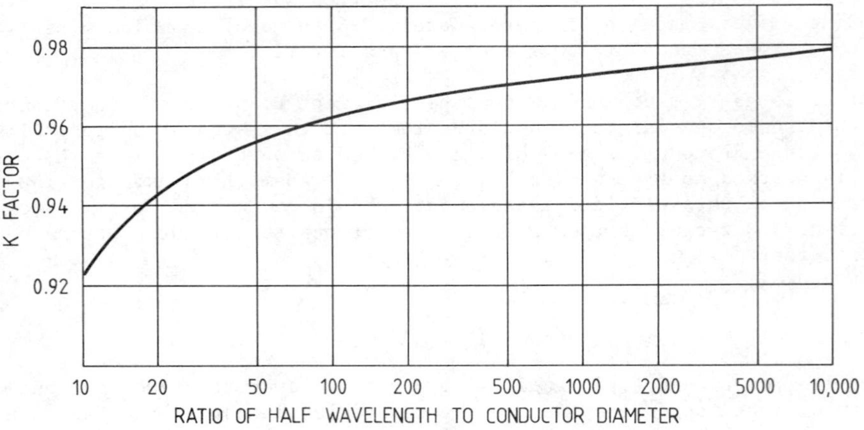

Fig. 1. The curve shows the constant k, used in eq. 4 to give the physical length of a resonant half-wave antenna. The ratio (x-axis) is calculated from eq. 5.

antenna is fixed. For frequencies near 30 MHz and where the k value (see Fig. 1) is large, the circumference of the loop is:

$$C_m = \frac{306.3}{f_{MHz}} . \qquad (6)$$

A signal may be received from almost anywhere within the perimeter of a horizontally polarized loop. A vertically polarized loop has directional sensitivity when the transmitter is outside the perimeter. Maximum radiation is perpendicular to the plane of the loop and is shaped like a 'figure 8'. Directional accuracy is approximately 5°; but when the loop is horizontally polarized, directionality is severely reduced. The theoretical gain in comparison to a half wave dipole is approximately +3 dB. Loop antennas can be constructed for higher frequencies with the addition of a tuning capacitor and these are termed, tuned loops.

MULTI-ELEMENT DIRECTIONAL ANTENNA

Multi-element antennas are usually only suitable for frequencies above 30 MHz. These antennas offer the benefit that, at higher frequencies, added elements increase the gain. Also, directional antennas concentrate power received over a much narrower angle. In this case directionality is improved to within 0.5^o when the signal source is > 20 λ away.

Optimal design will take into account the gain and front-to-back ratio of the antenna. To obtain a narrow capture area (and hence good directionality), the front-to-back ratio is optimized in the direction of signal origin. In practice, there are various compromises between antenna gain and front-to-back ratio which may be determined by element spacing, number and length of elements, and these are dictated, to a large extent, by the specific application.

Element lengths are approximately ± 5 percent of 0.5 λ and spacing between elements may vary from 0.1 to 0.2 λ. The different elements (Fig. 2) are (1) the driven element which is supplied power from the transmitter, or in the case of reception of radio signals, is the one that actually supplies power to the receiver, (2) the parasitic element which obtains power through electrical coupling to other elements

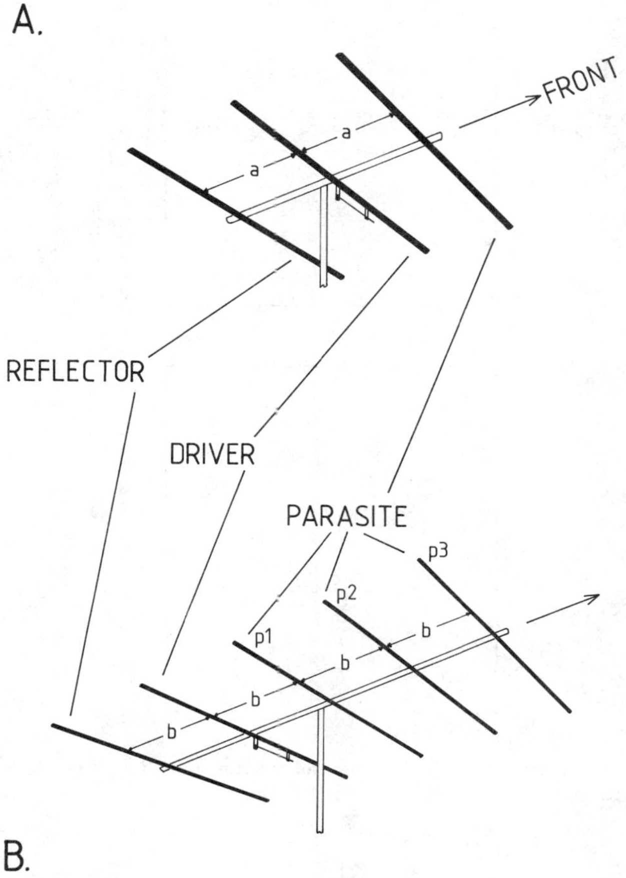

Fig. 2. A. A three element directional Yagi on the 102 MHz band has an element spacing (a) = 43.7 cm, reflector length = 145.5 cm, driver = 137.4 cm and parasite = 130.6 cm.
B. The five element Yagi designed for 173 MHz has an element spacing (b) = 35.5 cm, reflector length = 86.4 cm, driver = 81.3 cm, P1 = 76.2 cm, P2 = 75.6 cm, P3 = 74.9 cm.

in the array, and (3) a special category of parasite called the reflector, which is closely associated with the driven element. The reflector is located at the 'back' of the antenna. The Adcock and Yagi are special types of multi-element antennas.

Adcock or 'Lazy H' Antenna

The Adcock antenna (Fig. 3) is commonly used in radio tracking as it may be folded up when not in use; has excellent directional qualities (± 2°) and has a high gain compared to the half wave dipole (see Table 1). The disadvantage of an Adcock is that while in use, it is cumbersome, however, if the elements are hinged for folding, it can be stored in a small area. Table 1 shows the differences between fractional spacing of the two elements relative to antenna gain. In practice, 0.5 λ spacing of elements is optimum for frequencies in the range of 100 MHz.

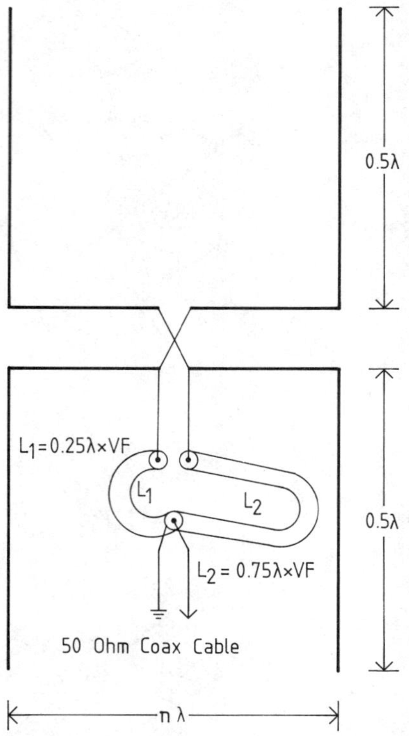

Fig. 3. An H-Adcock antenna and balun match. The dimensions of λ are given in Table 1.

Table 1.
Gain of Adcock antenna compared with dipole under different element spacing

3/8-wave spacing — 4.4 dB
1/2-wave spacing — 5.9 dB
5/8-wave spacing — 6.7 dB
3/4-wave spacing — 6.6 dB

The Adcock antenna may be connected to a receiver through a balun matching network (see Fig. 3 and matching the antenna to coaxial cable). It is critical that each loop length be precisely measured (0.25 λ and 0.75 λ) to maintain directional accuracy and impedance matching. A cable velocity factor (eq. 10) must also be taken into consideration (see electrical length of cables) when computing these loop lengths.

Yagi Antenna

A multi-element array with linear dipole elements is frequently called a 'Yagi' antenna. This antenna has extremely good directionality and has the highest gain (when many elements are utilized) of any antenna discussed here. Yagis are routinely used for long range biotelemetry and radio tracking because of these qualities. The simplest multi-element Yagi consists of two elements where one element is a driver and the other a reflector. Fold-up models are excellent for field usage. Fixed antenna systems may employ more than one Yagi stacked together and, if fed out of phase, will achieve higher quality directionality (± 0.5°). The null peak system described below is used for this purpose.

Effective Yagi design takes into account (1) element length, (2) spacing between elements, and (3) the total number of elements within the antenna. Figure 2b shows a typical five element Yagi antenna with a reflector, driven element and 3 parasites. Theoretically there is no limitation to the number of parasites which may be used, but the operating frequency is the practical limiting factor. At 30 MHz, antennas with more than 4 elements are large and impractical to build, whereas on VHF bands, size and wavelength allow use of many elements. Figure 4 shows the relationship between number of elements and antenna gain in decibels (dB) above the gain of a half wave dipole.

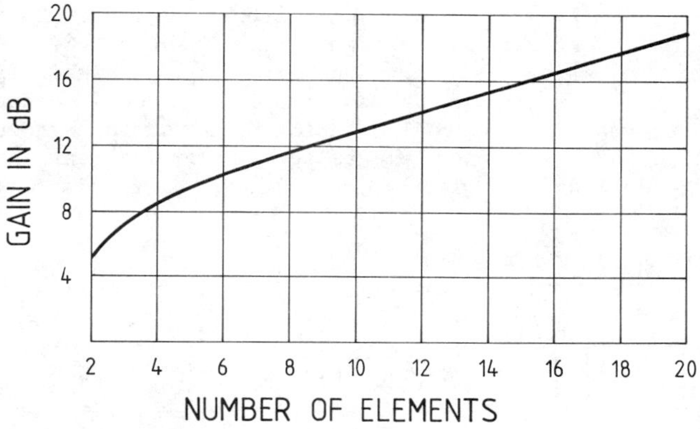

Fig. 4. Gain of a Yagi antenna compared to a half wave dipole versus number of elements on the Yagi.

Yagi Construction

Element lengths and spacings described in the following section will not necessarily be exactly those calculated using theoretical antenna design. The practical compromises are usually in terms of sacrificing gain for a greater band width (that is, what the antenna will receive above and below the designed frequency) or increasing front-to-back ratio for more gain. This means that in general the element length and (to a less critical extent) spacing will not always conform to the values calculated to give maximum gain.

Figure 2A and B shows the dimensions for a 102 MHz and 173 MHz Yagi. These antennas can conveniently be made from aluminum tubing which is obtainable from most electrical or plumbing suppliers. For permanent applications or in corrosive marine environments, stainless steel is preferable to aluminum tubing. Element diameter should be no more than 1.5 cm (except the driver element) and the boom 2.5 to 4 cm. The driven element should be close to 0.5 λ in length and 2-3 times the diameter of the other elements. Parasitic elements should be approximately 5 percent less in length than the driver and the reflector should be approximately 5 percent longer.

Wavelength (λ) calculations should take into account the k factor (Fig. 1) which essentially decreases the overall physical length of each element. The lengths of the different elements (gain optimized) are calculated by using the following equations:

$$\text{driver}_m = \frac{146.0}{f_{MHz}} \tag{7}$$

$$\text{reflector}_m = \frac{152.4}{f_{MHz}} \tag{8}$$

$$\text{first parasite}_m = \frac{138.7}{f_{MHz}} \quad . \tag{9}$$

These are average lengths determined using a length/diameter ratio of 200-400 (k = 0.96-0.97) and element spacing of 0.1 to 0.2 wavelengths.

The gamma match (trombone slide, Fig. 5) for the two Yagis in Fig. 3A and B should be made out of the same material as the elements. The antenna should be tuned for a minimum voltage standing wave ratio (VSWR) either using a VSWR bridge or by actual field testing (using a continuous tone transmitter located > 20 wavelengths away) by tuning the 'trombone slide' for a maximum signal. For higher frequencies (above the 200 MHz range) the elements may be made from solid rod. Conductors placed near elements of the array may cause severe detuning effects, which lower antenna gain and directional accuracy. Antenna elements should be no closer than 0.5 λ to the nearest conductor. Holding or mounting the antenna by its elements also causes detuning.

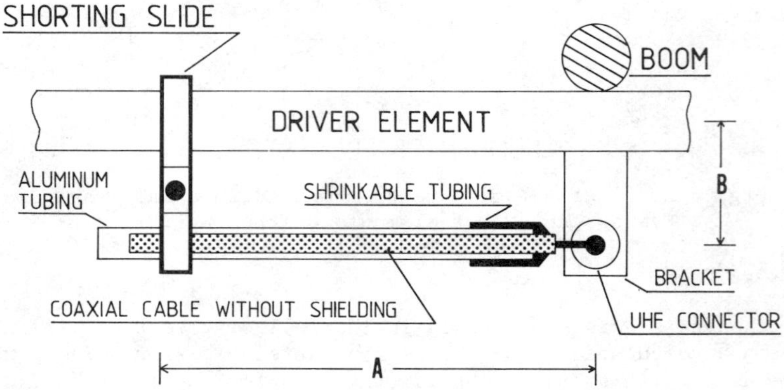

Fig. 5. The gamma match used on Yagis of Fig. 2, to match 50 or 75 coax cable. See text for explanations of dimensions.

OMNI-DIRECTIONAL ANTENNAS

Where animals are moving within a confined area, loss of signal from a transmitter is minimized by using an omni-directional antenna. These antennas can also be useful in radio tracking, for example, if an animal has been lost, an omni-directional antenna may be used on a vehicle to quickly locate the signal. It may also be used for studies of visits to a particular place, such as nest attendance of birds. The omni-directional antenna has a capture area of 360°, and uniform gain in all directions. Four equally spaced elements act as an artificial ground plane and are approximately 5 percent longer than the quarter wavelength radiator (Fig. 6). This antenna may be fed directly with 50 Ω coaxial cable with only a slight impedance mismatch.

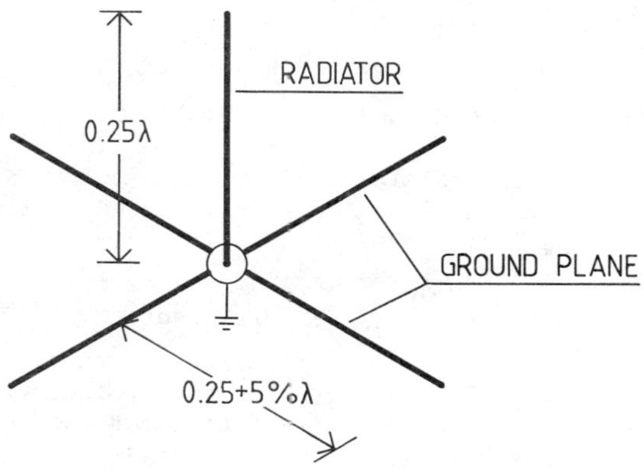

Fig. 6. An omni-directional ground plane antenna.

MATCHING THE ANTENNA TO COAXIAL CABLE

Techniques used for matching cable impedance and minimizing VSWR on receiving antennas include the gamma match (better known as the trombone slide), the T match (a special case of the gamma match), and balun matches. The balun and gamma matching methods adapt nicely to the use of coaxial cable which is readily available.

The reason for matching the cable to the antenna, is that, at various frequencies the antenna exhibits a different impedance which the cable 'sees' as it receives the signal. The compromise is between matching impedance (minimizing the standing wave ratio) or allowing the antenna to operate over a wide frequency band. If we minimize the VSWR this tends usually to minimize the band width of the antenna but increases the gain. If the VSWR is not minimized then gain on the antenna is greatly reduced and also the radiation pattern of the antenna is changed. Matching of impedance is achieved by changing the characteristics of the matching network, either by changing the length and spacing of the 'trombone slide' (Fig. 5), or by changing the length of the loop size (balun matching, Fig. 3).

Gamma Match Construction

The gamma match (Fig. 5) used on Yagis should have a capacitor (50 pf) placed in series with the feed line for 50 MHz and it should be 30 cm long. At 150 MHz the gamma match should be approximately 13 cm long and should use a 14 pf series capacitor. Antennas using these measurements can be fed directly with 75 Ω coax cable. For matching a multi-element array to a 50 Ω line the length of the rod should be 0.04-0.05 λ (Fig. 5, dimension A), outside diameter 1/3 that of driver diameter and spacing (center to center of driven element), approximately 0.007 λ (Fig. 5, dimension B).

CABLE TYPES

Common types of receiving cable are RG58U, RG59U, RG11 and RG8. These coaxial cables have a resistance of either 50 Ω (RG58 and RG8) or 75 Ω (RG11 and RG8). At frequencies above 30 MHz and approaching 200, the attenuation through 30 m of cable is approximately 2-3 dB. On this basis alone the length of cable between receiver and antenna should always be kept to an absolute minimum so as not to nullify the advantages of a high gain antenna. Antenna cable should be frequently checked for abrasion or wear due to rough field use and the effects of weather on the insulation. Frayed or worn cable should be replaced immediately as increased signal attenuation occurs when the insulation breaks down.

ELECTRICAL LENGTH OF CABLES

Whenever reference is made to a coaxial cable being a multiple of λ it refers to its electrical length and not its physical length calculated from eq. 2. Electrical length is given by the equation:

$$\text{Electrical length}_m = \frac{300}{f_{MHz}} \times \text{Velocity factor} . \tag{10}$$

This velocity factor is a ratio of the actual signal velocity along the coax to the signal velocity in free space. The velocity factor for the four different types of cable mentioned above is approximately 0.66. For other cable velocity factors, consult the manufacturers. Note that a velocity correction was made for the balun matching loop on Fig. 3.

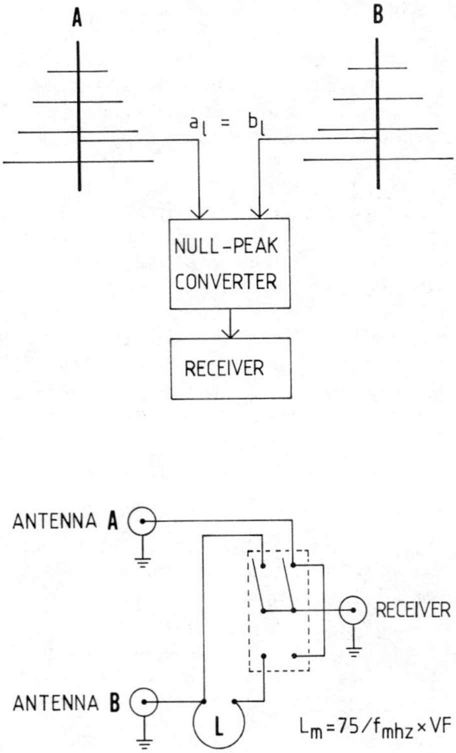

Fig. 7. The block diagram and schematic of a null-peak dual Yagi arrangement. The two coax cables a_L and b_L must be identical in length. Dashed lines on the schematic represent a double pole, double throw switch. VF is the velocity factor of the coax used in the loop (L).

NULL-PEAK MULTI-ANTENNA ARRAYS

Several Yagis fed in parallel boost forward gain, but at a cost of an overall increase in size. Two Yagis mounted side-by-side can be used conveniently for radio tracking purposes. Sibly and McCleery (this volume, page 347, Fig. 2) used these vertically polarized Yagis for tracking seagull movements over long distances (5-20 km). The base of this antenna could be rotated from inside a Landrover. The system utilized a null-peak device (Fig. 7). When the antennas were aimed directly at a distant transmitter (>> 20 λ) and both Yagis were in phase, the receiver gave a peak signal and when they were switched out of phase, a null was received. The null peak system created a very accurate method by which triangulation could be achieved with a precision of ± 0.5°.

Acknowledgements — I would like to thank N. Ball, N. Hough, D. Macdonald and J. Shaffery for reading and commenting on earlier drafts of this paper. Many thanks are also due to my wife, Beverly, for her continued support. Figures 1 and 4 are reproduced with the kind permission of the American Radio Relay League.

REFERENCES

Sibly R. M. and McCleery R. H. (1980) Continuous observation of individual Herring gulls. In C. J. Amlaner, Jr. and D. W. Macdonald (eds.), *A Handbook on Biotelemetry and Radio Tracking*, pp. 345 - 351. Oxford: Pergamon Press.

The American Radio Relay League Antenna Book. (1974) Newington, Connecticut: The American Radio Relay League, Inc.

A Collapsible Dipole Antenna for Radio Tracking on 102 MHz

T. PARISH

Institute of Terrestrial Ecology, Hill of Brathens, Banchory, Kincardineshire AB3 4BY, U.K.

Abstract – This paper describes with diagrams the method of construction and use of a collapsible dipole antenna for radio tracking of badgers fitted with 102 MHz transmitters. Construction is from readily available materials and methods of calculating dimensions are described with appropriate references. Its performance compares favorably with similar rigid antennas.

The method of use is described together with a modification which under ideal conditions can determine not only the bearing of the transmitter but also its direction. Disadvantages of this antenna are that it is relatively fragile when extended and that it lacks the precision and gain of larger H-Adcock and Yagi arrays, however these are outweighed for short range work and eventual visual contact by its compactness and the degree to which it is protected when folded.

INTRODUCTION

Studies of the European badger using radio tracking equipment have yielded many practical problems in the use of such equipment under difficult field conditions. There are problems of attacking the transmitter and of protecting it from moisture, physical damage by the animal and impact damage on rocks and tunnel walls.

Another field problem has been to make a receiver antenna which was sufficiently accurate to locate an animal when active or resting and to 'home in' to find its resting site or to make visual contact; and yet was compact and easy to carry when not in use.

Transmitters for this type of radio location are likely to operate on 102 MHz in the U.K. (or 104.6-105 MHz for new equipment) (Taylor and Lloyd, 1978). Each element of an antenna designed for this frequency must be 70 cm long: 95 percent of the theoretical ¼ wavelength of 73.5 cm (Laidlaw, 1957; Amlaner, 1980, this volume). A dipole for this frequency is therefore going to be at least 1.4 m long and even if it is made of sturdy aluminum tubing, it is liable to damage when used over rough terrain, walls and fences or in dense coniferous plantations particularly at night. It is also very cumbersome to carry while using other equipment such as an image intensifier or infra red binoculars.

To resolve these difficulties a hand held dipole antenna was made such that the elements could be extended in use and collapsed into the protection of the handle, when not required.

MATERIALS AND METHOD OF CONSTRUCTION

Basic materials required are (1) two telescopic transistor radio antennas with a hinge above the lower section which is not included in the minimum extended length of 70 cm; (2) 20 cm of 25 mm internal diameter Class C Alkathene water pipe; (3) three 25 mm rubber bungs and coaxial cable, clamp and plug to suit receiver.

The method of construction is shown in Fig. 1. The rubber bungs are trimmed to a tight push fit in the water pipe and two of them are bored with two holes each to support the antennas side by side. The base segments of the antennas are sleeved with thin walled rubber gas tubing or wrapped with Polyvinyl chloride (PVC) tape. The remaining bung has one hole to take the 75 Ω co-axial lead which is then attached to the antenna with solder or screws. The antenna which is connected to the center wire of the co-axial cable should be marked in some way (e.g. red paint) and will be termed the 'collector' in the following description of use of this device.

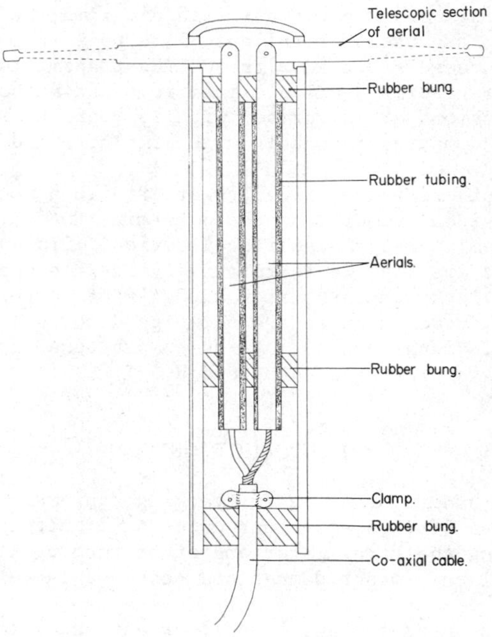

Fig. 1. Diagramatic section of antenna partially extended showing relative positions of components.

Two 8 cm deep notches are cut in opposite sides of one end of the water pipe, shaped like inverted keyholes so that they will grip the antennas firmly. The bungs and antennas are then pushed into the tube to about 1 cm below the top. A cable clamp is fitted above the lower bung which is threaded on the cable and pushed into the bottom of the tube.

Finally the cable is trimmed to an appropriate length, which for practical purposes should be 0.66 of the theoretical $\frac{1}{4}$ wavelength, or multiples of this unit (Laidlaw, 1957; see also Amlaner, 1980, this volume). Fine trimming ($\frac{1}{2}$ cm at a time) of the free end of the lead can locate a peak of sensitivity and another lead of identical wire can then be cut exactly to this length.

METHOD OF USE

When not in use the antenna can be collapsed and put in the receiver bag or in the pocket. In use the antennas are extended and folded through 90° to rest in the notches in the outer tube thus forming a simple dipole shape.

The following description of the use and performance of this antenna was obtained under optimal conditions with few reflected signals on a gentle slope with the observer approximately 5 m above the transmitter at a range of 250 m. The antenna is held above the head (Fig. 2) with its axis in a horizontal plane and is rotated through 360°. Nulls are obtained when the axis of the antenna is in line with the signal source and the strongest signal is received when the antenna is more or less at 90° to the source (Fig. 3). The polar diagrams relate strength of signal received to the direction of the 'collector' element of the dipole with a stationary transmitter. Figure 3 shows that there are two identical nulls indicating that the transmitter is in either of two opposite directions from the observer and only by repeating the readings from other locations or having a second observer at another location, can the signal source be found by triangulation.

Fig. 2. Holding the antenna horizontally for field testing.

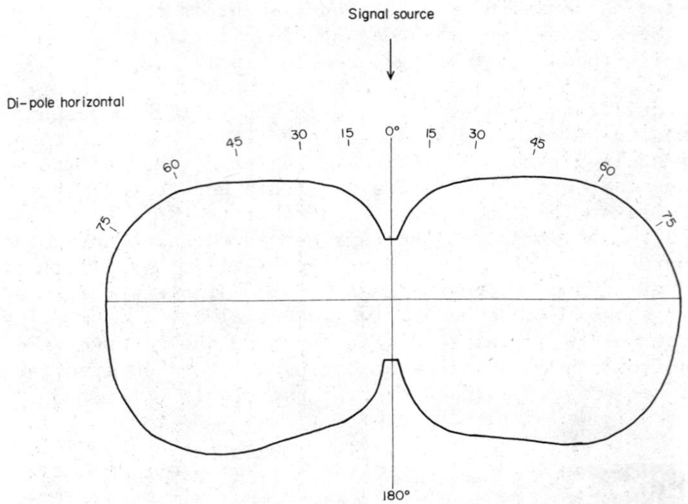

Fig. 3. Polar diagram of signal strength received in relation to direction of 'collector' with axis of antenna horizontal.

Field experiments have shown that if the antenna is tilted 15° from the horizontal with the 'collector' element uppermost (Fig. 4), the polar diagram is no longer symmetrical about the 90° axis (Fig. 5). When the antenna is rotated through 360°, one of the nulls is audibly narrower and more pronounced indicating not only the bearing but also the direction of the transmitter.

If the animal is moving (signal strength varying) the antenna must be rotated at first through 360° to obtain an approximate direction and then alternated rapidly through about 90° so that rapid comparisons of signal strength can be made by ear to detect the best null. This procedure should be repeated regularly when following or 'homing-in' on an animal to eliminate spurious effects of topography and reflected signals. When no longer required the antenna is simply collapsed into the handle for easy carrying (Fig. 6).

DISCUSSION

This antenna compares favorably in performance with similar rigid tubular aluminum dipoles. Disadvantages of this antenna are that it is fragile when opened and must be opened and collapsed with some care. It also lacks the precision and high gain of the vertically polarized H-Adcock double dipole (Taylor and Lloyd, 1978), or Yagi and double Yagi arrays (Hallberg, Janza and Trapp, 1974) which limits its use for long range precise triangulation. A further disadvantage is that its precision is frequently affected by reflected signals but the observer can usually overcome this by moving, often only a few meters. Its compactness and the degree to which it is protected when folded however make it ideal for 'homing-in' on an animal to make visual contact or to discover where the animal is resting by day. It is particularly useful for locating badgers and subsequently following them visually using infra red binoculars or an image intensifier, as an extended antenna is a great disadvantage when trying to move quietly over rough terrain and fences, or through dense vegetation.

Fig. 4. Dipole tilted 15° from horizontal.

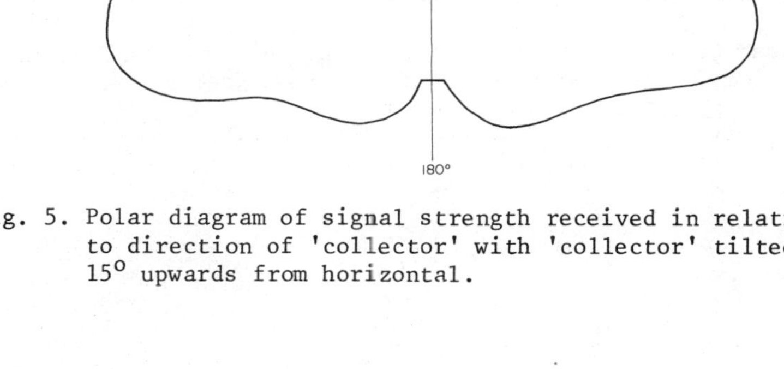

Fig. 5. Polar diagram of signal strength received in relation to direction of 'collector' with 'collector' tilted 15° upwards from horizontal.

Fig. 6. Collapsing the dipole antenna.

REFERENCES

Amlaner C. J., Jr. (1980) Telemetry antenna design. In C. J. Amlaner, Jr. and D. W. Macdonald (eds.), *A Handbook on Biotelemetry and Radio Tracking*, pp. 251-261. Oxford: Pergamon Press.

Hallberg D. L., Janza F. J. and Trapp G. R. (1974) A vehicle-mounted directional antenna system for biotelemetry monitoring. *Calif. Fish Game* 60, 172-177.

Laidlaw R. (1957) *Practical TV Aerial Manual for Bands I and III.* 80 pp. London: Norman Price.

Taylor K. D. and Lloyd H. G. (1978) The design, construction and use of a radio-tracking system for some British mammals. *Mammal Rev.* 8, 117-141.

A Practical, Remotely-controlled, Portable Radio Telemetry Receiving Apparatus

R. M. SMITH and B. TREVOR-DEUTSCH

Department of Biology, Laurentian University, Sudbury, Ontario P3E 2C6, Canada

Abstract — A light weight, two-tower receiving system was developed in response to the need for accurate positional fixes of tagged animals in the field, without introducing the problem of bias in the form of observer presence in or near the study site. The system can be remotely operated by either one or two workers, and can operate for approximately twelve hours on a single battery charge. It accommodates a wide variety of receivers and antennas, and can be assembled from easily obtainable items at modest cost.

INTRODUCTION

Radio tracking has demonstrated its merits as a useful system for the location and monitoring of animals in the field. Current limitations center around two major problems. The first is the availability of small, reliable transmitters with suitable range, while the second concerns the flexibility, sensitivity and accuracy of the receiver system. While many improvements to the former have occurred, we feel that the versatility of available reception systems leaves much to be desired. This is particularly so in the case of localized, intensive surveys in which accurate fixes must be taken often and where the presence of an observer would be likely to modify the behavior of the species under study.

We have developed a two-station receiving system which:

1. can be remotely controlled by one operator,
2. can be quickly relocated and re-referenced
3. can be used to take reliable locational fixes within a short period of time.

Single operator fixes can usually be completed within 30 s, once the approximate bearings are known, and if the period between readings is reasonably short. There is always a larger inherent error in locations made, which is directly proportional with time which has elapsed between readings.

SYSTEM LAYOUT

Two yagi antennas are supported on each of the two telescope masts by a reversible, antenna rotation device. This device consists of a 12 V planetary gearhead motor, rotation limit switches and a linear variable potentiometer (Fig. 1). The potentiometer is used to reference the position of the antenna mast, which is belt driven by the motor. The mast is solid ½ in steel, 40 cm long, and is supported at each end by an aluminum bushing, bolted to the mast by means of welded steel plates. The

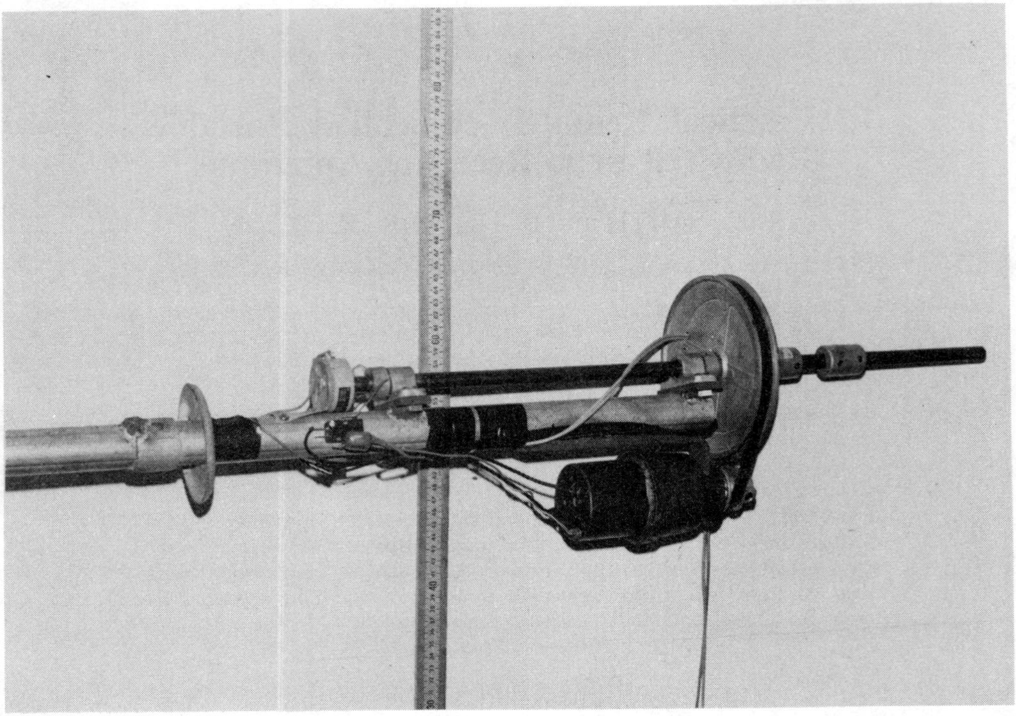

Fig. 1. Mast position device. Motor is spring loaded to tension belt. Limit switches are set to engage after a 185° sweep. The antenna cross-boom is clamped to the large aluminum collars situated to the right (above) of the pulley. The large ring located at the extreme left is used for anchoring ropes. A three or four rope configuration may be used.

resulting configuration has great rigidity. The potentiometer shaft is inserted at the lower end of the mast shaft and secured by a set screw. There is virtually no possibility of inaccurate positioning as a result of longitudinal twisting when this method of assembly is used. The whole unit has an extremely narrow profile to the wind. Control and reception is achieved remotely using solid state switching devices, external preamplifiers and a central control unit (Fig. 2). A 12 V lead acid battery supplies power to each tower. Since the heaviest power demand is at the tower site (each motor requires 8 A during operation), it is convenient and economical to place them there, thus eliminating the necessity of converters, transformers and bulky power lines. Connections are made from the towers to the central control unit via the cable controlling antenna rotation and coaxial cable. Null peak switching and signal preamplification is performed at each tower by remote control. The central control unit connects the receiver(s) and a digital ohm meter to the towers (Fig. 3). The ohm meter is the device that displays mast position. To calibrate the meter, a resistance is arbitrarily set for a bearing of 180°. Resistance change then occurs linearily over the 180° sweep as an exact function of 4.94 Ω degree^{-1}. A digital

Fig. 2. Central control unit 2. Unit measurements are 26 × 13 × 6 cm overall. Unit weight is 772 g. Central control unit 1 is not illustrated, but is approximately one-half the overall size of the unit illustrated above. Refer to Fig. 3 for circuit details.

ohm meter must therefore be used to develop this maximum accuracy. There are two ammeters incorporated into the control unit which show current drain resulting from the gearhead motors. This allows the operator to estimate remaining charge in the batteries as well as indicating load upon each motor. They also point out motor failure or jamming.

OPERATION

The system makes use of the null peak theory of radio location developed by Cochran (Banks *et al.*, 1975) and shows the accuracy typical of the receivers and antennas being utilized. On the 148 MHz band, this accuracy is approximately 0.5°, and thus generates the 'error polygon' formulated and described by Heezen and Tester (1967). If higher resolution is required, then the position resistance change must be recalibrated to a higher resistance change per degree.

In field practice, the towers are placed in optimum positions for signal reception and then connected to the power supplies and central control unit. They are then referenced with respect to either each other or to magnetic north, using transect or compass, respectively. This is accomplished by first rotating the mast until it stops at the limit switch, then manually rotating the whole tower until it is exactly aligned along the antenna reference bearing. Once in position, the ohm meter is set at its reference point setting by the use of variable potentiometers incorporated into the control unit.

Synchronous fixes are not possible unless two operators are present to control each tower separately, but accurate readings can be taken by one operator provided that the subject is not moving rapidly across the radio field.

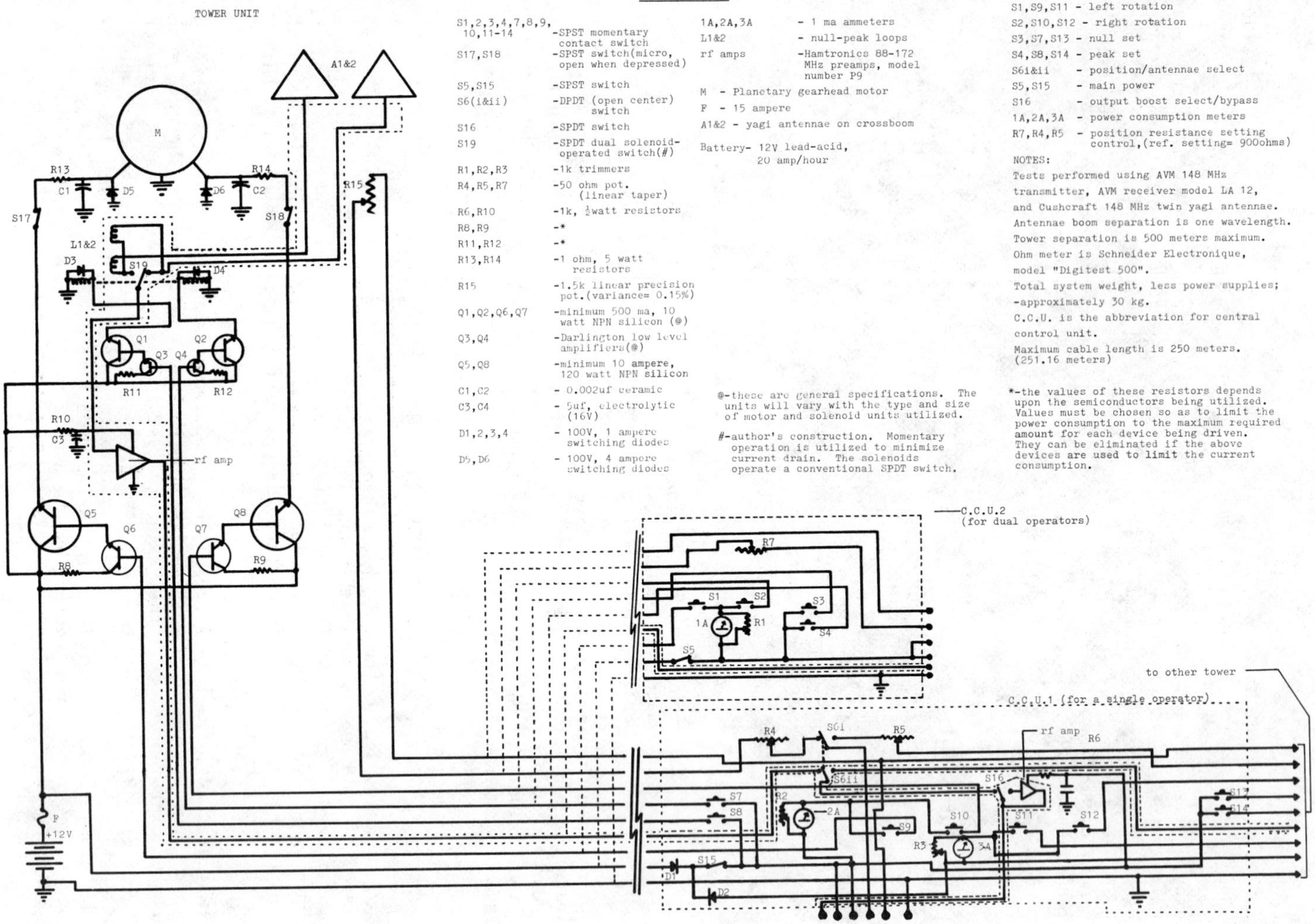

COMPONENT LIST

S1,2,3,4,7,8,9, 10,11-14	-SPST momentary contact switch
S17,S18	-SPST switch(micro, open when depressed)
S5,S15	-SPST switch
S6(i&ii)	-DPDT (open center) switch
S16	-SPDT switch
S19	-SPDT dual solenoid-operated switch(#)
R1,R2,R3	-1k trimmers
R4,R5,R7	-50 ohm pot. (linear taper)
R6,R10	-1k, ½watt resistors
R8,R9	-*
R11,R12	-*
R13,R14	-1 ohm, 5 watt resistors
R15	-1.5k linear precision pot.(variance= 0.15%)
Q1,Q2,Q6,Q7	-minimum 500 ma, 10 watt NPN silicon (@)
Q3,Q4	-Darlington low level amplifiers(@)
Q5,Q8	-minimum 10 ampere, 120 watt NPN silicon
C1,C2	- 0.002uf ceramic
C3,C4	- 5uf, electrolytic (16V)
D1,2,3,4	- 100V, 1 ampere switching diodes
D5,D6	- 100V, 4 ampere switching diodes

1A,2A,3A	- 1 ma ammeters
L1&2	- null-peak loops
rf amps	-Hamtronics 88-172 MHz preamps, model number P9

M - Planetary gearhead motor

F - 15 ampere

A1&2 - yagi antennae on crossboom

Battery- 12V lead-acid, 20 amp/hour

@-these are general specifications. The units will vary with the type and size of motor and solenoid units utilized.

#-author's construction. Momentary operation is utilized to minimize current drain. The solenoids operate a conventional SPDT switch.

FUNCTIONS

S1,S9,S11	- left rotation
S2,S10,S12	- right rotation
S3,S7,S13	- null set
S4,S8,S14	- peak set
S6i&ii	- position/antennae select
S5,S15	- main power
S16	- output boost select/bypass
1A,2A,3A	- power consumption meters
R7,R4,R5	- position resistance setting control,(ref. setting= 900ohms)

NOTES:

Tests performed using AVM 148 MHz transmitter, AVM receiver model LA 12, and Cushcraft 148 MHz twin yagi antennae.

Antennae boom separation is one wavelength.

Tower separation is 500 meters maximum.

Ohm meter is Schneider Electronique, model "Digitest 500".

Total system weight, less power supplies; -approximately 30 kg.

C.C.U. is the abbreviation for central control unit.

Maximum cable length is 250 meters. (251.16 meters)

*-the values of these resistors depends upon the semiconductors being utilized. Values must be chosen so as to limit the power consumption to the maximum required amount for each device being driven. They can be eliminated if the above devices are used to limit the current consumption.

Fig. 3. Schematic.

REFERENCES

Banks E. M., Brooks R. J. and Schnell J. (1975) A radio-tracking study of home range and activity of the brown lemming (*Lemmus trimucronatus*). *J. Mammal.* 56, 888-901.

Cochran W. W. and Brander R. B. (1971) Radio-location telemetry. *Wildlife Management Techniques*, Third Edition, Vol. 3, pp. 95-103.

Cochran W. W., Warner D. W., Tester J. R. and Kuechle V. B. (1965) Automatic radio-tracking system for monitoring animal movements. *BioScience* 15, 98-100.

Cochran W. W. and Lord R. D., Jr. (1963) A radio-tracking system for wild animals. *J. Wildl. Mngmt.* 27, 9-24.

Heezen K. L. and Tester J. R. (1967) Evaluation of radio-tracking by triangulation with special reference to deer movements. *J. Wildl. Mngmt.* 31, 124-141.

Construction of an Acoustic Receiver and Hydrophone

L. C. IRELAND* and K. D. LAWSON, JR.**

*Bermuda Biological Station for Research, St. George's West, 1-15, Bermuda
**Neil Brown Instruments, Box 498, Cautaumet, Massachusetts 02543, U.S.A.

Abstract — The construction of an acoustic receiver and hydrophone suitable for use in underwater tracking experiments is described. The receiver is a beat frequency oscillator design and the hydrophone features a frequency selective preamplifier. The equipment detects frequencies of from 40 to 60 KHz and is most sensitive to 50 KHz signals. A 1 μV, 50 KHz signal at the hydrophone is clearly audible at the headphones. Both the receiver and the preamplifier may be easily adjusted for the detection of either higher or lower frequencies. The hydrophone may be used for either omnidirectional or directional signal detection.

INTRODUCTION

Ever increasing use is being made of telemetry to obtain information about the behavior and physiology of animals in the wild. Most work in the area has involved radio tracking of terrestrial animals. Until recently, few studies with aquatic animals were performed because of the difficulty of propagating suitable energy through water. Over ranges of more than a few meters, sound is the most practical form of energy for underwater telemetry. Advances in acoustic telemetry make it possible to obtain data on many species of aquatic animals over periods of time and in locations where it is impossible to obtain information by means of any other technique (Stasko and Pincock, 1977; Ireland and Kanwisher, 1978). Unfortunately for the investigator, commercial acoustic telemetry equipment is expensive. Here we present all the information necessary to construct a simple, inexpensive, and compact acoustic receiver and hydrophone suitable for use in a variety of different experimental situations.

MATERIALS AND METHODS

Figure 1 is a circuit diagram for a beat frequency oscillator receiver and a frequency selective hydrophone preamplifier. Amplification within the hydrophone is important to eliminate electrical interference from boat motors and other sources. Frequency selective preamplifiers partially eliminate background noise. The circuit diagram is a modification of one shown in Kanwisher, Lawson and Sundnes (1974). Changes to the circuit reflect experience gained over several years in the field with this equipment.

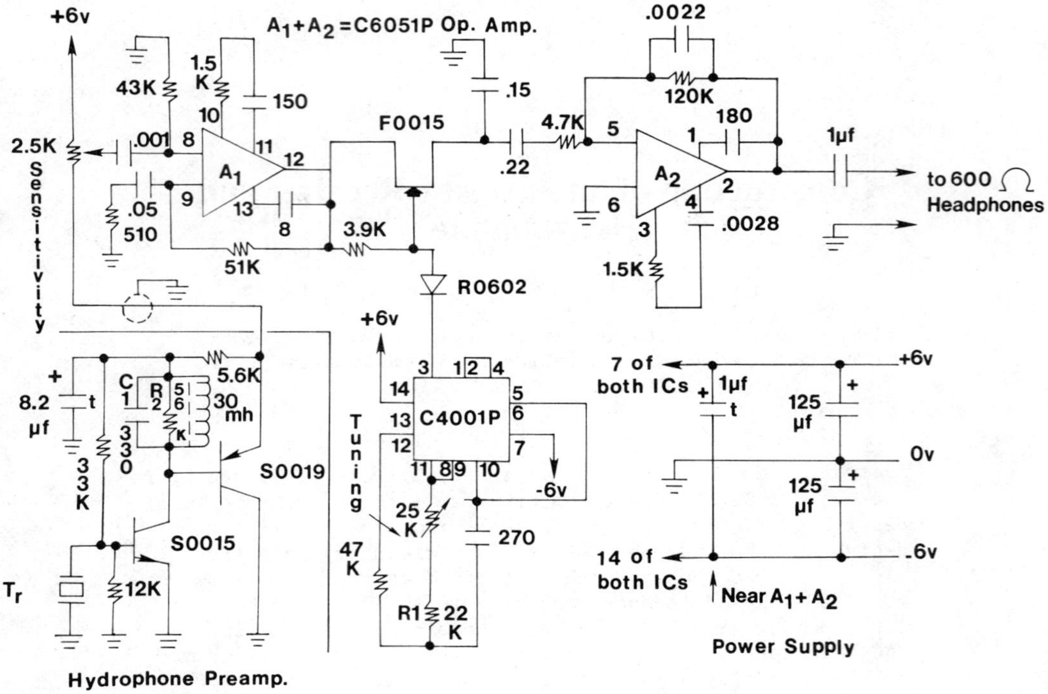

Fig. 1. Circuit diagram for an acoustic receiver and a frequency selective hydrophone preamplifier. All semiconductors are from the HEP® series by Motorola. The miniature 30 mH inductor is a Tortel TG-20-7518 D. The transducer, 1.65 cm o.d., 1.49 cm i.d., 0.89 cm long, TCD-5H® material, is manufactured by Marine Resources Incorporated. (Adapted from Kanwisher *et al.*, 1974. Courtesy of the U.S. National Marine Fisheries Service.)

The receiver may be assembled on a 10 × 15 cm perforated board. Except where indicated in Fig. 1, location of parts is not critical. Care should be taken when handling and soldering the CMOS C4001P integrated circuit as static electricity can damage it. With the component values shown in Fig. 1, the receiver detects frequencies of 40 to 60 KHz. Use of a precision 10-turn potentiometer (Clarostat 62JA) and a 10-turn dial with 1000:1 indexing accuracy (Spectrol 15-1-11) permits tuning to a given frequency with good accuracy. The receiver may be easily reset to detect either a higher or lower range of frequencies. Decreasing the value of resistor R1 tunes the receiver for higher frequencies; increasing it has the opposite effect. We use 1.5 V alkaline AA batteries to power the receiver. Carbon-zinc 1.5 V AA batteries may also be used and are available in all but the most remote corners of the world. A waterproof metal case houses the receiver. The case is connected to receiver ground.

The preamplifier is assembled on a thin, 1.5 × 4.0 cm perforated board. The location of parts is not critical. With the components shown in Fig. 1, the preamplifier is most sensitive to signals of 50 KHz. Sensitivity is approximately 3 dB down at 45 and 55 KHz. The preamplifier can be made maximally sensitive to another frequency by changing the value of capacitor C1. The formula for the proper value of C1 is:

$$C1\ (pF) = [\,330 \times (50/\text{desired frequency in KHz})^2\,].$$

The value of resistor R2 must then be adjusted so that the preamplifier once again has a gain of 100. The preamplifier is connected to the receiver by means of waterproof, 2-conductor shielded cable (Belden 8412). The shield is connected to ground at the receiver only. Care should be taken when soldering the leads from the preamplifier to the inside and outside of the cylindrical transducer; excessive heat can damage it. After the leads are in place, each end of the transducer is sealed with a 0.08 cm thick piece of Lexan (polycarbonate). The pieces of Lexan® are the same diameter as the transducer and held in place with cyanoacrylate adhesive. The transducer is now 'air backed'.

The transducer and preamplifier are both potted inside a thin walled glass epoxy tube. Tube and transducer should be a close fit if high acoustic-to-electric efficiency is to be realized. We used 7 cm lengths of 1.77 cm inner diameter tubing with a 0.08 cm wall, sealed at one end with a circular piece of glass epoxy, 0.15 cm thick (Stevens Tubing Corp.). Hysol® R9-2039 epoxy resin and H2-3651 hardener (Dexter Corp.) have proven excellent for potting hydrophones.

If an omnidirectional hydrophone is sufficient for one's experiments, the equipment is now ready to use. Sensitivity is remarkable for such simple devices. A 1 μV, 50 KHz signal at the hydrophone is clearly audible in the headphones. If a directional hydrophone is required, the potted transducer and preamplifier may be fitted inside a 90°, foam neoprene lined, stainless steel cone (see Stasko and Polar, 1973). The cone can be mounted on one end of a rotating shaft fixed to a boat. Brackets of the kind supplied with small electric fishing motors are ideal for holding the shaft. We used an aluminum shaft made up of several separate sections. Receiver, hydrophone, shaft, and mounting bracket all fit readily into an average sized suitcase.

Acknowledgements — This material is based upon work partially supported by the National Science Foundation under Grant No. BNS 78-18196. Contribution No. 807 of the Bermuda Biological Station for Research.

REFERENCES

Ireland L. C. and Kanwisher J. W. (1978) Underwater acoustic biotelemetry: procedures for obtaining information on the behavior and physiology of free-swimming aquatic animals in their natural environments. In D. I. Mostofsky (ed.), *The Behavior of Fish and Other Aquatic Animals*, pp. 341-379. New York: Academic Press.

Kanwisher J. W., Lawson K. D. and Sundnes G. (1974) Acoustic telemetry from fish. *Fish. Bull (U.S.)* 72, 251-255.

Stasko A. B. and Pincock D. G. (1977) Review of underwater biotelemetry, with emphasis on ultrasonic techniques. *J. Fish. Res. Bd Can.* 34, 1261-1285.

Stasko A. B. and Polar S. M. (1973) Hydrophone and bow-mount for tracking fish by ultrasonic telemetry. *J. Fish. Res. Bd Can.* 30, 119-121.

Heart Rate Telemetry Using Micropower Integrated Circuits

T. G. WOLCOTT

Department of Marine Science and Engineering, North Carolina State University, Raleigh, North Carolina, U.S.A.

Abstract — Two heart rate/temperature telemetry transmitter circuits have been developed for estimating activity, physiological state or metabolic rate. Use of programmable micropower operational amplifiers permits total quiescent supply currents below 25 μA while retaining sufficient sensitivity for use on invertebrates (crabs). The first circuit responds to electrical activity of the myocardium or cardiac ganglion. It has the advantage of sensing an unambiguous all-or-nothing electrical signal which has an exact one-to-one correspondence with heart beat. Its disadvantages are low noise immunity due to high input impedance, sensitivity of other muscle potentials, possible signal loss due to electrode motion or degradation, and a requirement for critical electrode placement. The second circuit, slightly more complex, responds to a.c. impedance changes associated with heart movements. The signal waveform changes with heart rate and degree of filling and therefore is not as unambiguous as ECG. This circuit is, however, immune to both muscle potentials and external noise fields, shows very long electrode life, and does not require precise electrode placement. Both transmitters can be constructed with package weights below 5 g including batteries for two weeks operation.

INTRODUCTION

The intensity, pattern, and metabolic cost of various activities are important variables in physiological and behavioral ecology. Of the various correlates of metabolism, heart rate appears to be the most suitable indicator of short-term variations (Gessaman, 1973). This paper presents two transmitters developed for monitoring the heart rate of free ranging ghost crabs (*Ocypode quadrata*) and some preliminary results which indicate that heart rate measurements will be a useful field tool.

CIRCUIT DESIGN

The requirements for monitoring the heart rate of invertebrates were somewhat more stringent than those for working on vertebrates. As with any small animal, low carrying capacity dictated minimal power drain to maximize life from small batteries. The transmitters themselves had to be small; nevertheless they had to be extraordinarily sensitive by vertebrate standards. Crab hearts are thin walled little bags of muscle and produce a signal orders of magnitude smaller than that obtainable from a mouse heart. The solutions to these requirements were essentially two-fold. First, programmable micropower operational amplifiers ('op amps') were used as high gain functional blocks. These integrated circuits have high input impedance and common mode rejection, are inherently stable, draw minimal supply current, and require a minimum of external components. Secondly, a pulsed output was used rather than transmitting a continuous analog signal of the ECG. This approach, pioneered by Baldwin (1973), reduces the duty cycle of the power consuming radio circuitry to a few percent, thus drastically lowering total power requirements for a given RF output and range. A third design requirement was that proper operation should not depend on the maintenance of any resistances in excess of 2 MΩ. Higher resistances were deemed likely to change as a result of inevitable exposure to soil moisture and salt water. (This criterion might well apply to implantable transmitters as well.) Two transmitter types have been designed using these approaches: the first senses the ECG, and the second senses variations in cardiac a.c. impedance.

The ECG sensing heart rate transmitter (Fig. 1) has a quiescent current drain of less than 10 μA and can be constructed to weigh less than 4 g (less batteries). The

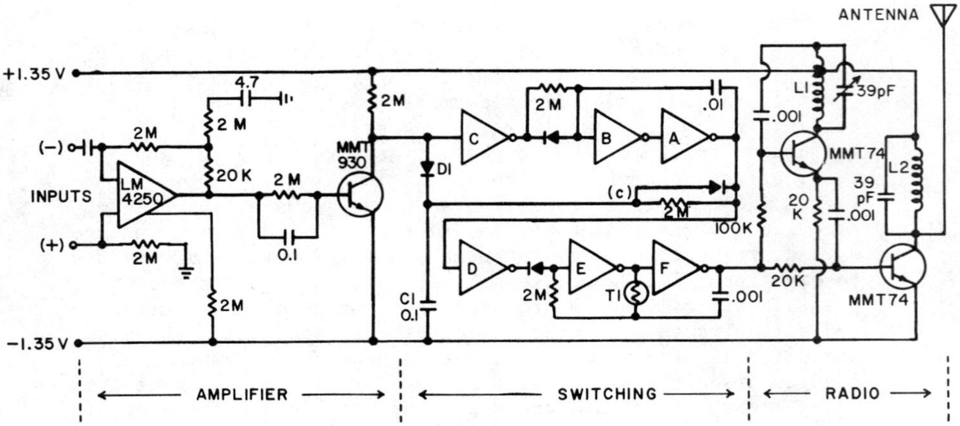

Fig. 1. ECG sensing heart rate/temperature transmitter. A-F = CMOS 4049 hex inverter buffer. T1 = 250 K thermistor. L1, L2: 5T, No. 28 A.W.G., 5 mm diameter; L1 tapped at 1T.

feedback loop of the operational amplifier passes d.c., resulting in an op amp output centered at ground potential. A.C. gain is controlled by a voltage divider coupled through a 4.7 μF capacitor to ground. It is set, for use on ghost crabs, at 1000 — about the maximum for the LM4250 at the required slew rate. The output voltage of the op amp biases the MMT930 transistor into saturation except when negative going swings of the op amp output withdraw base drive through the 0.1 μF capacitor. When this occurs the transistor is shut off and the collector voltage rises rapidly from nearly zero to the supply voltage.

This high amplitude rectangular wave signal is fed to switching circuitry built around a 4049 CMOS hex inverter buffer. Each time the transistor shuts off and the input of inverter 'C' goes high, several things happen. First, the 'oneshot' comprising inverters 'B' and 'A' is triggered; its output goes low for about 15 ms. The output of inverter 'D' goes high for the 15 ms period, permitting the multivibrator comprising inverters 'E' and 'F' to oscillate at the frequency determined by thermistor T1. The output of 'F' provides base drive for the radio transmitter, in this example a 2-stage 100 MHz FM circuit. Simultaneously, the output of 'A' discharges C1, which will hold the input of 'C' low for about 100 ms until C1 is recharged and D1 is again reverse biased. This limits output pulse duration to 15 ms regardless of input pulse length (up to 100 ms) and prevents retriggering of the high impedance amplifier circuitry by switching transients from the high current RF section. The net result is a 15 ms, temperature modulated radio pulse at each beat of the heart. Transmission of temperature permits the observer to distinguish heart rate changes due to activity from those due to temperature variation. Total current consumption at 100 bpm is about 100 μA.

This circuit has the advantage of sensing an unambiguous, all-or-none electrical event which has a one-to-one correlation with heart contractions. Its disadvantages include a susceptibility to triggering by muscle potentials during locomotion or other movement and a moderate sensitivity to external noise fields (e.g. power line hum). Due to the common mode rejection of the operational amplifier, even in the unbalanced configuration shown, hum is less of a problem than with traditional discrete designs. This circuit has been used in the laboratory under fluorescent lights without a Faraday cage. In vertebrates, large signal amplitudes would probably permit balanced operation by deletion of the voltage divider and grounding capacitor in the negative feedback loop, and noise rejection should be excellent. More serious limitations to the application of this circuit on crabs involve electrode placement. Apparently the myocardial potentials are so small and diffuse that proper operation depends on placing the electrodes precisely over the cardiac ganglion, which produces a clear, fast spike. It is, however, a small target and problems have arisen in achieving correct initial placement and in maintaining signal strength after electrode implantation. Apparently there can be relative movement between heart and carapace with changes in blood volume, or the electrodes themselves may be degraded over time by scarring over, polarization, or other processes which increase their impedance. These difficulties led me to explore alternative methods of detecting cardiac movements.

Circuits sensitive to variations in a.c. impedance ('impedance pneumographs') avoid the problems of detecting low level biopotentials. However, because they are complex and rely on their own power to induce an a.c. current in the animal rather than deriving signal power from the animal itself, they have been used almost exclusively in laboratory studies (Hoggarth and Trueman, 1967; Ansell, 1973; Cumberledge and Uglow, 1977; McDonald, McMahon and Wood, 1977).* The design worked out for an impedance sensing transmitter which would remain within the size and power consumption requirements for small animal telemetry (Fig. 2) again made use of programmable micropower op amps to minimize complexity and to hold quiescent current drain below 15 μA.

The a.c. signal to the heart electrodes is applied, and impedance changes are sensed, by a multivibrator constructed from the op amp OA1 and its associated components. The frequency of the multivibrator is determined (other components being held constant) by the a.c. impedance across the heart electrodes, which are in positive feedback path. At an impedance of 1 kΩ the circuit oscillates at about 2 KHz. Positive feedback resistor R1 is adjusted to about 1000 times the cardiac impedance for maximum sensitivity of the circuit. A.C. current across the heart is less than 5 μA. The period when the op amp rectangle wave output is high is short and essentially independent of cardiac impedance due to D1; the period when the output is low is proportional to cardiac impedance. Thus, differences in frequency caused by variations in cardiac impedance cause variations in average signal level which can be rectified and detected by diode D2 and its associated RC network. The detected signal is

---

*Dr. Adrain Bottoms informs me that he developed an impedance pneumograph type transmitter at the University of Hull, England, but has not yet published his work.

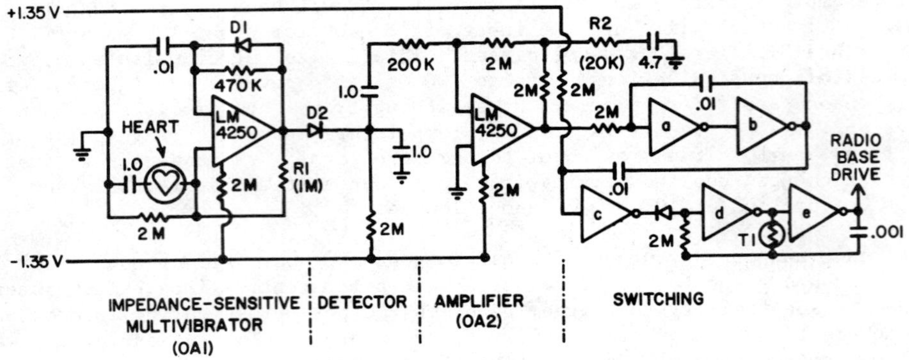

Fig. 2. Impedance sensing heart rate/temperature transmitter. a-f = CMOS 4049 hex inverter buffer. T1 = 250 K thermistor.

amplified by the second op amp (OA2) configured as in the ECG amplifier; gain is set by R2.

The cardiac impedance signal is a slow wave rather than a fast spike, and different switching circuitry must be employed. Inverters 'a' and 'b' act as a high gain amplifier, converting the signal into a rectangular wave; the positive feedback capacitor assures nonlinear operation and fast switching. Each time the output of 'b' goes low, the output of 'c' is driven high for about 15 ms and the multivibrator comprising 'd' and 'e' oscillates at a frequency determined by thermistor T1. The output of 'e' provides base drive for the radio circuitry. The circuit also has a total current drain of about 100 μA at 100 bpm, and can be built to weigh less than 5 g.

The impedance sensing circuit has several unique advantages: it is virtually immune to both muscle potentials and external noise fields, has indefinite electrode life, does not demand precise electrode placement, and is adaptable to detecting other movements not associated with electrical activity. It is, however, somewhat more complex than ECG sensing circuits and measures a less unequivocal phenomenon than ECG.

Both the ECG sensing and impedance sensing circuits have been tested on a captive crab in the laboratory; both produced acceptable signals for more than two months (a single pair of electrodes fortuitously served in both modes). The ECG sensing circuit, as expected, gave occasional false beats due to extraneous muscle potentials. The impedance sensing circuit also gave false beats at extremely low (below 30 bpm) and extremely high (above 150 bpm) heart rates. These errors were due to the impedance variation becoming multiphasic at the extreme rates (Fig. 3). The fundamental rhythm could usually be detected by ear, although automatic recording devices would not distinguish artifacts from heartbeats. Future experimentation will reveal whether the artifacts can be avoided by using other electrode orientations.

A system for obtaining continuous recordings of heart rates from animals in the field was developed to complement the transmitters. A clockdrive recorder with a 1 mA meter movement was driven by the circuit shown in Fig. 4. T1, a 555 timer i.c. wired as a one-shot, is triggered by the amplified audio output of the radio receiver. Its voltage output pulse is converted to a current pulse by the Howland circuit incorporating op amp OA1. Thus, each heartbeat results in the injection of a fixed charge into capacitor C1, and the voltage appearing on C1 (or at the output of OA1) is proportional to heart rate. This voltage is converted into meter drive current by OA2,

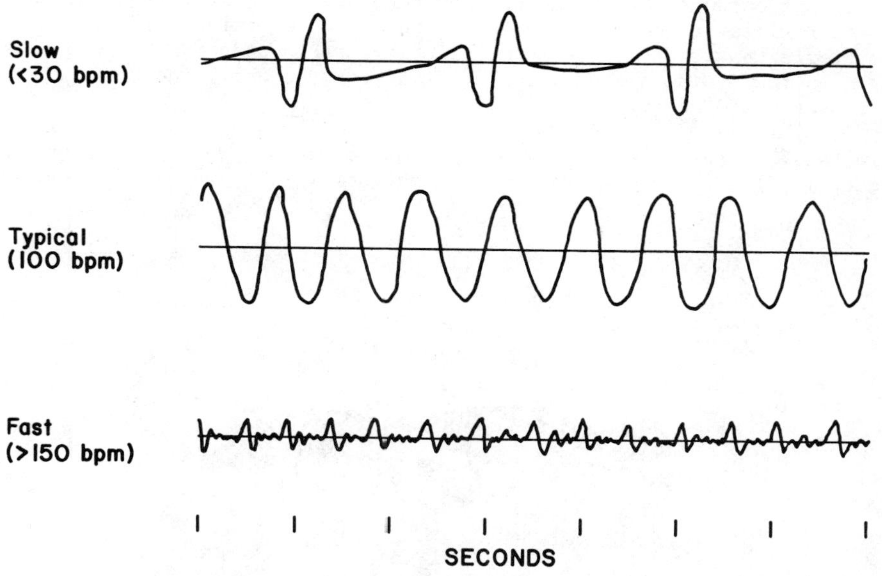

Fig. 3. Cardiac impedance signals (diagrammatic) occurring at the amplifier (CAZ) output of the impedance sensing circuit (Fig. 2). Triggering of an output pulse occurs at each negative going zero crossing; multiphasic signals at extremely low and high heart rates therefore give 'extra beats.'

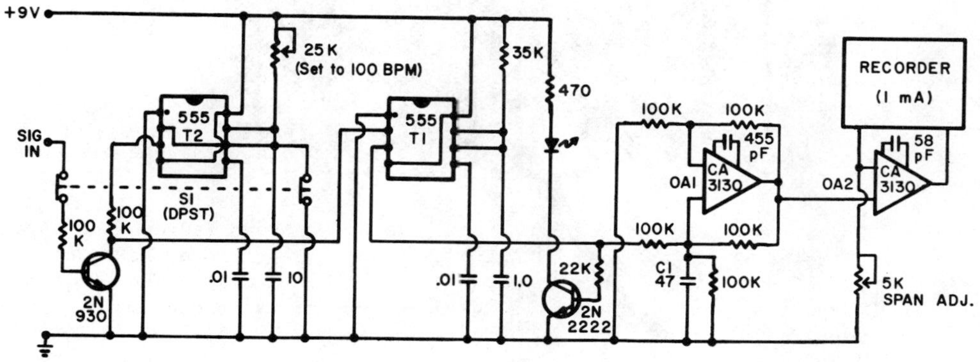

Fig. 4. Circuit which converts audio pulses from the radio receiver to heart rate proportional recorder drive current (0-1 mA). CA 3130 op amps are used because their common mode and output ranges include ground. Opening S2 applies 100 bpm calibration signal to recorder drive.

a voltage-to-current converter. Since the meter is current driven, variations in meter resistance due to temperature variation have no effect on readings. A second 555 timer, T2, is used to produce a 100 bpm calibration signal; the recorder span is adjusted in the field to the calibration value with a 5 K potentiometer. A light emitting diode provides a visual correlation between audio output and timer output.

This permits setting radio volume to a level where triggering occurs on all heartbeats but not on noise. Temperature was recorded by constructing and calibrating a CMOS multivibrator identical to the temperature sensitive multivibrator in the transmitter except that a logarithmic (audio taper) potentiometer was substituted for T1. In the field, the potentiometer was rotated until the pitch it produced matched that heard on the radio receiver.

A field trial of this system, utilizing a miniaturized prototype of an earlier (Wolcott, 1977) ECG sensing transmitter design (Figs. 5 and 6) was conducted with

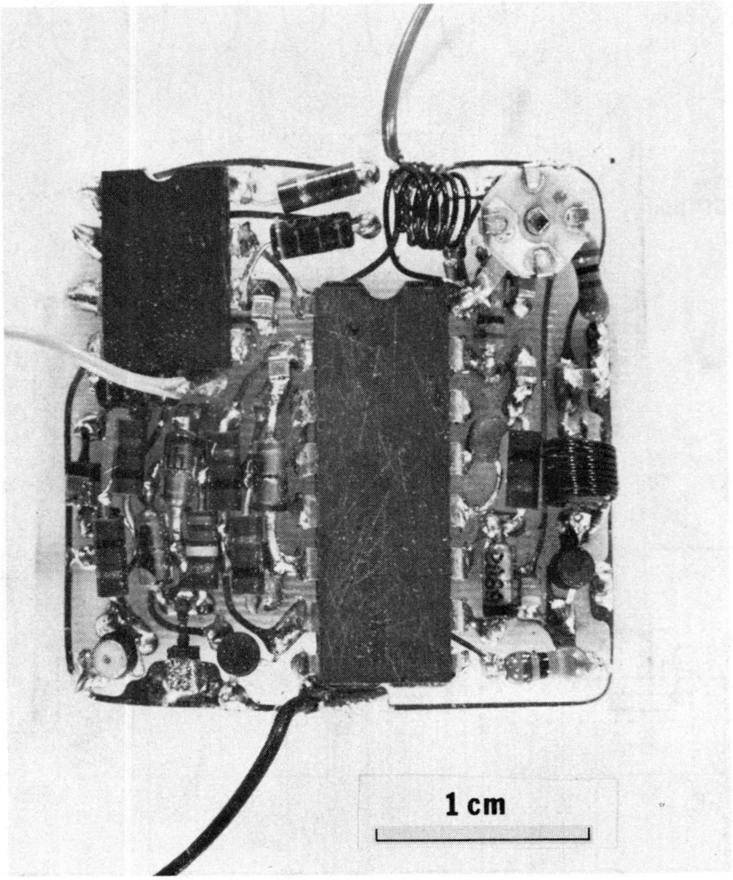

Fig. 5. Miniaturized prototype of ECG sensing transmitter (this circuit differs slightly from that in Fig. 1). Plastic DIPs are ground to 2 mm thickness to reduce weight and volume. Encapsulated weight = 4 g.

a crab which was later found to be in poor condition. Nevertheless, heart rate not only reflected changes in temperature (and presumably metabolic rate) over the long term, but also provided a very detailed record of activity (Fig. 7). Responses of heart rate to changes in activity in other crabs (tested before construction of the recorder) were extremely rapid; heart rate rose from 30 to 150 bpm within 5 s upon commencement of locomotion, and declined to resting levels within 1-2 min after cessation of activity. Sudden increases in heart rates of crabs 20-50 cm deep in their

Fig. 6. Heart rate/temperature transmitter on ghost crab (*Ocypode quadrata*). Trailing Teflon insulated wire is transmitting antenna.

burrows occurred when investigators approached within 10 m of the burrows, indicating the sensitivity of these animals to sand borne vibrations. Preliminary studies in a treadmill respirometer (C. F. Herreid, State University of New York at Buffalo, personal communication) have indicated a close correlation between exercise metabolism and heart rate. This heart rate telemetry system thus shows considerable promise as a method of estimating activity and its metabolic cost in ghost crabs and possibly in other highly mobile small animals.

Acknowledgements — These devices were developed with support from the N.C.S.U. Department of Marine Sciences and Engineering and grant DEB-77 from the National Science Foundation. The cooperation of the National Marine Fisheries Service Laboratory at Beaufort, N.C. and Fort Macon State Park on Bogue Banks, N.C., during the field work is gratefully acknowledged. Contribution 79-03 from the Department of Marine Science and Engineering, North Carolina State University.

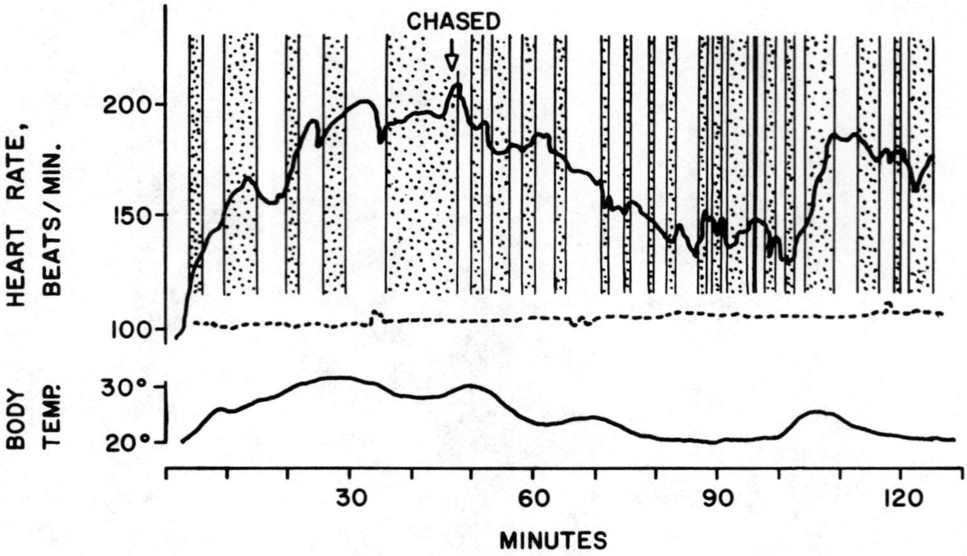

Fig. 7. Heart rate of a free ranging ghost crab (*Ocypode quadrata*). Stippled bars are periods of active locomotion which tend to be followed by increases in heart rate; open bars represent quiescent periods which tend to be associated with declining heart rates. Dotted line = heart rate of quiescent crab in burrow.

REFERENCES

Ansell A. D. (1973) Changes in oxygen consumption, heart rate, and ventilation accompanying starvation in the decapod crustacean *Cancer pagurus*. *Neth. J. Sea Res.* 7, 455-479.

Baldwin H. A. (1973) Instrumentation for remote observation of physiology and behavior. In J. A. Gessaman (ed.), *Ecological Energetics of Homeotherms*, Chap. 6, pp. 67-76. Utah State Univ. Monogr. Ser.

Cumberlidge N. and Uglow R. F. (1977) Size, temperature, and scaphognathite frequency-dependent variations of ventilation volumes in *Carcinus maenas* (L.) *J. exp. mar. Biol. Ecol.* 30, 85-93.

Gessaman J. A. (1973) Final comments. In J. A. Gessaman (ed.), *Ecological Energetics of Homeotherms*, Chap. 14, pp. 148-155. Utah State Univ. Monogr. Ser.

Hoggarth K. R. and Trueman E. R. (1967) Techniques for recording the activity of aquatic invertebrates. *Nature, Lond.* 213, 1050-1051.

McDonald D. G., McMahon B. R. and Wood C. M. (1977) Patterns of heart and scaphognathite activity in the crab *Cancer magister*. *J. exp. Zool.* 202, 33-44.

Wolcott T. G. (1977) Radiotelemetry of heart rates: a tool for field studies of activity and energetics, *Am. Soc. Zool.*, December 1977, Toronto. *Am. Zool.* 17, 895.

An Implantable Transducer for the Measurement of Respiratory Air Flow

A. J. WOAKES and P. J. BUTLER

Department of Zoology and Comparative Physiology, University of Birmingham, Birmingham B15 2TT, U.K.

Abstract — A transducer has been developed that is implantable, has a low power requirement and produces a linear directional measurement of respiratory air flow. The air velocity is sensed by the phase shift that it produces in a beam of ultrasound passed diagonally across the trachea between two identical piezoelectric transducers. Preliminary results indicate a linearity of better than 2 percent over a flow range of ± 20 l min^{-1}, with a response time of approximately 5 ms. The transducer can be fitted to tracheae of 1 cm and greater diameter, and constricts the tracheal cross section by 3 mm^2. The driver, receiver and phase measurement functions are provided by three standard CMOS digital integrated circuits.

Many attempts have been made to measure respiratory air flow in unrestrained animals (Berger, Hart and Roy, 1970; Kimmich, 1972, 1976; Hörnicke *et al.*, 1974). Most workers have used masks placed over the mouth and nostrils, with a consequent interference with the natural behavior of the subject. Implantation of the flow transducer is therefore essential if this major benefit of telemetry systems, particularly if used for longterm studies of behavior, physiology and energetics, is not to be lost. This is especially important if the flow transducer is used in conjunction with an oxygen sensor in order to measure the oxygen consumption of a free ranging animal. An implantable transducer and associated electronics have therefore been developed as a modular section of a multichannel transmitter. This report is of a preliminary nature as little experience has yet been gained of the operation of the device. However, the results that have been obtained are encouraging and it is hoped that this early publication will help other researchers who as yet have no means available for studying this important parameter.

The principle of operation of the transducer is similar to that of a number of instruments designed to measure blood flow (Zarnstorff, Castillo and Crumpton, 1962; Rayder, Meehan and Henriksen, 1973) in that the flow is sensed by the phase shift that it produces in a beam of ultrasound passed diagonally across the trachea between two identical piezoelectric transducers, T1 and T2 (Fig. 1). If one transducer, say T2, is driven at frequency f, then the time, t, taken for the sound to reach the other transducer, T1, d cm away is:

$$t_{2-1} = \frac{d}{C + V_{\cos}\theta}, \qquad (1)$$

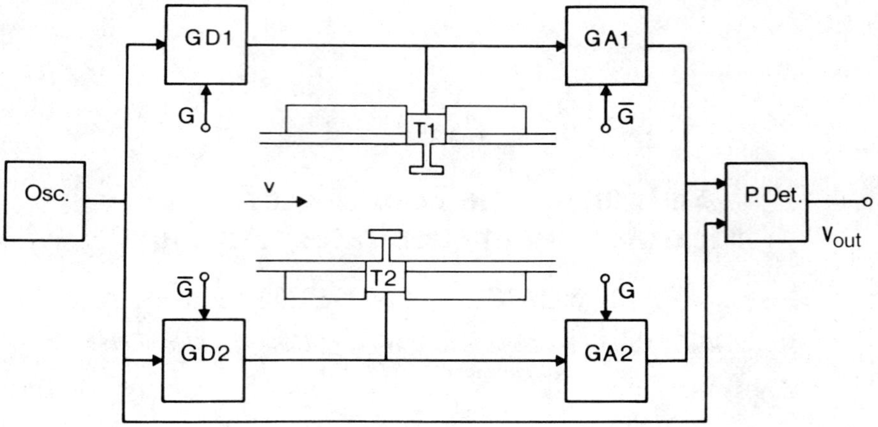

Fig. 1. Major elements of the flow meter. The oscillator (Osc.) provides a stable frequency that is used to drive one of the transducers (T1,2) through a gated driver circuit (GD1,2). The transducer generates a beam of ultrasound that crosses the trachea and is shifted in phase by a time delay that is partially dependent on the mean velocity (v) of the air flow in the trachea. The received signal is amplified by a gated amplifier (GA1,2) and compared with the master oscillator by the phase detector (P. Det.) which produces a voltage output proportional to the phase difference between the two signals.

where $V_{\cos}\theta$ is the component of mean air flow along the path, d, and c is the velocity of sound in air. This represents a phase shift, ϕ_{2-1}, of the transmitted signal, where

$$\phi_{2-1} = \frac{2\pi\, fd}{C + V_{\cos}\theta} \quad . \tag{2}$$

If the functions of the two transducers are reversed then the phase shift is:

$$\phi_{1-2} = \frac{2\pi\, fd}{C - V_{\cos}\theta} \quad . \tag{3}$$

The difference between these two phase shifts, $\Delta\phi$, is therefore given by:

$$\Delta\phi = 2\pi\, fd \quad \frac{1}{C - V_{\cos}\theta} - \frac{1}{C + V_{\cos}\theta} \tag{4}$$

$$= 2\pi\, fd \quad \frac{2V_{\cos}}{C^2 - V^2\cos^2\theta} \tag{5}$$

if $C^2 \gg V^2 \cos^2\theta$, this simplifies to:

$$\Delta\phi = \frac{4\pi\, fd\, \cos\theta \cdot V}{C^2} \quad . \tag{6}$$

Therefore:

$$\Delta\phi \propto V \ . \tag{7}$$

The difference between the phase shifts found with the ultrasonic beam travelling first in one direction, and then reversed, is therefore linearly related to the mean velocity of the air flowing between the two transducers.

In practice a stable, crystal controlled oscillator feeds two gated driver stages at the resonant frequency of two matched piezoelectric transducers T1 and T2 (Fig. 1). Only one driver at a time is active. They are controlled by the gate signal *G*, and its inverse $\bar{G}$, which is derived from the channel timing signals of the transmitter. When one transducer is operating as a sound transmitter, its associated gated receiver is switched off. When a transducer is receiving an ultrasonic signal, typically 2-5 mV in amplitude, its amplifier is switched on. The output of the amplifier is a clipped, square wave signal that retains the phase information of the input (Fig. 2).

The phase of this signal is compared with that of the main oscillator, and a voltage proportional to the difference is generated and forms the input to one channel of the multichannel transmitter. This one channel alternately carries the information

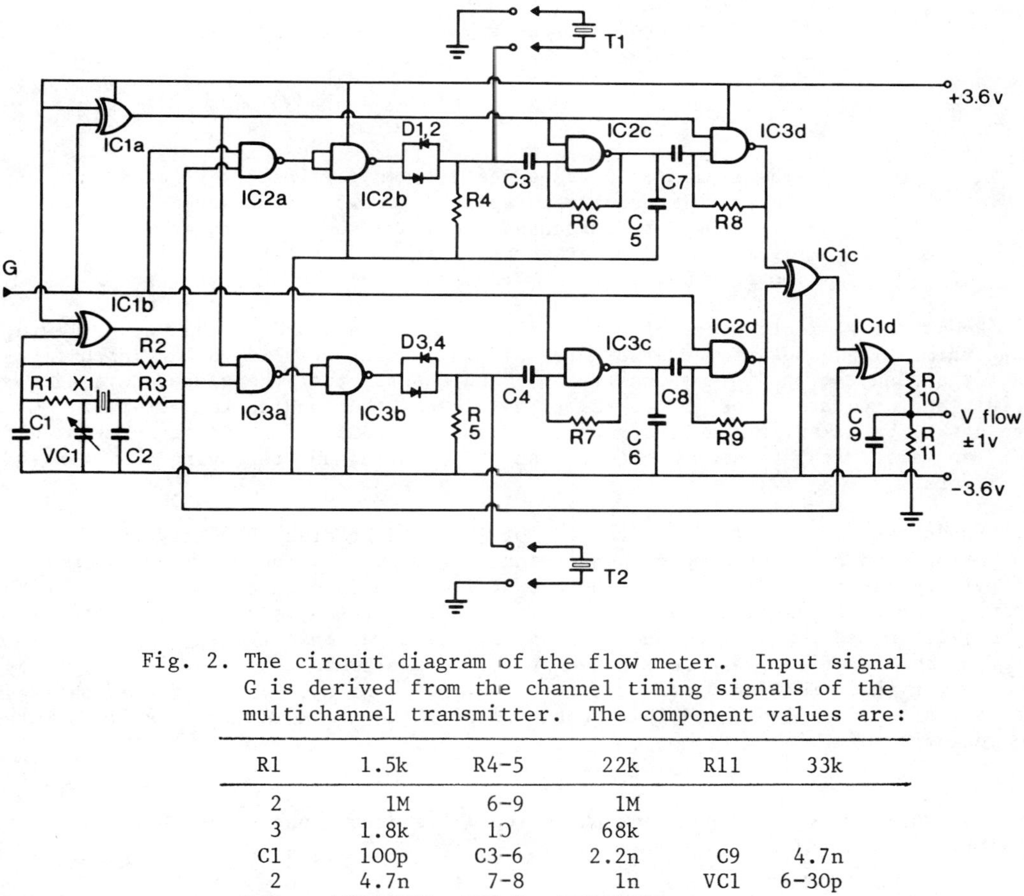

Fig. 2. The circuit diagram of the flow meter. Input signal G is derived from the channel timing signals of the multichannel transmitter. The component values are:

R1	1.5k	R4-5	22k	R11	33k
2	1M	6-9	1M		
3	1.8k	10	68k		
C1	100p	C3-6	2.2n	C9	4.7n
2	4.7n	7-8	1n	VC1	6-30p
IC1	MC14070B	IC2-3	MC14011B		
D1-4	OA202				

Resonating element X1 and transducers T1,2 are constructed from Mullard bimorph flexure elements MB7010.

derived from the two sequential measurements of phase shift. On reception, they are separated and subtracted in order to obtain the velocity measurement. Their sum gives information about the constant of proportionality, and allows corrections for changes in this factor to be made to the velocity readings. The oscillator, drivers, amplifiers and phase detector are built from three standard CMOS digital integrated circuits. The transducers are matched piezoelectric ceramic elements, 3 mm in diameter and have a resonant frequency of between 120 and 160 KHz. They are mounted on studs in order to project them into the lumen of the trachea (Fig. 3).

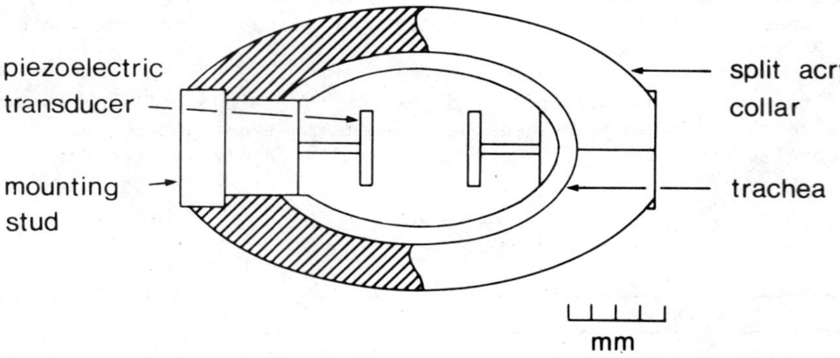

Fig. 3. The implantable transducer, shown partially sectioned. The acrylic collar is split into two sections which fit around the trachea. They are then held together with 1 mm diameter screws which also retain the mounting studs.

This keeps the elements clear of the walls of the trachea and of the mucus flowing along them. The bases of the studs are held in a split acrylic collar which fits tightly around the outside of the trachea, and ensures that the transducers are rigidly and accurately positioned relative to each other and to the air flow. Great care has to be taken in the construction of the instrument in order to prevent ultrasonic and electrical crosstalk between the transmitting and receiving elements and amplifiers.

The transducer may be fitted to tracheae of 1 cm and greater diameter, and each projecting stud restricts the cross section of the lumen by approximately 3 mm^2. Preliminary results indicate that the output from the transducer is linear to within 2 percent up to flows of at least 20 $l\ min^{-1}$ (Fig. 4), and is relatively insensitive to changes in gas composition and temperature. The response time is less than 10 ms (Fig. 5) and is effectively determined by the rate at which the flow is sampled by alternating the direction of the ultrasonic signal. Figure 6 shows traces obtained with the transducer implanted in a Canada goose (*Branta canadensis*) and illustrates the potential of this instrument.

Acknowledgement — We wish to thank the Science Research Council for their financial support.

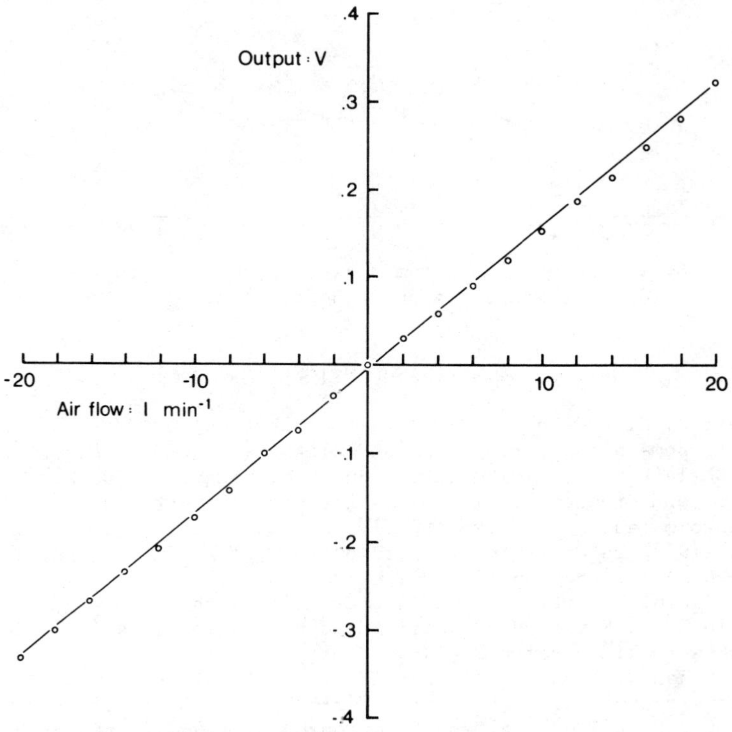

Fig. 4. Relationship between air flow and the output voltage of the flow meter. Air flow was measured with a Rotameter flow meter, 2-20 l min^{-1}.

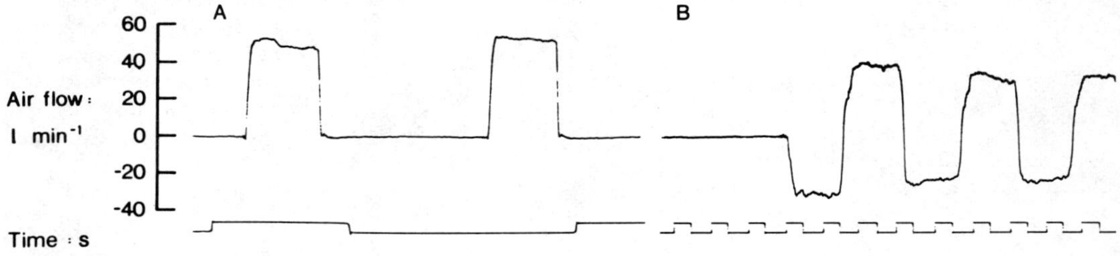

Fig. 5. Traces showing: (A) the response of the transducer to pulsed air flow; (B) respiration of a human, with the transducer used as an external flowmeter.

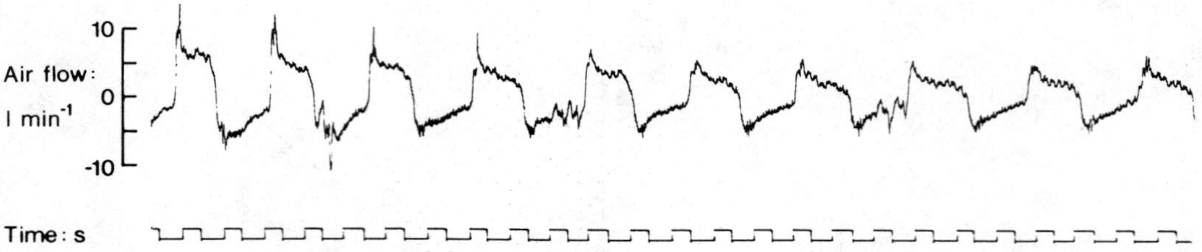

Fig. 6. Traces obtained from a resting goose, with the transducer implanted around the trachea.

REFERENCES

Berger M., Hart J. S. and Roy O. Z. (1970) Respiration, oxygen consumption and heart rate in some birds during rest and flight. *Zt. Vergl. Physiol.* 66, 201-214.

Hörnicke H., Ehrlein H. J., Tolkmitt G., Husch H., Nagel M., Decker D., Epple E., Kimmich H. P. and Kreuzer F. (1974) Respiratory telemetry in exercising horses. In P. A. Neukomm (ed.), *Biotelemetry II*, pp. 146-148. Basel: Karger.

Kimmich H. P. (1972) Telemetry of respiratory air flow. In H. P. Kimmich and J. A. Vos (eds.), *Biotelemetry*, pp. 11-120. Leiden: Meander NV.

Kimmich H. P. (1976) Telemetry of respiratory flow with a new magnetic type of flowmeter. In T. B. Fryer, H. A. Miller and H. Sandler (eds.), *Biotelemetry III*, pp. 25-28. New York: Academic Press.

Rayder R. D., Mehan J. P. and Henriksen J. K. C. (1973) An implantable blood pressure and flow transmitter. *IEEE Trans. Bio-Med. Eng.* BME-20, 37-43.

Zarnstorff W. C., Castillo C. A. and Crumpton C. W. (1962) A phase-shift ultrasonic flowmeter. *IRE Trans. Bio-Med. Electron.* BME-9, 199-203.

Implantable Telemetry System for Strain Measurements

N. H. NOWAIRA, F. MORRIS and S. RAKOWSKY

Bio-Medical Engineering Unit and Physics Department, North Staffordshire Polytechnic, Stoke-on-Trent, Staffordshire, U.K.

Abstract — In this paper we present design aspects of an inductively powered, implantable telemetry system which can be used in monitoring strains in orthopedic implants, in designing and testing of new devices and materials for internal fixation. As strain is usually measured by the electrical resistance of a gauge, attention has been given to telemetry systems which use resistive transducers. The continuous wave oscillator circuit chosen consists of a single field effect transistor (FET) with a feedback resistance in the source lead simulating the strain gauge. As this oscillator will be inductively powered, there will be variations in supply voltage (due to relative movements). Hence the changes in gauge resistance must produce greater changes in the oscillator frequency than those produced by supply voltage variations. The use of semiconductor strain gauges was found desirable.

INTRODUCTION

Because internal fixation of fractured bone depends on the production of local pressure areas between implant and bone and between bone fragments, assessing the stresses in fracture plates is of paramount importance in following the course of bone healing and in the design of new devices for internal fixation.

In 1969 Perren *et al.*, pioneered *in vivo* measurements of strain in fracture plates. They modified a four hole AO (Association for the Study of Internal Fixation) compression plate to incorporate strain gauges and thus to act, at the same time, as compression plate and load cell. They used this system in preliminary experiments using sheep by bringing the electrical leads through the skin. They emphasized that many preliminary experiments were required because of difficulties which arose from failure of insulation, retrograde infections, bending deformation and broken leads. In spite of these difficulties they could conclude:

(1) The system was sensitive to very small changes in length and pressure would have decreased to zero if there had been a loss of substance of 10-20 μ due to resorption. Minute changes of length due to instability would also cause sudden changes of pressure. (2) The pressure within the range studied (21 mn m^{-2}) lead to a slow loss of load which they attributed to Haversian remodeling. (3) The fact that pressure

recordings permit conclusions about longterm stability of fixation had potential in testing new devices for internal fixation.

The aim of the present project is to improve the technique of strain measurement in fracture plates by arranging an implant system which eliminates transcutaneous leads and which can enable the monitoring of strain over extended periods. To achieve the latter requirement, the telemeter is inductively powered.

Hence the complete system consists of four main parts (Fig. 1): (a) The power induction system; consisting of a 500 KHz power oscillator and primary induction coil. (b) The implanted pick up coil and rectification unit. (c) The implanted telemetry transmitter; this is a 4 MHz oscillator whose frequency will be frequency modulated (FM) by changes in the resistance of strain gauges which are attached to the fracture plate. (d) The external receiver in which the 4 MHz FM strain signal is detected and measured.

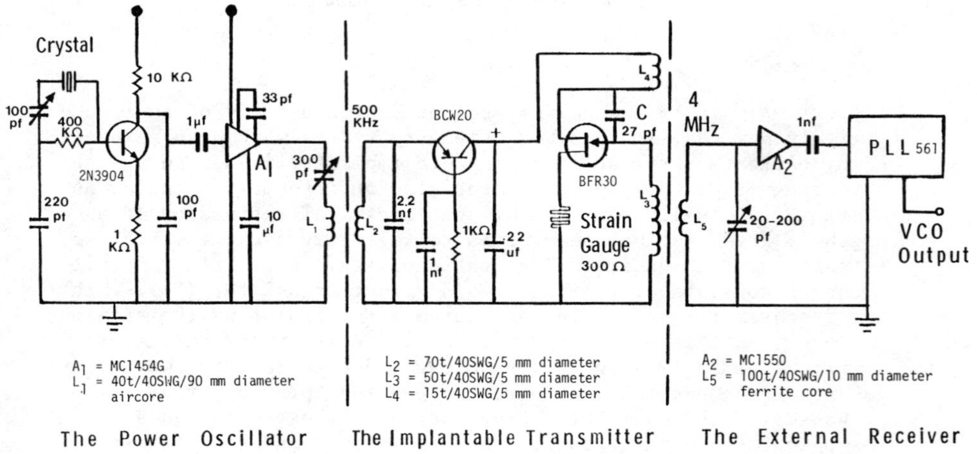

Fig. 1. Implantable telemetry system for strain measurement.

THE IMPLANTED TELEMETRY TRANSMITTER

As strain is usually measured by the electrical resistance of a gauge, attention is given to telemeters which use a resistive transducer. A commonly used circuit is the Hartley blocking oscillator (e.g. Fryer, Deboo and Winget, 1965; Clarke, 1966), in which the period of the Squeging oscillator is dependent on resistance. However, in an analysis by Lin and Ko (1968), they pointed out that the period of oscillation is also heavily dependent upon supply voltage. In an inductively powered system, the induced supply voltages will be subject to variation primarily due to small relative movements between the primary power coil and the implanted power pickup coil. Hence the Hartley blocking oscillator is considered inadequate for the current application.

Consequently, attention was centered on the continuous wave LC oscillator which consists of a single field effect transistor (FET) T3 with the strain gauge resistance R in the source lead. The drain and gate coil system acts as the radiating element.

The oscillator frequency (F) is dependent upon the resistance (R) of the strain gauge such that $\Delta F/(\Delta R/R)$ is approximately 6 KHz per 1 percent resistance. This relatively large variation is due to the FET being driven into saturation. Further, the variation of frequency with supply voltage (V) is approximately parabolic with a frequency minimum at V ≈ 2 V. However, the minimum may be broadened by the appropriate addition of drain gate capacitance. In Fig. 2 it is shown that the oscillator frequency only varies by 2 KHz for a variation of ± 0.5 V in the supply voltage.

Now the strains calculated from Perren's (1969a,b) results range up to 200 μ strain. With a metal foil gauge (resistance = 120 Ω and gauge factor = 2) the corresponding resistance range is up to 0.048 Ω. This change is too small to give a detectable frequency variation. Hence a semiconductor strain gauge with a gauge factor > 100 is desirable. In fact two gauges are used (one on each side of the neutral axis of the plate) to eliminate bending moments and consequently, the estimated variation in resistance is up to 10 Ω which would give an oscillator frequency variation up to 30 KHz.

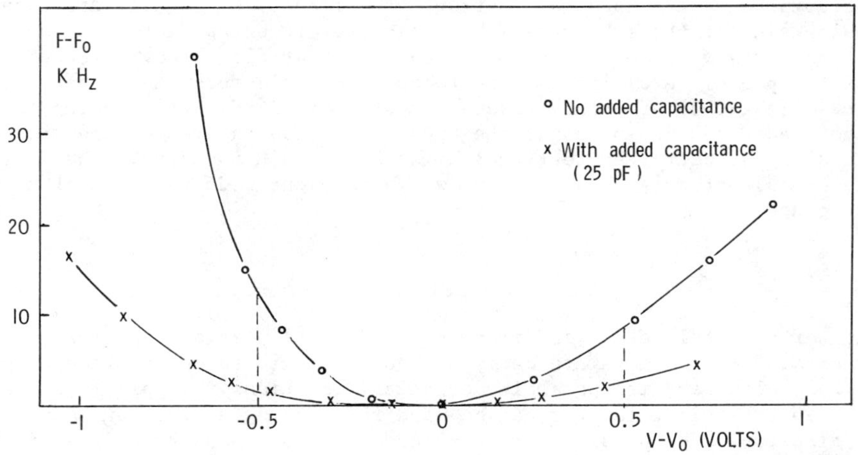

Fig. 2. The effect of adding capacitance between the drain and the gate.

THE EXTERNAL POWER SOURCE

The telemetry device does not contain batteries, but is powered by electromagnetic coupling. The external power induction system consists of a crystal oscillator, a power amplifier (A1) with low output impedence, a primary coil approximately 9 cm in diameter, and a variable capacitance in series to tune the resonant circuit at the required induction frequency.

For efficient power transmission, a high Q-factor coil is required. It was found that the optimum power was induced at 500 KHz by a 40-turn wire (18 SWG) coil having a Q = 180. The primary power coil was 3 cm distant along the common axis from the implantable secondary coil.

THE IMPLANTED PICKUP COIL AND RECTIFICATION UNIT

The secondary coil has a diameter of 5 mm, length of 4 mm and is wound on a ferrite core to increase the coupling coefficient. In order to minimize the interference between power induction frequency and the oscillator frequency, the power pickup coil is placed at right angle relative to the oscillator coils. The induced voltage is rectified by T2 which is operating in saturation so that it has a very low voltage drop (V_{ce} ≈ 50 mV).

THE EXTERNAL RECEIVER

Although the 4 MHz signal can be detected by a conventional radio receiver, the 500 KHz induction signal can cause interference. Hence the detection system consists of a RF amplifier A2 (bandwidth 730 KHz) feeding a phase locked loop (PLL). The voltage controlled oscillator (VCO) of the PLL is synchronized to the transmitter frequency. The VCO frequency is then measured by a digital frequency counter.

THE TELEMETRY CONSTRUCTION

A thick film hybrid technique is used for the production of the implantable telemeter. The conduction circuit is printed on a ceramic substrate (10 × 19 × 1 mm) and microminiature components are used to construct the circuit. The coils are then fixed on the opposite side of the substrate. The telemeter is encapsulated at the end of the fracture plate with two leads connected to the strain gauges attached at the center of the plate. Araldite is used to pot the whole telemeter, then it is given a layer of Silastic 382 for body tissue compatibility. Silastic tubing is used to sleeve the leads and the joints at the ends of the tubing are secured with a medical grade adhesive for better protection against body fluid invasion. The size of the encapsulated telemeter is about 1.5 cm^3 which is about 0.25 the size of a standard 6-hole AO plate.

CONCLUSION

The design of this telemeter was directed towards its possible application in orthopedic studies. *In vivo* data are badly needed to elucidate our knowledge about bone healing, remodeling and its possible mechanical reactions to implants of several types. Moreover, this telemeter can be used for testing devices and materials for internal fixation. The present work has succeeded to produce a telemeter which can fulfil the requisites of such applications. The telemeter, however, is not temperature compensated and therefore must only be used in constant temperature environments.

REFERENCES

Clarke K. K. (1966) Design of self limiting transistor sine wave oscillators. *IEEE Trans. Circuit Theory* CT 13, No. 1, 58-63.

Fryer T. B., Deboo G. J. and Winget C. M. (1965) Miniature long-life temperature telemetry system. *J. appl. Physiol.* 295-298.

Lin W. C. and Ko W. H. (1968) A study of microwatt-power pulsed carrier transmitter circuits. *Bio-Med. Engng.* 6, 309.

Perren S. M., Huggler A., Russenberger M., Straumann F., Müller M. E. and Allgöwer M. (1969) A method of measuring the change in compression applied to living cortical bone. *Acta. Orthop. Scand. Suppl.* 125, 6.

Perren S. M., Huggler A., Russenberger M., Allgöwer M., Mathys R., Shenk R., Willenegger H. and Müller M. E. (1969) The reaction of cortical bone to compression *Acta. Orthop. Scand. Suppl.* 125, 17.

An Implantable FM Telemetry System for Measuring Forces on Prosthetic Hip Joints

R. M. F. GOODMAN,* T. A. ENGLISH** and M. KILVINGTON***

**Department of Electronic Engineering, University of Hull, Cottingham, Road, Hull, U.K.*
***Hull Royal Infirmary, Anlaby Road, Hull, U.K.*
****Bioengineering Unit, Hull Royal Infirmary, Salthouse Road, Hull, U.K.*

Abstract — This paper is concerned with a totally implantable single channel FM telemetry system that has been designed to operate with a strain gauged version of an English low friction low offset style prosthetic hip joint. This enables measurements of the compressive forces acting on the prosthesis to be carried out via the telemetry system *in vivo*. The need for further investigations into the forces exerted on hip prostheses arose during the development of the English implant, after fatigue and fracture observations had been prevalent in other types of implant.

The paper describes the mechanical and electronic constraints imposed at the design stage of the system, and the final solution adopted. In essence, the system at present consists of a bridged gauge mounted on a carrier inside the prosthesis, which drives a miniature FM transmitter. The FM signal is received, demodulated, processed and displayed on a U.V. recorder in such a way that a direct print out of load patterns during selected activities is obtained. The system has been successfully implanted and operated, and many novel results have been obtained.

In addition to the system description the paper discusses the accuracy of the present system; also ways of improving this accuracy by means of further signal processing, and/or changes in the modulation scheme are outlined.

INTRODUCTION

The need for further investigations into the forces exerted on hip prostheses arose during the development of the English implant (English, 1975), after fatigue and fracture observations had been prevalent in other types of implant. Extensive mechanical testing of the implant had been carried out, but actual data on prosthetic hip forces was not available. Indeed, the published literature for walking forces in the femur varied from 1.8 to 7 times body weight, thereby making meaningful fatigue testing of the implant impossible. Most of the previous work on force measurement had been based on 'indirect' calculations, the only *in vivo* study being that of

Rydell, 1966. This study was limited in that measurements were started 6 months after operation, and restricted to one week because of the need to bring signal wires out through the skin.

It was therefore decided to measure directly the forces acting on the head of an English hip prosthesis using strain gauges, and to transmit this information out of the body via an implanted radio transmitter. The system has been designed and implanted, and the results obtained are the first ever reported results of *in vivo* measurement of human hip loads using a telemetric output.

DESIGN CONSTRAINTS

Several constraints have been elemental in producing the final system design. These are as follows:

(a) The hip prosthesis should be an English model C and the strain gauges should be mounted on the implant. These gauges will therefore be physically small, and give a low output.
(b) The transmission system should be totally implantable. The transmitter module must therefore be physically removed from the implant and strain gauges.
(c) Measurements should be made over as long a period of the patient's recovery as possible. Thus the transmitter requires high capacity batteries, and some means of switching them on and off externally, in order to conserve power.
(d) The transmitter must be physically small enough to be easily buried in body fat.
(e) Forces in the range 0 → 9000N, with a resolution of ≃ 50N are to be measured.

TRANSMITTER DESIGN

STRAIN GAUGES

The main factor governing the type of transmission scheme used is the type of strain gauge employed. The English prosthesis is manufactured in 316 stainless steel which yields at a strain level of approximately 0.04 percent. At a force of 2224N the strain expected is 0.0112 percent. This means that the maximum possible change in gauge resistance ΔR, relative to original resistance R, is given by:

$$\frac{\Delta R}{R} = eK \tag{1}$$

where e is the strain and K is the 'gauge' or K factor of the gauge which lies between 2 and 6 for metal gauges. Thus, for example, with a high output platinum-tungsten gauge which has $K = 4.5$, the maximum ratio $\Delta R/R$ is $\simeq$ 0.2 percent. In addition, this value is the maximum $\Delta R/R$, and in normal operation the loads expected would cause a change in resistance considerably less than this (say 100 times less). It is therefore necessary to operate several strain gauges in a bridge arrangement in order to produce a suitable output signal.

Figure 1 illustrates the solution adopted and shows the prosthesis which is different from the normal clinical model in that it is manufactured in two parts. The head part which forms a removable piston and fits into a cylinder machined out of the specially thickened neck of the femoral component. The removable neck is machined to allow for four miniature Dentronic 1800 series platinum tungsten strain gauges (R = 350 Ω) to be mounted around its circumference thus forming a simple load cell. Figure 2 shows a typical strain gauge, and Fig. 3 shows the bridge arrangement used for the gauges. The gauges are placed 90° apart around the circumference of the load cell. Each one has its gauge length axis at right angles to its neighbor. In axial

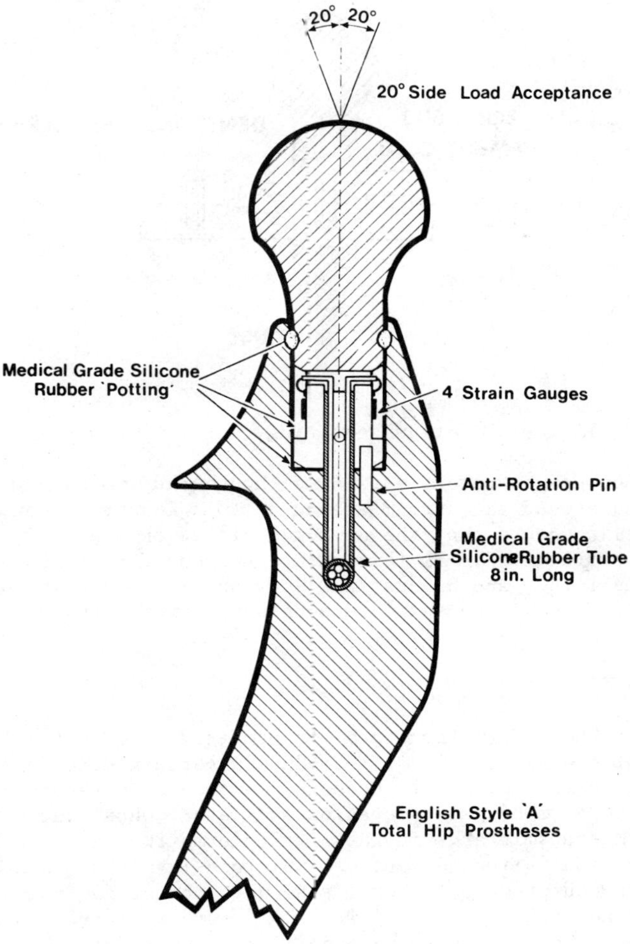

Fig. 1. English hip prosthesis.

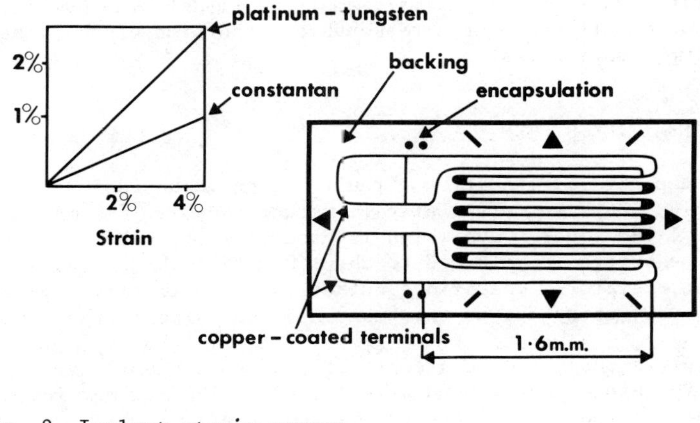

Fig. 2. Implant strain gauges.

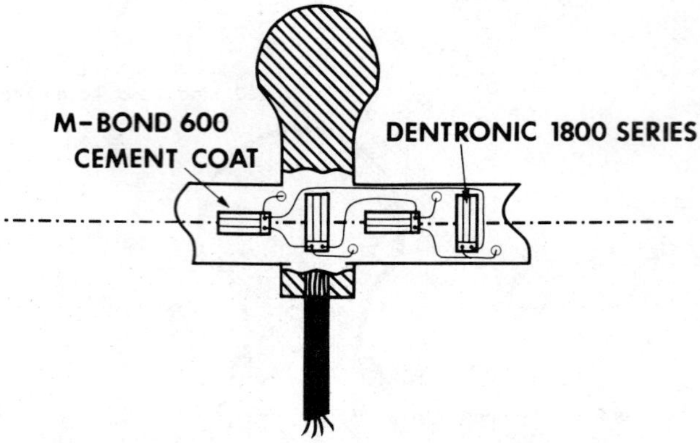

Fig. 3. Implant inter gauge wiring.

compression all four gauges detect strain, although the two having their gauge length axis aligned around the circumference do to a lesser extent. These circumferential gauges also ensure that the unwanted effect on the bridge of front/back off axis loading is minimized. No temperature compensation is included in the bridge for reasons of simplicity, and because the body should remain at sufficiently constant temperature to make compensation unnecessary for reasons of accuracy.

MODULATION SCHEME

In this application the information to be transmitted is essentially dynamic d.c., that is, loading of the prosthesis will produce a certain d.c. output from the gauge bridge. Following normal telemetric practice we should therefore use the bridge output to drive some form of pulse duration modulator, whose output pulse train can then modulate an FM transmitter. In this way information is in the mark/space ratio of the pulse train, and is independent of amplitude levels and hence battery life. This solution was not adopted for several reasons. The bridge output from no load to maximum is only about 0 to 3 mV V^{-1} input. In order to drive a pulse duration modulator this signal would have to be amplified, thus requiring more transmitter circuitry and giving lower battery life. Alternatively a high bridge input voltage would have to be used, and this would require a physically larger transmitter.

The final solution was to adopt a simple amplitude modulation system, and to check that accuracy was maintained over a reasonable proportion of the battery life by extensive pre-implantation testing.

TRANSMITTER SYSTEM

Figure 4 shows the final transmitter circuit design. The LM3909 oscillator chip runs off a single 1.35 V battery (Duracell mercury WH 3T2, 220 mA h) and produces a 1 V amplitude 1 KHz square wave which is used to excite the strain gauge bridge. The output of the bridge is balanced to about 0.5 mV peak-to-peak by means of the parallel balance resistor. The bridge output is approximately 3 mV at full load (9000N), and is very noisy due to the inherent chip noise in the excitation signal.

The bridge output is used to modulate directly an FM transmitter chip (SN102F) which operates on the VHF biotelemetry band at 102.3 MHz. The maximum input amplitude is 4 mV, so that bridge and transmitter are well matched.

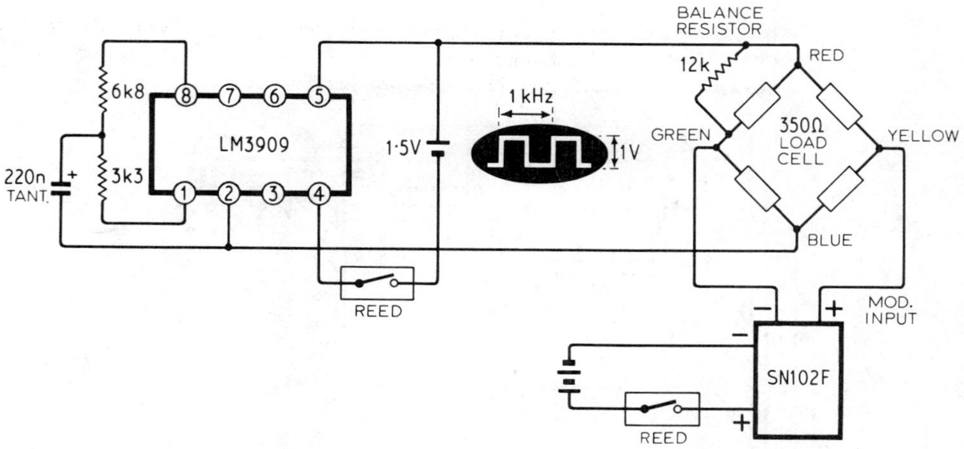

Fig. 4. Transmitter circuit.

Both the oscillator (1.5 V) and transmitter (2 × 1.35 V) battery circuits are fitted with subminiature reed switches which enable power to be switched on and off via a magnet at distances of between 25 to 50 mm from the transmitter. Battery life is approximately 70 h.

Figure 5 shows the mechanical construction of the transmitter which is approximately 25 × 35 × 10 mm and Fig. 6 shows the implanted system. Encapsulation is of paramount importance in any human implant and the required medical standard is achieved as follows. The gauges are sealed from the body fluids by coating with type A silicone adhesive. The cavity is allowed to cure and is applied slightly proud of the piston diameter. A further seal is provided by the addition of a groove in the neck and piston components, which are machined to line up on assembly. These grooves are filled with medical grade silicone rubber and allowed to cure, again slightly proud of the piston diameter. The polyvinyl chloride (PVC) wires attached to the strain gauges are covered by a thin walled silicone rubber sleeve which is sealed into the piston center hole with silicone adhesive. The cable is passed through the implant stem with the silicone sleeve protruding about ½ in. The main length of the PVC wires is then coated with Silastic adhesive type 'A' and a thick walled silicone rubber tube passed over them to bolt up to the implant stem and over the short length of thin walled tube. The signal cable is then attached to the transmitter and the Perspex box is filled with epoxy resin. The whole unit and cable are then covered in a moulded Silastic (382) covering.

THE RECEIVING SYSTEM

Figure 7 shows the receiving system. The FM signal from the transmitter is received on a simple 3 in loop antenna taped to the patient's skin. The antenna cable is long enough to allow complete freedom of movement when the patient is undergoing tests. The antenna signal is received and demodulated using standard Mullard modules. The noisy square wave receiver output is then passed to the main filtering circuits and to an audio amplifier/loudspeaker system. The signal is processed in the following way to provide a noise free d.c. signal whose amplitude varies in sympathy with the amplitude of the receiver output square wave. This d.c. signal is required to drive the chart recorder.

The noisy square wave is first passed through a 1 KHz Q = 10 bandpass active filter

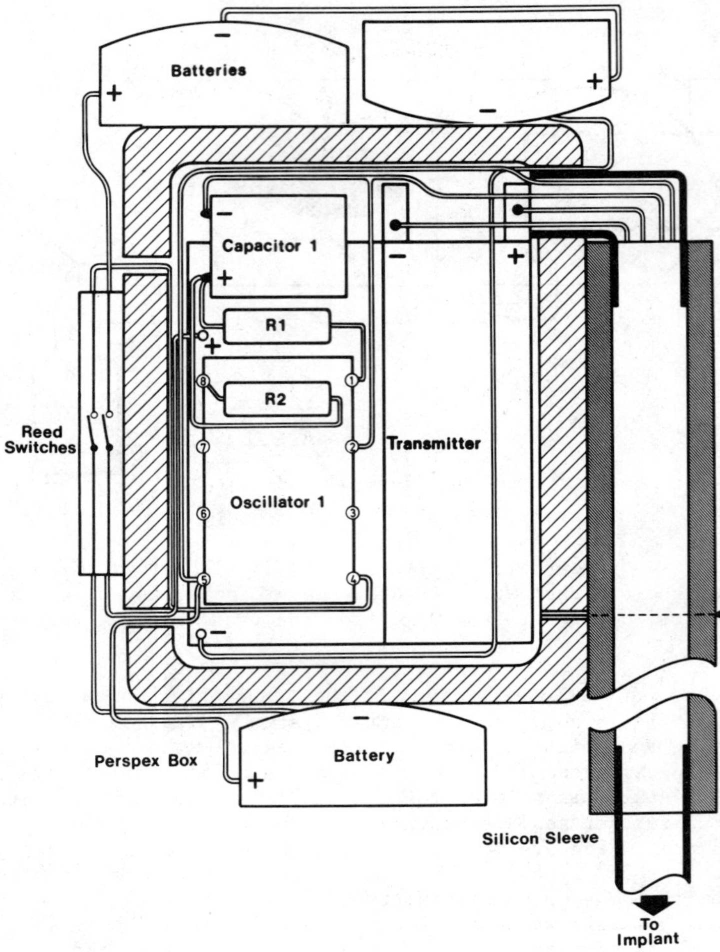

Fig. 5. Transmitter construction.

to remove all high frequency noise. The resulting 1 KHz output is buffered and fed to a precision full wave rectifier. The rectifier output is therefore a d.c. signal with a large amount of 2 KHz ripple. This signal is sufficient to drive the tuning meter. The signal is then smoothed by the integrating action of the 15 Hz low pass filter. The output is now d.c. with low ripple. The next stage performs level shifting and attenuation to provide a signal suitable for driving the U.V. recorder.

CALIBRATION AND ACCURACY

The implant and telemetry system were extensively calibrated and tested before implantation. A modified fatigue testing machine was used for calibration. The machine was capable of applying known static or dynamic loads of up to 9000N. Four ranges of load were recorded using a force applied axially down the implant head and neck. The ranges were: 0-2224N, 0-4448N, 0-6672N, 0-8896N. In addition, loads were applied at 20° angles to the axis within the range 0-4448N. This 20° off axis loading was felt to be the greatest direction at which a force would be applied to the implant head during normal walking or high load activities, and gave rise to an apparent fall in force of 7 percent when compared with axial loading.

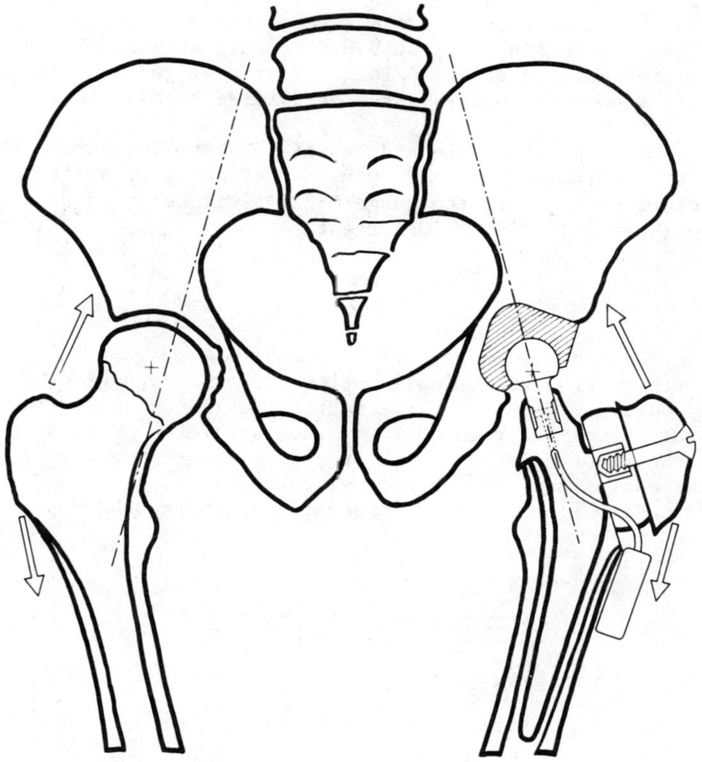

Fig. 6. Implanted system.

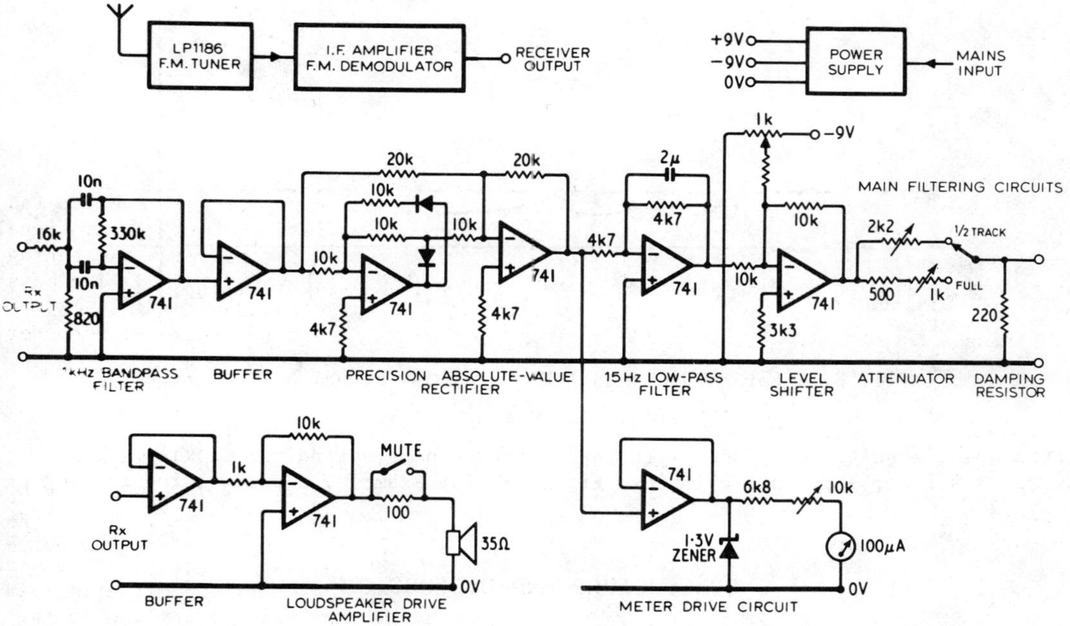

Fig. 7. Receiver circuits.

The system accuracy is affected by several factors. Firstly, the strain gauges are not temperature compensated, and the transmitter frequency varies with temperature. Secondly, because of the amplitude system used, receiver tuning (and hence transmitter temperature) affects the output U.V. recorder signal. Thirdly, off axis forces cause an apparent reduction in load. Fourthly, metal deformation and gauge 'creep' will affect results in high force activities such as running.

However, the system gave dependable and repeatable results when tested at body temperature, over several months, and over the period of transmitter battery life. An overall resolution of 40N in the range 0-2224N was obtained, and the overall system accuracy is estimated at ± 10 percent.

RESULTS

The patient selected for the operation was female and weighed 79.8 kg. The system provided hip force records at the implantation operation and at regular intervals during a 40 day period. Because of the factors mentioned above it was important to obtain an accurate base line recording for the zero force level at operation time. Figure 8 shows the recorded signal (in terms of patient body weight) at various times during the operation. The zero level was obtained by allowing the implant and transmitter to stabilize to body temperature but with the hip still dislocated.

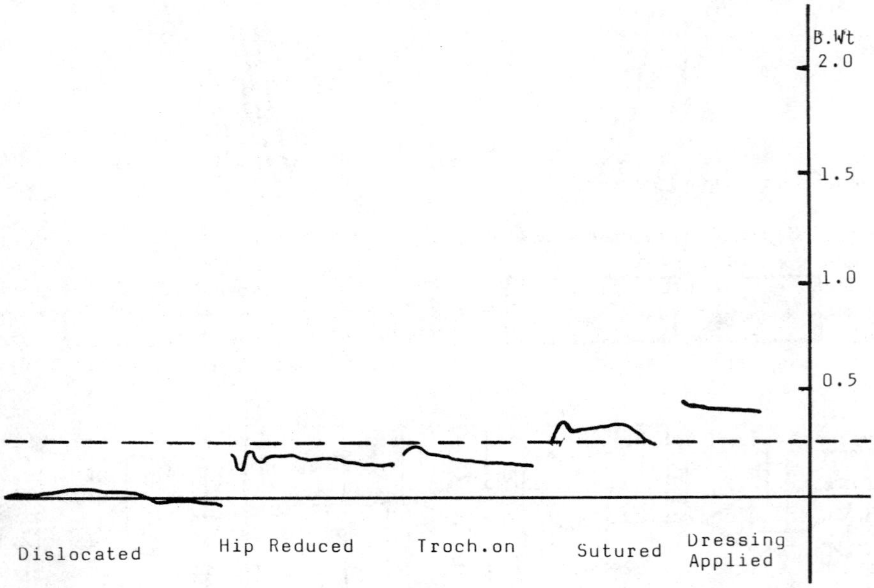

Fig. 8. Recorded hip load (AW). Day 0 — signal height at operation.

Subsequent results were taken with the patient supine, standing, walking with and without sticks, and stair climbing. Figure 9, for example, shows walking results on day 12.

CONCLUSIONS AND DISCUSSION

After day 40 no further signals were received. A possible cause of failure could have been the choice of silicone rubber for encapsulating the strain gauges as it

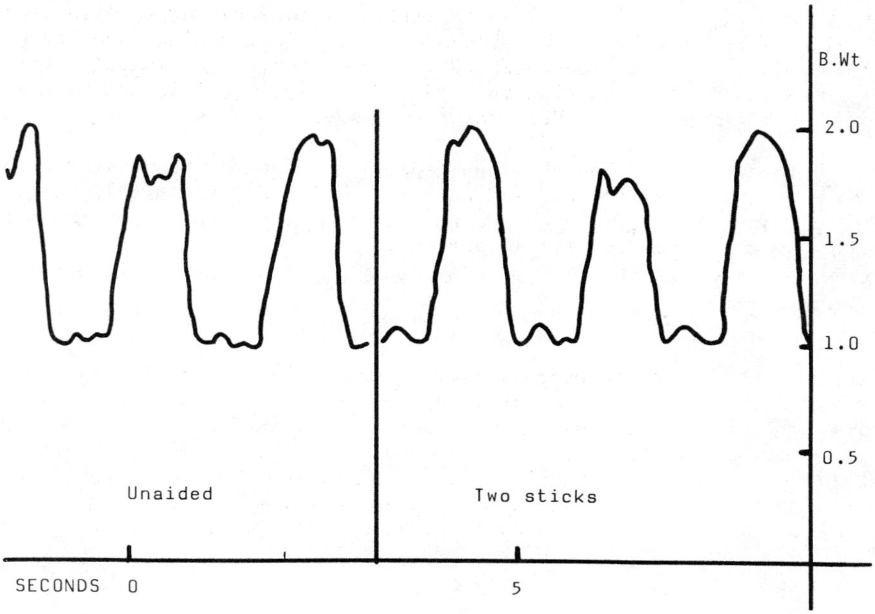

Fig. 9. Recorded hip loads (AW). Day 12 — walking with and without sticks.

has since come to our notice that the acetic acid present in uncured silicone could have corroded the gauges or solder connections.

The records obtained have given valuable information regarding forces acting on the head of the type of implant used. The design of the implant was intended to reduce bending loads on the stem and the operative technique (English, 1975) was intended to reduce the overall force acting across the hip joint by offsetting the trochanter and displacing the center for rotation towards the midline. The risk of dislocation, fatigue failure, and loosening is lower using this technique than with more standard implants.

The results obtained can be summarized as follows. Stance phase force, and swing phase force, were found to be 2.0 times and 1.0 times body weight respectively. The one-legged stance caused forces of 2.54 times and 2.18 times body weight at 12 and 40 days after operation.

The telemetry system has operated satisfactorily and given useful results. It is, however, possible to improve the system, and this is under consideration for any future clinical trials. Firstly, we can consider improvements to the existing design. These could take the form of temperature compensation on the strain gauges, and automatic receiver tuning via feedback from the filtered 1 KHz signal. This would alleviate transmitter temperature effects considerably, and speed up the determination of zero force at operation time. Secondly, we can consider redesigning the system to operate with a pulse duration modulation scheme. This would require a more complex transmitter, but in conjunction with strain gauge temperature compensation, would provide an absolute zero force level, in which system accuracy would be almost completely independent of transmitter temperature, amplitude, and frequency drifts. One possible way of achieving this is to consider the use of semiconductor strain gauges which have K values of up to 200. Such gauges, however, are extremely temperature dependent and would require complex compensation circuitry to be effective.

Finally, we can assert that the system has established the forces acting on this style of hip prostheses to within ± 10 percent, and that future clinical tests with improved electronics and gauge protection will be directed to confirming these results at greater accuracy, over longer periods of time, and with more varied activities.

REFERENCES

English T. A. (1975) Trochanteric approach to the hip for prosthetic replacement. *J. Bone Joint Surg.* 57-A, 1128-1133.

Rydell N. W. and Trychere A. B. (1966) *Forces Acting Upon the Femeral Head Prosthesis.* Gothenberg: Litotyp.

A Monitoring Chamber* for Trout Behavior and Physiology

W. F. MILLER, F. W. SANDWELL and P. A. GORTON

Water Research Centre, Stevenage Laboratory, Elder Way, Stevenage, Hertfordshire SG1 1TH, U.K.

Abstract — A new electrode chamber is described from which the locomotor activity, opercular movement, coughs, and heartbeat of free-swimming trout can be recorded without attachments to the fish. The chamber is more spacious than those previously described, and clear signals can be detected whether the fish is held in static water or in water flowing at several cm s^{-1}. The composite signal detected by the electrodes is amplified and then either filtered to provide separate respiratory and heartbeat signals or integrated to provide an index of locomotor activity.

The use of submerged metal electrodes to detect bioelectric signals from fish swimming freely in a tank is not new, but this paper describes considerable advances in the swimming space available for the fish, in the rate of flow of water that can be used without creating electrical interference, and in the number of biological parameters that can be recorded regardless of the position of the fish in the chamber. The electronic system developed for retrieving and processing the biological signals derived from the electrode chamber is also described.

INTRODUCTION

The respiratory and cardiac activity of fish can be monitored very precisely using catheters or implanted electrodes (Holeton and Randall, 1967; Saunders, 1961; Shelton and Randall, 1962). These methods are, however, complicated to use and affect the behavior of the test animal. An alternative approach is to use a pair of electrodes placed on either side of, but not in contact with, the fish. The electrophysiological signals produced by the fish, and the water movements caused by swimming, can be detected by the electrodes and displayed, after amplification, on a chart recorder. Spoor, Neiheisel and Drummond (1971), who reviewed the early work in this field, described an electrode chamber which could detect the respiratory movements of the opercula (gill covers). With no attachments to the fish the

*UK Patent applications 40052/76 and 28607/77.

technique was simpler to use and had less effect on behavior than methods involving catheters or implanted electrodes.

Electrode chambers of this type have been used in automatic water-pollution alarm systems (Cairns, Dickson and Westlake, 1975; Morgan, 1977; Sloof, 1977) and in sub-lethal toxicological studies (Drummond, Spoor and Olson, 1973; Thomas and Rice, 1975). The potential for movement within the chamber is, however, small because the electrodes, which are placed near the end walls, need to be close to the head and tail of the fish and the necessary alignment of the fish, perpendicular to the electrodes, is maintained by a narrow chamber width.

Marshall (1972) demonstrated that a signal related to heart beat could also be detected using essentially the same technique, and that a partial separation of the superimposed respiratory and cardiac components could be achieved by using simple electronic filters. An improvement in the filtering was shown by Yakimovich, Bergman and Weeks (1977), though their results required a tank shielded from electrical interference with a Faraday cage. For clear reception of heart beat signals the electrodes of both Marshall (1972) and Yakimovich *et al.* (1977) had to be nearer to the fish than were the electrodes of Spoor *et al.* (1971), and so the problem of continuous monitoring without physically restricting the fish remained.

This paper describes a new electrode chamber and electronic system in which some improvements on previous designs have been achieved. The electronic system was developed with assistance from Sands-Whiteley Research and Development Ltd., Cambridge, U.K.

THE MONITORING CHAMBER

The chamber (Fig. 1) differs from that of Spoor *et al.* (1971) in that the sensing electrodes are placed on its floor and roof rather than at its ends. The fish compartment is 610 mm long, 165 mm wide, and 76 mm deep and allows a trout of 150 mm to swim forwards for four fish-lengths and to turn comfortably through 360°. Water enters the chamber through a flow-straightening baffle similar to that of Poels (1975) and exits through a perforated plate. A standpipe maintains the water surface above the level of the top electrode which forms the roof of the chamber. The test fish can be placed in the chamber after removing the upper electrode.

With this electrode configuration the fish has considerably more freedom of movement yet the sensing electrodes are nearer to the fish than in the design of Spoor *et al.* (1971). This enables very clear respiratory and heart beat signals to be detected regardless of the position or attitude of the fish in the chamber. The extra space also makes the continuous monitoring of locomotor activity more meaningful and generally lessens the effect of the monitoring technique on behavior. An additional, unexpected advantage of this design is the fact that water velocities of several cm s^{-1} along the chamber can be used if required, whereas with electrodes placed across the flow (Spoor *et al.*, 1971) moving water creates strong interference which masks the biological signals. The explanation for this is not known, but it is thought to be related to the manner in which the common mode rejection characteristics of the amplifier respond to the two designs.

SIGNAL RETRIEVAL AND PROCESSING

The monitoring chamber was developed using a Brookdeal Electronics transformer (Model 9433) and differential ac amplifier (Model 9454) to amplify the biological signals derived from the electrodes. Subsequently these units were replaced with a less expensive, specially designed amplifier built up on a matched pair of transistors (type 2N3811) with second-stage amplification provided by an instrumentation

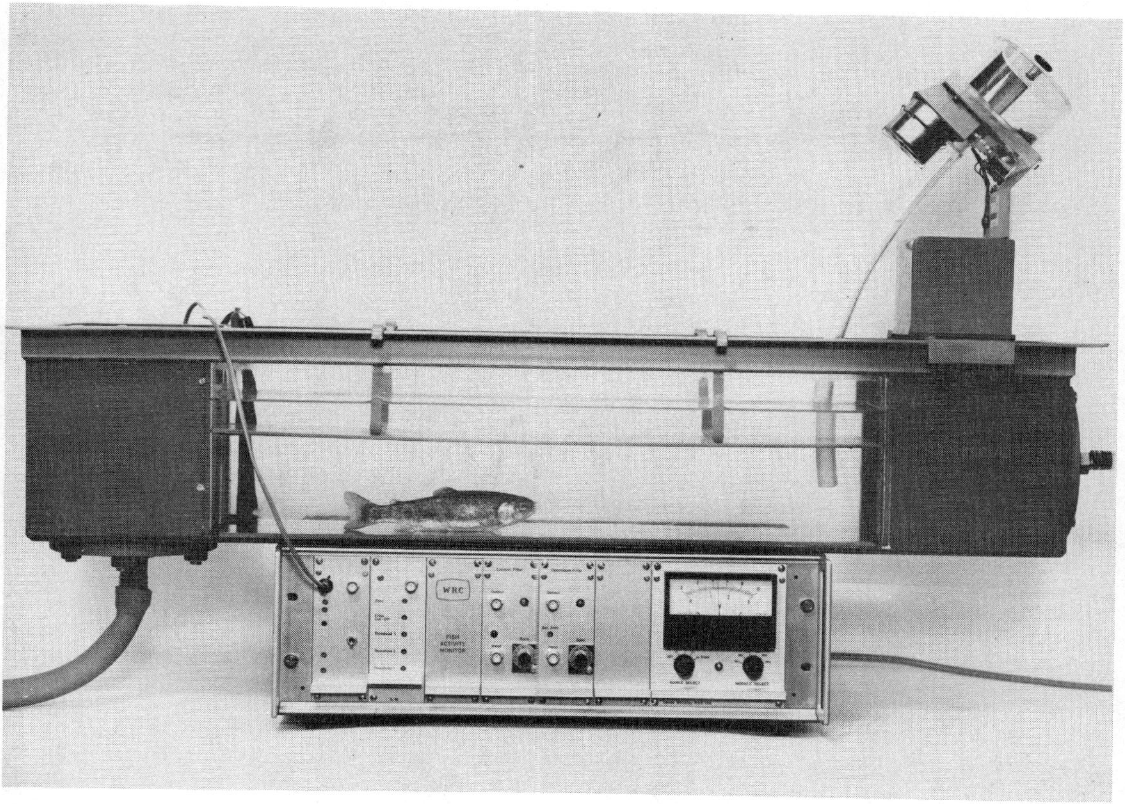

Fig. 1. Monitoring chamber and electronic unit.

amplifier (type AD521). Filters were used to limit the frequency response of the amplification system to approximately 1-40 Hz, and a dc feedback loop was incorporated to neutralize the charge which tended to build up spontaneously on either electrode.

The respiratory and cardiac components of the amplified signal were originally separated by using a Barr and Stroud variable filter (Model EF3). Lowpass filtering at 5 Hz was used to isolate the respiratory component while bandpass filtering at 20-40 Hz was used to isolate the electrocardioagram (ECG). This unit was replaced by a pair of fixed-frequency filters made with Barr and Stroud active filter modules (series EF 10/20). The two units comprised 6-pole Butterworth filters cutting off at the frequencies mentioned above at the rate of 36 dB/octave. Figure 2 shows the signals that can be obtained with the amplifier and the two filters, namely, (a) the electrocardioagram and (b) the movements of the opercula (including coughing). Clear opercular signals and the R waves of the ECG are almost always detectable. P waves, however, can be recorded from about one fish in two, while T waves are only seen in fish that give a particularly strong signal. Signals are displayed on a Devices (MX212) chart recorder.

A disadvantage of the technique is that these signals tend to become confused during swimming, this being more the case with the opercular signal than with the ECG. This fact has been turned to advantage, however, in the design of a system for monitoring locomotor activity. The system relies on the observation that the amplitude and duration of the interference which occurs during swimming increase with the intensity and duration of the locomotor activity which caused it. By rectifying and integrating

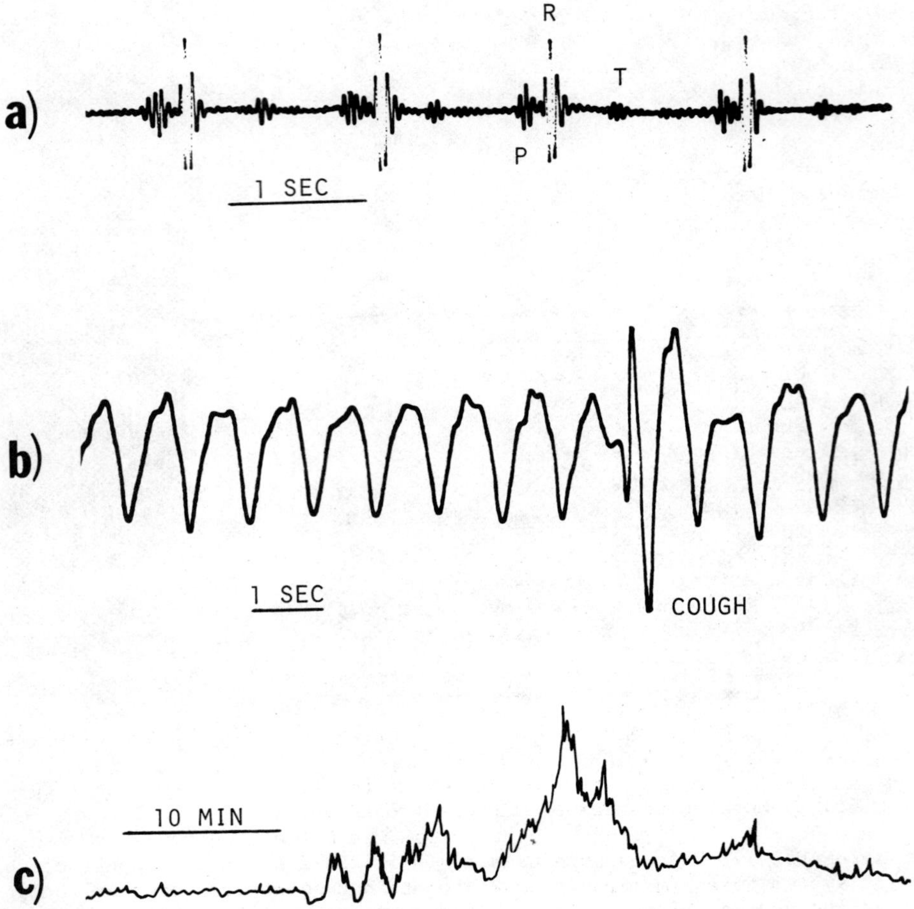

Fig. 2. a. Electrocardiogram showing P, R, and T waves.
b. Movements of the opercula, including coughing.
c. Locomotor activity.

the unfiltered output from the amplifier a measure of the amplitude and duration of this interference can be obtained. Figure 2 (c) shows a chart record of locomotor activity produced by an electronic module consisting of a fullwave rectifier and a simple capacitor integrator.

Figure 1 shows the modular electronic system designed at the Water Research Center to monitor fish activity. From left to right the modules are the amplifier, the locomotor activity monitor, an inscribed blanking-off panel, the ECG filter, the respiration filter, a blanking-off panel, and a power supply with a meter which displays the level of locomotor activity.

DISCUSSION

The monitoring system was developed for use in an automatic water pollution alarm system (Miller, 1977) but it is considered to be potentially useful in studies of fish physiology, behavior and sub-lethal toxicology. The degree of freedom allowed to the fish, the number of parameters that can be recorded, and the lack of attachments to the fish are seen as major advantages. The technique also works well with other species of fish, for example salmon smolts (*Salmo salar*), stickleback (*Gasterosteus aculeatus*), roach (*Rutilus rutilus*), tench (*Tinca tinca*), and goldfish (*Carassius auratus*).

Acknowledgement — This paper is published by permission of the Director of the Stevenage Laboratory, Water Research Centre.

REFERENCES

Cairns J., Dickson K. L. and Westlake G. F. (1975) Continuous biological monitoring to establish parameters for water pollution control. *Prog. Wat. Technol.* 7, 829-841.

Drummond R. A., Spoor W. A. and Olson G. F. (1973) Some short-term indicators of sublethal effects of copper on brook trout, *Salvelinus fontinalis*. *J. Fish. Res. Bd Can.* 30, 698-701.

Holeton G. F. and Randall D. J. (1967) Changes in blood pressure in the rainbow trout during hypoxia. *J. exp. Biol.* 46, 297-305.

Marshall W. C. (1972) Non-contact sensing of electric fields surrounding trout. MSc thesis, University of Wyoming, Laramie, Wyoming.

Miller W. F. (1977) Development of the WRC water quality monitor using fish. Paper presented at WRC Seminar on Practical Aspects of Water Quality Monitoring Systems. Stevenage, December 1977.

Morgan W. S. G. (1977) Biomonitoring with fish: an aid to industrial effluent and surface water quality control. *Prog. Wat. Technol.* 9, 703-711.

Poels C. L. M. (1975) Continuous automatic monitoring of surface water with fish. *Wat. Treat. Exam.* 24, 46-56.

Saunders R. L. (1961) The irrigation of the gills in fishes. 1. Studies of the mechanism of branchial irrigation. *Can. J. Zool.* 39, 637-653.

Shelton G. and Randall D. J. (1962) The relationship between heartbeat and respiration in teleost fish. *Comp. Biochem. Physiol.* 3, 237-250.

Sloof W. (1977) Biological monitoring based on fish respiration for continuous water quality control. National Institute for Water Supply, the Netherlands, Quarterly Report No. 11, December.

Spoor W. A., Neiheisel T. W. and Drummond R. A. (1971) An electrode chamber for recording respiratory and other movements of free-swimming animals. *Trans. Am. Fish. Soc.* 100, 22-28.

Thomas R. E. and Rice S. D. (1975) Increased opercular rates of pink salmon (*Oncorhynchus gorbuscha*) fry after exposure to the water-soluble fraction of Prudhoe Bay crude oil. *J. Fish. Res. Bd Can.* 32, 2221-2224.

Yakimovich M. J., Bergman H. L. and Weeks R. W. (1977) A noise-free method for separately recording opercular rate and heart rate from free-swimming fish. Paper presented at a Symposium, University of Wyoming, Laramie, Wyoming.

The Capture of Deer for Radio Tagging

R. H. SMITH

Zoology Department, The University, Whiteknights, Reading, U.K.

Abstract — Capture precedes many studies of large mammals in the wild. Three methods are available: stockades, netting and capture drugs. Stockades are expensive but can be effective. Netting is labor intensive and versatile, but requires considerable skill. Both methods are suitable for catching large numbers of unselected animals. Capture drugs are best administered by a syringe projectile and are suitable for catching small numbers of carefully selected animals. A recent development is the dart transmitter which enables the use of syringe projectors in dense forest conditions. Problems associated with each of these methods are reviewed; the health of the animals is a major consideration.

INTRODUCTION

A major problem in work on wild mammals is how to capture them safely and selectively. Three categories of capture techniques are available. In choosing the most suitable method the main consideration is whether a few, carefully selected animals or a large, unselected sample is required. There is also a choice between expensive versus labor intensive methods. Description of the different methods is biased by the author's experience with fallow deer, but as these are a difficult species to work with, this bias uncovers problems likely to be encountered with other deer species. Snares are unacceptable because of the risk to captured animals and so they are not considered here.

STOCKADES

The construction of a permanent or semipermanent stockade, where practical, is the most satisfactory way of capturing large numbers of deer. Animals are attracted or herded through a large funnel into enclosures from where they can be led singly through a passage into a crate or a handling crush. Once set up, a well designed stockade system can be operated by only a few people. However, the cost is likely to be prohibitively high for most scientific studies other than longterm ones. Permanent stockades are, of course, immobile and must be designed and positioned very carefully. Animal attractants may overcome the problem of avoiding the area around the stockade which may be associated with disturbance. A variety of attractants may be effective under different circumstances (e.g. winter feed, mineral licks, or even captive females). Stockades are unlikely to work in open country.

A semi permanent stockade with woven polypropylene walls (Oelofse, 1970) provides a sound, basic design for any stockade system and is described by Harthoorn (1976, p. 107). Animals are herded into a funnel, the mouth of which is at least 200 m wide and the depth even longer to avoid damaging the animals. The funnel leads through a gate into a straight sided enclosure, which in turn leads through a second gate into a circular corral. A circular final enclosure is highly recommended since it greatly reduces the risk of damage to the animal. The two gate system reduces the chances that an animal will double back and escape before entering the final enclosure. Oelofse used 3 m high plastic walls and has safely captured thousands of animals including wildebeest, zebra, impala and giraffe. A minimum height of 2.5 m would be required for larger deer species. Provided that the walls are high and opaque, the animals will run around rather than at the walls and are unlikely to attempt to jump out.

A simple stockade system has been used in conjunction with forest enclosures to capture fallow deer for controlled genetic crosses (Smith *et al.*, 1978). Two pairs of adjoining 5 ha enclosures are separated by a central capture area into which animals are attracted by winter feeding. The outer gates are shut when the deer are inside, and selected individuals can be split off and guided through a gate into the appropriate enclosure. Our fallow deer study has also used the very old device known as a saltory or deer leap (Shirley, 1867), which is effectively a oneway valve. Earth is removed from the inside of a fence to form a depression and banked up on the outside so that deer can jump into but not out of an enclosure. Once inside, animals can be funnelled into a handling crush or tranquillized using a syringe projector. A saltory does not need constant attention when in use and is easily fenced off when not required.

Corrals and traps have been used with varying degrees of success for capturing antelope (King, 1966), white-tailed deer (Hawkins, Autry and Klimstra, 1967), mule deer (Sauer, Gorman and Boyd, 1969), kangaroos (Keep and Fox, 1971) and moose (Le Resche and Lynch, 1973). The likelihood of success is enhanced if the walls are opaque (e.g. hessian should be draped over netting) and if the final enclosure is circular. Some quoted mortality figures of 20-35 percent are quite unacceptable and indicate that considerable thought must be put into the design and use of these systems to make them as effective and humane as possible.

NETTING

The use of nets for the capture and management of deer is a contentious issue (Dansie, 1977). However, in the absence of a stockade there is often very little choice if large numbers of animals are to be captured, e.g. Chapman and Chapman (1975). The use of cannon nets in bird ringing operations is well established, and the same technique has been used to capture white-tailed and fallow deer (Hawkins, Martoglio and Montgomery, 1968). An attractant is required to bring sufficient animals within range to make the exercise worthwhile. The necessity for strong, heavy nets imposes severe restrictions on range and usefulness. The technique is likely to be most effective on open ground where other netting methods are ineffectual.

Drop nets are most useful in seminatural conditions where animals regularly congregate in feeding areas. Ramsey (1968) reported the successful capture of several hundred white-tailed and axis deer, and Pienaar (1969) has used drop nets to capture many medium sized species of African game. All netting techniques require rapid handling of captives so that damage is minimized.

Long nets used in woodland can be efficient and humane, provided they are used by competent personnel. The principle is that animals are moved quickly, but not stampeded, into a long line of netting strung loosely on stakes, bushes or small trees. The net must suffer local collapse as an animal hits it, and the animal should very quickly be restrained, blindfolded and removed from the net. Blindfolding has an immediate calming effect. A hazard is that a number of animals may hit the same part of the net together, inevitably causing broken limbs, goring and suffering and involving risks to the handlers. Skilled beaters are required to bring a herd in slowly and split them off towards the net one or two at a time.

It is important to position the net well. Generally a funnel is required (Riney and Kettlitz, 1964) so that the animals cannot break away when they see the net. Ideally, the sides or wings should be brightly colored and the main catching net black or dark brown in forest, or beige in more open, drier country.

Controversy about the use of long nets has arisen because of unfortunate incidents and criticism from hearsay. Unnecessary mortality has occurred because of incompetent personnel who have damaged ungulates by kneeling on the thorax, and because of over zealous chasing. There is no doubt that capture in a net causes great stress, and so netting should not be undertaken lightly.

CAPTURE DRUGS

The capture of animals by the administration of stupefying drugs should be the least stressful and most satisfactory method available and it allows animals to be handled for some considerable time without physical restraint. However, problems of delivery, dosage and supervision generally make chemical capture less suitable except when only a few, selected animals are to be captured. A fairly recent book covers the subject excellently (Harthoorn, 1976).

Administration of the drug is either oral or hypodermic. Hypodermic injection of wild animals usually requires some sort of syringe projectile and an appropriate projector (e.g. a blow pipe, crossbow or a rifle or pistol using compressed gas or an explosive charge for propulsion). Harthoorn (1976) reviews the development of syringe projectors since Hayes *et al.* (1957), and Jones (1976) gives an up to date assessment of available weapons and syringe projectiles. A recent innovation not included in Jones' article is the Conservator range of tranquilizing equipment manufactured to order by H. V. Ilsley in Liverpool, U.K. The only weapon feasible for use in the field is a rifle type projector, and effective range is limited to 40 or 50 m.

The range of drugs suitable for use with different species is covered in detail by Harthoorn (1976). Muscle relaxants such as succinylcholine are unnecessarily cruel (Low, 1973; Lees and Done, 1973). The effects of morphine like analgesics such as etorphine (Reckitt and Colman) can be reversed by antagonists such as cyprenorphine. Commercially available formulations such as Large Animal Immobilon (Reckitt and Colman) may be unsatisfactory for particular species such as fallow deer which react adversely to the sedative acepromazine incorporated in the mixture (Jones and Manton, 1973). However, mixtures of pure etorphine with xylazine (Bayer) have been found to give a satisfactory response (Harrington and Wilson, 1974). Unfortunately the morphine like compounds and their antagonists can be extremely dangerous to man; all are listed as dangerous drugs and some as narcotics. Sedatives such as xylazine may be used for the safety of the worker but animal recovery may take several hours. There are reports that cyprenophine will reverse the effects of xylazine in some animals.

Hypodermic injection may be difficult, but oral administration is even more so. Because of the complex structure of the ruminant stomach, the bait must be chosen carefully if the drug is to be carried quickly through the rumen without risk of bacterial degradation (Harthoorn, 1976, p. 148). The tranquilizer diazepam (Roche) has been used to capture six species of deer with 3-4 percent mortality (Thomas, Robinson and Marburger, 1967). Other users have also encountered mortality in fallow deer (Murry and Dennett, 1963) and white-tailed deer (Montgomery and Hawkins, 1967), although smaller doses have proven effective and safe (Lees and Done, 1973). It is not possible to regulate the dose in individual animals. Some species (e.g. fallow deer) require such extraordinarily high doses that hundreds of grams of pure drug may be needed at a time (Thomas, Robinson and Marburger, 1967).

Whatever method is used to administer a drug, it is essential that a drugged animal be found and attended quickly. One way of finding drugged animals is to incorporate a transmitter into a syringe projectile, the needle of which must have either a barb or a cuff so that it stays in the animal. At present, one such dart transmitter is manufactured in the U.S.A. and is a modification of a Palmer Cap-Chur Syringe available from Wildlife Materials Inc. Although the designers of the dart transmitter report satisfactory performance (Lovett and Hill, 1977), the drastic modifications to the syringe are likely to reduce the accurate range of 30 m reported for the unmodified syringe by Jones (1976). I am currently collaborating with the manufacturer of the Conservator syringe projector to design and produce a dart transmitter with the same flight characteristics as an unmodified syringe projectile and hence a maximum accurate range of 40 to 50 m.

REFERENCES

Chapman D. and Chapman N. (1975) *Fallow Deer*. Lavenham: Dalton.

Dansie O. (1977) The case for catching deer. *Deer* 12, 140-142.

Harrington R. and Wilson P. (1974) Immobilon-Rompun in deer. *Vet. Rec.* 94, 362-363.

Harthoorn A. M. (1976) *The Chemical Capture of Animals*. London: Balliere-Tindall.

Hawkins R. E., Autry D. C. and Klimstra W. D. (1967) Comparison of methods used to capture white-tailed deer. *J. Wildl. Mngmt.* 31, 460-464.

Hawkins R. E., Martoglio L. D. and Montgomery G. G. (1968) Cannon-netting deer. *J. Wildl. Mngmt.* 32, 191-195.

Hayes F. A., Jenkins J. H., Feurt S. D. and Crockford J. A. (1957) Observations on the use of nicotine for immobilising semi-wild goats. *J. Am. vet. med. Ass.* 130, 479-483.

Jones D. M. (1976) An assessment of weapons and projectile syringes used for capturing mammals. *Vet. Rec.* 99, 250-253.

Jones D. M. and Manton V. J. A. (1973) Immobilon in deer. *Vet. Rec.* 93, 26-27.

Keep J. M. and Fox A. M. (1971) The capture, restraint and translocation of kangaroos in the wild. *Aust. vet. J.* 47, 141-145.

King J. M. (1966) The short step to extinction. *Zoologica Africana* 2, 9-11.

Lees P. and Done S. H. (1973) Immobilisation of fallow deer. *Vet. Rec.* 93, 455.

Le Resche R. E. and Lynch G. M. (1973) A trap for free-ranging moose. *J. Wildl. Mngmt.* 37, 87-89.

Lovett J. W. and Hill E. P. (1977) A transmitter syringe for recovery of immobilised deer. *J. Wildl. Mngmt.* 41, 313-315.

Low R. J. (1973) Immobilisation of fallow deer. *Vet. Rec.* 93, 454-455.

Montgomery G. G. and Hawkins R. E. (1967) Diazepam bait for capture of white-tailed deer. *J. Wildl. Mngmt.* 31, 464-468.

Murry R. E. and Dennett D. (1963) A preliminary report on the use of tranquilising compounds in handling wildlife. *Proc. S.E. Ass. Game Fish Commiss. Conf.*, Arkansas.

Oelofse J. (1970) Plastic for game catching. *Oryx* 10, 306-308.

Pienaar U. de V. (1969) Capture and immobilisation techniques currently employed in South African National Parks and Reserves. In F. B. Golley and H. K. Buechner (eds.) *A Practical Guide to the Study of the Productivity of Large Herbivores*. Oxford: Blackwell.

Ramsey C. W. (1968) A drop-net deer trap. *J. Wildl. Mngmt.* 32, 187-190.

Riney T. and Kettlitz W. L. (1964) Management of large mammals in the Transvaal. *Extrait Mammalia* 28, 189-248.

Sauer B. W., Gorman H. A. and Boyd R. J. (1969) A new technique for restraining antlerless mule deer. *J. Am. vet. med. Ass.* 155, 1080-1084.

Shirley E. P. (1867) *Some Account of English Deer Parks with Notes on the Management of Deer*. London: Murray.

Smith R. H., Johnson E., Hornby J. E. and Holt A. C. E. (1978) Long-haired fallow deer. *Rep. Forest Res., Lond.* 1978, 57-58.

Thomas J. W., Robinson R. M. and Marburger R. G. (1967) Use of diazepam in the capture and handling of cervids. *J. Wildl. Mngmt.* 31, 686-692.

A Simplified Technique for Mobile Radio Tracking

G. CEDERLUND and P. A. LEMNELL

Statens Naturvårdsverk, The National Swedish Environment, Protection Board, Grimsö Wildlife Research Station, S-770 31 Riddarhyttan, Sweden

Abstract — Radio tracking is an important tool in several studies on mammals and birds at Grimsö Research Station. Mobile tracking is essential to cover the 14,000 ha research area. The system recommended simplifies and shortens the time of this operation. A 6-element, horizontally polarized Yagi antenna, on a 3.5 m mast is operated from inside a vehicle. In working position the antenna is raised another 1.0 m. Bearings are read using a repeater compass system with a compass mounted at the top of the mast and a heading indicator mounted on a panel inside the vehicle (angular dependent electrical signals are transmitted through a cable to the heading indicator). The system eliminates aligning of the vehicle at each stop. Operation time is reduced to 1 to 2 min when tracking one animal. A test of the reliability of the repeater compass system shows a small systematic error related to the orientation of the antenna boom. Tests of the accuracy of the system within normal range (1000 m) showed a mean angular error of 3.3 ± 3.2°. Mean distance of location was 570 m.

INTRODUCTION

Radio tracking is an important tool in behavioral and ecological studies of wildlife. Techniques and equipment ultimately depend on the biological questions to be answered. Commonly, optimal accuracy requires bearings from at least 2 locations using a stationary antenna system or a mobile unit at predetermined sites.

Methods of aligning vehicles and recording bearings were suggested by Verts (1963); Proud (1969); Anderson and De Moor (1971); Haacke *et al.* (1973) and Hallberg *et al.* (1974), but the operator must get out of the vehicle and erect the antenna at each stop. Hutton *et al.* (1976) described alignment on a permanent system, but this could be troublesome under winter conditions. Kolz and Johnson (1975) required no orientation of the vehicle, but the equipment was heavy and did not permit elevation of the antenna more than 2 m above the vehicle roof.

At the Grimsö Wildlife Research Station, situated in the Taiga region of central Sweden, radio tracking is used for research on moose (*Alces alces*), roe deer

(*Capreolus capreolus*), mountain hare (*Lepus timidus*), fox (*Vulpes vulpes*), goshawk (*Accipiter gentilis*) and black grouse (*Lyrurus tetrix*). The mobile system recommended in this paper is used in the 14,000 ha research area. All operations are conveniently performed within the vehicle.

METHOD

Signals received with a 6-element, horizontally polarized yagi antenna, provided the best directionality. Bearings were recorded in the strongest signal position or by rotating the antenna on both sides of the peak lobe until equal strength was obtained, normally at the null-points. The final direction was then calculated by bisecting the angle formed by the null-points.

To reduce deflection caused by ground and vegetation, the antenna in working position was 4.5 m above the ground on a mast attached to a van-type vehicle. The 3.5 m mast was mounted through a hole in the center of the roof, and could be raised another 1.0 m when operating (Fig. 1). Antenna extension was manually accomplished from inside the cab using a handle attached to the lower end of the mast. A ball-bearing on the roof mounting facilitated rotation and stabilized the mast. Removal of the antenna from the vehicle could be accomplished within 5 min.

Fig. 1. Mobile tracking unit with the antenna elevated for operating. The antenna lowers from inside the vehicle to rest in the rack for driving.

Prior to moving the vehicle, the mast was lowered from inside and locked to prevent hitting tree branches and transmissions lines or bending the mast. The antenna was then oriented in the long axis of the van and the boom was attached to 2 aluminum tubes that were coated with polyvinyl chloride (PVC) material and supported by an aluminum frame on car-top carriers (Fig. 1). When the mast was lowered, the vehicle could be driven at highway speed.

Bearings were read using a repeater compass system (NM3 Neco Repeater Compass manufactured by Neco Marine Ltd., England) originally designed for sailing boats and consisting of a master compass, heading indicator, and power pack (Fig. 2). Power was provided by the vehicle battery. The master compass had a built in sensor coil that transmitted angular-dependent electrical signals through a cable to the heading

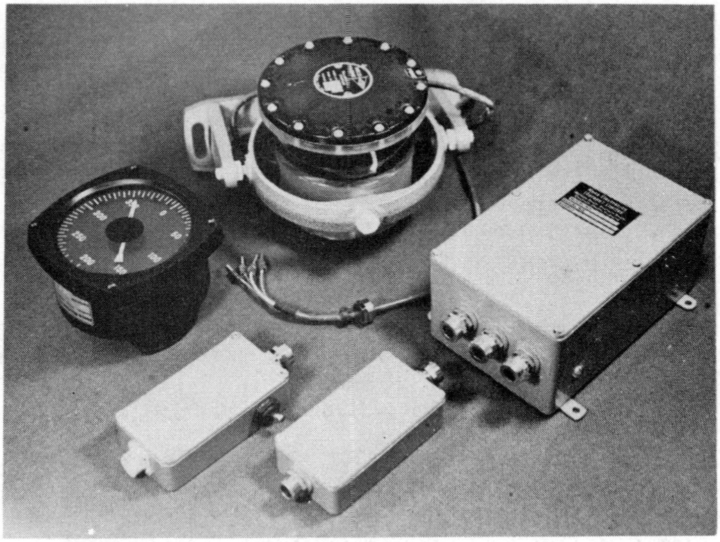

Fig. 2. The repeater compass system.

indicator (the magnesyn principle). The compass was mounted at the top of the mast, just above the antenna plane, and oriented along the boom toward the front of the antenna, to provide the zero bearing when the antenna was directed true north.

Bearings were read on the heading indicator, mounted on a panel inside the vehicle and faced so that the operator could simultaneously rotate the mast and read the direction of the antenna. The coaxial cable connecting the antenna with controls in the vehicle ran inside the mast tube, and permitted the antenna to rotate 4 to 5 rounds without difficulty.

Tracking was conducted by 1 or 2 persons. The repeater compass eliminated the necessity of alignment of the vehicle at each station. If 1 person drove and another operated the radio equipment, the time lapse between successive readings was reduced. Mast elevation, reading and mast lowering took 1 to 2 min when tracking 1 animal.

A test of reliability of the repeater compass against a handheld compass (Suunto Fast accuracy Compass) revealed a system error related to the orientation of the antenna boom. The pattern of the deviation was almost identical for all fixed stations, showing a sinusoidal function on a 400° round with ranges of $\pm 3^{\circ}$. This enabled general correction for each recorded bearing. The cause of the deviation was not understood fully. Further technical investigations rejected the suggestion that there was a magnetic deflection from the vehicle or ground.

Tests were conducted to estimate the accuracy of the system. A radio equipped dog was placed in known locations within normal receiving distance from fixed stations. Bearings were taken from at least 2 stations. A total of 39 bearings were obtained on 13 different locations of the transmitted dog. Corrected bearings gave greater accuracy than did uncorrected (sign test, $P < 0.05$). The angular errors were $3.3^{\circ} \pm 3.2^{\circ}$ and $4.2^{\circ} \pm 3.2^{\circ}$, respectively. However, within normal working range (1000 m) correction decreased mean error of location by about 8 m ($\bar{x}$36.5 m versus $\bar{x}$ 28.3 m) at a mean distance of 570 m.

A double yagi antenna provided better directionality than did a single yagi, but was more difficult to mount semi-permanently on a vehicle. Hallberg *et al.* (1974), using such an arrangement, recorded an accuracy of $\pm 1^{\circ}$ in over 2000 trials. They did not mention deflection from objects in the terrain, but trees, topography, and other structures may increase errors. Initial tests at Grimsö with a double yagi placed a few meters above ground in dense vegetation supports this belief.

Acknowledgements — We are indebted to T. Dreyfert for suggesting the technique and for valuable discussions. We thank the field personnel for assistance during the tests, L. Norling for statistical treatment of data, and L. Keith for reviewing the manuscript. This work was supported by O. Engkvist Foundation and The National Swedish Environment Protection Board.

REFERENCES

Anderson F. and De Moor P. P. (1971) A system for radio-tracking monkeys in dense bush and forest. *J. Wildl. Mngmt* 35, 636-643.

Haacke H., Hirneiss R., Hommel R., Kraft H. and Landauer G. (1973) Funkortung freilebender Füchse. *Z. Jagdwiss.* 19, 90-98.

Hallberg D. L., Janza F. J. and Trapp G. R. (1974) A vehicle-mounted directional antenna system for biotelemetry monitoring. *Calif. Fish Game* 60, 172-177.

Hutton T. A., Hatfield R. E. and Watt C. C. (1976) A method for orienting a mobile radiotracking unit. *J. Wildl. Mngmt* 40, 192-193.

Kolz A. L. and Johnson R. E. (1975) An elevating mechanism for mobile receiving antennas. *J. Wildl. Mngmt* 39, 819-820.

Proud J. F. (1969) Wild turkey studies in New York by radio-telemetry. *N.Y. Fish Game J.* 16, 46-83.

Verts B. J. (1963) Equipment and techniques for radio-tracking striped skunks. *J. Wildl. Mngmt* 27, 325-339.

Time-sharing for Biological Signals in Radio Telemetry

M. D. CRAGGS and C. M. ANDREW

MRC Neurological Protheses Unit and Department of Physiology, Institute of Psychiatry, London, U.K.

Abstract — Using inexpensive CMOS integrated circuits we have developed a simple multi-channel system for continuously monitoring low level biological signals by time sharing the transmission of a conventional single channel FM telemetry link. This device does not require additional circuits for preamplification on every channel and as a consequence uses relatively little power. Time sharing is achieved by switching the original low level signals through a standard analog multiplexor and feeding the output, together with synchronizing pulses derived from digital clocking circuits, directly into a commercially available biotelemetry transmitter.

This system has been successfully used to record simultaneously, eight channels of physiological information in freely moving subjects, and there is the potential for being extended to sixteen channels.

Ultimately it is hoped to develop a totally implantable transmitter based on this design that has low power consumption and is adequately protected from body fluids.

INTRODUCTION

The need to record many physiological parameters from unrestrained animals has led to the development of multi-channel radio telemetry systems (Mackay, 1970; Voegeli and Kraft, 1972). For some animals, such as monkeys, the ideal device should be designed for total implantation, thus avoiding infection prone open wounds and tampering by the animal. An implantable device must be small, resistant to body fluids and above all, operate at very low currents particularly if many channels are required.

Recently a four channel radio telemetry system (Voegeli and Kraft, 1972) has been tried and tested on small animals (Borbely, Baumann and Waser, 1972; Gottesmann *et al.*, 1977). For power conservation this device used pulse interval modulation (PIM), rather than continuous amplitude or frequency modulation (AM/FM), of an HF transmitter and was constructed from discrete components in conventional transistor circuits. The main disadvantage of this system was the need to provide continuous amplification for low level biological signals, such as the electroencephalogram (EEG), on every channel, thus multiplying quite dramatically the power requirements as the number of channels increased. Ideally it should be possible to time share a single amplification stage by switching rapidly between all the low level inputs, the output of the amplifier then being used to modulate the carrier frequency of a radio transmitter in one of the ways mentioned above. Since 1972 there has been a rapid development in the field of low power integrated circuits using Complementary symmetrical Metal Oxide Semiconductor (CMOS) technology (RCA Corporation, 1977). This significant advance has enabled implantable circuits to be designed, using CMOS multiplexors and few other external components, which can switch stimulation pulses to a number of nerves in prosthetic devices (Gheewala *et al.*, 1975; Clark *et al.*, 1977; Donaldson, 1979, this volume). The concept of an electronically controlled switch for low level biological signals was not necessarily considered during the development of CMOS multiplexors, but we have found that by using such integrated circuits an inexpensive system could be developed for continuously monitoring these signals on many channels by time sharing the transmission of a conventional single channel FM radio telemetry link. This device does not require additional circuits for preamplification on every channel and as a consequence uses relatively little power. To extend the link to sixteen channels would not require much extra power save that to switch channels more quickly.

METHOD

Figure 1 shows schematically the complete radio telemetry system. Time sharing in the transmitter was achieved by switching the original low level signals through a standard 4051B eight channel analog multiplexor and feeding the output directly into

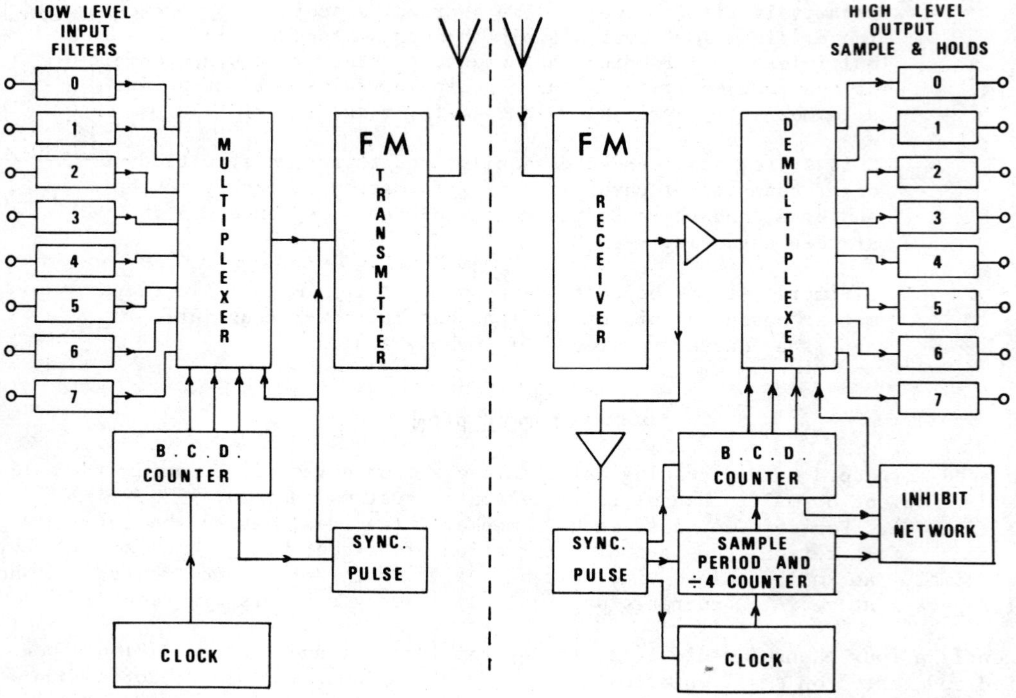

Fig. 1. Block diagram for an eight channel radio telemetry system using CMOS analog multiplexors to switch low level biological signals.

a commercially available FM biotelemetry transmitter (SNR102F, Dynamic Electronics Ltd.) tuned to the medical band 102.2-102.4 MHz (Fig 1 – left). A 4510 B binary coded decimal (BCD) counter, driven by a 4528B dual monostable wired as a 1:50 μs mark space ratio pulse generator, clocks the multiplexor for eight sequential counts, the remaining two being used as a synchronizing pulse for reception (Fig. 2 – left).

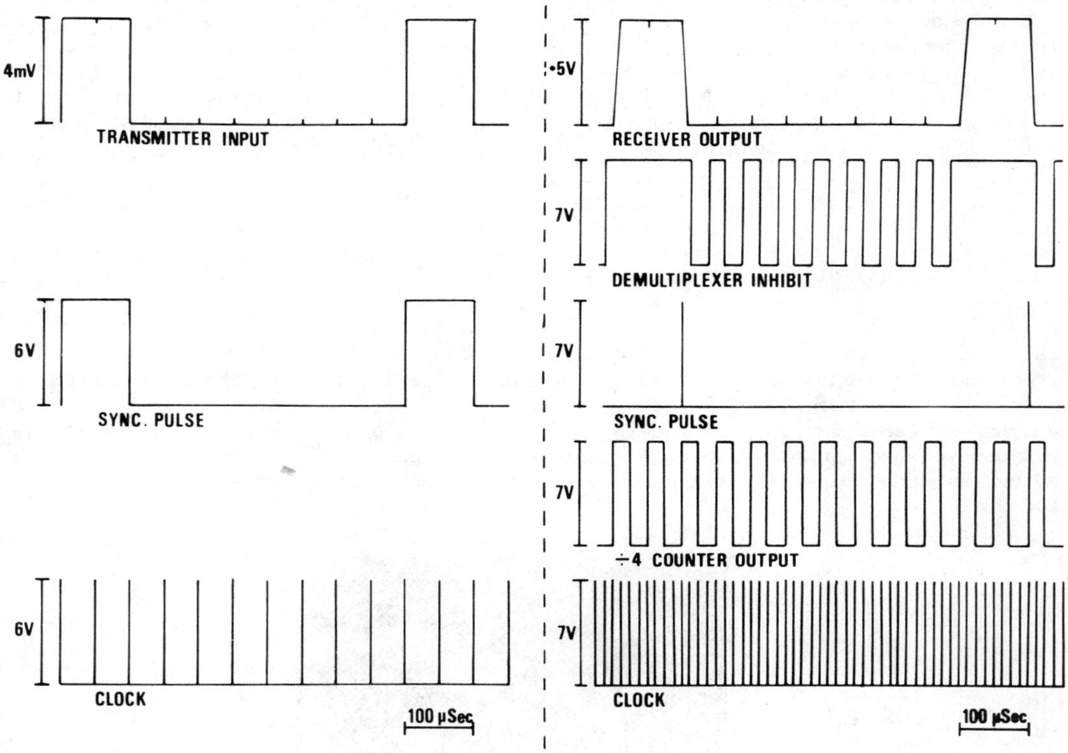

Fig. 2. Sequence of coding and decoding pulses to operate CMOS analog multiplexors.

Each channel was sampled for 50 μs every 500 μs. To avoid excessive cross channel coupling, particularly on sequential channels, a small low pass filter comprising a 0.1 μF capacitor and 100 kΩ resistor in parallel was necessary on each gate of the multiplexor. This undesirable coupling appears to be an inherent characteristic of these CMOS gates and arises because the gate is capacitively coupled to the control switching at power rail level. Empirically, the filters were arranged as a compromise of the conflicting requirements of low cross talk and useful high frequency response. The low frequency response was limited by a 1.0 μF blocking capacitor in series with the gate. The input impedance of the gates, when open, was mainly determined by the input characteristics of the commercial transmitter (approximately 75 kΩ). By operating the multiplexor at 6 V a negligible additional resistance of about 150 Ω was placed in series with the low level signal. The extra noise generated by this small resistance could not be detected. A decoupling capacitor (330 pF) was placed on the multiplexor output. The 3.0 mV synchronizing pulse was derived from the 6 V pulses of the unused fourth output of the BCD counter by a resistor and diode network and, with the low level multiplexed signals, fed directly to the input of the transmitter. This BCD pulse was also used to inhibit the multiplexor during the synchronization period. The power consumption of the switching circuits was approximately 250 μA but the total requirement was increased by the commercial transmitter which itself used 3 mA on a separate 3 V supply. Four standard 1.5 V watch cells (Mallory Duracell 'Pacemaker' type 10L125B2) with a capacity of 40 mAh were used to power the switching circuits giving a life expectancy of about 160 h.

The transmitted signals were received by a standard commercially available FM tuner (Sinclair Project 80) and fed into a demultiplexing (4051 B) circuit (Fig. 1 – right). The synchronizing pulse was detected and used to reset the decoding counters (4516B Binary) and an oscillator (CMOS 7555) running at four times the clock frequency of the transmitter; this enabled a sample and hold circuit to select only the second and third quarters (25 μs) of each 50 μs sample period for each channel (Fig. 2 – right). By this technique it was possible to (a) accommodate small fluctuations in the frequency of sampling and (b) remove the slight irregularities, caused by poor settling time, in the leading and trailing edges of the sample.

RESULTS

TRANSMISSION OF TEST SIGNALS

Figure 3 shows the reproducibility of a 70 μV sine wave at 10 Hz applied to one input of the eight channel telemetry system. The transmitted signal had the high frequency noise component removed by low-pass filtering in both the transmitter and receiver. The overall pass band of the radio telemetry system was 1 Hz-150 Hz (3dB), a range suitable for physiological signals from brain, heart and muscle. In this frequency range the noise on each channel did not exceed 5 μV with the inputs open circuited, and the cross talk between adjacent channels was better than -60 dB at 100 Hz.

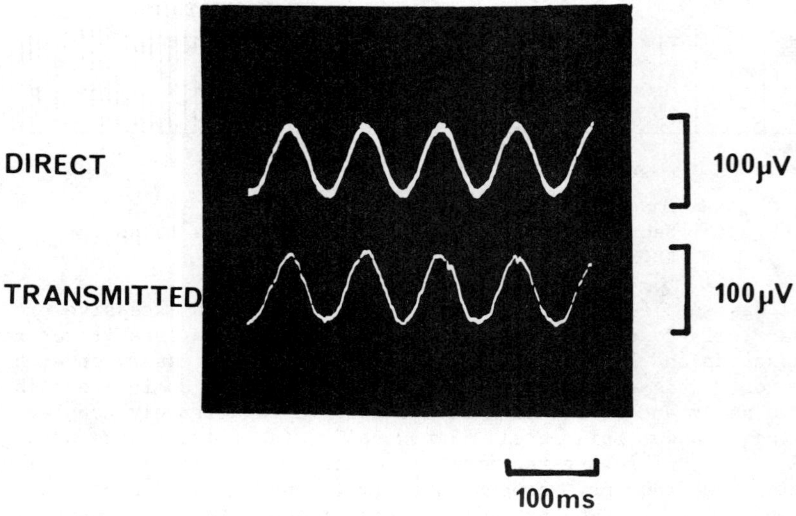

Fig. 3. Reproducibility of a 70 μV sine wave at 10 Hz applied to one input of the eight channel radio telemetry system.

TRANSMISSION OF BIOLOGICAL SIGNALS

Various physiological signals on channels 1 to 4, recorded between silver/silver chloride disc electrodes and a common reference, are shown in Fig. 4. The source impedance was approximately 5 kΩ. The low voltage EEG on channel 2 showed no sign

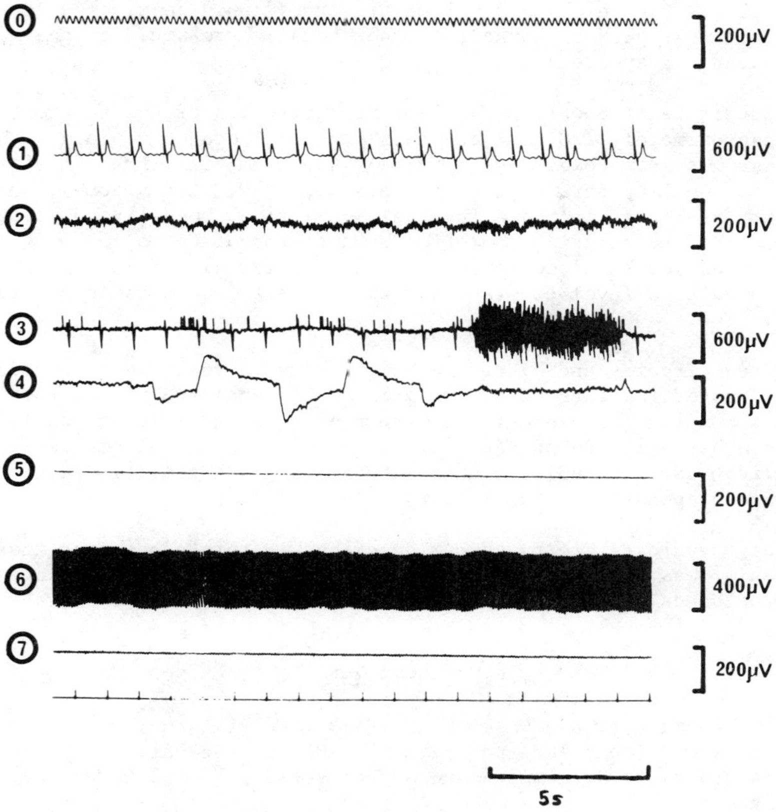

Fig. 4. Transmission of biological signals.
Channel:
0 — 50 µV 5 Hz sine wave.
1 — ECG.
2 — EEG — vertex.
3 — EMG — wrist extensor muscle.
4 — Eye movements.
5 — Short-circuited input.
6 — 400 µV 10 Hz sine wave.
7 — Open circuited input (100 kΩ).

of cross talk from the high amplitude ECG on the preceding channel. A comparison of channels 6 (400 µV 10 Hz sine wave) and 7 (open circuit) also reveals very little evidence of cross talk.

DISCUSSION

The purpose of the work described in this paper was to establish the principle of using CMOS analog multiplexors for switching low level biological signals before amplification, and then transmitting them by conventional radio telemetry. At the cost of some high frequency response, dictated by clocking rate and the need for low pass filters on each input, we have shown that it is possible to record eight channels of physiological information by this method. The system lends itself to extension to sixteen channels with little extra power consumption, and for differential recordings, two, eight channel multiplexors could be clocked together. By allocating more multiplexor channels to one signal there would be an improvement in upper frequency response.

The main expenditure of power in the switching circuits of our transmitter is in the clock (approximately 120 μA), but this might be reduced tenfold with a crystal controlled oscillator. Power consumption of the commerical FM transmitter was prohibitively high but, by careful design (Perkins, 1980, this volume), and the use of a shorter duty cycle, as in pulse interval modulation (Voegeli and Kraft, 1972), it should be possible to reduce consumption by a factor of up to 100. By combining the advantages of low level switching and pulse interval modulation power needs could be reduced to a level consistent with a totally implantable multichannel transmitter.

CMOS integrated circuits and hybrid techniques have already been introduced into hermetically sealed implants (Clark *et al.*, 1977; Donaldson, 1980, this volume). Miniaturization using the present generation of CMOS multiplexors would be limited by the need to have filters on each input, but with new developments such as self-aligning silicon gate technology, control gate coupling effects could be minimized, allowing us to dispense with the filters.

The short range radio telemetry system described in this paper used inexpensive standard components, was suitable for monitoring physiological signals, and could be mounted on the heads of small animals.

REFERENCES

Borbely A. A., Baumann I. and Waser N. M. (1972) Multichannel telemetry of physiological parameters (Body Temperature, EKG, EEG) in the rat. II Applications in Neuropharmacology. In H. Kimmich and J. Vos (eds.), *Biotelemetry*, pp. 318-388. Leiden: Meander.

Clark G. M., Black R., Dewhurst D. J., Forster I. C., Patrick J. F. and Tong Y. C. (1977) A multiple-electrode hearing prosthesis for cochlear implantation. *Med. Progr. Technol.* 5, 127-140.

Donaldson N. de N. (1980) Encapsulation and packaging of implanted components. In C. J. Amlaner and D. W. Macdonald (eds.), *Handbook on Biotelemetry and Radio Tracking*, pp. 217-224. Oxford: Pergamon Press.

Gheewala T., Melen R. and White R. (1975) C-MOS implant to aid deaf. *Electronics*. February 20, p. 38.

Gottesmann Cl., Rodi M., Rebelle J. and Mailett B. (1977) Polygraphic recording of the rat using miniaturised telemetry equipment. *Physiol. Behav.* 18, 337-340.

Mackay R. S. (1970) *Bio-Medical Telemetry*. Chapter 1, 2nd Edn. London: John Wiley & Sons.

Perkins T. A. (1980) Low power VHF transmitter for multiplexed radio tracking. In C. J. Amlaner and D. W. Macdonald (eds.), *Handbook on Biotelemetry and Radio Tracking*, pp. 201-204. Oxford: Pergamon Press.

RCA (1977) COS/MOS Integrated Circuits Databook, Somerville, U.S.A. RCA Corporation.

Voegeli F. and Kraft W. (1972) Multichannel telemetry of physiological parameters (Body Temperature, EKG, EEG) in the rat. I Design and Methods. In H. Kimmich and J. Vos (eds.), *Biotelemetry*, pp. 371-380. Leiden: Meander.

A New Automatic Food Dispenser

L. J. SHIELDS

Department of Biology, University of California at Los Angeles, Los Angeles, California 90024, U.S.A.

Abstract — An easily constructed and reliable automatic seed dispenser is described. Various small radio tagged and unmarked animals have learned to obtain seeds from it in the wild. Precise measurements of feeding and hoarding behavior can be made using this device.

INTRODUCTION

During radio tracking field studies of hoarding and feeding behavior of wild woodrats, *Neotoma fuscipes*, I have found that the animals will learn to obtain millet seeds at a field station by pressing a bar. Incorporated in the field system is an inexpensive and reliable automatic feeder of new design with several features that make it useful for both field and laboratory: (1) it can be built by moderately skilled persons with simple tools, (2) it has no parts that require close tolerances, and (3) it can be scaled up or down in size. I have used small feeders of this design to train goldfish and larger ones to train birds and small mammals. The feeder will dispense any particulate food that does not pack, e.g., certain kinds of commercial fish foods, cracked corn, or various small seeds, such as millet. It has proved to be extremely reliable. A feeder of this design has been working continuously under field conditions for the last 6 months, performing over 100,000 deliveries with few failures.

DESIGN

As can be seen in Fig. 1, the moving parts consist of (1) a lever with a food scoop on one end, (2) a solenoid plunger supported by a weak spring, and (3) a linkage between the lever and the solenoid. When the solenoid is energized the scoop end of the lever flips upwards rapidly. The food contained in the scoop continues to move on a line tangent to the circle described by the scoop, and its trajectory takes it over the edge of the food reservoir. The scoop end of the lever is weighted so that the scoop falls back into the food and refills at the end of each stroke.

Although the feeder will work with only a momentary switch closure to energize the solenoid, it is recommended that a device such as the electronic monostable multivibrator (one-shot) shown in Fig. 2 be used to deliver a short pulse to the solenoid. With a pulse of about 250 ms, the solenoid is de-energized by the time the lever

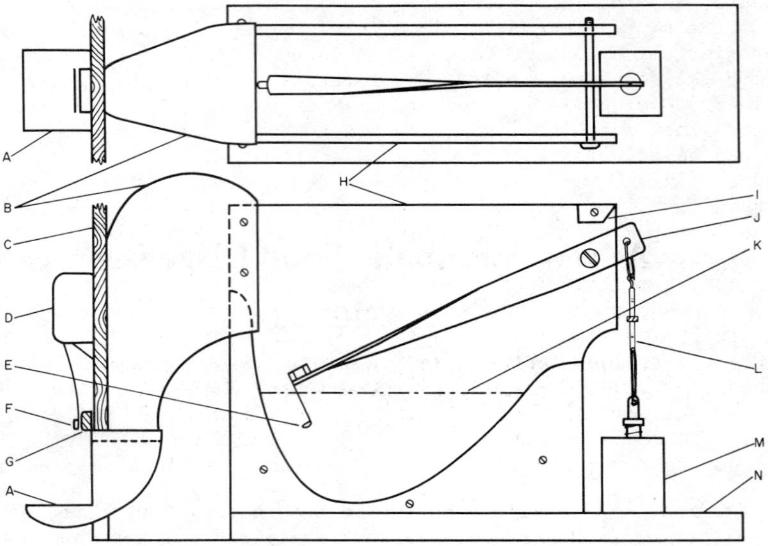

Fig. 1. Diagram of seed dispenser. (A) Food cup; (B) food chute; (C) plywood box; (D) micro-switch; (E) food scoop; (F) pressing bar; (G) fine screened package for training inducement; (H) Plexiglas sides; (I) wooden spacer to separate sides; (J) lever; (K) fill line; (L) linkage; (M) solenoid; (N) feeder base. The dimensions of the Plexiglas sides are 28 cm long by 22 cm high; the spacing between the sides is 7 cm.

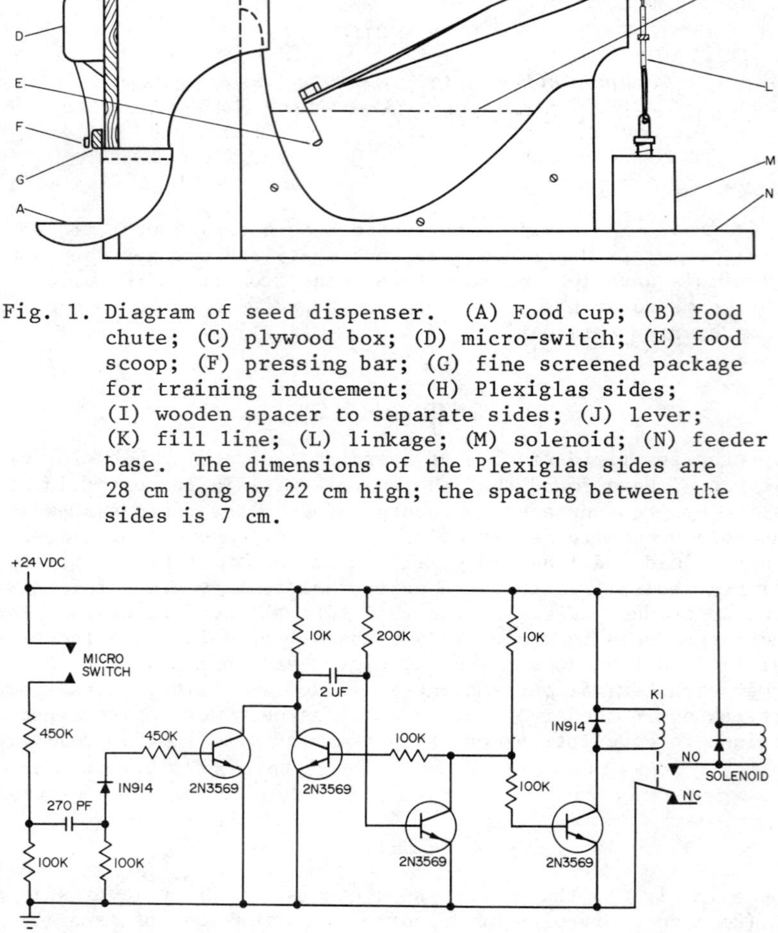

Fig. 2. Schematic diagram of solenoid control circuit. K1 should be a multi-pole relay to operate recording devices as well as the solenoid.

reaches the top of its swing. The solenoid plunger then is returned to its original position, 1.5 cm above the solenoid, by its supporting spring, and the weighted end falls unencumbered from its maximum height. This ensures that the scoop is buried in the food and becomes refilled completely.

The lever is made of 0.3 cm thick aluminum stock. The scoop is made of 1 mm thick aluminum sheet. Several interchangeable scoops delivering different amounts of seeds can be made for a single feeder. A scoop 6 mm long by 4 mm wide by 2 mm deep will deliver approximately 0.15 g per stroke. A large nut bolted or glued to the lever near the scoop is used as a weight. The lever is pivoted on a 6 mm bolt. A small hole is drilled 3.4 cm behind the pivot hole for the attachment of the solenoid-lever linkage. The length of the linkage is the only critical dimension. However, this length can be adjusted easily by using a universal Sturmey-Archer bicycle-gear cable adjuster. This device is tied to both the solenoid and the lever by several loops of strong fish line or waxed linen string. In addition to being an easy way to adjust the linkage length, the cable adjuster also acts as a swivel, preventing a change in linkage length due to twisting caused by rotation of the solenoid plunger. As previously mentioned, the plunger should be adjusted so that, when the scoop is well buried in the food, there is about 6 mm of slack on a solenoid whose travel is 1.5 cm. The food reservoir has 6 mm Plexiglas sides. The bottom and the ends are made of a single piece of 7 cm pine. It is important that the food reservoir has adequate width and depth. This keeps the food from packing and allows several hundred deliveries between refills. The reservoir is filled with seeds to a 6 cm depth. The feeder will not work well if it is over-filled. A chute to deliver the food to a cup can be fashioned readily by soldering together pieces of tin cans.

In the field this feeder is contained in a plywood box with the food chute leading to an outside food cup (6 cm wide, 4 cm long and 1.6 cm deep). A bar operating a microswitch is placed 7.5 cm above the food cup and 6 mm away from the plywood box.

DISCUSSION

Wild California ground squirrels (*Citellus beechyi*), brown towhees (*Papilio fuscus*) and California thrashers (*Toxostoma redivivium*), as well as woodrats, have learned to obtain millet seeds by pressing the bar. A bar pressing response can be elicited by placing peanut butter and seeds in a small fine screened package just above and behind the bar. These inducements generally can be removed after a single night. Any of these animals may spend up to 2 h per day feeding and/or hoarding. Woodrats will take as much as 80 g of food per night and also will feed occasionally in daylight. Ground squirrels may press as many as 1500 times per day.

I have identified individual animals by means of radio tagging (Shields, 1980, this volume) and have recorded feeding events automatically on a 24 h basis. By this means it is possible to make precise measurements of feeding times and amounts of food collected, and to determine inter and intra-specific dominance at the feeder. Important insights concerning the ecological relationships between various animals may emerge from data obtained in this manner (report in preparation).

It should be emphasized that the dimensions given are not critical. Adjustments in the location of the linkage attachment to the lever, the length of the linkage, and the depth to which the feeder is filled will allow the construction of functional feeders with widely varying dimensions.

Acknowledgements — This research was supported by NSF grant GB-27553 to Dr. J. L. Kavanau. I appreciate his considerable help and advice while preparing this report.

REFERENCE

Shields L. J. (1980) The determination of free ranging rodent activity by telemetry. In C. J. Amlaner, Jr. and D. W. Macdonald (eds.), *A Handbook on Biotelemetry and Radio Tracking*, pp. 667-672. Oxford: Pergamon Press.

Optical and Radio Optical Techniques for Tracking Nocturnal Animals

T. G. WOLCOTT

Department of Marine Science and Engineering, North Carolina State University, Raleigh, North Carolina, U.S.A.

Abstract — Optical and radio optical transmitters permit localization and tracking of nocturnal animals with reliability and precision that cannot otherwise be achieved. Radio tracking seldom offers resolution of better than a wavelength, and use of shorter wavelengths increases problems of absorption and reflection. Visual tracking with night vision scopes is often impossible due to animals' cryptic coloration. This paper describes high efficiency LED flasher circuitry which permits visual localization of subjects in open terrain to 300 m with a night vision scope. Addition of an RF section permits homing to visual range by radio from greater distances or in visually cluttered habitats. Schematics for radio optical transmitters in several frequency bands, some capable of transmitting one or two channels of data, are presented.

INTRODUCTION

Optical and radio optical devices offer considerable advantages over traditional telemetry techniques for obtaining precise fixes and activity budgets of nocturnal animals. Radio tracking alone, although it does give long range, provides only approximate tag location. Use of shorter wavelengths can increase the precision of transmitter localization but may exacerbate problems due to reflection or absorption of the radio signal. Fixes obtained by triangulation become increasingly approximate at greater distances due to angular errors, while at close range the animal is likely to be disturbed by the maneuvering of the tracker trying to get a second bearing before the subject moves. Even fixed antenna grids used with captive populations display considerable error (Chute *et al.*, 1974). In the case of diurnal animals the uncertainties can often be eliminated by visually locating and following the animal with a telescope or binoculars. This does not work with nocturnal animals unless one can see in the dark.

Fortunately, seeing in the dark has become possible with the development of image intensifying night vision scopes ('starlight scopes'). Unfortunately, these devices are monochromatic and cannot 'see' color contrasts, nor can they create light-dark contrast where none exists. Nocturnal animals, most of which are cryptically colored and patterned, remain exceedingly difficult to see against their natural backgrounds, rendering the starlight scope nearly useless. Attachment of an optical transmitter

(Wolcott, 1977) to the animal overcomes this problem, extending the range of the starlight scope to several hundred meters.

CIRCUITS AND APPLICATIONS

The simplest way to produce flashes suitable for tracking is to use LED's which incorporate both the LED chip and a flasher circuit chip in a single T-1 3/4 package (e.g. Litronix FLR 4403). These units have a duty cycle of about 50 percent, an average current drain of about 15 mA, and require a 5 V supply. These power requirements, which would restrict the use of such devices to very large animals and would certainly negate any advantages gained by the small size of the self flashing LED, have precluded the use of these devices for nocturnal field work.

Optical transmitters with short duty cycles and moderate power consumption (0.5-1.0 mA at 1 Hz repetition rate) may be simply constructed with the LM 3909 LED flasher I.C. (National Semiconductor), a LED, and a 300 μF capacitor. The flasher can be powered by a single cell (1.35-1.5 V) battery. These circuits have been found useful on medium sized animals such as beavers (Brooks and Dodge, 1978), where the size of the 8-pin DIP I.C. and large capacitor are acceptable. Ample battery capacity must be carried; the LM 3909 is rather inefficient for a telemetry device. It has leakage current in excess of 40 μA at 1.5 V, and more importantly incurs substantial power losses by passing all of the LED drive current through charging and current-limiting resistors.

Smaller, more efficient transmitters are required for use on small animals. Only low capacity batteries can be carried, and power losses must be minimized so that most of that capacity will be available for producing useful signals. Highly efficient, simple devices which flash a LED can be constructed utilizing a 3-transistor switching oscillator (Fig. 1). Such a circuit may have leakage currents in the order of 2 μA, and minimizes resistive losses by applying short, high current square wave pulses to a LED without current limiting resistors. Oscillation of the circuit begins as the emitter of Q2 approaches 0V due to C1 discharging through R1. Since the base of Q2 is held one diode voltage drop above ground by the base-emitter junction of Q3, leakage current begins to flow through Q2. High gain positive feedback between Q1 and Q2 quickly leads to saturation of Q1 and of Q3 until C1 is recharged through Q2. When Q2's emitter voltage rises high enough to cut off current through Q2, all three transistors revert to the 'off' state. Operation of Q3 in the saturation mode minimizes power losses in the load circuits; this is particularly important when the load is a high current device like a LED. Pulse duration is set by R2, which limits current flowing through Q2 to recharge C1 and reset the circuit. Repetition rate is set by R1.

The advantages of a LED flasher based on this circuit are its simplicity, small size and weight (less than 1 g), and efficiency (life of up to 7 weeks with 4 g of batteries). Details of construction are given in an earlier paper (Wolcott, 1977). The circuit has been successfully used for field tracking of beach crabs (*Ocypode quadrata*, Wolcott, 1978), can be used to transmit one data channel (e.g. temperature by substituting a thermistor for R1), and may be useful for obtaining photographically time budgets of captive animals in the dark by simply counting dots on a time exposure. Disturbance of subjects can be minimized by using LEDs emitting at red or infra red wavelengths, to which most animals are relatively insensitive. The principal disadvantage of optical tracking alone is that it works only at moderate ranges over a clear line of sight. Its usefulness is therefore limited to animals which occupy reasonably predictable home ranges in open habitats. Tracking nocturnal animals which move about a great deal, or which occupy visually cluttered habitats, requires the addition of a radio beacon to the transmitter to permit homing in to visual range.

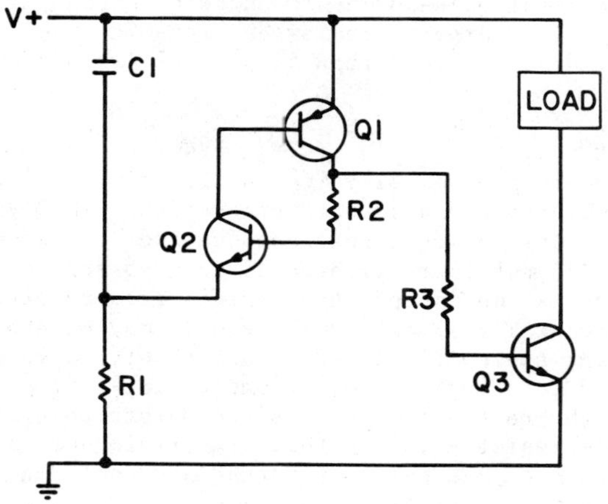

Fig. 1. Three transistor switching oscillator suitable for driving flashing-LED optical transmitters. V^+ = 2.7 V minimum.

Radio optical transmitter designs must take into account the differing requirements of optical and radio receivers. Optical receivers (eyes or starlight scopes) are sensitive to the total light quanta intercepted. Applying a very short (1 ms), high current pulse to the LED is therefore the most efficient means of producing the light output. In contrast, radio receivers and particularly the human ear require longer pulses to distinguish signals from noise. The radio transmitter must therefore be driven in longer, lower current pulses than the LED.

A crystal stabilized radio optical transmitter which meets these criteria and is compatible with commercially available telemetry receivers is shown schematically in Fig. 2. The switching oscillator is essentially the same as in Fig. 1 with the addition of a second PNP transistor (Q4), which saturates at the same time as Q1 and charges C2. C2 provides base drive for the 15 ms RF pulse, which begins

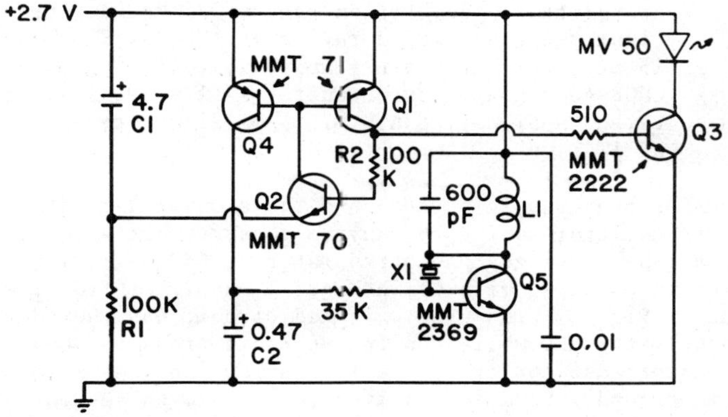

Fig. 2. Crystal stabilized DW radio optical tracking transmitter. Repetition rate = 1 Hz; duration of light pulse = 1 ms; duration of pulse = 15 ms. Current drain = approximately 30 μA. L1, a 22 mm loop, resonates at 30 MHz with 600 pF; x1 = 30 MHz. Other frequencies may be used by proper choice of crystal, resonant circuit and RF transistor.

simultaneously with the light pulse. The RF output frequency is determined by the resonant circuit and by X1. Temperature may be transmitted by substituting a thermistor for R1. Current drain is less than 50 µA; the circuit can be constructed to weigh less than 2 g.

A CMOS-switched circuit suitable for AM or FM operation and transmission of one or two data channels is shown schematically in Fig. 3. A multivibrator comprising inverters 'B' and 'C' alternately drives inverters 'A' and 'D', producing alternating light and radio pulses. The radio output is modulated by the multivibrator made from inverters 'E' and 'F', which oscillates as long as the output of inverter 'A' is high. The result is a 15 ms 'beep' at the receiver, the pitch of which is determined by the temperature of T1. Temperature readout may be accomplished by constructing and calibrating an identical multivibrator with a logarithmic potentiometer in place of T1 and rotating the potentiometer until it produces a pitch identical to that heard on the radio receiver. A second variable may be transmitted by incorporating a variable resistor in the timing multivibrator and thus varying repetition rate. This circuit may be built to weigh less than 3 g.

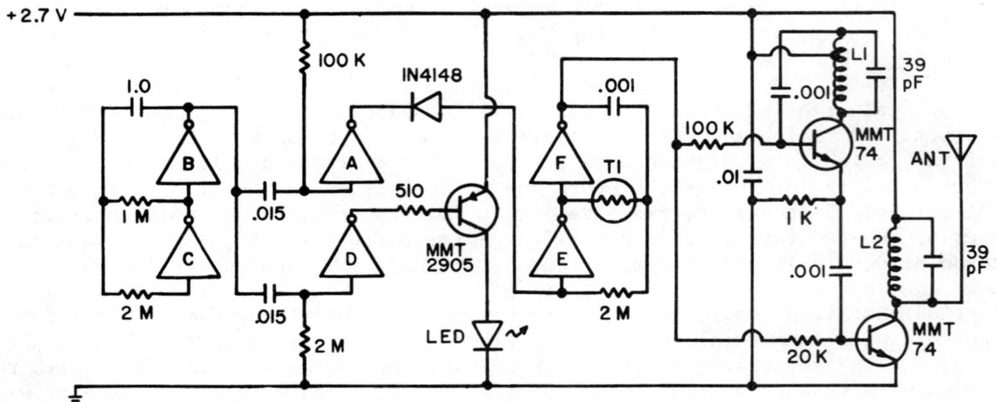

Fig. 3. CMOS switched radio optical tracking/temperature transmitter. Repetition rate = 0.5 Hz; duration of light pulse = 1 ms; duration of 100 MHz FM RF pulse = 15 ms. Current drain = approximately 25 µA. A-F = CMOS 4049 hex inverter buffer. T1 = 250 K thermistor. L1, L2 = 5T No. 28 A.W.G., 5 mm diameter; L1 tapped at 1T.

Several variations of these basic optical and radio optical transmitters are possible. Shutting off the transmitter during the day to conserve batteries is possible in nonburrowing animals by placing a phototransistor or CdS cell with a dark resistance of several megohms in parallel with C1 in Figs. 1 and 2, or from the input of inverter 'C' to ground in Fig. 3. The CMOS switched circuit has the additional option of shutting down only the LED while continuing radio transmission during the day, by connecting a phototransistor or CdS cell with low 'on' resistance from the input of inverter 'D' to ground. Duration of light and radio pulses may be tailored to specific requirements by adjusting the appropriate RC time constants. The LEDs, which normally direct most of the light along an axis parallel with the leads, may be modified according to the viewing requirements. For instance, the application on ghost crabs required that the light be directed toward the horizon throughout 360°, rather than in a beam straight up. This was accomplished by using an air bubble as a reflector. A small hole was drilled into the LED directly over the semiconductor die; the hole was then refilled with clear epoxy resin containing a single bubble (Fig. 4). This type of device was very useful in determining activity budgets

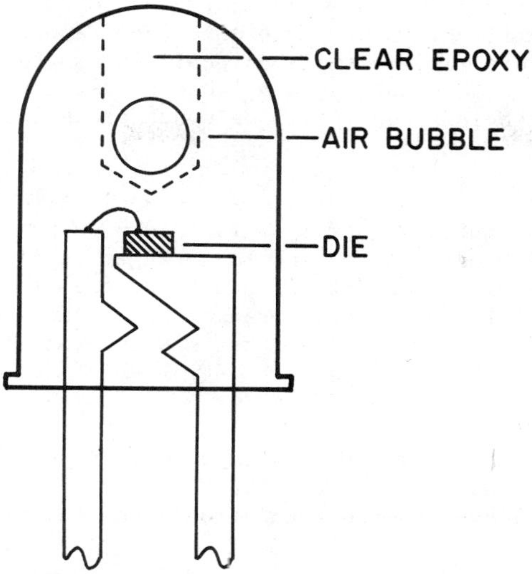

Fig. 4. LED modified for higher visibility perpendicular to the axis. Air bubble trapped in clear epoxy serves as reflector.

of ghost crabs (*Ocypode quadrata*). It gave satisfactory outputs when the crabs were roaming the beach or entering the seawater swash, and gave radio ranges of 0.5 km even when the animals were 0.5 m deep in their burrows (see Fig. 5).

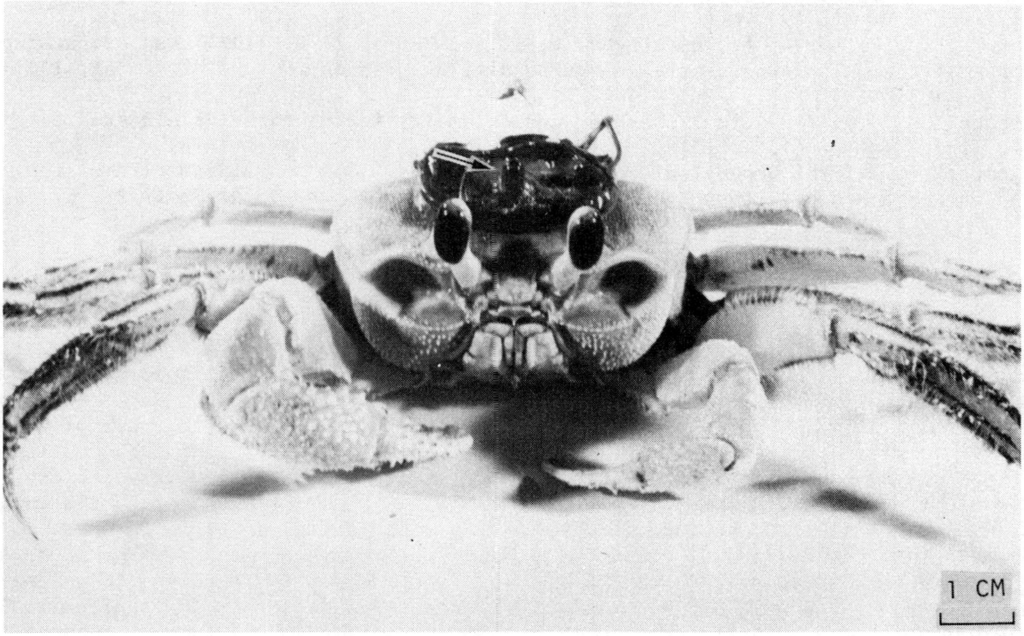

Fig. 5. Radio optical transmitter on a ghost crab (*Ocypode quadrata*). Trailing Teflon insulated wire is transmitting antenna. Arrow = flashing LED.

Individual coding of optical transmitters is possible by using different colored LEDs, different repetition rates or, with additional CMOS digital circuitry, pulse codes (Batchelor and McMillan, in preparation). Most applications will require radio optical transmitters rather simple optical transmitters; in those instances individual identification is simply accomplished by using different radio channels. The LED flashes from each individual will be synchronous (or alternate) with radio pulses on an identified channel.

Radio optical transmitters optimized for a given application can combine the advantages of both radio and optical tracking. The radio output provides low precision of localization, but high penetration and long range; it permits homing into visual range from considerable distances without hindrance from interposed obstacles. The light output is useful only over a short range, but confers high precision; it thus permits exact location and facilitates direct observation of undisturbed nocturnal animals in the field.

Acknowledgements — These devices were developed with support from N.C.S.U. Department of Marine Sciences and Engineering, the N.C. Marine Science Council, and Grant No. DEB-77 from the National Science Foundation. The cooperation of the National Marine Fisheries Service Laboratory, Beaufort, N.C., and Fort Macon State Park, Bogue Banks, N.C., during field work is gratefully acknowledged. Contribution No. 79-02 from the Department of Marine Science and Engineering, North Carolina State University.

REFERENCES

Batchelor T. A. and McMillan J. R. (Zoology Dept., Auckland University, New Zealand). A visual marking system for nocturnal animals. (In preparation).

Brooks R. P. and Dodge W. E. (1978) A night identification collar for beavers. *J. Wildl. Mngmt.* 42, 448-452.

Chute F. S., Fuller W. A., Harding P. R. J. and Herman T. B. (1974) Radio tracking of small animals using a grid of overhead wire antennas. *Can. J. Zool.* 52, 1481-1488.

Wolcott T. G. (1977) Optical tracking and telemetry for nocturnal field studies. *J. Wildl. Mngmt.* 41, 309-312.

Wolcott T. G. (1978) Ecological role of ghost crabs, *Ocypode quadrata* (Fabricius) on an ocean beach: scavengers or predators? *J. exp. mar. Biol. Ecol.* 31, 67-82.

Automatic Monitoring of Nest Attentiveness in Incubating Birds

E. N. SMITH* and E. G. BOLEN**

*Division of Natural Science and Mathematics, Northeastern State University, Tahlequah, Oklahoma 74464, U.S.A.
**Department of Range and Wildlife Management, Texas Tech University, Lubbock, Texas 79409, U.S.A.

Abstract — A radio telemetry system is described that automatically monitors and reccrds nest occupancy during incubation of two pairs of black-bellied whistling ducks, *Dendrocygna autumnalis*, occupying two nearby nesting sites. Each incubating bird carries a short range radio collar. Each pair of radio transmitters operate on different frequencies and the two transmitters of each pair are pulsed at different pulse rates. Demodulation provides time division multiplexing between nests and yields a unique output to a battery operated chart recorder for each possible incubation condition of the two nest boxes. Actual field data are provided. Transmitters operate at 150 MHz and weigh about 15 g. Range is limited to a few meters and a battery life of several weeks is attainable.

INTRODUCTION

Use of biotelemetry has been well established for wildlife studies (Schladweiler and Ball, 1968; Will and Patric, 1972), and the effects of radio packs on waterfowl have been described (Greenwood and Sargeant, 1973; Gilmer *et al.*, 1974). Attempts to monitor nest attentiveness have employed mechanical devices (Kendeigh, 1952); such studies have been limited to those species where only one member, usually the female, attends the nest (cf. Stewart, 1962).

This report describes telemetry equipment used to automatically monitor nest attentiveness for birds when both sexes alternately share incubation. The system was successfully used to monitor incubation behavior of black-bellied whistling ducks, *Dendrocygna autumnalis*, nesting at the Welder Wildlife Foundation near Sinton, Texas, in July, 1978. The long tenure of the pair-bond, including shared incubation duties, in this species is described elsewhere (Bolen, 1971).

The system consists of four 150 MHz transmitters, 2 VHF receivers and a multiplexed demodulator circuit capable of recording nest attentiveness of 2 pairs of birds incubating in 2 separate nests. The next bcxes erected for black-bellied whistling ducks are predator proof and readily utilized by the birds in lieu of natural cavities (Bolen, 1967). Data are continuously recorded on an unattended single channel strip chart recorder.

SYSTEM BLOCK DIAGRAM

The complete system consists of four transmitters, 2 fast pulse rate transmitters mounted on the females of each pair and 2 slow pulse rate transmitters attached to the males of each pair (Fig. 1). Two pairs of incubating birds in nearby nests can be simultaneously monitored with this system. Each pair is on a different radio frequency so that each can be recognized by the incoming signal. Two antennas (one at each nest) provide signals to each of 2 VHF receivers. The audio signal from each receiver is demodulated and a voltage is derived as a function of the pulse rate. Gated dual comparators detect the absence of a signal (nest empty), a low rate transmitter (male present), or a high rate signal (female present). Multiplexing enables monitoring of 2 receivers by alternately disabling each dual gated comparator. A unique current output for each condition indicates the condition within each nest. Table 1 shows the relation between recorded meter reading and nest attentiveness of 2 nests.

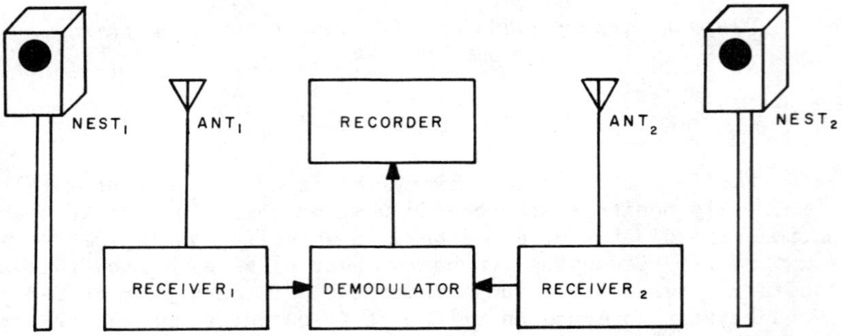

Fig. 1. System block diagram. By using time division multiplexing, one recorder automatically monitors 2 incubating pairs of birds. Each antenna may be up to 100 m from the recording site. The antennas consist of a 46 cm wire attached to the rear of the wooden nest boxes connected to the receiver by coax cable.

Table 1.
Recorder meter readings indicating simultaneous conditions of attentiveness at each of 2 nests during incubation. The telemetry system is designed for species where both sexes share incubation duties

Meter reading	Nest	Attentiveness
0.00 mA	Nest 1	No bird present
0.20 mA	Nest 1	Male present
0.40 mA	Nest 1	Female present
0.60 mA	Nest 2	No bird present
0.80 mA	Nest 2	Male present
1.00 mA	Nest 2	Female present

TRANSMITTERS

The radio transmitter consists of a free running multivibrator, differentiating circuit, electronic switch, and RF oscillator (Fig. 2). The RF oscillator, similar to

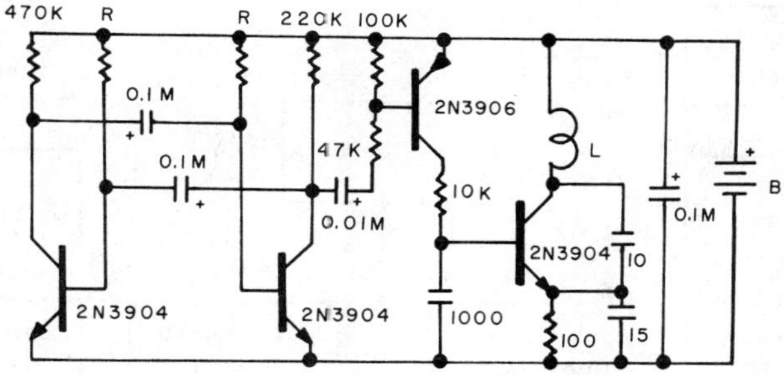

Fig. 2. Transmitter schematic diagram. Resistors in Ω unless designated otherwise. Capacitance is picofarad except M indicates microfarad. R determines the pulse rates; male transmitter 10M, female transmitter 4.7M. The coil L determines the radio frequency. Approximately 3 turns 26 gauge enameled wire operates at 150 MHz. B consists of two mercury cells, Hg-630 in series.

that described by Smith and Salb (1975), operates at 150 MHz and is turned on briefly each time the multivibrator output falls to zero volts. This provides a low duty cycle for low current drain and long battery life. Multivibrator base resistors are selected to provide the desired pulse rate. A transmitted range of several meters is attainable with an average current drain of less than 100 μA, extending battery life to several weeks. Increasing the 0.01 mfd coupling capacitor increases duty cycle and range at the expense of battery life.

The transmitters weighed about 15 g, or less than 2 percent of the average weight (800-890 g) of adult black-bellied whistling ducks (Bolen, 1964). The transmitter was attached to a plastic collar adapted from those designed for geese (Sherwood, 1966); the collars were color-coded for visual identification of each bird during subsequent inspections of the nest box.

MULTIPLEXED DEMODULATOR BLOCK DIAGRAM

The audio signal from each receiver is amplified by a saturating amplifier and each pulse from the transmitter triggers a one-shot multivibrator (Fig. 3). Since the duration of the one-shot is constant, the duty cycle is a function of the pulse rate. The one-shot output is filtered and a voltage is developed that is proportional to the pulse rate. Two gated comparators compare the filtered voltage with 2 references and respond when the voltage exceeds the reference voltage. Each comparator effects the output current in a unique way so the output current describes 4 of the 6 possibilities (Table 1).

DEMODULATOR SCHEMATIC DIAGRAM

Input audio signals are amplified by the operational amplifiers (Fig. 4). Negative going outputs trigger the 555 timers used as one-shot multivibrators. The output is filtered and applied to 2 comparators. The 3rd 555 timer is operated as an astable multivibrator and provides the multiplexing signal. When the output is positive, 2 comparators are inhibited. The multiplexing signal is inverted by IC-1a, which enables the other 2 comparators when the 555 timer output is high. Alternately, the output of IC-1a provides the 0.6 mA output signal to the recorder if no birds are present in the second nest. Output currents may vary slightly depending on

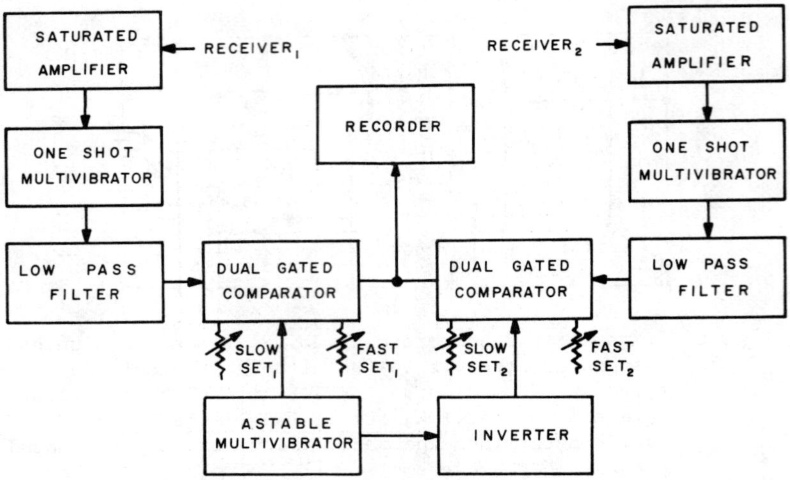

Fig. 3. Block diagram of the demodulator. Time multiplexing was achieved by the use of 2 gated dual comparators. Each nest is monitored every 30 s.

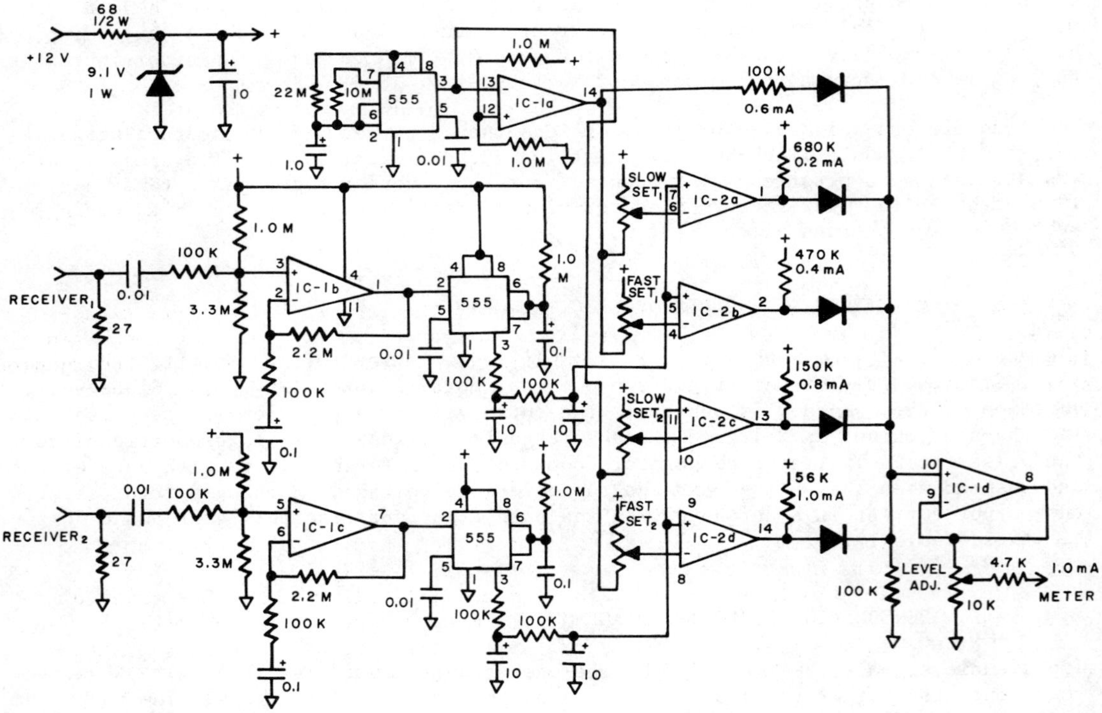

Fig. 4. Demodulator schematic diagram. Resistors in Ω, capacitance is in microfarads. Unlabeled pots are 100 K trimpots. IC-1 is LM 324N quad operational amplifier and IC-2 is LM 339N quad comparator.

resistor tolerance and substitutions may be made for different values. Output amplifier IC-1d operates as a voltage-follower to provide current to drive the 0-1 mA meter of the recorder. A slow chart speed (12 in 24 h^{-1}) was selected so a continuous record was obtained (Fig. 5).

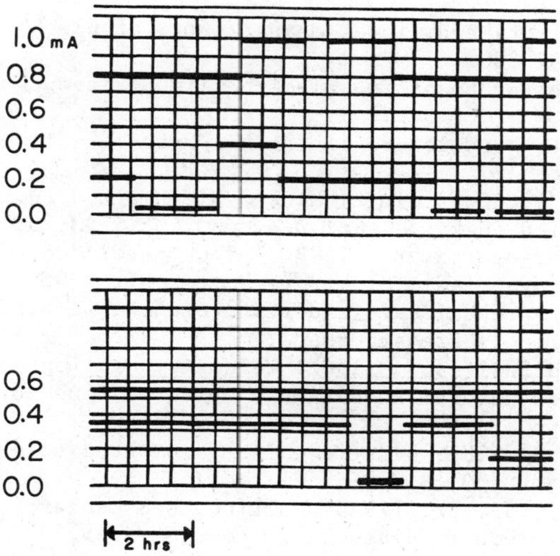

Fig. 5. Attentive behavior of incubating black-bellied whistling ducks. Upper panel was simulated in the laboratory; the lower panel contains data recorded in the field. A Rustrak 12 V DC recorder Model 288/F2146 (available from Gulton Industries, Inc., Manchester, N.H., U.S.A.) was used.

ADJUSTMENTS

Each low rate comparator is adjusted to respond to the respective low rate transmitter, but not to the high rate transmitter. Conversely, the high rate comparator is adjusted to respond to the high rate transmitter. Receiver volume must be adjusted so a weak signal (null orientation of the radio collar) operates the demodulator. Too high a volume setting results in response to static or ignition noise.

APPLICATION

The system described should find wide application for nesting birds where both parents incubate the eggs. It could also be used with altricial species for a record of feeding behavior by the parent. The system automatically records the time each parent is on the nest, and if shift changes are made at the nest or if the nest is vacated during shift change. Finally, this system could be used with mammals in behavioral studies dealing with social relationships.

Acknowledgements — Portions of this research were supported by Welder Wildlife Foundation and Caesar Kelberg Foundation. Additional support was provided by NIH (MBS) Grant 5 S06 RR08123 and MARC Grant 5 T32 GM07694. We thank W. C. Glazener, Director of the Welder Wildlife Foundation, for his support of our work. Travel funds (for E. N. Smith) were supplied by the College of Arts and Sciences and by a Faculty

Research Grant from Northeastern State University. We gratefully acknowledge Charles Amlaner and David Macdonald for organizing this conference and for the invitation to participate.

REFERENCES

Bolen E. G. (1964) Weights and linear measurements of black-bellied tree ducks. *Texas J. Sci.* 16, 257-260.

Bolen E. G. (1967) Nesting boxes for black-bellied tree ducks. *J. Wildl. Mngmt.* 31, 794-797.

Bolen E. G. (1971) Pair-bond tenure in the black-bellied tree duck. *J. Wildl. Mngmt.* 35, 385-388.

Gilmer D. S., Hall I. J., Cowardin L. M. and Riechmann J. H. (1974) Effects of radio packages on wild ducks. *J. Wildl. Mngmt.* 38, 243-252.

Greenwood R. J. and Sargeant A. B. (1973) Influence of radio packs on captive mallards and blue-winged teal. *J. Wildl. Mngmt.* 37, 3-9.

Kendeigh S. C. (1952) Parental care and its evolution in birds. *Illinois Bio. Monogr.* 22, 1-358. Univ. Illinois Press, Urbana.

Schladweiler J. L. and Ball I. J. (1968) Telemetry bibliography emphasizing studies of wild animals under natural conditions. Univ. of Minnesota, Bell Mus. Nat. Hist. Tech. Rept. 15. 31 pp.

Sherwood G. A. (1966) Flexible plastic collars compared to nasal discs for marking geese. *J. Wildl. Mngmt.* 30, 853-855.

Smith E. N. and Salb T. J., Jr. (1975) Multichannel subcarrier ECG, respiration, and temperature biotelemetry system. *J. Appl. Physiol.* 39, 331-334.

Stewart P. A. (1962) Nesting attentiveness and incubation period of a wood duck. *Bird-Banding* 33, 85-89.

Will G. B. and Patric E. F. (1972) A contribution toward a bibliography on wildlife telemetry and radio tracking. New York Fed. Aid in Fish and Wildl. Rest. Proj. W-123-R and Rhode Island Agric. Exp. Stn. Contrib. 1439. 56 pp.

Continuous Observation of Individual Herring Gulls during the Incubation Season Using Radio Tags: An Evaluation of the Technique and a Cost-Benefit Analysis of Transmitter Power

R. M. SIBLY* and R. H. McCLEERY

Animal Behaviour Research Group, Department of Zoology, University of Oxford, South Parks Road, Oxford OX1 3PS, U.K.

Abstract — We aimed to observe mated pairs continuously when they were off their territories during the incubation season. Important gaps in our observations were of 3 types, when (1) we lost a bird (3 percent); (2) we were catching up to a bird that had flown away (1 percent); and (3) we were at the site but could not distinguish the bird's behavior (4 percent). More powerful radio tags reduce type 1 gaps, and have less effect on the bird's breeding success than being caught, provided that the weight of the package is less than 30 g. We discuss the biases introduced by these gaps in our observations and try to allow for them.

INTRODUCTION

A prerequisite of any behavior study is observation of behavior in natural conditions. Since we wished to study the behavior off territory of parent herring gulls during the incubation season, we began by trying to watch mated pairs as closely as possible. Without a tag of some kind, such an enterprise would be impossible, especially in a large colony such as that in which we worked on Walney Island, Cumbria, U.K., which we now estimate to be 20,000-30,000 pairs of herring gulls (*Larus argentatus*) nesting with a similar number of lesser black-backed gulls (*Larus fuscus*). Special advantages of herring gulls as study animals are (1) large size, (2) accessible nests, (3) relative tolerance of man, (4) relative accessibility of sites utilized, and (5) diurnality.

The questions we shall pose here are (1) How near to continuous are the observations we made? (2) What biases are introduced because of gaps in our observations? (3) What is the best trade-off of transmitter power versus weight?

MATERIALS

We attached radio tags to herring gulls breeding at the gull colony at South Walney as follows: in 1975 (pilot observations) 3 pairs and 2 singletons; in 1976, 5 pairs and 4 singletons; and in 1977, 7 pairs and 4 singletons. 39 birds were caught on the

---

*Present address: Department of Zoology, University of Glasgow.

nest within a week of laying using a walk-in trap (Amlaner *et al.*, 1978) (see Fig. 1) and 1 was caught before laying by baiting (1 g Avertin in a 10 cm^2 bread 'sandwich') under Nature Conservancy license. Each bird was measured, tagged, dyed yellow on the neck feathers with picric acid, and released within a few hours of catching: we then protected the clutch until the bird returned. In 13 cases the birds were later recaught to recover the transmitters.

We used 2 types of transmitter operating in the 173.2-173.5 MHz band. 15 two-stage transmitters (weighing 8 g) supplied by AVM Instrument Co. (model SB2) with mercury batteries supplied by Mallory (TR-112N, 8 g); 13 one-stage transmitters (2 g) built in our workshops with mercury batteries (RM-675, 2 g).

To the transmitter was attached a 30 cm whip antenna made of 18-gauge stainless steel spring wire and a 30 cm Teflon coated multistrand ground plane wire. In our early packages Araldite completely enclosed the transmitter and battery and the harness was made from two-strand plastic coated wire: this packaging weighed about 40 g. In a later version Araldite did not entirely cover all surfaces but was used to provide structural strength, and the harness was made from nylon coated elastic webbing 3 mm wide: this packaging weighed 7-10 g. The harness was wrapped firmly around the bird in a 'figure of eight' pattern crossing the breast in such a way that the radio pack lay on the gull's back at the base of the neck, the whip antenna lay parallel to the tail, and the earth wire was sewn to the harness.

For tracking we used two Landrovers equipped with null-peak antenna systems consisting of twin four-element Yagis mounted on a rotating boom (Fig. 2.), and AVM type LA12 receivers. For most of the time the vehicle served as a mobile hide; for tracking on foot handheld three-element Yagis were used. Another receiver was used in the field station at the colony, with 2 alarm systems both based on a type 741 operational amplifier used as a voltage comparator (Fig. 3). One was triggered by the disappearance of a radio signal, the other by a nest relief monitored by a balance set in the ground under the nest. We received additional field help from Dr. M. I. Webber in 1976 and Dr. R. Ashcroft in 1977.

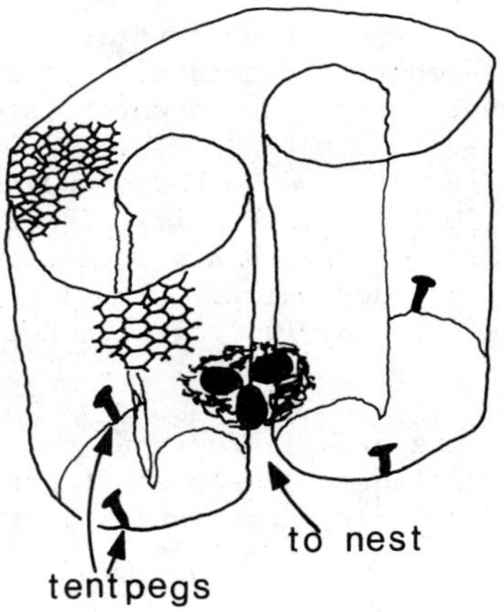

Fig. 1. The walk-in trap to catch birds on the nest made of 'chicken-wire' roughly 1 m high. When the catcher appears from the direction of the entrance the bird attempts to escape through one of the 'false exits' (Amlaner *et al.*, 1978).

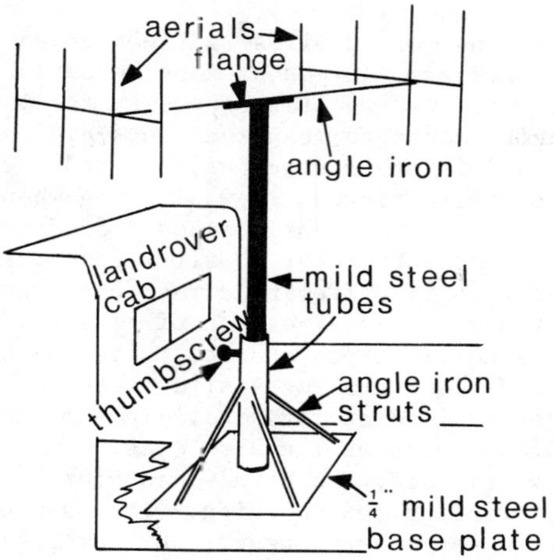

Fig. 2. The null-peak antenna system as mounted in our Landrovers. Joints were welded and bearings were fitted in base tube. The thumbscrew had a very heavy thread and its hole was reinforced; angle iron was 1" × 1" × 1/8".

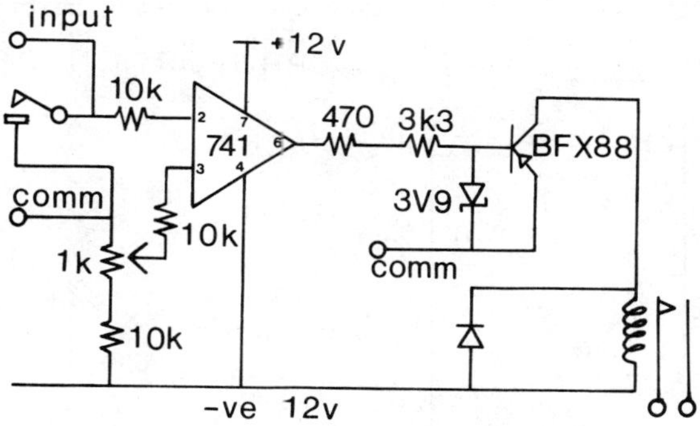

Fig. 3. The alarm circuit. When the bird moves the radio signal fades, the input (from the receiver) becomes less negative than the threshold voltage (set by using the variable resistor, 1 kΩ wire wound, ten turn) and the relay closes ringing a domestic doorbell. The circuit will also operate with two 9 V radio batteries (PP9 or equivalent).

METHOD

Our plan of action was to note the direction the bird went when the alarm sounded, to try to find the bird, and then to note (using a tape recorder) the times at which it performed feeding, preening, resting, drinking, and bathing. Time resolution was 1 min and short gaps in the observers' attention were ignored if they lasted less than this time. Brief notes were also made about the food taken, the feeding site, the number and behavior of other gulls at the site, and the weather. No observations were made of the behavior of the bird on territory.

RESULTS

We made observations on 14 out of 22 birds that continued incubation in a normal fashion when released. There are 4 types of gaps in our observations of birds off the territory (Fig. 4). Firstly, we sometimes lost the bird. For present purposes we use *missing* in a special sense to mean that we could not make radio contact for a period of more than 2 h — in other cases of lost radio contact we inferred the location of the bird from the direction it was flying when contact was broken or from the site on which it was subsequently found. 5 of the 9 birds of Fig. 4 only used local sites. We use *local* to refer to sites less than 25 km by road from the colony. Non-local sites were in different directions from local sites and these destinations were inferred from the direction of flight. No attempt was made to follow birds to non-local destinations and the criterion of missing, above, was not then applied. When a bird flew to a local site there was typically a delay, *catch-up time*, before the observer arrived at the site. In some cases this gap seriously biases our assessment of behavior at the site (see below). When at the site the observer could not always classify the bird's behavior into the categories of feeding, preening, resting, drinking, and bathing. The bulk of this discrepancy (Fig. 4) arose during night observation when, however, we know from the nature of the site that the bird could not have fed. The observer was able to make inferences about the bird's behavior without seeing it if (1) the frequency of the radio signal changed in a way he knew to be characteristic of bathing, (2) the bird was known to be in a group all members of which were behaving identically, or (3) the site precluded feeding and use of water.

Thus there are 3 major sources of bias in Fig. 4, arising from the time the bird was missing, catch-up time, and time that behavior was not classfied when the observer was at the site. We shall use these measures to assess the benefits of stronger radio signals.

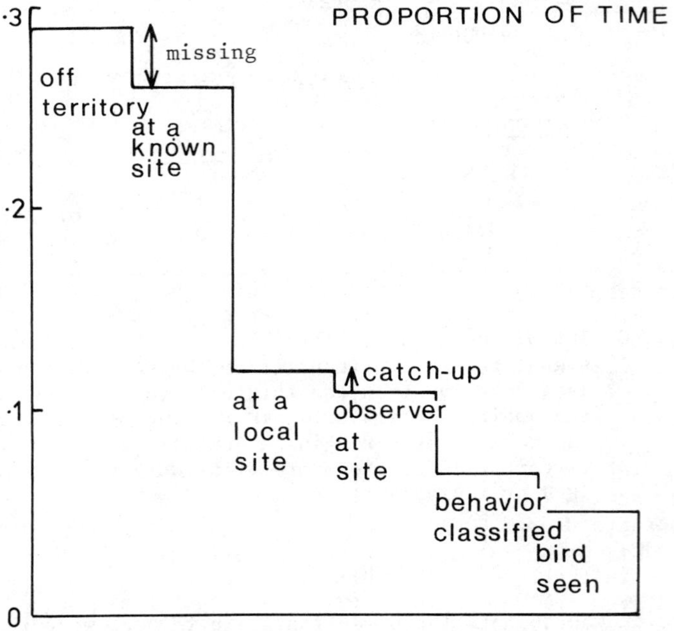

Fig. 4. Overview of results of tracking 9 birds for 1530 h (strong signals). The most serious gaps in our observations are those when birds were missing and those when the observer was catching-up (see Table 4) with the animal.

COST-BENEFIT ANALYSIS OF TRANSMITTER POWER

We classified our one- and two-stage transmitters as producing *strong signals* (9 birds) if the range when the bird was flying was about 15 km or *weak signals* (5 birds) if the range was 5 km. It can be seen from Table 1 that radio tags producing stronger signals went missing considerably less than weak signals. On the other hand catch-up times at local sites (Table 2) were not consistently lower. These figures are medians; for comparison we provide also the average visit durations. Note that catch-up time is only a substantial proportion of visit duration when the destination was 'worms' or 'within colony' (see below). The *tip* (4 km road distance) is the local municipal refuse tip; *worms* (15-25 km) refers to agricultural land, mainly pasture fields, on the mainland covering an area of about 50 km^2; *nearby mussels* (1 km) refers to a mussel skear on the island's beach, *bay* (5 km) refers to feeding sites exposed at low tide in Morecambe Bay; *beach* (1 km) is the island's beach; *within colony* (1 km) refers to (probably watering) sites within the colony.

Table 1.
The benefits of stronger radio signals: 1. Less often missing

Signals	% Time missing	(Hours followed)
Strong	3	(1530)
Weak	14	(1140)

Table 2.
The benefits of stronger radio signals: 2. No help in catching-up with birds at local sites (catch-up time in min.)

Signals	Tip	Worms	Nearby mussels	Bay	Beach	Within colony
Strong	11	85	15	4	10	10
Weak	8	90	8	28	4	7
Visit duration (min)						
	212	108	132	158	69	17

Stronger signals can only be produced by increasing the weight of the package attached to the bird. Package characteristics might affect the bird's behavior in a number of ways: increasing the amount of preening, affecting the choice of feeding site, decreasing the amount of time spent on territory, or increasing the amount of time spent resting on the beach. It is quite likely that any of these would lower the breeding success of a pair because clutch predation is high in the Walney colony.

We have used the survivorship of at least one member of the clutch as a measure of the effect of a package (Amlaner *et al.*, 1978). Figure 5a shows the effects of catching both parents and attaching packages of different weight. The effects can be compared directly since all pairs were caught in the same area at the same time, and differed only in their treatment which was chosen at random. The radio tags in Fig. 5b differed from the dummy tags of Fig. 5a, in that a 30 cm antenna protruded from them, and a groundplane wire was sewn into the harness. In addition the neck feathers of birds receiving radio tags were dyed yellow and their nests were disturbed by being placed on balances buried in the ground. These birds nested in a different area of the colony from those of Fig. 5a at various times over 3 seasons so that, amongst other things, the amount of disturbance near their nests may have differed from that of Fig. 5a.

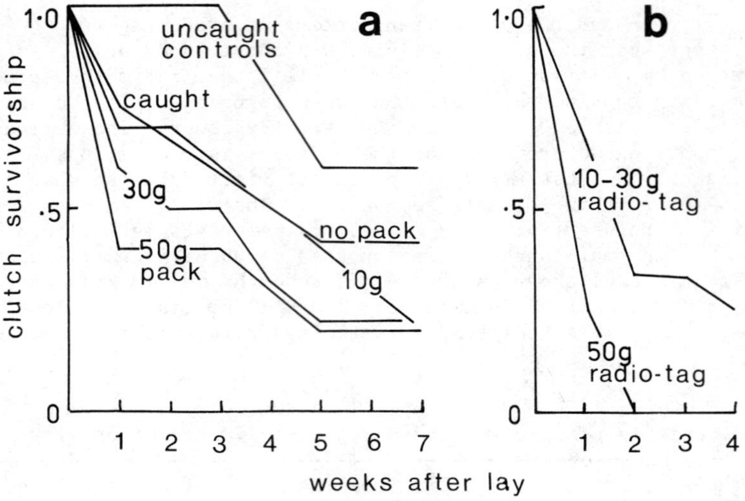

Fig. 5. The cost of heavier radio tags: clutch survivorship was lower when heavier tags were attached to both parents just after the clutch was completed. 10 pairs in each treatment, except that only 5 received 50 g radio tags. (a) The trauma of catching the bird had more effect than the attachment of dummy tags weighing up to 50 g (Walney herring gulls weigh 800-1000 g) (Amlaner *et al.*, 1978). (b) The effect of radio tags was slightly greater, possibly because a 30 cm antenna and a groundplane wire protruded from them (see text for details).

DISCUSSION

Although stronger signals did not increase the proportion of time that a bird's behavior could be classified (Table 3) or catch-up time (Table 2) it did substantially reduce the time that birds went missing (Table 1) and we therefore conclude that stronger radio signals are well worth having, provided that the weight of the package does not exceed 30 g (Fig. 5), about 3-4 percent of body weight.

Table 3.
The benefits of stronger radio signals: 3. No help to observers at local sites in classifying behavior of birds

	% time behavior classified					
Signals	Tip	Worms*	Nearby mussels	Bay	Beach	Within colony
Strong	82	61*	79	0	34	100
Weak	75	85*	68	0	79	100

*Birds were observed for a much lower percent of time.

Our success in tracking with strong signals is underestimated in Fig. 4, since over half of the birds spent all their time at local sites. Nevertheless there are 3 serious sources of bias (Table 4) in attempting to reconstruct behavior sequences we missed. The accuracy of such a reconstruction will depend on what it is required for, but in some cases it would serve for an evaluation of, for example, the amount of time spent feeding, resting, or preening at each site. The effect of bias must be considered separately in every case. Our general preliminary assessment is given in Table 5. The birds followed spent most of their time feeding, preening or resting so that we can estimate the bias and allow for it for each individual. On the other hand drinking and bathing occur much more rapidly, often within seconds of arriving at a site, before the observer had arrived. We certainly cannot estimate the bias in our observations of bathing and drinking, except at the tip.

Table 4.
Can we reconstruct unobserved behavior sequences?

Source of bias	Can bias be allowed for?
1. Bird missing	No
2. Catch-up delay	Yes if behavior at destination is stereotyped
3. Visibility of bird depends on its behavior	Yes if behavior at 'microsites' is stereotyped

Table 5.
Can we reconstruct unobserved behavior sequences?

Behavior	Can bias be allowed for?
Feeding	Yes
Preening	Yes
Resting	Yes
Drinking } Bathing	Only by using observations on other randomly chosen gulls to construct an 'expected' behavior sequence

Could the sources of bias (Table 4) be eliminated? Certainly increasing the number of observers would help. Nevertheless we estimate that at least 6 observers would be required to watch simultaneously at different sites (i.e. a team of 6-12) in order to drastically reduce these biases. At least some of them would require Landrovers and/or boats, and all would need radio tracking equipment and two-way radio communication.

What would be the result of longer-term tracking? The weak points of our technique would include breakage of whip antenna, harness, battery and water-proofing. Few of our harnesses lasted more than a year. We are indebted to N. Ball for the following information on the fates of 7 single birds caught with Avertin bait in the colony in February, 1978, to which were attached 45 g transmitter packages powered by solar-panel band rechargeable NICAD batteries supplied by Telemetry Systems, Inc. These packages had inferior 30 cm whip antennas and were attached as above by 1 cm nylon-coated elastic webbing. One harness broke in April, 1978, and one male died in May, 1978. One male was not heard after June, 1978. In April, 1979 two males and two females were still heard, but all of their antennas had broken off. Only one of the four (a male) was defending a territory. Transmitter performance was much lower than expected: at most 3 km when the bird was flying, 0-1 km when on the ground, but not reduced overnight.

This model was obviously of limited value, but we suspect that most of the problems will be overcome in time. Most worrying to us was the low reproductive success, which we ascribe to the heavy potting material and bulky solar panel.

CONCLUSION

We consider that, given our resources, our method and technique worked well, and gave us some idea of the continuous behavior sequences of individual herring gulls. The limitations of the method are demonstrated in Table 5, but we ought to conclude by pointing out its unique strengths: the data gathered are qualitatively different and new, and could not be obtained in any other way.

Acknowledgements — This project was financed by NERC grant to D. J. McFarland while R. M. Sibly was Ernest Cook Trust Research Fellow of the Royal Society. Workshop facilities were provided by Department of Zoology, University of Oxford. We are very grateful to R. Cheney, A. Price, M. Dolan and T. Barker for constructing equipment, to Mike Webber and Ruth Ashcroft for their generous and enthusiastic support in the field, to the Cumbria Naturalist's Trust for allowing us access to sites, and to all those who helped us in other ways.

REFERENCES

Amlaner C. J., Sibly R. M. and McCleery R. H. (1978) The effects of telemetry transmitter weight on breeding success in Herring gulls. *Biotelemetry Patient Monitg.* 5, 154-163.

III. Evaluation, Data Acqusition and Analysis

A Comment on the Use of Radio Tracking in Ecological Research

A. N. LANCE* and A. WATSON**

**Nature Conservancy Council, 12 Hope Terrace, Edinburgh, U.K.*
***Institute of Terrestrial Ecology, Blackhall, Banchory, U.K.*

Abstract — Radio tracking was widely heralded when it first appeared in ecology two decades ago, but in our opinion its use has seldom exploited its full potential. We suggest that this has been due less to drawbacks in the equipment than to how the technique has tended to be used. Perusal of the ecological literature suggests that many studies based on radio tracking have been mere descriptions of movements and activity, with no explicit hypothesis testing.

Better transmitter design and construction have overcome several of the equipment's early failings. An example of improvements in performance is given from our own field experience. Developments in microcircuitry will bring further improvements, but these alone will not make radio tracking's contribution to ecology more effective unless users look beyond the temptation of studies which just exploit the technique's efficiency. Some varied ecological topics well suited to study by telemetry are listed.

INTRODUCTION

Ever since radio telemetry was first adapted for biological research, ecologists have realized that tracking a radio tagged animal might be more efficient and predictable than observations from sightings or trapping. The ability to locate and follow free ranging animals is essential to most wildlife studies. Potentially, radio tracking can show the whereabouts of an animal continuously, and with minimal disruption of its behavior. However, as an advantage to research, a technique is only as useful as the solutions to the problems to which it is applied. In the 16 years since the first major symposium on biotelemetry (Slater, 1963), radio tracking has become well established in ecology. It seems appropriate in this present symposium to ask whether the technique's achievements have lived up to expectations.

There are two main aspects to this question: (i) technical performance records and (ii) radio tracking's contribution to the study of important ecological problems. We are not concerned here with the technical principles of radio tracking, with the merits of different types of equipment, or with biotelemetry (which is more the province of environmental physiology). Our concern instead is with the usefulness of radio tracking as a tool in ecological research. Our suggestion is that this has been less satisfying thus far than might have been expected, not so much because of deficiencies in the equipment, but because of how the technique has often been employed.

TECHNICAL PERFORMANCE

The main theoretical constraints on the performance of radio tracking equipment are signal range and life span. The actual constraints on the user are more numerous, however, and include: (i) the cost and availability of equipment compared with that of more conventional techniques; (ii) breakages and other malfunctions; (iii) the difficulties of effecting repairs or replacements while in the field; (iv) possible adverse effects on the animal; (v) the restricted size, weight, (and hence performance) of transmitters necessary to minimize iv; (vi) the ways in which the decision to use radio tracking can affect the logistics of fieldwork. The limitations in vi are often the most serious of all.

Premature transmitter failure can be frustrating if it interrupts a valuable field experiment, but better design and structure have overcome several of the commoner causes of failure. The improvement in transmitter performance shown in Table 1 is an example from our own experience. Also, because more commercial suppliers are

Table 1.
Improvements in signal-life of transmitters of similar basic design, during three successive ecological studies*

	Study 1 (1966)		Study 2 (1968-70)		Study 3 (1971-72)	
	No.	%	No.	%	No.	%
Transmitters fitted+	9	-	34	-	8	-
Transmitters achieving full expected life^x	0	0	9	38	4	67
Transmitters achieving 75 percent of expected life	2	22	11	44	5	83
Transmitters achieving 50 percent of expected life	4	44	24	89	5	83
Sources or premature failure‡						
early battery exhaustion	6		6		1	
antenna breakage	2		4		1	
damage by predator attack	0		4		2	

*Study 1: Lance (1970). Study 2: Lance (1978). Study 3: Lance (1971).
+Some of the transmitters fitted during studies 2 and 3 were removed while they were still functioning.
^xMean expected transmitting life was 60 days with 1 mW signal output from a 1.4 V 800 mAh mercury cell.
‡Source of failure could not be determined for all transmitters. The differences between studies in the proportions of transmitters achieving full expected transmitting life are significant at $P < 0.025$ in a 2 × 3 contingency table. The differences in proportions achieving 75 percent of expected life are significant at $P < 0.10$.

now in the market, radio tracking equipment is easier to obtain than before, and replacements and repairs at short notice are easier to arrange. As judged by the little published evidence to the contrary, effects on the animals have been less problematic than once feared. It is surprising, however, that so few published studies involving radio tracking have included evidence such as that provided by Leuze (1980, this volume) and Smith (1980, this volume). Admittedly, testing may be difficult when radio tracking is the only effective way of getting the information, but this does not absolve the basic need to any research project, to ensure that the technique gives unbiased results.

Of the consideration in vi, the following is perhaps the most significant. Even when catching and radio marking pose no practical difficulties, one person using a portable receiver (the most common situation) can seldom track more than 10-12 animals at frequent intervals, and still have time for other kinds of fieldwork. With all his time and effort committed to this one method, the success of his study may depend entirely on the data which radio tracking can given him. Accidents, predation, and transmitter failure may reduce his already small sample enough to leave him with difficulties in statistical analysis, or he may have to restrict the scope of his study—perhaps both. Two possible tactics are to increase the number of field staff, or to resort to automated equipment (Cochran, Warner and Tester, 1964), but these increase the study's costs. Brander and Cochran (1971) rightly stress cost-effectiveness as a primary requisite for radio tracking studies. Cost and logistics may partly explain why so many telemetry based studies have been confined to the movements, habitat selection, or home range of a few individuals for which large amounts of information can be obtained easily. We suggest, however, that this is not the complete explanation.

RADIO TRACKING'S UNFULFILLED POTENTIAL

In one bibliography of radio tracking during the technique's first decade in ecology (Schladweiler and Ball, 1968), more than three-quarters of the entries appear to have concerned studies primarily on activity or movements (Table 2). If studies of this kind are so common in the literature, a prospective user of radio tracking

Table 2.
Types of studies based on radio tracking, during the period up to 1968[+]

	No.
Total number of studies based on radio tracking[x]	71
Movements or activity as the subject[‡]	53
Other types of behavior as the subject	4
Breeding or mortality as the subject	7
General ecological studies	7

[+]Data from part I of Schladweiler and Ball (1968).
[x]As inferred from study title, excluding papers on telemetry systems development, migration and navigation.
[‡]Combined total for studies on home range, habitat use, and movement and activity patterns.

might be forgiven for thinking that they were the technique's most important application. But activity and movements are just two relatively simple kinds of information which can be derived from radio tracking, and should not be confused with its essential function which is merely to relocate animals efficiently. Accordingly, the technique's potential usefulness is limited less by its performance than by the type of research question that is asked. Radio location offers a means of making many kinds of observations more efficiently, and should not be just an end in itself (Macdonald, 1978). Furthermore, while aiding other observations, radio tracking provides data on locations and movement as an incidental bonus anyway.

Some varied topics for which radio tracking can be used to advantage are: (a) Survival studies of secretive animals (Cook *et al.*, 1967); (b) finding animals at specific times, such as feeding periods (Robel, 1969a); (c) longterm studies of, for example, growth or social behavior (Lance, 1978); (d) finding animals at successive phases of their performance, in order to compare the phases (Lance, 1978); (e) comparing movements or activity with any of a-d (Robel, 1969b; Newton and Marquiss, in press); (f) comparison of individual differences with conspecifics (Lance, 1978).

Since radio tracking is done with individual animals, it is well suited to the study of differences between individuals. Differences in individual performance have been rather neglected by ecologists generally, and not just by those who use radio tracking. Opposing assumptions about the status of different individuals in a population have been implicit in the rival views of some population ecologists for many years. Do differences amongst the individual members determine their population's performance, or is individual performance no more than a random sample of the population's performance? This is a question requiring the study of individuals in depth, a virtually impossible task if it is difficult to keep in touch with them. Radio tracking offers one of the best means available for doing so. Statistical obstacles may occur (Dunn and Gipson, 1977; Macdonald *et al.*, 1980, this volume; Voigt and Tinline, 1979, this volume), but these are not sufficient reason for ignoring the question. If they are anticipated at the outset, at least some statistical difficulties can be reduced by stratified sampling, or by manipulating individuals' performance in an experiment.

Radio tracking used in the foregoing ways would contrast sharply with the temptation t use it indiscriminately; e.g. with radios fitted unselectively and information merely gathered *ad hoc*. As the interpretation of *ad hoc* data must also be *ad hoc*, such a use of radio tracking is unscientific.

RADIO TRACKING IN THE FUTURE

During its growing popularity in recent years, radio tracking equipment has been made smaller, cheaper, more dependable, and more efficient. Microcircuit technology may bring further improvements of this kind, and the use of satellites, solar powered transmitters, and computerized receivers may become more practicable than at present. Sophistications like these may make radio tracking more efficient as a research tool, and perhaps more prestigious to the user, but we do not believe that they will make it more effective to ecology. To fulfil its real potential, the technique must be applied more often to the more important problems of ecology, and less often to those which merely capitalize on the technique's convenience. Many of the studies reported elsewhere in this volume provide good examples of the way ahead.

Acknowledgements — The National Research Council of Canada, the Natural Environment Research Council in Britain, and the Forest and Wildlife Service in the Republic of Ireland gave grants for the radio tracking equipment used in Table 1. We thank Dr. Ian Newton for useful discussion.

REFERENCES

Brander R. B. and Cochran W. W. (1971) Radio-location telemetry. In R. H. Giles (ed.), *Wildlife Management Techniques*, 3rd ed., Chapter 9, pp. 95-103. Washington, D.C.: The Wildlife Society.

Cochran W. W., Warner D. W. and Tester J. R. (1964) The Cedar Creek automatic radio-tracking system. Univ. Minnesota Nat. Hist. Mus. Tech. Rep. 7.

Cook R. S., White M., Trainer D. O. and Glazener W. C. (1967) Radiotelemetry for fawn mortality studies. *Bull. Wildl. Diseases Ass.* 3, 160-165.

Dunn J. and Gipson P. (1977) Analysis of radio telemetry data in studies of home range. *Biometrics* 33, 85-101.

Lance A. N. (1970) Movements of blue grouse on the summer range. *Condor* 72, 437-444.

Lance A. N. (1971) Research on red grouse. Animal Production; Research Rep. for 1971, pp. 115-116. Dublin: The Agricultural Institute.

Lance A. N. (1978) Survival and recruitment success of individual young cock red grouse *Lagopus l. scoticus* tracked by radio-telemetry. *Ibis* 120, 369-378.

Leuze C. C. K. (1980) The application of radio tracking and its effect on the behavioral ecology of the water vole, *Arvicola terrestris* (Lacépède). In C. J. Amlaner, Jr. and D. W. Macdonald (eds.), *A Handbook on Biotelemetry and Radio Tracking*, pp. 361 - 366. Oxford: Pergamon Press.

Macdonald D. W. (1978) Radio-tracking: some applications and limitations. In B. Stonehouse (ed.), *Animal Marking: Recognition Marking of Animals in Research*, pp. 192-204. London: Macmillan.

Macdonald D. W., Ball F. and Hough N. G. (1980) The evaluation of home range size and configuration from radio tracking data. In C. J. Amlaner, Jr. and D. W. Macdonald (eds.), *A Handbook on Biotelemetry and Radio Tracking*, pp. 405-424. Oxford: Pergamon Press.

Newton I. and Marquiss M. Research on sparrowhawks. Inst. of Terrestrial Ecol. Ann. Rep. for 1978. Cambridge: Institute of Terrestrial Ecology, in press.

Robel R. J. (1969a) Nesting activities and brood movements of black grouse in Scotland. *Ibis* 111, 395-399.

Robel R. J. (1969b) Movements and flock stratification within a population of blackcocks in Scotland. *J. Anim. Ecol.* 38, 755-763.

Schladweiler J. L. and Ball I. J. (1968) Telemetry bibliography emphasizing studies on wild animals under natural conditions. Univ. Minnesota Nat. Hist. Mus. Tech. Rep. 15.

Slater L. (ed.). (1963) *Bio-telemetry*. New York: Pergamon Press.

Smith H. R. (1980) Growth, reproduction and survival in *Peromyscus* carrying intraperitoneally implanted transmitters. In C. J. Amlaner, Jr. and D. W. Macdonald (eds.), *A Handbook on Biotelemetry and Radio Tracking*, pp. 367-374. Oxford: Pergamon Press.

Voigt D. R. and Tinline R. R. (1980) Strategies for analyzing radio tracking data. In C. J. Amlaner, Jr. and D. W. Macdonald, *A Handbook on Biotelemetry and Radio Tracking*, pp. 387-404. Oxford: Pergamon Press.

The Application of Radio Tracking and Its Effect on the Behavioral Ecology of the Water Vole, *Arvicola terrestris* (Lacépède)

C. C. K. LEUZE

Culterty Field Station, Newburgh, Ellon, Aberdeenshire AB4 0AA, U.K.

Abstract – Radio tracking was used as the main technique in a study of dispersion and social organization of the water vole, and the casual mechanisms relating these parameters. 217 water voles were radio tracked, for at least 2 months each, including 87 juveniles for at least 7 months each, and 40 adults for at least 12 months each. A simple, low cost, and home made tracking system is described. The linear movements of water voles along the edges of waterways permitted signal intensity alone to be used for location. The effect of the radio collar on radio tagged individuals was measured in the field by direct observation. Performance in five parameters was compared before and after the attachment of the collar: activity above ground was reduced for 24 h but thereafter was normal. Range, territory size, social status, reproductive success in females, and bodyweight in juveniles and adult males were unaffected. Radio tracking was used to measure dispersion and nearest neighbor distances. A most important aspect of the technique was that it permitted field experiments of a complexity comparable to laboratory approaches.

INTRODUCTION

In the present paper I shall describe the logistics of the radio tracking technique used to study dispersion and social organization and the casual mechanisms relating these two parameters. I will briefly describe firstly, the radio collar and Ashwell's circuit, secondly, the effect of this collar on five parameters of the individual's performance relevant to the scientific aims of this study and, thirdly, how I applied the technique in the field and sampled radio locations. The results of the main study will not be dealt with, but rather how the logistics were developed in relation to the questions asked will be discussed.

RADIO TRACKING TECHNIQUES

The low cost and easily home made radio collar is based on a simple oscillator circuit without crystal control of its frequency (Ashwell-Transmitter, Fig. 1a). It had a band width of 3 MHz and transmitted on the commercial VHF band (88-168 MHz Bambino

Challenger). Varying the frequency output and click rate allowed me to operate, on average, 43 transmitters concurrently within the same area. To describe population dispersion of small mammals it was an essential requirement that many transmitters be operated concurrently. The coil antenna of the transmitter (Fig. 1b) provided good vertical polarization which improved reception of the signal in dense vegetation, at the expense of signal range (maximally 180 m).

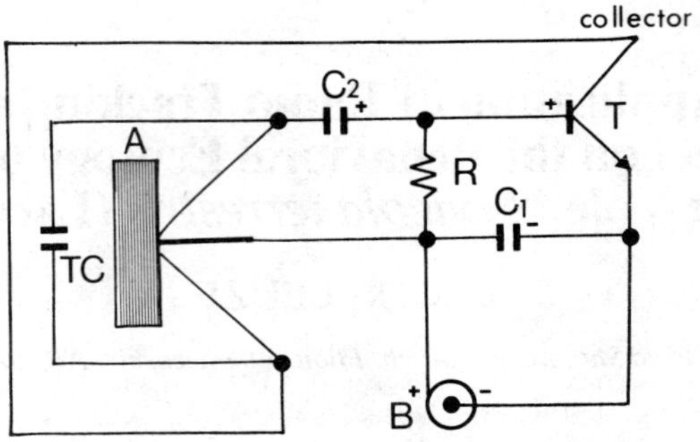

Fig. 1a. Ashwell transmitter circuit. (TC) Tuning capacitor (5-39 p*F*), (R) resistor (220-330 kΩ), (C_1) capacitor (1000 p*F*), (C_2) capacitor (click rate control together with resistor, 0.47-4.7 μF), (T) Transistor, NPN, ZTX 320, (A) antenna (see Fig. 1b), (B) 2 g Mallory Battery (1.5 V; due to leakage usually 1.35 V).

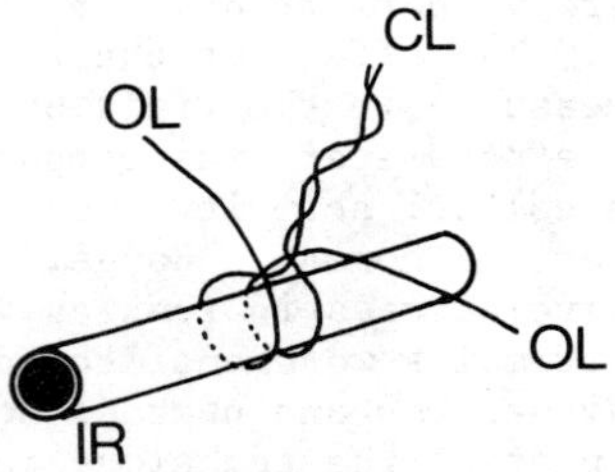

Fig. 1b. Diagram of the windings of the coil antenna of the transmitter. (IR) Iron resin compound cylinder, (CL) center lead (286 enamel coated wire), (OL) outer lead.

To assemble the circuit, I glued the antenna, C_2, and the transistor in a row on the collar (plastic cable tie with a nonrelease ratchet). The remaining components were positioned in the resulting gaps and were held by their solder joints. The battery (2 g Mallory, field life 23 days) was glued to one end of the row to permit easy exchange and was connected with conductive epoxy. I encased the unit in several thin layers of silicone rubber ('Bathtub Caulk'). The total weight was 5.6-6.1 g, being maximally 6 percent and on the average 3 percent, of the bodyweight of the water voles.

Since water voles move in linear ranges along the banks of waterways, I could use the intensity of the signal for its location. The signal intensity dropped drastically about 2 m from its source and the location error was within 1 m. The antenna provided with the commercial receiver was sufficient for my purposes. A detailed manual for the construction of the unit is available on request.

THE EFFECT OF THE RADIO COLLAR ON WATER VOLES

INTRODUCTION

An assessment of the effect of the technique on the parameters to be studied is very rarely given. Tester (1971) found no published reports on small mammals, but noted 'abnormal' behavior for 2 days after attachment. Hamely and Falls (1975) showed that the activity of *Microtus* was reduced for up to 10 days after attachment. I have measured the impact of the radio collar on the individual's performance in five parameters.

MATERIAL AND METHODS

The measurement of the effect of the radio collar was based on the comparison of individuals in the field marked for identification prior to radio tracking by fur clipping. Their behavior was measured before and after the attachment of the collar. Since radio packages make individuals more obvious than does fur clipping, I have also tested the parameters described below in individuals wearing color coded collars (0.3 g). These were as visible as radio packages and I have found no observer bias between the fur clipped and color tagged animals. The parameters selected for the assessment were crucial to the analysis of the behavioral ecology of this species. They were also chosen to be representative of the effects on: firstly, short and longterm activity (activity above ground, range and territory size over two months), secondly, on physiological capability (reproductive success in adult females, body-weight in adult males and juveniles) and thirdly, on social status (interactions won and lost when carrying and not carrying a collar). For the field techniques used to measure these parameters see Leuze (1976).

RESULTS

SHORT AND LONGTERM ACTIVITY

The activity shown above ground by 16 fur clipped water voles was reduced for 24 h after the attachment of the radio collar (Fig. 2), but was thereafter the same as before attachment. I have lumped the results for the four sex and age classes because differences between individuals were greater than between classes.

Territory size per 2 months in 14 adult females was similar when carrying (241.7 m, SD 59.5) and not carrying a transmitter (243.6 m, SD 57.9) (Mann-Whitney U-Test: $U = 104$, $P > 0.05$). The same result was obtained for range sizes in adult males and juvenile females. Corresponding results from juvenile males are lacking since their long distance movements necessitated continuous radio tracking.

PHYSIOLOGICAL CAPABILITY

Reproductive success in fur clipped adult females, when carrying and not carrying a radio collar over the whole of the gestation and lactation period (35 days, Leuze, 1976), was measured in number of weanlings first seen with each adult female (Leuze, 1976). Females carrying a collar produced 7.03 weanlings (SD 1.26, $n = 24$) and those not carrying a collar produce 7.42 (SD 1.68, $n = 13$). Reproductive success thus did not differ significantly between these categories (Mann-Whitney U-Test: $U = 155$, $z = 0.032$, $P = 0.49$).

Changes in body weight (± g) over one month did not differ significantly between animals with collars and those without in adult males (Mann-Whitney U-Test: $U = 105$, $P > 0.05$), in juvenile males ($U = 56$, $P > 0.05$) and in juvenile females ($U = 58$, $P > 0.05$).

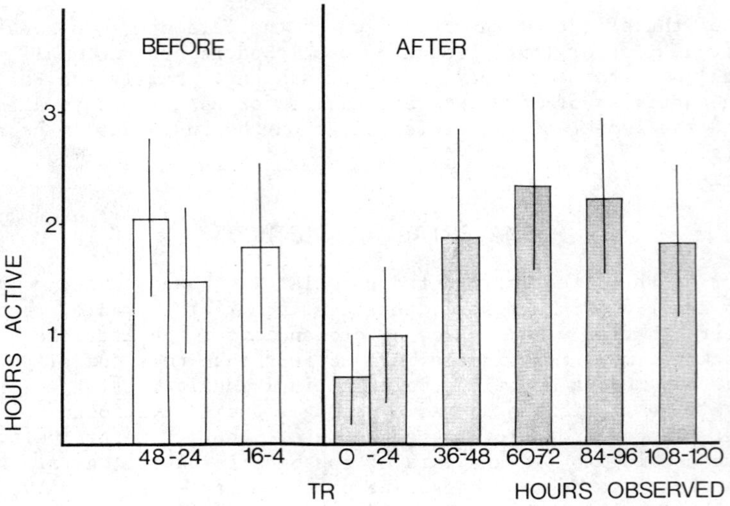

Fig. 2. The activity shown above ground by 16 water voles before and after the attachment of the radio collar recorded throughout by direct observation. (TR) Trapping (within 4 h) and attachment of collar.

SOCIAL STATUS

For the analysis of the impact of the radio collar on social status I selected 7 pairs of adult females in which the members were of equal social status and not wearing a radio collar. After attachment of a radio collar to one member of the pair I observed whether or not equality was maintained in all pairs. I have selected pairs of females with equal status since it must be expected that observable changes due to the impact of the collar would be more difficult to detect in females of greatly different social status. Within all 7 pairs equality was maintained after the attachment of the collar to one individual. The same results were obtained for adult males and juvenile females. A corresponding analysis was impossible for juvenile males because they were never seen to be involved in agonistic interactions.

DISCUSSION

None of the parameters measured gave indications of longterm debilitating effects of the radio collar on the animal carrying it. Although the activity shown above ground was significantly reduced for the first 24 h after the attachment of the collar, the activity levels were normal within 48 h. This indicates that no long-term adjustment was needed for normal activity to proceed. Reproductive success, range and territory size, body weight and social status were unaffected by this technique. These favorable results, in comparison to Hamley and Falls' (1975), may be due to the small relative size of the radio collar. The only aberrant behavior I noticed was the pushing of the collar around the neck so that the radio lay on the back of the neck.

THE APPLICATION OF RADIO TRACKING

MEASUREMENT OF DISPERSION AND DISPERSAL

To reveal range size and location in those individuals which maintained a stable monthly range I used discontinuous radio tracking, with equal tracking times spent in each 10 m interval of waterway, until the increase of range size with increasing tracking time had levelled off (Odum and Kuenzler, 1955). Radio locations were plotted on large scale maps.

Dispersing juvenile females swam 3-4 times as much as resident ones and moved, usually without interruption, until they entered marginal unoccupied habitat. I followed dispersing juvenile females continuously by radio tracking from a canoe at a distance of about 100 m. Predatory attacks by herons, *Ardea cinerea*, barn owls, *Tyto alba*, and pike, *Esox lucius*, were observed frequently while following these females and about 50 percent of all dispersing females were preyed upon. The maximum distance moved by these females up to the point of predator contact was 1257 m, calculated from the center of the previous home range. Resident juvenile females, wearing the same radio collars as dispersing ones, had 86× less chance of being preyed upon. In contrast, dispersing juvenile males moved slowly, spending days with different mother/daughter groups and the radio tracking of their movement was basically as for residents.

I required, on average, 62.4 h of radio tracking per month to reveal 87 ranges. During the study I radio tracked 217 individuals each for at least 2 months, including 87 juveniles for at least 7 months, and 40 adults for at least 12 months.

NEAREST NEIGHBOR DISTANCES

In adult males, whose ranges overlapped completely, I employed nearest neighbor techniques to analyze spacing. The technique required that all adult males were radio tagged and that radio locations were taken within the shortest possible time span (on average 6 min between nearest neighbors). The observed distribution of nearest neighbor distances was compared to expected distributions calculated from computer simulations. The statistical comparison between observed and expected nearest neighbor distances showed that for the greatest part of each month, adult males spaced themselves out within their overlapping ranges. Field experiments were used to determine the mechanism of this spacing.

EXPERIMENTAL APPROACHES

One of these experiments, for example, involved the alteration of scent fields along the run system of adult males and the measurement of responses of the males to the altered scent fields. Such complex field experiments were only feasible when the test individual and all its neighbors were radio tagged. The technical problems involved in guaranteeing standardized field experiments of the above kind are immense: the time that scent fields were exposed to the air before being contacted by the test individual had to be the same between experiments and the test animal had to be the first to enter the experimental site. With the help of radio tracking I could monitor location and movements of the test animal and its neighbors prior to the experiment. Thus, radio tracking can permit field workers to conduct experiments comparable in complexity to laboratory approaches.

Acknowledgements — I am greatly indebted to the late Gordon Ashwell who designed the transmitter circuit. Dr. Ian J. Patterson carried out the computer simulations, read the manuscript and provided invaluable criticism, Mr. R. Elliott and Mr. C. Hancock are gratefully acknowledged for correcting the English in parts of the manuscript. The study was supported by grants from the Dr. C. Duisberg Stiftung, Deutscher Akademischer Austauschdienst, and the Mainzer Akademie für Wissenschaften und Literatur.

REFERENCES

Caruthers A. D. (1973) Capture-recapture methods applied to a population of known parameters. *J. Appl. Ecol.* 42, 125-146.

Hamley J. M. and Falls J. B. (1975) Reduced activity in transmitter carrying voles. *Can. J. Zool.* 53, 1476-1478.

Leuze C. (1976) Social behavior and dispersion in the water vole, *Arvicola terrestris* (Lacépède). Unpl. Ph.D. thesis. University of Aberdeen.

Odum E. P. and Kuenzler E. J. (1955) The measurement of territory and home range size in birds. *Auk* 72, 128-137.

Tester Y. R. (1971) Interpretation of ecological and behavioral data on wild animals obtained by telemetry with special reference to errors and uncertainties. *Proc. Symp. Biotelemetry*, Pretoria, South Africa: 385-408.

Growth, Reproduction and Survival in *Peromyscus leucopus* Carrying Intraperitoneally Implanted Transmitters*

H. R. SMITH

USDA, Forest Service, Northeastern Forest Experiment Station, Hamden, Connecticut 06514, U.S.A.

Abstract — The effect of an intraperitoneal transmitter implant on the rates of growth, reproduction and survival of *Peromyscus leucopus* was studied. Growth was measured in weaned juveniles for 21 days following implantation. Observed growth rates based on average daily weight gain (g) per 24 hours were 0.32 and 0.31 for implanted and control mice respectively. Reproductive performance of females in a wild population was based on their observed breeding condition during censusing and the number and size of litters found in nest boxes. Of the implanted females 37 percent produced litters compared to 33 percent for controls. Litter size averaged 4.3 and 5.0 respectively for implanted and control mice. Neither of these differences was significant. Survival rates were determined for juveniles, sub-adults, and adults in a wild population. No significant differences were found in monthly disappearance or overwintering survival between implanted and control mice for any age or sex category. This study failed to reveal any deleterious effects caused by intraperitoneal implantation and it was concluded that implanted mice behaved normally.

INTRODUCTION

Biotelemetry technology and application have increased significantly during the past decade. Despite widespread use of telemetry to study a variety of wild animals under natural conditions it remains uncertain as to whether radio tagged animals behave normally. It has been generally assumed that placement of a transmitter in or on the animal does not cause significant behavioral or physiological change (Rawson and Hartline, 1964; Marten, 1972; Corner and Pearson, 1972). Tester (1971) found no published reports of adverse behavioral effects in mammals but warned that both short and long term effects should be anticipated. Reduced activity in radio tagged voles (*Microtus pennsylvanicus*) was reported by Hamley and Falls (1975) whereas Madison (1978),

---

*The use of trade, firm, or corporation names in this publication is for the convenience and information of the reader. Such use does not constitute an official endorsement or approval by the U.S. Department of Agriculture or the Forest Service of any product or service to the exclusion of others that may be suitable.

working with the same species, found no adverse effects. There is little quantitative data available to support the premise that radio tagged animals behave normally.

The utilization of the white-footed mouse, *Peromyscus leucopus* (an important predator of forest insects), in an integrated pest management program for gypsy moth, *Lymantria dispar* L., prompted the development of an intraperitoneal transmitter to quantify mouse mortality (Neely and Campbell, 1973). Determination of their radio transmitter's feasibility for use in *Peromyscus* made it imperative to examine its effects on the behavior and physiology of the animal. It was hypothesized that implanted *P. leucopus* can be said to behave normally if their growth rates, reproductive performance, and survival rates do not differ significantly from *P. leucopus* without implant. The results of a study to test this hypothesis are presented in this paper.

MATERIALS AND METHODS

STUDY AREA

The study area consisted of 2.8 ha of mixed deciduous-hardwood forest on the Lake Saltonstall Reservoir watershed located in Branford, Connecticut.

TRAPPING OF ANIMALS

Live trapping with Sherman® traps was used to monitor reproductive success and survival (i.e. disappearance from population which included mortality and emigration). A 9 × 14 trapping grid was established with one trap (7.5 × 7.5 × 30 cm) placed at each station. Trap stations were at 15 m intervals. Trapping was conducted during a 16-day period in July, 1975 and 5-day periods in August through November, 1975 and March and April, 1976. Trapping was not attempted during the winter months because of the high risk of mortality to mice caught in live traps at that time. In addition to trapping, mice were observed semimonthly in nest boxes located on the study area. Animals were numbered by specific toe removal when first captured and that number along with the following data were recorded: sex, age, place of capture, general condition, and reproductive condition (males: testes scrotal or non-scrotal; females: teats not visible or visible — enlarged from suckling young, lactating or gravid). Only those in advanced stages of pregnancy could be determined. Mice were categorized by age class using the following criteria: predominantly gray: juvenile; gray dorsal band and brown lateral bands: sub-adult; predominantly brown: adult.

The study area was divided into quadrants to provide a more even distribution of test and control animals. Each animal was listed by sex and age within the quadrant in which its activity was centered. The center of activity was determined after day 5 of the July trapping period using the exclusive boundary strip method described by Stickel (1954). Animals in each quadrant and within each category were selected at random to receive intraperitoneal implants. Those not implanted served as controls.

Implantation of transmitter capsules began on day 7 of the July trapping period. Capsules were similar to the transmitter package developed by Neely and Campbell (1973) with one major modification: the capsules were coated with a paraffin-polyvinyl compound instead of silicone rubber. The total implant package measured less than 16 mm and weighed an average of 0.84 ± 0.004 g.

Selected animals were captured and taken in the trap to the field laboratory where surgery was performed. Animals were released at their point of capture as soon as the last animal had recovered from surgery. No mice were in captivity long than 4 hours. Implantation continued through day 16 of the July trapping period and each trapping period through November, 1975.

Growth was measured in 15 juvenile *P. leucopus* born in captivity from wild parents. Four litters were divided (implant:control) as follows: (2♀ , 1♂:2♀, 1♂), (1♀, 1♂:2♀), (1♀, 1♂:1♂), and (1♂:1♀). All animals were weaned at time of surgery and the age of the animals ranged from 23 to 25 days. Their weights were measured daily (nearest 0.01 g) for 21 days. Animals were individually caged and maintained in a field laboratory. Conditions in the field laboratory closely approximated daylength and ambient temperature (ca 22°C night, 26°C day). Water consumption (measured to the nearest ml) and food consumption (commercial laboratory chow for mice) measured to the nearest 0.01 g were recorded daily.

SURGICAL PROCEDURE

Methoxyflurane inhalation anesthetic was used by introducing approximately 2 cm^3 on a cotton swab inside a wire container within a tight wooden box, measuring 7.5 × 13.5 × 21.5 cm. The lid of the box was equipped with a window for observation. The mouse was placed in the box and observed until it was anesthetized and was then removed and placed on its back on a sterile drape. A small cone fitted with a cotton plug was placed over the mouse's head and additional methoxyflurane was introduced on the cotton as necessary. Sterile instruments and sterile rubber gloves were used. The surgery was clean but not sterile; hair was not clipped and drapes were not used. The abdominal midline of the mouse was scrubbed with a solution of Zephiran® chloride, glycerine, and alcohol. A midline incision about 2 cm long was made with a No. 12 surgical blade and the opening in the skin was elongated with scissors. A midline musculature and peritoneum stab incision was enlarged with mosquito forceps. The encapsulated transmitter was placed in an antiseptic solution (Zephiran chloride) and then rinsed in sterile water before being inserted.

One suture was required to close the peritoneum and muscle and a second to close the skin incision. The suture material was No. 0000 chromic opthalmic surgical gut with a swedged-on needle. The length of time required for anesthesia, surgery, and recovery was recorded for each animal.

RESULTS

Nine juvenile (average weight 8.95 g ± 0.52 g) and 57 wild trapped *P. leucopus* (representing all age classes) were implanted. Anesthesia time (time from first exposure to anesthetic in the box until the animal lost consciousness at which time it was removed) averaged 2.0 ± 0.4 min for juveniles and 4.8 ± 0.2 min for the wild mice. Anesthesia time varied with the strength of anesthesia remaining in the box. Depth of anesthesia was determined by respiration rate. Surgery time (time required for surgical implantation procedure) averaged 12.9 ± 0.8 min for juveniles and 9.8 ± 0.3 min for the wild mice. Recovery from anesthesia was rapid and without any apparent post-operative effects except in two cases mentioned below. Implanted juveniles required an average of 18.7 ± 9.4 min to regain sufficient use of their limbs so as to move about in the recovery chamber. However, one animal required 71 min. Recovery time for mice in the wild population averaged 13.3 ± 0.9 min. One animal succumbed to an apparent overdose of anesthesia. Overdose by methoxyflurane is unusual if sufficient oxygen is provided (Barry, 1972). No obvious complications such as paralysis were observed; however, one juvenile male removed its sutures and disemboweled itself the first night following surgery.

FOOD AND WATER CONSUMPTION

Over a 21-day period eight implanted juvenile *P. leucopus* on a diet of Purina Laboratory Chow® consumed a daily average of 0.21 g g^{-1} body weight (weight of capsule included) as compared to a daily consumption of 0.22 g g^{-1} body weight for the control animals (Fig. 1). The difference in food consumption was not significant ($P > 0.05$). Daily water consumption was significantly different ($P < 0.05$) between implants and controls, being 0.44 and 0.51 ml g^{-1} body weight respectively. Capsule weight is included in body weight.

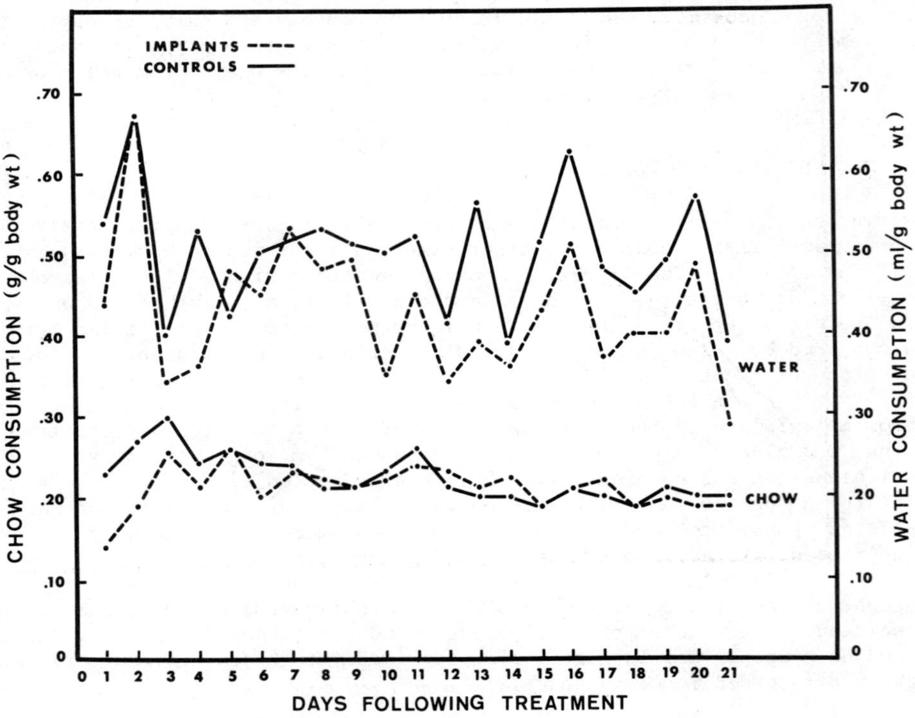

Fig. 1. Average daily consumption of water (ml g^{-1} body weight) and food (g g^{-1} body weight) for eight implanted and seven control juvenile *P. leucopus*.

SOMATIC GROWTH

Analyses of variance (ANOVA) were used to evaluate the effects of implantation (accounting for differences between litters) on both short term and long term changes in body weight. Body weight is probably the best single criterion of growth of the whole organism (Layne, 1968). No long term differences in body weight 21 days after surgery were attributable to implantation ($P > 0.05$). The comparative growth curves (Fig. 2) are nearly identical. Implanted mice showed an average daily weight gain of 0.32 g compared to 0.31 g for controls. However, a significantly greater variation in weight ($P < 0.01$) was found between individuals from different litters as compared to differences between siblings. The standard deviation between individuals in the same litter was $\pm$ 0.76 but $\pm$ 2.33 for mice from different litters.

Significant ($P < 0.01$) short term effects (24 hours after surgery) on weight change were attributable to implantation. There was no evidence of dissimilarity between litters in short term weight change. Eight implanted juveniles lost an average of 0.32 $\pm$ 0.13 g during the 24 h period following surgery, while control mice gained an average of 0.56 $\pm$ 0.09 g during the same time period. Implanted mice required 2.0 $\pm$ 0.3 days for their real body weight (capsule weight not included) to exceed their

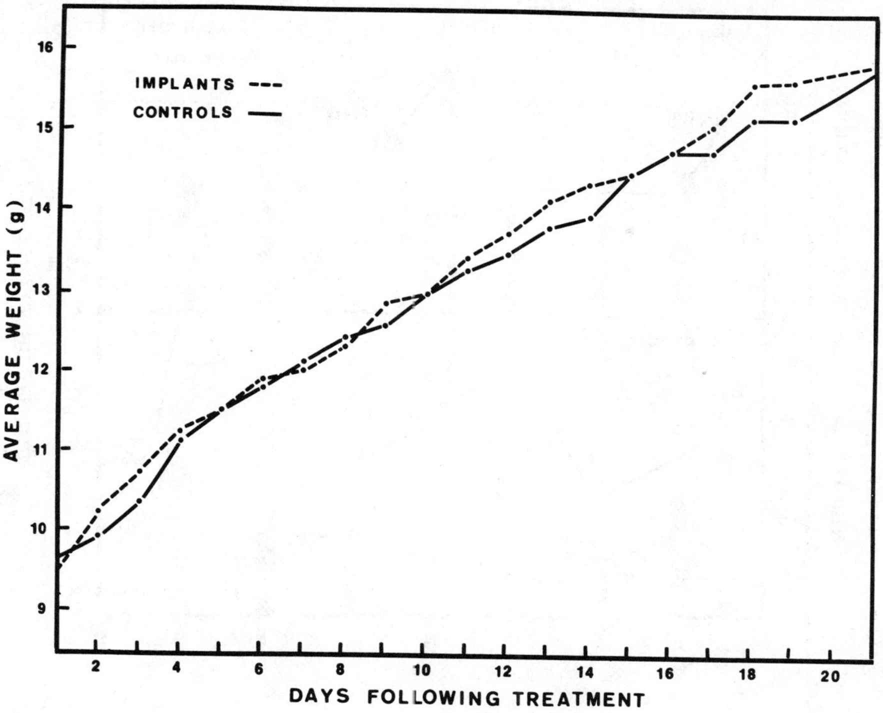

Fig. 2. Growth curves for implanted and control juvenile *P. leucopus* based in average daily weight change in g.

weight recorded prior to surgery. Initial mean weight was 9.0 ± 0.5 g for both groups of mice.

REPRODUCTION

Comparison of reproductive success (Fig. 3) between implanted and control *P. leucopus* revealed no significant differences. Reproductive success was defined as parturition having taken place. On the basis of observed reproductive condition indices (described in Methods) taken during each trapping period, and beginning one month after treatment for each animal, females were assigned to three categories: (a) definitely reproduced, (b) probably reproduced, and (c) did not reproduce. Twenty-one implanted females were observed; of those, eight (38 percent) definitely reproduced; three (14 percent) probably reproduced; and 10 (48 percent) did not reproduce. Comparable reproductive percentages for 24 control female mice were eight (33 percent), three (13 percent), and 15 (54 percent), respectively. Only six litters were actually observed (nest boxes) during the study. Litter size averaged 4.3 (three litters) and 5.0 (three litters) for implanted and control mice respectively.

SURVIVAL

One hundred twenty-three wild *P. leucopus* were live trapped (57 received transmitter implants and 66 remained as controls). Age distribution for implanted mice was: 12 juveniles, 12 sub-adults, and 33 adults; for controls: 23, 14, and 34 respectively.

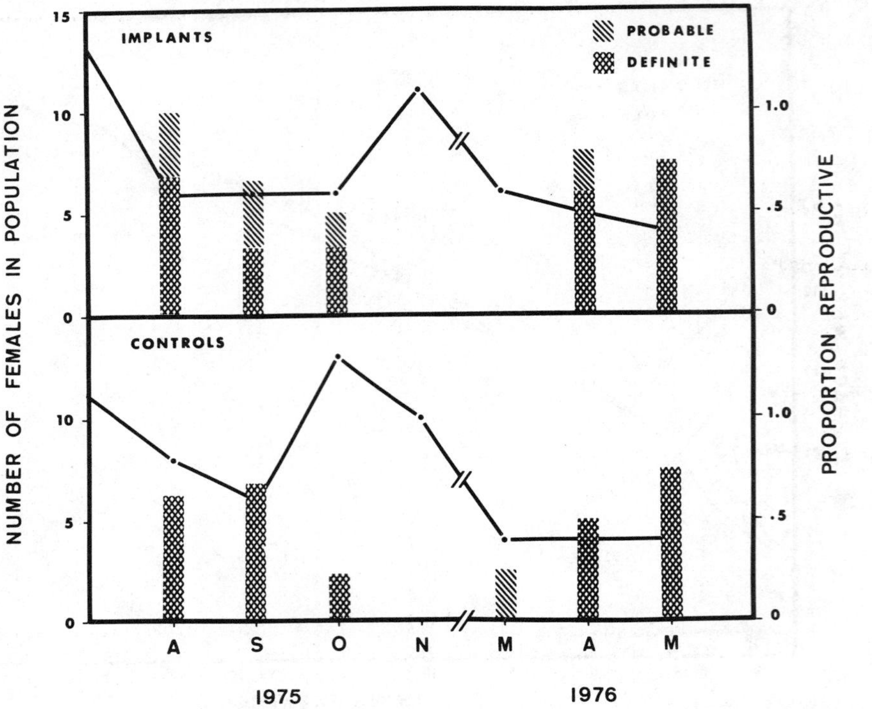

Fig. 3. Proportion of reproductive female *P. leucopus* (bar graphs) in relation to the number of females in the implant and control population (line graph).

Survivorship curves by sex for implanted and control are shown in Fig. 4. Disappearance from the population included mortality and emigration. Differences in survival due to implanting for either sex were not significant ($P > 0.05$, t-test). Neither were any significant differences ($P > 0.05$, χ^2) found between any age classes. The difference in survival shown in Fig. 4 between males in the second month was due largely to the transients that were caught during the extended July trapping period. Despite this, implanted animals tended to be less transitory. Analysis of the March, 1976 census indicated a winter survival rate of 19 percent of the previous fall implanted population and a 12 percent survival rate for controls. This difference was not significant ($P > 0.05$, χ^2).

DISCUSSION

The objective of this study was to determine whether intraperitoneally implanted *P. leucopus* behaved normally. Tester (1971) warned that both short and long term effects should be anticipated. It was assumed that any abnormal behavior due to implantation would be reflected in one or more of the following parameters: growth rate of juveniles, reproductive performance in the wild, and survival in the wild. This study failed to reveal any deleterious effects due to implantation and concluded that implanted mice behaved normally. Despite this conclusion it should not be assumed that *P. leucopus* carrying a different transmitter package or any other species carrying an intraperitoneal transmitter will behave normally.

This study has provided fundamental information useful in biotelemetry studies on

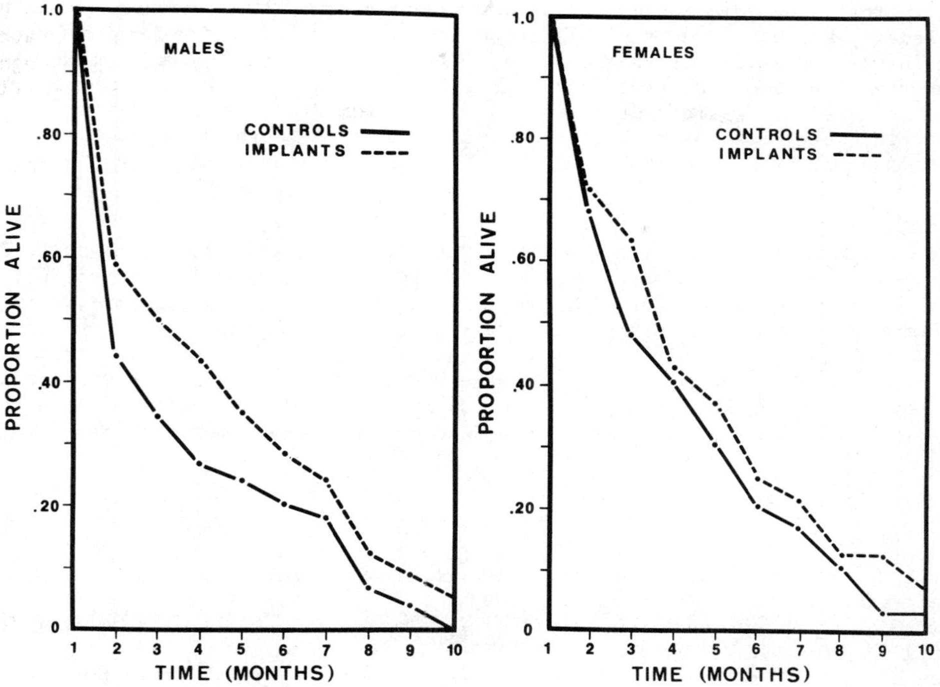

Fig. 4. Survivorship curves for implanted and control male and female *P. leucopus* based on disappearance after first capture.

animal behavior in the wild. It is notable that implanted juveniles weighing 6 to 9 g grew at a normal rate and that implanted mice in a wild population would not be handicapped in relation to survival, especially during the winter. Any loss or impairment of agility or motor coordination would have certainly reduced survival in this heavily preyed upon species.

The short term effect in weight loss in implanted juveniles was attributed to a reduction in food intake. The animals did not return to a normal food consumption rate until the third day after surgery. Conversely, there was no reduction in water consumption during this time. The long term difference in water consumption may be questioned because water bottles with glass sipper tubes were used. Mice often 'play' with these bottles causing them to leak. Therefore, measurement of water consumed was less precise than measurement of food consumption.

There is further difficulty in assuming that certain differences that were not shown to be statistically significant ($P < 0.05$) were therefore not biologically significant. For example, Wecker (1962) found that *Peromyscus* parasitized by *Cuterebra* sp. were less likely to emigrate than non-parasitized *Peromyscus* when only one larva was present but not subject to higher mortality rates. It is possible the higher survival rate of implanted mice (19 to 12 percent) could reflect a similar phenomenon even though it was not statistically significant.

The long term difference in growth rate between litters emphasizes the importance of dividing siblings between test groups in order to minimize bias. Future studies are necessary to provide comparative data relative to transmitter collars and subcutaneous implants. Some consequences of these techniques have been reported (Smith and Whitney,

1977). Hamley and Falls (1975) cautioned researchers against continuing to report casually that the radio tagged animal was behaving normally. The widespread use of biotelemetry has permitted the collection of a variety of data which previously was unobtainable. However, it is the 'uniqueness' of this technique that makes testing of adverse effects on behavior difficult because control groups in the wild are not easily followed. A variety of mammalian species have the potential for evaluating various telemetry systems. This can provide a more thorough understanding of stress factors imposed by telemetry and their influence upon behavior.

Acknowledgements — I thank the New Haven Water Company, especially Bob Hart, for assistance in locating and use of the study area. The assistance of the following were also greatly appreciated: Bob Neely, Michelle Wall, David Emerson, Wendy Sabol and Kathy McManus (technical); Gerald Walton (statistical, and manuscript review); Roger Zerillo (photographic); and R. A. Lautenschlager and John Podgwaite, Northeastern Forest Experiment Station, for their critical review of this manuscript. The work reported was funded in part by a U.S. Department of Agriculture sponsored program entitled, "The Expanded Gypsy Moth Research and Development Program."

REFERENCES

Barry W. J. (1972) Methoxyflurane: An anesthetic for field and laboratory use on squirrels. *J. Wildl. Mngmt.* 36, 992-993.

Corner G. W. and Pearson E. W. (1972) A miniature 30-MH_2 collar transmitter for small mammals. *J. Wildl. Mngmt.* 36, 657-661.

Hamley J. M. and Falls J. B. (1975) Reduced activity in transmitter-carrying voles. *Can. J. Zool.* 53, 1476-1478.

Layne J. N. (1968) *Biology of Peromyscus*, Vol. 2, Chap. 6, pp. 148-253. Spec. Publ., Amer. Soc. Mammal.

Madison D. M. (1978) Movement indicators of reproductive events among female meadow voles as revealed by radio telemetry. *J. Mammal.* 59, 835-843.

Marten G. G. (1972) The remote sensing approach to censusing. *Res. Popul. Ecol.* 14, 36-57.

Neely R. D. and Campbell R. W. (1973) A system for gathering small mammal mortality data. U.S. Dept. Agric. For. Serv. Res. Paper NE-280. 6 pp.

Rawson K. S. and Hartline P. H. (1964) Telemetry of homing behavior by the deermouse, *Peromyscus*. *Science* 46, 1596-1597.

Smith H. R. and Whitney G. (1977) Intraperitoneal transmitter implants — their biological feasibility for studying small mammals. In *Proc. First Intnl. Conf. Wildl. Biotelem.*, pp. 109-117. Laramie, Wyoming.

Stickel L. F. (1954) A comparison of certain methods of measuring ranges of small mammals. *J. Mammal.* 35, 1-15.

Tester J. R. (1971) Interpretation of ecological and behavioral data on wild animals obtained by telemetry with special reference to errors and uncertainties. *Proc. Symp. Biotelem.*, pp. 385-408. Pretoria, So. Africa.

Wecker S. C. (1962) The effects of botfly parasitism on a local population of the white-footed mouse. *Ecology* 43, 561-565.

An Evaluation of Several Grid Trapping Methods by Comparison with Radio Telemetry in a Home Range Study of the Eastern Chipmunk (*Tamias striatus* L.)

B. TREVOR-DEUTSCH and D. F. HACKETT

Biology Department, Laurentian University, Sudbury, Ontario P3E 2C6, Canada

Abstract — Grid trapping has been used to determine small mammal home range for almost 50 years. In the absence of a more precise standard against which to calibrate, at least nine trap data interpretation methods have been developed, each yielding different values. Home ranges generated from these methods were compared to those calculated using much more accurate radio telemetry, to determine the formers' relative precisions. The Standard Circle and Koeppl's 70 percent Ellipse generated home ranges closest to those determined with radio telemetry. Assumptions concerning the accuracy of peripheral successful traps in defining boundaries of home ranges were tested, as were those concerning spatial utilization near unsuccessful traps within home range.

INTRODUCTION

The home range of the eastern chipmunk (*Tamias striatus* L.) has been described by a number of authors including Allen (1938); Bole (1939); Burt (1940); Blair (1941, 1942); Smith (1942); Manville (1949) and Yerger (1953). In all of these studies, grid trapping and one of several methods of data interpretation have been employed to determine home range. There is however, no general agreement on which method of estimating home range from trap data is most precise. Since several have been in common use, comparability among published home range studies has been tenuous (Mohr, 1943, 1947; Hayne, 1949, 1950; Yerger, 1953; Stickel, 1954; Van Winkle, 1975; Koeppl, Slade and Hoffman, 1975). Also it has long been suspected that home range as determined by grid trapping may differ from true home range (Chitty, 1937; Chitty and Dempson, 1949; Dice, 1938; Burt, 1940; Manville, 1949; Hayne, 1949, 1950, Stickel, 1954, 1968; Justice, 1961; Brooks and Banks, 1975), but this has never been adequately tested or quantified (Ambrose, 1969).

Radio tracking with directional antennas provides a means of gathering locational data on free ranging animals which occupy large areas. This technique is not subject to the biases and uncertainties or grid trapping and has been successfully applied to the study of home range since the early 1960s (Marshall, Gullion and Schwab, 1962; Cochran and Lord, 1963; Verts, 1963; Ellis, 1964; Storm, 1965; Tester and Siniff, 1965; Craighead, 1972). Unfortunately, difficulties with transmitter weight, short broadcast range, and unreliable battery life hampered early attempts at miniaturizing equipment for use with rodent-sized animals (Rawson and Hartline, 1964; Beal, 1967; Kolz, Corner and Teitsen, 1972; Corner and Pearson, 1972). It was not until the recent development of powerful light weight transmitter packages (Banks, Brooks and Schnell, 1975; Mineau and Madison, 1977; Madison, 1977, 1978) that radio tracking with directional antennas became practical for the study of small, highly mobile forest rodents such as chipmunks.

The present study employed two techniques: the conventional grid trapping approach patterned after Burt (1940) and Manville (1949), and radio tracking using equipment similar to that of Banks *et al.* (1975). The aim of the investigation was to use telemetry as a basis on which to evaluate the accuracy of grid trapping.

MATERIALS AND METHODS

STUDY AREA

The study site was 3.6 ha of forested slope with southern aspect, in the Lake Laurentian Conservation Area, Sudbury, Ontario, Canada (47° 10' N, 82° 00' W).

FIELD METHOD

Field work was undertaken between 26 May and 2 September, 1976. As many as 20 chipmunks were radio tracked hourly during the day throughout this period. A 'conventional' trapping grid was operated within range of the radio towers for 3-day trapping periods separated by 3-day intervals during which all traps had been removed from the grid. Thus, trapping and non-trapping treatments were alternated, while radio surveillance was continued daily. No trapping or radio tracking was done on rainy days.

Grid Trapping

One hundred and eighty-seven Sherman live traps (7.5 × 7.5 × 30 cm) were laid out in eleven rows and seventeen columns. Intertrap spacing was 15 m, resulting in a grid area of 3.6 ha.

Traps were baited with peanut butter and sunflower seeds, and were checked three times a day. Captured animals were examined to determine their sex, general health, and reproductive condition. Chipmunks were earmarked using a Number 1 leather punch, according to the system devised by Burt (1940). Radio collars were attached, checked, or changed at these times.

Radio Tracking

Tracking of radio tagged animals was accomplished by triangulation from two towers that were located 220 m apart, and about 30 m beyond the northern corners of the trap grid (Fig. 1). Two simultaneously taken bearings, one from each tower, were required to radio locate a chipmunk. Operators synchronized their activities by walkie-talkie contact, or by prearranged scheduling.

Bearings were taken nine times per day on radio tagged animals and on a 'reference' transmitter, the latter to ensure that the directional antennas and compass rosette pointers were aligned. Telemetry equipment was manufactured by AVM Instrument Co., Champaign, Illinois, U.S.A.

A FORTRAN program intitled 'CHIP' was created to convert telemetry bearings to computer drawn maps of chipmunk locations. The program also calculated the center of activity for each animal (Hayne, 1949), the distance of each radio location from

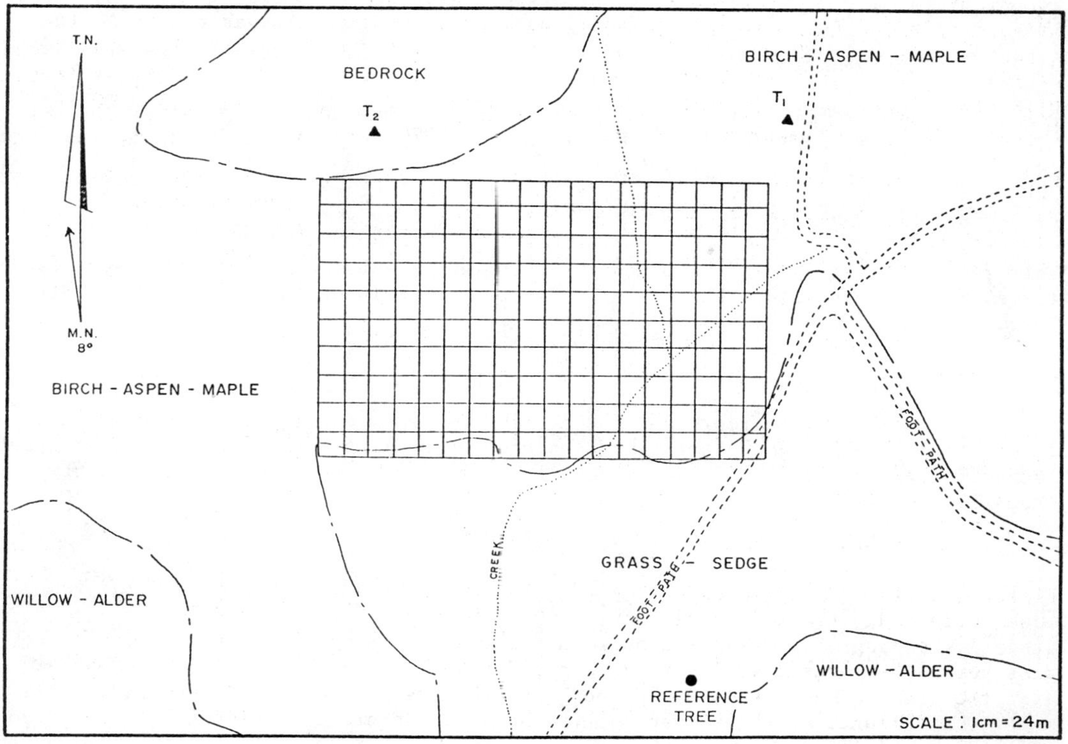

Fig. 1. Layout of the study site. The rectangle consisting of horizontal and vertical lines represents 11 rows and 17 columns of traps which comprised the trap grid. Locations of radio towers, T1 and T2, are marked by the symbol ▲. A reference transmitter was placed at tree ● for alignment and mapping purposes (see text).

the center of activity, distances between consecutive locations, the 'Standard Circle' (Harrison, 1958), an elliptical model of probabilistic home range (Koeppl *et al.*, 1975), and errors in estimation.

Minimum home range values (Dalke, 1942; Mohr, 1943) for radio tracking data were derived using the program 'POLY', designed by Dr. M. Herman of the Department of Mathematics, Laurentian University. Both programs are available from the authors at cost prices.

RESULTS AND DISCUSSION

A COMPARISON OF METHODS FOR CALCULATING HOME RANGE SIZE

A number of different methods have been used to calculate home range size from trap data. Conventional methods are based on the summing of areas around capture sites according to specific sets of rules. The most widely used of these are:

Minimum Home Range Method

The most peripheral capture points are joined by straight lines. Home range is considered to be the area contained within the polygon thus formed (Dalke, 1942; Mohr, 1943, 1947).

Inclusive Boundary Strip

Home range is determined in the same way as in 'Minimum Home Range', with the addition of a peripheral 'boundary strip' equal in width to half the intertrap distance (Blair, 1940; Stickel, 1946, 1954).

Exclusive Boundary Strip

This is similar to 'Inclusive Boundary Strip' except that areas around unsuccessful traps are excluded unless they fall in direct line between successful traps (Stickel, 1946, 1954).

Added Squares

Each trap is assumed to represent the center of a square with sides equal to the intertrap distance. Home range is calculated by summing the areas of squares around successful traps and those unsuccessful traps that join disjunct capture points (Manville, 1949).

Observed Range Length Circle and Ellipse

Half the distance between the most widely separated capture points is used as the radius of a circular or elliptical home range representation, depending on the scatter of trap data (Stickel, 1946, 1954).

Adjusted Range Length Circle and Ellipse

Half the intertrap distance is added to the radius described for 'Observed Range Length', and a circular or elliptical home range representation is calculated (Stickel, 1946, 1954).

Assumptions and relative merits of these methods have been reviewed by Hayne (1949) and Stickel (1954).

There have also been a number of attempts at describing home range in terms of probability rings or ellipses surrounding the geometric mean of captures. These have been summarized by Keoppl *et al.* (1975) and Van Winkle (1975).

Two probalistic models considered in this study are:

The Standard Circle

This model defines a 'standard diameter' as twice the square root of the mean square of all distances between capture points, and the center of activity. A circle based on this diameter is called the 'Standard Circle' and contains 68.26 percent of all captures (Harrison, 1958; Doebel and McGinnes, 1974).

Koeppl's Model

This model shapes probability ellipses of various confidence intervals to best fit the data. Major and minor axes pass through the center of activity (geometric mean) at an angle determined by the scatter of points. Then the data are rotated until the major and minor axes parallel the X,Y axes. In this position, the standard deviations of all X's and Y's become the radii of a rotated 'Standard Ellipse'. This ellipse is then adjusted for sample size and desired confidence interval (Koeppl *et al.*, 1975).

Each of the above methods of calculation results in a unique home range estimate. Figure 2 illustrates how differently these methods would interpret the same three hypothetical capture points. Table 1 presents home range estimates for chipmunks captured five or more times in the present study, as determined using the above nine methods of calculation. Average home range estimates varied from 3,925 m^2 to 28,882 m^2 (a seven-fold difference) using conventional analyses, and from 3,925 m^2 to 89,919 m^2 (a twenty-two fold difference) if Koeppl's 95 percent probabilistic ellipse is considered. Clearly therefore, comparability among studies in which different home range calculation methods have been used, is tenuous.

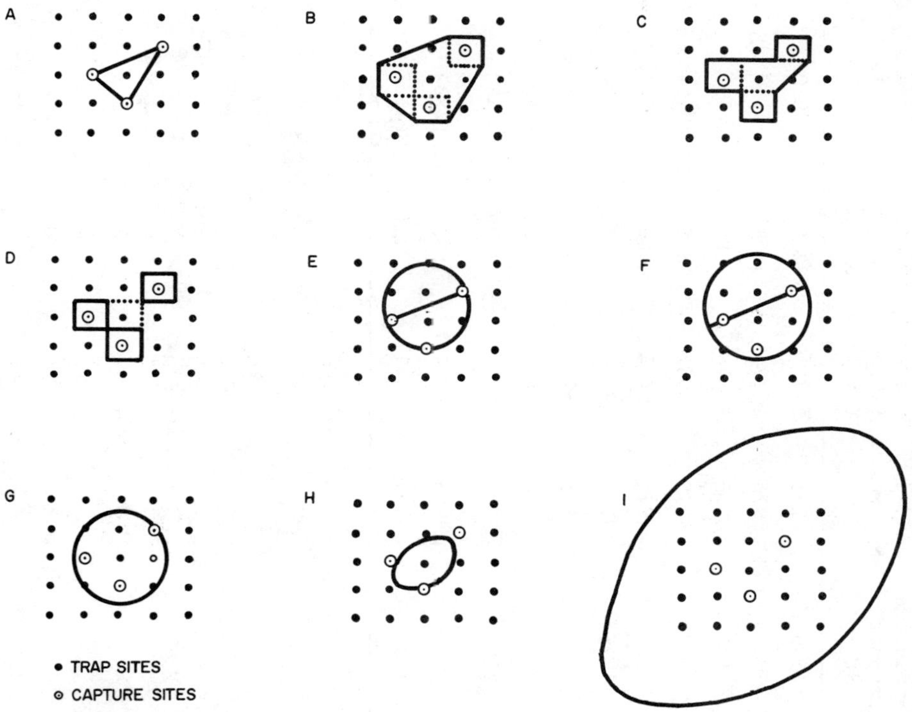

Fig. 2. An illustration of nine different interpretations of home range, based on the same three capture points.* (a) Minimum Home Range; (b) Inclusive Boundary Strip; (c) Exclusive Boundary Strip; (d) Added Squares; (e) Observed Range Length Circle; (f) Adjusted Range Length Circle; (g) Standard Circle; (h) Standard Ellipse; (i) Koeppl's 70 percent Ellipse.

COMPARISONS BETWEEN TRAP AND RADIO DETERMINED HOME RANGE ESTIMATES

In the present context, we refer to Trap Determined Home Range (an estimate based on trap data), Radio Home Range (an estimate based on telemetry data) and Actual Home Range which is in fact the area occupied by an animal. Until recently, it has not been possible to determine which method of trap data calculation yields values closest to 'Actual Home Range'.

*This figure is patterned after the work of Stickel (1954), who illustrated the Inclusive and Exclusive Boundary Strip Methods in a similar fashion.

Table 1.
Home range estimates (m^2) resulting from nine different methods of calculation

Chipmunk Field Number	Captures	Minimum	Inclusive Boundary Strip	Exclusive Boundary Strip	Added Squares	Observed Range Circle	Adjusted Range Circle	Standard Circle	Koeppl's 95% Ellipse	Koeppl's 70% Ellipse
E6	14	5963	8775	4556	4275	17203	20867	7860	23453	8040
E14	7	20925	25988	7088	6975	52279	58535	43150	202038	54086
E15	5	7425	11588	3600	3375	22698	26880	17599	179598	34791
E24	12	7650	10800	4613	4275	14957	18385	8454	32726	10856
E32	6	10575	15638	4388	4275	39408	40863	3444	168897	38848
E33	9	4500	7538	3825	3150	18627	22432	8991	27848	8460
E35	6	2138	4388	2138	2025	5945	8171	3861	27353	6503
E43	7	10125	14625	4950	4950	42638	48305	25783	115904	31028
E70	6	2363	4905	2250	2250	8825	11499	5463	36951	8785
Mean		7963	11583	4156	3925	24731	28882	17289	89919	22377
		*(1.97)	(2.86)	(1.03)	(0.97)	(6.11)	(7.14)	(4.27)	(22.22)	(5.53)

*( ) mean area in acres.

At the present state of the art, Actual Home Range is most closely approximated by 'Radio Home Range' because radio tracking using light weight transmitters (≈ 5 percent of body weight) does not appear to interfere with movements of study animals except during a brief period of adjustment after collaring (Brander and Cochran, 1971; Brooks and Banks, 1971; Lindsey *et al.*, 1973; Hamley and Falls, 1975; Cranford, 1976; Madison, 1978) and when transmitter batteries are changed. Also, radio tracking permits frequent location of animals anywhere within the home ranges whereas trap based location points are restricted to trap sites and historically, capture frequency has been limited to one or several per day. On these premises, a comparison has been made between trap revealed home ranges determined using eight methods of calculation, and radio estimated home ranges using the Minimum Method (Dalke, 1942; Mohr, 1943, 1947) on peripheral radio locations. Only those three animals whose spatial activities were restricted to the area of the trap grid and which were well sampled by radio locations, were included.

Table 2 gives individual radio and trap determined home ranges. Trapping estimates within rows are based on identical data: therefore the differences observed are entirely due to the method of calculation.

Based on the premise that Minimum Radio Estimates most precisely approximated Actual Home Range, trapping estimates compared as follows:

1. The Standard Circle (11 percent too small),
2. Inclusive Boundary Strip (22 percent too small),
3. Minimum, and Exclusive Boundary Strip (58 percent too small),
4. Added Squares (62 percent too small),
5. Observed Range Length Circle (75 percent too large),
6. Adjusted Range Length Circle (113 percent too large),
7. Koeppl's 95 percent Probabilistic Ellipse (266 percent too large).

Thus, the Standard Circle gave the 'best' trap based estimate of Minimum Radio Home Range.

Our results differed from those of Stickel (1954). This is not surprising since her testing procedures were entirely threoetical and based on unlikely assumptions about home range shape and the success of traps.

A number of assumptions have been made by workers using trapping methods and several of these lent themselves to being scrutinized using the present study. The assumption that all traps within a home range will be successful is false. Of the 20 home ranges completely and partly sampled by the trap grid, all contained unsuccessful traps. This may have reflected differential utilization of home range or simply that some traps were in less attractive locations. The assumption that experimental animals do not travel beyond peripheral unsuccessful traps is also false. If animals can evade intervening traps, then it seems reasonable that they could also range some distance past peripheral capture sites without being detected by grid trapping. Indeed, on several occasions study animals were visually identified outside their trap determined home ranges. Superimposed radio tracking and grid trapping maps (Fig. 3) also indicated chipmunk movement beyond capture points.

There were in each case several captures outside the radio revealed home range. Apparently, these occurred while the operators were servicing the grid; these movements would undoubtably have been detected if radio tracking had been conducted during these periods. This raises the possibility that animals were 'chased' off their home range by human presence, encountered traps, and were captured, thus biasing trapping results.

The observation that chipmunks ranged beyond peripheral unsuccessful traps may explain the relative efficiencies of some of the trap based methods of estimating

Table 2.
A comparison of home range estimates (m^2) of eight trap based methods of calculation, and Minimum Radio Home Range

Chipmunk Field Number	Days monitored by telemetry and trapping	No. of captures	No. of radio locations	Minimum Radio Home Range	Minimum Method	Inclusive Boundary Strip	Exclusive Boundary Strip	Observed Range Length Circle	Adjusted Range Length Circle	Added Squares	Standard Circle	Koeppl's 95% Ellipse
E6	5	10	14	9842	4613	7763	3600	15949	19483	3600	7389	26493
E24	5	12	19	9994	4500	8325	4725	14957	18385	4275	8454	32717
E33	6	6	32	8316	2700	5850	3375	18358	22167	2925	9280	43732
Average home range				9384	3938	7313	3900	16430	20012	3600	8374	34314
Percentage of Minimum Radio Home Range				100	42	78	42	175	213	38	89	366
Relative comparison of various estimates with the Minimum trapping estimate				2.4	1.0	1.9	1.0	4.2	5.1	0.9	2.1	8.7

Table 3.
Trap based home range estimates (m^2) of Koeppl's Probability Ellipses and 'Rotated Standard Ellipse', compared with Minimum Radio Home Range (m^2)

Chipmunk Field Number	Days monitored by telemetry and trapping	No. of captures	No. of radio locations	Minimum radio Home Range	Trap based method of calculation: Confidence Interval of Keoppl's Model						Rotated Standard Ellipse
					99%	95%	90%	75%	70%	50%	
E6	5	10	14	9842	51383	26493	18474	9861	8316	4497	2640
E24	5	12	19	9994	60327	32717	23301	12768	10853	5929	3627
E33	6	6	32	8316	113427	43732	27223	12603	10398	5218	2521
Average home range				9384	75046	34314	22999	11744	9856	5215	2929
Percentage of Minimum Radio Home Range				100	800	366	245	125	105	55.6	31.2

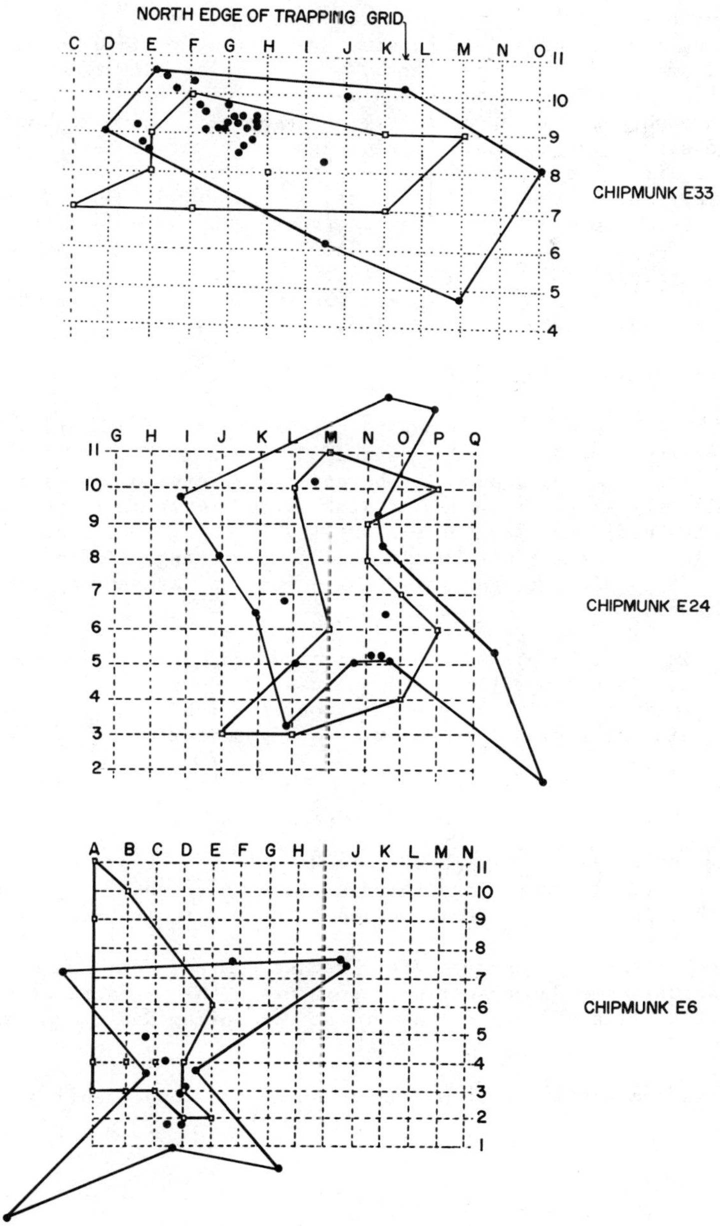

Fig. 3. Radio and trap determined home ranges of three eastern chipmunks. Lettered columns and numbered rows represent the trap grid. Lines joining peripheral trap captures enclose Minimum Trap determined Home Ranges. Lines joining peripheral radio locations enclose Minimum Radio Home Ranges. ● radio location, ■ trap capture, 1 burrow, 2 trap determined center of activity, 3 radio determined center of activity.

home range size. Minimum, Added Squares and Boundary Strip methods either deal directly with capture points or grant an extra margin of half the intertrap distance. By underestimating travels beyond successful traps, they yielded conservative estimates. Observed and Adjusted Range Length Circles employ only the two most widely separated captures and assume these to represent the diameter of a circle. Yet the scatter of capture points was not circular. As a result, these methods over estimated home range size. The Standard Circle also interprets home range in a circular way. However, it centers itself around the geometric mean of all the data, and encloses (68.26 percent of the capture points) (Harrison, 1958). Its particular properties combined fortuitously to provide the 'best' trap based estimate among those methods examined, despite the fact that it incorporated no data on movements beyond successful traps. Koeppl's Probability Ellipse approximated Actual Home Range poorly because the 95 percent confidence ellipse was much too large when based on the low capture numbers available from a live trapping study.

A Modified Version of Koeppl's Probabilistic Ellipse

Koeppl's Model is used to generate probabilistic ellipses representing home range in confidence intervals of 99, 95, 90, 75 and 50 percent. Koeppl *et al.* (1975) favored the 95 percent ellipse, possibly because this level of significance is commonly used in statistical treatments, but since it over estimated Actual Home Range by 266 percent, the efficacy of other probability levels were tested. The 70 percent ellipse was in close agreement with Minimum Radio Home Range (Table 3). At 105 percent of the latter value, it generated the best trap based probabilistic approximation of home range.

It is not surprising that Koeppl's Model can be adpated to estimate home range well from trapping data. Each probability ellipse centers, shapes, and orients itself to the scatter of points. It is only a question of which size of ellipse best encompasses the data and includes extra travels beyond peripheral successful traps.

CONCLUSIONS

1. Grid trapping reveals Trap Determined Home Range, as distinguished from Actual Home Range because: (i) animals range through areas where they are not detected by traps and (ii) the presence of workers on the trap grid influences animal movement.

2. The results of this study indicate that Trap Determined Home Range can be used to approximate Actual Home Range most precisely using the Standard Circle and Koeppl's 70 percent Ellipse (89 and 105 percent of Minimum Radio Home Range, respectively).

3. Further studies are required to determine the general applicability of these findings.

REFERENCES

Allen E. G. (1938) The habits and life history of the eastern chipmunk. *N.Y. State Mus. Bull.* 122 pp.

Ambrose H. W. (1969) A comparison of *Microtus pennsylvanicus* home ranges as determined by isotope and live trap methods. *Am. Mid. Nat.* 81, 535-555.

Banks E. M., Brooks R. J. and Schnell J. (1975) A radiotracking study of home range and activity of the brown lemming (*Lemmus trimucronatus*). *J. Mammal.* 56, 888-901.

Beal R. O. (1967) Radio transmitter-collars for squirrels. *J. Wildl. Mngmt.* 31, 373-374.

Blair W. F. (1941) The small mammal population of hardwood forest in northern Michigan. Contr. Lab. Vert. Genetics, University of Mich., No. 17, 1-10.

Blair W. F. (1942) Size of home range and notes on the life history of the woodland deer-mouse and the eastern chipmunk in northern Michigan. *J. Mammal.* 23, 27-36.
Bole B. P., Jr. (1939) The quadrat method of studying small mammal populations. *Cleveland Mus. Natur. Hist. Sci. Publi.* 5, 15-77.
Brander R. B. and Cochran W. W. (1971) Radio-location telemetry. In R. H. Giles (ed.), *Wildlife Management Techniques*, pp. 95-103. The Wildlife Society, Washington, D.C.: 633 p.
Brooks R. J. and Banks E. M. (1971) Radio-tracking study of lemming home range. *Comm. Behav. Biol.* 6, 1-5.
Burt W. H. (1940) Territorial behaviour and populations of some small mammals in southern Michigan. *Misc. Publ. Mus. Zool.* University of Michigan, No. 45, 58 pp.
Chitty D. (1937) A ringing technique for small mammals. *J. Animal Ecol.* 6, 36-53.
Chitty D. and Dempson D. A. (1949) Prebaiting small mammals and a new design of live trap. *Ecology* 30, 536-542.
Cochran W. W. and Lord R. D., Jr. (1963) A radio-tracking system for wild animals. *J. Wildl. Mngmt.* 27, 9-24.
Corner G. W. and Pearson E. W. (1972) A miniature 30-HMz collar transmitter for small animals. *J. Wildl. Mngmt.* 36, 657-661.
Craighead F. C., Jr. (1972) Grizzly bear prehibernation and denning activities as determined by radio-tracking. *Wildl. Monogr.* 32, 6-35.
Cranford J. A. (1976) Home range and habitat utilization by *Neotoma fuscipes* as determined by radiotelemetry. *J. Mammal.* 58, 165-172.
Dalke P. D. (1942) The cottontail rabbits in Connecticut. *Connecticut State Goel. Natur. Hist. Surv. Bull.* 65, 1-97.
Dice L. R. (1938) Some census methods for mammals. *J. Wildl. Mngmt.* 2, 119-130.
Doebel J. H. and McGinnes B. S. (1974) Home range and activity of a gray squirrel population. *J. Wildl. Mngmt.* 38, 860-867.
Ellis R. J. (1964) Tracking raccoons by radio. *J. Wildl. Mngmt.* 28, 363-368.
Hamley J. M. and Falls J. B. (1975) Reduced activity in transmitter-carrying voles. *Can. J. Zool.* 53, 1476-1478.
Harrison J. L. (1958) Range of movement of some Malayan rats. *J. Mammal.* 39, 190-206.
Hayne D. W. (1949) Calculation of size of home range. *J. Mammal.* 30, 1-18.
Hayne D. W. (1950) Apparent home range of *Microtus* in relation to distance between traps. *J. Mammal.* 31, 26-39.
Justice K. E. (1961) A new method for measuring home ranges of small mammals. *J. Mammal.* 42, 462-470.
Koeppl J. W., Slade N. S. and Hoffman R. S. (1975) A bivariate home range model with possible application to ethological data analysis. *J. Mammal.* 56, 81-90.
Kolz A. L., Corner C. W. and Tietjen H. P. (1972) A radio frequency beacon transmitter for small mammals. *J. Wildl. Mngmt.* 36, 177-179.
Lindsey G. D., Nass R. D., Hood G. A. and Hirata D. N. (1973) Movement patterns of Polynesian rats (*Rattus exulans*) in sugarcane. *Pacific Sci.* 27, 239-246.
Madison D. M. (1977) Movements and habitat use among interacting *Peromyscus leucopus* as revealed by radiotelemetry. *Can. Field-Nat.* 91, 273-281.
Madison D. M. (1978) Movements and habitat use among interacting *Peromyscus leucopus* revealed by radiotelemetry. *Can. Field-Nat.* 91, 273-281.
Manville R. H. (1949) A study of small mammal populations in northern Michigan. University of Mich. Press, Ann Arbor, Michigan.
Marshall W. H., Gullion G. W. and Schwab R. G. (1962) Early summer activities of porcupines as determined by radio-positioning techniques. *J. Wildl. Mngmt.* 26, 75-79.
Mineau P. and Madison E. M. (1977) Radio-tracking of *Peromyscus leucopus*. *Can. J. Zool.* 55, 465-468.
Mohr C. O. (1943) A summary of North American small mammal censuses. *Am. Mid. Nat.* 29, 545-587.
Mohr C. O. (1947) Table of equivalent populations of North American small mammals. *Am. Mid. Nat.* 37, 223-249.
Rawson K. S. and Hartline P. H. (1964) Telemetry of homing behaviour by the deer mouse (*Peromyscus*). *Science* 146, 1596-1598.

Smith R. A. (1942) The biology of a small mammal community in a central New York woodlot. Doctorate thesis, Cornell University, 218 p.

Stickel L. F. (1946) Experimental analysis of methods for measuring small mammal populations. *J. Wildl. Mngmt.* 10, 150-159.

Stickel L. F. (1954) A comparison of certain methods of measuring ranges of small mammals. *J. Mammal.* 35, 1-15.

Stickel L. F. (1968) Home range and travels. In J. A. King (ed.), *Biology of Peromyscus*, pp. 373-411. spec. Publ. Amer. Soc. Mamm., 2, xiii + 1-593.

Storm G. L. (1965) Movements and activities of foxes as determined by radio-tracking. *J. Wildl. Mngmt.* 29, 1-13.

Tester J. R. and Siniff D. B. (1965) Aspects of animal movement and home range data obtained by telemetry. *Trans. N. Am. Wildl. and Nat. Resources Conf.* 30, 379-393.

Van Winkle W. (1975) Comparison of several probabilistic home-range models. *J. Wildl. Mngmt.* 39, 118-123.

Verts B. J. (1963) Equipment and techniques for radio-tracking striped skunks. *J. Wildl. Mngmt.* 27, 325-339.

Yerger R. W. (1953) Home range, territoriality, and populations of the chipmunk in central New York. *J. Mammal.* 34, 448-458.

Strategies for Analyzing Radio Tracking Data

D. R. VOIGT* and R. R. TINLINE**

**Ministry of Natural Resources, Box 50, Maple, Ontario, Canada*
***Department of Geography, Queen's University, Kingston, Ontario, Canada*

Abstract — Previously used methods to analyze radio tracking data have not generally taken advantage of all the information associated with location fixes. Most early analyses were preoccupied with size and shape of home range. Analysis strategies with options determined by study objectives, type of data and species-specific variables are described. Descriptions can include cartographical, graphic and tabular depictions of distributions for spatial, habitat, and activity type utilization, direction and orientation as well as interaction potential measures. A sample set of data from a red fox radio tracking study was used to demonstrate methods of analysis. Analysis was performed interactively, which permitted the researcher to interrogate the data base with few inherent assumptions. Computer algorithms for measuring area and depicting borders of home ranges were compared for consistency, sensitivity and richness for biological implications. Two methods employing cell frequency counts were found to be useful for investigating the internal anatomy of home ranges. Emphasis is placed on describing approaches to answering questions on how animals use their home areas rather than the computer program package developed. A review of home range analysis and literature is also included.

INTRODUCTION

Radio transmitters have been used to track animals for over 20 years and their potential for home range studies was described early (Adams, 1965). Recently, a great many studies have employed radio tracking to learn more about the biology of wildlife species, but a review of the literature indicates early promises of application have not been fully realized. Many studies employing telemetry still present data on movements only in terms of size or shape of home ranges. Frequently, analyses only involve determination of the area occupied by the minimum area method (Mohr, 1947) with limited use of other information inherent in the data. Thus, it appears there has not been a concomitant evolution of methods of analysis with methods of data acquisition.

A possible cause of failure could be linked to the ways in which analyses of home range data have evolved over the past 40 years. Much of the original data was obtained from trap-capture studies. Generally, analyses have been directed at determining perimeters, enclosing areas or determining probabilistic models of the data distribution. In most cases far less consideration was given to the internal anatomy of home range and significance of movements of animals. Sanderson (1966) recognized the need to explore the reasons for animals' movements and Adams and Davis (1967) concluded that radio tracking could permit such investigations.

Although a radio tracking approach is often far superior technically, numerically and biologically to trap-capture methods it has caused new problems (Siniff and Tester, 1965). In many cases, a great quantity of data is amassed and manual manipulation during analysis is no longer feasible. Continuous data from fixed tower automatic tracking systems requires the use of computers (Siniff and Tester, 1965; Deat *et al.*, 1980, this volume). A practical solution which also minimizes resolution problems of such systems has been to accept discontinuous tracking, employ small mobile receivers, and calculate and code fixes by hand. Often other constraints such as time, finances, staff or study objectives also result in discontinuous data. In such situations, problems similar to trap-capture studies may occur since the data set is a sample which may be biased as a result of acquisition methods, especially in wide ranging, elusive species which are difficult to track (Mech, 1974). In comprehensive, longterm studies, increased care in data collection, recording and handling is necessary to ensure consistent accuracy.

In addition to these problems, investigations of movements and ranges from telemetry data still present the two major problems inherent in live-trapping data: (i) what assumptions about the data distribution is the researcher willing to make in order to interpret animal behavior; (ii) how is the sample design (or lack of one) related to those assumptions. For example, recent attempts to bound distributions of fixes with confidence limits based on standard ellipses assume that the fixes are distributed in a bivariate normal fashion and are independent (Jennrich and Turner, 1969; Mazurkiewicz, 1971; Koeppl, Slade and Hoffmann, 1975). Initially, researchers may be unwilling to make such assumptions although they recognize the utility of probabilistic statements (Van Winkle, 1975). Furthermore, the use of discontinuous data may result in a poor relationship between a sample design and assumptions inherent in analysis methods.

This paper is an attempt to provide useful means of utilizing discontinuous radio tracking data to explore wider implications of home range use. Since our approach evolved from previous home range analyses, an experimental data set was tested in relation to previous concepts. From this, the implications of animal movement data were further studied which resulted in a general strategy for investigating the data in an interactive fashion.

OBJECTIVES OF MOVEMENT STUDIES

Telemetry now provides the tool to investigate complete socio-ecological relationships. Movement data can be supplemented with telemetry of physiological measures such as body temperature, heart and respiration rates. The objective of most movement studies is to measure features of an animal's activity and relate them to environmental factors. Primary among these features are spatial characteristics of movements such as shape, size, orientation and dispersion, resulting in measures of environmental use. Studies considering frequency of use permit determination of intensive use areas, preferred areas and travel routes. They also provide base data for energetic requirements as well as inter- and intra-specific relationships including potential interaction.

To meet these objectives a strategy for analyzing movement data should: (i) permit investigation of activity with respect to temporal and spatial distributions of fixes; (ii) make explicit the assumptions of a given methodology; (iii) use a minimum of assumptions, and (iv) deal with discontinuous data.

AN ANALYSIS STRATEGY

DATA PROCESSING

Experimental Data Set

A subset was chosen from a large data collection involving red foxes (*Vulpes vulpes*) and three other species, radio tracked over a 4-year period in southern Ontario, Canada. These data were collected to describe movements, activity, home ranges and interaction of these species as part of a wildlife rabies research program. From a sample of 84 different foxes, 12 were chosen which were genetically related (3 generations). Consideration was given to representing variation in: season (spring, summer, fall, winter), year (1975, 1976, 1977), sex (7 females: 5 males), age (pup, yearling, adult), family, behavior (as residents, transients), habitat type, number of fixes/fox (27-701) and length of time monitored (less than 3 months — over 3 years). The data were discontinuous for reasons described earlier. The telemetry system evolved from one described by Cochran and Lord (1963) to that refined by Lotimer (1979).

Access in the study area permitted locating all radio tagged foxes at any time in the study area. Ability to get within standard distances (0.15-0.45 km) from foxes permitted consistent accuracy in the radio fixes to the level of 100 m square grid cells. These were coded to a 9-digit number describing the X, Y coordinates as per the Universal Transverse Mercator grid. Other information collected concurrently included fox number, date, hour, sex, age and family as well as activity status. Activity status was coded stationary/active, as determined from the variation in signal intensity of successive pulses, and day/night extended from sunrise/sunset. Table 1 shows data organization as stored on computer files. Included are other useful variables (*) which were calculated from raw field data during data base encoding.

System Components

The software that permitted the analysis discussed above consists of two major components. The first component, a set of data base building algorithms, included: (i) a CRUNCH package to extend the raw data by adding seasons, converting alphanumeric dates to numeric dates and calculating distances, times and rates between successive fixes, and (ii) a CODING package to convert raw data to binary form for use with the query language described below.

The second software component, the set of data analysis procedures, included: (i) a text editor; (ii) a query language, and (iii) a package of area determination algorithms (HOME). The editor operated via a CRT screen display and gave the researcher an easy means of editing, updating, retrieving, sorting, transferring and searching files. Both the query language and the HOME package were manipulated through the editor.

The query language was designed to allow the researcher to request cross tabulations and prepare subsets of the data set using English-like statements. For example, the query in Table 2 gives distances between successive fixes during four seasons for an adult female fox. Increased movement in the fall (column d) is readily apparent. In practice, the query language has proven to be an excellent means of exploring a data set and setting up other analyses. Finally, HOME was used interactively to select parts of a data file and the appropriate analysis techniques.

The software components were designed for an inexpensive mini-computer, the Digital PDP11V03 with two floppy drives, a Lektromedia screen, a Diablo printer and a small Tektronix flatbed plotter. The major advantages of this configuration were: (i) the machine was dedicated to a single user permitting the reliable, fast response necessary

Table 1.
Data organization as stored on floppy disc files (* indicates variables calculated from field data)

1	2	3	4	5	6	7	8	9	10	11	12	13	14	15	16	17	18	19
F	3	05	F	P	1	JY	15	75	1020	1657.431	5948	49041	7	3	0	0	0	1
F	3	05	F	P	2	JY	15	75	2135	1657.899	5943	49040	7	3	673	50	5	2
F	3	05	F	P	3	JY	16	75	1140	1658.486	5940	49037	7	3	845	42	3	3
F	3	05	F	P	2	JY	16	75	2147	1658.908	5941	49042	7	3	607	50	5	4
F	3	05	F	P	2	JY	16	75	2215	1658.927	5945	49040	7	3	27	44	98	5
F	3	05	F	P	3	JY	17	75	1054	1659.454	5944	49042	7	3	758	22	2	6
F	3	05	F	P	2	JY	17	75	2145	1659.906	5949	49037	7	3	650	70	7	7
F	3	05	F	P	2	JY	18	75	540	1660.236	5951	49043	7	3	475	63	8	8
F	3	05	F	P	1	JY	18	75	600	1660.250	5946	49040	7	3	20	58	173	9
F	3	05	F	P	3	JY	18	75	840	1660.361	5945	49043	7	3	159	31	12	10

1	SPECIES	F=fox,C=coyote,R=raccoon,S=skunk
2	NUMBER	animal identifying number
3	FAMILY	family number
4	SEX	m= male or f= female
5	AGE	p= pup , y= yearling , a=adult
6	STATUS	1=active day,2=active night,3=stationary day, 4=stationary night, o= unknown
7	MONTH	month of year, alphabetic
8	DAY	day of the month
9	YEAR	75,76 or 77
10	HOUR	hour of day
11	*BASEDAY	days since Jan.1 1971
12	FIXX	X co-ordinate (UTM system)
13	FIXY	Y co-ordinate (UTM system)
14	*MONTH2	numeric month
15	*SEASON	winter=1, spring=2,summer=3,fall=4
16	*SPAN	time between successive fixes (minutes)
17	*DIST	distance between successive fixes (tens of km)
18	*RATE	speed between successive fixes (tens of km)
19	*SEQUENC	sequential number of record

for easy interactive computing; (ii) the secondary storage medium: floppy discs which hold 250 k bytes, were cheap (US $5.00) and compact (size of 45 rpm record); (iii) the combination of a screen, plotter and incremental printer permitted a variety of graphics output, and (iv) the operating cost was low. The system was developed by the Department of Geography, Queen's University, Ontario (Moore and Tinline, 1978; Stevens, Tinline and Moore, 1979).

SPATIAL INVESTIGATIONS

AREA DETERMINATIONS

Literature Review

Methods used previously to measure home ranges can be broadly classified as: (i) minimum area methods (Dalke, 1942; Mohr, 1947); (ii) center of activity methods (Hayne, 1949), and (iii) grid cell methods (Siniff and Tester, 1965).

The first and simplest of these methods arose from the analyses of trap-capture data and is the most commonly used measure of area in the literature. A basic assumption is that the observed animal will reveal the boundaries of its home range. Thus, a

Table 2.
A query and output showing distribution of distances between successive fixes for an adult female fox (F02) during 4 seasons in 1977

NUMBER=2 and YEAR=77

Winter = 1	Spring = 2	Summer = 3	Fall = 4	
	1	2	3	4	sums
DIST = 0 to 49	8	18	13	8	47
DIST = 50 to 99	8	5	4	16	33
DIST = 100 to 149	5	7	7	10	29
DIST = 150 to 199	3	1	3	10	17
DIST = 200 to 249	4	0	3	9	16
DIST = 250 to 299	0	2	6	7	15
DIST = 300 to 349	0	0	1	6	7
DIST = 350 to 399	0	0	0	7	7
DIST = 400 to 449	0	0	0	4	4
DIST = 450 to 499	0	0	0	1	1
DIST = 500 to 549	0	0	1	2	3
DIST = 550 to 599	0	0	0	0	0
DIST = 600	0	0	0	0	0
sums	28	33	38	80	179

Comment: in this query the distance between successive movements for fox number two is examined by season for 1977. Note that all distances are in XXX E-2 format. Hence 600 means 6.00 km and 50 means 0.50 km.

```
Count every dist from 0 to 600 group 50
by every season
by number =2 and year=77
including marginals
```

line connecting the outermost points of capture will form a convex polygon. Stickel (1954) used concave figures but that procedure was not defined until Jennrich and Turner (1969). The convex polygon may not represent the minimum area of a home range as commonly assumed, since it may include large unused areas. It does, however, represent the minimum perimeter of a home range.

Since large holes may exist within the polygon, the figure represents an area over which the animal may roam but there is little guarantee, except in the case of continuous tracking, that the animal did not move beyond these sample boundary points (Brant, 1962). Attempts to correct the empty area problem included restricting the length of the line joining boundary points such as Harvey and Barbour's (1965) modified minimum area method or excluding areas by experience such as Able's (1969) atypical habitat elimination method. Efforts to correct the boundary problem included a variety of methods which add a boundary strip (Burt, 1943; Blair, 1940). While these variations meet the objectives of some researchers, the results from different methods are not comparable. Furthermore, no assessment of the relative use of portions of the bounded areas is given.

The second general method also evolved from the analyses of trap-capture data and assumed that the set of observations could be used to reveal a center of activity and that home range boundaries could be expressed in terms of a relationship with distance from this center. Methods employed to date include the geographic center (Hayne, 1949) and the frequency distribution of distances to define probabilities of capture from the center for circular home range (Calhoun, 1955). In most applications the result is not a discretely bounded home range but a statistical concept defining activity radii leading to concentric boundary zones (Dice and Clark, 1953; Calhoun and Casby, 1958). Noncircular home ranges led to the use of ellipses (Fitch, 1947; Mazurkiewicz, 1971) and later, elliptical boundary zones were used (Koeppl *et al.*, 1975; Jennrich and Turner, 1969).

A third general strategy was suggested by Siniff and Tester (1965) as a means of dealing with fixes obtained from continuous tracking data. Given such information, it would be possible to infer area directly by assigning each fix to a small grid cell on a gridded map of the study area. This meant that home range area could be calculated by counting used cells. Furthermore, "... boundary, centers of intensive use, islands of disuse, and the like" (Adams, 1965) could be determined.

Since most studies deal with discontinuous data, important considerations become: (i) determining the point (asymptote) at which increasing sample size gives little or no new knowledge about home range size (Stickel, 1954; Odum and Kuenzler, 1955; Metzgar and Sheldon, 1974), and (ii) inferring as much as possible about home range use with a limited data set. A solution is to join empty grid cells with used cells (Maxson, 1978) but the results are necessarily a function of the biological significance of the joining rules. Furthermore, the problem of range expansion or range shifts has often been ignored (Cooper, 1978a). Despite these difficulties, grid cell methods have potential to investigate internal aspects and size of home ranges including relationships with other ecological variables (Cooper, 1978b).

Methods

Three algorithms were devised to represent the three general methods of determining home range described above. The computer algorithm developed for the minimum area method (MAM) began by choosing the most northerly fix, then rotating a perimeter line clockwise until it intersected another fix. This new point was taken as the next perimeter point and the procedure repeated until the original starting point was reached. The result was a unique convex polygon enclosing the outermost data points It should be noted that attempts were made to modify the algorithm to use the restrictions on the length of the line joining perimeter points as suggested by Harvey and Barbour (1965) but these modifications only produce unique polygons in special circumstances. The area of the resulting polygon (Fig. 1a) was calculated using Simpson's Rule.

The second algorithm, the sorted polygon method (SP), assumed that a center of activity can be described as the mean X coordinate and mean Y coordinate of the set of fixes. Once this center was determined, the direction (azimuth) of each fix was calculated relative to this center. Azimuths were then sorted from smallest to largest and the points were joined in this order. The area of the resulting polygon (Fig. 1b) was calculated using Simpson's Rule.

The final algorithm depends on the assignment of fixes to grid cells on a map of the study area. The size of the cells was chosen to reflect the error inherent in calculating each fix. For the data set described in this paper, each grid cell measured 100 × 100 m.

The objective of this algorithm was to allow cells with fixes (marked cells) to influence cells around and between fixes. This was done in two ways. In the

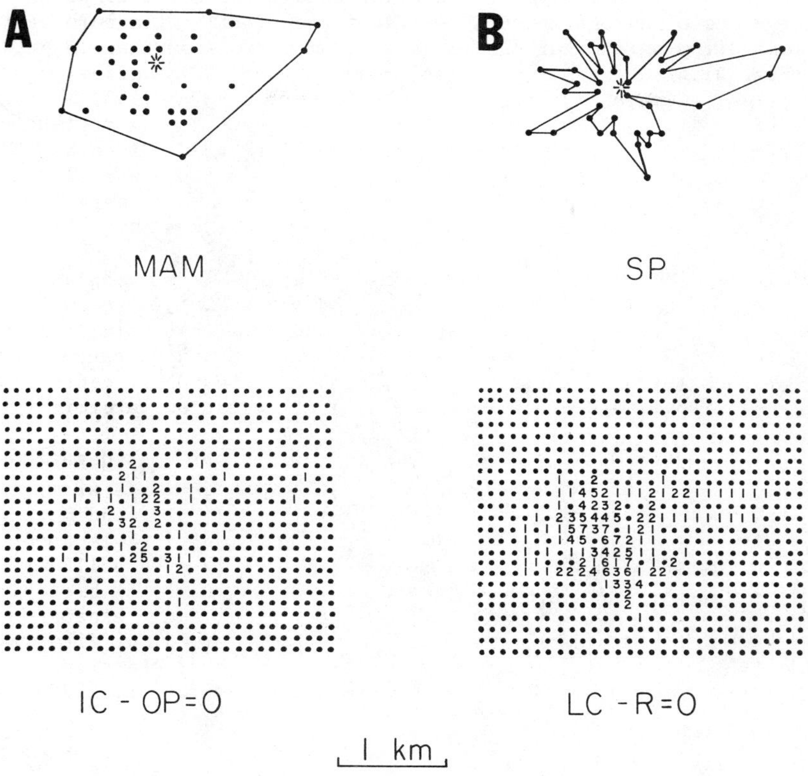

Fig. 1. Area determination methods for a female fox pup.

'influence cell' (IC) procedure, the grid cells immediately surrounding marked cells were recorded as 'used' cells. A count of used cells then became a measure of home range area. A variety of definitions of contiguity allowed the area of influence to be altered by the researcher as options. For instance, the definition 'queen's case, operator (OP)=1' implied that the 8 cells surrounding a marked cell were recorded as used cells. In this study, the definition 'queen's case, operator (OP)=0' (Fig. 1a) was also used to confine attention to the marked cells only. Other options may use: (i) the rook's or bishop's case to examine directional biases; (ii) larger operator sizes (where number of cells influenced equals $\{(2 \times \text{operator size}) + 1\}^2$). An option used in this study was to allow only fixes with 'active' status to influence an area larger than the marked cell.

To make further use of the data set a 'linked cell' (LC) method was employed. Where successive fixes were closely related in time and far enough apart for the animal to be moving in a straight line, cells along the probable travel path were recorded as used cells. Furthermore, at the end of such a travel path, fixes coded with an active status were allowed to influence an area as defined by 'queen's case, operator (OP)=1'. An investigation of travel rates for this data set suggested that rates greater than 0.1 km h^{-1} between subsequent fixes provided a useful joining rule. Using this restriction, less than 20 percent of all fixes were joined. Results are given for both LC with rate R = 0.1 km h^{-1} and rate R = 0. The latter definition joins all fixes in chronological order (Fig. 1b).

These methods were supported by either plotter or printer graphics as shown in Fig. 1. In addition, two other plotting algorithms allowed either point plots of the fixes or plots of chronologically linked fixes (CLF). CLF's proved to be very useful in separating periods of stable range from periods of range expansion and dispersal (Fig. 2).

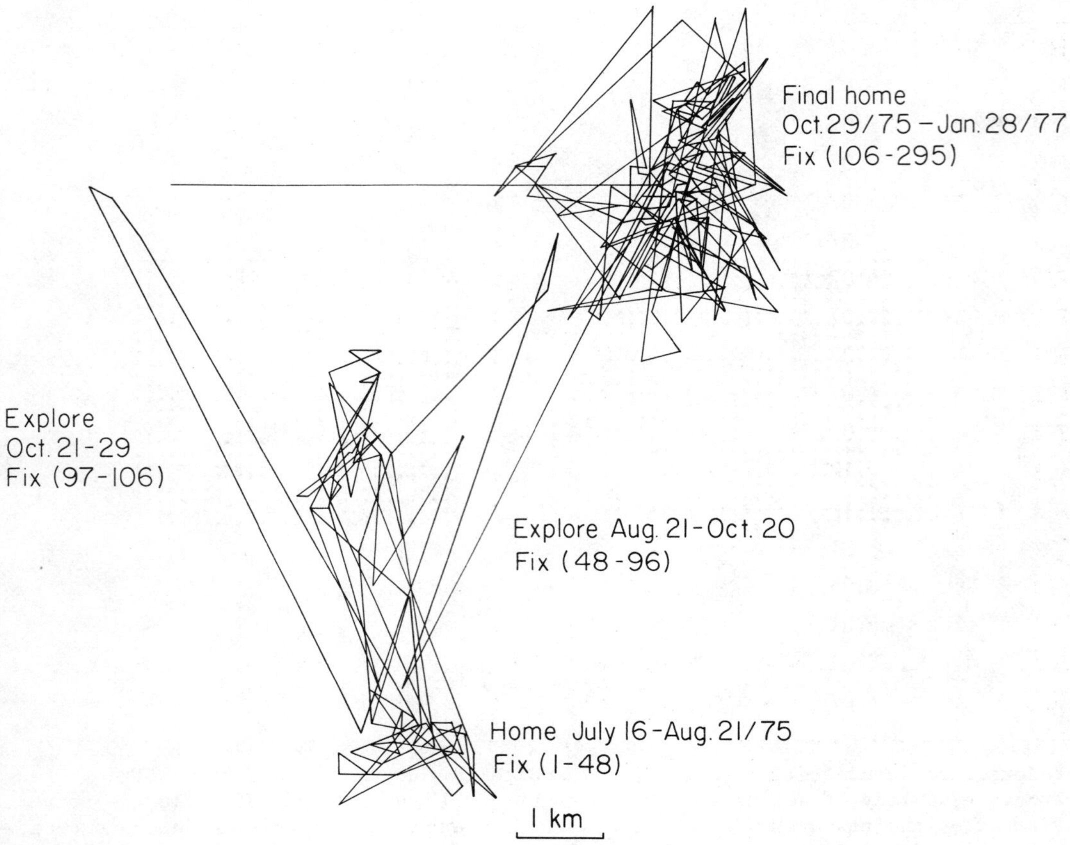

Fig. 2. Chronologically linked fixes of a male fox showing pup range, exploration, dispersal and final residency. Fixes before Oct. 29 were excluded for the analysis of final home range (F04).

RESULTS AND DISCUSSION

Table 3 lists home range areas calculated by six different methods, two representing operator size options for IC and two representing rate of travel options for LC. All fox periods on this table represent periods of relatively stable range as determined from CLF plots.

Absolute estimates of home range vary greatly among methods. However, the relative values of the estimates are closely correlated (Table 4). Specifically, the four IC/LC methods are highly correlated with each other. SP and MAM formed another less marked grouping. Pairwise regression analyses between the four IC/LC methods suggest that any one of these methods could be used to provide consistent estimates of the

Table 3.
Comparison of 6 area determination methods for 18 fox periods

Animal identification			Tracking period	N	Area (km^2) IC OP=0	OP=1	LC R=0	R=0.10	SP	MAM
F02[1]	F[2]	P[3]	15.07.75 - 16.11.75	106	0.82	2.49	2.24	0.91	0.95	4.08
F03	F	P	15.07.75 - 01.09.75	62	0.41	1.60	1.07	0.58	0.90	3.08
F04	M	P	16.07.75 - 21.08.75	48	0.36	1.22	0.87	0.44	0.50	1.15
F05	F	P	17.07.75 - 01.09.75	59	0.45	1.27	0.75	0.50	0.40	1.39
F15	F	P	16.07.76 - 23.08.76	27	0.17	0.56	0.39	0.17	0.46	0.58
F29	F	P	22.07.76 - 07.10.76	53	0.34	1.06	0.70	0.35	0.33	0.94
F30	F	P-Y	24.07.76 - 27.07.77	171	0.81	2.47	2.15	1.31	1.00	4.29
F31	M	P-Y	24.07.76 - 09.02.77	131	0.86	2.92	3.67	2.33	2.51	9.60
F34	M	P	29.07.76 - 30.10.76	55	0.43	1.67	1.66	1.07	1.00	3.25
F36	M	P	29.07.76 - 16.09.76	40	0.32	1.21	1.05	0.49	1.45	2.73
F37	F	P	30.07.76 - 18.11.76	87	0.60	1.93	2.23	1.06	1.87	4.11
F03	F	Y	22.12.76 - 09.08.77	203	1.05	3.86	3.49	2.63	1.47	6.53
F05	F	Y	31.12.75 - 17.07.76	144	0.97	3.95	4.68	3.23	5.83	18.67
F02	F	Y	05.01.76 - 18.05.76	300	1.72	6.76	6.34	5.67	4.50	11.93
F02	F	Y	18.05.76 - 10.09.76	71	0.51	2.36	3.12	2.01	3.23	4.35
F02	F	A	05.01.77 - 31.12.77	179	1.39	4.85	7.41	5.87	4.44	8.63
F02	F	A	02.01.78 - 31.03.78	70	0.41	1.79	2.18	1.07	3.68	6.09
F01	M	A	15.07.75 - 01.08.75	64	0.53	2.44	2.65	2.48	2.42	5.42

[1]F=fox, with identifying number
[2]F=female, M=male
[3]P=pup, Y=yearling, A=adult

Table 4.
Correlations among area determination methods

	IC OP=0 1	OP=1 2	LC R=0 3	R=0.1 4	SP 5	MAM 6
1	1.00	0.977	0.914	0.911	0.639	0.712
2	—	1.00	0.934	0.945	0.735	0.770
3	—	—	1.00	0.979	0.830	0.780
4	—	—	—	1.00	0.786	0.731
5	—	—	—	—	1.00	0.887
6	—	—	—	—	—	1.00

other IC/LC methods. Since this is not the case for the relationship between SP and MAM, it appears that, subject to further testing, we have identified three reasonably distinctive methods of measuring home range.

Choosing the method to estimate absolute values of home range is more difficult since there are no standards of accuracy. As Table 3 illustrates, middle range estimates are provided by IC (OP=1), LC (R=0) and SP. By definition, IC (OP=0) and MAM give low and high ranges respectively, although LC with a high rate value can provide arbitrarily low estimates but the logical lower limit is set by IC (OP=0). It is a count of all fixes and is some proportion of the actual movement area determined by sampling effort.

The problem of choosing methods is compounded by the fact that areal measures are a function of sample size (Fig. 3). Beyond some ceiling level increased sampling may not provide new areal information if range expansion, contraction or dispersal do not occur. To determine this level, areal measures were calculated for all animals in Table 3 for sample sizes increasing in steps as shown in Fig. 4. Note that areal estimates stabilized after 20 records so that the areal values in Table 3 for F04 should be indicative of its home range as a pup. Exploration and dispersal are shown by a sudden increase in area, especially for MAM at 50 records. This expansion is also indicated in the CLF plots for the same fox (Fig. 2). All areal estimates in Table 3 had sufficient records to level off in this way. In each instance, MAM was found to be the most sensitive to range expansion while SP and IC were better indicators of the ceiling level for a particular period. In every case, MAM estimates included large empty areas. Only SP areas can decrease with larger samples.

Until more extensive comparison work is done, several methods should be used (Metzgar, 1972). A combination of IC (OP=0,1), SP and MAM for increasing sample sizes would reveal a range of areal estimates and indicate whether these estimates approached their ceiling values. Work is currently in progress to identify the minimum sampling effort required to estimate these ceiling values under a variety of circumstances. This effort should prove useful even if continuous sampling is used, as continuous recording creates data storage problems.

Subsequent sections demonstrate that IC/LC methods also provide useful information on the internal anatomy of the home range. Future technological changes permitting inexpensive continuous tracking (large sample sizes) will further increase their usefulness. However, even with continuous tracking, the problem of inferring area from points or lines remains. At one level, the problem is solved by scaling grid cells and travel corridors to the level of accuracy of the fixes. Ultimately, however, the researcher will have to set the scale in both time and space on the basis of its biological significance.

UTILIZATION DISTRIBUTION

Knowledge of the frequency of movements in different portions of an individual's range is important in assessing environmental influences. Both IC and LC readily provide such analysis with their various options (Fig. 5). For individuals, pairs, or large groups of radio tracked animals, used cells can be plotted (Fig. 6) to depict effectively space partitioning for chosen time intervals. An option, to determine the relative importance of areas, involves graphing distance from home site, center of activity or high-count cell versus the density or number of used cells at unit distances (Fig. 7). Comparison over time may illustrate stability or shifts in use of areas. The relative importance of different directions is viewed by choosing sectors or quadrants as in Fig. 8. Alternatively, contouring of used cells can be done through computer algorithms but requires contouring (joining) rules.

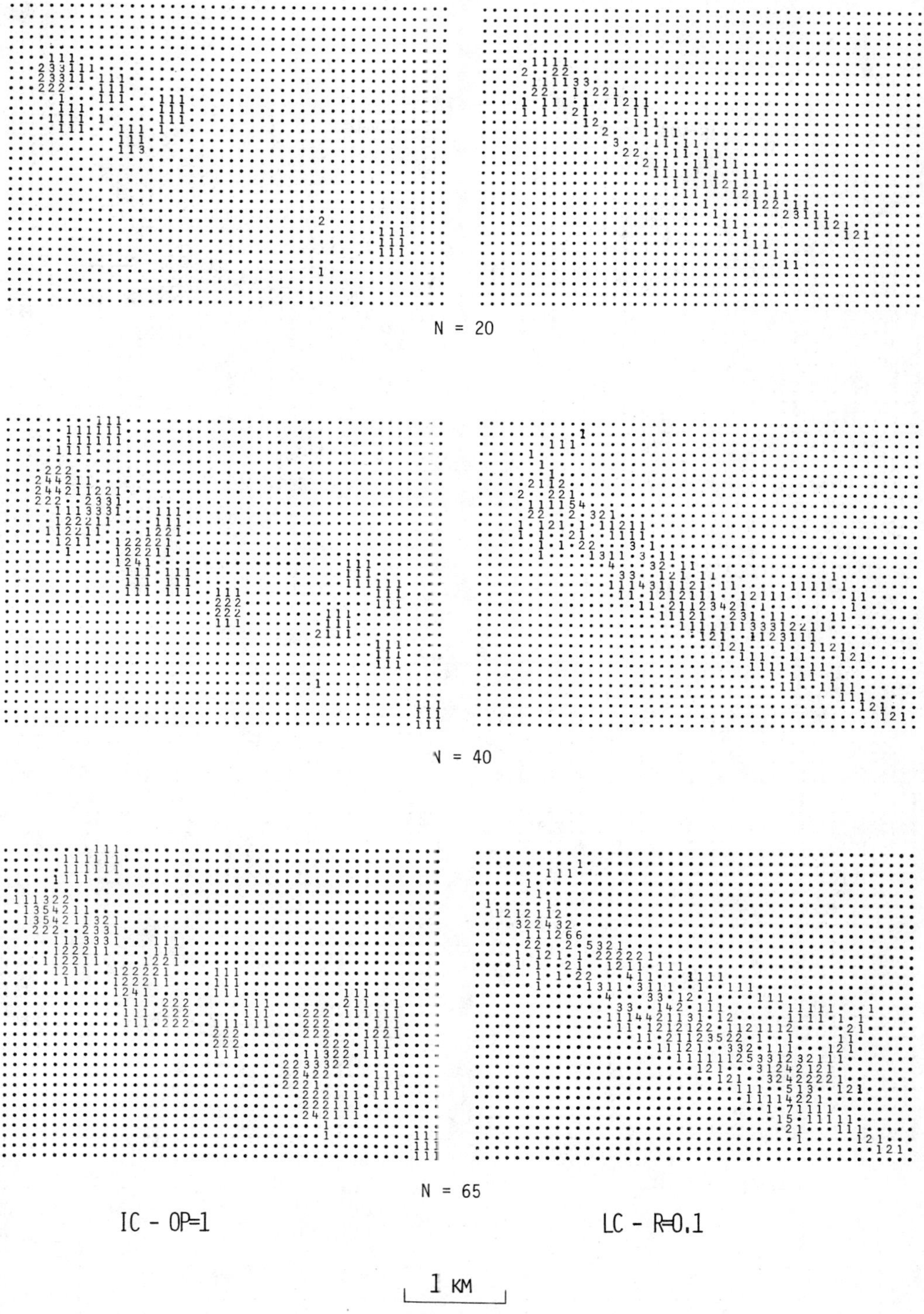

Fig. 3. A comparison between the home range of a male fox using LC and IC as sample size increases from N=20 to N=65.

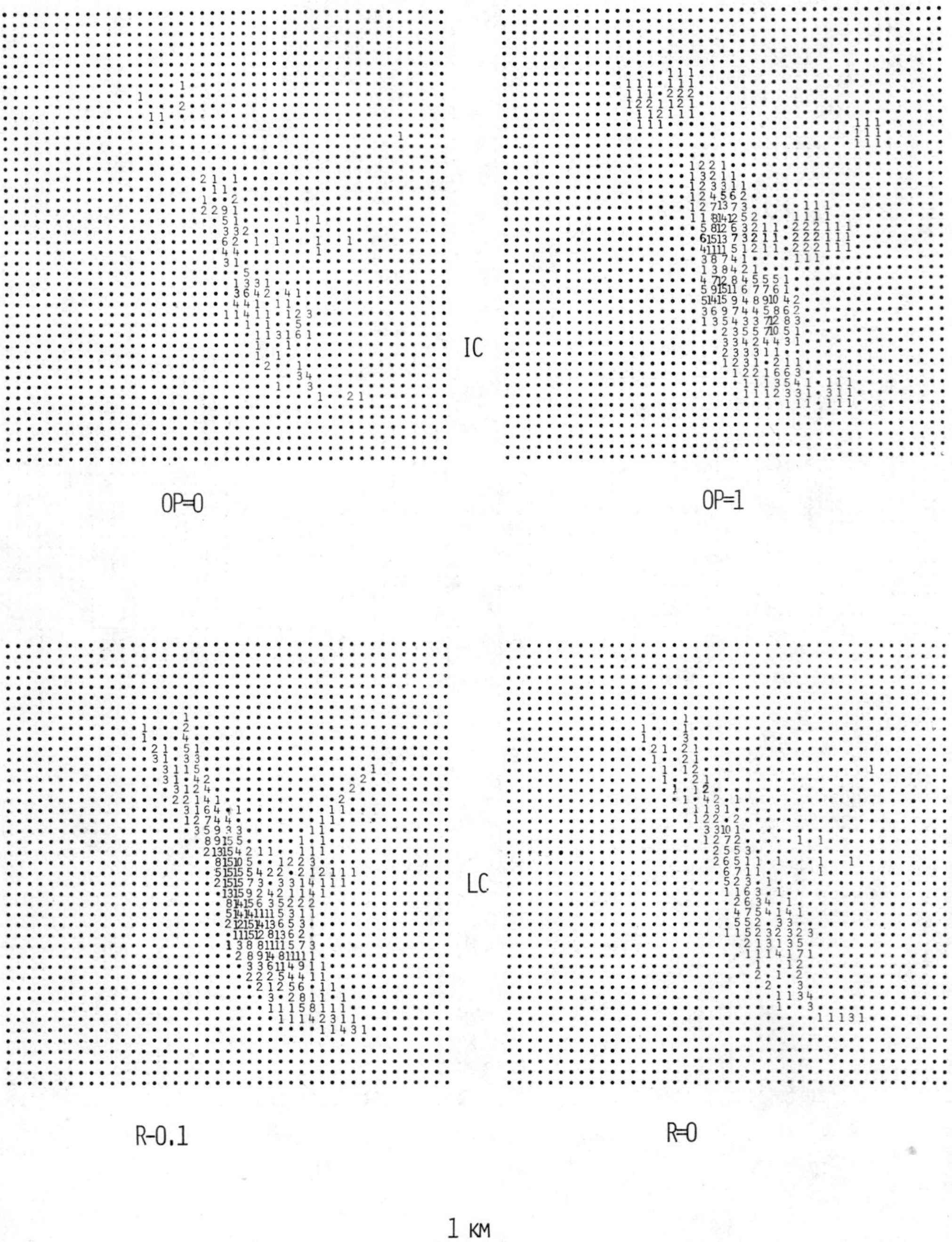

Fig. 4. Relationship between area and sample size for a male fox pup (FO4) using 4 methods.

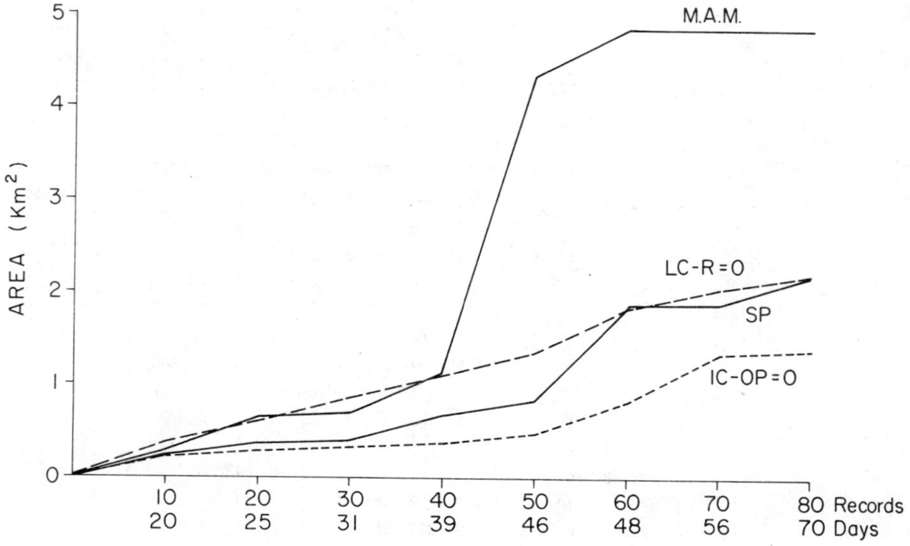

Fig. 5. A comparison of the home range of a female fox using LC (R=0 and R=0.1 km h^{-1}) and IC (OP=0 and OP=1).

HABITAT UTILIZATION

Through the use of IC with case operator options, high-count cells (or cell clusters) can be isolated for comparisons with features such as topography, habitat type, prey density, other attractants etc. assuming knowledge of the occurrence of these features temporally and spatially. Digitizing habitat units as polygons (Gilmer, Miller and Cowardin, 1973) provides base maps which are superimposed with IC to depict habitat use frequency. Interactive querying assists in evaluating temporal changes and sex class, age class or species specific differences. This two-dimensional depiction can be modified for arboreal species to three dimensions (Koeppl *et al.*, 1977). If IC is viewed as an X, Y matrix on the ground plane the Z dimension is a set of matrices at chosen units above the ground. Instead of determining areas or frequencies on the ground, volumetric measures can be related to three-dimensional habitat. Similarly, terrestrial species' movements can be related to land elevation, i.e. measuring use of area as a function of land form.

Despite the fact that insufficient information on habitat may be available, or accuracy of fixes results in cells which are too large to measure use of habitat, the value of being able to group large blocks of cells may still be important, as for example, in aerial survey assessment or aircraft baiting strategies.

BEHAVIORAL ANALYSIS

ACTIVITY TYPE DISTRIBUTION

A simple option is to modify IC to record activity type such as active or stationary, day or night (or other determined time periods) and plot such activity instead of used cells. The distribution of a type of status can then be investigated in relation

-	-	-	-	-	-	-	-	-	-	-	-	-	-	-	-	-	-	-
-	-	-	-	-	-	-	-	-	-	-	-	-	-	-	-	-	-	-
-	-	-	-	-	-	-	-	-	-	-	1	-	-	-	-	-	-	-
-	-	-	-	-	-	-	-	-	1	-	-	-	-	-	-	-	-	-
-	-	-	-	1	-	-	-	-	1	10	2	-	-	-	-	-	-	-
-	-	1	-	-	2	-	-	-	8	41	4	23	14	-	-	-	-	-
-	-	-	-	-	-	1	-	10	28	60	4	22	160	2	1	-	2	1
-	-	-	-	-	-	4	53	72	175	99	69	157	131	10	3	22	23	-
-	-	-	-	-	2	12	119	166	418	84	16	50	24	9	2	1	1	-
-	-	-	-	-	-	179	294	56	56	16	7	3	3	1	-	1	-	-
-	-	-	-	1	2	40	205	41	25	7	5	2	1	-	3	8	5	-
-	-	-	-	-	-	-	15	23	1	10	13	5	2	1	15	60	41	-
-	-	-	-	-	-	-	4	10	20	178	24	11	1	10	23	167	87	32
-	-	-	-	-	-	-	1	10	37	80	33	24	6	2	13	66	29	9
-	-	-	-	-	-	-	-	12	99	34	2	17	14	1	-	6	4	2
-	-	-	-	-	1	-	3	1	27	47	7	2	8	5	1	-	12	-
-	-	-	-	-	-	1	-	-	5	58	17	10	22	4	1	1	3	1
-	-	-	-	-	1	-	-	1	52	88	18	18	16	25	4	-	2	1
-	-	-	-	9	3	4	6	8	74	101	30	7	11	4	4	-	-	-
-	-	-	-	-	1	10	13	12	105	82	46	21	12	6	3	-	-	-
-	-	-	-	2	1	6	15	11	90	340	76	20	19	10	3	1	-	-
-	-	-	-	3	-	2	6	3	7	153	130	8	2	1	4	1	-	-
-	-	-	-	1	1	5	4	2	1	4	25	12	-	-	1	5	-	-
2	1	-	10	1	-	2	3	3	-	-	2	1	2	1	1	21	13	1
3	4	2	5	2	9	1	-	2	-	3	-	-	9	8	-	-	22	3
12	23	1	-	17	36	4	2	2	-	3	2	-	7	1	-	-	-	-
3	32	19	2	77	1	2	4	3	4	11	72	4	1	-	-	-	-	-
31	60	31	1	2	3	-	6	9	1	34	79	7	4	13	8	-	-	-
30	19	19	-	-	-	-	1	-	6	23	3	5	23	63	81	5	-	4
7	4	1	-	-	-	-	-	2	3	11	25	17	120	72	63	5	13	9
-	-	-	-	-	2	-	4	7	4	-	4	16	137	23	53	3	10	3
-	-	-	-	-	-	-	2	1	-	-	-	7	3	-	-	-	-	-
-	-	-	-	-	-	-	-	-	-	-	-	3	4	1	-	-	-	-

1 km

Fig. 6. Count of all fixes in core study area obtainable from IC (OP=0) or query. Circled numbers denote known fox dens.

to habitat and distance from home site, for example. At this point a suitable analysis may involve comparisons between siblings, adults and their young, or species. This method allows use of physiological measures or specific activities (hunting, grooming, foraging) to be correlated with locational data as discussed by Covich (1976).

Such correlations are further investigated with IC through the direction and orientation distribution. An origin of choice such as home site, highest frequency cell or geographic center is used. Through conversion of either high-count cells or the X, Y of a particular cluster of cells to polar coordinates, a vector can be drawn, the width being determined by the relative frequency rank of the cell. Comparisons among members of families or any other group chosen by day, season, year or home site are easily plotted with the aid of interactive querying. LC is used to compare vectors with probable travel corridors.

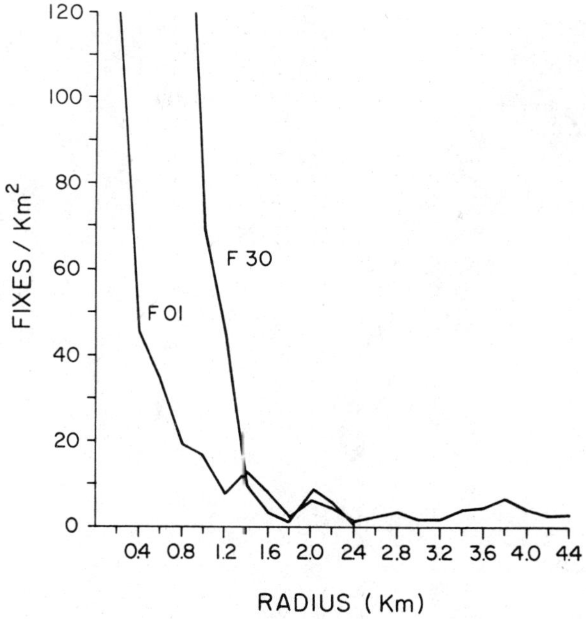

Fig. 7. Relationship between density of used cells and distance from the den for an adult male fox (F01) and an adult female fox (F30).

Interaction Measures

This determination derives from the utilization distribution using IC and/or LC. Comparing locations of two individuals within specified time intervals, say 3 weeks, provides interaction probability measures. Actual interaction can be determined from matching temporally paired observations. A modification employed during our study of family interactions of foxes involved the simple construction of a table comparing the mean distance between family members during designated time intervals and compared throughout a season. Wildlife managers may wish to test hunter versus game interaction potential in this way where, for example, hunter occurrence is a probability dependent on access zone (similarly, prey occurrence versus predator occurrence). In fact, all of the area determination methods can be used to obtain interaction measures with the groups to be compared determined by querying.

CONCLUSIONS

During our telemetry study several species were radio tracked over long periods. Although most of the data were discontinuous, over 14,000 fixes were obtained. Recently, many other studies have been involved in radio tracking a large number of animals, and like continuous tracking studies, have amassed a large amount of data. Such quantities of data invariably require computer analysis. When a data base file is prepared for the computer, a relatively small amount of additional effort can provide a large number of analysis options.

The basis of the strategy developed in this paper is that the interactive use of several methods with various options permit an exploration of data characteristics, assumptions and biological implications. Researchers may wish to utilize methods

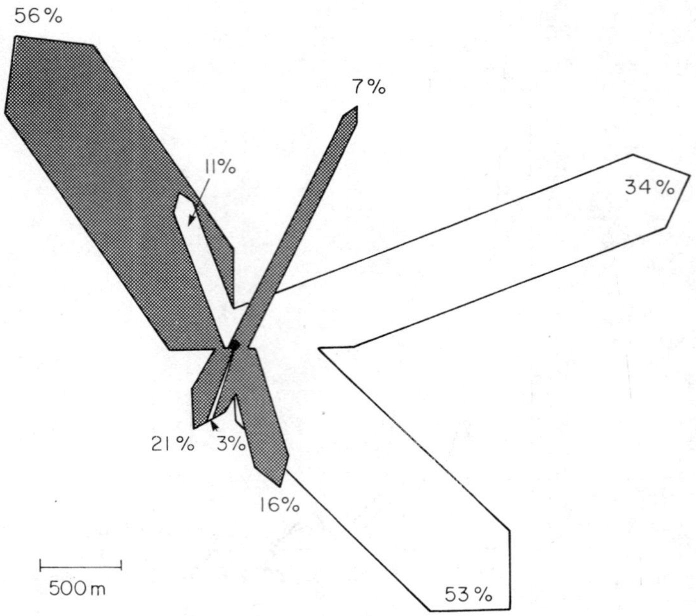

Fig. 8. Differences in use of 4 quadrants by an adult vixen in 1976 (white) and her yearling daughter in 1977 (black) in relation to the same den used for pup-rearing. (Vector width is proportional to percent of all fixes; vector length equals distance to farthest fix; vector bearing is the vector resolution of all fixes in a quadrant). Developed from an LC map.

described earlier in the literature for comparative purposes. Alternatively, most will wish to develop analyses to provide information for the specific objectives of their study. Rather than trying to develop a universal statistic, our approach has been to develop a fundamental package from which a great number of diverse measures may be readily calculated. This need frequently develops during early analyses when the objectives of the study and the type of data are known but the characteristics and suitability of the data set for a specific analysis is unknown.

The description of the computer software, particularly the query system, has been general since, to some extent, it is hardware dependent. This is not felt to be a major problem as hardware costs represent an increasingly small proportion of software development costs and since the strategy maximizes the interaction between the researchers' experience and the data set, a neglected aspect of most analyses strategies. A simple example is observation of the computer incrementally plotting chronologically linked fixes determined from radio tracking. This essentially relives both the data collection and movements of the animal and is especially important during longterm studies. Portions of the general strategy package described here are being investigated in a more detailed fashion and will be reported later.

Acknowledgements — We wish to express our gratitude to B. Pond, J. Litalien, N. Maxwell and B. Stevens of the Department of Geography at Queen's for their efforts in the development of the computer techniques described in this paper. We also wish to thank D. Johnston, B. Earle, I. Watt and F. Matejka of the Rabies Research Unit,

Ontario Ministry of Natural Resources for their dedication in assisting with the data. Finally, we wish to note that the paper could never have been completed without the indefatigable assistance of E. Harrison at Maple. This is Ontario Ministry of Natural Resources, Wildlife Br. Contribution No. 79-6.

REFERENCES

Ables E. D. (1969) Home-range studies of red foxes (*Vulpes vulpes*). *J. Mammal.* 50, 108-120.

Adams L. (1965) Progress in ecological biotelemetry. *Bioscience.* 15, 83-86, 155-157.

Adams L. and Davis S. D. (1967) The internal anatomy of home range. *J. Mammal.* 48, 529-536.

Blair W. F. (1940) Home ranges and populations of the meadow vole in southern Michigan. *J. Wildl. Mngmt.* 4, 149-161.

Brant D. H. (1962) Measures of the movements and population densities of small rodents. *Univ. Calif. Publ. in Zool.* 62, 105-184.

Burt W. H. (1943) Territoriality and home range concepts as applied to mammals. *J. Mammal.* 24, 346-352.

Calhoun J. B. (1955) A technique for investigating the distance parameter of home range. *J. Mammal.* 36, 45-52.

Calhoun J. B. and Casby J. U. (1958) Calculation of home range and density of small mammals. U.S. Public Health Serv. Public Health Monogr. 55. 24 pp.

Cochran W. W. and Lord R. D., Jr. (1963) A radio-tracking system for wild animals. *J. Wildl. Mngmt.* 27, 9-24.

Cooper W. E., Jr. (1978a) Home range criteria based on temporal stability of areal occupation. *J. theoret. Biol.* 73, 687-695.

Cooper W. E., Jr. (1978b) Home range size and population dynamics. *J. theoret. Biol.* 75, 327-337.

Covich A. P. (1976) Analyzing shapes of foraging areas: some ecological and economic theories. *Ann. Rev. Ecol. Syst.* 7, 235-257.

Dalke P. D. (1942) The cottontail rabbit in Connecticut. *State Geol. Nat. Hist. Survey Bull.* 65, 1-97.

Deat A., Mauget C., Mauget R., Maurel D. and Sempéré A. (1979) The automatic continuous and fixed radio tracking system of the Chizé Forest: Theoretical and practical analyses. In C. J. Amlaner, Jr. and D. W. Macdonald (eds.), *A Handbook on Biotelemetry and Radio Tracking*, pp. 439-451. Oxford: Pergamon Press.

Dice L. R. and Clark P. J. (1953) The statistical concept of home range as applied to the recapture of the deermouse (*Peromyscus*). Univ. Mich. Lab. Vert. Biol. Contrib. 62. 15 pp.

Fitch H. S. (1947) Ecology of a cottontail rabbit (*Sylvilagus audubon*) population in central California. *Calif. Fish Game* 33, 159-184.

Gilmer D. S., Miller S. E. and Cowardin L. M. (1973) Analysis of radiotracking data using digitized habitat maps. *J. Wildl. Mngmt.* 37, 404-409.

Harvey M. J. and Barbour R. W. (1965) Home range of *Microtus ochrogaster* as determined by a modified minimum area method. *J. Mammal.* 46, 398-402.

Hayne D. W. (1949) Calculation of size of home range. *J. Mammal.* 30, 1-18.

Jennrich R. I. and Turner F. B. (1969) Measurement of non-circular home range. *J. theoret. Biol.* 22, 227-237.

Koeppl J. W., Slade N. A. and Hoffmann R. S. (1975) A bivariate home range model with possible application to ethological data analysis. *J. Mammal.* 56, 81-90.

Koeppl J. W., Slade N. A., Harris K. S. and Hoffmann R. S. (1977) A three-dimensional home range model. *J. Mammal.* 58, 213-220

Lotimer J. S. (1979) A telemetry system for wildlife in Ontario. Ont. Min. Nat. Res., Wildl. Br., Maple, Ont. (In preparation).

Maxson S. J. (1978) Spring home range and habitat use by female ruffed grouse. *J. Wildl. Mngmt.* 42, 61-71.

Mazurkieicz M. (1971) Shape, size and distribution of home ranges of *Clethrionomys glareolus*. *Act Theriologica* XVI, 23-60.

Mech L. D. (1974) Current techniques in the study of elusive wilderness carnivores. *Proc. Int. Cong. Biol.* 11, 315-322.

Metzgar L. H. (1972) The measurement of home range shape. *J. Wildl. Mngmt.* 36, 643-645.

Metzgar L. H. and Sheldon A. L. (1974) An index of home range size. *J. Wildl. Mngmt.* 38, 546-551.

Mohr C. O. (1947) Table of equivalent populations of North American small mammals. *Am. Midland Nat.* 37, 223-249.

Moore E. G. and Tinline R. R. (1978) The PRISSA system. Population dynamics information systems and education planning at the local level. Report prepared for Ont. Min. Educ. Dept. of Geog., Queen's Univ., Kingston, Ont. 102 pp.

Odum E. P. and Kuenzler E. J. (1955) Measurement of territory and home range size in birds. *Auk* 72, 128-137.

Sanderson G. C. (1966) The study of mammal movements — a review. *J. Wildl. Mngmt.* 30, 215-235.

Siniff D. B. and Tester J. R. (1965) Computer analysis of animal movement data obtained by telemetry. *BioScience* 15, 104-108.

Stevens R., Tinline R. and Moore E. (1979) An interactive query system. Tech. Rept. No. 5, PRISSA. Dept. of Geog., Queen's Univ., Kingston, Ont. 112 pp.

Stickel F. (1954) A comparison of certain methods of measuring ranges of small mammals. *J. Mammal.* 35, 1-15.

Van Winkle W. (1975) Comparison of several probabilistic home-range models. *J. Wildl. Mngmt.* 39, 118-123.

The Evaluation of Home Range Size and Configuration Using Radio Tracking Data

D. W. MACDONALD,* F. G. BALL** and N. G. HOUGH*

*Department of Zoology, South Parks Road, Oxford, U.K.
**Department of Biomathematics, Pusey Street, Oxford, U.K.

Abstract — Using specimen data we compare conclusions based on different techniques for treating movement data. We believe that grid methods are particularly useful but stress that there is no substitute for field observations. Radio tracking used in conjunction with field craft can greatly increase the possibilities of watching animals and so reduce the need for reliance on the models discussed in this paper. We also explore methods of assessing the extent of overlap between ranges and both the dynamic interaction and probability of meeting of two animals. Applying these techniques to fox movement yields some information relevant to models of rabies transmission.

INTRODUCTION

By stalking a radio tagged animal a biologist can greatly increase his chances of watching its behavior and so discovering not only where it is, but what it is doing, and with whom it is doing it. In this way a description of the animal's movements and observations on its behavior can complement each other in the interpretation of the species' social system (Macdonald, 1978). In some cases even the most skillful fieldcraft used in conjunction with radio tracking is likely to be rewarded by only rare sightings. In such cases information on the animal's position is gathered with little knowledge of its behavior. How can this information be interpreted?

Considerable effort has been devoted to the analysis of location data from live trapping (Hayne, 1949; Brown, 1962; Mazurkiewicz, 1971; Randolph, 1977) and radio tracking (Tester and Siniff, 1965; Storm, 1965; Siniff and Jesson, 1969; Dunn and Gipson, 1977). Sanderson (1966) pointed out the inadequacy of the interpretative techniques in this field, especially in terms of the concept of home range proposed by Burt (1943). We hope to advise the would-be radio tracker on how he can analyze his results and to warn that the more mathematically complicated models do not necessarily give greater biological insight.

The first difficulty arises where there is doubt about the relationship between the animal's real use of space and the distribution of fragmentary 'glimpses' of this usage provided by sporadic radio fixes, sightings or trappings. For those studying social systems this is not a trivial problem; a slight difference in the position of home range boundaries of neighboring animals may have considerable implications for the interpretation of their social organization. This is illustrated by the work of Owings *et al.* (1977) who present data on the locations of the ground squirrel, *Spermophilus beechei*. These authors drew boundaries around sightings of individual squirrels (equivalent to radio 'fixes' in this context) and then measured the degree of overlap between the neighboring ranges. Part of their data is redrawn in Fig. 1 from which they conclude that the home range of male 'A' overlaps those of females 'C' and 'E'. The home range of squirrel 'E' is drawn by the 'convex polygon' method (Mohr, 1947; Southwood, 1966, p. 262; see also Odum and Kuenzler, 1955). A different protocol has been used for drawing the boundaries of squirrel ranges 'A' and 'C'; Stickel (1954) and Southwood (1966) describe graphical methods of treating radio fixes without the restriction of convexity, and Jennrich and Turner (1969) quote the 'map maker's' formula for computing the area of such polygons.

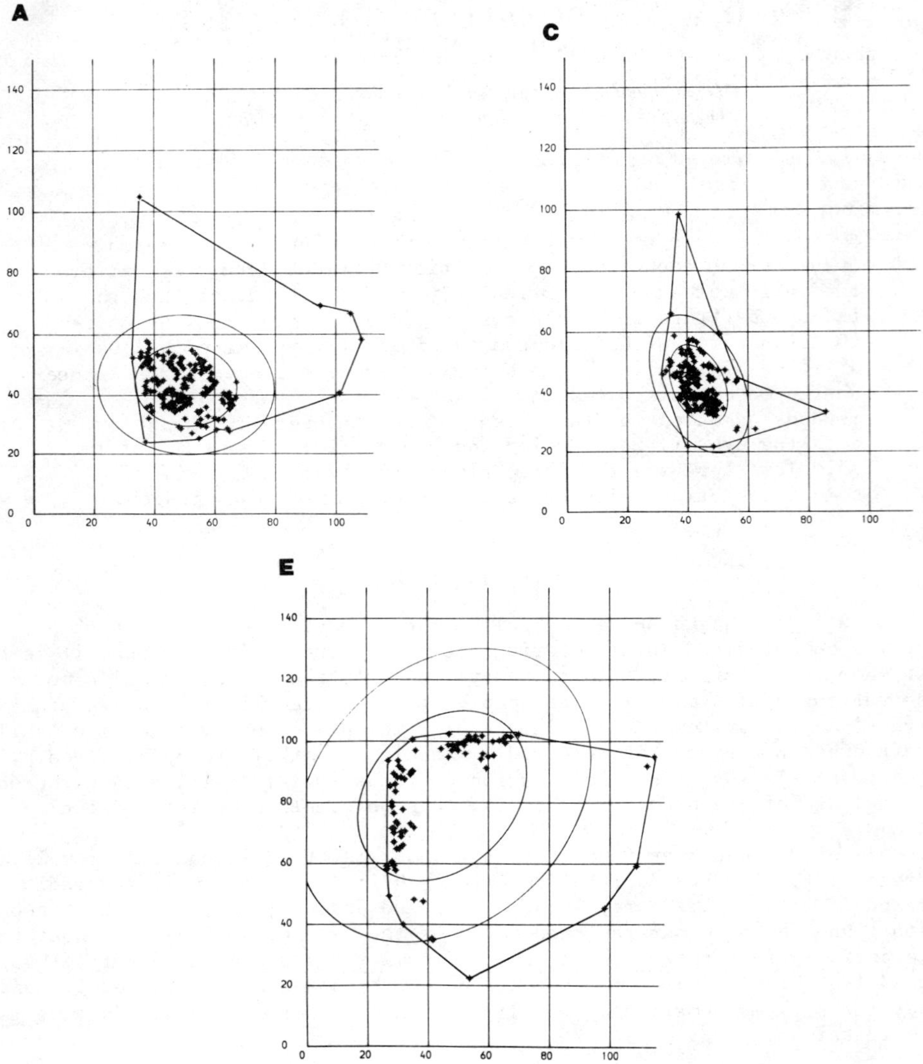

Fig. 1. Individual sightings of three ground squirrels (A male, C female, E female) showing estimates of their range size and configuration based on the observations of Owings *et al.* (1977) and using polygonal shapes and contrasted with 62.3 and 95.0 percent probability ellipses.

Various models of the probability of recapture or location in given areas or annuli have been proposed. The debate over which are most appropriate has revolved around the legitimacy of assuming either circular home ranges or the bivariate normality of animals' movements (see reviews by Jennrich and Turner, 1969; van Winckle, 1975). Both these reviews favor the covariance technique proposed by Jennrich and Turner (1969) and Mazurkiewicz (1969) which enables ellipses including fixed proportions of the variance in the X, Y coordinates of the locations to be drawn. This technique assumes a bivariate normal distribution, and is fully explained in Mazurkiewicz (1971, p. 57). A reanalysis of the squirrel locations using this covariance technique is also presented in Fig. 1. Table 1 shows that while Owings *et al.* (1977) calculated that about 60.1 percent of the home range of squirrel 'E' was overlapped by that of 'A', 95 percent probability ellipses for 'E' is only overlapped by that of 'A' by 17.1 percent. Furthermore, the 62.3 percent ellipses show that only 0.1 percent of 'E' is overlapped by 'A' in a 'core area'. Using this method one might conclude that male 'A' and female 'E' rather strictly avoided each other, as opposed to largely cohabiting. Without observations on what the animals were doing at their wayward locations, it would have been impossible to know which conclusion to draw and in this respect the sightings reported in this study are considerably more useful than an equivalent number of radio fixes. The covariance ellipses also provide parameters (area of ellipse), its orientation (radians) and the length of its semi-major and semi-minor axes (e.g. Table 2) which can be used for statistical comparison (Mazurkiewicz, 1971). These interesting data highlight two questions, (i) how can home range size be assessed, and (ii) to what extent can movements of two animals be analyzed to illuminate the relationship between them?

Table 1.
Estimates of the overlap between the home ranges of three ground squirrels, showing the percentage of home range (X) which is overlapped by the other home range (Y)

	Owings *et al.*				Convex polygon		
Y	X			Y	X		
	A	C	E		A	C	E
A	—	97.7	60.1	A	—	97.3	69.0
C	39.2	—	24.8	C	51.4	—	35.8
E	85.9	88.4	—	E	94.7	94.7	—
	62.3 percent ellipse				95 percent ellipse		
Y	X			Y	X		
	A	C	E		A	C	E
A	—	96.4	0.1	A	—	98.7	17.1
C	50.3	—	0.2	C	50.8	—	10.7
E	0.3	1.29	—	E	56.2	68.5	—

The size and shape of home ranges and hence the form of spatial relationships within a community are often a starting point for the analysis of a social system. Where location data are not supplemented adequately with direct observations, biologists often rely on mathematical models to generate estimates of home range size and configuration. In comparing some models we hope to show ways in which they can assist and ways in which no mathematics, however sophisticated, can remedy inadequate observation.

Table 2.
Various home range parameters for the ground squirrels (Fig. 1)

Squirrel	n	Owings *et al.* m^2	Convex polygon ha	62.3% ha	95% ha	semi-major m	semi-minor m	θ radians
A	158	0.36	0.38	0.07	0.21	29.2	23.2	-0.11
C	119	0.15	0.20	0.037	0.11	24.3	14.4	-1.16
E	81	0.52	0.52	0.23	0.70	54.3	41.0	0.79

SOME MODELS

The following models can be applied to bursts of movement data derived from radio tracking, in the form:

$$Z_1, Z_2, \ldots\ldots\ldots, Z_N$$

where $Z_i = (x_i, y_i)$ is the location of the ith observation of the animal with reference to a fixed x, y coordinate system.

CONVEX POLYGON

The home range is defined as the smallest convex polygon enclosing *all* the points $Z_1, Z_2 \ldots\ldots\ldots, Z_N$. (A polygon E is convex if the straight line between any pair of points inside E lies totally within E.)

BIVARIATE NORMAL AND CIRCULAR NORMAL MODELS

The home range is estimated as the area falling within an ellipse the axes of which have a length proportional to a specified percentage of the variance of the x and y coordinates in the sample $Z_1, Z_2 \ldots, Z_N$ and assuming that these locations are an independent sample from a bivariate normal distribution. The means of the x and y coordinates of the fixes are denoted by μ_1 and μ_2, the variances by σ_1^2 and σ_2^2 and the covariance by σ_{12}. The circular normal model (Calhoun and Casby, 1958) imposes the unrealistic restrictions that $\sigma_{12} = 0$ and $\sigma_1^2 = \sigma_2^2$.

The computation of the dimensions of these ellipses is explained in Appendix 1. The sizes of the ellipses are determined by the 'inclusion probability' $P(r)$, which depends on the values of r^2 which relates to the square of the length of the ellipses major and minor axes and arbitrarily sets the proportion of fixes to be included within the calculated ellipse. The choice of a value for r^2 and hence the 'inclusion probability' rests with the researcher: somebody interested in core areas of home ranges might wish to use $r^2 = 2$, $P(r) = 0.62$, in order to discover the smallest ellipse that includes the majority (about 62 percent) of the observations. Jennrich and Turner (1969) define a home range, in this context, as the area of the smallest subregion (set by r^2) which accounts for a specified proportion of the total utilization (Koeppel *et al.*, 1977; consider the special case of animals using the third dimension). Mazurkiewicz (1969) selects a value of $r^2 = 6$, ($P(r) = 0.95$), while Calhoun and Casby (1958) used $r^2 = 9$ ($P(r) = 0.99$). We use $r^2 = 2$ and $r^2 = 6$ throughout this paper.

It is helpful to remember that one standard deviation embraces 68 percent of the distribution about a mean of a normal distribution and thus the 62.3 percent probability ellipse is not dissimilar to one standard deviation about the geometric center of the fixes. The boundaries of these ellipses do not necessarily have any biological meaning, and do not shed more than a statistical light on what should, or should not qualify as part of the individual's home range, as defined on the basis of its 'routine' movements (Jewell, 1966).

BIVARIATE ORNSTEIN-UHLENBECK DIFFUSION MODEL

In this model the utilization of space by the animal is still assumed to be bivariate normal, but recognition is made of the correlation between the successive fixes along the animal's route (which is assumed to be a continuous path). The model assumes that the configuration of the home range is stable in the longterm, and that the animal's route is Markovian, i.e. a movement at time t depends only on the animal's position at time t and *not* on the history of its path before time t. It can be shown that the Ornstein-Uhlenbeck (O-U) Diffusion model is the only model which satisfies these assumptions. The critical difference between this model (proposed by Dunn, 1977a; Dunn and Gipson, 1977) and the Jennrich and Turner (1969) model is the assumption of continuous paths. Thus, two successive fixes are likely to be close together if the sampling interval is short. The potential bias inherent in computing ellipses from distributions derived from relatively short series of data (with obvious dependence between successive fixes) is compensated for in the O-U model but not with Jennrich and Turner's assumption of independent fixes. The extent of this compensation (e.g. in size and orientation of probability ellipses) depends on the pattern of the animal's movements within its presumed stable home range during the sampling period (e.g. the pattern of day range use in the context of total home range use). It is a consequence of the model that the animal's movements have a centralizing tendency (i.e. it is inclined to return towards the center of its stable home range). Thus the magnitude and orientation of the centralizing tendency calculated from the particular sample data is implicitly used to adjust the size and orientation of the Dunn and Gipson probability ellipses with respect to those of Jennrich and Turner.

As with the Jennrich and Turner models, we cannot calculate the true values of the means, variances and covariance which govern the center, shape, and size of the home range or the parameter measuring its centralizing tendency, but if the assumptions of the O-U model are valid then this model can be used to produce estimates from the distributions of fixes and these estimates are used to draw the ellipses.

GRID METHODS

Rather than drawing a single contour around the boundary of a home range, the animal's movements are plotted in terms of the frequency of locations in each of a number of squares or polygons (Adams and Davis, 1967). Siniff and Tester (1965) used this technique to describe fox movements gathered continuously from an automated station and it is widely used by primatologists (e.g. Clutton-Brock, 1974).

THE 'FIELD-WORKER'S ESTIMATE'

In the following sections we contrast the features of each of these methods with the measurement of home range size and configuration drawn by hand on the basis of direct observation, interpretation of field sign and radio tracking.

COMPARISON OF MODELS

Each of the above home range models has been applied to two samples, one of 174 radio fixes on a female red fox followed in a suburban habitat and the second — fixes on a male fox followed over moorland. These data are displayed in Fig. 2 together with various estimates of home range size and configuration. (For other comparisons, see Mohr and Stumpf, 1966; Trevor-Deutsch and Hackett, 1980, this volume). The models have the following characteristics:

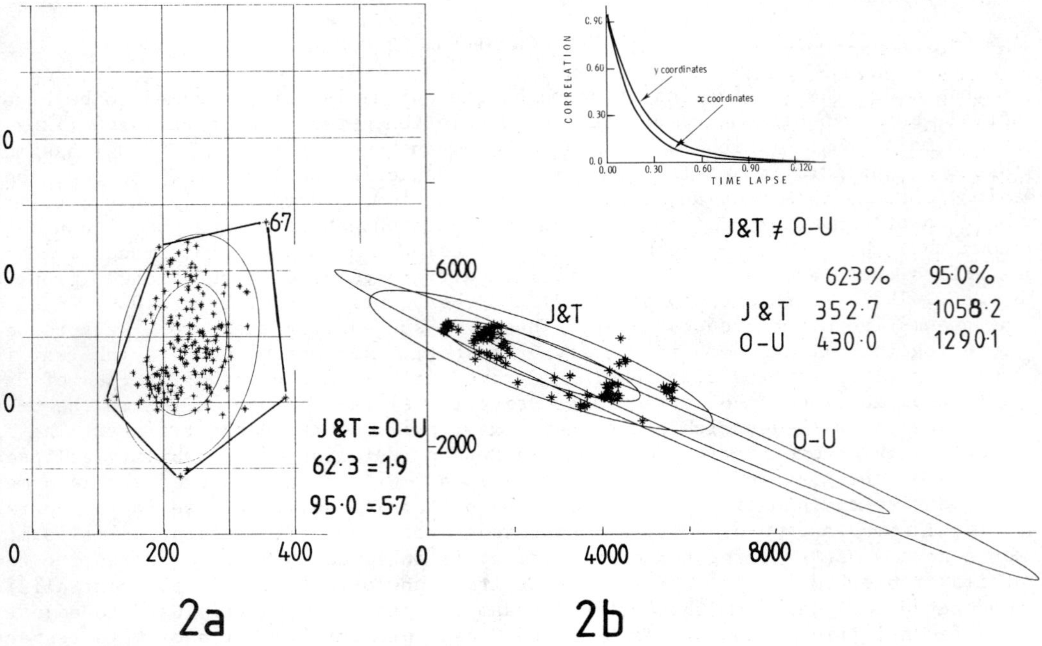

Fig. 2. Estimates of home ranges (scale in meters) of two foxes radio tracked by night in different habitats. (a) A vixen radio tracked in the suburbs of Oxford in the vicinity of a well provisioned feeding site. (b) A dog fox radio tracked on the moors of Cumbria feeding mainly on rabbits and birds. Convex polygons, 62.3 and 95 percent probability ellipses as calculated by both the simple bivariate normal (J and T) and the O-U processes are shown. The inset graph presents the serial correlation between successive *x*, *y* coordinates of the vixen's positions (areas in hectares).

CONVEX POLYGON

(i) This method is very sensitive to movements on the periphery of the animal's home range, irrespective of the frequency with which that area is visited. Individuals of most species occasionally travel to 'unusual' places, perhaps on an excursion from their normal range (e.g. Niewold, 1974). The effect of including peripheral locations in a convex polygon on interpretation of social behavior is illustrated in Fig. 1.

(ii) The area of the home range increases with sample size using this method and under the assumptions of the bivariate normal O-U process, will tend to infinity as sample size tends to infinity.

(iii) Large areas of land which are never visited can be included within the polygon if an animal's range is an irregular shape, perhaps delineated by topographical features.

CIRCULAR NORMAL AND BIVARIATE NORMAL

(i) A circular normal estimate of home range will usually embrace a larger area than a bivariate one. This is elaborated in Appendix 2 using a sphericity test (Anderson, 1958, p. 259) of the circular normal versus bivariate normal models.

(ii) The assumption of bivariate normality is unlikely to be realistic since the resources governing the pattern of animal movement are in many cases not normally distributed in space. For example mates, food supplies, and enemies are all likely to be clumped in distribution. This model does not allow for heterogeneous use of space either because of resource availability or social factors. This constraint has been emphasized by Metzgar (1972, 1973) and Fig. 3 shows that fox movement data may depart from normality. A sample of 36 such analyses performed on the x, y coordinates of movement data gathered from 18 resident red foxes in four different habitats showed that 21/36 departed from a normal distribution at greater than $P < 0.01$.

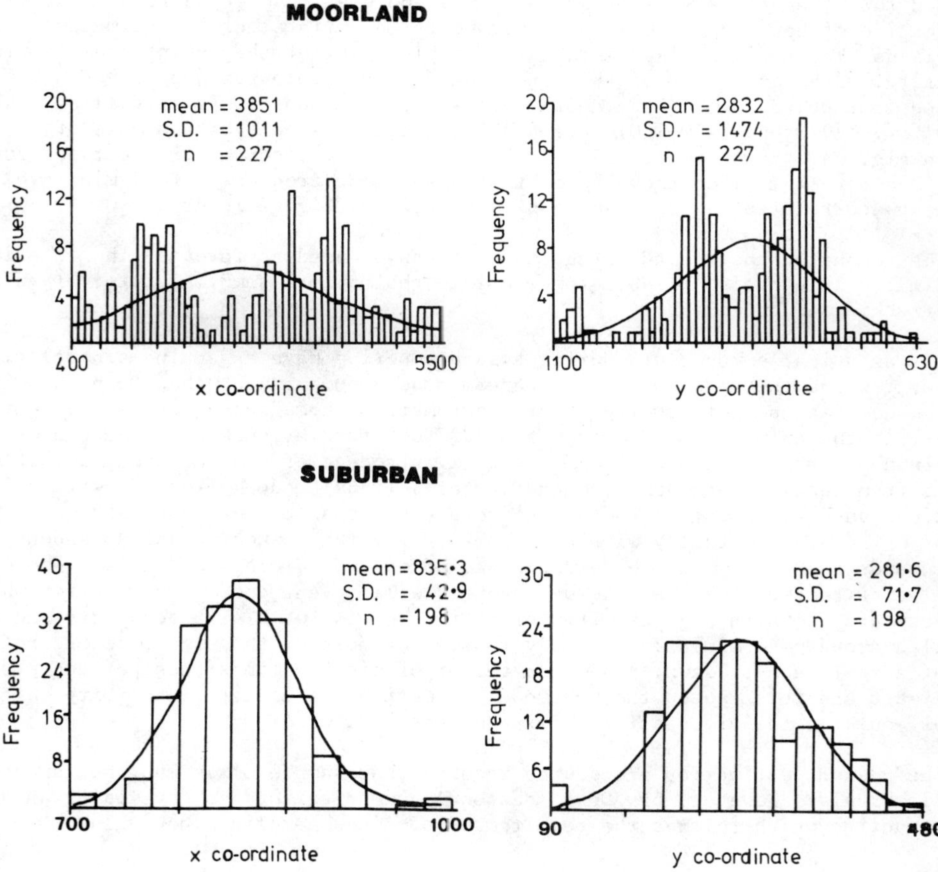

Fig. 3. Frequency distribution of the x, y coordinates (meters) of the foxes in Fig. 2 and the normal distribution which would underlie such a mean and variance. Testing the observed and predicted distribution with a χ^2 test revealed that while the suburban fox loosely approximated a normal distribution (ordinate $P < 0.8$, abscissa $P < 0.02$) the moorland fox did not (ordinate and abscissa $P < 0.001$).

(iii) In spite of its unrealistic assumptions Jennrich and Turner (1969), van Winkle (1975), and Koeppel *et al.* (1975) favor this technique which they consider to be statistically robust.

(iv) The bivariate normal probability ellipses reduce the disproportionate weighting bestowed on wayward points by the convex polygon.

(v) The bivariate normal estimate of home range may also include areas of land which are never visited, as in (iii) of the convex polygon and all other parametric methods.

ORSTEIN-UHLENBECK DIFFUSION PROCESS

(i) The tiny difference (0.2 ha) between the bivariate normal and O-U estimates of the home range (hence only one set of ellipses drawn) on Fig. 2a suggests that the assumption of independence along sample paths made in the bivariate normal model does not necessarily seriously invalidate its estimates of mean vectors and covariance provided that the sample size is quite large. However, a precise rule for answering the question of how large the sample must be for bivariate normal estimates to be valid is not known. The disadvantage to the biologist of the O-U process is its complexity, although a copy of this and other related computer programs can be obtained from Dr. J. Dunn (Dept. Mathematics, Univ. Arkansas, Fayetteville, Arkansas). Dunn *et al.* (1978, pp. 150-160) have published other aspects of analyses of the data in Fig. 2a. For example, by calculating the serial correlation between successive fixes for that vixen they showed that a sampling frequency of 30 min involves only a low correlation between successive locations (Fig. 2 inset).

(ii) The technique has the advantage that it can be used to forecast the location of the animal. It can also be used to calculate the optimal sampling interval while radio tracking (Dunn, 1977b).

(iii) If an animal's movements during a sample period have a genuine centralizing tendency towards the middle of its longterm stable home range, then Dunn and Gipson's (1977) model can use this to improve the estimate of home range size and orientation. However, if the animal's movements bear a different relationship to the center of its longterm home range, then the O-U model will introduce errors which may depart even farther from reality than Jennrich and Turner's (1969) model. Indeed, with a few exceptions such as repeatedly returning to a nest area to feed young, which anyway may not be located centrally within a home range, the assumption of a tendency to move towards the center of a home range has no obvious biological basis. Where no centralizing tendency exists, the O-U process will invent one. Thus the fox whose movements are shown in Fig. 2b often slept in the west of his range by day and then travelled along predominantly easterly routes for most of the night, before returning home at a fast pace at dawn. The mean vector of his movements thus pointed strongly to the east and this caused the O-U model to estimate a center of activity which was well outside the area that the fox ever visited.

(iv) The present estimation procedures require that sample fixes must be taken at fixed intervals. However, Dr. Dunn (personal communication) is developing an estimation routine which relaxes the requirement of fixed sampling intervals.

GRID METHODS

(i) Home range size can be computed by adding the area of all squares containing fixes; however this gives no measure of the distribution of these fixes, and is influenced by the size of grid square chosen.

(ii) Empty squares within the home range may result from sampling strategy or from biologically relevant factors. Voigt and Tinline (1980, this volume) describe systems that allow used squares to weight their neighbors and for linking squares through which an animal must logically have travelled between two sightings, even if it was not followed while doing so.

(iii) The size of the grid squares should be no smaller than the limitations in accuracy of the tracking system, which may be insufficient to resolve movements in terms of detailed habitat types (see Cooper, 1978).

(iv) For comparison of the pattern of use of home ranges using frequency values in different cells we use cell sizes described by *proportions* of total range size (in Fig. 4, squares of 550 m^2) since comparisons can otherwise only be made between ranges of identical sizes as the number of cells influences the frequency distribution of movement data. Two types of comparison may be made between these gridded representations of home range use: (1) the area used, and (2) the pattern of range use. Measures of variability (e.g. standard deviation, variance) and of diversity (e.g. Shannon-Weiner index, as used to compute species diversity), allow statistical comparisons between different patterns of range use, but are influenced by the number of sample cells and by the total number of observations, but the coefficient of variation (standard deviation divided by the mean) is unaffected by sample size.

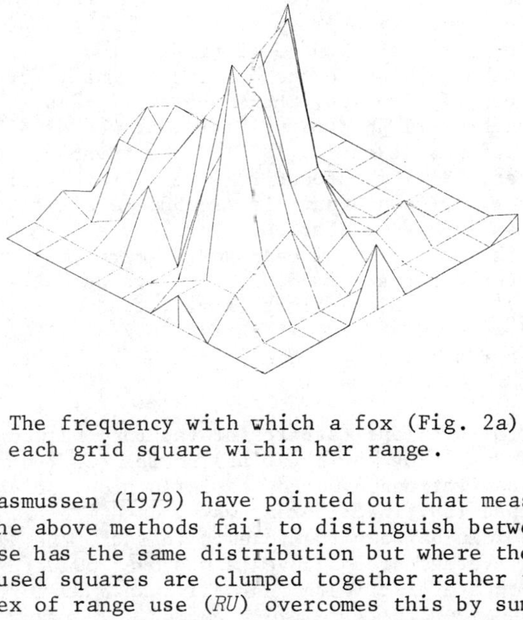

Fig. 4. The frequency with which a fox (Fig. 2a) was found in each grid square within her range.

(v) Rasmussen and Rasmussen (1979) have pointed out that measures of the intensity of home range use by the above methods fail to distinguish between ranges where the frequency of cell use has the same distribution but where the pattern of use differs, e.g. where heavily used squares are clumped together rather than regularly spaced. The Rasmussens' index of range use (*RU*) overcomes this by summing the number of fixes in a pair of cells and dividing this by the distance between them to give a value (d) which is similarly computed for each of the $N(N - 1)/2$ pairs of cells:

$$RU = \frac{\sum_{i=1}^{P} (d_i - \bar{d})^2/P}{\bar{d}}$$

where d_i is the summed total of fixes in the ith pair of cells, divided by the distance between their centers and $P = N(N - 1)/2$. *RU* takes greater values when intensively used cells (ideally hexagons, to equalize distances between neighboring cell centers) are clumped together.

INTERACTION AND OVERLAP

The second problem posed in the introduction concerned the analysis of interactions between the movements of two animals.

STATIC INTERACTION

The spatial overlap, or static interaction, of one animal on another is defined as the proportion of the latter's home range which is overlapped by the former's home range. More precisely, if A_1 and A_2 represent the areas of the home ranges of animals 1 and 2 and A the area of overlap, then the static interaction (S) is:

$$S_{12} = \frac{A}{A_1} \text{ and } S_{21} = \frac{A}{A_2}$$

where S_{12} is the static interaction of animal 2 on animal 1 etc. Static interaction is not reflexive (i.e. $S_{12} \neq S_{21}$) except in the improbable event of the two home ranges having equal area.

The problems of interpreting overlaps between home range configurations derived by different methods is illustrated by a study of the red fox's social system on Boar's Hill, Oxford, U.K. Radio tracking was used to predict the foxes' positions and so to facilitate observations which led to the conclusion that many resident foxes in that habitat lived in group territories, where 4-5 adult members of a social group shared a common range which overlapped very little, or not at all, with similar neighboring group territories (Macdonald, 1977). The bivariate normal model of home range (Jennrich and Turner, 1969) was found to give a useful comparative index of range size for these foxes, but would have led to hopeless misinterpretation of their social biology if applied in isolation. Territorial boundaries often followed topographical features like roads and ditches and so were neatly tesselated in a way which home range models could not accommodate. The average size of 5 group territories on the basis of all types of observation was 41.44 ha (S.D. 20.6 ha) while the elliptical estimates were 70.45 ± 30.72 ha and 20.79 ± 11.84 for the 95 percent and 62.3 percent ellipses respectively. The convex polygon estimates were 70.5 ± 56.7 ha.

Table 3 shows the extent of overlaps (static interaction) between and within the members of two social groups using the bivariate normal confidence ellipses. Differences in the pattern of habitat use by each fox influenced the configuration and orientation of each ellipse (or, in different ways, each minimum polygon) to disguise the real spatial relationship between the foxes. In fact, the 95 percent confidence ellipses suggest that on average group I overlapped over 50 percent of group II ranges. Even if movements are distributed normally these distortions would still occur as different group members utilize their common home range differently and hence have different variances on their x, y coordinates. Small sample size would also alter ellipse size.

AN INDEX OF STATIC INTERACTION

Computed areas of overlap based on probability ellipses may have useful applications in some models concerned with probabilities of overlap and relative chances of encounter, but they should be interpreted with the caution (cf. Table 3). Figure 5 shows night time radio fixes of four adult foxes over 19 nights during January, 1979 on Boar's Hill, Oxford. The overlap between the two neighboring ranges was generally thought to be small on the basis of our fieldwork although several probable excursions were recorded, and yet the overlap of ellipses Si (female) and Bl (female) with RW (female) and SN (male) is considerable (Fig. 5 inset).

Table 3.
The percentage of overlap between estimated ranges of eight foxes belonging to two social groups whose two group ranges were believed to overlap very little in reality

Y	X							
	Group A				Group B			
	SN male	TP female	FT female	HU female	RO male	DF female	JO female	WT female
				62.3 percent				
SN male	—	83.1	51.4	39.0	3.3	0.0	1.3	7.2
TP female	68.0	—	60.0	43.5	11.6	0.4	11.3	11.0
FT female	54.3	77.5	—	81.2	3.3	0.0	9.6	28.2
HU female	32.5	44.5	64.3	—	9.9	0.0	7.5	23.6
RO male	1.9	8.3	6.0	7.0	—	86.3	100.0	85.8
DF female	0.0	0.2	0.0	0.0	52.5	—	74.1	62.3
JO female	0.5	5.2	3.4	3.4	64.5	78.5	—	77.6
WT female	2.4	9.3	9.1	9.7	50.0	60.0	70.1	—
				95.0 percent				
SN male	—	94.0	71.1	62.5	44.6	42.2	53.1	66.0
TP female	77.0	—	67.5	62.3	50.4	50.8	61.1	76.4
FT female	75.1	87.3	—	92.0	46.9	42.7	51.9	60.4
HU female	52.3	63.8	72.9	—	40.2	36.3	44.0	54.6
RO male	26.3	36.4	26.6	28.3	—	91.8	100.0	93.1
DF female	15.1	22.3	14.4	15.5	56.0	—	80.7	73.5
JO female	20.7	29.1	19.1	20.4	67.0	87.8	—	83.8
WT female	22.6	32.1	19.6	22.4	54.2	70.6	73.0	—

In order to compute static interaction, given any two ellipses, new coordinate axes can be defined so that one ellipse is in standard form with center at the origin and major and minor axes along the coordinate axes. A numerical algorithm is then used to determine the points of intersection (0, 2, or 4). If the ellipses do intersect, the area of intersection is calculated by using standard integration theory to evaluate the areas below each of the (2 or 4) sections of elliptical arc. Details of the method and of a computer program may be obtained from the second author.

DYNAMIC INTERACTION

By dynamic interaction we mean the way in which movements of two animals are related, e.g. through attraction or avoidance. This is not necessarily equivalent to their tendency to physically encounter each other, as described below. To illustrate the problem consider the home ranges of two animals, 1 and 2, of which both are circular, centered on the origin and with equal radii. Hence, static interaction is total. They share the same home range but do their movements influence each other? Figure 6 shows the simulated marginal distributions of the distances between the two animals (i.e. the distance between them at any one time), depending on whether ρ is high, low

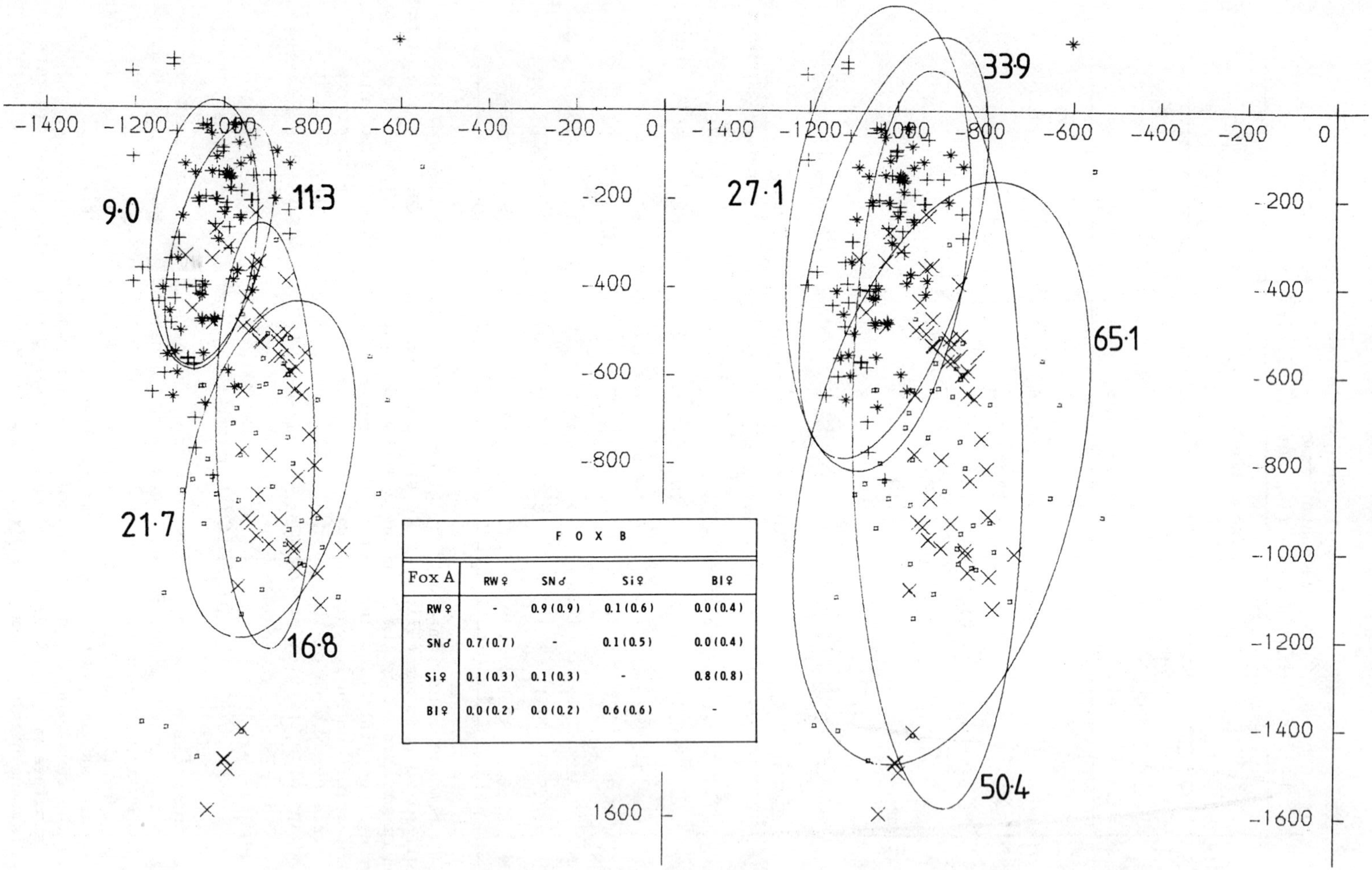

Fox A	FOX B RW♀	SN♂	Si♀	Bl♀
RW♀	-	0.9 (0.9)	0.1 (0.6)	0.0 (0.4)
SN♂	0.7 (0.7)	-	0.1 (0.5)	0.0 (0.4)
Si♀	0.1 (0.3)	0.1 (0.3)	-	0.8 (0.8)
Bl♀	0.0 (0.2)	0.0 (0.2)	0.6 (0.6)	-

Fig. 5. Radio fixes taken during January, 1979 on four adult foxes on the outskirts of Oxford, RW female (*), SN male (+), Si female (X) and BL female (□). 62.3 and 95 percent bivariate normal probability ellipses are shown and the inset table shows the computed static interaction (fox A on fox B) between these foxes (95 percent in brackets). Home range sizes given in hectares.

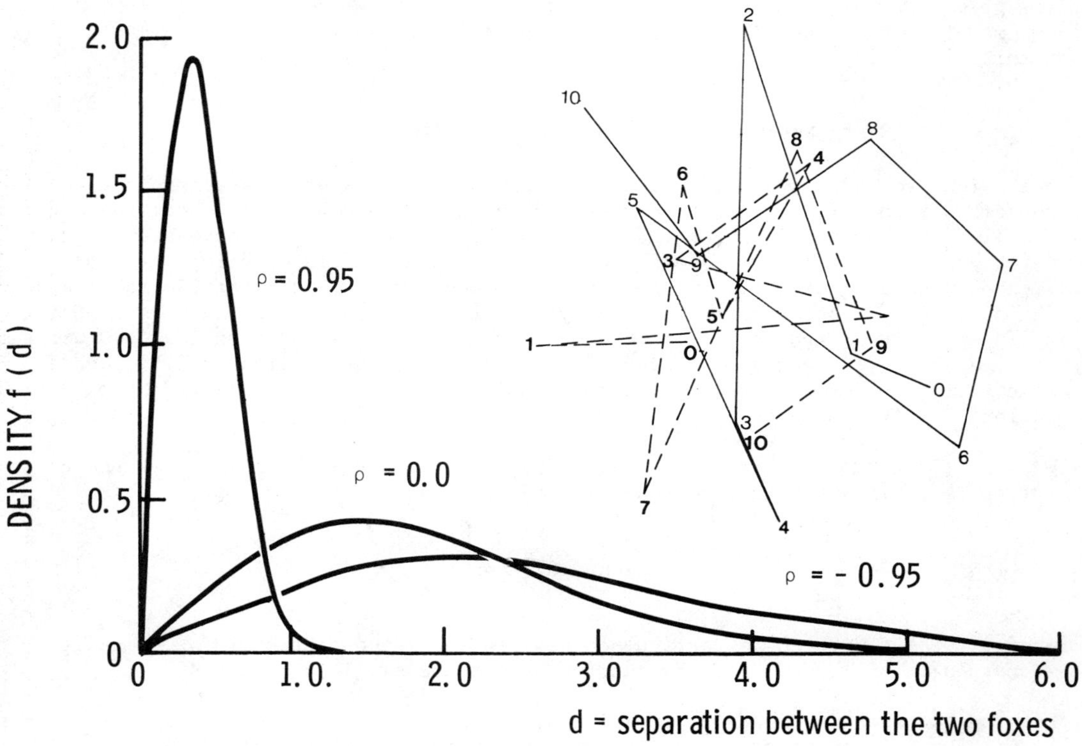

Fig. 6. The marginal distribution of the separation between two simulated foxes whose movements were correlated in different ways. The inset shows the paths of two simulated foxes whose movements were negatively correlated at ρ = −0.3, numbers referring to the 1st, 2nd, *n*th move of each fox.

or neutral. ρ is the correlation between the coordinates x_1 and x_2 (and equally y_1 and y_2) of the two animals.

It can be shown that $d(t)$, i.e. the distance between the two animals at time t, has a marginal distribution with density

$$f(d) = 2\lambda t e^{-\lambda t^2} \quad \text{where } \lambda = \tfrac{1}{4}(1 - \rho).$$

Graphs of $f(d)$ for ρ = −0.95, 0 and +0.95 are given in Fig. 6. As can be seen the graphs are very peaked for high positive values of ρ and relatively flat for large negative values of ρ. It is noteworthy that the difference between neutrality and attraction is more obvious than that between neutrality and avoidance, since there are so many places to avoid another animal and only one to meet him. Hence avoidance by two animals of each other is harder to detect in practice (see inset Fig. 6, where ρ = −0.3). The animals are avoiding each other, but this would not be obvious at first sight, so a test is needed.

By simplifying the O-U model, in dropping the assumption of dependent fixes but still keeping the possibility of dynamic interaction between the two animals, we provide in Appendix 3 a test of whether the animals behave neutrally to each other or otherwise. Copies of the program for this test are available from F. Ball. J. Dunn (in press) has a more sophisticated, but mathematically complex test, which uses his O-U model rather than an extension of the Bivariate Normal model, for testing dynamic interaction.

INTERACTION BETWEEN FOXES

The extent to which foxes interact in their movements is of interest both to those concerned with interpretation of their social system and to those modelling and controlling the spread of rabies (Macdonald, 1980). Few real data have been published which enable a modeller to assess the likely frequency of interaction between foxes. Figure 5 shows fixes of four foxes in two neighboring groups. For SN (male) and RW (female) the W statistic for testing dynamic interaction between them is calculated as in Appendix 3; and equals 1.146. Now, W is distributed as F on 4 and 90 degrees of freedom ($F_{4,90}$) and the 5 percent level of $F_{4,90}$ distribution is 2.5. Hence since $1.146 < 2.5$ we cannot reject the hypothesis H_0 that there is no dynamic interaction.

We can illustrate that there is little dynamic interaction by looking at the correlation matrix *between* the two foxes' coordinates

$$\rho = \begin{bmatrix} \rho_{xx} & \rho_{xy} \\ \rho_{yx} & \rho_{yy} \end{bmatrix} \quad \text{which are, } \rho = \begin{bmatrix} 0.175 & -0.028 \\ 0.041 & 0.154 \end{bmatrix}$$

where, for instance ρ_{xy} = correlation between the x coordinate of fox RW and the y coordinate of fox SN. All these correlations are small as expected.

For the two females Si and Bl, $W = 1.256$, and from the distribution we again cannot reject the null hypothesis that there is no dynamic interaction.

The correlation matrix ρ as defined before is

$$\rho = \begin{bmatrix} -0.075 & 0.189 \\ -0.211 & 0.219 \end{bmatrix}$$

the correlations again being small.

In reality these foxes could avoid each other in much more subtle ways than we have tested for, but this model could readily be adapted to test for different patterns of avoidance, or attraction. However, these are likely to be beyond the resolution of most radio tracking studies (especially where data are gathered discontinuously) and will only be soluble by direct observation.

The chance of two animals meeting depends on the distance over which they can detect each other, and how each of them reacts to the detection of the other. It is important for models of rabies transmission to know how frequently foxes come into contact. By calculating the parameters of probability ellipses of two foxes and working out the underlying distributions for each ellipse (taking account of the covariance of their movements) it is then possible to calculate the probabilities that the two animals will be within a given distance of each other (see Dunn, 1979). The data on Fig. 5 have been treated in this way to generate the probability curves in Fig. 7, which show the chances of RW female and SN male being within given distances of each other, and similarly for Bl female and Si female (bearing in mind that the accuracy of radio fixes in this habitat is probably within ±30 m).

The curves on Fig. 7 show that, for instance, Bl (female) and Si (female) pass within 10 m of each other with a probability of 7.58×10^{-4}, so that within the 10 h darkness

	RW	SN	Si	Bl
RW	-	0.02	0.0	0.0
SN	0.06	-	0.0	0.0
Si	0.04	0.0	-	0.02
Bl	0.0	0.0	0.04	-

20 m

50 m

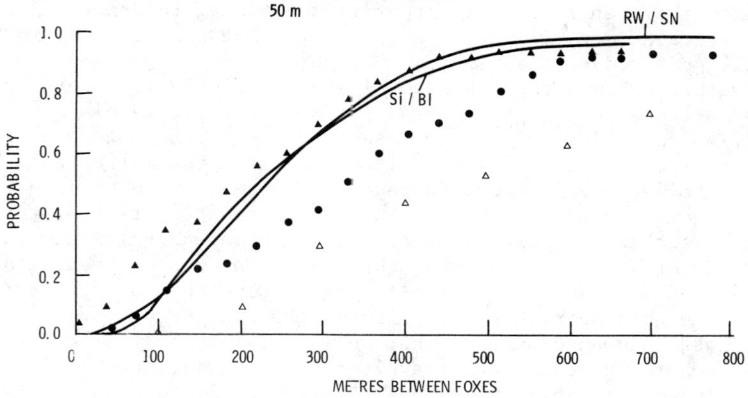

Fig. 7. The probability of the foxes whose movements were depicted in Fig. 5 meeting each other. Lines show probabilities based on the Jennrich and Turner (1969) ellipses. Symbols show probabilities based on the raw data: RW female/SN male (▲), Si female/Bl female (●), SN male/Si female (△).

of a winter's night they would be that close for an average of 0.45 min. Depending greatly on the habitat, foxes can sense each other at least 30-50 m apart and considering a separation of 30 m (approximately the resolution of the original tracking data) the probability increases to 6.79×10^{-3} or just over 4 min 10 h^{-1}. Of course, the way each of two foxes will react when they sense each other depends on many circumstances (e.g. season, status, proximity of food and, presumably also, whether one of them is rabid), as will the duration of each meeting.

Figure 7 also presents probabilities of separation of RW (female)/SN (male), Si (female)/bl (female) and, by contrast, SN (male)/Si (female) based empirically on the data. These probabilties, for distances of 20 m and 50 m, are given on the inset table. In fact, these suggest slightly higher probabilities of meeting for foxes within a group than do the calculated curves; the probability of RW (female)/SN (male) being within 10 m is 2×10^{-2}, (12 min/10 h) and for Si (female)/Bl (female) is 1.9^{-2}. In the absence of adequate direct observations, this technique provides a starting point for generating contact probabilities for models of rabies transmission.

CONCLUSIONS

Where they can be supported with adequate observations we advocate drawing home range boundaries by hand and measuring their areas simply with a planimeter. Only in this way can interpretations of spatial relationships be as finely tuned as are animal societies in reality. Of course this leaves more room for 'wishful thinking' by the biologist, but any undesirable consequences of this can be countered by reporting in slightly more detail the criteria used for drawing given boundaries.

Other estimates of home range can also be included to facilitate comparison within the literature.

The drawback of all quantitative estimates of spatial organization is that they do not necessarily distinguish between the animals' behavior at different locations. From movement data alone there is no hope of interpreting the social behavior, e.g. of an individual which is territorial but whose territory does not correspond to its home range. Biotelemetry may soon help to solve this type of problem (Macdonald and Amlaner, 1980, this volume). Grid square (or hexagon) methods seem to us to be preferable since they can readily incorporate additional relevant information on a relatively fine scale (e.g. Gilmer *et al.*, 1973; Voigt and Tinline, 1980, this volume). Probability ellipses reduce the hazard, inherent in convex polygons, of including 'excursions' within a home range, but may leave out highly frequented areas. Convex polygons are unaffected by large numbers of fixes in a single place, e.g. while the animal sleeps, which could for most purposes, make complete nonsense of probability ellipses. In some cases probability ellipses can give realistic range borders (e.g. Fig. 2a, where the data were uncomplicated anyway) and have been used to great effect in some studies (e.g. Randolph, 1977). The added sophistication of the O-U process would seem to be either an improvement or a mistake depending on how the animal moves. In other words, without knowing the answer one cannot predict how appropriate a given estimate of its behavior will be! Trevor-Deutsch and Hackett (1980, this volume) illustrate this point by their finding that 70 percent probability ellipses happen to fit most chipmunk ranges rather well, but they only discovered this after knowing exactly where the chipmunks travelled. Of course, once a particularly appropriate inclusion probability has been discovered in this way it can be used, with reservations, with greater confidence than an arbitrarily chosen inclusion probability.

Acknowledgements — D. W. Macdonald is grateful to Dr. D. Rogers for first introducing him to the problems of computing confidence ellipses and to Dr. J. Dunn for conducting a specimen analysis of some of the data presented here. Professor P. Armitage, C. Amlaner and M. Newdick kindly read the manuscript and G. Carr helped with some of the fieldwork.

REFERENCES

Adams L. and Davis S. D. (1967) The internal anatomy of home range. *J. Mammal.* 48, 529-536.

Anderson T. W. (1958) *An Introduction to Multivariate Statistical Analysis*. London: J. Wiley, 374 p.

Brown L. E. (1962) Home range in small mammal communities. In B. Glass (ed.), *Survey of Biological Progress 4*, pp. 131-179. London: Academic Press.

Burt W. H. (1943) Territoriality and home-range concepts as applied to mammals. *J. Mammal.* 24, 346-352.

Calhoun J. B. and Casby J. U. (1958) The calculation of home-range and density of small mammals. Public Health Manag. No. 55. U.S. Dept. of Health, Educ. and Welfare.

Clutton-Brock T. H. (1974) Primate social organisation and ecology. *Nature, Lond.* 250, 539-542.

Cooper W. E. (1978) Home range size and population dynamics. *J. theoret. Biol.* 75, 327-337.

Dunn J. E. (1977a) The multivariate Ornstein-Uhlenbeck process in studies of home range. Univ. of Arkansas, Stats. Lab. Tech. Rept. No. 2 (July 15, 1977).

Dunn J. E. (1977b) Optimal sampling in radio telemetry studies of home range. Univ. of Arkansas, Stats. Lab. Tech. Rept. No. 5 (June 27, 1977).

Dunn J. E. and Gipson P. (1977) Analysis of radio telemetry data in studies of home range. *Biometrics* 33, 85-101.

Dunn J. E., Karpoff G., Hirsch M. and Cappy G. (1978) A characterization data of the home ranges of fifteen animals in terms of the multivariate Ornstein-Uhlenbeck diffusion process. Univ. of Arkansas, Stats. Lab. Tech. Rept. No. 8 (January 15, 1978).

Dunn J. E. (1979) A complete test for dynamic territorial interaction. Proc. 2nd Int. Conf. on Wildl. Biotelem. Univ. Wyoming, Laramie, pp. 159-169.

Gilmer D. S., Miller S. E. and Cowardin L. M. (1973) Analysis of radio tracking data using digitized habitat maps. *J. Wildl. Mngmt.* 37, 404-409.

Hayne D. W. (1949) Calculation of size of home range. *J. Mammal.* 30, 1-18.

Jennrich R. I. and Turner F. B. (1969) Measurement of non-circular home range. *J. theoret. Biol.* 22, 227-237.

Jewell P. A. (1966) The concept of home range in mammals. *Symp. Zool. Soc. Lond.* 18, 85-109.

Koeppel J. W., Slade N. A. and Hoffmann R. S. (1975) A bivariate home range model with possible application to ethological data analysis. *J. Mammal.* 56, 81-90.

Koeppel J. W., Slade N. M., Harris N. S. and Hoffmann R. S. (1977) A three-dimensional home range model. *J. Mammal.* 58, 213-220.

Macdonald D. W. (1977) The behavioural ecology of the red fox, *Vulpes vulpes:* a study of social organisation and resource exploitation. D.Phil. thesis, Oxford University.

Macdonald D. W. (1978) Radio tracking: some applications and limitations. In B. Stonehouse (ed.), *Animal Marking: Recognition Marking of Animals in Research*, pp. 192-204. London: Macmillan.

Macdonald D. W. (1980) *Rabies and Wildlife: A Biologist's Perspective*. Oxford: Oxford University Press, in press.

Macdonald D. W. and Amlaner C. J. (1980) A practical guide to radio tracking. In C. J. Amlaner, Jr. and D. W. Macdonald (eds.), *A Handbook on Biotelemetry and Radio Tracking*, pp. 143 - 159. Oxford: Pergamon Press.

Mazurkiewicz M. (1969) Elliptical modification of the home-range pattern. *Bull. Acad. Pol. Sci.* Ch. II, 17, 7, 427-431.

Mazurkiewicz M. (1971) Shape, size and distribution of home ranges of *Clethrionomys glareolus* (Schreber 1780). *Acta Theriol.* 16, 23-60.

Metzgar L. H. (1972) The measurement of home range shape. *J. Wildl. Mngmt.* 36, 643-645.

Metzgar L. H. (1973) Home range shape and activity in *Peromiscus leucopus*. *J. Mammal.* 54, 383-390.

Mohr C. O. (1947) Table of equivalent populations of North America small mammals. *Am. Midland Nat.* 37, 223-249.

Mohr C. O. and Stumpf W. A. (1966) Comparison of methods of calculating areas of animal activity. *J. Wildl. Mngmt.* 30, 293-309.

Niewold F. J. J. (1974) Irregular movements of the red fox (*Vulpes vulpes*) determined by radio-tracking. *XI Int. Congr. of Game Biol. Stockholm*, 331-337.

Odum E. P. and Kuenzler E. J. (1955) The measurement of territory and home-range size in birds. *Auk* 72, 128-137.

Owings D. H., Borchert M. and Ross V. (1977) The behaviour of California ground squirrels. *Anim. Behav.* 25, 221-230.

Randolph S. E. (1977) Changing spatial relationships in a population of *Apodemus sylvaticus*, with the onset of breeding. *J. Anim. Ecol.* 46, 653-676.

Rasmussen D. R. and Rasmussen K. L. (1979) Social ecology of adult males in a confined troop of Japanese macaques (*Macaca fuscata*). *Anim. Behav.* 27, 434-445.

Sanderson G. C. (1966) The study of mammal movements — a review. *J. Wildl. Mngmt.* 30, 215-235.

Siniff D. B. and Jesson C. R. (1969) A simulation model of animal movement patterns. In J. B. Cragg (ed.), *Advances in Ecological Research 6*, pp. 185-219. London: Academic Press.

Siniff D. B. and Tester J. R. (1965) Computer analysis of animal movement data obtained by telemetry. *Bioscience* 15, 104-108.

Southwood T. R. B. (1966) *Ecological Methods, with Particular Reference to the Study of Insect Population*. London: Methuen. 391 p.

Stickel L. F. (1954) A comparison of certain methods of measuring ranges of small mammals. *J. Mammal.* 35, 1-15.
Storm G. L. (1965) Movements and activities of foxes as determined by radio tracking. *J. Wildl. Mngmt.* 29, 1-13.
Tester J. R. and Siniff D. B. (1965) Aspects of animal movement and home range data obtained by telemetry. *Trans. XXXth N. Am. Wildl. Nat. Res. Conf.* 352-379.
Trevor-Deutsch B. and Hackett D. F. (1980) An evaluation of several grid-trapping methods of comparison with radio tracking in a home range study of the Eastern chipmunk (*Tamias striatus* L.). In C. J. Amlaner, Jr. and D. W. Macdonald (eds.), *A Handbook on Biotelemetry and Radio Tracking*, pp. 375-386. Oxford: Pergamon Press.
Voigt D. R. and Tinline R. R. (1980) Strategies for analyzing radio tracking movement data. In C. J. Amlaner, Jr. and D. W. Macdonald (eds.), *A Handbook on Biotelemetry and Radio Tracking*, pp. 387-404. Oxford: Pergamon Press.
van Winkle W. (1975) Comparisons of several probabilistic home range models. *J. Wildl. Mngmt.* 39, 118-123.

APPENDIX 1

ESTIMATION OF PARAMETERS AND CALCULATION OF PROBABILITY ELLIPSES FOR BIVARIATE AND CIRCULAR NORMAL MODELS

Assume that we have a series of radio fixes, $Z_1, Z_2, \ldots, Z_N$, where $Z_i = (x_i, y_i)$ is the location of the animal at the ith fix with reference to a fixed x, y coordinate system. In both models the mean locations on the x and y coordinate axes is estimated by the sample means $\bar{x}$ and $\bar{y}$ respectively. Estimates of the variances, $s_1{}^2$ and $s_2{}^2$, and the covariance, s_{12}, of the x, y coordinates are calculated as follows:

$$s_1^2 = \left(\sum_{n=1}^{N} (x_n - \bar{x})^2\right)/(N-2) \quad \text{and} \quad s_2^2 = \left(\sum_{n=1}^{N} (y_n - \bar{y})^2\right)/(N-2)$$

$$s_{12} = \left(\sum_{n=1}^{N} (x_n - \bar{x})(y_n - \bar{y})\right)/(N-2) .$$

From these values the angle (θ) between the major axis of the probability ellipse and the x axis of the grid, is given by:

$$\text{Tan } 2\theta = 2s_{12}/(s_1^2 - s_2^2).$$

The whole coordinate system can be translated so that the origin becomes the center of activity, and then rotated clockwise through an angle θ about the new origin so that the covariance between the animal's locations on the new coordinate system becomes zero.

If the new coordinates measured from the center of the ellipse are u and v, then estimates of the variances in this system will be:

$$s_u^2 = s_1^2 + s_{12} \tan\theta \text{ and } s_v^2 = s_2^2 - s_{12} \tan\theta \text{ with } s_{uv} = 0.$$

The lengths of the semi-major (M) and semi-minor (m) axes of the probability ellipse are: $M = r\, s_u$ and $m = r\, s_v$ where the value of r determines the probability level of the ellipse and thus the probability of the animal being found in an elliptical estimate of home range, i.e. the 'inclusion probability' given by: $P(r) = 1 - e^{-r^2/2}$ (see Mazurkiewicz, 1971, p. 58, and Jennrich and Turner, 1969, p. 232, where the same function is called p).

The corresponding confidence ellipse is given by: $(u^2/s_u^2) + (v^2/s_v^2) = r^2$ enclosing an area of $\pi r^2 s_u s_v$ square units.

In the circular normal model we estimate the common variance, σ^2, by:

$$\hat{\sigma}^2 = (N - 2)(s_1^2 + s_2^2)/(2(N - 1)) \ .$$

The estimate of home range, with inclusion probability $1 - \alpha$, is then a circle center $(\bar{x}, \bar{y})$ and radius $\sqrt{\{2 \ln(1/\alpha)\}}\ \hat{\sigma}$.

APPENDIX 2

A TEST FOR VALIDITY OF THE CIRCULAR NORMAL MODEL

Calculate the matrix

$$s = \begin{bmatrix} s_1^2 & s_{12} \\ s_{12} & s_2^2 \end{bmatrix}$$

as before. Write $|s| = s_1^2 \times s_2^2 - s_{12}^2$ (the 'determinant') and $tr(s) = s_1^2 + s_2^2$ (the 'trace'). Calculate $W = |s|/\{\frac{1}{2}tr(s)\}^2$. We reject the circular normal model of $W < \beta^{2/(N-1)}$ where $100(1 - \beta)$ is the significance level of the test. It is possible to relate this test and in particular the statistic W to the two estimated areas of home range A_0 and A_1 calculated respectively by the circular and bivariate normal models from the expression $W = (N-1)A_0/((N - 2)A_1)$ we get an intuitive interpretation of the test in that the circular normal model is rejected if the ratio of home range areas A_0/A_1 is above a fixed value; inevitably $A_0/A_1 > N-2/N - 1$.

APPENDIX 3

A hypothesis test for the presence of dynamic interaction, i.e. whether the animals behave independently of each other or not, assuming an extension of the bivariate normal home range model. (This is a particular case of a test given in Anderson, 1958, p. 230 *et seq.*)

We assume that we have series of data $Z_1, Z_2, \ldots, Z_N$ where $Z_i = (Z_{i1}, Z_{i2})$ and $Z_{ij} = (x_{ij}, y_{ij})$ are the coordinates of the jth animal at the ith data point. As in the bivariate normal model we assume that $Z_1, Z_2, \ldots Z_N$ is an independent sample from a multivariate normal distribution, $N(\mu, \Lambda)$ where

$$\mu = \begin{bmatrix} \underline{\mu}_1 \\ \underline{\mu}_2 \end{bmatrix}$$

$$\mu_i = \begin{bmatrix} \mu_{i1} \\ \mu_{i2} \end{bmatrix} \qquad i = 1, 2$$

is the mean location vector of the ith animal, and

$$\Lambda = \begin{bmatrix} \Lambda_{11} & \Lambda_{12} \\ \Lambda_{21} & \Lambda_{22} \end{bmatrix}$$

where Λ_{ij} is a 2×2 covariance matrix between animals i and j. Any dynamic interaction is contained in the off diagonal sub-matrix $\Lambda_{12} = \Lambda_{21}^T$ of Λ. Hence we wish to test H_0: $\Lambda_{12} = \Lambda_{21}^T = \underline{0}$ i.e. independence between the two animals versus H_1: $\Lambda_{12} \neq \underline{0}$ i.e. some dynamic interaction.

We define a 4 × 4 matrix A by

$$A = \sum_{n=1}^{N} (Z_n - \bar{Z}_n)(Z_n - \bar{Z})^T$$

where

$$Z = \frac{1}{N} \sum_{n=1}^{N} Z_n .$$

We partition A into four 2 × 2 matrices as follows

$$A = \begin{bmatrix} A_{11} & A_{12} \\ A_{22} & A_{22} \end{bmatrix} .$$

If we define V and W by

$$V = \frac{|A|}{|A_{11}|\,|A_{22}|}$$

$$W = \frac{1 - \sqrt{V}}{\sqrt{V}} \quad \frac{N - 4}{2}$$

then we reject the hypothesis H_0, that there is no dynamic interaction, if

$$W > F_{4,2(N - 4)}(1 - \alpha)$$

where 100 α percent is the size of the test. ($F_{4,2\,(N-4)}\,(1-\alpha)$) is the 100(1 - α) percent level of the F distribution with degrees of freedom $\nu_1 = 4$, $\nu_2 = 2(N - 4)$.

Admittedly this test is based on the unrealistic assumption of the data points Z_0, Z_1, ... Z_N being independent but this is unlikely to be critical unless either the sample size or the sample interval is small.

A Computer Based Data File Management System for Telemetry Studies on Free Ranging Animals

J. C. KROLL and P. C. JOHNSON

Stephen F. Austin State University, Nacogdoches, Texas 75962, U.S.A.

Abstract — Telemetry studies on free ranging animals quickly generate large quantities of location data which are often difficult to analyze. A computer based data file management system, based on a modification of the Statistical Package for the Social Sciences, was developed for acquisition, update and analyses of such data. Field data are recorded directly on a miniature remote data terminal, eliminating need for field data forms and verification. Location coordinates are subsequently plotted by a computer program interfaced with the Tectronix remote terminal. Time elapsed from data acquisition to update and/or analyses is less than 24 h.

INTRODUCTION

In radio telemetry studies on free ranging animals, considerable thought and planning are usually expended in development of monitoring systems; however, such studies, by their nature, quickly generate large amounts of location or physiological data. Little work has been done in regard to manipulation of these data. Upon completion of a study, the researcher is often faced with analyses of a virtually unmanageable data set. This paper briefly discusses a simple, efficient and inexpensive data file management and mapping system, developed by researchers at Stephen F. Austin State University, which allows for weekly or even daily updating of telemetry data. Our system differs from computer-interfaced continuous radio tracking systems in that cost is held to a minimum. Many researchers cannot afford to use such expensive procedures.

COLLECTION OF DATA

The Statistical Package for the Social Sciences (Nie *et al.*, 1975) was modified for use in telemetry studies. Prior to initiation of field research, a command file is established to process the data set into usable form. Variable formats and names, as well as, coded values are established *a priori*. An example of such a command file we are now using for white-tailed deer (*Odocoileus virginianus*) research follows:

```
RUN NAME            INITIAL RUN OF DEER TELEMETRY DATA
FILE NAME           C:DEER, DEER TELEMETRY COMMAND FILE
VARIABLE LSIT       DTYPE, DATE, NUMBER, SEX, STA1,STA2,SIGNAL
                    B1,B2,T1,T2,XCO,YCO,RAIN,SNOW,TEMP,
                    HUM,WINDD,WINDV,MOON,CLOUD
INPUT FORMAT        FIXED(F2.0,F5.0,F2.0,F1.0,2F3.0,F1.0,
                    2F3.0,2F4.0,2F5.0,2F1.0,F3.1,F3.1,F1.0,
                    F3.0,F1.0,F2.0,7X)
N OF CASES          UNKNOWN
INPUT MEDIUM        DC/D:DEER
VAR LABELS          DTYPE, DATA TYPE/
                    DATE, YEAR AND JULIAN DATE/
                    NUMBER, DEER NUMBER/
                    SEX, SEX OF ANIMAL/
                    STA1, NUMBER OF STATION ONE/
                    STA2, NUMBER OF STATION TWO/
                    SIGNAL, SIGNAL STRENGTH/
                    B1, BEARING ONE/
                    B2, BEARING TWO/
                    T1, TIME ONE/
                    T2, TIME TWO/
                    XCO, X-AXIS COORDINATE/
                    YCO, Y-AXIS COORDINATE/
                    RAIN, HAS THERE BEEN RAIN?/
                    SNOW, HAS THERE BEEN SNOW?/
                    TEMP, TEMPERATURE IN DEGREES CELCIUS/
                    HUM, RELATIVE HUMIDITY/
                    WINDD, WIND DIRECTION/
                    WINDV, WIND VELOCITY/
                    MOON, MOON PHASE/
                    CLOUD, PERCENT CLOUD COVER/
VALUE LABELS        DTYPE (1) TELEMETRY DATA/
                    SEX (1) MALE (2) FEMALE (3) UNKNOWN/
                    SIGNAL (1) STRONG (2) MODERATE (3) WEAK
                    (4) VERY WEAK (5) VARIABLE/
                    RAIN TO SNOW (1) YES (2) NO/
                    WINDD (1) SOUTH (2) NORTH (3) WEST
                    (4) EAST (5) SOUTHWEST (6) SOUTHEAST
                    (7) NORTHWEST (8) NORTHEAST/
                    MOON (1) FULL (0) NO MOON (2) THREE-QUARTERS
                    (3) HALF (4) QUARTER (5) NEW/
MISSING VALUES      TEMP TO WINDV (0)/
READ INPUT DATA
FREQUENCIES         GENERAL=NUMBER
SAVE FILE           DC/P:DEER
FINISH
```

This command file identifies all variables, retrieves the data from disc storage and combines variable names and value codes with the data, storing the new file on disc. Three standardized suffixes are used for ease in identification of files: *C* denotes a command file, *D* a data file and *P* an SPSS file. Data can also be stored for backup in card form. Disc storage is most economical for our data files.

Animals are located using standard single or double Yagi antennas, null-peak box and/or hand-held double Yagi. Bearings (degrees from true north) and X-Y coordinates (from a computer generated grid map) are obtained at two or more fixed stations. Location data are recorded by two methods. First, a standardized, coded data form is completed at the time of each location. Second, data can be recorded on a miniature field data terminal with 8 K memory capacity. On completion of each tracking period (usually 24 h), data are communicated to the computer via a telephone and acoustic coupler. Coded data forms serve as a backup data set and for verification of data files.

DATA FILE MANAGEMENT

Data files are updated on a weekly or daily basis. Prior to analyses, data are double verified against the coded data forms. SPSS command files are used to extract data subsets. Subset structure significantly reduces computing costs. For example, data pertaining only to a single animal or a particular data or habitat are selected from the overall data set. This can be accomplished with a few simple SPSS commands.

COMPUTER PROGRAM *SHADE*

The general utility of computer graphics is well recognized, but frequently involves expensive graphics terminals. Shaded maps which utilize overprinting to produce visual density effects, however, are relatively easy to produce at standard terminals or line printers and represent an intermediate technology available to the large majority of users. The program presented here, *SHADE** is used to produce shaded probability distributions about radio telemetry fixes on individual deer, against a background of various map features (roads, streams, etc.), but is flexible enough to provide a variety of mapping options including maps or uniformly shaded quadrats where shading density represents population quadrat density.

General characteristics of the program include:

1. Selection of mapping symbols.
2. Specification of map ranges.
3. Choice of two map sizes.
4. Choice of output device.
5. Choice of a logarithmic decay model or square quadrat model in determining shading about an X,Y-coordinate.
6. Scaling of the shading to provide weighting relative to the observed population density or equal weighting of all population densities.

DATA PREPARATION

SHADE utilizes two user supplied data sets generated by SPSS command files, one containing background mapping information, while the other provides individual population fixes (i.e. an X,Y-coordinate and population density estimate). The format for the two data sets is identical:

X	Y	Density	Symbol
F8.3	F8.3	F5.0	A1

Each data record consists of the X,Y-coordinates of a map feature or animal location, a density estimate (zero for map features) and a map symbol.

Mapping features are kept separate from population radio fix data since it is frequently desirable to use a consistent background map while selecting various population data sets for mapping. With population data stored in SPSS files, output of

**SHADE* was developed initially to provide probability density distributions for southern pine beetles dispersing from infestations and was funded by the Expanded Southern Pine Beetle Research and Applications Program.

selected radio fix data is easily obtained through a combination of OUTPUT MEDIUM, *SELECT IF and WRITE CASES commands. Thus population data can be changed without altering the background map data.

COMMANDS

SHADE is set up as a load module on the Honeywell CP-V computer at Stephen F. Austin State University and may be accessed through remote time-sharing terminals as an interactive routine, or in a batch processing mode using a prepared command file or card deck.

SHADE recognizes 12 commands, consisting of a command word followed by a blank and a string of format-free parameter specifications separated by commas. Table 1 provides a brief description of the commands and their parameters. Two commands, MODE and SCALE will be discussed in detail below.

Table 1.
Commands recognized by *SHADE*, including parameters supplied by the user. Brackets enclose optional input. Parentheses enclose alternate input

Command	Parameters	Comment
TITLE	Character String (-72)	Labels output maps
LABEL	Character String (-72)	Labels X-axis
MODE	A, B, (C)	See text for explanation
SIZE	$\binom{L}{S}$	Map size: L = 100 char. wide S = 50 char. wide
SCALE	N	N = Minimum recognizable density. See text for explanation
MIN-X	X_1	Sets X and Y axis range; $X_2 < X_1$, $Y_2 < Y_1$
MAX-X	X_2	
MIN-Y	Y_1	
MAX-Y	Y_2	
SHOW		Provides list of current parameter values
PRINT	$\binom{T}{L}$	Triggers printout and specifies device: T = Terminal; L = Line Printer
BYE		Terminates mapping session

MODE

The MODE command is followed by three parameters, A, B and C, which determine the relationship between the shading value of a map location (*SHADE*), and its distance from an X,Y-coordinate of an animal fix. Specifying A and B sets the parameters of the logarithmic decay model describing dispersal about the fix. The model used is:

$$SHADE = A + B \ln X, \tag{1}$$

where X is the distance from the fix location to the limits of location accuracy in map units.

If B = 0, a third parameter must be specified, fix accuracy, C, which determines the limits about the fix within which shading will occur. If B ≠ 0, the fix accuracy is calculated from A and B as the X-axis intercept. When B = 0 and C is specified (in map units), a *square* quadrat of dimension 2C centered on the X,Y-coordinate of the fix is shaded uniformly with the shade value determined from the fix density and current scale factor. Two examples are provided below (the computer prompts with a '?'):

I.	?MODE 1.0,-1.5	... This produces a fix accuracy of 1.94 map units; shading is maximum at the fix and declines in relation to the natural log of the distance from the fix.
II.	?MODE 1.0,0.0,0.25	... This produces a shaded quadrat 0.50 map units on a side centered on the fix.

SCALE

The SCALE command is followed by a numeric parameter which determines the effect of fix density on shading. The default value is 1.0, which provides shading in direct proportion to relative density, while setting the scale parameter equal to the maximum density causes all fixes to receive equal weighting regardless of density (with the exception of a density of 0, which receives no shading ... a weighting factor of 0). The weighting factor, W, is computed from density (N), maximum density (N_m) and the scale factor (SC) as follows:

$$W = (N/N_m)\ (SC), \tag{2}$$

and is multiplied by the shading value computed from equation 1. If W exceeds 1.0, it is adjusted to a value of 1.0. Therefore, W ranges from 0.0 to 1.0, with values of N greater than SC being weighted according to their ratio to N_m. SC is termed the minimum recognizable density, with all values of N greater than 0 and less than or equal to SC being considered equal (a value of N = 0 produces a point with no shading).

MIN-X, MAX-X, MIN-Y, MAX-Y

The four minimum and maximum range commands are used to limit the X and Y axes of the map produced. Since the physical size of the map remains constant, this 'blows-up' specified areas for detailed examination. If the Y-axis is truncated to maintain the 54 line limit, an appropriate message is included in the output. The maximum Y-axis range is effectively 90 percent of the specified X-axis range. Two examples are provided below:

I.	?MAX-X 5.0 ?MAX-Y 7.25 ?MIN-X 0 ?MIN-Y 1	... This series of commands will produce an X-axis range of 5 map units (0.0-5.0) and Y-axis range of 4.5 map units (1.0-5.5) — 54 lines.
II.	?MAX-X 12.25 ?MAX-Y 81.32 ?MIN-X 1.72 ?MIN-Y 75.00	... This series of commands will provide an X-axis range of 10.53 map units (1.72 to 12.25) and a Y-axis range of 6.32 map units (75.00-81.32) — 36 lines

(Note: The MODE command readjusts the X and Y range by the addition of the critical distance to each end to allow complete mapping of the shaded area about the maximum and minimum fixes. If specific range limits are required, specify them *after* entering the MODE command.)

SAMPLE RUNS

Figures 1a–d provide examples of maps produced with *SHADE*. Mapping data and population fixes are from a study of white-tailed deer in East Texas. Each fix represents the location of a single deer through radio telemetry, and each map represents a time series on that deer through a season (the codes 1-4 represent spring through winter).

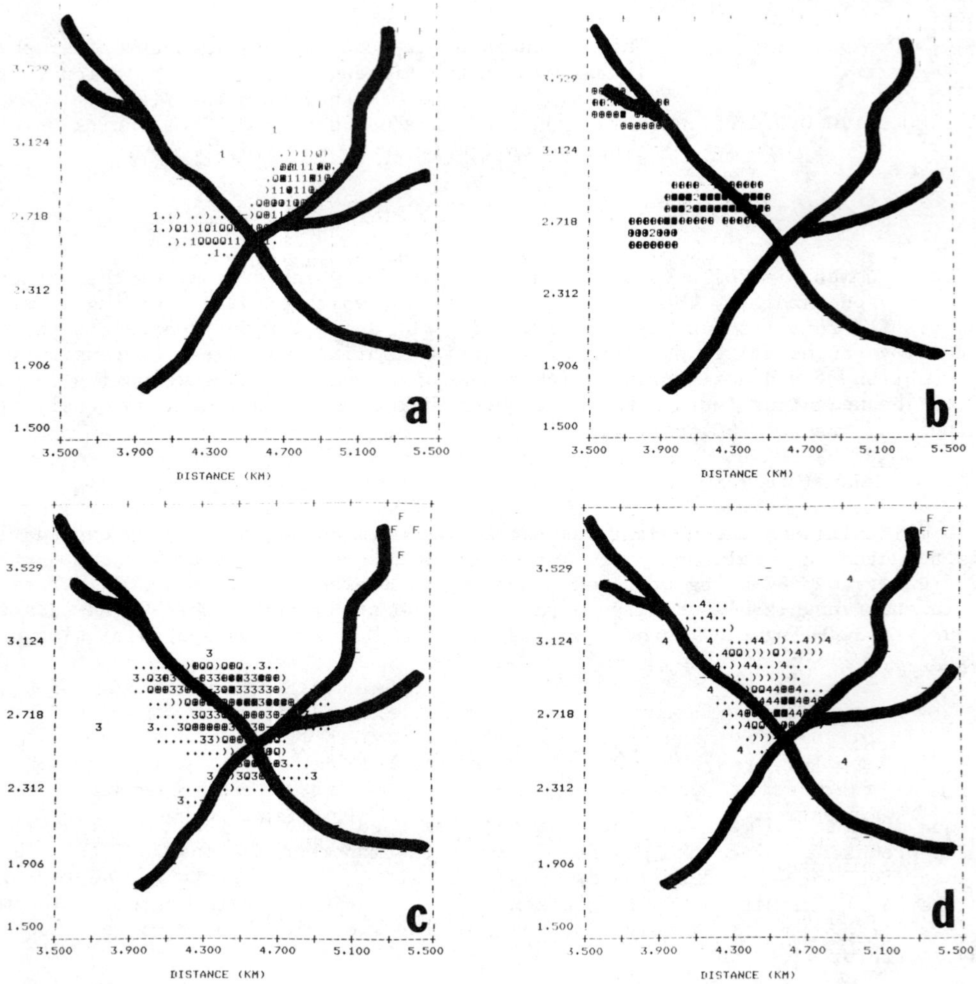

Fig. 1. (a) Spring home range for a white-tailed deer showing center of activity (shaded area). The black line represents a road system within the study area. (b) Summer home range. (c) Fall home range shows a shift as demonstrated by the darker overscoring for multiple fixes. (d) Winter home range shifted again.

REFERENCE

Nie N. H., Hull C. H., Jenkins J. G., Steinbrenner K. and Bent D. H. (1975) *Statistical Package for the Social Sciences*, 675 p. London: McGraw-Hill.

Remote Data-Acquisition in Rehabilitation

J. MACGREGOR, R. F. LEVY and T. G. GRASSIE

Bioengineering Unit, University of Strathcylde, Glasgow, U.K.

Abstract — Poliomyelitis, peripheral vascular disease, 'strokes' and serious traumatic injury all contribute to the relatively large population of physically handicapped members of our society. In an attempt to restore function, at least in part, the rehabilitation physician or surgeon is called upon to prescribe orthotic or prosthetic devices and to modify them from time to time in response to patient performance. Assessment can be undertaken (a) by questioning the patient about his mobility at home and work and (b) by inspection of gait in the severely limited confines and atypical environment of the outpatient clinic. Both approaches are notoriously fallible. In a pilot attempt to introduce a degree of objectivity a group of lower limb amputees were fitted with miniature FM radio transmitters to telemeter physical activity and physiological cost parameters over the immediate environs of a city center campus. In the end the results were extremely disappointing, main problems were, interference, signal loss, limited range etc. The added requirement of acquisition of 3 channels of data over 24 h periods in unrestricted patients travelling up to 30 miles distance to their home led to the examination of a miniature analog tape recorder system carried by the patient. The effects of the move were so far-reaching as to alter the entire philosophical approach to the investigation. Perhaps the only real limitation is the non-availability of data in real-time.

Using modified 'Medilog' recorders with an integral crystal clock signal on one channel, 24 h records of ECG, trunk acceleration and body postures are obtained. On replay through purpose-built analyzers a dramatic picture of the 'life-style' of the patient emerges with temporal relations of changes in posture etc. along with quantitative measures of mobility and, of course, physiological cost. Once the transducers are fitted, the cables taped down and the recorder secured to a broad waist-band worn next to the skin, the patient is dressed and able then to undertake all his normal daily and overnight activities, immersion only excepted. Current experience suggests that only very rarely does the patient feel restricted to any degree. Acquisition of high-quality data in the domiciliary or working environment can bring a new and valuable dimension to the problems faced by the handicapped in the real-life situation.

INTRODUCTION

Society now recognizes the term 'handicapped' in a general and non-specific way. In clinical circles and in the social services there are occasions when some assessment of the degree of handicap is essential for legal or clinical reasons. Recent legislation for instance has provided mobility allowances for people who are considered to have a severe physical impairment reducing their opportunities of normal employment or social life in their community. In all these instances the criteria applied are quite subjective and seldom amenable to impartial measurement.

It is obviously desirable that reliable methods be developed which give a reasonable picture of the abilities of an individual patient. Essentially there are two factors which must be identified. Firstly, the actual mobility achieved by the patient and secondly, the physiological cost required to produce that degree of mobility. Mobility may be reduced if the patient is 'lazy' while on the other hand it may be increased if the subject is determined to prove his suitability for a particular post. Motivation is undoubtedly a most important factor in the degree of physical independence achieved by the subject. Any system therefore which attempts to define the physical activity of the patient in quantitative terms must also take account of the effort required by the patient to achieve that result.

A second factor which must be borne in mind is the need to extend any assessment from the strictly artificial environment of a hospital, clinic, day center or surgery to the normal environment in which the patient lives i.e. his domiciliary or employment environment. In addition, if his performance has to be adequately measured then the sample must be a large one and representative of the different aspects of his daily living pattern which are important to him. We have thus developed a concept of 'life style' in which quantitative measurement must take place over a period of at least 24 h in a typical day in which all normal activities are undertaken while unobtrusive and non-invasive assessment techniques are action.

When first attracted to this problem, it was thought that a radio telemetry method might well be suitable for transmitting information from appropriate transducers towards a recording system. It rapidly became clear, however, that our patients were living within an area which extended to 50 km from our laboratory and that the volume of information likely to be transmitted and stored was very substantial indeed. Initial trials within the laboratory brought to light a large number of technical problems related to interference, the fugitive nature of signals in a steel reinforced concrete building with many electrical services, and we were forced to consider whether immediate access to the data in real-time was essential. Our conclusion was that in this situation there was little merit in immediate access once the system had been adequately set up and calibrated and signal acquisition verified. We therefore turned to a data storage system utilizing miniature analog tape recorders.

We have since developed a Longterm, Ambulatory, Physiological, Surveillance Equipment (LAPSE) which in essence is a variety of biomedical transducers fed through signal conditioning units into a miniature analog tape recorder. The recorder has a 24 h endurance and in practice an ECG signal, a trunk acceleration signal, a posture signal from position sensors and an internal clock signal with a basic frequency of 60 Hz, can all be recorded for a 24 h period on to a high quality C120 standard tape cassette.

The tape cassette is replayed and the signals demodulated before being passed to special purpose analyzers which, among other things, identify (a) beat-to-beat interval from successive R wave detection on the ECG and (b) successive step intervals by detection of characteristic oscillations on the trunk acceleration signal. Both intervals are measured against the recorded clock signal on tape, making the entire operation independent of error introduced through variation in capstan speed, friction, tape stretch, loss of battery power etc. The availability of a posture indication permits the analysis to be performed on portions of the data related strictly to sitting, standing, lying and also permits statistical operations relating speed of walking to instantaneous heart rate etc.

The impact of such a detailed view of an individual patient's 'life style' proved far reaching and it became necessary to alter the philosophy of the approach to mobility investigations of the handicapped patient. The volume of data to be stored is self-evident and the only practical limitation in the system is that the information is not available in real-time.

METHOD

For the purpose of this communication, the LAPSE system employed utilized three chest leads yielding an ECG signal (reference, manubrium; negative, xiphisternum, positive, mid-acillary line, level of the negative electrode). A Pye Electrodynamics accelerometer (BLA 2) was mounted on the mid-sternum between reference and negative electrodes. The transducers were applied to carefully prepared skin sites and the cables taped to the body surface to reduce disturbance artifacts during daily activities. Transducers were connected to a modified Medilog miniature analog tape recorder (Oxford Medical Systems Ltd.) which was worn in a soft leather pouch attached to a webbing belt around the waist. The entire assembly was worn next to the skin so there is no need for the patient to disturb the recorder when changing clothes etc.

Part of the calibration procedure for the system requires that ambulatory patients walk on a figure-of-eight walkpath in our laboratory. The time to cover 200 m is recorded along with the number of steps taken, thus providing information on 'step length' and 'step interval' for the particular cadence employed.

Normal walking is such a deep-rooted activity that most subjects, when asked to walk at their own speed (and when it is pointed out to them that they are not in a hurry nor are they loitering) tend to adopt a 'preferred gait' which is extremely reproducible. This is true not only of normal individuals but of patients with physical handicaps including amputees, poliomyelitis patients and those with orthotic braces etc. In our experience any departure from this general rule is usually due to a failure in the investigator to distract the subject from his or her usual environment during the course of the test so that auronomic conditions may prevail.

The figure-of-eight walkpath is employed as, in our view, walking in a circle is not representative of normal locomotion. In our laboratory the figure-of-eight crossover is used between normal circuits to allow the patients to experience turns in both directions.

In the laboratory it is easy to line up the replayed locomotion data with the individual circuits at different cadences and thus validate the 'step length' versus 'step interval' relationship. We also establish their correlation with the patients' heart rate at that time.

Once the patients are fitted up with the transducers and recorder they are dressed and returned to their usual environment. They are instructed to continue with their normal way of life in all respects with the exception that the equipment should not be immersed in water. At the conclusion of the investigation period the patient returns for further calibration tests in the laboratory, if required, and then the equipment is carefully removed, the skin washed, dried and dusted with talc. It is only rarely that a mild erythema is found and it is unusual for the subject to report any adverse comments after a 24 h period.

The C120 tape cassette is replayed and the analyzers determine heart beat and step intervals. Much of the processing within the analyzers is digital but among other features the series of intervals are numerically reciprocated and digital/analog converted to develop voltages which are used to drive a conventional chart recorder so that we have analog output of instantaneous heart rate and step rate. Alternatively the trunk accelerometer signal may be output to indicate the general level of activity of the individual and an indication of posture is also given. An example of this form of output is shown in Fig. 1.

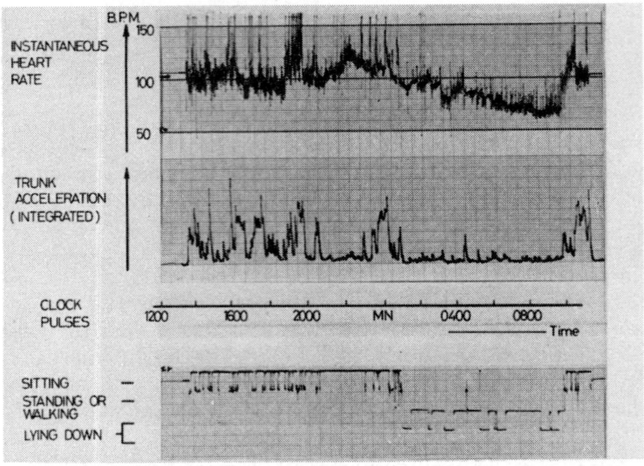

Fig. 1. Activity and heart rate output.

In the present investigation we are confining our attention to the relationship between *physiological cost index* and *walking speed*. The physiological cost index is an attempt to correct instantnaeous heart rate for factors related to differences in stroke volume between individual patients (a function of fitness). The Index is a measure of performance achieved. The resting heart rate is subtracted from the observed heart rate, averaged over the period of interest. The heart rate difference in beats per minute is divided by the speed of the subject in m min^{-1} thus the 'physiological cost index' has the dimensions of 'additional heart beats m^{-1} travelled'. The speed is taken either from walkpath data or by computing from the observed step interval using our knowledge of the patients' steps interval versus step length relationships.

RESULTS

LAPSE studies have been applied to a wide range of normal individuals and some of the results are shown in Fig. 2. Here we have the physiological cost index plotted against speed for normal men and women in an age range from 14 to 50 years. The data shown concern only the preferred cadence, self selected by the subject, in level walking and does not demonstrate the range of speeds achievable by the subject. It has been previously shown that the preferred cadence is the optimal physiological cost point for the individual subject and at lesser and higher cadences physiological cost indices rise. The figure has been drawn to illustrate the boundaries of physiological cost index as being at 0.5 net beats m^{-1} with a preferred speed range of from 60 to 85 m^{-1}. Figure 3 shows the results for a group of individuals who are quite severely handicapped where physiological costs are high when walking and the achieved speed is low. When these individuals are placed in wheelchairs their performance improves dramatically and it may take them into the normal range of performance. The figure also shows a number of normal individuals undertaking mobility tests in wheelchairs.

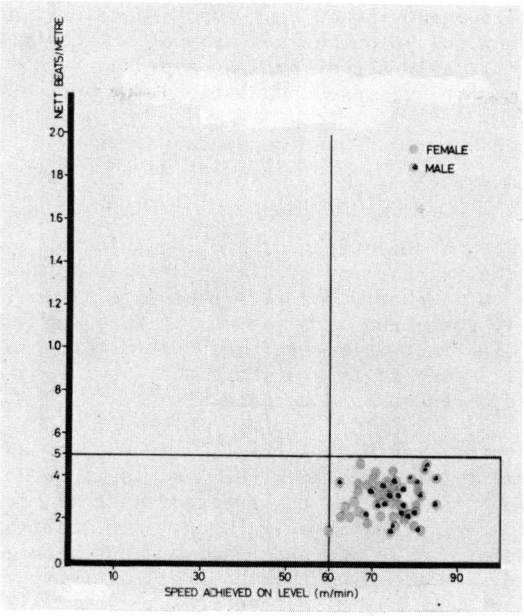

Fig. 2. Normal mobility performance at 'preferred speed'.

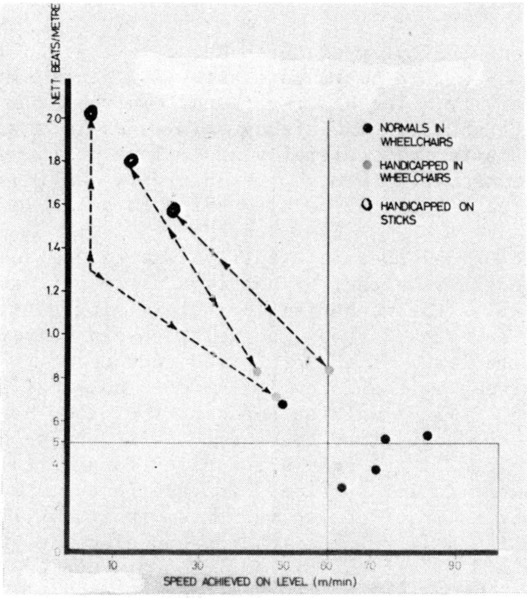

Fig. 3. Wheelchair mobility performances at 'preferred speed'.

DISCUSSION

The traditional physiological view of cost in the metabolic sense is firmly related to the concept of oxygen uptake and the measurement of respiration parameters in the ambulant subject. It is obvious that such an approach is quite unsuited for unobtrusive monitoring of the handicapped population, for periods of up to 24 h.

The correlation between heart rate and oxygen uptake is well known and well understood and undoubtedly is a very useful index of metabolic cost particularly when applied over extended periods of time.

A common question raised is the effect of emotional factors on heart rate as a disturbing influence on the reliability of heart rate *per se* as a measure of external work. It seems clear to us that faced with real-life situations in a real-life world we are not overly concerned with the way in which an increase in physiological cost is produced, our interest is solely in the fact that the increased cost occurs. Thus any tachycardia, however it is brought about, is a genuine additional load to the patient and should be taken into account.

In the record from a normal individual (Fig. 1) we can see an example of this problem. The subject, a TV commentator, undertook the test during a normal working day. From the record it is obvious that by far the greatest part of her day was spent sitting with only small episodes of standing or walking. She did in fact feel she had experienced a physically trying day as she was rushing about between film locations, television studios etc. In the event her physical activity was interspersed with many periods of driving, possibly also stressful. Total walking time proved to be as little as 37 min, a fact which surprised her considerably.

At about 9 p.m. (2100 h) the lady was embarking upon a fairly momentous dinner engagement which involved matters of personal import. The heart rate *rises* in the *absence* of physical activity as demonstrated by the trunk acceleration signal and reaches a peak around 10.30 p.m. Presumably the fall in heart rate thereafer is an indication of the success of the interview.

All in all a very reasonable view of this subject emerges. Here is someone who spends most of the working day with a heart rate above 100 beats min^{-1} irrespective of the level of physical activity. The heart rate pattern when the subject retires to bed shortly after midnight shows a slow steady fall which is *atypical*. Normally subjects would have a more dramatic fall initially and a long period of a steady heart rate averaging 50 to 60 beats min^{-1} throughout the night. Obviously much more detailed examination of these physiological relationships is possible.

The concept of physiological cost is a fairly complex one and the only justification for any particular index is whether or not it adds to our understanding of the problems facing the patient. The mechanism for determining physiological cost index has been described above and Fig. 2 shows the relationship of physiological cost index to walking speed on the level at the preferred cadence of individual subjects. There is a large overlap between male and female subjects but the limits of 0.5 net beats m^{-1} on the one hand and 60 m min^{-1} walking speed on the other have been set quite empirically in relation to the population observed. This data should be compared with that shown in Fig. 3 where one of the possible applications of this approach to domiciliary or ambulatory assessment is illustrated. In this instance we are considering the role of the wheelchair as an effective mechanism in rehabilitation. A number of patients were tested in their own wheelchairs and also when wearing long leg braces and walking with the aid of sticks. In addition a group of normal individuals were encouraged to undertake a series of tests in borrowed wheelchairs.

The implications are obvious. Patients who are normally confined to chairs are extremely handicapped without them. Physiological costs are extremely high and the mobility achieved is very poor. When placed in chairs their performance improves dramatically but it is fair to say it does not quite come into the normal range except for one subject.

The normal subjects do rather less well than one might have expected and the interpretation to be placed on this result is that the patient who appears in the normal population range has developed a quite remarkable upper body fitness and strength in the shoulders and arms.

The merit of the method is not necessarily in demonstrating differences between patients in and out of chairs but in providing means whereby any further alteration in prescription or in the management of the patient can be measured. It now remains for clinic teams in various areas of rehabilitation to consider how they may best employ a method of domiciliary cost evaluation in the routine management of the problems with which they are undoubtedly faced.

The Automatic, Continuous and Fixed Radio Tracking System of the Chizé Forest: Theoretical and Practical Analysis

A. DEAT, C. MAUGET, R. MAUGET, D. MAUREL and A. SEMPERE

Centre d'Etudes Biologiques des Animaux Sauvages, 79360 Beauvoir-sur-Niort, France

Abstract — The eco-ethological research based on radio tracking of European wild mammals in the Chizé forest, focuses on social structure, the utilization of space, and activity rhythms. A fully automated 10-channel system has been specially adapted for the Chizé forest. Transmitted signals are received by pairs of Yagi antennas on top of three 30 m pylons located on high land in the Reserve. After initial amplification, the signals are fed via coaxial cables to the receiving unit in the laboratory. Each antenna rotates through 360° every four minutes, during which it picks up signals from all ten transmitters. After reception, the angular positions of the animals in relation to magnetic North are displayed numerically and simultaneously recorded on magnetic tape. A theoretical analysis gave an estimate of overall system accuracy of ± 1°, which has been verified by an experimental analysis of the influence of different parameters (vegetation, topography, transmitting direction, etc.). Data are processed after storage on a PDP 11-40 computer by means of a specially designed interface. An initial program makes it possible to structure the recorded data on the basis of one data file per channel per day, including the computed angular positions and times, from which the position of the animal is calculated in Cartesian coordinates. Structured data files, in times and coordinates, are obtained and stored in the memory. These generate hourly displays of the animal's movements superimposed on an outline of the forest, from which behavioral characteristics such as movement speed, home range, territory and rhythms of activity/rest are computed. This system has enabled us to obtain original results from boar, deer, fox and badgers regarding their utilization of space and its implication for their social relationships.

INTRODUCTION

A great number of direct and indirect techniques have been developed to study mammalian movements and activity patterns (reviewed by Sanderson, 1966). These different techniques give only an approximation to quantifiable parameters: home range, movements (migration, immigration, emigration) and territory, but a new technique, namely radio tracking provides a better approach (Lemunyan *et al.*, 1959).

The description (Cochran and Lord, 1963) of a mobile radio tracking system (Cochran *et al.*, 1965) opened new ways of investigation. Location data which were difficult to obtain on big mammals by any other technique, became available. Since then, this method has been widely used on a great number of animal species. The major scope of these studies is in wildlife management, requiring a precise knowledge of home range and movements (migrations, emigrations). These problems led us to develop an automatic, fixed and continuous radio tracking system.

STUDY AREA

The Chizé forest is in mid-western France (46° 10 North latitude, 0° 30 East longitude), near the Atlantic coast (Fig. 1). The average elevation is 60 m, ranging from 50 to 90 m, and has a temperate, oceanic climate. Its major features are: average maximum temperature — $20^{\circ}C$ in summer; average minimum temperature — $2^{\circ}C$ in winter; annual rainfall — 800 to 900 mm in 140-150 days, most rain falls during October-November (local data). The forest is an European-temperate-deciduous biome type of the paleartic zone. The major tree species are represented by oak (*Quercus sp.*) and beech (*Fagus sylvatica*) with conifers (*Pinus maritima, P. sylvestris*) in isolated plantations. The Government Forest occupies 5000 ha of which 2620 ha is a fenced National Reserve. The natural fauna is abundant and diverse: wild boar (*Sus scrofa* L.), roe deer (*Capreolus capreolus* L.), fox (*Vulpes vulpes* L.), badger (*Meles meles* L.) and other mustelids. In this Reserve the CNRS (Centre National de la Recherche Scientifique) established the CEBAS (Centre d'Etudes Biologiques des Animaux Sauvages) in 1968 to specialize in ecological, ethological and physiological studies of wild mammals in their natural environment.

DESCRIPTION OF THE RADIO TRACKING SYSTEM

The radio tracking system is modeled on the one described by Cochran *et al.* (1965) and adapted for the Chizé forest. The principle is of localization by radio triangulation, and different steps in the system are presented in Fig. 2 and are described below.

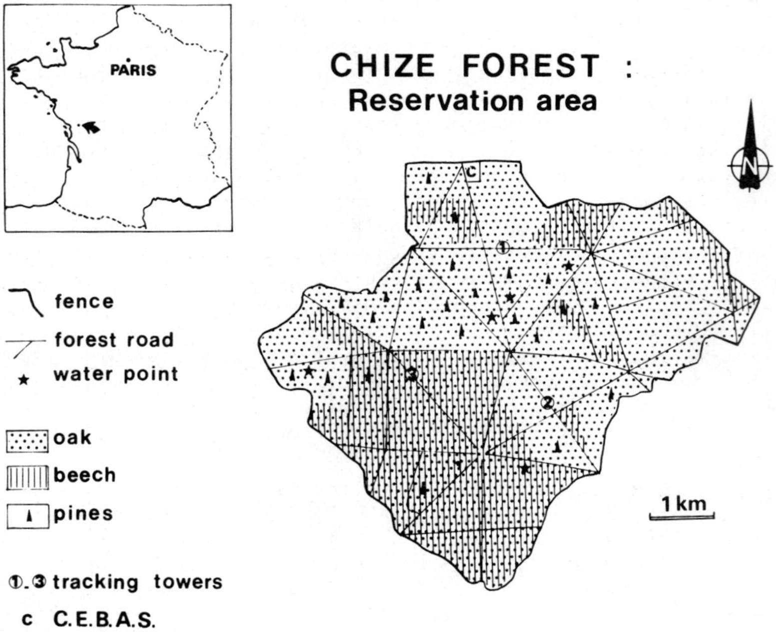

Fig. 1. Study area.

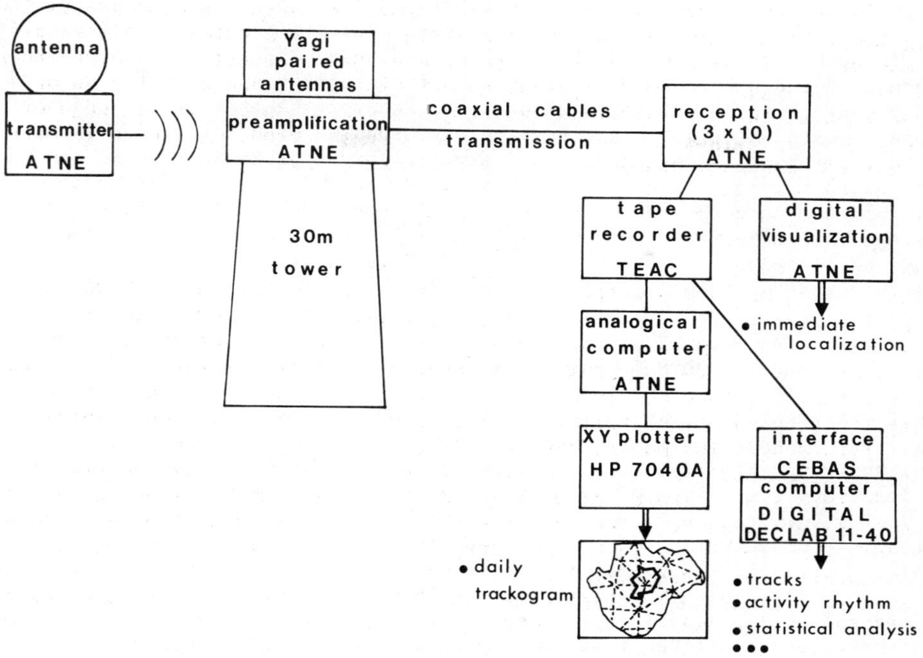

Fig. 2. Block diagram of the system.

A crystal controlled transmitter (ATNE*) transmits modulated pulses in the 72 MHz frequency range. The transmitting antenna is a loop. Signals are received by three paired Yagi antennas (Marquès, 1972) which rotate. Antennas are placed on the top of three 30 m towers located on high points within the Reserve (see map, Fig. 1). After an initial amplification stage, the signals are conducted by coaxial cables to an automatic receiving system in the laboratory. The antenna rotates at a constant speed of 360° per 4 min. During a single revolution, one antenna tower can receive signals from 10 transmitters. Thirty receivers (3 × 10) collect the data. Angular locations, measured with reference to magnetic north are simultaneously displayed on each receiver. This permits immediate localization of each tracked animal. They are also tape recorded (TEAC Data-pack recorder), such that, every four minutes date, time and the three angles measured by the three towers are recorded for each animal.

Two processing routines are available. Data processing by an analog computer (ATNE* prototype) allows the position of each animal to be reconstructed in xy coordinates with a plotter (Hewlett Packard HP 7040 A) on a 1/25000 map of the forest. For each animal this provides rough daily 'trackograms'. After storage in computer memory, the other routine allows processing on a Digital Declab 11-40 Computer.

TRANSMITTERS

The first experiments with radio tracking in the Chizé forest involved roe deer whose size and morphology made it a good test-species for radio collar design. The behavior of other animal species in the forest required the transmitter to be more robust.

---

*Applications Techniques Nouvelles en Electronique, ORSAY (France).

The first prototype transmitter was encapsulated with epoxy resin. A flexible plastic collar protecting a loop antenna was used to attach the transmitters to the animals' necks. However, this method, modelled on an AVM Instrument Co. design, was soon rejected. Because the animals are frequently trapped again, we needed a method of attachment which would allow us to recover the transmitter undamaged and to replace the battery quickly. For these reasons we manufactured an aluminum block to house the transmitter and battery (Mallory type RM 1H for the roe deer, badger and fox, type RM 12R for wild boar). So that the various elements did not move within the case they were potted with silicone rubber and a cover was fixed by screws and then made waterproof by silicone (Fig. 6). The use of lithium batteries provided an increased transmitter lifetime. Recently, we have replaced the aluminum with lightweight plastic (Ertalon 6 SA).

THEORETICAL ANALYSIS OF THE SYSTEM

Transmitters are located by three angular measures, from three receiving towers delimiting a triangle within the Reserve. Every direction is measured with a theoretical accuracy $\varepsilon = \pm 1$ degree. For one antenna revolution with an animal in a fixed location, the system gives three dots corresponding to the three possible pairs ($\theta_1\ \theta_2\ \theta_3$). An area, representing the smallest surface in which one can be sure to find the animal, is delimited from these three triangulations. The most probable location is the geometric center of the polygon delimited by: $\theta_1 \pm \varepsilon$, $\theta_2 \pm \varepsilon$, $\theta_3 \pm \varepsilon$. Each one of the three possible pairs, has an area which indicates where the animal is. This area, which looks like a parallelogram, has two properties: the further the transmitter from the two towers, the larger the area and the further the angle of two directions from 90°, the less likely the localization along the largest diagonal. In practice, for every point in the forest, one can select two triangulations such that a third does not improve the accuracy of the animal's location.

The directions are chosen on the basis of two criteria: (a) smallest distance to the towers and (b) angle of the two directions closest to 90°. This raises the problem of the angle at which we should stop using the bearings of the two nearest towers and start using a tower further off but with an angle closer to 90° to one of the first two towers (Fig. 3). In this case the two nearest towers are T_2 and T_3. The dot A nearing the limit $T_2\ T_3$ ($\theta_2\ \theta_3$) will remain the best pair as long as:

$$AB < AD$$

$$\text{but } AB = AC \times \text{TAN}\ \frac{\alpha_1}{2}$$

$$\text{and } AC \simeq AE = l_1 \times \text{TAN}\ l^o$$

$$\text{so } AB \simeq l_1\ \text{TAN}\ l^o \times \text{TAN}\ \frac{\alpha_1}{2}$$

$$\text{in the same way: } AD \simeq l_2 \times \text{TAN}\ l^o \times \text{TAN}\ \frac{\alpha_2}{2}$$

$$\text{the condition } AB < AD \text{ equivalent to: } \quad l_1 \times \text{TAN}\ \frac{\alpha_1}{2} < l_2 \times \text{TAN}\ \frac{\alpha_2}{2}$$

$$\text{and is realized for: } \alpha_1 < 130^o.$$

The limit of validity of the pair ($\theta_2\ \theta_3$) is extended to the geometric locus where T_2 and T_3 are seen from an angle of 130°. Likening triangle $T_1\ T_2\ T_3$ to an equilatera triangle, the upper limit of the angle α will be 130° for the three pairs ($\theta_1\ \theta_2\ \theta_3$). Taking into account the forest's dimensions, no lower limit of the angle will be determined.

These considerations led us to divide the forest into 'best pair' zones (Fig. 4).

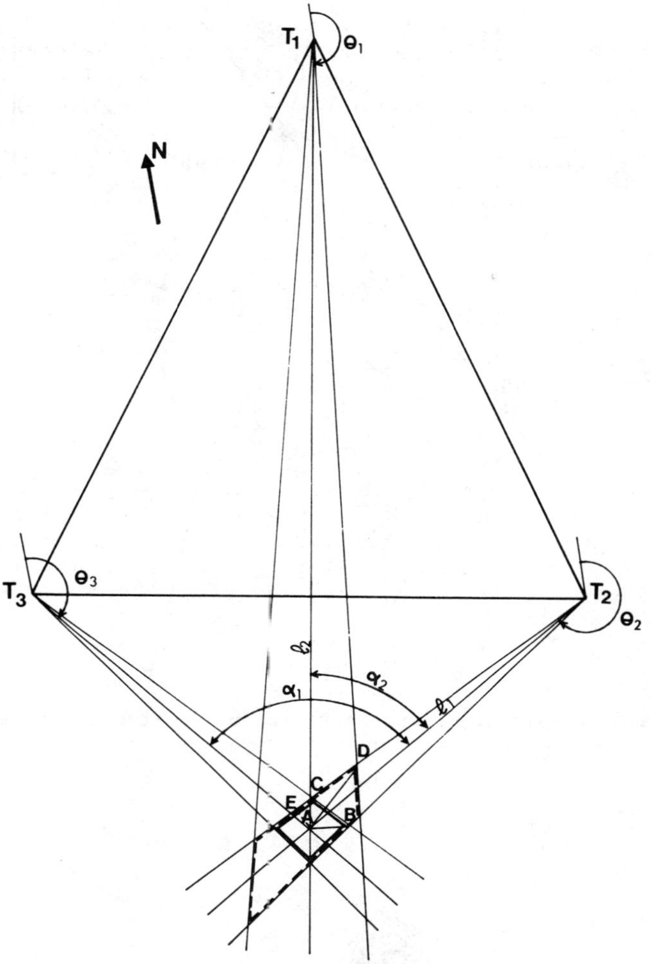

Fig. 3. Choice of the 'best-pair' (see text).

Angular boundaries are fixed to each zone to allow direct sorting of the recorded data θ_1 θ_2 θ_3. It is possible to delimit a curvilinear triangle in the middle of the T_1 T_2 T_3 triangle, in which an animal's location would be accurate by taking into account the three pairs (θ_1 θ_2 θ_3); but since this area would be small the delineation of the certainty polygon and the calculation of the geometric center are complicated, we chose instead to select one of the three pairs inside the triangle.

To this static accuracy problem one must add the dynamic one. The three receiver antennas turn 360° within 4 min and face North at the same time. So, even if every bearing is correct, since it may take 4 min between the pointing of two receiver antennas, an error in the localization of a moving animal may occur. The faster the animal and the larger the lapse between two triangulations, the greater the error. Because of this lapse, a linear movement will be recorded as an ellipse of which the major axis will be in the direction of movement while its minor axis will be proportional to the animal's speed and to the lapse between the two triangulations. Consequently, during an animal's movements, one must choose the pair θ_i, θ_j, that will give both good static accuracy and good dynamic accuracy (given by the smallest difference $\theta_i - \theta_j$).

Figure 5 shows forest zones where, for every pair (θ_i, θ_j), the two bearings θ_i

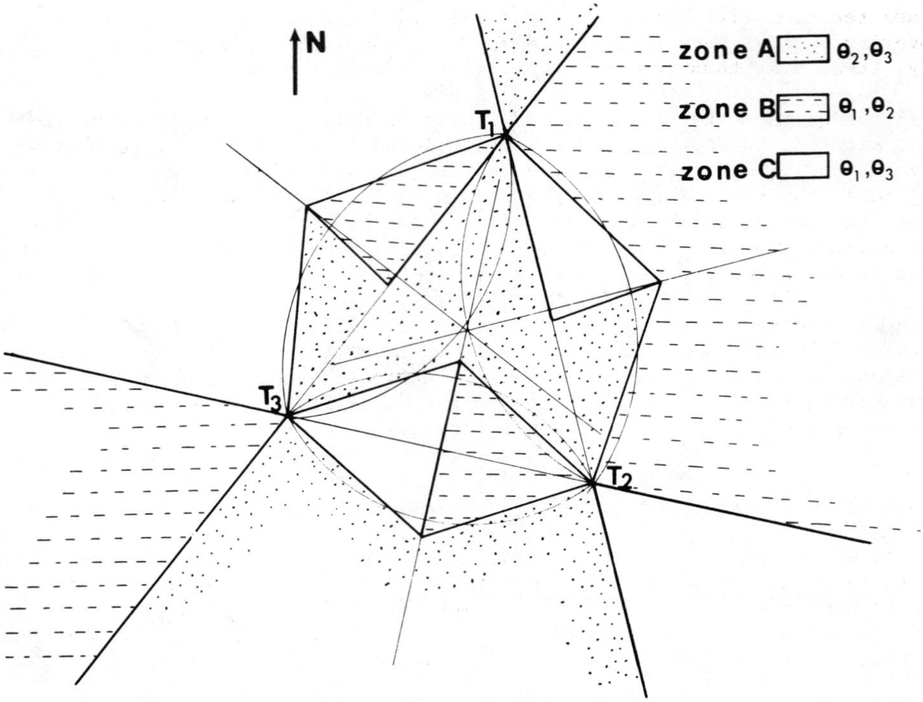

Fig. 4. Theoretical representation of 'best-pair' zones.

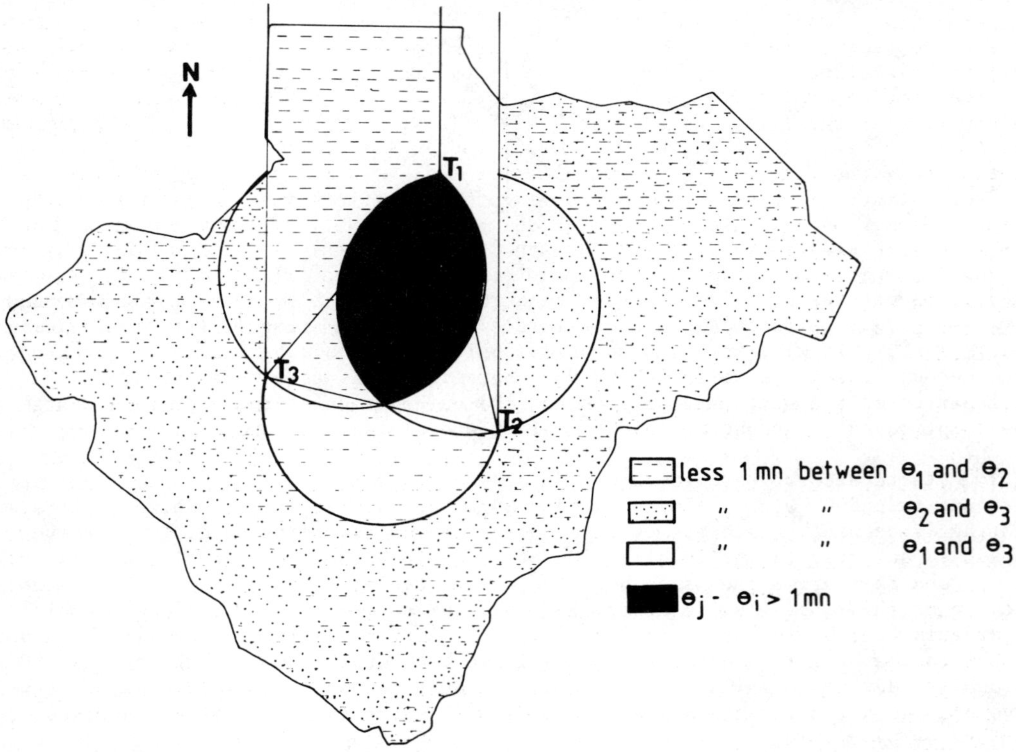

Fig. 5. Zones of time displacing for each pair of angles.

and θ_j are recorded with a time interval less than one minute. There is a central zone covering a great part of triangle $T_1 T_2 T_3$ where, whatever the pair (θ_i, θ_j), it always takes more than one minute between the pointing of θ_i and θ_j.

Superposition of these two theoretical studies (static and dynamic) shows that there are three kinds of zones: (a) good static and dynamic accuracy; (b) good dynamic accuracy (θ_i and θ_j bearings separated by less than one minute) but with static accuracy less than that defined by the choice of the 'best-pair' and (c) other zones with more than one minute between θ_i and θ_j bearings. It is possible to change the receiver antenna synchrony to shift these zones and possibly to obtain a better recording of animal movements in a selected sector.

A bearing is recorded by the receiving system only if it can integrate about 20 transmitting 'bleeps' corresponding to 20 s of reception. The double Yagi antennas are rather directional so the system cannot integrate enough signal if the animal moves quickly in the opposite direction to the receiving antenna. In practice bearing rate decreases as the animal's speed increases.

EXPERIMENTAL ANALYSIS OF THE SYSTEM

To test system performance and the influence of various parameters, we have performed systematic tests in the forest. To test the theoretical precision of system bearings, a badger transmitter (antenna length = 33 cm) was placed in a fixed location within the central zone of the triangle delimited by the towers. Testing the system precision statistically over ten months, revealed an accuracy of ± 1 degree.

To test the effects of other parameters we varied each of the following, one at a time: (1) Transmitting direction: circular antennas have a maximum transmitting direction which is in the antenna plan. In the fenced area, there is no unnecessary dependence on the transmitting direction, as long as the antenna plane remains vertical; if this plane is horizontal, the accuracy decreases. (2) Transmitter-receiver distance: this parameter is of little importance. (3) Transmitter-ground distance: this parameter has little effect for heights between 0 and 75 cm. (4) Topography: the forest relief lies between 50 and 90 m above sea level. The bearing rate decreases if the transmitter is not in a direct line with receiving antennas. (5) Vegetation: it appears from these tests that results are better in winter than summer (dense cover). (6) Fence: a three meter wire netting fence encloses the forest. When the transmitter collar is close to the fence, no direction is recorded if the maximum transmitting direction is parallel to the fence, but if it is perpendicular to the fence, bearings exist but are inaccurate. As it is placed further from the fence, accuracy and bearing rate get better and beyond 10 m, no influence was observed.

Parameters described as unimportant, may, in particular circumstances cause problems, particularly in combination. In real conditions, local and standing patterns (topography, distance to the towers, fence effect) or local and transitory patterns (unfavorable climatic conditions, dense cover, obstacles) lead to bad signal reception.

We conducted other experiments to investigate the effect of an animal's movements. An operator walked along a road with a transmitter collar around his arm. He traveled at about 4 km h^{-1} (speed arbitrarily chosen) and noted where he was every 4 min in step with the antenna bearings. Records were then compared with the journey.

The first thing to note is that bearing rate decreased by about 30 percent during movement. Accuracy decreased too; a rectilinear journey there and back was reconstructed as an ellipse, and the minor axis is larger as time between the two bearings taken into account is larger. On the other hand, as soon as the operator stopped for long enough to permit a bearing in a fixed location, accuracy was regained.

System accuracy may be given at ± 1 degree for fixed location records. Travel at low speed may be reconstructed reasonably well, in spite of some missing records; but the faster the animal moves, the less easily defined the journey. Short travels (≤ 4 min) from a particular location and ending at the same place, may go unrecorded.

AUTOMATIC DATA PROCESSING (Fig. 6)

The data is recorded on a cassette at the rate of one message every 4 min. Every message begins with a 'clock' word followed by 30 'receiver words' (all 36 bit words composed are shown in Fig. 7a). Processing is done on a Digital Declab 11-40 (16 k, 16 bit words). An interface built at CEBAS makes it possible to differentiate the useful signal from the 'clock' word and to transform each recorded word into two 16 bit words which are sent to a standard serial interface (Digital DR-11 L) (Fig. 7b).

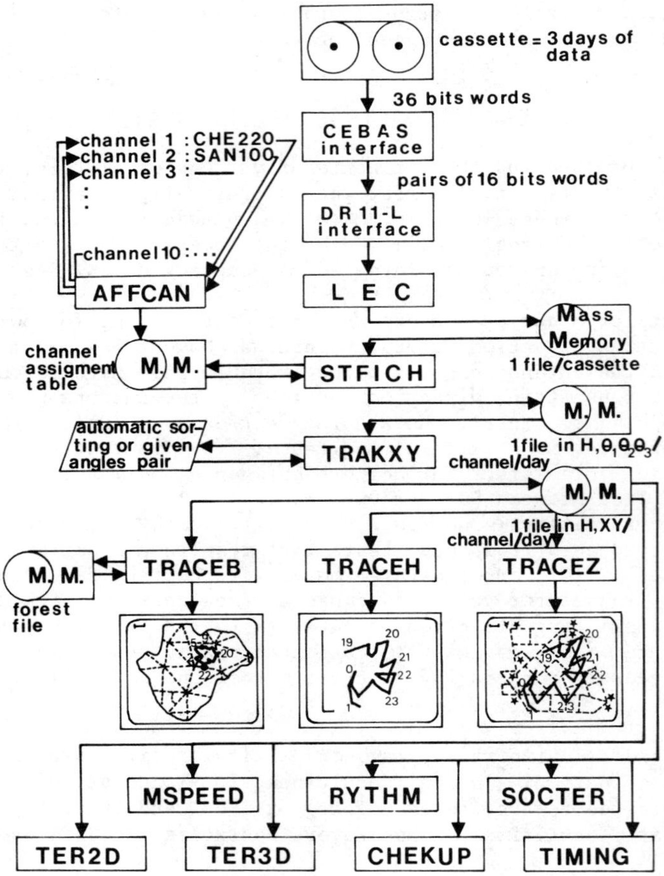

Fig. 6. Different steps of automatic data processing.

The program 'LEC' written in assembly language transfers data contained on cassette to computer memory. A Fortran program 'AFFCAN' fills a channel assignment table which holds the name (three letters) and the number (three figures) of the animal occupying the channel for each of the 10 channels. These six characters will form part of the title of the files created by the file-structure program 'STFICH'. This program establishes files for each channel and day, containing the time and the three measured angles θ_1 θ_2 θ_3. Because of the necessity of frequent addressing within each word (transcoding, swapping, etc.) this program has been written in Assembly-language for better computer efficiency.

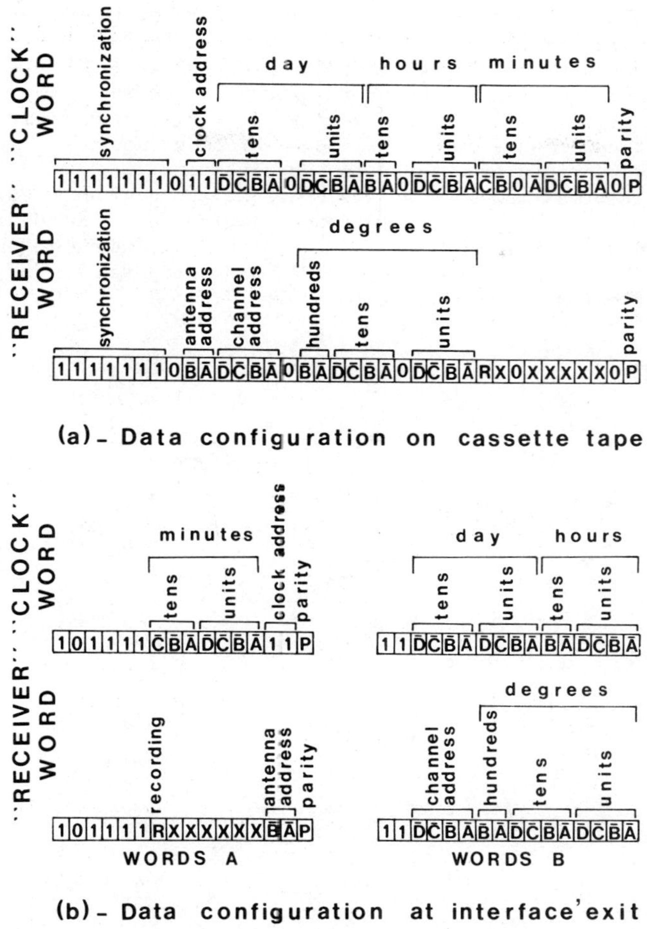

Fig. 7. Data configuration.

CALCULATIONS

In the section on theoretical analysis of the system, we saw that it was possible to distinguish 'best-pair' zones in the forest. The TRAKXY program, written in FORTRAN calculates the position of an animal, either by making calculations with a given pair of angles (for example to eliminate records of an inaccurate tower) or by running an automatic sorting of the zones, as follows: if three angular records exist, the program finds the zone where the animal is and calculates X, Y coordinates from the pair (θ_1 θ_2), (θ_2 θ_3), or (θ_1 θ_3), which theoretically gives the best static accuracy in that zone; if only two angles have been recorded, calculations are made with this pair. These files (structured in X, Y) obtained at a rate of one per animal and day, are used for all later computing.

DISPLAY PROGRAMS

An acquisition program of the forest coordinates (fence and roads) and a drawing program of the map make it possible to store on a disc unit a forest picture file which will be superimposed over the animal's movement tracking. A display of the animal's travels can be obtained on the display screen by the program TRACEB (Fig. 8). Moreover, this program can suppress, if necessary, all points corresponding to a speed, given by the operator and considered as impossible (for instance, records obtained from secondary antenna lobes, or a mistaken record). The same program also generates a display on a digital plotter, with different colors for night and day. The operator can then examine individual areas by use of the program TRACEH. One can choose a new origin of coordinates, scale, time interval between the plotting of two points (with a maximum of 3 s) and hours at the beginning and end of handling with an inscription on the screen of the hour, every hour. Moreover, this program is interactive and waits for questions on the light-pen; using it one can obtain the exact time of any point on the teletype. TRACEZ displays an extended movement with superposition of the forest zone where the animal is.

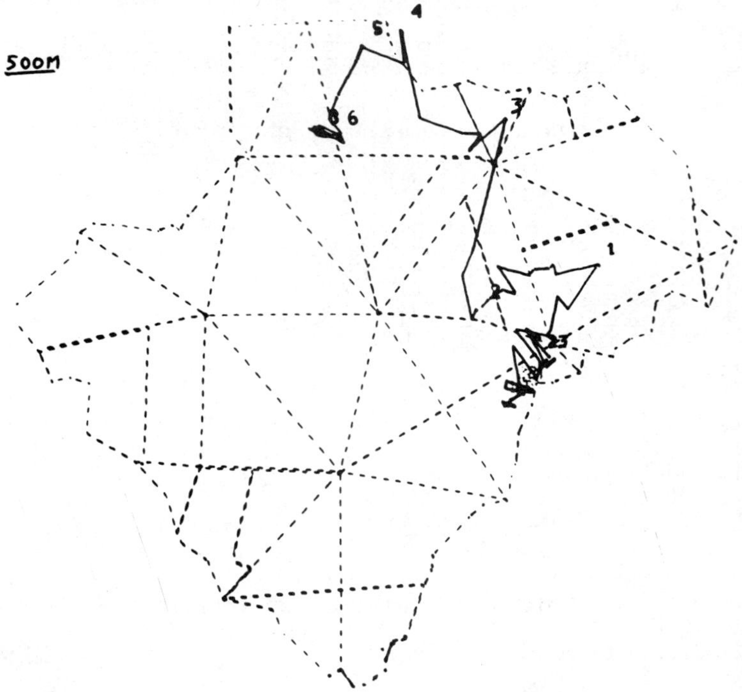

Fig. 8. Travel drawn by the program TRACEB.

EXPLOITING PROGRAMS

The program MSPEED computes the length and mean speed of the animal's travel between any two hours. This program creates a new file which represents the smoothed travel which may be displayed on the screen with the programs TRACEB or TRACEH.

The program RYTHM computes times of entry that exist inside a circle of a given radius, around any point of travel, if inactivity lasts for more than a specified time. This allows a rather good representation of activity rhythms and determination of resting places and activity zones. The programs SOCTER and SEQTER divide the animal's territory into a matrix and compute, for the whole tracking period or for some interesting sequences, the number of times the animal was in each area unit. The files obtained can be examined visually by program TER2D to have a 2-dimensional ground plan representation of territory occupation or by program TER3D in a 3-dimensional drawing that gives a good idea of occupation density (Fig. 9). If superimposed with a vegetation map, one can interpret occupation in terms of food zones, resting zones and so on. Next we study different activity sequences. A process that

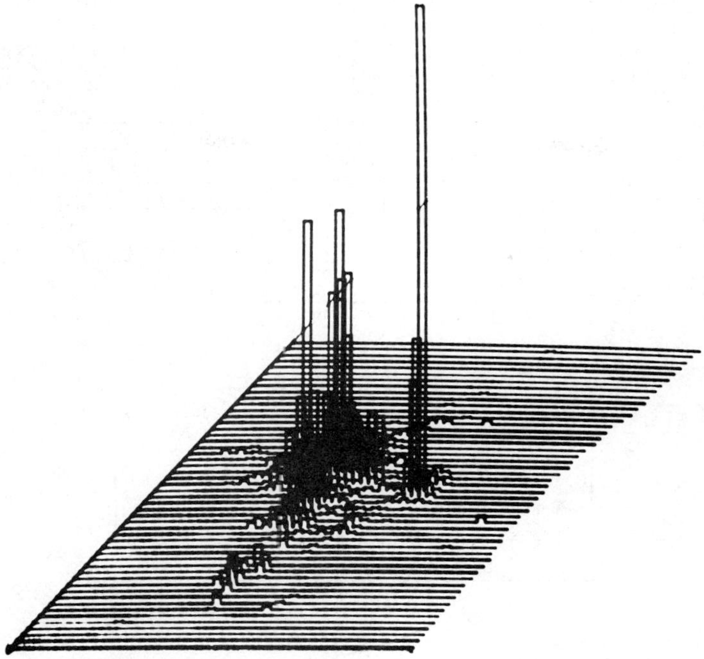

Fig. 9. Three dimensional territory occupation.

has been developed for wild boar. The researcher, with the aid of drawings and the program RHTHM delimits sequences: resting, activity zones, standing zones (which are zones of little activity and short duration) and movements. The program CHEKUP sums up the duration of each kind of activity, the activity area, the travel length and locomotion speed; it edits all these data, stores them and draws an 'integrated trackogram'. The file created will then be processed by a statistical method for time series data. Another approach to this problem has been developed for foxes and badgers: TIMING is an entirely automatic program which computes sequences of rest and activity from data files and from several criteria which define rest and activity and which dictate what is to be done with missing or unreliable data (Fig. 10). This program is rather sophisticated because (a) of the difficulty of identifying all kinds of activity, (b) the great number of data patterns and (c) the sequential aspect of the processing. This program edits the different sequences of activity with similar characteristics and stores them in a single file for the whole track period of the animal. This file may then be processed by statistical methods for time series data.

CONCLUSION

After the description and the analysis of the working of the radio tracking system of the Chizé forest, it is necessary to criticize the characteristics of its parts: transmission, reception and recording exploitation.

Transmission: the frequency that was selected (72 MHz) seems to be the best for a forest area such as the Chizé forest. All the transmitters used so far had a range of better than 4 km. Another important characteristic is the reliability of the ATNE transmitters, especially the frequency stability, and the possibility of reuse after the capture of the animal.

Reception is characterized by its precision. For an animal that is resting or moving slowly, the precision of the systems is ± 1 degree. It decreases when the speed of movement rises.

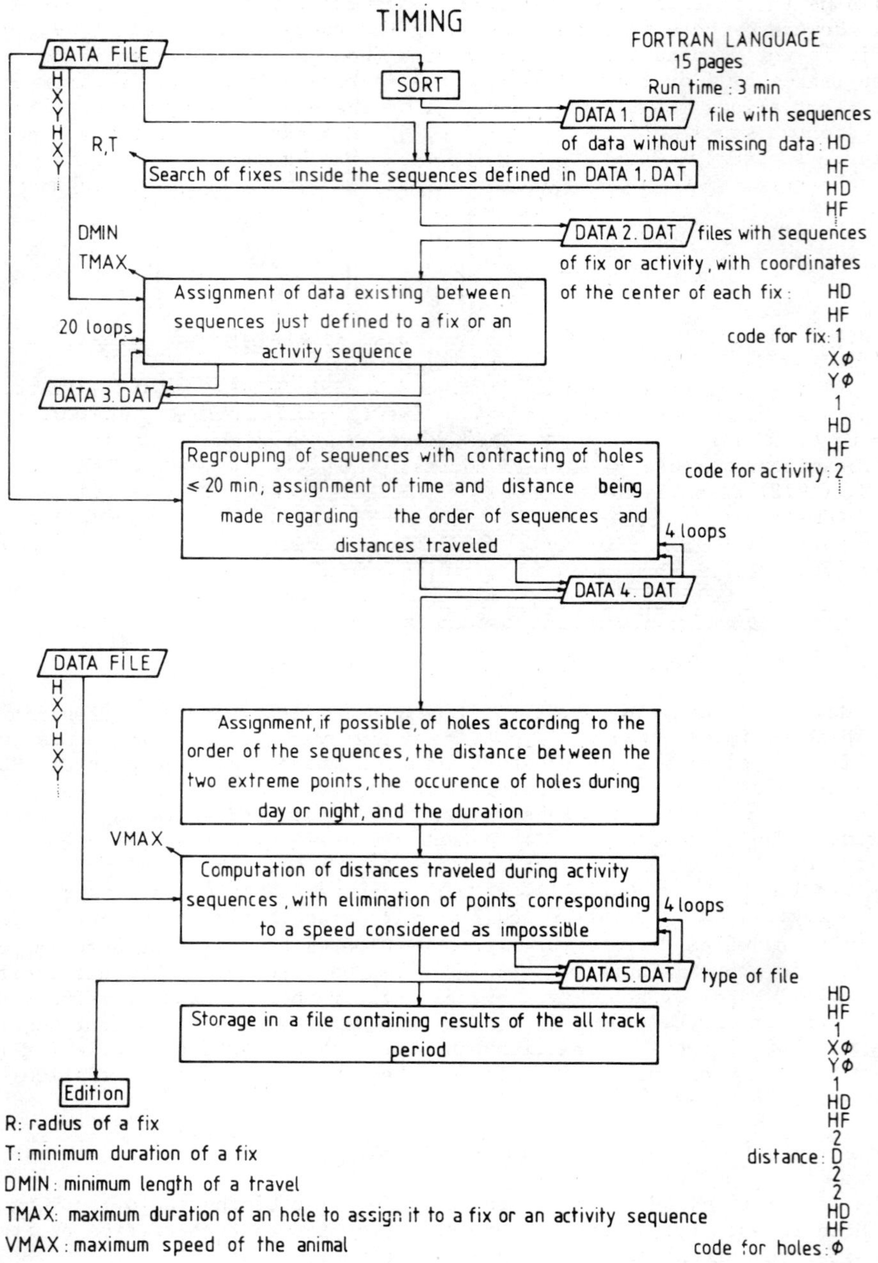

Fig. 10. Chart of the program TIMING.

Recording-exploitation: the original features of our system are versatility and automation. Indeed, it is possible to follow ten animals at the same time and to record automatically, every 4 min their coordinates for several months. Then, their tracks are displayed and analyzed by the computer. The large number of radio locations allow us statistically to refine the precision of the system during the sequences of resting or slow movements of the animals. This sophisticated radio tracking system gathers raw-data of the same kind as those given by a mobile radio tracking system. It permits us to determine the home range, movements, and activity rhythm of large mammals. However, there is a restriction in the range of utilization of our system. So far, it is not adapted for precise studies (homing, locomotor activity) on mammals with home ranges of 1 or 2 ha. The original feature of our data is their continuous character that allows longitudinal analysis of the behavior of free ranging animals. However, the two systems (fixed or mobile), are not exclusive, and may complement each other. In fact, the mobile system may confirm, by short term analysis on several animals, a particular behavior pattern suggested during a longterm study on one animal.

REFERENCES

Cochran W. W. and Lord R. D., Jr. (1963) A radiotracking system for wild animals. *J. Wildl. Mngmt.* 27, 9-24.

Cochran W. W., Warner D. W., Tester J. R. and Kuechle W. B. (1965) Automatic radio-tracking system for monitoring animal movements. *BioScience* 15, 98-100.

Lemunyan C. D., White W., Nyberg E. and Christian J. J. (1959) Design of miniature radiotransmitter for use in animal studies. *J. Wildl. Mngmt.* 23, 107-110.

Marquès M. (1972) Ensemble de réception 72 MHz à deux aériens permettant la localisation d'émetteurs destinés au radiotracking d'animaux. *Mammalia* 36, 299-304.

Sanderson G. C. (1966) The study of mammal movements. A review. *J. Wildl. Mngmt.* 30, 215-235.

An Automatic Telemetry System for Tracking and Physiology

P. A. LEMNELL

Statens Naturvardsverk, The National Swedish Environment Protection Board, Grimsö Wildlife Research Station, S-770 31 Riddarhyttan, Sweden

Abstract – An automatic radio tracking and biotelemetry system has been designed for moose. Initial tests indicate a tracking accuracy of ± 15 m in a 4000 ha space using hyperbolic triangulation. The system can handle up to 60 animals with a sample of each animal every 2 min.

INTRODUCTION

At the Grimsö Wildlife Research Station in central Sweden an automatic telemetry system for simultaneous tracking and transponding of physiological data from moose has recently been set up. The system has been developed for the Environmental Protection Board by Sensitron AB in cooperation with the staff of the station. The project is funded by the O. Engkvist foundation.

OBJECTIVE

The objectives have been to construct a flexible automatic system which combines high precision and accuracy of tracking with transmission of coded information. This would be an ideal tool for projects with a combination of ecological and physiological problems as well as for studies on behavioral interactions within populations. Home range and habitat utilization studies will achieve more detailed information with less man power than with ordinary methods.

NEW APPROACH

Positioning is made by the hyperbolic principle using the measurement of time delay from received signals, which gives high accuracy over large areas. The transmitter on the animal is selectively activated by remote control (Fig. 1). Physiological information is transmitted from an internal transmitter (Fig. 2) to the external transponder (Fig. 3) and relayed to a central unit by a frequency shift technique simultaneously along with position data. The system is automatic. A microcomputer (Fig. 4) and a digital taperecorder handle timing of individuals, data reduction and storing. The stored information can be displayed on a TV monitor for manual observation. All stored data are relayed to a central computer by a terminal. At the central computer the data can be treated in several ways by help of a library of programs.

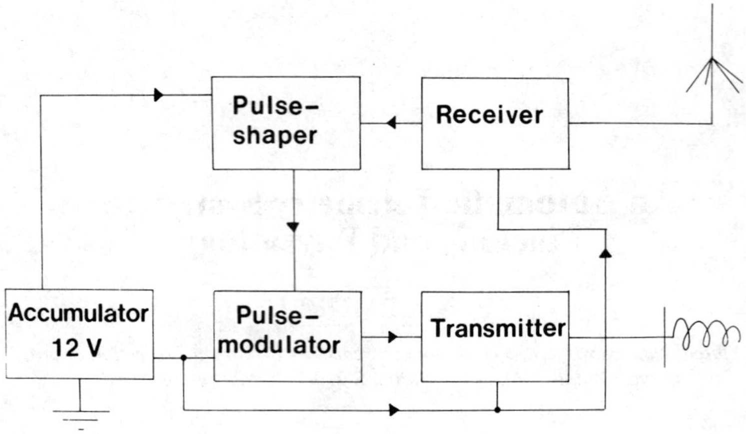

Fig. 1. Block diagram of repeater station.

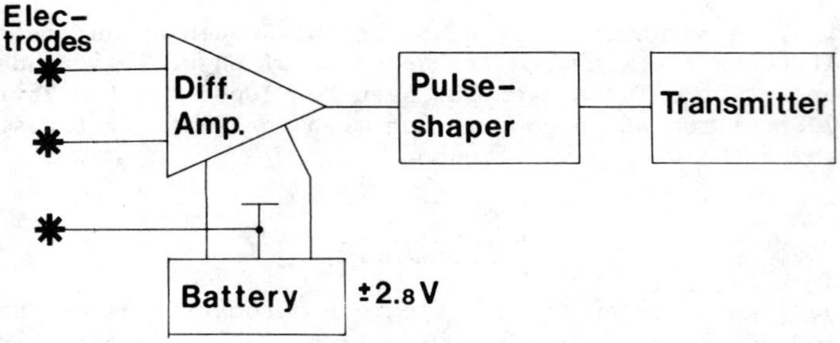

Fig. 2. Block diagram of physiological transmitter.

UP TO DATE INFORMATION

At the beginning of March, 1979 the first free roaming moose was instrumented (after immobilization from a helicopter). The moose will be intensely studied in May during a cross-country run in the area with about 2000 people participating.

CAPACITY

The prototype transmitter system at Grimsö has been designed for moose, but the goal for the future is to include smaller animals by miniaturization of the radio package. The system covers about 4000 ha with a mean accuracy of ± 15 m. By quality sorting of an abundance of data, accuracy can be increased and the usable range extended.

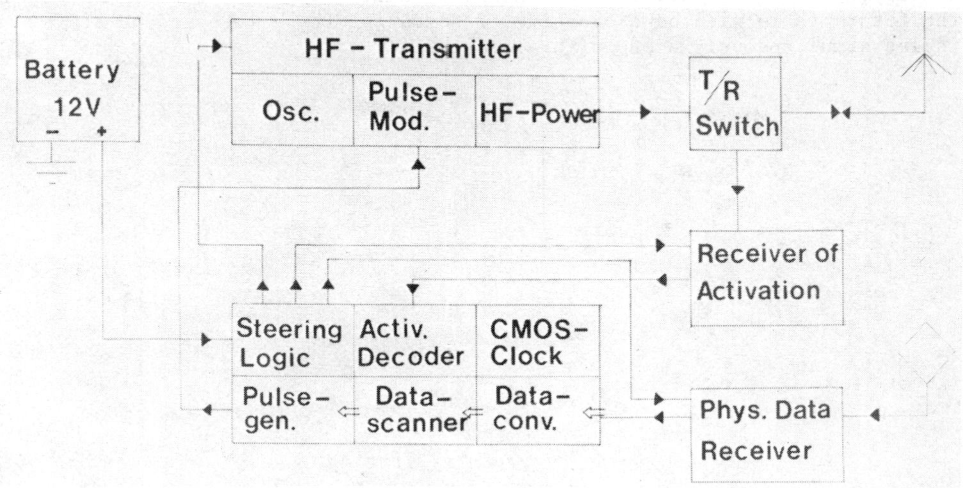

Fig. 3. Block diagram of transponder and HF transmitter on animal.

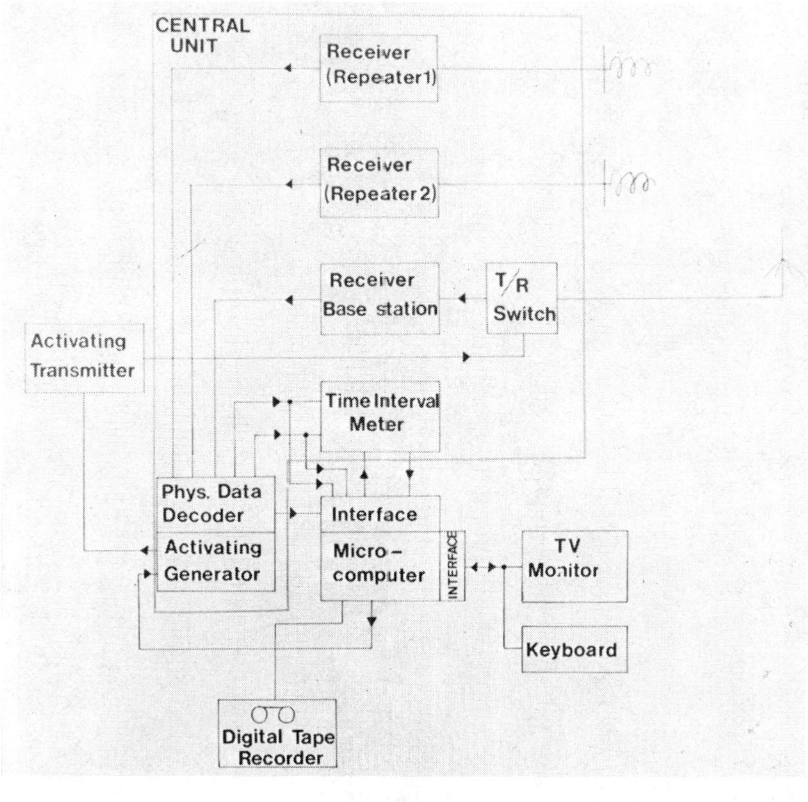

Fig. 4. Block diagram of base station with data acquisition system.

The maximum capacity of the test system is 60 animals with a sample rate per animal of 1 every 2 min. Together with a radio fix, heart rate of the moose is given. In the future there will be a capacity of conveying three independent binary coded variables simultaneously along with a radio fix.

Animal Activity Recorded by Radio Tracking and an Audio Time Lapse Recorder

G. GÖRANSSON

Ecology Building, Helgonav 5, S-223 62 Lund, Sweden

Abstract — A simple radio tracking recording system was developed to register diel activity patterns of resident or semi-resident animals such as pheasants, badgers, stoats, and foxes.

The system consisted of an omnidirectional antenna, a receiver, a tape recorder, and a timer. The receiver and tape recorder were powered by drycell batteries. The circuit could be turned on by a timer, adjustable from 1 min to 12 h. Normally recording periods of 10-15 s every 5 min were used.

This system detects signals from a transmitter in spite of disturbing noise. This may otherwise be difficult using conventional recorders. Normally one diel period is compressed to one hour on the tape.

When using compact cassettes and recording periods as mentioned above, the cassette has to be changed every day. Large tape rolls and a slower tape speed will allow longer periods without service need.

Activity data are given for incubating pheasants, and resident badgers.

INTRODUCTION

Measuring an animal's activity manually by radio tracking is relatively easy. Short measurements with longer intervals in between will often register periods of activity and rest in a satisfactory way. However, long periods in the field are necessary to obtain these registrations. Automatic recording of the signals received will save much time. Conventional techniques involving chart recorders have, however, several disadvantages: (1) it may be difficult to distinguish signals from noise, (2) recorders often need a high voltage, and (3) equipment is expensive. I will give the details of a simple and inexpensive system suitable for field work.

EQUIPMENT

The system is depicted in Fig. 1. The signals from 27 MHz receiver are recorded on a commerical tape recorder. Both receiver and recorder are controlled by a timer.

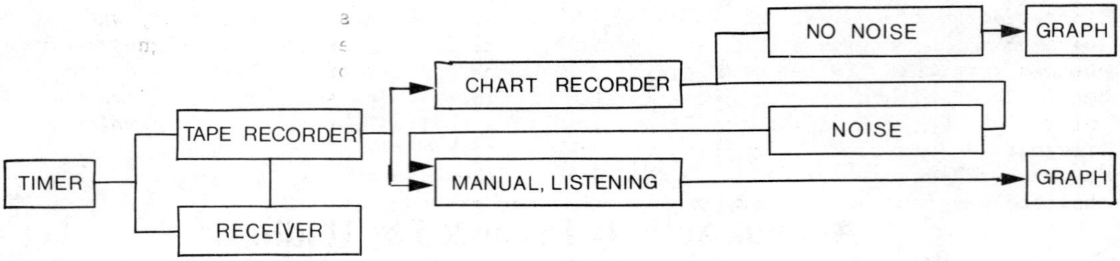

Fig. 1. Equipment and operation of the method.

The signals can be recorded on a chart recorder connected to either the receiver or tape recorder. The chart may be difficult to interpret on account of disturbing noise; when this happens the signal can often be traced by a person listening to the tape (since the ear will usually distinguish between signal and noise). The signal pattern is then plotted by hand.

An omnidirectional antenna makes signal strength independent of the direction towards the transmitter. Any receiver is suitable if sensitive enough; a cheap walkie-talkie, if provided with a beat frequency oscillator works well. A simple (mono or stereo) cassette recorder can be used. Recorder and receiver are powered by drycell batteries. A stereo recorder is able to record activities of two animals, or two kinds of activities of a single animal (a twin transmitter with two different channels is required), provided that two receivers are used, which are operated at different frequencies. A multichannel system may be constructed where recordings from several channels are registered in sequence.

I constructed two different types of timers. One type consisted of a battery clock with a Teflon disc fastened to the minute hand shaft (thus the disc made one turn in 1 h; if fastened to the hour hand shaft, one turn would be made in 12 h). Rectangular openings were cut near the rim of the disc, and oriented with their main axes parallel to the radii of the disc. A silver contact was mounted immediately above the disc in such a way that it slipped into the openings and made contact with another silver contact immediately below the Teflon disc. Time of closure could be adjusted by changing the width of the openings in the Teflon disc or by moving the contacts nearer to the center of the disc.

The second type was modified by using ceramic magnets instead of openings in the rim of the disc, and magnetic switches mounted below the rim. One advantage of magnetic switches is that they do not corrode. Two switches were soldered together in such a way that both had to be closed to close the circuit. When the magnet passed over the switches the circuit was closed for a time due to the distance between the switches. The number of switches placed below the disc determines the number of closures of the circuit.

The timer closes a relay, which in turn starts the receiver and the recorder. A sequence of time lapse recordings is then registered. Recording 10 s every 5 min, a 60-min tape will last about 30 h. These recordings may then be transferred to a chart recorder. Doubtful recordings on the chart may be checked by listening to the actual recording on the tape.

APPLICATIONS

I have used this method for studying the activity of pheasant and badger, and others in The Wildlife Ecology Group at Lund, have used it for stoat and fox.

The incubation rhythm of pheasant hens was recorded in May and June, 1976. The antenna was constructed of coated copper wire placed in a loop around the nest. Timer, receiver and recorder were placed 10 m from the nest. Sensitivity was adjusted so as to give a signal when the hen was in the nest, but no signal when she was more than 5 m away. Figure 2 shows some records of the activity of three hens. The diel activity of hen A is shown. Time-off was spent feeding (two hours per day). The activity of hen C was similar to that of hen A, but hen B was more irregular. This might be explained by the fact that her activity was recorded in May when the vegetation was poorly developed (the others were recorded in June); thus she may have been disturbed more than the others.

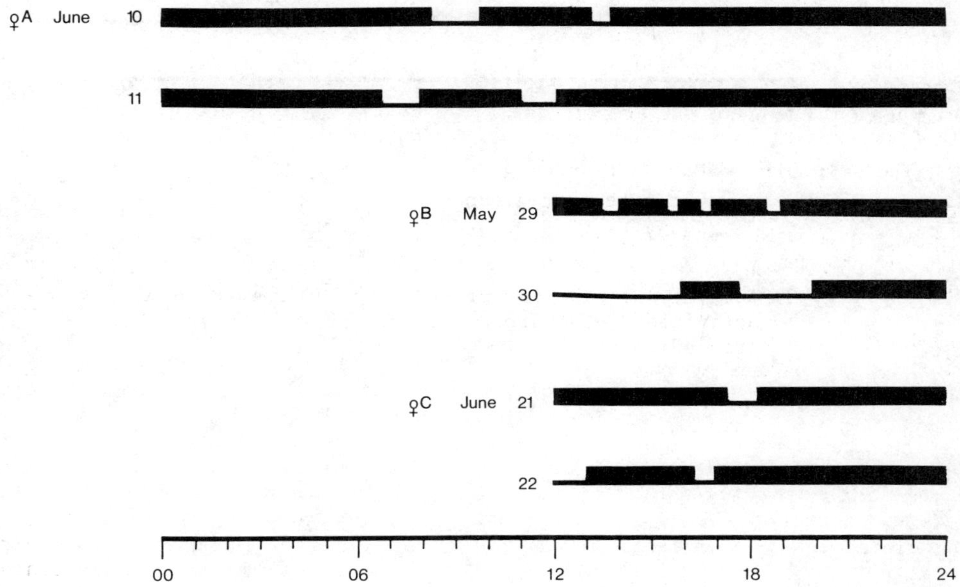

Fig. 2. Incubating activity of three pheasant hens recorded at 5 min intervals for 10 s.

The activity of a resident badger was recorded in March (Fig. 3). The strength of the signal indicated whether the badger was below or on the ground. The badger regularly became active in the evening and left the sett for about 6 h. At midnight 16/17 March, however, the badger spent 2.5 h in the sett. It normally returned some time between 2 and 4 a.m., but sometimes left the sett for one hour during early morning.

CONCLUSION

This method is inexpensive, it saves time and electrical power which makes it suitable for field conditions, and is applicable to many studies concerned with the activity of mainly resident animals. The possibility of using a multichannel device makes it suitable for following a number of individuals simultaneously, thus facilitating studies of strongly seasonal activities, where otherwise a single or at most a few individuals could have been studied. The possibility of listening to the tape recordings, thus ensuring that interpretation of the signals is correct, further enhances the usefulness of the method.

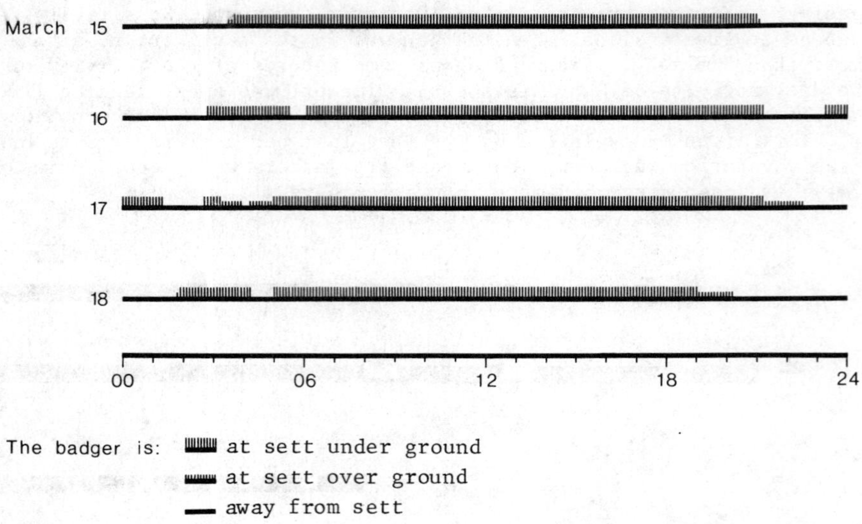

Fig. 3. Activity of a badger recorded for 10 s every 6 min. Activities below ground, on ground and off the nest are indicated.

Ultrasonic Tracking of Juvenile Cod by Means of a Large Spaced Hydrophone Array

A. D. HAWKINS, G. G. URQUHART and G. W. SMITH

D.A.F.S. Marine Laboratory, Aberdeen, U.K.

Abstract — An ultrasonic transmitter inside a cod is tracked by measuring differences in the time of arrival of acoustic pulses at an array of hydrophones spaced several hundred meters apart. Each hydrophone is tuned, with a bandwidth of 5 KHz, to the transmitter center frequency of 77.5 KHz, and is omnidirectional in the horizontal plane.

The received signals are processed by a multichannel superheterodyne receiver with a single tunable local oscillator common to all channels. The demodulated receiver outputs are fed to comparators and the resultant logic level pulses to a minicomputer.

The interrupt facility of the computer triggers off the program to determine the times of arrival of the pulses, while a parallel input identifies the hydrophones. The order of arrival of the pulses and the time of arrival differences are displayed on a visual display unit (VDU). The position of the fish, in rectangular coordinates, is computed using data from 3 selected hydrophones and can be obtained every 1.5 s on a fast printer.

Juvenile cod have been tracked continuously for up to 11 days. Data on the distance moved by the fish has been shown by periodogram analysis, fast Fourier transform and autocorrelation to have a clear diurnal periodicity.

Techniques have been developed for analyzing the pattern of use of the fish's home range.

INTRODUCTION

The ecology and behavior of even the more common species of marine fish are only poorly understood. These animals, living in an environment which is foreign to man, can only be studied indirectly, mainly by sampling catches. Direct observation is rarely possible, since the visual range of a diver, or underwater television, is limited to a few meters especially in murky North Atlantic waters. Active sonar equipment operates over greater ranges but cannot readily resolve individual fish close to the sea bed. To examine the detailed movements of single fish, some form of tracking technique must be used where the fish is labelled with a transmitter.

The position of an underwater object is not easily determined by means of radio transmission as electromagnetic waves are highly attenuated by salt water. Sound waves have much more favorable propagation characteristics.

The location of fish or other marine organisms labelled with small ultrasonic transmitters or transponders is now commonplace, a wide variety of tracking systems being employed to locate and record the position of the underwater acoustic beacon. Perhaps the simplest and commonest is the use of one or more directional sonar receivers to locate the transmitter from a series of bearings. A more sophisticated sonar which both transmits and receives can locate a transponder in terms of a bearing and range measurement. These methods are excellent for tracking highly mobile fish in the open sea, but good and rapid bearings can only be achieved at great expense with highly sophisticated receivers.

For following fish in inshore or enclosed waters, fixed installations can perform the tracking. One such method, with a large spaced hydrophone array laid on the sea bed and connected to a minicomputer has been developed by us and is employed to follow the movements of young fish and invertebrates in a Scottish sea loch.

OPERATING PRINCIPLE

The system operates on the hyperbolic navigation principle similar to that used by radio navigation systems for ships and aircraft. For underwater tracking, sound pulses are sent out by a transmitter attached to the fish and these are detected by hydrophones laid on the sea bed at known locations. Each pulse transmitted from the fish propagates through the water at a known velocity, the time of arrival at each of the hydrophones being dependent upon the distance between the fish and hydrophone. By measuring the relative times of arrival of the pulses transmitted by the fish at three hydrophones (t_0, t_1, t_2), two independent time intervals are obtained (t_1-t_0, t_2-t_0) which are sufficient to determine the position of the transmitter. The locus of points corresponding to any particular time interval is a hyperbola with the two hydrophones as its foci. The two independent time delays define two hyperbolae which intersect at the position of the fish.

DESCRIPTION OF THE SYSTEM

TRANSMITTERS

The ultrasonic transmitters were designed and constructed by the MAFF Fisheries Laboratory, Lowestoft. They are pingers, producing a continuous train of pulses or pings. Each pulse is a tone burst of about 75 KHz of 1.5 ms duration, with an interval between pulses of 1.5 s. The peak acoustic power output is about 0.75 W. The complete device totally encapsulated, including batteries is 14.7 mm dia. and 58 mm long, weighing 20 g in air and about 10 g in water.

The acoustic pulse is produced by a single, ring-shaped piezoelectric transducer which is closely coupled to the water. The transducer is driven electrically through a matching transformer at the frequency of its lowest mechanical resonance (i.e. in the circumferential mode) to produce the required output as efficiently as possible.

HYDROPHONES

The tracking hydrophones have a stacked array of transducer elements, each element being similar to the single element ring transducer of the transmitter. The array is arranged to be omnidirectional in the horizontal plane and to have a beamwidth of 30° in the vertical plane. This means that the hydrophones are most sensitive looking horizontally along the sea bed and are relatively insensitive to surface noise and reflections.

The transducers, operating at or near resonance, can be regarded as current generators with a large parallel capacitance. The preamplifier, which is an integral part of the hydrophone, was therefore designed with a low input impedance differential current amplifier input stage, which effectively shunts the parallel capacitance. This is coupled to the main amplifying and driving stages by a tuned transformer. The center frequency is set at 77.5 KHz with a 3 dB bandwidth of 5 KHz.

CABLES

The hydrophones are connected to the shore equipment by individual cables each with a pair of twisted conductors and a conducting screen, with an overall sheath of polyurethane. The cables, which are resistant to attack by marine organisms and show no significant seawater leakage even after several months immersion, are protected from abrasion in the intertidal zone by passing them through Polyvinyl chloride (PVC) tubing. The characteristic impedance of the cable is 120 Ω at 75 KHz, and though the signal attenuation over a 600 m length is about 3 dB, the signal to noise ratio is not significantly degraded.

HYDROPHONE TERMINATION UNIT AND MAIN AMPLIFIERS

The hydrophones are plugged into a seven channel termination unit which houses impedance matching and power supply decoupling circuits. This unit also contains power supply and signal monitoring facilities. The signal from each hydrophone channel is then connected to a low noise amplifier of variable gain and fixed passband where most of the amplification of the signals takes place.

RECEIVERS

The heart of the receiving hardward is a seven channel, amplitude modulated, superheterodyne receiver with a common local oscillator and digital outputs. Each of the seven identical channels has a mixer stage where the transmitter signal, in the 75 KHz range, is combined with the locally generated frequency of about 530 KHz to produce a fixed intermediate frequency (IF) of 455 KHz. A narrow band filter of either 750 Hz or 2 KHz can be selected to pass only the 455 KHz signal to the IF amplifiers. The detected transmitter signals are thus transformed to equivalent signals at the IF with a large improvement in signal to noise ratio. The signal is then demodulated and operates a voltage comparator so that signals exceeding the preset threshold voltage are converted to digital (logic level) pulses.

Logic circuitry rejects very short spurious pulses caused by natural sea noise and accepts only those of a present minimum duration of 0.5 ms. Retriggering is inhibited for a further 15 ms after each accepted pulse so that the circuit responds only to the leading edge of the direct path pulse. Output pulses for each of the seven channels, corresponding to the acoustic pulses detected by the hydrophones, are then generated and fed to the computer.

THE COMPUTER

The system includes a C.A.I. Alpha LSI 2 minicomputer with a floppy disc storage unit, a Data Recording high speed printer and a Newbury Laboratories visual display unit (VDU). With the fish tracking program running, the machine times the incoming signals to the nearest 0.1 ms and the VDU displays a continuous list of pulse timings at the active hydrophones. The program checks the data for consistency, then computes the position of the transmitter relative to the hydrophone positions which have been previously determined and stored in the floppy disc unit. After further checks, the result of the computation is then sent to the printer which writes out the real time together with X and Y coordinates at meters relative to the chosen axes. Figure 1 shows a shcematic flow diagram of the fish tracking program.

THE SYSTEM IN PRACTICE

The complete system is mounted in a single standard 19" rack containing the various subunits. The fast printer and VDU are placed on a bench close by. In addition, a multichannel oscilloscope is used to monitor the system's operation and a multitrack tape recorder can be connected to record the signals. The equipment can readily be transported to a new location, or placed on board a ship, but is normally housed in a wooden hut on the sea shore. Power is supplied from a portable generator.

Each hydrophone, mounted upon a steel frame approximately 1 m high, is laid individually on the sea bed from a small boat working from the sea towards the shore. The initial placement is approximate, the instruments being laid in a triangular pattern about 400 m apart. The position of each relative to the others is subsequently determined by placing a pinger at each one in turn and measuring the time taken for the signal to reach the other hydrophones. After choosing coordinates arbitrarily for one of the hydrophones, the computer then calculates the coordinates of all other hydrophones in the range from these measurements. The set of coordinates, or range data, is then stored in the computer's disc unit ready for tracking. With the tracking program running, the coordinates of any transmitter in the range can be instantly determined. The position of the array relative to a chart of the area can be determined by placing a pinger at two or more known positions marked on the chart.

Fish, which are caught by hook and line at shallow depths, are tagged with transmitters either by insertion into the stomach or by a surgical implantation into the abdominal cavity. In both cases the insertion is performed under MS222 anesthesia. Fish are held overnight in a cage on the sea bed and are subsequently released by triggering a catch on the cage door with a 'messenger' weight. If the fish moves outside the range of the array, additional hydrophones may be laid to extend coverage.

Depth surveys of the area within the array are performed from a boat fitted with a pinger and echosounder. Inspection of parts of the area are also carried out by a SCUBA diver carrying a pinger.

ANALYSIS OF THE DATA

The frequency with which the fish coordinates are given can be specified before the tracking run, the time of each fix being printed simultaneously. A computer program extracts the following information for each pair of successive fixes generated during tracking, on the assumption that movement between the fixes has been in a straight line.

- distance moved
- mean speed
- angle of movement relative to the grid
- change in angle between one direction of movement and the last.

The program also averages the information for each hour of the run.

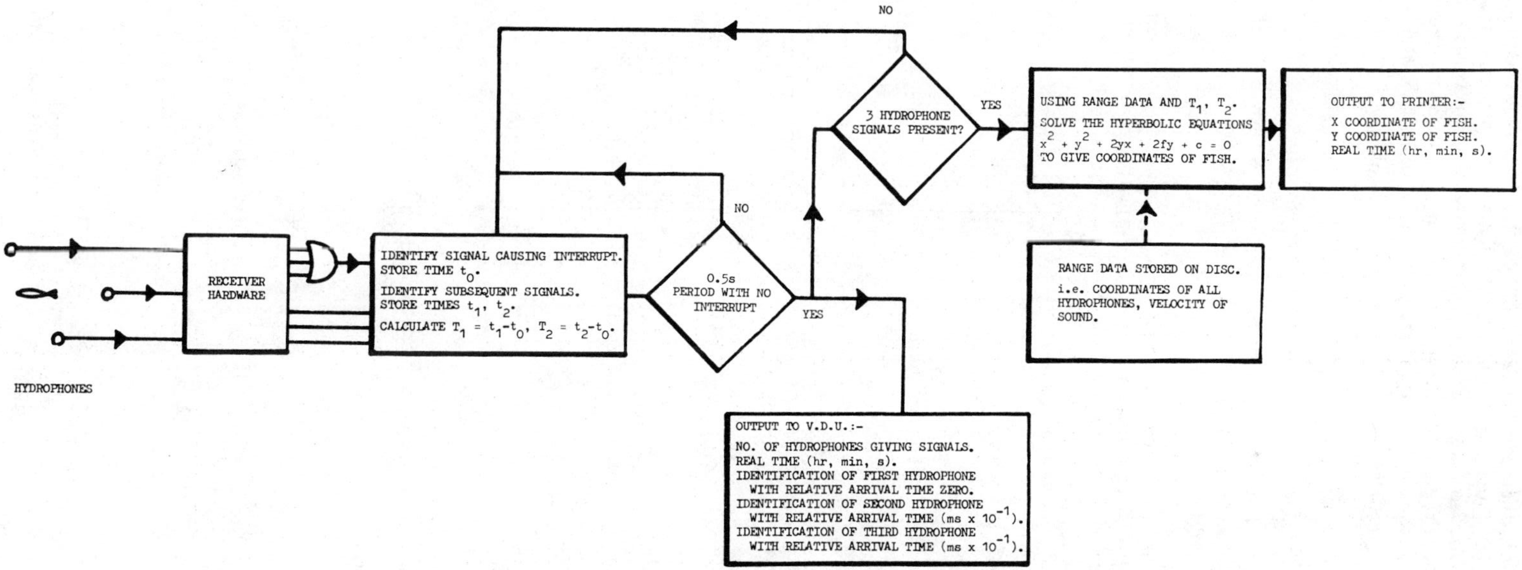

Fig. 1. A schematic flow diagram of the fish tracking computer program. For clarity, only three hydrophones are shown but any number can be used and up to seven at a time can be connected to the system.

In general, to reduce data handling problems the time between successive fixes is standardized at 5 min. Over this time, though the fish do not move in a straight line, the distance between successive fixes provides an index of fish activity. For more detailed tracking the time interval is reduced to one minute, with occasional fixes at even shorter intervals. Both the temporal and spatial patterns of fish activity are examined.

TEMPORAL PATTERNING

The hourly mean value for distance moved by the fish in a 5 min period is taken as an index of activity within that hour, and a time series of hourly activity values is produced for the whole run.

The time series is split into its component sine waves by means of a Fast Fourier Transform producing a power spectrum, an illustration of which is given in Fig. 2a.

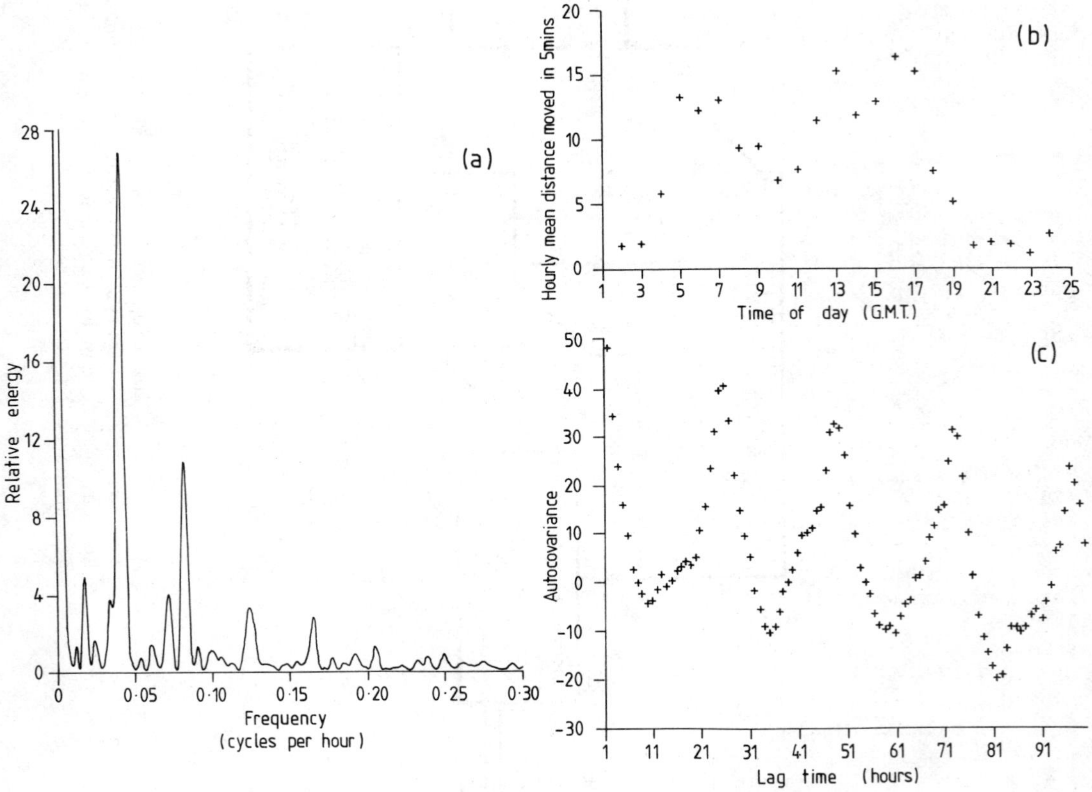

Fig. 2. The analysis of an activity time series. a) A Fourier spectrum showing a marked peak at 24.2 h period length. b) A form estimate of the waveform assuming a 24 h periodicity. c) An autocorrelation showing peaks at successive 24 h periods.

A periodogram analysis (Enright, 1965) is performed, based on the principle that, if a time series is grouped at a period length not present in the data, then the resulting averaged waveform would approach zero amplitude as the time series approaches infinite length. Estimates of the period lengths present in the data can thus be gained by plotting indices of the 'intensity' of the waveform against corresponding period lengths.

A 'form estimate' or averaged waveform for any period length is a necessary step in producing the periodogram, and helps the analysis of activity patterns, giving an indication of the shape of the waveform of rhythms present. For example Fig. 2b shows a form estimate of a cod activity pattern based on a 24 h period length. A marked bimodality is evident.

Autocorrelation, the technique of correlating a time series against an identical series progressively lagged further behind the first, gives a series of autocovariances, peaks in these values occurring at time lags corresponding to period lengths appearing in the data. Figure 2c gives a plot of autocovariances, showing a marked 24 h periodicity. Binkley (1976) has suggested that all the above methods are equally useful in the investigation of period length. Our time series are short in relation to the period lengths (e.g. a 10 day run showing 24 h periodicity), so that a few atypical values can greatly affect the final analysis. Our data is also very noisy and all three methods of analysis are employed to establish the temporal pattern of fish activity.

SPATIAL PATTERNING

Five minute fixes are plotted to determine the size of the home range occupied by the fish. Home ranges have been successfully described by statistical distributions of the spread of points (e.g. Zachs and Falls, 1978). Our data did not fit any single statistical distribution (see Fig. 3), and a more empirical approach was adopted.

A computer is used to speed calculation. The fixed grid is first split into areas of approximately 4 × 4 m to comply with limitations of computer storage, the total number of fixes occurring in each box over the time period for which the home range is being calculated is then found. Contours of 'equal fix number' are then drawn over the area, and the area between each contour calculated. Starting from areas of highest 'fix number', contours enclosing areas where a given percentage of the fish's time is spent can then be drawn to make this model more compatible with more refined statistical descriptions.

From this model, an index of the stability of the home range over time can be calculated, as suggested by Cooper (1978). The overlapping area for successively calculated home ranges can be found, and the total number of fixes within that area summed. An index of stability (S) can then be defined as S = No. of fixes in area of overlap/total no. of fixes in both home ranges. This method has an advantage over Cooper's method (which calculates a stability index merely in terms of area overlap) because it allows weighting for areas which the animal uses intensively. If such areas were continually being 'lost' from the home range this would suggest greater instability than if little used peripheral areas of the same size were lost.

The minimum observation time to provide an accurate picture of a stable home range can be established from the stability index. The home range is defined by a low number of successive fixes, and the stability index between this 'home range' and that described by the next equal number of fixes is calculated. If the number of fixes to define successive home ranges is progressively increased, the stability index will also increase before stabilizing. The number of fixes (and therefore observation time, if time between fixes has been standardized) at which the stability index increases by less than a standard value (e.g. 5 percent) can thus be defined as an index of the time taken for the animal to cover its whole home range.

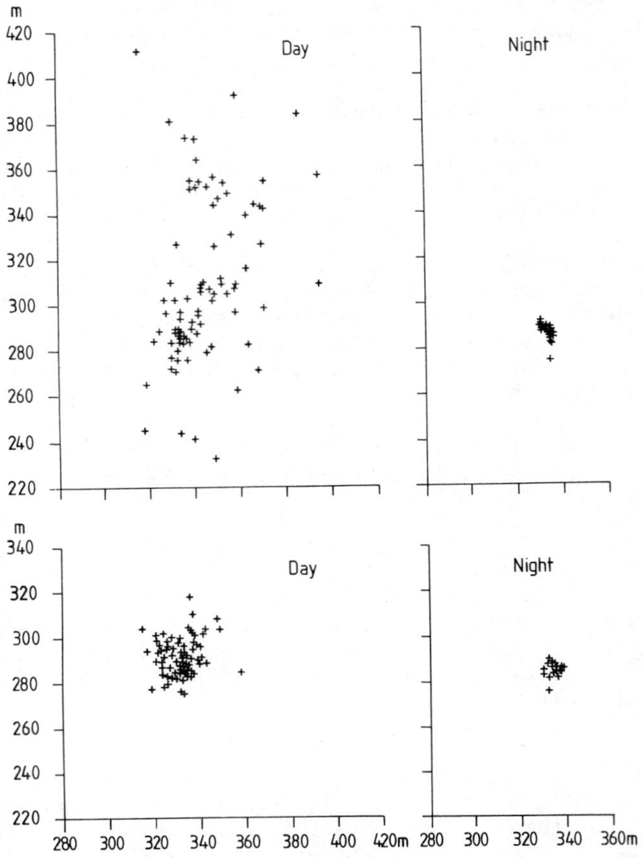

Fig. 3. Scatter diagrams of the day and night home ranges of a juvenile cod on successive days.

THE BEHAVIOR OF JUVENILE COD IN LOCH TORRIDON

The hydrophone array has been applied to a study of cod in a Scottish sea loch. Young cod settle in this area in their first year and may stay there for up to two years. Data from one fish, Christian, caught on the 22 August, 1978, released the next day and tracked continuously for 11 days is given below as an illustration of the behavior of the cod in late summer.

Fourier analysis and autocorrelation carried out on the activity time series both showed pronounced cyclic activity with a period length of approximately 24 h (Fig. 2a and c). No other periodicities of importance were found in the data. The form estimate of the same data (Fig. 2b) gives the activity within the 24 h cycle and clearly shows greater activity during the daylight hours with very low activity at night. A similar pattern is evident if the areas of the home range by day and by night are compared (Table 1).

Home range area, was calculated by two different methods. The first, proposed by Koeppl *et al.* (1975) calculates the area on the basis of a statistical model assuming a bivariate normal distribution of fixes. Values given in parenthesis are those calculated by a graphical method where extreme points in the range are joined by

Table 1.
Home range statistics for 'Christian' cod. Area calculated by graphical method (see text for details) appears in parenthesis. The difference between day and night areas calculated by the statistical model are significant at the 0.005 level (Mann Whitney 'U' test, U = 87.0)

Date		23/8	24/8	25/8	26/8	27/8	28/8	29/8	30/8	31/8	1/9	2/9	Study period
Home	Day		1967	303457	7304	960231	1388090	2345829	328116	1311936	2691562	1807103	
range		—	(299)	(4059)	(1321)	(9911)	(9882)	(14076)	(15886)	(10945)	(10227)	(10313)	
area													1774098
(m^2)	Night	21.6	2.2	7.6	7.7	265061	225265	4182	1046	17	427	—	
Center	Day	—	334.4	336.0	332.9	338.9	344.3	341.4	362.3	343.3	343.8	256.9	
of													341.1
activity	Night	334.1	334.4	334.1	334.7	342.6	348.7	336.1	336.4	335.5	334.8	—	
(x)													
Center	Day	—	287.2	291.8	289.0	297.5	295.3	306.0	319.5	295.2	299.4	344.7	
of													294.5
activity	Night	286.3	287.3	287.5	286.8	259.4	250.5	286.3	286.8	285.9	285.9	—	
(y)													

straight lines and the area of the resulting complex figure calculated. The areas calculated by the statistical model greatly overestimate the areas used by the cod, an indication that a bivariate normal distribution is not a good description of the spread of activity. This is especially true in the nights of 27 and 28 August where activity was centered in two areas some 60 m apart resulting in vastly inflated estimates of home range area.

Table 1 also gives the coordinates of the centers of activity (Hayne, 1949) which are very consistent over the period of study. From this it seems that not only do juvenile cod remain in the same general area over fairly long periods of time but that their distribution of activity around a central area would also appear to be maintained. Movement patterns of individuals at different times of the year and at different stages of their life in the loch are at present under study.

REFERENCES

Binkley S. (1976) Computer methods of analysis for biorhythm data. In P. J. Decoursey (ed.), *Biological Rhythms in the Marine Environment*. Columbia: University of South Carolina Press.

Cooper W. E. (1978) Home range criteria based on the temporal stability of areal occupation. *J. theoret. Biol.* 73, 687-695.

Enright J. T. (1965) The search for rhythmicity in biological time series. *J. theoret. Biol.* 8, 426-468.

Hayne D. W. (1949) Calculation of size of home range. *J. Mammal.* 30, 1-18.

Koeppl J. W., Slade N. A. and Hoffmann R. S. (1975) A bivariate home range model with possible application to ethological data analysis. *J. Mammal.* 56, 81-90.

Zach R. and Falls J. B. (1978) Bivariate normal song territories in Ovenbiards. *Can. J. Zool.* 56, 2088-2092.

Automation of Data Collection in Ultrasonic Biotelemetry

D. PINCOCK

Electrical Engineering Department, University of New Brunswick, Fredericton, N.B., Canada

Abstract — The validity of many studies using ultrasonic biotelemetry for determining fish behavior can be questioned because of the small number of fish involved. The reason for this small number is often the fact that the task of decoding and analyzing sensor data is extremely time consuming. For this reason, an on-going objective of our work at the University of New Brunswick has been to develop telemetry systems in which sensor data is produced in a format permitting immediate computer analysis of field data. This objective has now been met and the methods used are described in this paper.

In general, the data gathering system consists of three components — an ultrasonic receiver, a decoder and a digital recorder. The receiver, on detecting a transmitted pulse should produce a logical indication for processing by the decoder. This task is complicated by highly variable noise levels and echoes which can be delayed even hundreds of milliseconds. Receiving techniques and detection methods which work satisfactorily in this environment are discussed. The decoder accepts signals from one or a number of receivers, extracts sensor data and passes it (usually along with a time of day indication) to the recorder. To accomplish these tasks, while achieving a package suitable for use on small boats, microprocessor-based logic is used. The paper discusses several special purpose systems which have been used as well as a subsequent system which should be adaptable to most applications. Finally, the paper describes the cassette recording and playback system which has been developed.

INTRODUCTION

In many studies which use ultrasonic biotelemetry for determining fish behavior, the results are somewhat inconclusive because of the small number of fish involved. The cost of transmitters, boat time, etc. is certainly a factor contributing to these small samples but a more serious limitation is encountered in cases in which data from sensors is being transmitted. Because of the characteristics of the medium and the requirement for simple transmitter electronics, transmission is usually pulsed with data encoded either by pulse repetition rate or, in the case of several sensors, by a pulse train containing several pulse intervals (Stasko and Pincock, 1977). Usually, decoding of this information is done either manually in the field or by subsequent interpretation of recorded (e.g. strip charts) signals. The former method makes field work very tedious and, in many cases, leads to an inadequate sampling frequency while the latter often leads to an accumulation of 'data' whose interpretation is so time consuming that it often lags the field work by months or years. A continuing objective of our work at the University of New Brunswick, Canada, has been the development of telemetry systems which overcome the problems described above and permit the true potential of ultrasonic biotelemetry to be realized. It is the purpose of this paper to give an *overview* of this work and to present some of the *systems aspects* entering into the design of an automated system.

In concept, the automated system is very straightforward consisting of a hydrophone, receiver, decoder and a digital recorder. The receiver, on detecting a transmitted pulse, should produce a logical indication for processing by the decoder. The decoder accepts signals from one or a number of receivers, extracts sensor data and passes it (usually along with a time of day indication) to the recorder. To provide interaction in the field, the decoder would have some sort of display and would possibly produce an analog recreation of the sensor signal suitable for strip chart recording. It should be noted that such chart recordings do not eliminate the need for digital recording since it is usually necessary to convert this information to computer-readable form for analysis. This task is extremely time consuming while the digitally recorded data is directly readable by a computer. The following sections give details on major system components.

RECEIVER

Because of the highly variable noise conditions and severe multipath which exist, the design of a suitable receiver is the most challenging aspect of the entire system. The signals shown in Figs. 1 and 2 are instructive. As can be seen, the main effect of multipath in shallow water (Fig. 1) is to broaden the transmitted pulse. In other situations, such as in relatively deep water (Fig. 2), distinct echoes can occur even hundreds of milliseconds after reception of the pulse. Transmitter coding and signal detection method must be chosen with these propagation conditions in mind.

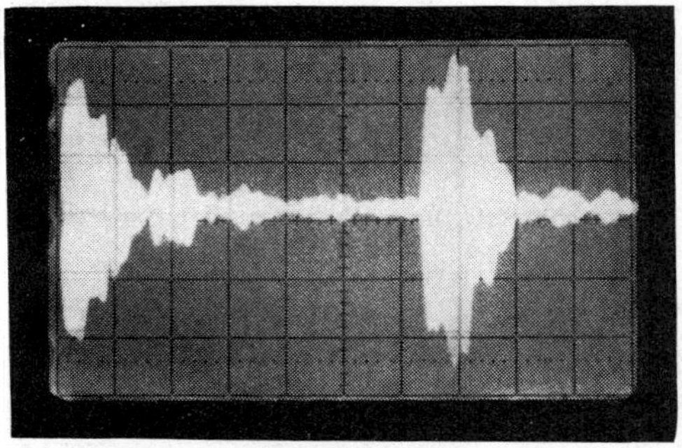

Fig. 1. Effect of multipath in shallow water (water depth 20 m; range 100 m; transmitted pulse width 16 ms; horizontal scale 20 ms div^{-1}).

For most of our work we have used a superheterodyne type of receiver which is now commercially available (CR-40, Communication Associates, Huntington, N.Y.) and have processed the output of an audio detector to obtain a logical signal indication. The basis of this processing is to say that a signal or echo is present if the detector output exceeds some multiple of the average level for at least a few milliseconds

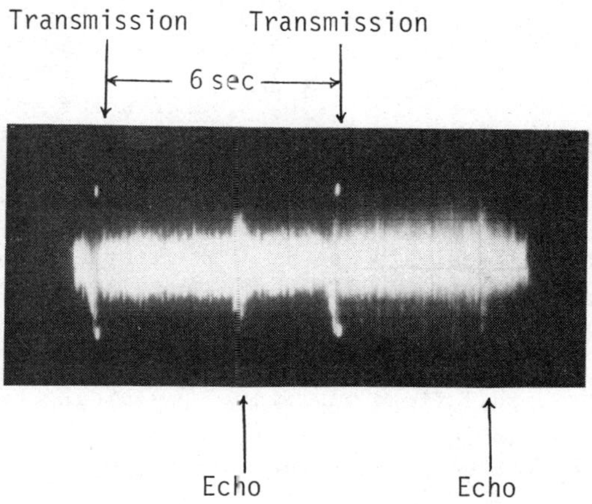

Fig. 2. Effect of multipath in deeper water (water depth 500 m; range 1000 m; transmitted pulse separation 0.6 s).

(typical transmitter pulse durations are at least 10 ms). Once this is done it is only necessary to reject the echoes. The techniques used to accomplish this are outlined below.

For situations in which multipath produces the effects typified by Fig. 1, the rejection of echoes is straightforward. All that is necessary is a disabling of the signal detection logic for an appropriate period (typically 150 ms) after each detection. At the end of this period echoes will have vanished. Of course, in order for this approach to be successful, the minimum interval between transmissions must be longer than the period during which the receiver is effectively disabled. Therefore, our transmitters are designed not to transmit pulses less than 300 ms apart.

When the effects typified by Fig. 2 exist, the situation is somewhat more difficult. While it is possible to use the same approach as outlined above, the interval that the receiver would have to be disabled would be much longer. This would impose a requirement on the minimum transmitter pulse spacing which would be so long that, for many applications, the data rate would be too low. In addition, the long silences between transmitted pulses would cause difficulties in applications in which the transmitter must be tracked. Therefore, these echoes are rejected on the basis of their amplitude which, because of increased propagation distance and reflection loss, is smaller than that of direct signals. This technique has worked fairly well but is not foolproof. In particular, for areas in which reflection losses are small (e.g. rocky bottom) and the range is large, the amplitude difference between signal and echo can be small and reliable detection can only occur by decreasing the range. Also, saturation of the receiver (such as is evidenced by the brightening at the signal peaks of Fig. 2) must be avoided or amplitude information is lost. Of course, receivers which destroy amplitude information (for instance, those based on phaselocked loops or those possessing a fast acting AGC) are not suitable.

DECODER

Two approaches are possible for design of the decoder. The first involves the construction of a special purpose instrument while the second involves the use of a general purpose computer or calculator. To date, we have tended to use the former approach but for certain applications we are implementing the latter for reasons which are explained below.

Figure 3 shows a typical decoder developed at the University of New Brunswick. This particular decoder is for single sensor transmitters (temperature, pressure or swimming speed) whose pulse repetition rate is linearly proportional to the sensor voltage (Pincock and Luke, 1975; Pincock *et al.*, 1978; Pincock *et al.*, 1979). The decoder measures the period between received signals, checks for validity and then applies the linear calibration specified by the front panel switches and displays the data. In addition, at user-selectable intervals, data is presented along with an elapsed time indication at the output connector for recording. Decoders of this type are microprocessor-based (either SC/MP or 8085 because of their hardware simplicity) with software resident in Read Only Memory (ROM). Other similar decoders incorporate such features as analog outputs, multisensor capability, time of day entry, etc. One of the main problems with this approach to decoder design is its inflexibility. Each new application requires redesign of the front panel and circuitry before the software can be written. Furthermore, the ability to communicate with the program is limited by the data entry method (switches) and, therefore, many functions which could easily be implemented in the program (e.g. calculations involving nonlinear calibration curves) are not because of the relatively large amount of operator input required.

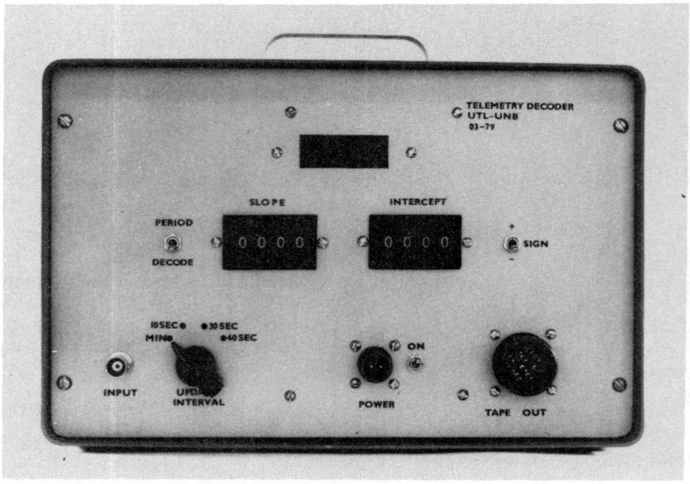

Fig. 3. A typical special purpose decoder.

Because of the difficulties mentioned above, we have undertaken the design of a decoder whose hardware is more general. Data entry is accomplished by means of a calculator-type keyboard, and eight digits of display are available. The entry of such information as calibration data, time of day, update interval, display data, etc. is accomplished by an interractive dialog involving the keyboard and display. Thus, not only is the ability to pass data to the program greatly enhanced, but new applications are implemented through software modifications only.

The decoders described above are most suitable in applications where portability is very important, but in other applications where more bulky equipment is acceptable their disadvantages become limiting. In particular, the decoder performs basically a data collection function and, therefore, there is little possibility of analyzing the data as it is being collected. This is certainly unsatisfactory for long field experiments. Also, the development of the ROM — based program can only be carried out in a laboratory which has the proper development tools (PROM programmer, logic analyzer, cross-assemblers, etc.).

The disadvantages mentioned above are overcome through the use of a general purpose computer which examines the receiver output, does the required decoding, displays results on associated peripherals (printer, screen, plotter, etc.) and stores data on some mass storage device (tape, floppy disc, etc.). With the possible exception of an interface to the receiver, this approach involves no hardware design, but, rather, a selection of computer and peripherals with the desired input, display and storage capabilities. Software is structured so that the basic decoding and data storage functions are initialized through a dialog with the operator. This implements the function of the special purpose decoders described above but with more power because of more flexible input/output devices (e.g. video terminal). In addition to this data collection function, the software can be structured to permit analysis and display of collected data through programs written in higher level languages (e.g. FORTRAN, BASIC). Thus, for example, the progress of a long experiment can be checked and different analysis techniques tried as the experiment is being conducted. This permits modification of the experiment or data collection if results are not totally satisfactory. It should be re-emphasized that the size, power requirements and cost of the general purpose computer rule out its use for a large number of applications.

RECORDER

Cassette recording offers the advantages of relatively large capacity at low cost and power consumption, and was used for this application. Experience gained with audio cassette units indicated that reliable recording was best achieved through bypassing the record electronics and using saturation recording. This being the case, it was decided to base the recording system around one of the several digital cassette units now available at a cost comparable to that of a good audio unit (we use the CM-600 from Braeman Computer Devices, Burnsville Minnesota).

The logic of the recorder unit is also microprocessor-based. The main functions performed by the microprocessor are blocking of data (to obtain a reasonable recording density), tape motion control and phase encoding on recording and unblocking and decoding on playback. In order that decoders and recorders can be easily tested in the laboratory and can communicate easily with computers which will analyze data, all communication between them is according to the RS-232C standard using ASCII characters.

FUTURE DEVELOPMENTS

There are always situations in which the echo rejection techniques described above could fail. Greatly improved echo rejection will be achieved by means of software running either on the decoder or on a microprocessor integrated into the receiver. Another shortcoming of the present method is that the threshold test used for signal detection is inefficient in that it does not make full use of the available signal energy. Implementation of a matched filter will permit a given range to be achieved with less transmitter power.

Acknowledgement — The design philosophy and system concepts described in this paper have evolved during the development of a number of biotelemetry systems over the past few years. David Luke, Don Church and Fred Voegeli have contributed greatly to this work. The contributions and interest of a number of biologists, in particular Aivars Stasko, Frank Carey, Mike Butler and Peter Hurley, have also been invaluable. Financial support has been received from the World Wildlife Fund and the National Research Council of Canada. Financial support and boat time were provided by D. Merten and Communication Associates.

REFERENCES

Luke D. M., Pincock D. G. and Stasko A. B. (1973) Pressure-sensing ultrasonic transmitter for tracking aquatic animals. *J. Fish. Res. Bd Can.* 30, 1402-1404.

Pincock D. G. and Luke D. M. (1975) Systems for telemetry from free-swimming fish. *Proc. IERE Conf. Instrum Oceanogr.*, Bangor, Wales, pp. 175-186.

Pincock D. G., Luke D. M., Voegeli F. A. and Butler M. J. A. (1978) A multichannel ultrasonic telemetry system for physiological and behaviour studies of bluefin tuna. International Symposium on Biotelemetry, Garmisch, Germany.

Pincock D. G., Luke D. M. and Voegeli F. A. (1979) An automated multisensor underwater telemetry system for biotelemetry. To be presented at Ultrasonics International, Graz, Austria.

Stasko A. B. and Pincock D. G. (1977) Review of underwater biotelemetry, with emphasis on ultrasonic techniques. *J. Fish. Res. Bd Can.* 34, 1261-1285.

IV. Applications in Biotelemetry

Telemetric Measurements of Hemodynamic Responses to Car Driving in Coronary Patients

R. ZERZAWY, K. BACHMANN and K. SCHEBELLE

Medizinische Poliklinik, Ostliche Stadtmauerstrafze 29, D-852, Erlangen, Germany

Abstract — In 15 patients with coronary heart disease, hemodynamic response to car driving was analyzed by telemetry utilizing magnetic tape recording storage of direct arterial blood pressure, pulmonary artery pressure, heart rate and rate-pressure-product. The effect of isosorbide dinitrate on these parameters was measured in 8 patients. Driving causes a moderate increase of all parameters immediately after starting into city traffic, with a rise of systemic blood pressure to 162/89 mm Hg, of pulmonary artery pressure to 34/16 mm Hg, of heart rate to 103 beats min^{-1} and of rate-pressure-product to 167 × 10^2 mm Hg min^{-1}. Continued driving in city traffic reduces all parameters, and driving on a highway shows no significant difference to resting values. In one patient with left ventricular wall aneurysm, an extreme increase of pulmonary artery pressure to 85/30 mm Hg was measured during the first few minutes of driving. Isosorbide dinitrate is effective in lowering both pulmonary artery and systemic blood pressure. Out of 8 patients, only one developed arterial hypotension with systolic blood pressure below 100 mm Hg.

INTRODUCTION

Driving a car is one of our most common activities. It is therefore of particular importance to have knowledge about the hemodynamic reactions in the coronary patient to this kind of stress. Because of the mainly chronic nature of coronary heart disease, the consequences concern a wide range of the patient's activities. The decision of whether or not a patient is able to drive in traffic is of considerable social weight. The two major criteria for this decision are the functional impairment of the heart and the cardiovascular stress induced by driving.

There is little knowledge about the hemodynamic response to driving in coronary patients. In this group of patients, only telemetric recordings of the electrocardiogram as an indicator of myocardial ischemia and cardiac rhythm have been performed (Bellet *et al.*, 1968; Simonson *et al.*, 1968; Taggart, Gibbons and Somerville, 1969).

Measurements of arterial blood pressure have until now been performed only in healthy persons. The cuff method was used by Hoffmann *et al.*, 1971; and direct continuous blood pressure recordings have been reported previously from our laboratory (Bachmann and Zerzawy, 1972; Zerzawy, Fleischer and Bachmann, 1974).

The previous studies generally agreed that driving under usual traffic conditions is only a moderate stress factor to the heart (Fig. 1). Only in few coronary patients, ST-segment alterations and/or significant arrhythmias could be observed (Bellet *et al.*, 1968; Simonson *et al.*, 1968; Taggart, Gibbons and Somerville, 1969).

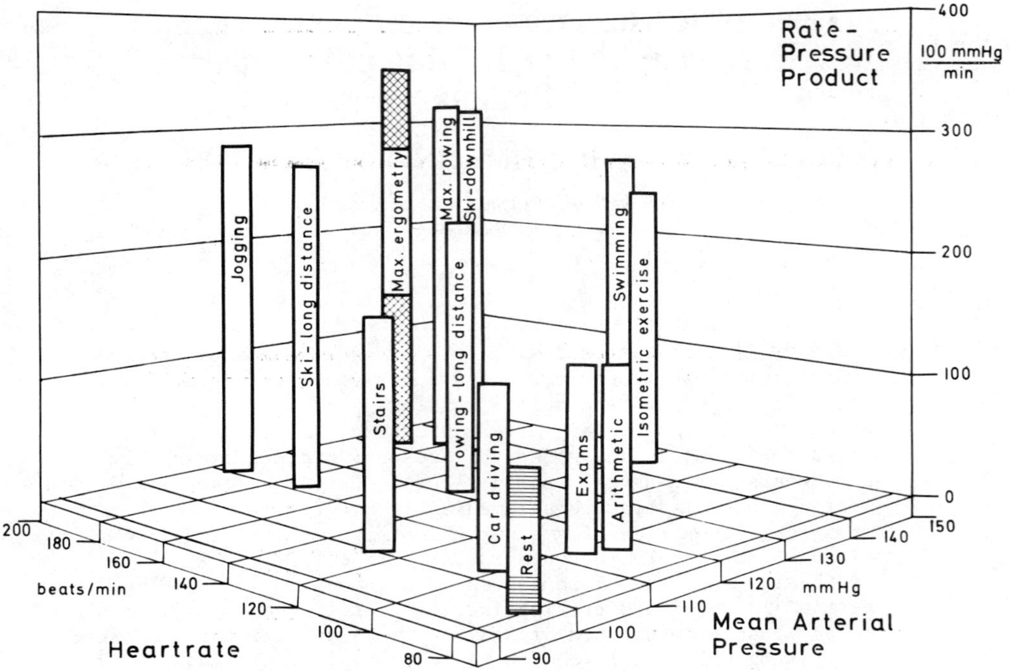

Fig. 1. Three-dimensional display of heart rate (on the left), mean arterial pressure (on the right) and systolic pressure-rate-product (vertical scale) in healthy persons during different types of mental stress and physical exercise. Car driving is located in the lower ranks of this scheme, indicating minor cardiovascular stress.

The purpose of this study was to measure central and peripheral hemodynamic parameters in coronary patients during actual driving conditions.

PATIENTS AND METHOD

In 15 patients with angiographically documented coronary heart disease (age 50 ± 5 years) direct arterial blood pressure was measured via a Teflon-microcatheter in the brachial artery, and pulmonary artery pressure was measured by a balloon-tipped floating catheter, which was introduced via a cubital vein. The pressure signals and two leads of the ECG (bipolar) were amplified and, after frequency-division multiplexed modulation, recorded on a portable stereo tape recorder. For the estimation of left ventricular stroke volume, intermittent records of the impedance cardiogram were made.

Each patient drove an identical route in a Volkswagen bus, in which they drove before the test. From these patients, 8 repeated the tour 5 min after sublingual intake of 5 mg of isosorbide dinitrate (ISDN). The length of the route was 35 km, which included city and highway traffic. The trip took an average time of 50 min.

RESULTS

HEMODYNAMICS DURING DRIVING

In the terms of heart rate, arterial pressure and systolic pressure-rate-product, the hemodynamic reaction to driving in coronary patients is very similar to the results obtained from 11 healthy young persons, as it is shown in Fig. 2 (Zerzawy, Fleischer and Bachmann, 1974). Whereas heart rate and diastolic arterial pressure

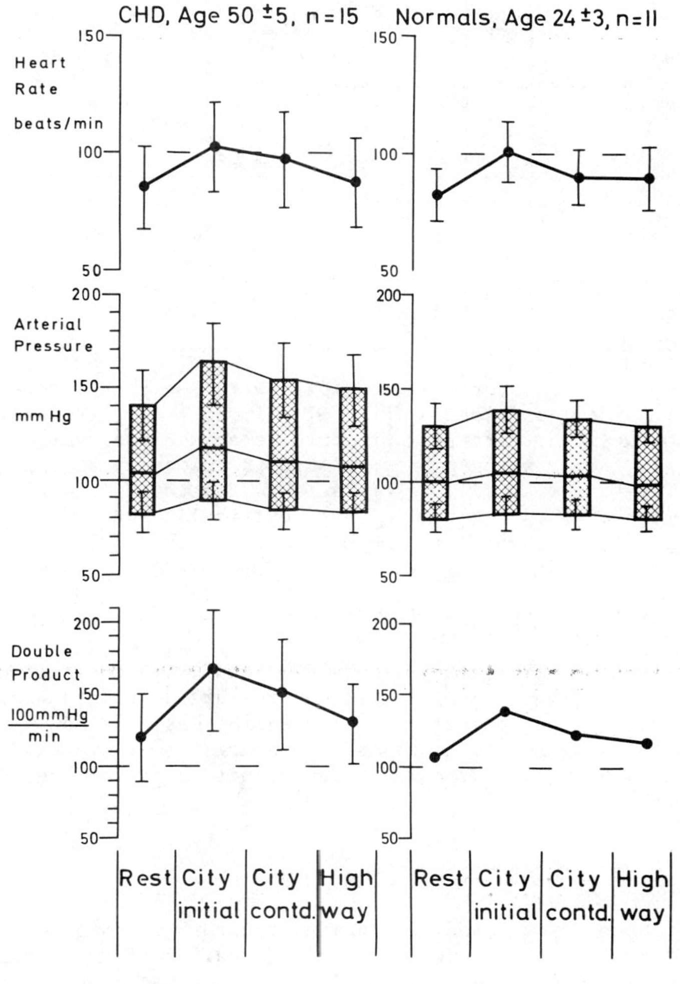

Fig. 2. Mean values and standard deviations of heart rate, arterial blood pressure (systolic/mean/diastolic) and systolic pressure-rate-product before and during car driving in 15 patients with coronary heart disease (CHD; left panel) and in 11 young healthy subjects (right panel).

show practically identical profiles under different traffic situations, the resting value and all driving values of systolic arterial pressure are 10 to 20 mm Hg higher than those of the young control group. Accordingly, the systolic pressure-rate-product is a similar amount higher in the coronary group (rest: 120×10^2 mm Hg min^{-1}, initially after start into city traffic: 167×10^2 mm Hg min^{-1}) than in the young control group (rest: 106×10^2 mm Hg min^{-1}, initially after start into city traffic: 140×10^2 mm Hg min^{-1}).

The increase of pulmonary arterial pressure during driving is also moderate. Only at the start into city traffic, pulmonary arterial pressure rises by an amount worth mentioning (Table 1). Driving on the highway has no effect at all to pulmonary arterial pressure.

Table 1.
Pulmonary artery pressure (mean values and standard deviations) before and while driving in 15 patients with coronary heart disease

	Syst. PAP mm Hg	Diast. PAP mm Hg	Mean PAP mm Hg
Rest before start	24.7 ± 10.5	12.1 ± 6.9	18.1 ± 7.8
City initial	33.9 ± 16.7	16.1 ± 10.7	24.4 ± 13.7
City continued	30.0 ± 13.3	14.2 ± 10.0	21.5 ± 11.3
Highway	25.2 ± 11.1	11.7 ± 6.6	18.6 ± 8.1

One exception from this behavior was seen in a patient with a left anterior wall aneurysm after myocardial infarction (Fig. 3). Immediately after the start into city traffic, a rise of pulmonary arterial pressure to an abnormal value of 85/34 mm Hg was recorded. This enormous increase in diastolic pulmonary arterial pressure as an indicator of left ventricular filling pressure is due to an extensive loss of left ventricular compliance. This reversible left ventricular pump failure is not accompanied by any significant changes in the ECG or in arterial blood pressure.

RESULTS AFTER ISDN

A single dose of sublingually administered ISDN results in a significant reduction of both arterial and pulmonary arterial blood pressure during the whole time of the drive (Fig. 4). Furthermore, the driving induced increases of both pressure profiles are largely reduced. The influence of ISDN on heart rate is nearly neglegible (Fig. 5), so that the systolic pressure-rate-product is prevailingly reduced by a significant amount.

DISCUSSION

As has been shown by previous studies, driving a car is not a greater stress factor for the heart than examinations, solving mathematical problems or climbing up stairs (Bachmann, Zerzawy and Schebelle, 1978). In terms of heart rate, arterial pressure and pressure-rate-product, driving is to be classified below most physical exercises, as displayed in Fig. 1. For this reason, the majority of coronary patients with exercise-induced angina and/or previous myocardial infarction are to be considered as hemodynamically 'normal' during driving.

The problem, which arises from our study is, that patients with extensive myocardial

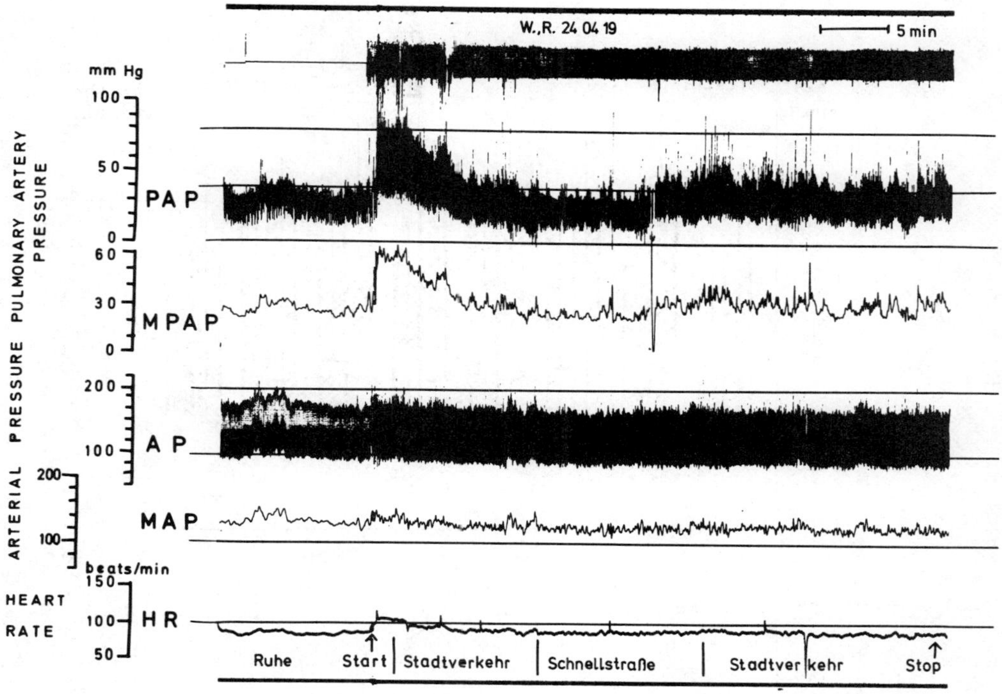

Fig. 3. Original recording from a 58 year old patient with left anterior wall aneurysm. From the top to the bottom: ECG, pulmonary artery pressure (PAP), mean pulmonary artery pressure (MPAP), arterial pressure (AP), mean arterial pressure (MAP), and heart rate (HR). Immediately at the start of driving in city traffic, there is an abnormal increase in pulmonary artery pressure, which after a few minutes returns to resting values. Driving on the highway (Schnellstrasse) causes no alteration of hemodynamic parameters compared to rest. Note, that arterial blood pressure and heart rate show only minimum changes during the whole drive.

scars due to infarctions may develop left ventricular pump failure. For this group of patients, the demonstrated pressure-lowering effects of ISDN are helpful.

ISDN usually causes a highly significant reduction of arterial pressure. Therefore one has to be aware of the fact, that patients with a low normal blood pressure in the untreated state may develop arterial hypotension with systolic values below 100 mm Hg. Out of the 8 patients under study, only 1 showed such a response. In this case, systolic blood pressure fell to a minimum of 80 mm Hg, with the average value being maintained at 90 mm Hg for the whole time of the drive.

In conclusion, magnetic storage biotelemetry has facilitated, for the first time, a study of the hemodynamic profile in the untreated and treated coronary patient under actual driving conditions. The results are helping to answer the urgent question about their ability to use their car. Thus, these findings are a step forward in advising both patient and physicians about the directives.

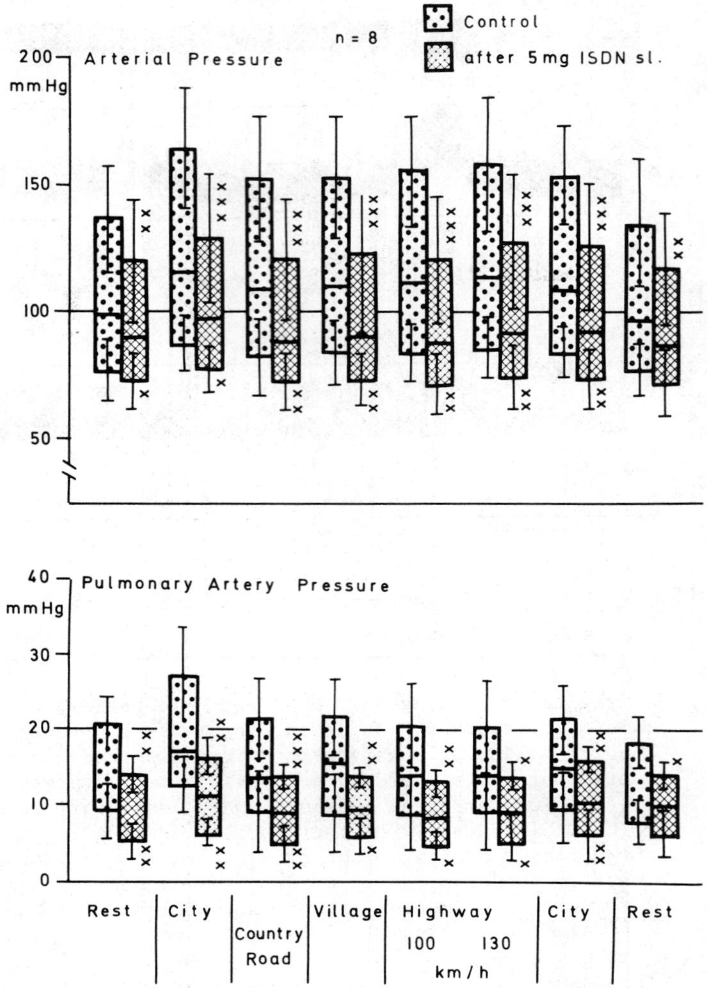

Fig. 4. Hemodynamic effect of 5 mg of isosorbide dinitrate (ISDN) in 8 coronary patients during driving under various traffic conditions. Mean values and a standard deviation of arterial blood pressure and of pulmonary artery pressure (+++: $P < 0.001$, ++: $P < 0.01$, +: $P < 0.05$).

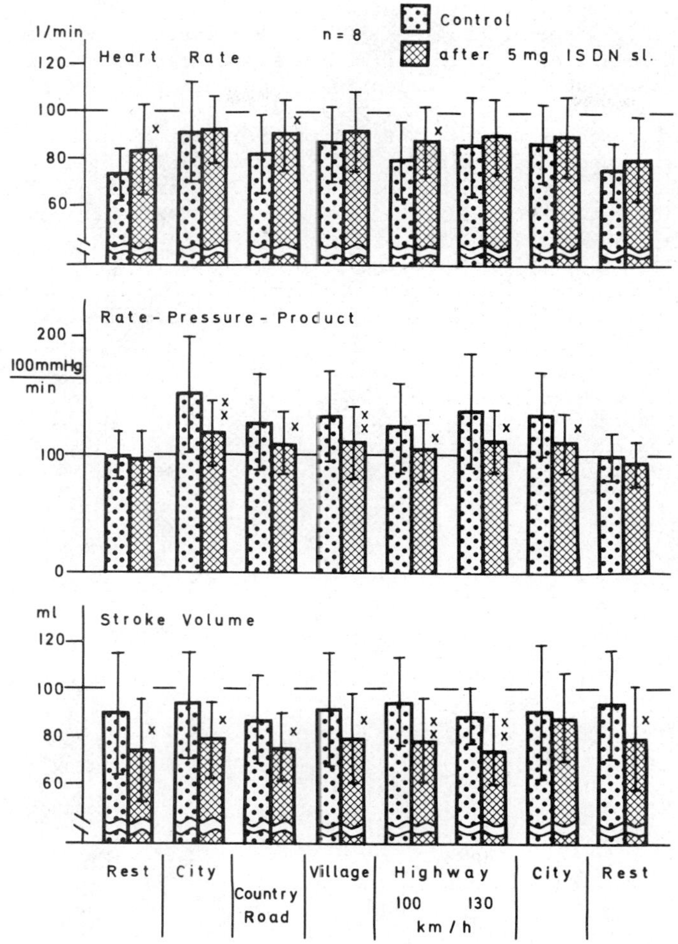

Fig. 5. Effect of 5 mg of ISDN on heart rate, systolic pressure-rate-product and on stroke volume (estimation by impedance cardiography) in 8 patients with coronary heart disease during driving under various traffic conditions. There is only a slight trend to an increase in heart rate. The rate-pressure-product is reduced by a significant amount in the majority of situations.

Acknowledgements — This work was supported by a grant from the Deutsche Forschungsgemeinschaft.

REFERENCES

Bachmann K. and Thebis J. (1967) Die drahtlose Übertragung kontinuierlicher direkter Blutdruckmessungen. *Zt. Kreislaufforsch.* 56, 188.

Bachmann K. and Zerzawy R. (1972) Radiotelemetry of direct blood pressure measurements in aorta, pulmonary artery and heart. In H. P. Kimmich and J. A. Vos (eds.), *Biotelemetry*, pp. 183-195. Leiden: Meander N.V.

Bachmann K., Zerzawy R. and Schebelle K. (1978) Die kardiale Belastung des Kraftfahrers: Radiotelemetrie von arteriellem Druck, Pulmonalarteriendruck und Elektrokardiogramm bei Gesunden und Patienten mit koronarer Herzkrankheit. *Unfall-Sicherheitsforschung im Straßenverkehr*, 19, 54.

Bellet S., Roman L., Kostis J. and Slater A. (1968) Continuous electrocardiographic monitoring during automobile driving. *Am. J. Cardiol.* 22, 856.

Hoffmann H., Mayer H., Grundei M. and Meier W. (1971) Verkehrsmedizinische Untersuchungen zur Frage des Kreislaufverhaltens, der körperlichen und seelischen Belastungen von Kraftfahrzeugführern in verschiedenen Fahrzeugtypen auf Fernfahrt. *Zentralbl. Verkehrsmed.* 17, 193 (Part I), 18, 1 (Part II).

Simonson E., Baker C., Barms N., Keiper C., Schmitt O. H. and Stackhouse S. (1968) Cardiovascular stress (electrocardiographic changes) produced by driving an automobile. *Am. Heart J.* 75, 125.

Taggart P., Gibbons D. and Somerville W. (1969) Some effects of motor-car driving on the normal and abnormal heart. *Br. Med. J.* 4, 130.

Zerzawy R., Fleischer H. and Bachmann K. (1974) Environmental studies on blood pressure, ECG and EEG using a programmable four channel telemetry system. In P. A. Neukomm (ed.), *Biotelemetry II*, pp. 134-136. Basel: Karger.

Practical Problems of Controlling Medical Implants via Radio Links

T. A. PERKINS and N. J. CHAFFEY

MRC Neurological Prostheses Unit, Institute of Psychiatry, London, U.K.

Abstract — An increasing number of stimulators implanted for medical purposes are powered and controlled via transdermal radio links. The implanted electronics may be kept relatively simple and hence reliable, while the more complex control circuits, being external, can remain accessible for servicing or alteration. The external circuitry needs little protection from moisture, unlike implants. However, some of the problems involved in implants are not entirely removed when the bulk of the electronics is external. The power supply remains a problem principally because the power transfer efficiency between two RF coils separated by 1-2 cm of flesh is generally in the range 10-30 percent. Pulse generators and other control circuits must consume current sparingly. With careful circuit design and NiCd rechargeable batteries it has been possible to produce stimulation controllers that are quite acceptably compact for patients to carry around. The devices must still be reliable — indeed — rugged. As it is impracticable to have transmitters and control circuits in the same box, highly durable cabling must connect between them. Some solutions to these problems are illustrated by the equipment designed to control implants for cerebeller stimulation for the treatment of some epileptics and also implants for the stimulation of the cauda equina of paraplegics, enabling them to stand on their own feet.

INTRODUCTION

The authors develop control equipment for the British Medical Research Council Neurological Prostheses Unit directed by Professor G. S. Brindley. The prostheses produced by this Unit follow the general scheme shown in Fig. 1. The work of the Unit includes the treatment of epilepsy by stimulation of the cerebellum as described by Cooper *et al.* (1974); Brindley *et al.* (1977) and Cooper (1978). The stimulation controllers in this case have to provide 1 ms pulses at 10 Hz on two channels each of whose amplitude may be adjusted. The two channels operate one at a time, alternating every 6 min 50 s. Femoral nerve stimulation, as described by Waters (1977) and Brindley, Polkey and Rushton (1979) is used by this Unit to enable paraplegics to regain some use of their legs. This nerve stimulation causes the patient's own leg muscles to straighten

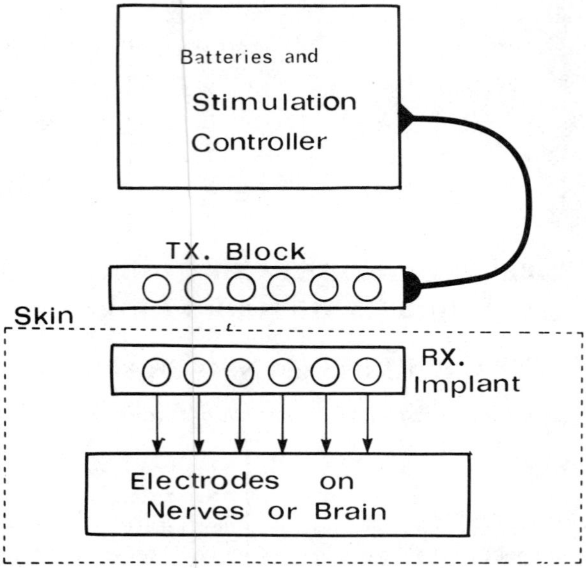

Fig. 1. General scheme of a neurological prosthesis.

the knees with sufficient strength to support the patient's weight. The paraplegic may then stand, with crutches used for balancing. He may also 'walk' by alternately placing both crutches forward and then drawing both stiff legs up to them. The stimulation controllers for the paraplegics have to provide up to six amplitude adjustable channels of 100-300 μs pulses at 10-15 Hz. A switch operated by the patient causes the stimulation gradually to turn on or decay so that the patient may arise from or descend to his wheelchair in a controlled manner.

RADIO LINKS

The passive implanted 19 mm coil receivers used for these projects, as shown in Fig. 2b, are developments of those described by Donaldson (1973). The receivers are set out in an array of 27 mm pitch with 7 and 9 MHz receivers alternating as shown in Fig. 2d so that cross-talk is kept reasonably low while the overall size of the implant is surgically acceptable. The common emitter Hartley circuit of Fig. 2a is simple enough for the components to fit inside the 19 mm transmitting coil. To maximize stability of the received signal, the turns ratios were chosen to give critical coupling for a transmitter-receiver range of about 1 cm (see Fig. 2c). R1 of Fig. 2a is selected during construction such that around 22 V appear across a 470 Ω load on the appropriate type of receiver at a 1 cm range, for a 34 V supply to the transmitter, power transfer efficiency being around 12 percent. The output is almost linear with supply voltage from 5 V upwards. The transmitters are set into an epoxy resin block in the same geometric configuration as the implanted receivers.

CONTROL CIRCUITS

POWER SUPPLY

Full circuit diagrams of the stimulation controllers are shown in Fig. 3 (femoral nerve) and Fig. 4 (cerebellar). As the current consumed by the logic is, in each case, small compared with the mean current used by the transmitters the PP3 size of

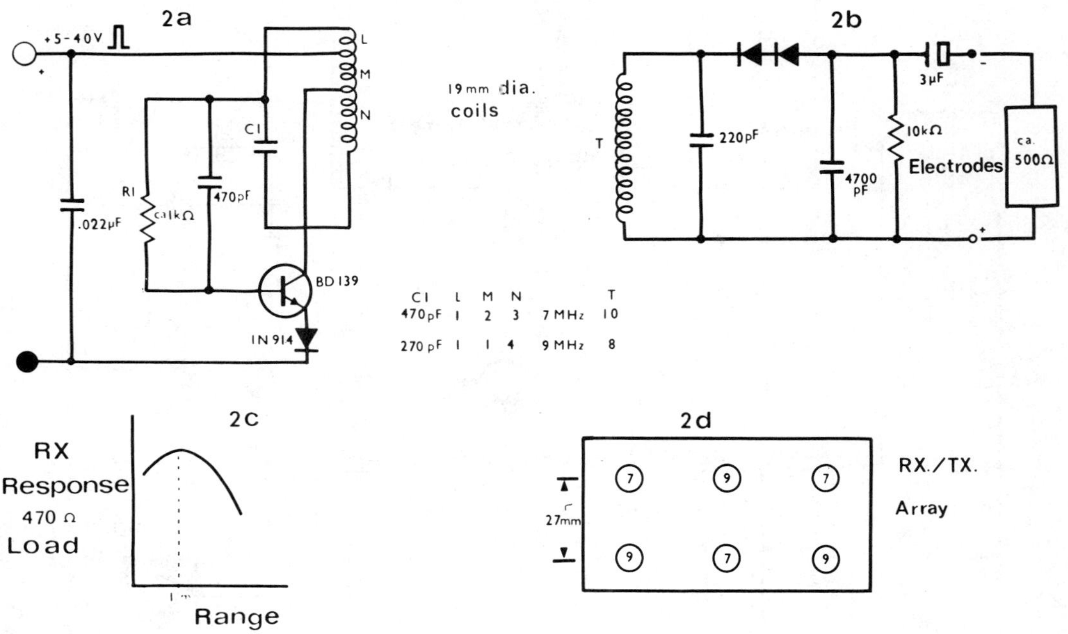

Fig. 2. The radio links.

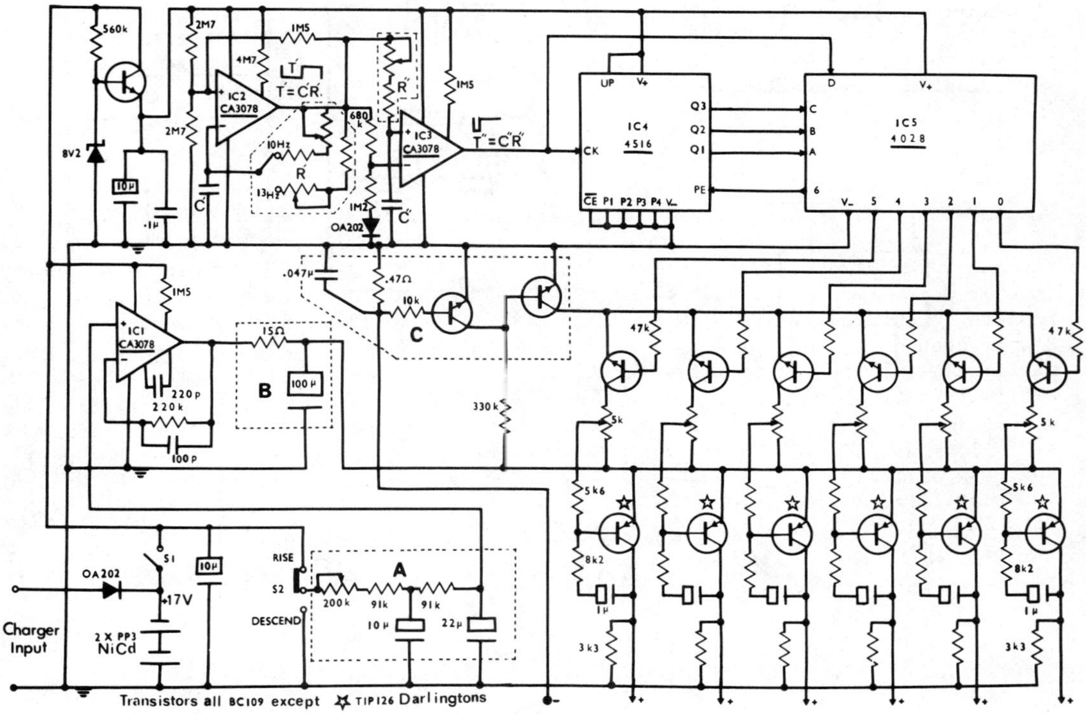

Fig. 3. Circuit of stimulation controller for paraplegics.

R 6	1 k
R 7	220 k
R 8	150 k
R 9	4 M7
R 10	3 M3
R 11	2 M2
R 12	2 k2
R 13	
R 14	R22
R 15	27 R
R 16	180 k
R 17	560 k
R 18	10 k
R 19	33 k
R 20	10 k
R 21	1 k8
R 22	10 k
R 23	4 k7
R 24	10 k
R 25	33 k
R 26	10 k
R 27	1 k8
R 28	10 k
R 29	4 k7
R 30	4 k7
VR1,2	2 k presets
C 1,2,3	100 µF
C 4	
C 5	4 µF
C 6	33 n
C 7	470 pF
C 8	47 µF
C 9	10 µF
C10	100 n
C11,12	4 µ7
D 1,4,5,6,7	OA202
D 2	8V2 Zener
D 3	1N4003
T 1,2,6,7,8	BC479
T 3,5	BC109
T 4	BC107
T 9,10	TIP121
ICI	CD4020
B 1	4 × PP3 NiCd

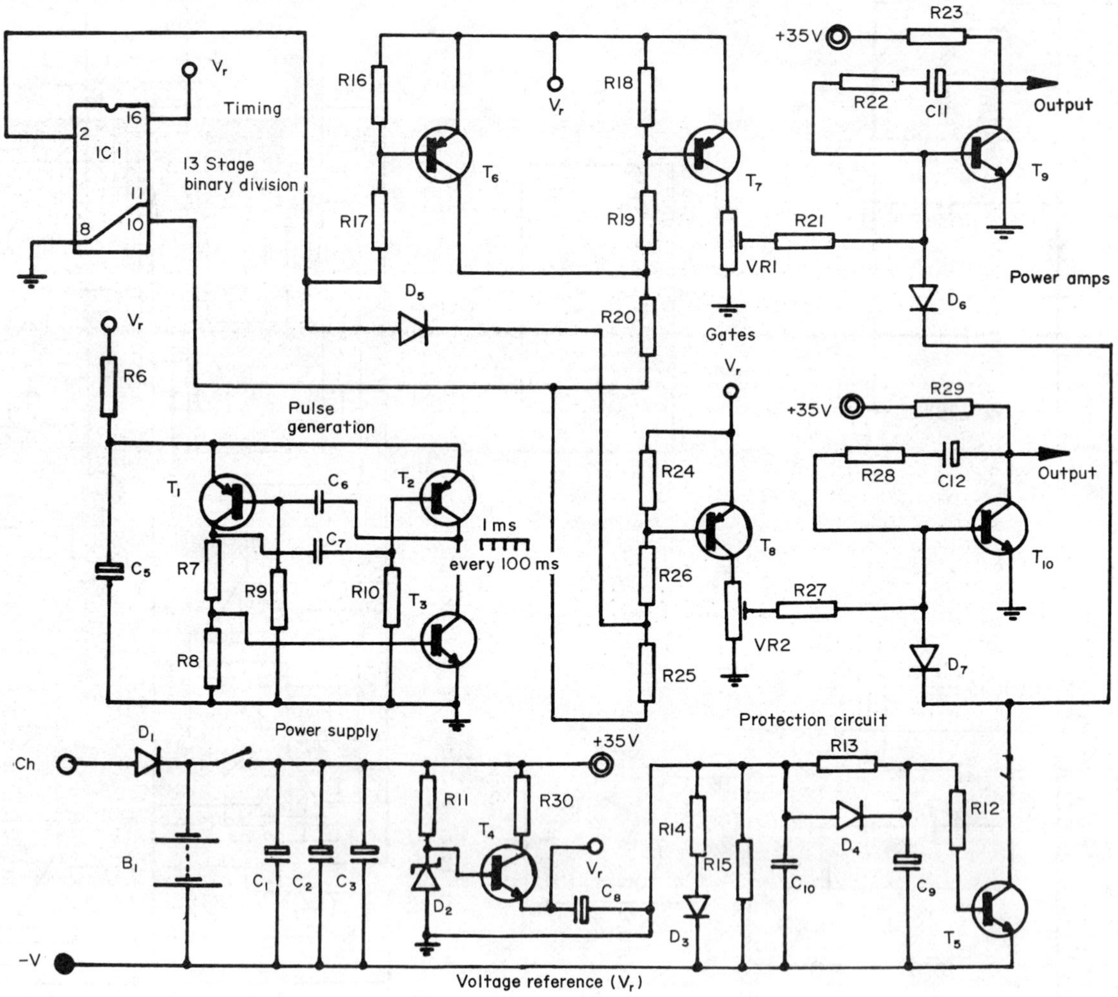

Fig. 4. Circuit of controller for cerebellar stimulation.

NiCd battery is more than sufficient in capacity for 12 h of continuous use before it is recharged by the patient. This makes it feasible for both circuits and batteries to be contained in pocket sized boxes as shown in Figs. 5 and 6.

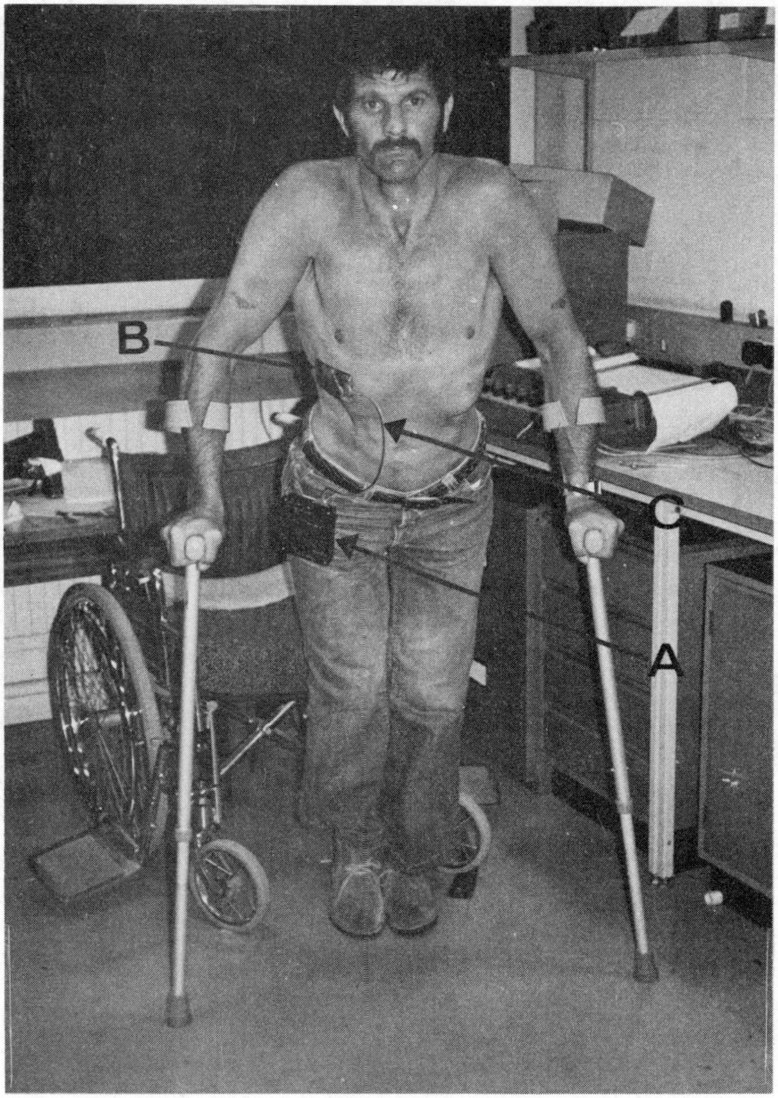

Fig. 5. Paraplegic Mr. G. L. standing with the aid of femoral nerve stimulation. (A) Rugged ABS box contains circuit of Fig. 3 and sufficient NiCd batteries for at least 12 h use between charges. (B) Araldite moulding containing six transmitters stuck to skin over receiver implant. (C) Cable and connectors between A and B.

PULSE GENERATION

Unfortunately CMOS pulse generators that were stable, power economical and compact were not available at the time of design — hence the use of operational amplifiers

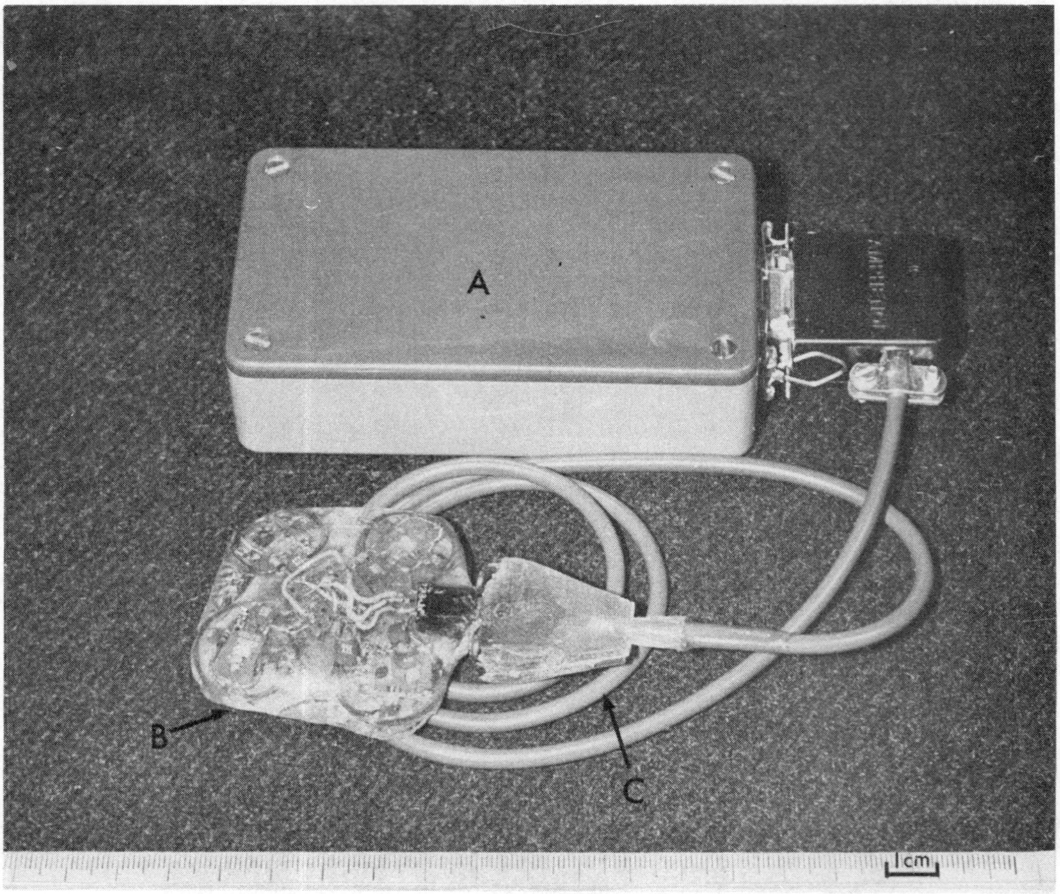

Fig. 6. (A) Rugged ABS box contains circuit of Fig. 4 and sufficient NiCd batteries for at least 12 h use between charges. (B) Araldite moulding containing four transmitters — diagonal pairs are connected in parallel. (C) Commercial connectors and tinsel lead.

(op. amps) IC2 and IC3 in Fig. 3 and the discrete component pulse generator in Fig. 4. These 'micropower' op. amps use approximately 100 μA but do give a stable output. The compact discrete component version gives a very snappy output, uses only 6 μA but does need good power supply decoupling.

PULSE DISTRIBUTION LOGIC

Except where current drive or voltage requirements are too great CMOS logic is used for distributing the pulses to the various channels because of its low power consumption. As the outputs of the femoral nerve controller are staggered in time neither device has more than one channel operating at any instant. This reduces interaction between transmitters, evens the load on the power supply and simplifies output circuit protection.

MODULATION

The transmitters are modulated by varying the power supply to them. This is achieved with common emitter Darlington amplifiers with parallel negative feedback – presets being used to adjust the stimulation strength. The paraplegics need to arise and descend gradually, which is catered for by the switch S2 and the ramp circuit 'A' controlling, via ICI, the power supply to the outputs (see Fig. 3).

PROTECTION CIRCUITS

Because the outputs are staggered in time the protection circuitry can safely be common to all channels. In Fig. 3 the circuit 'C' provides a 1 A limit. The smoothing circuit 'B' allows large output pulses while ICI provides a 14 mA limit on mean current. The protection in Fig. 4 contains a timing element allowing 1 A pulses of 1 ms duration but the current limit decays with time. Ultimately the mean current permitted is about 14 mA. This short circuit protection allows small output transistors to be used with confidence.

CONNECTOR CABLE

It is unfortunate that battery weight and RF interference make it impractical to have a stimulation controller and its transmitter block in the same box as this would avoid an 'unreliable' connector lead between the two (see Figs. 5 and 6). Tinsel lead, however, has been found satisfactory provided there is adequate strain relief between cable and connectors. Dead skin etc. occasionally necessitates cleaning of the connector contacts.

RESULTS

At present three epileptics and two paraplegics use prostheses as described here. Although all the epileptics have shown some apparent fit rate reduction since implantation, the efficacy of this treatment is still being investigated by others in this Unit (see Brindley *et al.*, 1977; cf. Cooper *et al.*, 1974).

Both paraplegics had complete spinal lesions about two years prior to implantation. Initially, endurance on standing amounted to only a few minutes. After 4 months of leg exercises for 1-2 h daily, the quadricepts femoris muscle's strength and fatigue resistance had improved such that Mr. A. F. was able to stand for 35 min or 'walk' for 15 min; Mr. G. L. was able to stand for 75 min or 'walk' for 5 min (see Brindley *et al.*, 1979). It was found that over-strong stimulation tended to cause hip flexor spasm. Obviously with insufficient stimulation the legs are unable to support the patient's weight. As the skin over the receiver implant is quite mobile, this makes it difficult to maintain adequate stability of stimulation strength, especially when walking.

The degree of mobility provided by these prostheses is as yet of little practical value to the paraplegics. As their physical health has improved as a result of the leg exercises, they are at least being kept in trim for any future improvements in technique which may increase their leg control and mobility to a useful level.

FURTHER DEVELOPMENT

As neurological prostheses require more channels the use of multiplexed radio links becomes highly desirable. One or two large coils instead of many small coils will improve the stability of stimulation strength which is a tiresome problem for the paraplegics using the present coils. Also there will be space to use M.O.P.A. transmitters as in Ko, Liang and Fung (1977), which will be less dependent on transistor parameter spread. The improvement in power transfer efficiency will, hopefully, cover the increase in power required by a 16 or 24 channel implant. Implants with I.C. demultiplexors and multiplexing stimulation controllers using micro-processors are already under development in this Unit. Injection moulding will aid production of connector lead fairings.

Acknowledgement — J. D. Cooper is gratefully acknowledged for his great contribution to receiver and transmitter design.

REFERENCES

Brindley G. S., Falconer M. A., Fenton G. W., Fenwick P. B. C., Polkey C. E. and Rushton D. N. (1977) Immediate effects of cerebellar stimulation in an epileptic patient. *EEG Clin. Neurophysiol.* 41, 539-540.

Brindley G. S., Polkey C. E. and Rushton D. N. (1979) Electrical splinting of the knee. *Paraplegia* 16, 428-435.

Cooper I. S. (1978) Some technical considerations of cerebellar stimulation. In I. S. Cooper (ed.), *Cerebellar Stimulation in Man*, pp. 13-17. New York: Raven Press.

Cooper I. S., Amin I., Gilman S. and Waltz J. M. (1974) The effect of chronic stimulation of cerebellar cortex on epilepsy in man. In I. S. Cooper, M. Riklan and R. S. Snider (eds.), *Cerebellum, Epilepsy and Behaviour*, pp. 119-171. New York: Plenum Press.

Donaldson P. E. K. (1973) Experimental visual prosthesis. *IEE Proceedings* 120, 281-298.

Ko W. H., Liang S. P. and Fung C. D. F. (1977) Design of RF powered coils for implant instruments. *Med. Biol. Engng. Comput.* 15, 634-640.

Waters R. (1977) Electrical stimulation of the peroneal and femoral nerves in man. In F. T. Hambrecht and J. B. Reswick (eds.), *Functional Electrical Stimulation: Applications in Neural Prostheses*, pp. 55-64. New York: Marcel Dekker.

Characterization of Diurnal Activity in Man, Using LAPSE

R. F. LEVY, T. G. GRASSIE and J. MACGREGOR

Bioengineering Unit, University of Strathcylde, Glasgow, U.K.

Abstract — Diurnal activity of man was studied using a miniature 4-channel tape recorder. Heart rate, step interval and acceleration of humans in natural circumstances were monitored. Examples are given of stressful conditions such as pregnancy and parachuting.

INTRODUCTION

Miniature 4-channel tape recorders for use in the measurement of physiological variables have been described by McKinnon (1974). Based on their use, LAPSE (longterm ambulatory physiological surveillance equipment) has been developed. The basic concepts have been described by Karagozoglu (1977) and further development and applications by MacGregor (1978). This communication is intended to illustrate the flexibility of the system, the range of information and the type of data which is obtained.

LIFE STYLE (24 HOUR RECORD)

Ergonomists tend to view human activities in terms of their energy cost. While it is no doubt interesting to be aware of the oxygen uptake of individuals during particular events it is obviously unrealistic to attempt to characterize a whole day's activity with such a parameter.

The posture sensors in LAPSE indicate those moments of the day when the subject is standing, sitting and lying down. Simultaneous information on heart beat and step interval give an indication of the physiological cost of mobility. The temporal pattern of these relationships is an illuminating view of a patient's lifestyle and for example would highlight the practical effect of any physical handicap.

BBC TV reporter Ms. Fidelma Cook was investigated (Fig. 1) during a typical working day with its episodes of creative desk work interspersed with car travel to locations for filming. In the evening a long dinner party involved matters of some personal importance. The stress builds up to a considerable emotional climax around 9 p.m., after which heart beat interval returned towards the normal daily level.

The record shows clearly variation in heart rate throughout the day and night, not all episodes of tachycardia being due to physical activity as indicated by the record

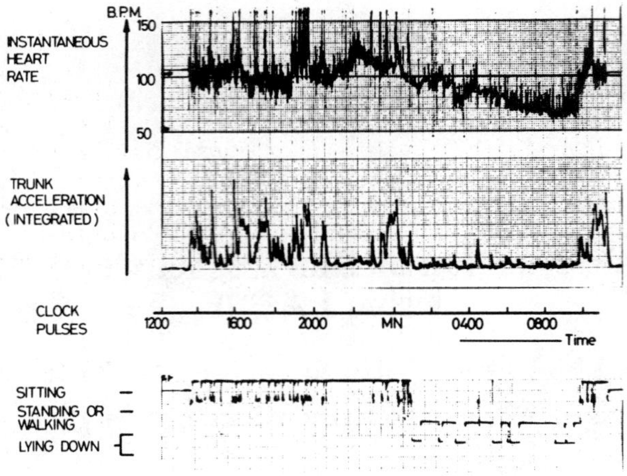

Fig. 1. Behavioral and physiological responses of a female reporter.

of trunk acceleration. All-in-all quite a stressful day. Automatic analysis of the posture signal demonstrates a division of the day in a way which the subject herself found rather surprising.

EVALUATION OF STRESS (EXTRACTION OF ACUTE EPISODE)

A 32 year old male parachutist (a highly experienced instructor) undertook a test. The extract shows intantaneous heart beat interval over a period of twenty minutes (Fig. 2).

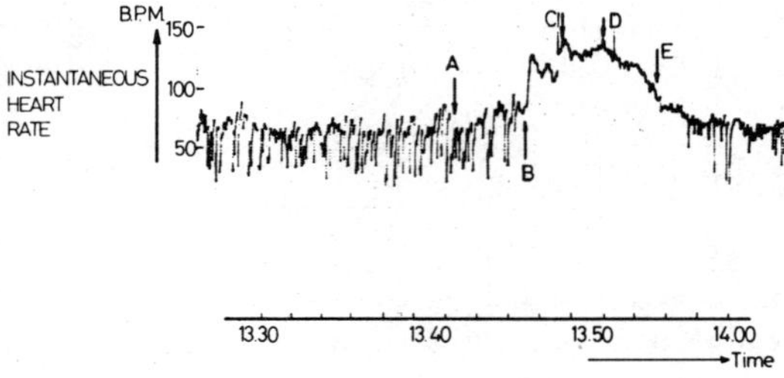

Fig. 2. Heart rate of a male parachutist.

The events indicated are:

(a) commitment of aircraft to approach for parachute drop;
(b) substantial response as instructor musters the 'delayed-drop party' to the exit hatches;
(c) the moment of truth — instructor leaves plane: falls and deploys main chute;
(d) heart rate — rises as ground aiming-point approaches;
(e) normal recovery curve on termination of stress period.

MEASUREMENT OF PHYSIOLOGICAL COST

As an example of the editing power of the method we illustrate (Fig. 3) a series of histograms compiled from the heart beat intervals of a patient with a normal pregnancy.

The portion of the record, when the patient is lying down at night, is processed in each instance and it is therefore possible to compare the distribution of instantaneous heart beat interval at different stages in the pregnancy. The patient was a primagravida aged 28 years and the first three records are at 26, 30 and 37 weeks respectively. A further two post-partum records are illustrated to show:

(a) the return of a relatively high heart rate to the normal subject, and
(b) the reappearance of a bimodal distribution of a 'shoulder' to the distribution, probably representing a return of periods of sinus arrythmia in the post-partum patients.

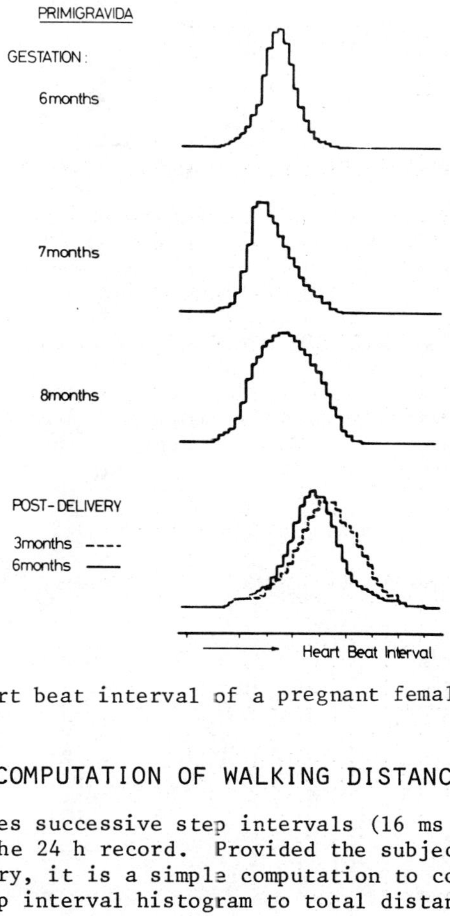

Fig. 3. Heart beat interval of a pregnant female.

COMPUTATION OF WALKING DISTANCE

A typical output includes successive step intervals (16 ms resolution) for all periods of walking in the 24 h record. Provided the subject is previously 'calibrated' in the laboratory, it is a simple computation to convert step interval to step length and the step interval histogram to total distance travelled.

As an example and in order to demonstrate the essential reliability of the method, two subjects of widely different physical characteristics undertook to walk together over varying terrain for a distance of around 4.8 km (Fig. 4). When the two records were individually processed the adult woman had travelled 4783 m and her eight year old son 4770 m.

Pairs of subjects walking together have been studied over measured distances up to

16 km and, in general repeatability is found to be better than 3 percent. This is more than adequate for use in practical investigation into the walking capabilities of patients.

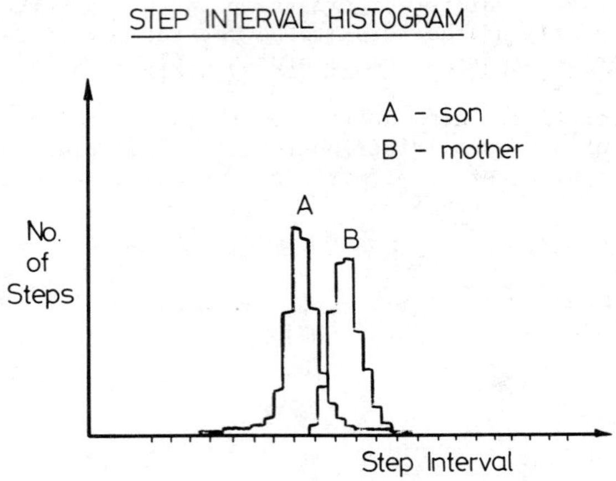

Fig. 4. Step interval of male and female.

REFERENCES

Karagozoglu G. (1977) A system of assessment of physiological cost of daily activities of unencumbered patients. Ph.D thesis, University of Strathclyde, Glasgow.

MacGregor J. (1979) Rehabilitation and ambulatory monitoring. In R. M. Kenedi (ed.), *Rehabilitation of the Disabled*. London: Macmillan, in press.

MacKinnon J. B. (1974) Miniature 4-channel cassette recorder for physiological and other variables. *Biotelemetry Proc. Int. Symp.*, Paris. Basel: Karger.

Longterm Monitoring of the Electroencephalogram in Epileptic Patients

A. KAMP,* N. J. I. MARS** and T. WISMAN***

*Biomedical Department, Institute of Medical Physics TNO, Utrecht, The Netherlands
**Department of Bio-informatics, Twente University of Technology, Enschede, The Netherlands
***Department of EEG, Epileptic Center Meer and Bosch, Heemstede, The Netherlands

Abstract — The duration of a normal electroencephalographic (EEG) examination is relatively short. The chance of obtaining clinically relevant information from a conventionally recorded EEG in epileptic patients is small due to the unpredictable timing of epileptiform EEG phenomena. Therefore, in order to facilitate the diagnosis and the treatment of epileptic patients, great clinical interest lies in carrying out longterm (24 h) EEG recording in epileptic patients, who preferably are able to perform normal daily activities. This results in technical problems of EEG recording and the generation of an enormous amount of EEG data which has to be stored and analyzed if no provision is made for selective registration of the clinically relevant parts of the record. We constructed a modular EEG-telemetry system which permits a longterm registration of 16 EEG signals from relatively free moving patients. The basic module of this system is formed by a small, portable 16 channel EEG-amplifier which feeds into a pulse code modulation (PCM) module. To date cable telemetry has been used for recording the EEG before, during and after an epileptic seizure. In order to obtain the desired data reduction, the system has been used in conjunction with a computerized buffer memory which provides selective registration and storage of the relevant parts of the record. The patient is under video monitoring (VTR) throughout, and elapsed time is registered both on the VTR and on the EEG.

INTRODUCTION

Electroencephalography is an important clinical neurophysiological method for the diagnosis and the treatment of epileptic patients. Besides the diagnostic value of an EEG recorded in patients suspected of epilepsy, the recording of epileptiform EEG activities is also of great interest for obtaining an insight into the nature of epileptic disease and for judging the effect of medication. Duration of the normal EEG recording procedure is relatively short as compared to a circadian period of 24 h. Due to the unpredictable timing of epileptiform EEG phenomena, the chance of obtaining diagnostic information from a conventionally recorded EEG in epileptic patients is slight. Furthermore a conventional record might be incomplete. The chance of recording the EEG during an epileptic seizure, which is of great importance for correct diagnosis, is even less. The effect of medication, particularly the effect of the way in which the anti-epileptic drugs are administered is difficult to judge on the basis of intermittent EEG examinations (Rowan *et al.*, 1977). Thus there exists a great clinical interest in carrying out longterm (24 h) EEG recording, in order to facilitate both the diagnosis and treatment of epileptic patients. Therefore a project was started with the aim of providing a means of longterm EEG monitoring and to solve related problems concerned with the application of non-standard EEG recording methods and the resulting necessity for data reduction.

RECORDING METHODS

Longterm EEG electroencephalography involves the technical problem of recording from subjects who preferably are able to perform their normal daily activities. Telemetry had to be applied which, to a certain extent, would allow the subject to move freely. From the point of view of EEG interpretation and clinical usefulness, it was important to provide longterm EEG registration facilities, the number of recorded channels and the electrode placement were in accordance with standard EEG techniques. So, we developed a modular telemetry system whereby the choice of any particular method of signal transmission could be easily implemented according to the requirements of the investigation concerned. A 16 channel portable EEG amplifier was constructed which could serve as a basic module for any necessary form of preprocessing and/or telemetry of EEG data. It consists of 16 identical differential amplifiers having characteristics which are largely in agreement with standard EEG requirements as regards technical performance. These characteristics are: differential input impedance 20 $M\Omega$, CMRR > 60 dB, equivalent input noise < 1.5 μV peak to peak, frequency range 0.16-100 Hz, gain 1000 ×.

Telemetry methods were straight forward using multilead cable transmission and line transmission of multiplexed (pulse code modulation) data over a single lead coaxial cable. A modulator module for wireless transmission of EEG data is being constructed for those instances when a longer range might be required than is acceptable with cable transmission. The construction of a module for pre-processing of EEG data, to be used in conjunction with a portable taperecorder, is being considered.

DATA REDUCTION

During longterm EEG monitoring of epileptic patients much irrelevant data is recorded and has to be stored and analyzed if no provisions are made for filtering those of relevance for a given examination. The clinically relevant parts of the record usually concern EEG segments of short duration, prior to, during and after the occurrence of epileptiform EEG phenomena. Selective registration of these EEG segments would achieve tremendous data reduction, and also save technician's time, storage capacity and recording paper.

Because the EEG recorded from a patient immediately prior to and during an epileptic seizure is of great importance for a correct diagnosis (Brazier, 1972, 1973) we wanted first to provide means for the selective registration of electro-clinical seizures. Therefore a computerized buffer-memory was constructed which resembles the seizure recording system described by Ives, Thompson and Gloor (1976).

Our system differs in that it offers a greater flexibility and higher performance (Fig. 1). High quality recordings of seizures were needed for research carried out using computer assisted analysis methods. A PDP 11/04 processor with 16K memory controls the whole system. The 16 channel EEG is lowpass filtered, sampled, held and by means of a multiplexer, digitized sequentially to 14 bits accuracy. Digital access provides a high speed data stream to the memory. Using double buffering the data is transferred to a disc. Normally a few minutes of EEG are recorded on disc

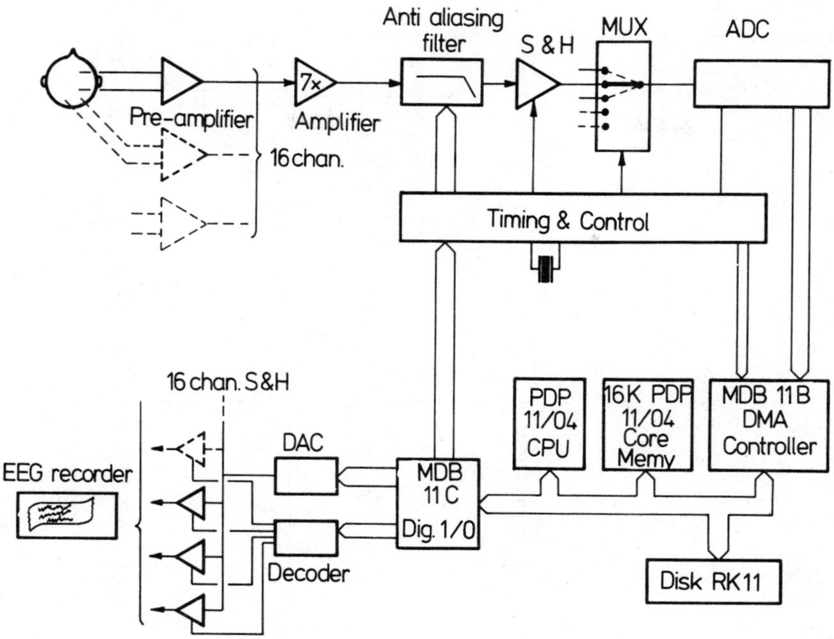

Fig. 1. Block diagram of seizure recording system.

whereby new information is recorded over the oldest information. When a seizure is recorded the recording process is stopped and EEG data prior to and during the seizure is stored on disc. A D/A-channel provides means for on-line or off-line print-outs of the seizure EEG. Some specifications are:

(a) sample frequency (per channel) can be chosen: 128, 256, 512, 1024 Hz;
(b) number of channels: 1 . . . 16;
(c) recording period: pre > 0 s; post ≥ 0 s;
(d) number of seizures to be recorded: 1 . . . 25;
(e) aperture uncertainty of sample and hold: 4 ns;
(f) resolution 14 bits (in view of large dynamic range of seizure EEG's).

SYSTEM FOR LONGTERM EEG MONITORING

The telemetry provisions and the seizure recording system described have been implemented in a clinical setup for carrying out routine longterm EEG monitoring. Besides the EEG information obtained during an epileptic seizure the behavior of the patient is also of great importance for a correct diagnosis. Therefore the patient is also monitored by means of a closed-loop TV circuit and behavioral information is stored on video tape. The block diagram of Fig. 2 exemplifies the complete monitoring system. The upper part of the drawing shows the personal equipment carried by the patient. The 16 channel EEG amplifier module is equipped with an input module providing a preselected electrode placement from a standard 10-20 electrode placement system. An output module performs preprocessing of the EEG data by means of low pass filters (cut-off frequency 40 Hz) before being applied to a portable PCM-encoder (manufactured by Kaiser), which by means of an optically coupled output circuit also provides an electrical isolation. The safety of the patient is a necessary provision considering the fact that he can move around and skin connection can be made to potentially uninsulated non-medical electronic instruments.

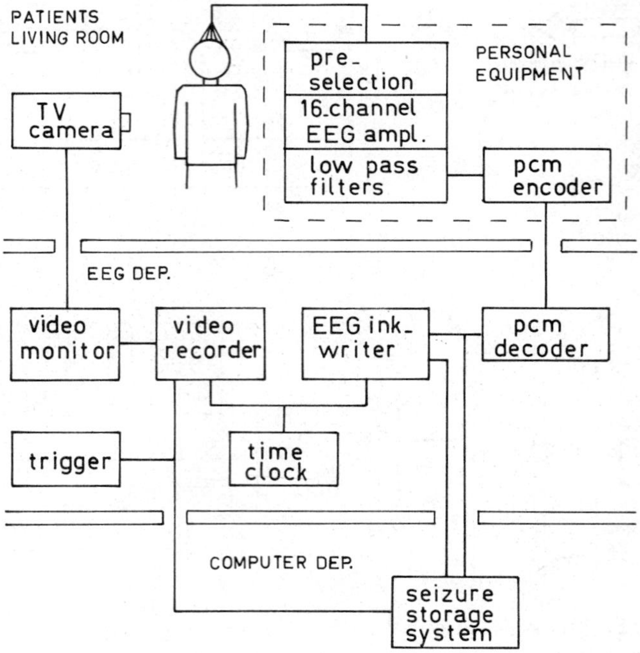

Fig. 2. Block diagram of a clinical set-up for longterm monitoring of EEG and behavior in epileptic patients.

By means of a single lead flexible cable the sequentially digitized output of the PCM-encoder is coupled to a PCM-decoder which restores in analog form the original 16 channels of EEG data. These data can be written out continuously by means of a 16 channel ink-writer or applied to the seizure buffer-memory for storage and be written out whenever an epileptic seizure is recorded.

The behavior of the patient is continuously monitored by means of a TV camera and displayed on a VTR; behavioral information is stored on video tape, whereby longterm video recording is provided by using a lowered tape speed and a frame frequency of 4 frames s^{-1}. In order to retain the time relation between EEG behavioral data, the elapsed time is alpha numerically displayed in the video picture and also on the paper record by means of a character generator. When, on the basis of behavioral and/or EEG data, the onset of a seizure has been recognized, a trigger signal is fed to the buffer memory in order to start the storage on disc of the seizure EEG and also to the video recorder which then is switched to the normal recording mode.

DISCUSSION

Systems for longterm monitoring in epileptic patients have been described by several authors. Using a different methodology, such as cable transmission (Ives *et al.*, 1976), wireless transmission (Willison, 1976) or storage of EEG data using a portable tape recorder (Ives and Woods, 1975; Sato, Penry and Dreifus, 1976) each of these systems provides longterm recording facilities which, to a certain extent, fulfils specific or general clinical requirements. The system described here (Fig. 2) uses a methodology similar to the one described by Ives *et al.* (1976). It differs in that it is more flexible because means are provided for recording from subjects, which, depending on the telemetry technology used, can move within a certain range. A modular design was chosen in order to be able to choose the method of signal transmission to any particular recording situation. So far it has been our experience that in a clinical environment line transmission offers satisfactory recording facilities for carrying out different types of examinations in epileptic patients. The ease of operation has allowed us to make longterm EEG recording as a routine clinical method. Future use and developments will also depend on the outcome of an evaluation study to determine the benefits of longterm EEG recording as compared to conventional EEG examination. Our aim of providing optimal EEG performance has been of great value, enabling the electroencephalographer to use standard methods of EEG interpretation.

REFERENCES

Brazier M. A. B. (1972) Spread of seizure discharges in epilepsy: anatomical and electrophysiological considerations. *Exp. Neurol.* 36, 253-272.

Brazier M. A. B. (1973) Electrical seizure discharges within the human brain: the problem of spread. In M. A. B. Brazier (ed.), *Epilepsy: Its Phenomena in Man,* pp. 153-170. New York: Academic Press.

Ives J. R. and Woods J. F. (1975) 4-channel 24-hour cassette recorder for long-term EEG monitoring of ambulatory patients. *Electroenceph. clin. Neurophysiol.* 39, 88-92.

Ives J. R., Thompson C. J. and Gloor P. (1976) Seizure monitoring: a new tool in electroencephalography. *Electroenceph. clin. Neurophysiol.* 41, 422-427.

Rowan A. J., Binnie C. D., DeBeer-Pawlikowski N. K. B., Goedhart D. M., Gutter Th., Van Parys J. A. D., Meinardi H. and Meyer J. W. A. (1977) 24-hour EEG studies with very frequent serum antiepileptic drug concentration determinations in the study of sodium valproate. In H. Meinardi and A. J. Rowan (eds.), *Advances in Epileptology,* pp. 255-260. Amsterdam: Swets and Zeitlinger B. V.

Willison R. G. (1976) Radio-telemetry and EEG analysis in the study of patients with seizures. In *Applications of Electronics in Medicine,* IERE Golden Jubilee Conference, the Institution of Electronic and Radio Engineers, 8-9 Bedford Square, London.

The Radio Pressure Pill—Its Use when Measuring Truncal Stress

D. A. STUBBS and G. C. DAVID

Department of Human Biology and Health, Materials Handling Research Unit, University of Surrey, Guildford, Surrey GU2 5XH, U.K.

Abstract — The work of the Materials Handling Research Unit is aimed at the prevention of back injuries and a methodology, using radio pressure pills, is reviewed which enables truncal stress to be quantified. Such measurements have been taken from several thousand volunteers using this method, and a variety of situations have been investigated. In industry, prevention is directed firstly at ameliorating existing handling problems, and secondly at providing safe lifting capacities for the design of future working environments. Where sporting or leisure pursuits are investigated, the main aim is not to deter people from participating, but to provide safeguards for young people embarking on such pursuits. Examples will be given of all three radio pressure pill applications, a method which is now an established, safe and acceptable research procedure which is being used to prevent people suffering from back injury.

INTRODUCTION

Many people today suffer from back pain as a result of injury at work or during leisure pursuits. The extent of their suffering can range from acute disability to chronic discomfort, and it is estimated that 13.2 million man days are lost each year in Britain through the painful back (Benn and Wood, 1975) at an estamated cost to the country of £350 million (US $700 million).

The work of the Materials Handling Research Unit is aimed at the prevention of back injuries. For this work, use is made of the observation that when rapid extension of the trunk is required, or when heavy weights are being lifted, there is increased intra-abdominal pressure caused by the simultaneous contraction of the diaphragm, levator ani, oblique and transverse abdominal muscles, sometimes supported by similar increase in pressure in the thoracic cavity. The operation of this pneumatic mechanism was postulated by Keith (1923) and investigated by Davis (1956, 1957, 1959a, b); Bartelink (1957); Morris, Lucas and Bresler (1961) and Davis and Troup (1964). Subsequent research by Stubbs (1975); Davis and Stubbs (1976, 1977, 1978) and Stubbs and Osborne (1979) has established that there is a close correlation between the magnitude of pressure induced within the trunk by heavy physical activity and the magnitudes of the forces acting on the spinal mechanism. Thus, these pressure measurements can be used as an indirect method of assessing truncal stress.

THE RADIO PILL

The original studies on intra-truncal pressure changes during physical activity and postural changes of the trunk were conducted using partially inflated rubber balloons in the body cavities connected to a pressure transducer via a catheter (vagina, Murphy and Mengert, 1933; rectum, Rushmer, 1946; rectum, stomach, perotoneum and bladder, Adno, 1956; rectum and stomach, Davis, 1956; stomach, Eie and Wehn, 1962; and oesophagus and stomach, Davis and Troup, 1964). More recently radio pressure pills developed by Watson, Ross and Kay (1962) (Fig. 1) have been used to study trunk stress during physical activities (Davis and Troup, 1966; Kumar, 1971; Davis and Stubbs, 1976, 1977, 1978; Stubbs and Osborne, 1979). Their main advantage over other methods is that they can be used in the field as well as in the laboratory, with minimal interference to the subjects under investigation.

With the present work Rigel* model 7014 pressure sensitive pills are used (Davis, Stubbs and Ridd, 1977) and are easily swallowed by volunteers. The pill (Fig. 1),

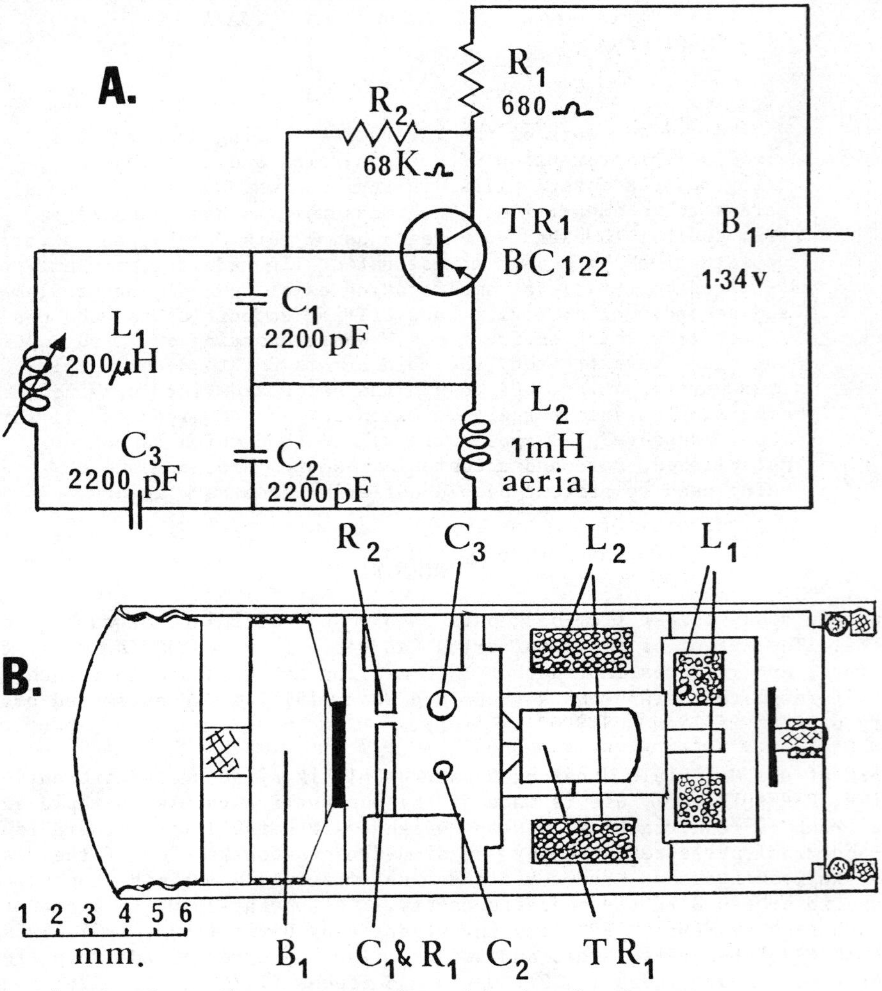

Fig. 1. The radio pressure pill. (a) Circuit diagram.
(b) Mechanical layout.

---

*Rigel Research Ltd., 99 Gander Green Lane, Sutton, Surrey, U.K.

which is powered by a miniature mercury battery, type RM312, consists of a transistor oscillator, the frequency of which is controlled by a diaphragm operated variable inductor. The signal from the pill is picked up by a unidirectional antenna, strapped about the subject's waist, which is connected to a portable Rigel receiver, model 7040, and a Smith's Servoscribe, model 501.2 (Fig. 2).

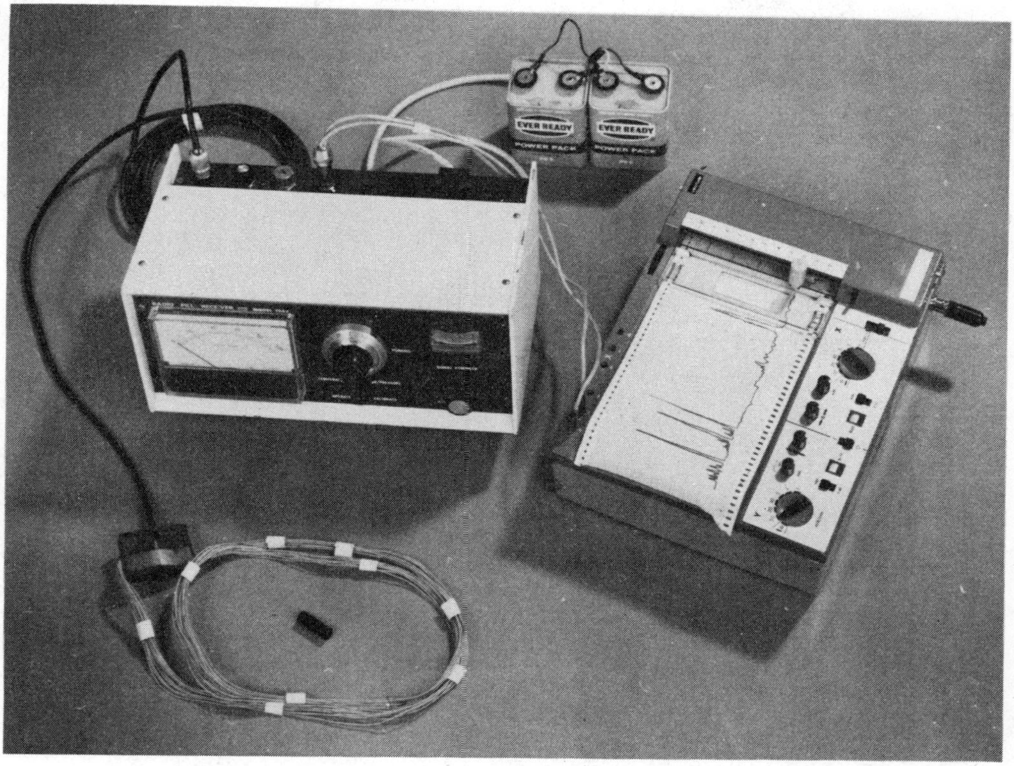

Fig. 2. View of the apparatus used in truncal stress evaluation showing the pill, antenna and portable receiver and recorder.

APPLICATIONS

Within industry, prevention of back pain is directed firstly at ameliorating existing handling problems and secondly at providing safe handling capacities for the design of future working environments (see Fig. 5). With the former approach, once the stress on the trunk, caused by a given activity is known, as indicated by intra-abdominal pressure, this can be compared with standard measurements (Davis and Stubbs, 1976) and the safety of any part of a particular procedure can be assessed. If the activity appears hazardous, alternative procedures are tested, and where a safe method is found this is recommended for adoption. Often, just a simple change in posture can make a given task safe as shown in Fig. 3. These results indicate that using the legs to provide the lifting force, with the trunk as erect as possible, causes least trunk stress.

Of recent interest, has been the truncal stresses imposed on nurses when lifting patients using the recommended handling techniques. The results, from one of several lifts using each recommended technique, for a male nurse are shown in Fig. 4, indicating the magnitudes of the peak pressure, which gives an indication of the truncal

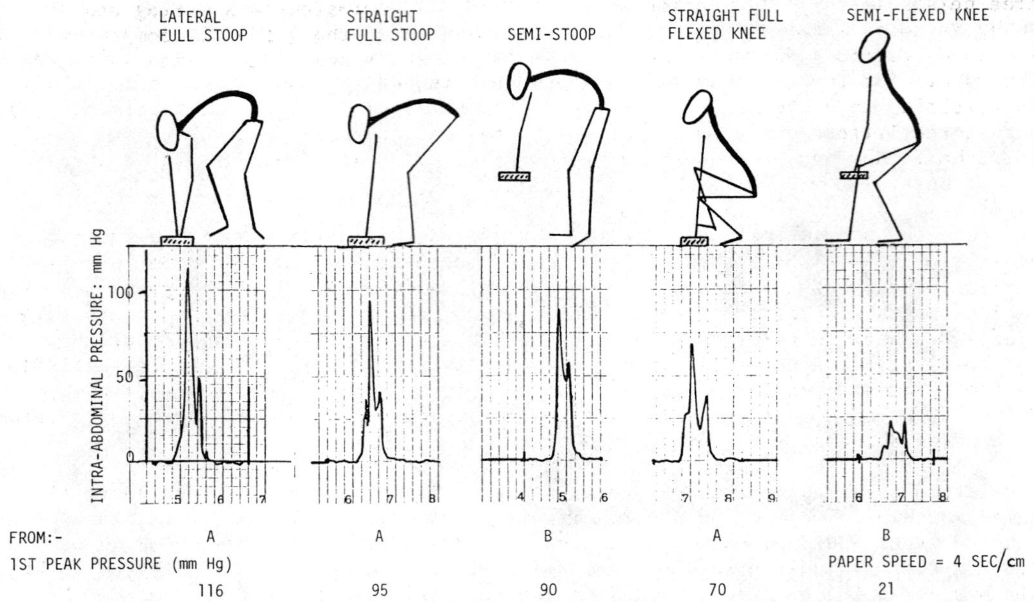

Fig. 3. Intra-abdominal pressure records of a male subject lifting 35 kg from (a) ground level, and (b) a raised surface 0.39 m high to a height of 1.11 m using five lifting techniques.

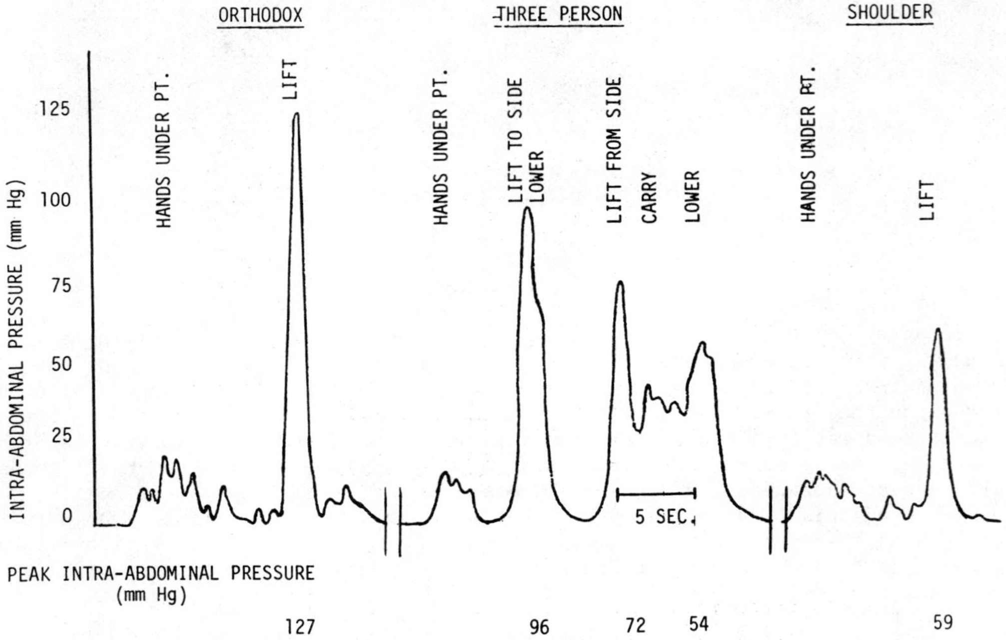

Fig. 4. Intra-abdominal pressure records for a male nurse (in training) assisting to move a 65 kg patient using three recommended techniques.

stresses involved in each maneuver. That the stresses are high in the orthodox and three person lifts is a function of the stooped or semi-stooped starting posture which, as noted earlier (Fig. 3), cause high stress on the lifter in comparison with an erect trunk posture. Further, in the orthodox and three person lift, the lifters are far removed from the patient, thus increasing the torque on the trunk, and resulting in higher spinal stress. With the shoulder lift, the forces are less because the lifters are much closer to the patient and the resultant torque much diminished. However, the question to be answered is 'are these levels of stress safe or unsafe?'

In a detailed study of manual handling hazards in the building and construction industry, Stubbs (1975) and Davis and Stubbs (1976) found that peaks of intra-abdominal pressures in excess of 100 mm Hg or 13.3 kPa were frequently induced with model tasks taken from occupations with a significantly higher incidence of back injuries. Thus, in relation to patient handling, for males one would conclude that the stresses encountered in the orthodox and three person lifts are sufficiently high to be considered unacceptable. Further, from biomechanical considerations, one would conclude that it was similarly undesirable to use female nurses for either of these techniques.

In relation to establishing safe handling capacities for equipment designers it seemed worthwhile to examine what magnitudes of loads in different relationships with the trunk would be expected to give pressure of the order of 100 mm Hg or 13.3 kPa during handling tasks. One such designer's contour is illustrated in Fig. 5, this example being for the single safe vertical lifting forces (kg) that can be

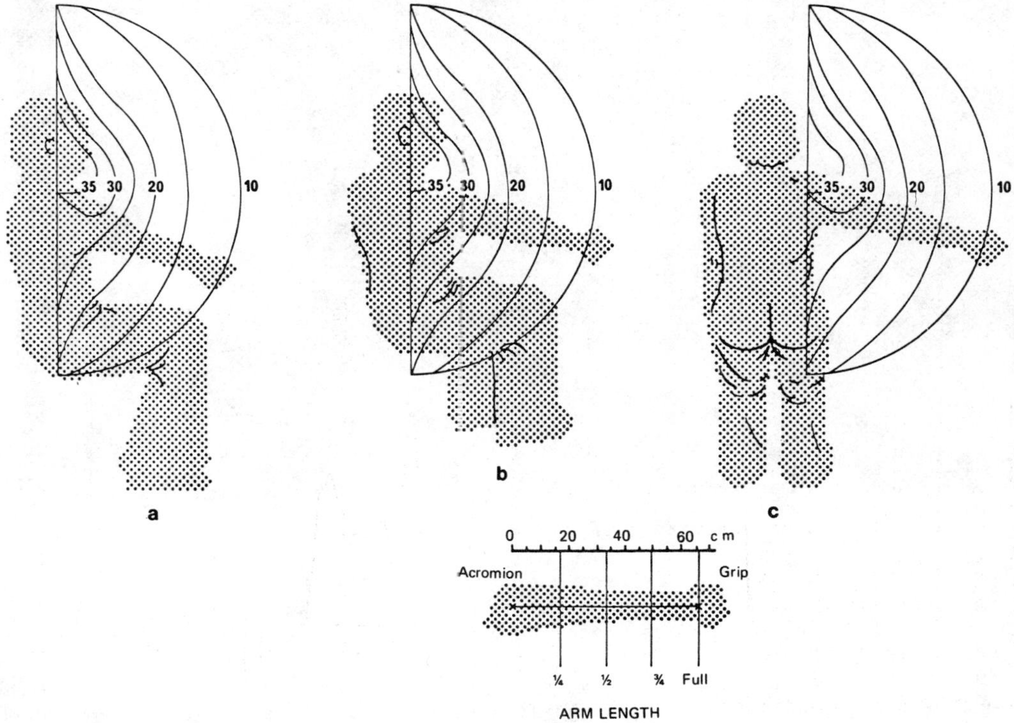

Fig. 5. Contours of safe values (kg) for one-handed vertical lifts in the planes shown below: (a) hand directly in front of the body (sagittal plane); (b) hand in a plane at 45° from the sagittal plane; (c) hand in a plane at 90° from the sagittal plane.

exerted using one hand by seated, fit young, adult males. If the lifting activity is carried out at a rate greater than one per minute, then the contour values must be reduced to 70 percent of those shown. Such contours can be used either to examine safety in existing handling tasks, or for designing a new environment, where one can build the system around the workforces' safe capabilities (Davis and Stubbs, 1977).

Where sporting or leisure pursuits are investigated the main aim is not to deter people from participating, but to provide safeguards for young people embarking on such pursuits, and to provide design evaluation facilities for existing equipment or new ones developed to, for example, take the backache out of gardening. Figure 6 illustrates such a trial where a new fork design is being compared with existing tools in relation to the spinal stresses involved in their use.

Fig. 6. An evaluation of a proposed new garden fork.

Another recent pilot study undertaken was to investigate the spinal forces incurred during water-ski jumping (Fig. 7). This was initiated because of the alarmingly high incidence of injury to both knees and backs of water-ski jumpers (Horne and Cockshott, 1977). To quantify the 'take-off' and 'landing' impacts at the ankle and base of the spine, accelerometers (Pye Dynamic, type BLA2) were attached to the water-skis and the skin overlying the spinous process of the fifth lumbar vertebra. The outputs were fed through a power supply and amplifier system into an Oxford Medilog Series 4.2. The radio pressure pill could not be used for obvious reasons and instead intra-abdominal pressure was monitored using an Entran Model No. EPA-125E-30AW miniature pressure transducer mounted in a Ryles catheter which was swallowed either via the nasal or oral cavity. Again, the output was recorded on the

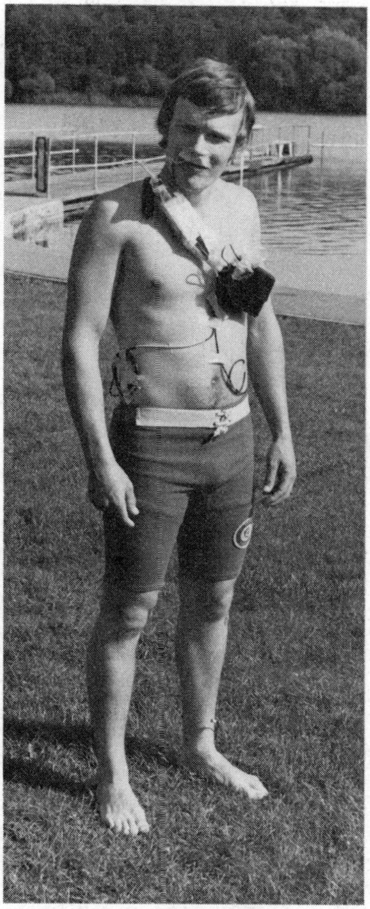

Fig. 7. Showing subject instrumentation used in the analysis of the spinal forces incurred during water-ski jumping.

Medilog. Each of four jumps was filmed using a film speed of 64 frames per second. Two subjects were investigated, the first being an experienced jumper using a 1.83 m ramp and a boat speed of 58 km h^{-1}, the second being a 13 year old junior using a 1.52 m ramp with a boat speed of 40 km h^{-1}. For the first subject 21 peak pressures in excess of 100 mm Hg were observed (maximum 196 mm Hg) and the mean spinal 'g' forces were 4.5 and 9.5 for 'take-off' and 'landing' impacts respectively. For the junior, mean pressures were 30 mm Hg and 75 mm Hg with mean 'g' forces of 2 and 4.6 for 'take-off' and 'landing' impacts respectively.

The 'g' force profiles did not approach the accepted level of 10 g for 500 m s^{-1}, for pilot/air crew ejections. However, pilots seldom eject 20 times a day which is approaching the number of jumps a ski-jumper may complete in a training session. Reid, Kopp and Verhoeven (1977) reach a similar conclusion in noting that the forces that occur duing 'take-off' and 'landing' appear not to be of an intensity to result in injury to a normal, healthy athlete. However, they continue, that frequent exposure to these impacts may well prove hazardous to the health of the jumper. This is reinforced by Horne and Cockshott (1977) who recommend that consideration be given to the elimination of ski-jumping in the junior division, that is those skiers 13 years of age and under.

DISCUSSION

The quantification of spinal stress by measurement of intra-abdominal pressure using radio pressure pills is now an established, safe and acceptable research procedure which is being used to prevent people suffering back injury in many parts of the world. Studies have been carried out in many industries and these have lead to the adoption of recommendations which will significantly reduce the risk of back injury. Often only a small change in posture or technique is needed. For example, intra-abdominal pressure measurements using radio pressure pills showed that back stress was unacceptably high when workers pushed hand trolleys, loaded with lengths of timber, in a factory manufacturing prefabricated sections for bungalows. They were advised to pull the loaded trolleys, which ensured they took up a more favorable posture when applying the necessary force, and subsequent measurements showed that stress on the back had been more than halved. Sometimes a redesign of the work area is called for. In a chemical engineering plant the hose connections on down pipes had been sited at or near ground level. Measurements on operators, making the connection between the flexible hose and the down pipe, showed that the combination of the weight of the hose and the stooping posture required to reach the connecting parts caused high levels of back stress. The solution recommended was to raise the mouth of the down pipe to the average elbow height of the workforce, thus allowing the operator to stand erect whilst making the connection and to handle the hose at waist level. The technique has also been used successfully to contribute to the design of new equipment. A box-type enclosure was proposed to contain and protect telecommunications equipment but the extent to which the complete unit could be safely handled was unknown. A mock-up was built and volunteers, who had swallowed radio pressure pills, lifted and handled the unit in environments similar to those in which its future use was planned. As a result of the study various modifications were suggested, such as the relocation of handles and the types to be used, and also limits were set to the height to which the unit could safely be stacked and also lifted in confined space. Thus workers were successfully prevented from injuring their backs whilst using the equipment right from its introduction.

There are several areas where improvements can be made. The present pills have to be recovered because of their cost (£55, US $110). In addition, the present system requires an umbilical cord between antenna and receiver. If a disposable pill could be developed that either interfaced via a miniature receiver/demodulator with, for example, a Medilog, or could be retransmitted via a secondary telemeter, then this would allow a much wider application of the methodology to prevent the often needless suffering imposed by back pain.

REFERENCES

Adno J. (1956) Some aspects of the mechanics of the abdomen. *South African Med. J.* 30, 535-539.

Bartelink D. L. (1957) The role of abdominal pressure in relieving the pressure on the lumbar intervertebral disc. *J. Bone Jt.Surg.* 39B, 718-725.

Benn R. T. and Wood P. H. N. (1975) Pain in the back: an attempt to estimate the size of the problem. *Rheum. Rehab.* XIV, 121-128.

Davis P. R. (1956) Variations of the intra-abdominal pressure during weight lifting in various postures. *J. Anat.* 90, 601(P).

Davis P. R. (1957) Studies on the functional anatomy of the human vertebral column, with special reference to the thoracic and lumbar region. Ph.D. Thesis, London.

Davis P. R. (1959a) Posture of the trunk during the lifting of weights. *Br. Med. J.* 1, 87-89.

Davis P. R. (1959b) The causation of hernia by weight-lifting. *Lancet* 2, 155-157.

Davis P. R. and Troup J. D. G. (1964) Pressures in the trunk cavities when pulling, pushing and lifting. *Ergonomics* 7, 465-474.

Davis P. R. and Troup J. D. G. (1966) Effects on the trunk of handling pit props at different working heights. *Ergonomics* 9, 475-484.

Davis P. R. and Stubbs D. A. (1976) Report of the International Symposium on Safety in Manual Materials Handling. DHEW, NIOSH, 78-185.

Davis P. R. and Stubbs D. A. (1977) A method of establishing safe handling forces in working situations. *Ergonomics* 20, 557(P).

Davis P. R. and Stubbs D. A. (1978) Safe levels of manual forces for young males, 1, 2, and 3. *Appl. Ergonomics* 8, 141-150; 8, 219-228; 9, 1, 33-37.

Davis P. R., Stubbs D. A. and Ridd R. E. (1977) Radio pills: their use in monitoring back stress. *J. Med. Engng. Tech.* 1, 209-212.

Eie N. and Wehn P. (1962) Measurement of the intra-abdominal pressure in relation to weight bearing of the lumbo-sacral spine. *J. Oslo City Hosp.* 12, 205-217.

Keith A. (1923) Hunterian lectures. Man's posture: its evolution and disorders. *Br. Med. J.* 1, 587-590.

Kumar S. (1971) Studies of the trunk mechanics during physical activity. Ph.D. Thesis, University of Surrey.

Morris J. M., Lucas D. B. and Bresler B. (1961) Role of the trunk in stability of the spine. *J. Bone Jt Surg.* 43A, 327-351.

Reid J. G., Ropp P. M. and Verhoeven A. P. (1977) Impact forces on water-ski jumpers. Biomechanics Lab., School of Physical and Health Education Queen's University, Ottawa, Canada.

Murphy D. F. and Mengert W. F. (1933) Intra-abdominal pressure created by muscular effort. *Surg. Gynaec. Obstet.* 57, 487-493.

Rushmer R. F. (1947) The changes in pressure in the peritoneal cavity produced by sudden deceleration of experimental animals. *J. Aviation Med.* 18, 199-206.

Stubbs D. A. (1975) Trunk stress in construction workers. Ph.D. Thesis, University of Surrey.

Stubbs D. A. and Osborne C. M. (1979) Back injuries: a comparison between the nursing profession and the construction industry, Part I and II. *Nursing*, in press.

Watson B. W., Ross B. and Kay A. W. (1962) Telemetering from within the body using a pressure sensitive radio pill. *Gut* 3, 181-186.

Radio Telemetric System for Medical Supervision during Swimming

V. E. MANOILOV, Y. V. STUPIN, V. I. LEBEDEV and N. A. ZOTOV

The Scientific-Research Biotelemetrical Laboratory of Kislovodsk Resort Council, Kislovodsk, U.S.S.R.

Abstract — A 2 channel EKG transmitter system has been designed and tested on humans. Sports swimmers have had their EKG monitored during exercise in an open-air pool. Further application to medically supervised activites are being made with the biotelemeter.

The strength of the cardiovascular system of a human organism is one of the important constituent parts of its functioning and adaptation to different loads. Lately much attention has been given to the problem of medical control of the cardiovascular system during swimming in the process of taking hydrotherapeutic treatment. Taking into account the importance of the problem we developed a 2 channel radio telemetry system, permitting the recording of an electrocardiogram from two bipolar chest leads on a human being or on an animal during swimming.

The main technical characteristics of the system are given below. The complete system includes: (1) A transmitting device, consisting of a crystal controlled transmitter with input signal transformation and dimensions of 80 × 50 × 25 mm; an autonomous feeding source assembled in a separate case with dimensions of 70 × 35 × 25 mm; and a set of electrodes complete the transmitter. The device is hermetically sealed and has a total weight of 3 kg. It is fastened to the waist of a swimmer. (2) The receiver was portable and had automatic feeding permitting field use with dimensions of 260 × 130 × 100 mm. The radio telemetry system is made according to a double frequency modulation scheme and secures transmitting of the tension 0.03-4 mV in the frequency range 0.1-100 Hz with an error not more than ± 10 percent. The input resistance is 2 MΩ and the system range is 100 m.

The electrodes for reading electrical activity of a heart are cuplike sensors with a surface diameter of 14 mm. The electrodes are fastened on the body by means of two-sided sticky tape that ensures sufficient encapsulation.

The system was successfully tested in the Kislovodsk open air swimming pool during the training process of sports swimmers and also during medical studies of patients with blood circulation pathology during their gymnastics in water. The system receives information about heart activity of a swimmer in and above water which extends the sphere of its application.

Prospects for the Use of Biotelemetry in Studies of the Differentiation of Mammalian Locomotor Systems

S. SALMONS

Department of Anatomy, The Medical School, University of Birmingham, Birmingham, U.K.

Abstract — The technology already exists for recording multiple channels of electromyographic (EMG) information via a radio frequency link, and has been used, for example, in the analysis of gait patterns. What is not generally appreciated is the contribution such techniques could make to understanding the development of the locomotor system. The differentiation of mammalian skeletal muscle into slow-contracting and fast-contracting types is usually completed in the first few weeks following birth. In this process the muscles become highly specialized, yet they retain into adult life a capacity for undergoing transformation of fiber type. Evidence from experiments using implanted electronic stimulators points to nerve impulse activity as the regulatory factor responsible. Biotelemetry of EMG would enable this hypothesis to be tested further by recording changes in endogenous patterns of muscle activity under conditions in which transformation of fiber types takes place: in cross-reinnervation, in altered thyroid status and during early postnatal development.

INTRODUCTION

The laser was once described as 'a solution in search of a problem'. The same comment could apply to biotelemetry, in so far as it has not yet realized its full potential for solving biomedical problems. The reasons for this are perhaps inherent in the technique itself. Most biotelemetry systems pose stringent specifications and their design involves a series of trade-offs which tend to be unique to each application. The resultant lack of standardization, particularly in regard to frequency allocations, has discouraged the commercial interest which might have made such systems more widely available, and has tended to confine biotelemetry to those laboratories which possess adequate engineering facilities and expertize. With rare exceptions, therefore, instrumentation tends to be developed in isolation from biomedical problems, while the scientists with those problems are either unaware of the possibilities offered by biotelemetry, or if aware are not equipped to make use of them. This paper is intended to draw attention to some of this unexploited potential.

ELECTROMYOGRAPHY

It has been known since the middle of the last century that contraction of skeletal muscle is accompanied by electrical activity, although it was only after the development of cathode ray oscillography that the nature of this rapidly varying signal, or electromyogram, could be demonstrated visually. Nowadays electromyography (EMG) belongs to the normal repertoire of diagnostic techniques available to the modern hospital.

The diagnostic use of EMG consists of an examination of the waveform generated under conditions of rest, minimal and maximal voluntary contraction of the muscle. Regardless of the manner in which the EMG signal is processed it is not a reliable measure of the force generated by the muscle except under tightly circumscribed, artificial conditions; for example a fatiguing muscle exhibits copious electrical activity although the corresponding force may be only a fraction of normal. The tension developed by an individual muscle is, however, technically difficult to measure, and with isolated exceptions (Barnes and Pinder, 1974; Walmsley, Hodgson and Burke, 1978) has not been measured at all under conditions of unrestricted movement. The electrical signal, or its integral or envelope, has therefore found application as a convenient indicator of the presence of contractile activity in a muscle. It has been used as an endpoint in studies of the central nervous mechanisms underlying the initiation of voluntary movement (Evaerts, 1968). It has been exploited by a number of groups as a signal for controlling a powered prosthesis, the so-called myoelectric arm. It can be used to examine the co-ordinated activity of different muscles in the posture and locomotion of animals, and to explore changes in the roles of the homologous muscles in animals adapted evolutionarily for a different way of life. Such techniques also find clinical application, providing a valuable adjunct to force plate and cinephotographic methods in the assessment of gait. In this way abnormalities of gait can be defined more closely, and their improvement or otherwise during a course of remedial treatment assessed quantitatively. In all these applications, and particularly in the case of locomotor studies, it is important to minimize encumbrance or restraint of the experimental subject by the measuring apparatus. The use of radio telemetric techniques eliminates trailing wires, an advantage which has not been overlooked (Baumann and Baumgartner, 1974; Rasmussen, Chan and Goslow, 1978; Amlaner, 1978). A logical and welcome extension of this philosophy is the incorporation of amplifying and transmitting circuitry in the surface mounted EMG electrodes, obviating the necessity even for the electrode leads which would normally be required for connection to a waist mounted transmitter (Westbrook, Fryer and Rositano, 1976).

The surface electrodes record somewhat generalized activity from superficially located muscles. For recording from less accessible muscles, electrodes must be placed within the muscle of interest. In clinical examination concentric needle electrodes are used. In freely moving subjects the electrodes take the form of fine wires, insulated except for a small active area at the tip. They are normally introduced into the muscle with the aid of a fine hypodermic needle, the tip of the wire being folded back to form a barb which retains the wire in the muscle. Such an arrangement has the further advantage that the tip of the electrode records from an area undamaged by the passage of the hypodermic needle (H. P. Clamann, personal communication). These electrodes permit the analysis of patterns of single motor unit activity during the normal use of the muscle. They are not without problems, however. The barbed electrodes have a tendency to migrate into the muscle, making it more difficult to fix the recording site or to localize it subsequently. The fine wires are fragile and if used in long runs prone to inductive interference. For animal studies these considerations point to a need for a completely implantable multichannel EMG biotelemeter, with the amplifiers and possibly the multiplexing circuitry located close to the electrodes, and power and signal links being conducted to a common central transmitter. With the advent of micropower integrated circuits such a system would be well within the scope of existing technology, and would enable the research, as opposed to the clinical, possibilities for EMG biotelemetry to be more fully explored.

MUSCLE DIFFERENTIATION AND THE ADAPTIVE RESPONSE

EMG as an indicator of the moment-to-moment contractile activity of a muscle is a well established concept. Very much less familiar are the longterm consequences of a given pattern of activity for the differentiation of the muscle. To understand the significance of this role it is necessary to consider briefly the normal development of muscle.

The skeletal muscles of most mammals undergo important changes in the first few weeks following birth. During this time the muscle fibers become specialized for their functional role in the adult animal. From this process, two distinct types of muscle fibers emerge. Slow, or 'red' muscle fibers are adpated for maintenance of posture and sustained activity; they contract and relax slowly, derive their energy from an aerobic metabolism, and are resistant to fatigue. Fast, or 'white' muscle fibers are suited to intermittent periods of intense activity; they contract and relax rapidly, derive their energy from the less efficient anaerobic pathway, and are readily fatigued. Whole muscles can be referred to as 'slow' or 'fast' when their overall characteristics are determined by a predominance of one or the other fiber type.

These muscle types are further distinguished by the patterns of impulse activity delivered to them by their motor nerves. The motoneurones innervating slow muscles generate a sustained low frequency pattern of activity; in contrast, motoneurones supplying fast muscles subject them to intermittent bursts of more intense activity. The characteristic physiological and biochemical properties of slow muscles appear to emerge in response to the establishment of the adult pattern of motoneurone activity. If the motoneurones are rendered quiescent by isolation of the spinal cord a few days after birth, muscles which would normally have acquired slow contractile characteristics develop as fast muscles (Buller, Eccles and Eccles, 1960a). Conversely an implantable electronic stimulator (Salmons, 1967) can be used to subject a fast muscle to a continuous low frequency pattern of activity of the type normally received by slow muscles. Under these conditions a fast muscle undergoes an orderly sequence of changes which ultimately bring about a complete transformation to a slow muscle by all the criteria which have so far been applied (Salmons and Vrbová, 1969; Sréter *et al.*, 1973; Pette *et al.*, 1973; Romanul *et al.*, 1974; Sréter *et al.*, 1974; Sréter *et al.*, 1975; Salmons and Sréter, 1976; Pette *et al.*, 1976; Hudlická, *et al.*, 1977; Salmons, Gale and Sréter, 1978; Rubinstein *et al.*, 1978; Heilmann and Pette, 1979). Two aspects of this response are worthy of emphasis. Firstly, the response comprises qualitative as well as quantitative changes in the complement of proteins synthesized by the muscle, evidence of a fundamental redirection of protein synthesis within the cells. Secondly, the response is elicited just as easily in fast muscles of an adult animal, and this suggests that the changes represent an adaptive response which enable a muscle to accommodate the changing functional requirements which may be encountered at any time during the animal's life. Such a concept can provide insight into previous experiments relating to the influence of nerve on muscle differentiation.

CROSS-INNERVATION

In 1960 Buller *et al.*, reported the results of experiments in which the motor nerves of a fast and a slow muscle had been cut and cross-anastomosed; as a result, the fast muscle had been reinnervated by the nerve which previously supplied the slow muscle, and *vice versa*. The contractile characteristics of the muscles had been found to alter, the fast muscle becoming slower and the slow muscle faster. These changes have since been found to extend to a variety of biochemical and histochemical properties, although the degree of transformation is not as great as that induced by electrical stimulation. One explanation of these phenomena which was widely accepted attributed the neural influence to chemical trophic factors transported by the nerve to the muscle. However, if cross-union of the nerves interchanges the characteristic patterns of impulse activity reaching the slow and fast muscles, the results would be fully consistent with the adaptive-response hypothesis. The fast muscles, subjected to continuous low frequency activity would tend to acquire slow muscle properties; the slow muscles, relieved of such a pattern, would tend to undergo reciprocal changes. Such an interpretation is supported by the work of Salmons and Sréter (1976) who used the chronic electrical stimulation technique to show that in the absence of an actual change in the pattern of impulse activity reaching a muscle, cross-innervation had no significant effect. Nevertheless, the evidence that cross-innervation transposes the patterns of activity reaching the muscles remains circumstantial. Biotelemetry of the EMG could furnish direct evidence on this point by recording in the freely moving animal the pattern of motor unit activity in muscles reinnervated by foreign nerves. Control data could be obtained from muscles whose nerves had been sectioned and self-anastomosed.

During reinnervation of a muscle the first axons to arrive tend to branch profusely and supply a disproportionate number of muscle fibers. These contacts, once formed, are not displaced by the arrival of subsequent axons. As a result the motor units in a reinnervated muscle are not dispersed as randomly as in a normal muscle. It would therefore be necessary to sample the activity at a variety of sites within the muscle in order to avoid possible bias. Such an experiment therefore calls for multichannel operation, but it is clear that a biologically significant result could be obtained with only a few days of recording.

THYROIDAL INFLUENCES ON SKELETAL MUSCLE

The thyroid gland plays an important part in a number of developmental processes, and in the adult animal its hormonal secretions regulate the level of metabolic activity in cells. It appears that, in addition to their metabolic effects, thyroid hormones also influence the fiber type composition of skeletal muscles. Ianuzzo *et al.* (1977) observed a significantly greater proportion of slow fibers in muscles of animals whose thyroid glands had been removed; conversely the proportion of slow fibers in animals to whom additional thyroid hormones had been administered was smaller. These effects could be attributed to a direct influence of thyroid hormones on muscle cells. The adaptive-response hypothesis offers an alternative interpretation which was not discussed by Ianuzzo *et al.*, and which is worth testing experimentally. According to this interpretation, thyroid hormones have an indirect influence on muscle differentiation, exerting their effects via an influence on the patterns of discharge of the corresponding motoneurones. One of the obvious manifestations of hypothyroidism is the lethargic state of the affected individual. Under these relatively static conditions postural patterns would be expected to prevail, producing an increase in the proportion of slow fibers. Conversely the hyperthyroid condition is characterized by continuous restless movement, in which phasic patterns of activity would predominate, with a corresponding decrease in slow fiber content.

To what extent is it reasonable to suggest that fiber type transformations can be produced by changes in motor patterns of activity which are nevertheless within physiological limits? Two lines of evidence illustrate the degree to which postural muscles can respond under such conditions. Firstly, changes similar to those reported by Ianuzzo *et al.*, for the hyperthyroid animals have been seen in rat soleus muscles which were partially relieved of postural activity by denervation of antagonistic muscles (Guth and Wells, 1972). Secondly, as the accompanying Fig. 1 shows, the physiological increase in postural activity brought about by exposure of the animal

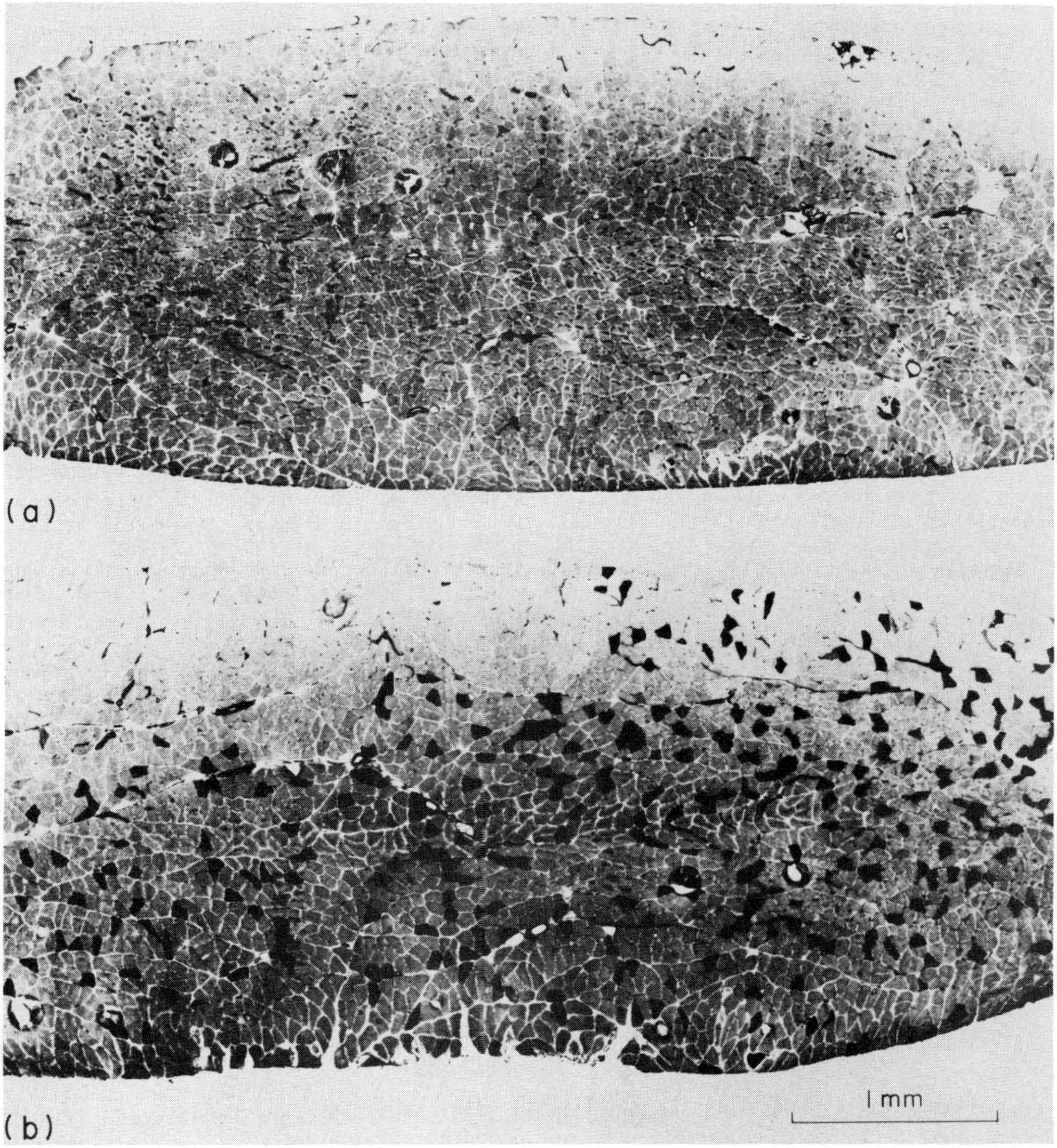

Fig. 1. Effects of increased postural activity on rat soleus muscles. In this experiment, part of a study in progress with J. Oyama of the N.A.S.A., Ames Space Research Center, rats were raised from weaning to 6 months of age in a specially designed centrifuge which subjected them to a constant gravitational field of 2.3 g. Cryostat sections from the soleus muscles of these animals (a) and those of normal gravity controls (b) were processed together for the demonstration of myofibrillar ATPase with alkali preincubation. Soleus muscles of rats raised under hypergravity conditions were composed entirely of alkali-labile fibers, unlike those of normal gravity controls which always contained a complement of alkali-stable fibers. Similar observations have been made by Martin and Romond (1975).

to hypergravity conditions produces changes even greater than those seen in the thyroidectomized animals. It is therefore possible that the thyroidal influence on muscle represents a response to a change in functional demands rather than to an excess or absence of the hormones *per se*. This proposition could be tested by analyzing EMG activity from muscles of normal, hypothyroid and hyperthyroid animals, an experiment to which biotelemetric techniques would be admirably suited.

CHANGES IN ACTIVITY DURING DEVELOPMENT

In many mammals, including the rat, rabbit, and cat, fast and slow fiber types appear in the muscles during the first few weeks after birth. The process is therefore accessible to experimental investigation. The adaptive-response hypothesis assigns a crucial role to activity in this postnatal phase of muscle differentiation, and there are two types of observation which are consistent with such a role. Firstly, muscles which in the adult animal are slow have some of the biochemical characteristics of fast muscles when examined at an early postnatal stage. In terms, for example, of the light chains of myosin the sequence of changes during early development is strongly reminiscent of that evoked in adult fast muscles by chronic low frequency stimulation (Sréter *et al.*, 1974; Margreth, 1975; Pelloni-Müller, Ermini and Jenny, 1976a, b; Syrový, 1976; Rubinstein and Kelly, 1978). Similarly, an increase in the proportion of slow fiber types has been observed to take place during the normal development of muscles in the guinea pig (Karpati and Engle, 1968), rat (Rubinstein and Kelly, 1978) and man (Keens *et al.*, 1978; Elder, 1978). These findings suggest that the emergence of slow fiber properties in the course of development is contingent on the establishment of adult patterns of motor activity in the maturing nervous system. Secondly, there is evidence that these properties can be neither established nor sustained in the absence of normal postural activity. Reduction or elimination of this activity by procedures such as spinal cord section and isolation, limb fixation or nerve section, arrests the normal process of histochemical maturation of the muscle (Karpati and Engél, 1968; Rubinstein and Kelly, 1978). Furthermore, the fiber type composition of lower limb muscles in patients paralyzed by spinal cord lesions show an abnormally low proportion of slow fibers; indeed in some cases slow fibers are completely absent from such muscles (Grimby *et al.*, 1976).

Considerable interest therefore attaches to the actual patterns of impulse activity received by limb muscles of animals during the first six weeks following birth. Because of the small size of a neonatal animal and the need to ensure that any monitoring procedure does not entail rejection by its mother, this type of experiment presents a considerable technical challenge to the bioengineer.

CONCLUSION

Three experiments have been described here in order to give substance to my contention that there is considerable unexploited potential for the application of biotelemetry to studies of the differentiation of the locomotor system. Naturally these examples do not exhaust the possibilities for research in this area, much of which could, in principle, be carried out with existing technology. Looking to the future, there are exciting prospects for studying the central nervous mechanisms underlying movement and posture, and the way in which the appropriate pathways are established. If biotelemetry is to play its part in these developments there must be a greater awareness both of the techniques and of their scientific potential. Further commitment will be needed to the erosion of traditional boundaries between the disciplines: in degree courses, in the constitution of research groups, and in the allocation of research funds. This, rather than any technical obstacle, will be the true determinant of progress in this area.

Acknowledgements — I gratefully acknowledge grant support from the Muscular Dystrophy Group of Great Britain for some of the studies referred to in this article.

REFERENCES

Amlaner C. J., Jr. (1978) Biotelemetry from free ranging animals. In B. Stonehouse, *Animal Marking: Recognition Marking of Animals in Research*, pp. 205-228. London: Macmillan.

Barnes G. R. G. and Pinder D. N. (1974) *In vivo* tendon tension and bone strain measurement and correlation. *J. Biomechanics* 7, 35-42.

Baumann J. U. and Baumgartner R. (1974) Synchronised photo-optical and EMG gait analysis with radiotelemetry. In P. A. Neukomm (ed.), *Biotelemetry II*, pp. 94-96. Basel: Karger.

Buller A. J., Eccles J. C. and Eccles R. M. (1960a) Differentiation of fast and slow muscles in the cat hindlimb. *J. Physiol.* 150, 399-416.

Buller A. J., Eccles J. C. and Eccles R. M. (1960b). Interactions between motoneurones and muscles in respect of the characteristic speeds of their responses. *J. Physiol.* 150, 417-439.

Elder G. C. B. (1978) Metabolic differentiation of human soleus muscle during development. In *Abstracts IVth International Congress on Neuromuscular Diseases*, Montreal.

Evaerts E. V. (1968) Relation of pyramidal tract activity to force exerted during voluntary movement. *J. Neurophysiol.* 31, 14-27.

Grimby A., Broberg C., Krotkiewska I. and Krotkiewski M. (1976) Muscle fiber composition in patients with traumatic corc lesion. *Scand. J. rehabil. Med.* 8, 37-42.

Guth L. and Wells J. B. (1972) Physiological and histochemical properties of the soleus muscle after denervation of its antagonists. *Exp. Neurol.* 36, 463-471.

Heilmann C. and Pette D. (1979) Molecular transformations in sarcoplasmic reticulum of fast-twitch muscle by electro-stimulation. *Eur. J. Biochem.* 93, 437-446.

Hudlická O., Brown M., Cotter M., Smith M. and Vrbová G. (1977) The effect of long-term stimulation of fast muscles on their blood flow, metabolism and ability to withstand fatigue. *Pflügers Arch.* 369, 141-149.

Ianuzzo D., Patel P., Chen V., O'Brien P. and Williams C. (1977) Thyroidal trophic influence on skeletal muscle myosin. *Nature* 270, 74-76.

Karpati G. and Engel W. K. (1968) Correlative histochemical study of skeletal muscle after suprasegmental denervation, peripheral nerve section, and skeletal fixation. *Neurol.* (Minneap). 18, 681-692.

Keens T. G., Bryan A. C., Levison H. and Ianuzzo C. D. (1978) Developmental pattern of muscle fiber types in human ventilatory muscles. *J. appl. Physiol.* 44, 909-913.

Margreth A. (1975) In W. G. Bradley, D. Gardner-Medwin and J. N. Walton (eds.), *Recent Advances in Myology*, Intl. Congress Ser. 360, pp. 307-309. Amsterdam: Excerpta Medica.

Martin W. D. and Romond E. M. (1975) Effects of chronic rotation and hypergravity on muscle fibers of soleus and plantaris muscles of the rat. *Exp. Neurol.* 49, 758-771.

Pelloni-Müller G., Ermini M. and Jenny E. (1976a) Myosin light chains of developing fast and slow rabbit skeletal muscle. *FEBS Lett.* 67, 68-74.

Pelloni-Müller G., Ermini M. and Jenny E. (1976b) Changes in myosin light and heavy chain stoichiometry during development of rabbit fast, slow and cardiac muscle. *FEBS Lett.* 70, 113-117.

Pette D., Müller W., Leisner E. and Vrbová G. (1976) Time dependent effects on contractile properties, fiber population, myosin light chains and enzymes of energy metabolism in intermittently and continuously stimulated fast twitch muscles of the rabbit. *Pflügers Arch.* 364, 103-112.

Pette D., Smith M. E., Staudte H. W. and Vrbová G. (1973) Effects of long-term electrical stimulation on some contractile and metabolic characteristics of fast rabbit muscles. *Pflügers Arch.* 338, 257-272.

Rasmussen S., Chan A. K. and Goslow G. E. (1978) The cat step cycle: electromyographic patterns for hindlimb muscles during posture and unrestrainted locomotion. *J. Morphol.* 155, 253-269.

Romanul F. C. A., Sréter F. A., Salmons S. and Gergely J. (1974) The effects of a changed pattern of activity on histochemical characteristics of muscle fibres. In A. T. Mihorat (ed.), *Exploratory Concepts in Muscular Dystrophy II*, Intl. Congress Ser. No. 333, 344-348.
Rubinstein N. A. and Kelly A. M. (1978) Myogenic and neurogenic contributions to the development of fast and slow twitch muscles in rat. *Devl. Biol.* 62, 473-485.
Rubinstein N., Mabuchi K., Pepe F., Salmons S., Gergely J. and Sréter F. A. (1978) Use of type specific antimyosins to demonstrate the transformation of individual fibres in chronically stimulated rabbit fast muscles. *J. Cell Biol.* 79, 252-261.
Salmons S. (1967) An implantable muscle stimulator. *J. Physiol.* 188, 13-14P.
Salmons S., Gale D. R. and Sréter F. A. (1978) Ultrastructural aspects of the transformation of muscle fibre type by long term stimulation: changes in Z-discs and mitochondria. *J. Anat.* 127, 17-31.
Salmons S. and Sréter F. A. (1976) Significance of impulse activity in the transformation of skeletal muscle type. *Nature* 263, 30-34.
Salmons S. and Vrbová G. (1969) The influence of activity on some contractile characteristics of mammalian fast and slow muscles. *J. Physiol.* 201, 535-549.
Sréter F. A., Elzinga M., Salmons S., Luff A. R. and Mabuchi K. (1975) The N^{τ}-methylhistidine content of myosin in stimulated and cross-reinnervated skeletal muscles of the rabbit. *FEBS Lett.* 57, 107-111.
Sréter F. A., Gergely J., Salmons S. and Romanul F. C. A. (1973) Synthesis by fast muscle of myosin light chains characteristic of slow muscle in response to long-term stimulation. *Nature* 241, 17-19.
Sréter F. A., Romanul F. C. A., Salmons S. and Gergely J. (1974) The effect of a changed pattern of activity on some biochemical characteristics of muscle. In A. T. Milhorat (ed.), *Exploratory Concepts in Muscular Dystrophy II*, Intl. Congress Ser. No. 333, 338-343.
Syrový I. (1976) The relation between ATPase activity and light chains of myosin in developing, adult and denervated muscles of several animal species. *Physiol. bohemoslov.* 25, 295-300.
Walmsley B., Hodgson J. A. and Burke R. E. (1978) Forces produced by medial gastrocnemius and soleus muscles during locomotion in freely moving cats. *J. Neurophysiol.* 41, 1203-1216.
Westbrook R. M., Fryer T. B. and Rositano S. A. (1976) A wideband EMG telemetry system. In T. B. Fryer, H. A. Miller and H. Sandler (eds.), *Biotelemetry III*, pp. 329-332. New York: Academic Press.

Temperature Sensitive Telemetry Applied to Studies of Small Mammal Activity Patterns

D. W. OSGOOD

Department of Zoology, Butler University, Indianapolis, Indiana 46208, U.S.A.

Abstract — The availability of small, temperature sensitive, radio transmitters has added a new dimension to the study of small mammal activity. Transmitter collars have been used with *Eutamias townsendii*, *Glaucomys sabrinus*, *Tamiasciurus hudsonicus*, *Sciurus niger*, and *Ochotona princeps*. These biotelemeters provide the same position and movement information that is available using ordinary radio tracking equipment. However, the addition of temperature sensitivity also makes it possible to determine whether the animal is in a nest or moving through its environment. Careful monitoring of temperature fluctuations allows inference of other activities, e.g., sleep, moving in the nest, foraging, and taking temporary shelter in some protected location. When correlated with the corresponding environmental information, these data allow new insights into the biology of small, secretive, homiotherms.

The use of a physiological telemetry transmitter in studying the behavior of nocturnal subterranean, or otherwise difficult to observe small mammals is discussed. The transmitter (Fig. 1) is a modification of the crystal controlled radio tracking transmitter described by Cochran and Lord, 1963; Osgood and Weigl, 1972. It is a simple, blocking oscillator with a pulse rate controlled by a thermistor. Heating the transmitter increases the pulse rate while cooling it decreases the rate. With careful calibration it can measure absolute temperatures to ± 0.5°C but, for this application the detection of small changes in temperature is more important than absolute accuracy. The receiver used is a modified walkie-talkie equipped with earphones, a beat-frequency oscillator (BFO), and a directional loop antenna.

Adding temperature sensitivity to a typical radio tag provides the means for obtaining considerable behavioral information (in addition to location) from its signal. If the transmitter is in the form of a collar (with the temperature sensing element outside the animal), the temperature sensed will be that of the environment but with an important contribution from the body heat of the tagged animal. When the ambient temperature (T_A) is at least 5°C lower than the subject's body temperature (T_B), the collar temperature (T_C) can provide clues to the activity of the animal.

If the subject is sleeping in a well insulated nest the T_C will be only a few degrees lower than T_B. As the animal awakens and moves within the nest, air flow

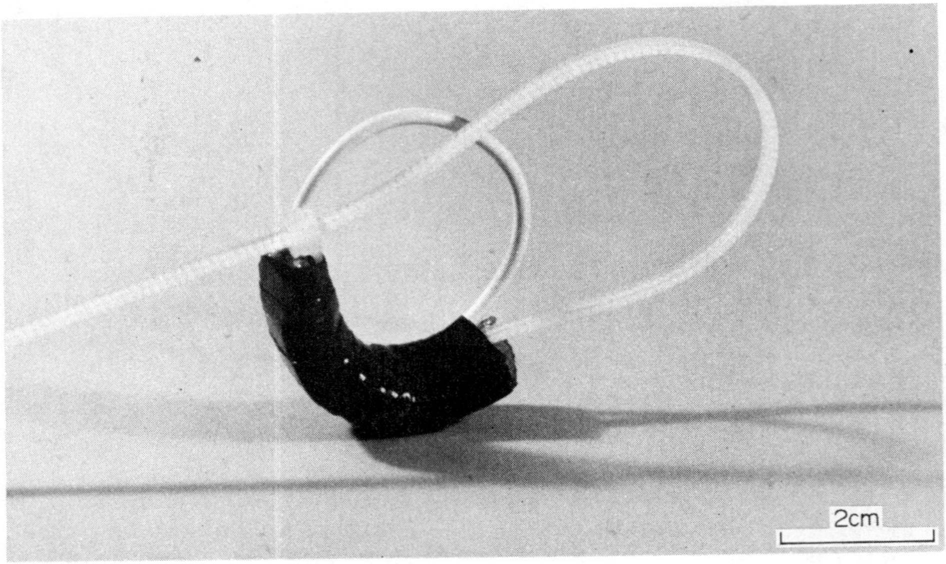

Fig. 1. The temperature sensitive collar transmitter with its cable-tie attachment device.

past the transmitter will cause the transmitter to cool by 1-3°C. Frequently the animal will resume its resting position for short periods, allowing the collar to rewarm. The result is a pattern of fluctuating, but relatively high temperature readings during the awakening phase of activity.

If the animal leaves the sheltered nest and enters cooler ambient air, activity is signaled by a sharp drop in T_C. The magnitude of this drop is largely a function of T_A, but it may also be influenced by wind or the movements of the subject. The collar temperature of the foraging animal will be more variable than that recorded from inside the nest, but if T_A remains relatively constant, variation in T_C can be interpreted behaviorally. Depending on the behavior patterns of the species, a brief increase in T_C of several degrees may indicate that the subject has stopped in the shade of a tree branch or that it has entered a nest for a period, too short to allow the transmitter to reach normal 'nest temperature'. Similarly, a brief T_C drop of several degrees may mean that the animal has passed through a windy location or that it emerged from a burrow system to forage for a short period. Clearly, it is critical to keep records of T_A in all possible microhabitats that the subject may be expected to enter. The following paragraphs provide examples of the use of this system in studies of two species with widely different habits and problems of observation.

Figure 2 is a sample collar temperature record from a female northern flying squirrel, *Glaucomys sabrinus*. Ambient temperature remained almost constant (14-15°C) for the entire observation period. There was a slight (4-8 km h^{-1}) wind throughout the night. This animal was using a well insulated nest located in a large oak tree trunk cavity. Full darkness (operationally defined as light level requiring artificial light for the biologist to reach the study area) occurred approximately 2100 h. At about this time fluctuating temperature readings from the collar indicated awakening of the subject. At 2130 h the subject left the nest for a foraging-feeding period of 160 min. During this time the animal entered protected locations on four occasions for periods of several minutes each. These temporary shelters may have been animal burrows on the ground, tree crotches, or other microhabitats sheltered from

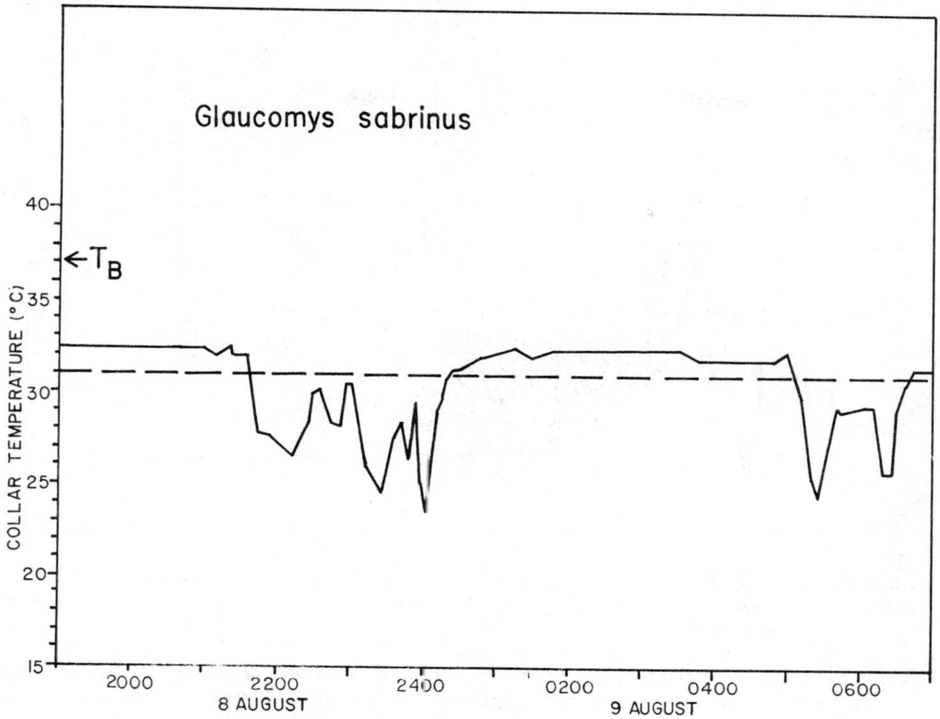

Fig. 2. Sample temperature record for a female flying squirrel. The broken line at 31°C is an arbitrary estimate of the division between collar temperatures indicative of activities occurring within the nest and those outside.

the wind. At 2410 h the animal re-entered its nest and went to sleep. The sleep period continued until 0500 h at which time the squirrel left the nest for a second feeding period of about 90 min just before daylight. During this period approximately 30 min was spent in some sheltered location. The pattern of two activity periods, one beginning just after dark and a second ending at daybreak, is typical for this species (Weigl and Osgood, 1974).

Pikas, *Ochotona princeps,* live in rock slides in alpine locations throughout western North America. In summer their behavior patterns are easily studied since they are diurnal and tolerate human observers within 50 m. Observation of pika activity during winter is much more difficult. This species lives under a continuous snow cover for up to eight months of the year yet remains active during this time. Food, in the form of 'hay', is stored during the snow-free summer months and is consumed when snow cover makes foraging difficult or impossible. The use of temperature sensitive transmitter collars has allowed me to investigate winter activity of this species.

Figure 3 is a sample collar temperature record for a pika under snow cover more than 1 m deep. The subnivean temperature remains near 0°C throughout the winter, but in a few places, channels in the snow remain open and carry air from the surface beneath the snow. In these channels air temperature is much more variable and closely matches that of the snow surface. The graph shows that, while this animal maintains a basically diurnal activity pattern, it is also active at night. Activity outside the

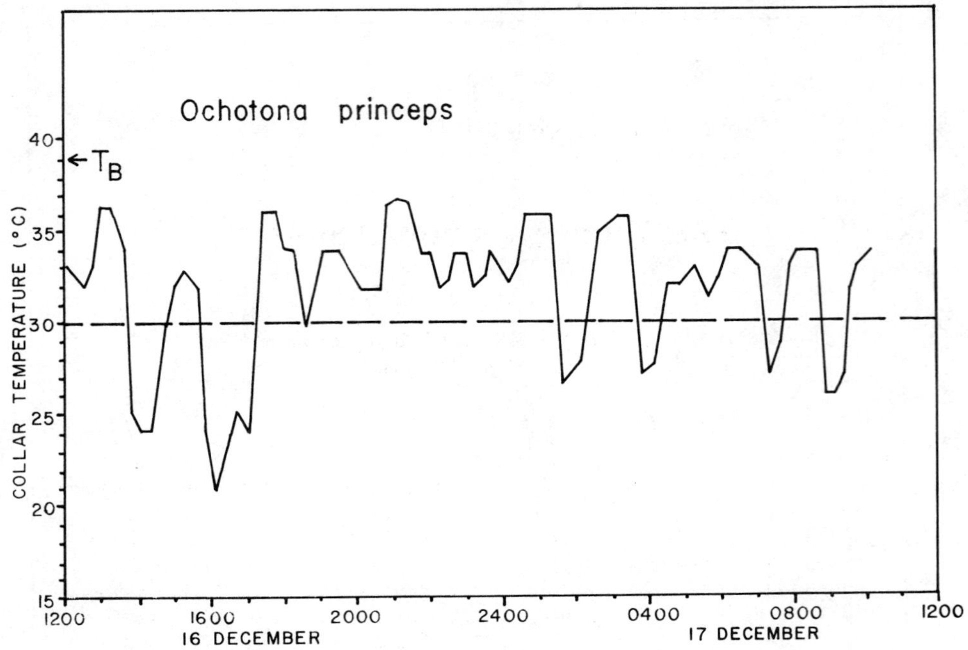

Fig. 3. Sample temperature record for a male pika. Snow depth on this date varied from 1 to 2 m. Subnivean temperature was constant at 0°C while air temperature varied from a high of 4°C to a low of -3°C over the observation period. Broken line at 31°C as in Fig. 2.

nest occurs primarily in the afternoon with shorter periods occurring at sunrise and again shortly afterwards. The two early morning excursions (0200 and 0400 h) were also of short duration. The sharp decrease in T_C at 1600 h may represent the animal moving into one of the subnivean air channels with a consequent increase in the rate of heat loss by the transmitter collar. Another noteworthy feature of this record is the variable nest temperature seen in this species. It would suggest that time spent in the nest is not solely occupied by sleeping, at least not in the 'tight ball' characteristic of rodents. One would predict from these data that pikas move extensively within their nests. The maximum T_C's observed (37°C) are within 2°C of the 39°C reported T_B for this species, indicating a very well insulated nest.

Acknowledgements — The field data reported here were collected in collaboration with Peter Weigl (*Glaucomys*) and Nancy Frazier (*Ochotona*). The pika study was supported by grants from the Shell Foundation and The Mazamas. A sabbatical leave and faculty fellowship from Butler University made the work possible.

REFERENCES

Cochran W. W. and Lord R. D., Jr. (1963) A radio-tracking system for wild animals. *J. Wildl. Mngmt.* 27, 9-24.

Osgood D. W. and Weigl P. D. (1972) Monitoring activity of small mammals by temperature telemetry. *Ecology* 53, 738-740.

Weigl P. D. and Osgood D. W. (1974) Study of the northern flying squirrel, *Glaucomys sabrinus*, by temperature telemetry. *Am. Midl. Nat.* 92, 482-486.

The Animal as a Sound Source

W. M. SCHLEIDT

Department of Zoology, University of Maryland, College Park, Maryland 20742, U.S.A.

Abstract — Many animals use specific calls as 'long distance communication signals' for the attraction of a mate or in territorial defense. Such calls are in many cases loud enough to carry beyond the caller's home range or territory, and therefore can be detected and recorded without disturbing the caller. Such sound signals have certain properties in common with the electromagnetic signals of comparable wavelength, used in radio telemetry (geometry of spread, attenuation in the natural environment), and can be measured with comparable accuracy. The advantage of 'acoustical telemetry' is that the animal under investigation remains undisturbed (no trapping or handling), and limitations of transmitters (weight, life expectancy of batteries) do not apply. The disadvantage is that the particular time of measurement is beyond the observer's control.

If, in a particular species, radio tracking is feasible, radio microphones can provide valuable information in addition to the location of the source: especially faint vocalizations used in close-contact communication can be observed continuously, and other selected activities, which are accompanied by sound (locomotion, breathing, feeding, digging, gnawing, etc.) can be recorded, yielding an intimate insight into behavior and behavioral time budgets.

INTRODUCTION

To listen to a sound and then determine the location of its source is presumably the most ancient form of telemetry. Some animals seem to impersonate living beacons, providing information on their identity and location to conspecifics, and allow us to listen in without any harmful effect to the ongoing behavioral process. Radio tracking has the disadvantage that the transmitter life is limited (mostly due to short battery life, but also because of environmental stress), and that the animal which carries the transmitter has to be incommoded to a varying degree when the unit is attached, and/or carried. In some respects acoustical telemetry and radio tracking have similar properties (e.g., the wavelength of the signal) and can yield similar results (change of location in time, change of behavioral state), and therefore a systematic comparison of these two methods is in order.

The wavelength of sound in air for the range of human hearing (20 Hz to 20 KHz) lies between 15 m and 15 mm, and is quite close to the radio frequency wavelengths currently used in telemetry. I am not aware of a systematic comparison of the propagation of sound and radio signals in the natural environment of animals, but it appears that the basic geometry of the field (at least the far field) of a point source is quite similar, and that, assuming spherical spread, the signal will decrease by 6 dB when doubling the distance from the source. Obstacles in the path of a sound or electromagnetic wave reflect depending on their size, relative to the wavelength; i.e., an isolated object will produce a shadow if it is several wavelengths wide, but creates no disturbance when it is much smaller than one wavelength. The electrical impedance of a surface will be different from the acoustical impedance, in the numerical value at least, and the absorption and diffraction in the medium is noticably different especially in the short wavelength end of the common spectrum (for a detailed discussion see Piercy, Embleton and Sutherland, 1977; Chessell, 1977). As long as transmitter and receiver are close to the ground, detection seems to be a problem for both forms of energy. Sound waves are bent when passing through gradients of temperature or wind, and reflected (or partially absorbed) by the ground, or by other planes (e.g., the surface of the edge of a forest, or, within the forest, by the canopy above). Within the range of roughly 500 Hz to 5 KHz (0.6 to 0.06 m) the effects of geometry of propagation, and absorption and diffraction are comparable over distances up to several hundred meters. for both forms of energy. Sound sources can be localized more precisely with currently available technology, when arrival time differences are used (see below), due to the relatively slow speed of sound in air (334 ms^{-1} at 18°C).

The amount of energy which is radiated in the form of sound in a typical long distance communication signal is considerable: at a distance of 1 m from a singing bird, sound pressure levels of approximately 100 dB (re 2.10^{-5} Nm^{-2}) can be measured. This corresponds to 130 mW at the source. To what extent this energy can be utilized for acoustical telemetry depends on a variety of variables of the signal itself (especially its band width), and of the environment. Ambient noise level and excess attenuation in a particular frequency band (which are both dependent on various meteorological conditions and on the sound production of other animals), vary with the time of day and the season. During prevailing environmental conditions distances up to 100 m can be handled easily, and at optimal conditions, measurements of up to 1 km are possible.

THE SOUND SIGNAL

Sound signals have developed under a variety of selection pressures, and it is next to impossible to abstract any common features, even if we restrict our considerations to 'typical' long distance communication signals, with the same precision applied in the technical specifications of a radio transmitter. In the absence of a comprehensive study of the physical variables of long distance communication signals, I take the liberty to compile here a few rough estimates, which should facilitate a comparison of the kind of radio signals which are currently used in telemetry.

In the acoustical signals of animals the most striking feature is the absence of a continuously present 'carrier', and the use of intermittent signaling in a form which is very similar to 'pulse interval modulation' (Schleidt, 1973). Most calls fit within the range of 0.3 to 3 s, and the intervening intervals range from seconds to minutes. The frequency is rarely maintained at a fixed value, but most commonly modulated in some form, while amplitude modulation is virtually absent. Repetitions and more or less complex alternations of elements are common. The frequency band which one individual uses is relatively wide (one or more octaves) and, since the available frequency range is limited, the selectivity is achieved by a characteristic 'signature' of the signal, which encodes most strikingly the species of the caller, but in many cases also the individual identity (Schleidt, 1976). Unlike in radio communication, where each transmitter works continuously in a particular frequency band, animals use a relatively wide band communally, in a mode similar to 'time sharing'.

In radio telemetry, the time at which a particular measurement is made is well under the control of the investigator, while in the case of acoustical telemetry the listening observer is at the mercy of his subject. At first glance, this seems to be a serious drawback of the latter method though the situation is not as bad, when we consider that the time when the animal under investigation emits a signal is a potentially significant data point. If we are mostly interested in locating the animal, in the context of space utilization, we already know that commonly a long distance signal is repeated several times before it moves to a new location. The review of the ornithological literature indicates that in many species the singing territorial male has a strong preference to call from a limited number of singing perches, but quantitative data to support these claims are rare and further studies are urgently needed.

The potential for the identification of individual callers with the help of unique characteristics of the signal are excellent, provided that a sound-spectrographic analysis is possible. In my laboratory, a considerable number of long distance communication signals of a variety of species were analyzed, and in all cases unique features have been found in sufficient numbers to identify several (5 to 20) individuals within a given population. The only difficulty at present is that the important features cannot be detected readily by the unaided ear, and a sound spectrograph for field application is not yet available.

The kind of information carried in a sound, and the form in which it is endoded appears too variable to be discussed in general terms. Nearly always it conveys 'I am alive and well'. In many species, calling is observed predominantly in one sex (mostly the male), and only in sexually mature (or active) individuals, which limits the use of acoustical telemetry. At the same time this can be used to the advantage of the investigator, because it excludes at least the immature and senile individuals, which do not contribute to the productivity of the population. In some bird species only those males sing consistently which are not (yet) mated. Since in these cases usually the males arrive first at a territory, and the females follow later, acoustical telemetry can be used to monitor the annual re-settling of an area, and the stability of the population.

THE SOUND RECORDING

In radio tracking, either one individual transmitter is tracked with a single, portable and directional antenna, or several transmitters are monitored by two or more stationary, rotating, and also directional antennas by means of triangulation. In acoustical telemetry, the first method (individual tracking) is also possible, but not recommended because of the potential disturbance: in the worst case, the animal is chased around by the investigator, and the resulting data are meaningless. An acoustical adaptation of the second method was tested by the author in a pilot study, with highly directional 'shotgun microphones' as indicators, or with microphone pairs mounted at an exaggerated 'interaural' distance, but neither system gave satisfactory results. The relatively short duration of the acoustical signals, and the long silence between make it very difficult to obtain unambiguous readings of the sources location.

For acoustical telemetry, the most reliable and precise monitoring of sound sources is possible with microphone arrays. This method takes advantage of the relatively slow speed of sound in air, which results in significant differences in the time at which a signal from a distant source arrives at each microphone. The basic principle is best explained when we first consider the case of two microphones only: the wave front from a distant sound source will arrive first at the microphone which is closer to the source (M1), and at the second microphone (M2) with a delay which is a function of microphone configuration and the speed of sound. If the signal arrives simultaneously at both M1 and M2, we know that the source is somewhere on a line exactly between the two microphones; if the arrival time difference between M1 and M2 is as long as the sound needs to travel from M1 to M2, we know that the source is somewhere on a line cutting through M1 and M2, but behind M1; if the arrival time difference is shorter, but not zero, we know that the source is somewhere on a hyperbola which has M1 and M2 as its focal points, and which satisfies the observed difference in the length of the path (which can be calculated from the speed of sound). With two microphones (or two ears) we can only determine the approximate direction, approximate because we do not know the distance. In order to estimate the actual location of the source we need at least a third microphone (M3), which allows us to measure two additional arrival time differences, and to construct two more hyperbolas. The resulting three hyperbolas intersect at several locations and if only one location exists at which all three intersect, we conclude that this is the location of the source. If more than one threefold intersection is found, a fourth microphone is necessary to resolve this ambiguity.

In the first applications of this method (finding the location of the enemy's cannons during WW I) simple paper chart recorders were used in the field, and the arrival time differences were converted, with the help of tables, to locations relative to the microphone array. Today, a portable storage oscilloscope and an electronic hand calculator are the logical and practical alternatives. With relatively faint signals, imbedded in the ambient noise, direct field measurements do not provide sufficient resolution, and a better method is to record the signals on a 4 channel magnetic tape recorder, and conduct the analysis (arrival time difference measurements, and conversion to locations) in the laboratory, where sonographic analysis allows maximal recovery of the signal, and suitable computer programs can convert the arrival time differences to locations (Magyar, Schleidt and Miller, 1978); an underwater system has been described by Watkins, 1976.

There are several technical problems remaining, which hamper a general application of this method. (1) Only a few portable 4 channel tape recorders are commercially available, and they are all rather expensive. (2) The microphones which are readily available have to be modified with a more effective wind screen, and to a toroidal directional response (maximize sensitivity in the horizontal plane, minimize sensitivity downward and upward, to exclude excessive ambient noise from nearby insects, and from aircrafts overhead). Until such recorders and microphones are available, simplified versions can be used with only minor sacrificing of accuracy. Microphones with a cardioid characteristic can be mounted vertically, and covered by a fiberglass umbrella (which excludes to some extent aircraft noise) and which can support additional wind screen material. If the animals are in a cluster (e.g., frogs at a pond), only three microphones are sufficient to obtain unambiguous hyperbola intersections. Provided that the individual signal is of sufficient duration, and/or repeated several times, during which the animal does not move significantly, a stereo tape recorder can be used. One microphone is connected permanently to one channel, and the other two microphones are alternately recorded on the second channel. The switching can be done by hand, e.g., at the midpoint of the signal, monitored through headphones on channel 1, or by means of an electronic multiplexing system.

RADIO MICROPHONES

The preceding discussion on acoustical telemetry is in no way intended to undercut the use of radio telemetry, but mostly to review potential alternatives. Provided that favorable financial conditions prevail, and the particular species under investigation is cooperative, various combinations of the two methods could be most beneficial. In view of the great advances in the miniaturization of circuits, more complex transmitters can be built, and recent developments in the production of hearing aids, especially of miniaturized condenser microphones of very small weight (less than 2 g) make it possible to build miniature radio microphones. Until recently, an inexpensive radio microphone, tunable within the FM range, was available on the commercial market, but the general license has been revoked, and the sale discontinued. We have converted such units and incorporated into a collar weighing 25 g (including batteries). This unit was readily accepted by domestic cats and allowed us to record vocalizations in an unobstructed environment over distances up to 65 m with a commercial FM radio-cassette recorder in the range of 500 Hz to 15 KHz (McKinley, Dowell and Schleidt, 1976).

The use of radio microphones can resolve several problems which are encountered when direct observation or other recording methods are unavailable. First of all, even very faint vocalizations, used in close-contact communication, are easily recorded. The alternatives, bringing a conventional microphone close to the animal results in excessive disturbance, and the use of highly directional microphones from a distance (e.g., parabolic reflectors) can easily result in distortions of the signal. In social encounters it is usually impossible to judge from a distance, which animal gives which vocalization, an ambiguity which the radio microphone can easily resolve. Because of the fixed distance between microphone and mouth the relative loudness of different vocalizations of one animal can be measured with great accuracy, and this constant distance is also advantageous, when the vocal output of an animal is to be monitored over an extended time period.

REFERENCES

Chessell C. I. (1977) Propagation of noise along a finite impedance boundary. *J. Acous. Soc. Amer.* 62, 825-834.

Magyar I., Schleidt W. M. and Miller B. (1978) Localization of sound producing animals using the arrival time difference of their signals at an array of microphones. *Experientia* 34, 676-677.

McKinley P., Dowell B. and Schleidt W. M. (1976) Wireless microphone for studies of animal vocalizations. *Experientia* 32, 132-133.

Piercy J. E., Embleton T. F. W. and Sutherland L. C. (1977) Review of noise propagation in the atmosphere. *J. Acous. Soc. Amer.* 61, 1403-1418.

Schleidt W. M. (1973) Tonic communication: continual effects of discrete signs in animal communication systems. *J. theoret. Biol.* 42, 359-386.

Schleidt W. M. (1976) On individuality: the constituents of distinctiveness. In P. H. Klopfer and P. P. G. Bateson (eds.), *Perspectives of Ethology*, Vol. 2, pp. 299-310.

Watkins W. A. (1976) Biological sound-source location by computer analysis of underwater array data. *Deep Sea Res.* 23, 175-180.

Biotelemetry of the Vocalizations of a Group of Monkeys

J. P. GAUTIER

C.N.R.S., Station Biologique de Paimpont, 35380 Plélan le Grande, France

Abstract — Five monkeys belonging to the same family were equipped with miniature VHF, FM transmitters. The calls of each monkey were transmitted to five FM receivers and recorded on a multitrack tape recorder. The animals released in their usual enclosure were recorded during different activities: awakening, falling asleep, resting, feeding, alarming, etc. I present (a) the technical apparatus and its performance; (b) nature of different sounds recorded; (c) qualitative and quantitative vocal expression of one animal; and (d) study of vocal exchanges between several individuals.

INTRODUCTION

A social group of monkeys is composed of different age, sex and status classes of individuals. In the same context the behavior of each class differs from that of the others. The same is true for vocal behavior. My study has two aims: (i) to learn the nature, number and chronology of vocal events which characterize a particular class of animal; (ii) to understand the interactions between the vocal patterns emitted by different individuals during exchanges. Consequently I must identify and isolate the vocal participation of each individual. That is easy where only two or three animals are concerned but becomes impossible when the group is larger. Biotelemetry has been used to solve this problem (Gautier and Deputte, 1975; Catherine, 1977). Here I present the essential technical apparatus and its potential to study vocal behavior. The details and results concerning this latter part will be published elsewhere.

Five monkeys belonging to the same family* (one adult female and her offspring) were equipped with miniature VHF, FM transmitters (Fig. 1). Their calls were transmitted to five FM receivers and recorded separately on a multitrack tape recorder. The male of the group was not radio tagged; his rare and loud vocalizations were recorded on

*The family observed is composed of one *Cercopithecus pogonias grayi* adult male, one *Cercopithecus ascanius katangae* adult female. They gave birth to four hybrids in 1974. (1) 'Valérie', adult female, five years old; (2) 'François', subadult male, four years old; (3) 'Anasthase', juvenile male, three years old; (4) 'Danièle, young female, two years old. This group was chosen so that the hybrid nature of the vocalizations of the young could be studied also.

Fig. 1. 3 animals of the group equipped with their harnesses — the male (second from the left) is not equipped.

the general sound track. The animals released in their usual enclosure were recorded continuously, twice for two weeks during their various activities: awakening, falling asleep, resting, feeding, playing, alarming, etc. Video recordings were made simultaneously to observe any behavioral context (spatial relationships, postures, mimics) within which the vocalizations took place. Approximately 20 h of observations have been recorded.

BIOTELEMETRY SYSTEM — DESCRIPTION

The system is composed of four parts. The three principal ones are: transmitter, receiver and magnetic tape recorder. The fourth element, a graphic recorder, was only used to display the data.

TRANSMITTER

The transmitter apparatus is carried by the animals. It is composed of a small microphone, an oscillator and a battery. These three components are assembled in an adjustable leather harness.

The circuit diagram of the transmitter is shown in Fig. 2. It is a VHF FM (varicap) oscillator built to emit from 90 to 130 MHz. The frequency of oscillation is tuned with an adjustable capacitor 'C7'. The different components of the oscillator are assembled between two circular printed circuit boards (ϕ = 15 mm, length = 21 mm, weight $\leq$ 3 g).

The microphone (Lem, EM 38 B) is a miniature capacitor type (3 × 6 × 8 mm) which possesses an internal preamplifier which modulates the frequency of the carrier. The microphone can be connected to the coil and varicaps either, through a single

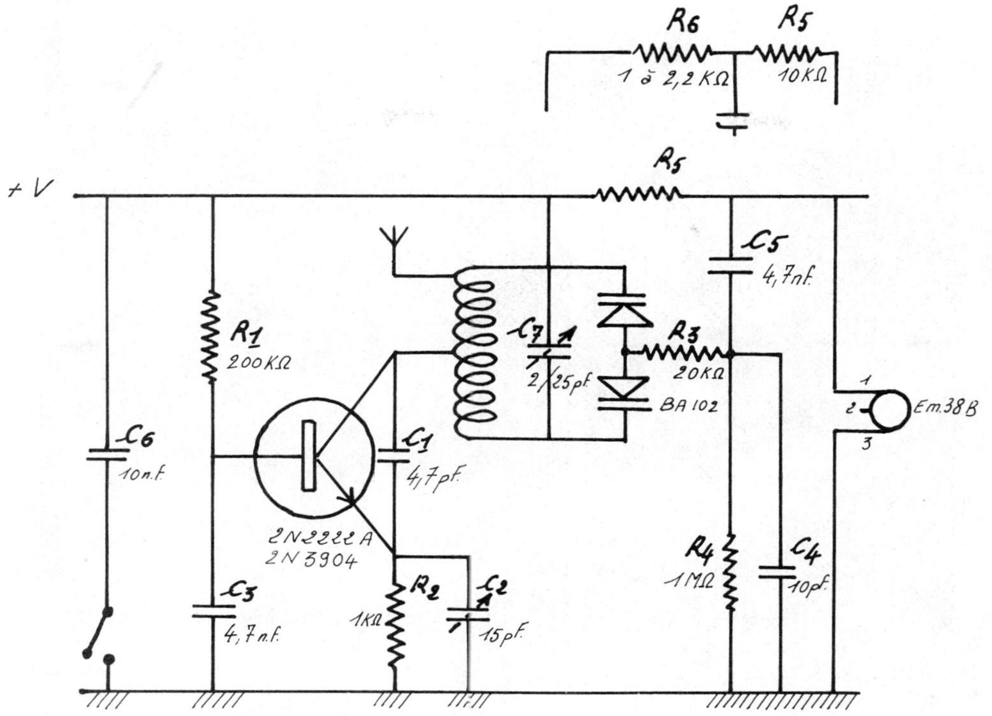

Fig. 2. Circuit diagram of the oscillator.

resistance 'R5' or a bridge of resistances 'R5/R6'. The second system constitutes an adjustable attenuator of the audio frequency (AF) signal. So the sensitivity of the microphone decreases but the amplitude of the frequency modulation of carrier is reduced and the receivers can keep in tune.

The microphone is fixed onto the middle of a rubber collar adjusted round the neck of each animal and is positioned over the larynx area. The rubber collar is protected by a leather one fixed with straps over the shoulders (see Fig. 3).

The Mallory-Lithium battery (LO 32) has a capacity of 1100 mAh and gives a nominal voltage of 3 V. This is the biggest element of the telemetry system (ϕ = 16 mm, length = 35 mm, weight = 12 g) and it is connected to the oscillator by a miniature switch in a plastic polyvinyl chloride (PVC) container (ϕ = 20 mm, length = 65 mm). This is rivetted onto one of the straps on the back position of the animal (see Fig. 3). The two wires of the microphone pass through a hole made in the PVC container and run inside one of the straps. The antenna runs inside the other strap.

The harness is composed of (i) the rubber and leather collars; (ii) the two straps made of two strips of leather sewn together (width 13 mm). One of them passes over the right shoulder and under the left arm-pit. The other follows a symmetrical path. They cross on the chest and on the back of the animal and can be adjusted by two small buckles. The weight range of the harness runs from 32 g to 42 g according to the animal's weight (1200 to 4000 g), i.e., between 2.7 to 1 percent of the monkey's weight.

With a 3 V battery, the current consumption of the five transmitters (oscillator + microphone) ranges from 760 to 1000 µA. The corresponding power output is 2.28 to

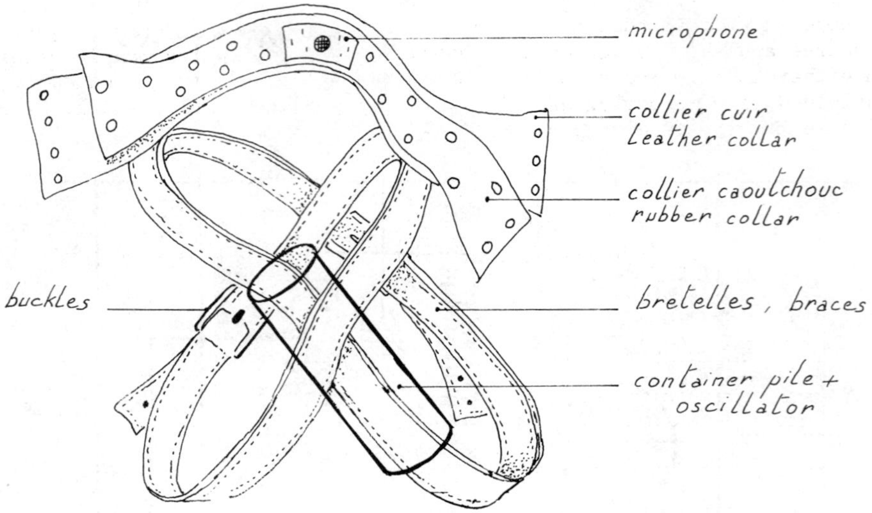

Fig. 3. Diagram of a harness.

3 mW and transmission range is approximately 50 m. Their lifetime varies from 45 to 60 days. The only problem with these simple transmitters is the variation of the carrier with temperature. The frequency increases when the temperature decreases and vice-versa.

RECEIVING/RECORDING SYSTEM

Five animals were radio tagged each with a transmitter of a separate carrier frequency. The five frequencies were individually received on five FM tuners. Each of the five AF signals was recorded on separate tracks of a multitrack tape recorder. Figure 4 shows the relationship between receivers and tape recorder.

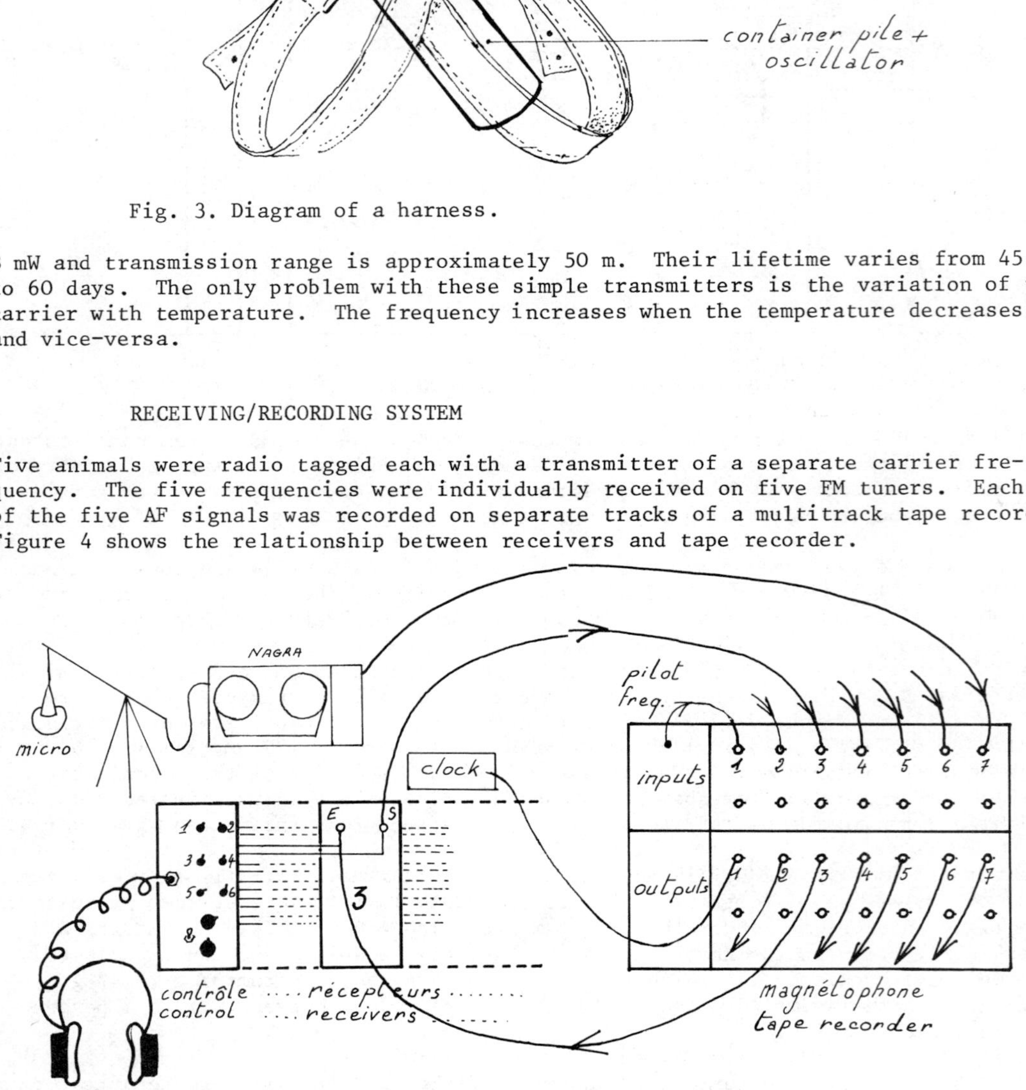

Fig. 4. Diagram of receiving/recording system.

Receiving System

It is composed of 6 FM tuners assembled together in a rack. They possess a common power source and share an antenna preamplifier. A monitoring module allows us to listen either to only one signal or to all the AF signals at the same time. That is possible during recording as well as during playback. Figure 5 gives the general organization of the receiving system.

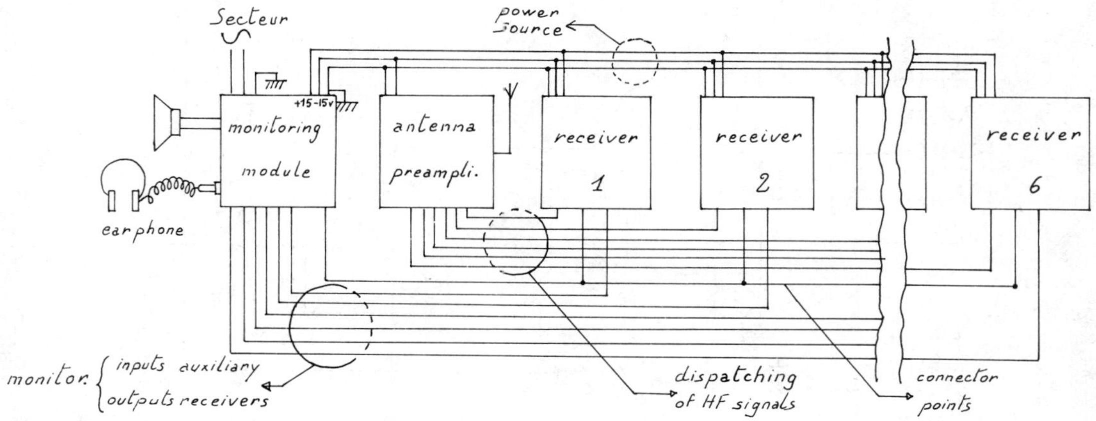

Fig. 5. Diagram of receiving system.

The commercial FM tuners (Görler) used for reception have been modified: (a) Their band of tuned frequencies has been raised from 88/108 to 96/126 MHz. The local FM stations are thus eliminated. (b) The automatic gain control is replaced by an adjustable AF amplfier. (c) The main modification concerns the automatic frequency control (AFC). The carrier of the oscillators is not very stable but the receiver must be able to follow it. An AFC with an adjustable high gain can follow a ± 1 MHz deviation around the carrier. Figure 6 shows these modifications.

The close proximity of the six tuners in the same rack can cause some disturbance phenomena. One receiver may become tuned onto the local oscillator of another one. To avoid this the total range of the 5/6 carriers of the transmitters must not exceed 10.7 MHz, the value of the intermediary frequency.

Magnetic Recorder

A multitrack tape recorder (Schlumberger MT 5530) was used at low speed (4.75 cm s^{-1}). Seven tracks were equipped with recording and playback amplifiers (direct). A pilot frequency (3.125 KHz) was recorded on the first track and feeds an electronic time keeper. The animals were recorded individually on tracks 2 to 6. The seventh was used for the male and surrounding sounds (see Fig. 4). The monitoring module of the receiving system allows control over the AF signals recorded on the different tracks. Their outputs were linked with the corresponding inputs of each receiving module.

DISPLAY SYSTEM

A four track graphic recorder (EEG, Reega IV, ALVAR) was used to display the data recorded on the magnetic tape. The graphic records reveal the amplitude modulations of the sounds which tells precisely: (i) the time distribution of vocal events of

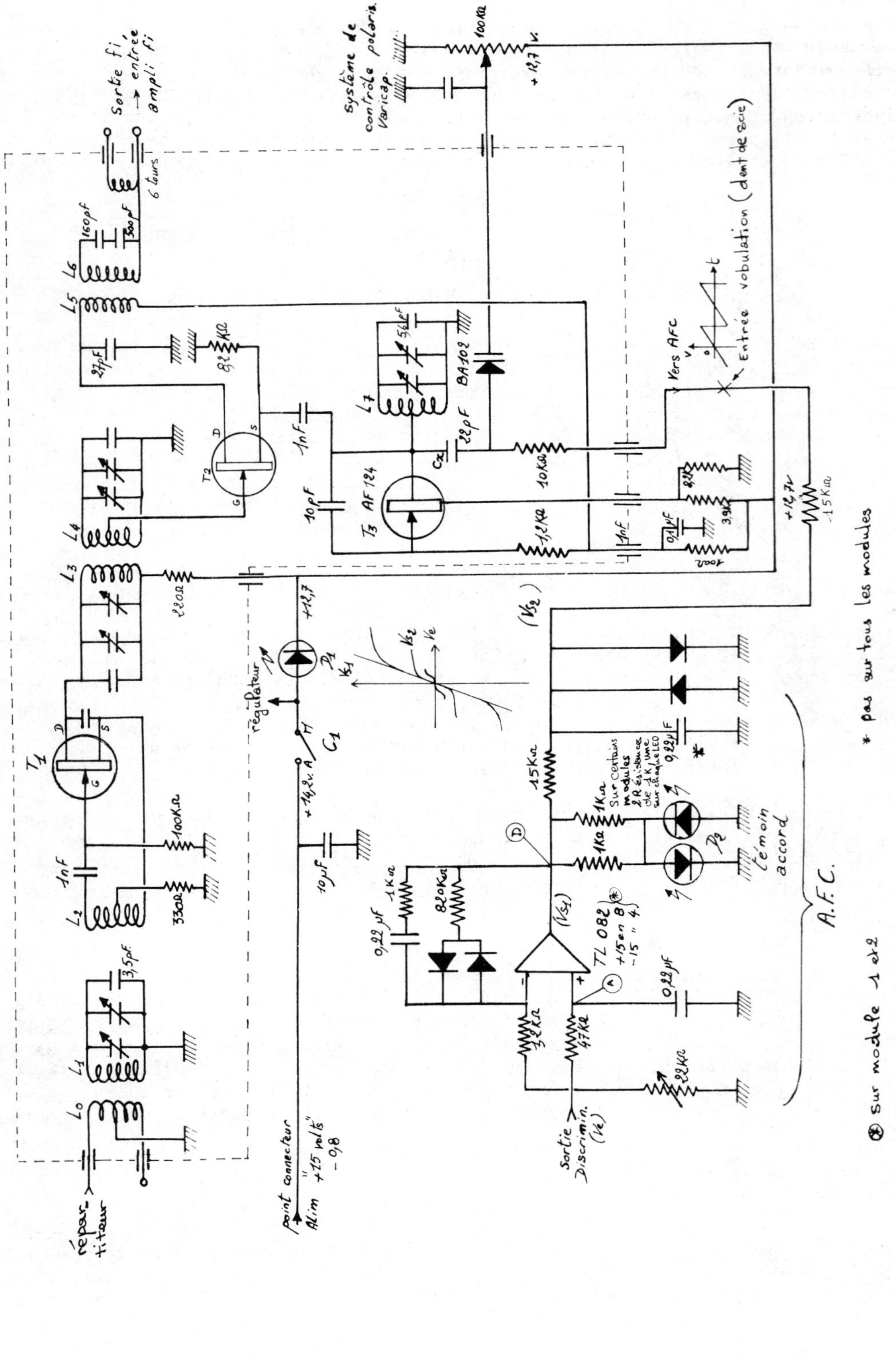

Fig. 6. Circuit diagram of the modified tuner and AFC.

each animal; (ii) the delay between the vocalizations of different animals in a vocal exchange (see Fig. 8).

The outputs of the magnetic tape recorder are linked to the inputs of the graphic one through the input/output of the receiving system. That is only used to control and modify the level of the AF signals. The graphic displays will be more or less recognizable according to the loudness of the sounds and/or the rhythmicity of their amplitude modulation. In general, graphic recording must be accompanied by simultaneous listening to analyze track by track the nature of the translated sounds.

RESULTS

NATURE AND GRAPHIC IDENTIFICATION OF TRANSMITTED SOUNDS

The position of the microphone over the larynx not only enables the transmission of calls but also of many other sounds, such as mastication, swallowing, and handling of the harness. Scratching behavior is easily recognized by its rhythm. Weaker sounds such as heart beats and breathing can be transmitted and recorded. When the young animals are sleeping, the head bends down onto the chest and presses the microphone onto the trachea. Breathing becomes sonorous and the close proximity of the carotids allows the detection of the heart beating. These phenomena have thus been regularly observed during the first experiment (microphone with weak attenuator). The adults do not transmit these parameters.

All the animals have been recorded during different activities. To characterize the vocal pattern of each animal, it is possible to listen to each track separately and compute the chronological succession of the different sounds. This is possible only when the number of calls is small. But when the frequency of vocalizations increases or when vocal exchanges are studied the graphic recorder must be used. However, operator identification of the sounds must be made while making the graphic records.

Graphic Records of the Nonvocal Sounds

Figure 7 shows different types of graphic records from nonvocal sounds. (i) *Heart rate* (Fig. 7A, B, C) can be calculated from collar or chest records. The latter are better and similar to a ECG taken on the same animal at the same time. (ii) *Breathing* (Fig. 7D) can be displayed but the recording conditions are too special to be of general use. (iii) *Scratching* is more apparent than mastication (Fig. 7E, F).

GRAPHIC RECORDS OF THE VOCALIZATIONS

Figure 8 compares the spectrographic analysis (a to e) with an actual AF sound from the graphic records of four types of calls (Gautier, 1975). Type 1 alarm call (Fig. 8, a/a') is a loud, sharp, brief and high pitched call. Its graphical representation is very bad and easily confused with an artifact. On the contrary aggressive type 4 call (Fig. 8, e/e') is easily identified by its rhythmic structure. The same is true of translations of the weak, low pitched, quavering type 2 cohesion call (Fig. 8, b/b') in comparison to those of the stronger, high pitched quavering type 6 contact calls (Fig. 8, c/c'; d/d').

The graphics can be 'cleaned up' by increasing the chart speed. However one is then quickly submerged by kilometers of paper. A compromise between good resolution of the sounds and a reasonable length of paper must be chosen by an experienced observer.

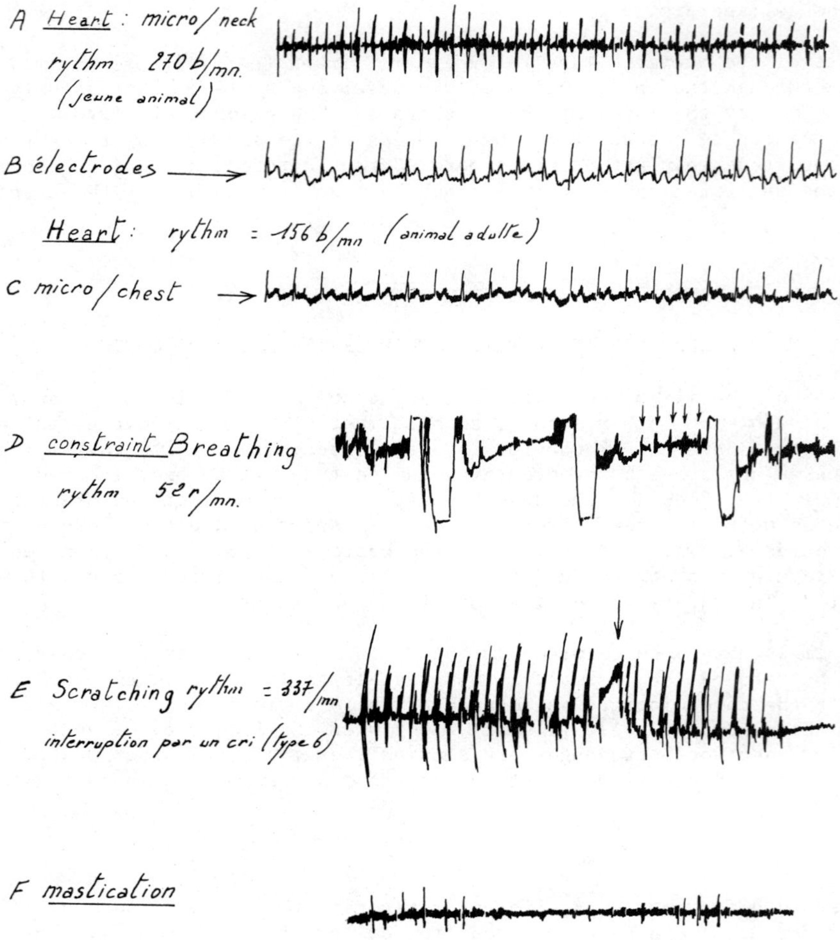

Fig. 7. Graphic records of the non-vocal sounds. A,C — heart beats transmitted respectively from a collar and chest microphone. B — Comparison with ECG record. D — Constraint breathing. E — Scratching display interrupted by a call (little arrow). F — Mastication display.

ANALYSIS OF VOCAL ACTIVITIES

The graphic records enable us to get a rigorous chronology of the sounds of each individual as well as interactions between animals.

Individual Vocal Activity: Vocograms

We already know that the succession of calls of one animal, without any stimulation from its conspecifics is not a random phenomenon. Cohesion and contact calls (type 2 and 6), for example, are clearly linked. On the other hand the comparison between the vocal patterns of different individuals allows us to establish relationships between vocal activity and age, sex or social status of the vocalizing animals.

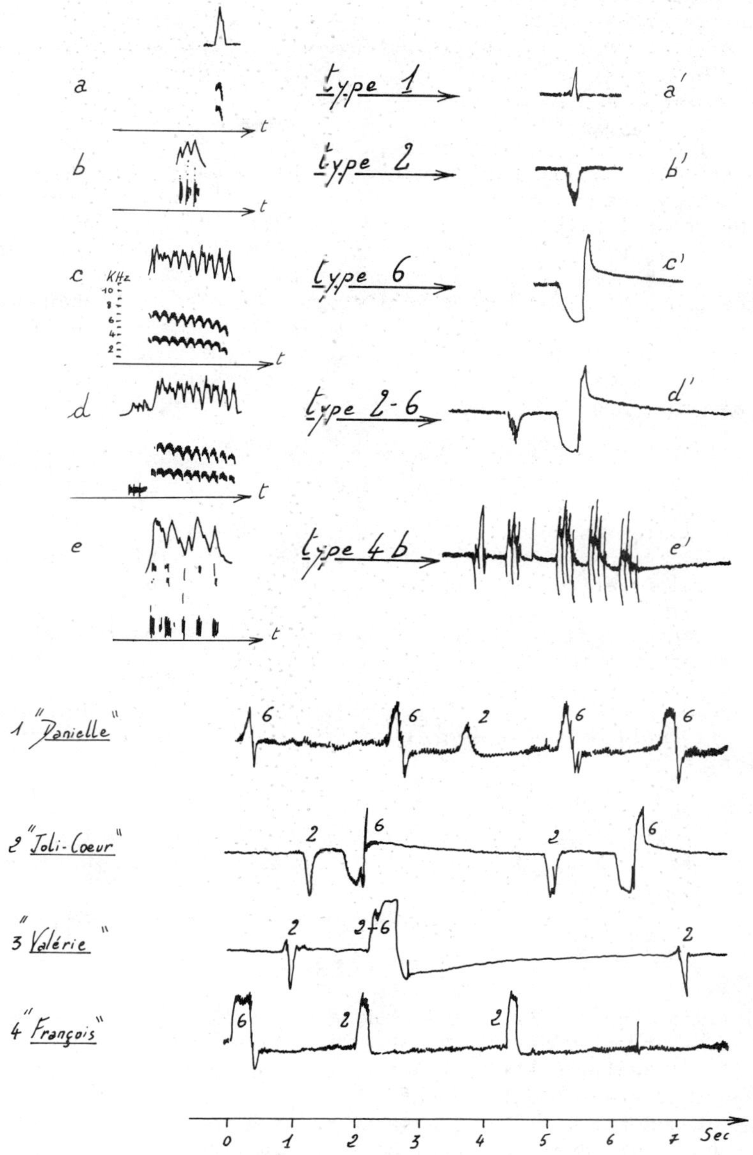

Fig. 8. Graphic records of the vocalizations. a to e: spectographic analysis of four type calls (Kay sonagraph). a' to e': amplitude modulation spectra of the same calls given by the graphic recorder. 1 to 4: Synchronous graphic displays of 4 animals in a vocal exchange. The numerals on the curves correspond to the different types of vocalizations.

In a given vocal sequence the nature and the number of sound types transmitted can be obtained by listening to the tape while temporal characteristics are given by the graphic display. Each type of call within the species repertoire will be transmitted with a characteristic frequency according to the given sequence and stimulating context. The vocal participation of an animal can thus be characterized by a 'vocogram' in which are presented the nature, number and mean frequency of the vocalized sounds and also their frequency distribution.

Inter Individual Vocal Activity: Vocal Exchanges

We already know that the sound of a call in a group of monkeys frequently releases a vocal response. It can be either immediate (phonoresponse) or delayed. So the 'vocogram' of each animal is continuously modified by those of other individuals. Consequently we generally find the calls of the different individuals, associated in a vocal exchange, of limited duration (see Fig. 8, 1 to 4). The measure of delay time between the successive calls must allow clarification of the phonoresponse concept and therefore allows chronological organization of each sequence of vocal exchanges. The comparison with an adequate sample of that series will give the probability of vocalization of one call according to the nature of preceding transmissions given by identified animals. These data could be subjected to stochastic Markovian process (see Altmann, 1965).

CONCLUSION

The biotelemetry system described allows accurate recording of all the vocal interactions in a social group of monkeys. It now becomes possible to begin a quantitative study of vocal inter communication processes which lead to causal problems.

The recording of scratching or mastication and swallowing enables us to monitor a few of the behaviors manifested by hidden animals in the field. With the help of proper transducers we could record and correlate locomotor and feeding activities according to age, sex and status in the same species as well as between different species. Thus it could be possible to quantify the cost/benefit of different adaptive strategies.

Acknowledgements — I particularly wish to thank L. Macé, technician at the CNET at Lannion, as well as J. Leray, J. Guerlais, B. Pommier from the IUT at Rennes for their technical assistance.

REFERENCES

Altmann S. A. (1965) Sociobiology of Rhesus monkeys. II Stochastics of social communication. *J. theoret. Biol.* 8, 490-522.

Catherine A. (1977) Réalisation d'une chaîne de Biotélémétrie (ensemble de réception) — Rapport de stage I.U.T. Rennes.

Gautier J. P. et Deputte B. (1975) Mise au point d'une méthode télémétrique d'enregistrement individuel des vocalisations: application à *Cercocebus albigena*. *La Terre et la Vie* 29, 298-306.

Gautier J. P. (1975) Etude comparée des systèmes d'intercommunication sonore chez quelques Cercopithecinés forestiers africains; mise en évidence de corrélations phylogénétiques et sociopsychologiques. Thèse Doct. Etat, Université de Rennes.

Feeding Dairy Cows by Using Radio Telemetry

K. A. COLLIS

Agricultural Research Council, Institute for Research on Animal Diseases, Compton, Newbury, Berkshire RG16 0NN, U.K.

Abstract – The Alfa Laval automatic feeder is designed to control the daily intake of concentrate by individual cows in a herd. It consists of a stall and manger, above which is a feed dispenser with a storage hopper. Signals from a collar worn around the cow's neck activate the dispenser to supply concentrate at a predetermined constant rate. The total feeding time ($min^{-24}h$) for each cow can be adjusted on a variable dial on the collar. Cows quickly adapted to using the feeders. Individuals made approximately the same number of visits to the feeder each day, but among cows the number of visits per day ranged from 4 to 46. Younger cows visited the feeder more often than older cows ($P < 0.01$). and initially wasted more time in the feeders when no feed was being dispensed ($P < 0.01$). There was a good correlation ($r = 0.96$) between the desired and actual concentrate intake of the cows.

INTRODUCTION

In order to achieve maximum production of milk in commercial dairy herds it is preferable to allow cows 24 h access to feed. However, some degree of control of intake is often necessary to prevent over-eating which can lead to digestive imbalance and economic waste. The Alfa Laval feeder was designed to control concentrate intake by individual cows.

MATERIALS AND METHODS

The feeder consists of a stall with a manger, and above this a feed dispenser with a storage hopper (Fig. 1). Signals from a collar worn around the cow's neck activate the dispenser to supply concentrate at a predetermined constant rate (Fig. 2).

The collar consists of a wire loop, which acts as an antenna, suspending a plastic case which encloses an energy cell. This energy cell discharges at a constant rate while the cow is not using the feeder. When a cow enters the feeding stall the collar receives a high frequency radio signal from the feeder which recharges the energy cell and drives a signal generator. Once fully charged the collar ceases to emit signals and no more feed is dispensed. The total feeding time ($min^{-24}h$) for each cow can be adjusted on a variable dial on the collar.

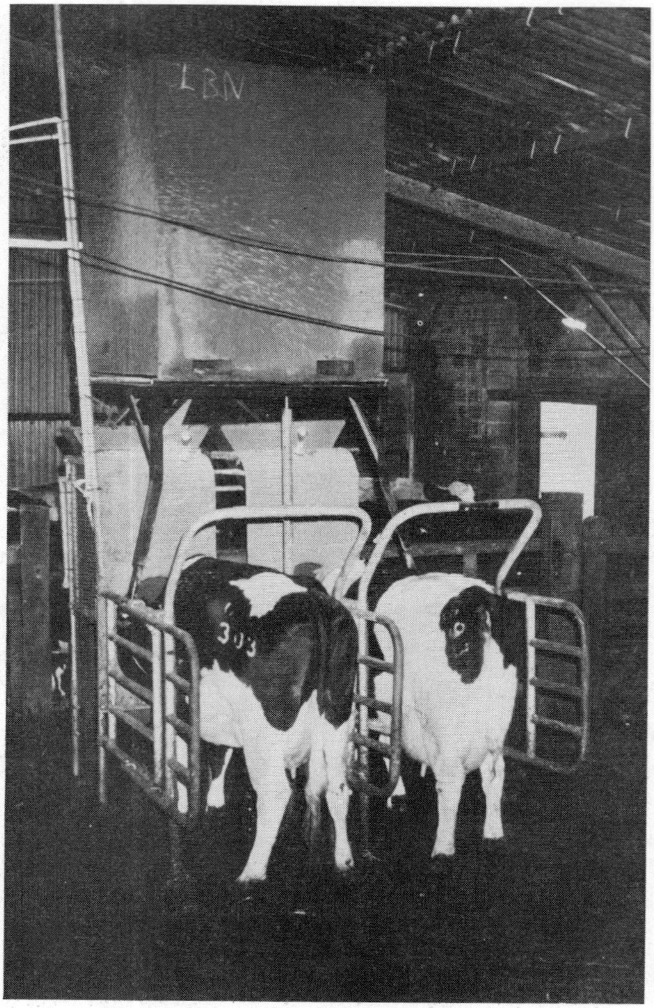

Fig. 1. The two Alfa Laval feeders, with cows in the normal feeding position.

Thirty-two cows which had not previously used the feeders were allowed access to two of them, situated at one end of their loose housing yard. Observations were made successively during two 72 h and three 48 h periods with intervals between the first day of each observation period of 1, 4, 4, and 4 weeks. Records were kept of the number of times each cow entered each feeder, the time for which concentrate was dispensed, and the time for which the cow remained in the feeder after concentrate was no longer dispensed.

RESULTS AND DISCUSSION

Within three days of being introduced to the feeders 28 of the 32 cows were using them regularly. The remaining 4 cows also used them regularly after a simple 5 min

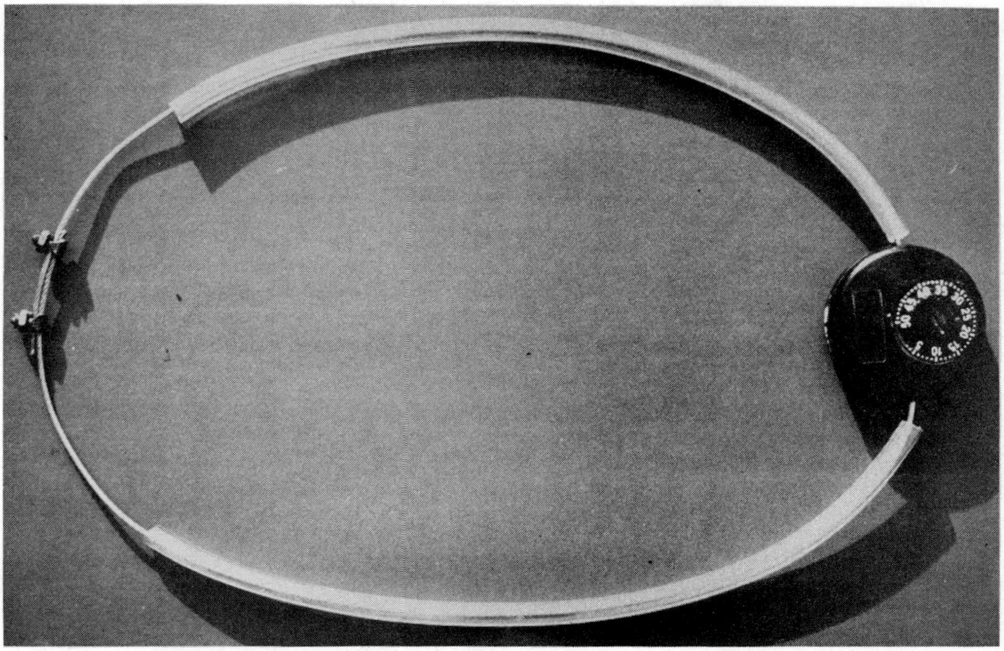

Fig. 2. A transponder collar.

training period which consisted of holding the cow in the feed stall with concentrate in the manger. This training never needed repeating and it has been adopted for all new cows introduced to the feeders.

Each cow quickly established a behavior pattern which remained fairly constant during the experiment. The daily number of visits each individual cow made to the feeder varied widely between cows. Table 1 gives the extremes of this variation (G370 and J425) together with a mean figure for all cows. Younger cows visited the feeder more often than older cows ($P < 0.01$).

Table 1.
Number of visits to both feeders in 24 hours

	First day of observation				
Cow no.	1	7	36	64	92
G370	6	7	8	8	6
J425	46	44	38	46	33
Mean no. of visits (all cows) (± S.E.)	16 (± 2.23)	18 (± 2.08)	17 (± 1.69)	18 (± 1.77)	13 (± 1.35)

The time each cow wasted in the feeders during 24 h, i.e., the time during which she was in the feeder but concentrate was not being dispensed, varied greatly and showed a negative correlation with age ($P < 0.01$). There was a steady decrease, greatest for the younger cows, in the time wasted during successive observation periods. Table 2 shows the extremes of this variation (H451 and J425), together with a mean figure for all cows. This reduction in wasted time is probably explained by a progressive modification of the cows' behavior with time. Initially there was much interest and strong competition. Cows would resist vigorous butts to the rear and remain in the feeding stall even though it may not have been dispensing. A similar 'blocking' effect was reported by Hyde *et al.* (1974). However, in this experiment, cows became increasingly willing to back out of the feeder, even when concentrate was being dispensed, and on most occasions the age of the cow had no effect on this response, cows backing out as readily for younger as for older cows.

Table 2.
Time wasted in both feeders (min^{-24h})

	First day of observation				
Cow no.	1	7	36	64	92
H451	5	2	2	1	3
J425	88	53	37	41	23
Mean time wasted (all cows) (± S.E.)	27 (± 4.54)	18 (± 2.44)	15 (± 1.90)	12 (± 1.80)	7 (± 0.99)

Once cows had established a pattern of visiting the dispenser, there was a good correlation (r = 0.96) between the desired and measured concentrate intakes. This agreed well with the results of Puckett *et al.* (1973).

Four cows which received only 74 percent, 74 percent, 82 percent and 86 percent of their intended ration were in early lactation and the settings on their transponders were especially high in order to encourage maximum milk production.

REFERENCES

Hyde G. M., Olver E. F., Hashbarger K. E., Puckett H. B., Spahr S. L. and Devore A. L. (1974) Factors affecting dispensing rate of an automatic dairy concentrate feeder. Proc. *IFAC Symposium on Automatic Control* for Agric. Univ. of Saskatchewan. pp. 1-11.

Puckett H. B., Hyde G. M., Olver E. F. and Hashbarger K. E. (1973) An automated individual feeding system for dairy cows. Nat. Dairy Housing Conf. Paper ASAE Special Publication Sp. -01-73, 272-280.

Atropine Effect on Fear Bradycardia of the Eastern Cottontail Rabbit, *Sylvilagus floridanus*

E. N. SMITH and D. J. WORTH

Division of Natural Science and Mathematics, Northeastern State University, Tahlequah, Oklahoma 74464, U.S.A.

Abstract — Multichannel transmitters were surgically implanted in the abdominal cavity of five mature eastern cottontail rabbits, *Sylvilagus floridanus*, and the heart rate response to fear investigated. Transmitted heart rate was obtained for 10 min prior to disturbance. The rabbit was then stimulated for 5 min by approach of the investigator. Ten minutes of recovery data were collected following disturbance. Fear bradycardia was observed in each of 17 experiments. Onset of bradycardia occurred within 5 s of the stimulus. Recovery required from 1 to 4 min. No recovery tachycardia was observed. Mean heart rates during prestimulus, stimulus and recovery were 209 ± 1.1, 156 ± 1.4 and 198 ± 0.7 bpm ($\bar{x}$ ± SE) respectively.

In subsequent experiments, atropine was injected to block vagal influence of the heart. Fear bradycardia was present in each of 17 experiments, however, the magnitude was decreased. Atropine treated heart rates during prestimulus, stimulus and recovery were 254 ± 0.8, 202 ± 1.4 and 238 ± 1.0 bpm ($\bar{x}$ ± SE) respectively. Mean heart rate reduction of 25.4 percent was observed in the experiments without treatment and 20.5 percent in atropine treated animals, indicating possible involvement of the sympathetic portion of the autonomic system.

INTRODUCTION

The classic response to fear is increased heart rate and general readiness for 'fight or flight'. Sympathetic activity increases and the animal is generally prepared for strenuous activity. Alternately, an animal may react to threat by passively 'freezing or hiding'. The latter response is often characterized by bradycardia and has been observed in reptiles, birds and mammals (Belkin, 1968). Telemetry measurements of heart rate from a free ranging submerged alligator dropped from 30 bpm to less than 2 bpm when frightened (Smith, Allison and Crowder, 1974). Free ranging Uinta ground squirrels showed a decrease of 100 bpm when approached by man (Ruff, 1971). E. Folk, Iowa Medical School, observed fear bradycardia in radio tagged woodchucks and snowshoe rabbits (unpublished observations). Smith (1978) reported marked fear bradycardia in free ranging woodchucks, *Marmota monax*, if the animals had access to a burrow. Captive woodchucks showed aggressive behavior and tachycardia when approached.

The eastern cottontail rabbit, *Sylvilagus floridanus*, is a widely distributed lagomorph inhabiting wooded and swampy areas of much of north and central America. When chased by a dog or other predator, rabbits run and show the typical 'fight or flight' response including marked tachycardia. The first response of cottontails to a threatening stimulus is 'freezing' and bradycardia (unpublished observations). This response can be repeated under laboratory conditions. The purpose of this study was to investigate fear bradycardia of the eastern cottontail rabbit and to determine the effect of the parasympathetic blocking agent, atropine, on the heart rate response.

METHODS

Five mature cottontail rabbits were captured in eastern Oklahoma (Cherokee County) by box trap (Havahart, No. 2). Rabbits were maintained in standard rabbit cages and fed a diet of Purina Rabbit Chow, alfalfa hay, and mixed fresh carrots, lettuce, apples and dandelion leaves. Drinking water was available. Incoming rabbits were tranquilized by the addition of 0.5 mg day^{-1} piper-acetazine (Pitman-Moore, Inc., Psymod) to the drinking water. Failure to tranquilize field caught rabbits occasionally resulted in rabbits injuring themselves while trying to escape. The use of the tranquilizer was discontinued for 48 h before each animal was used experimentally. Rabbits were anesthetized for surgery with ketamine hydrochloride (Parke-Davis, Vetalar 100 mg ml^{-1}). The multichannel ECG and temperature transmitter was inserted into the abdominal cavity (Essler and Folk, 1961) through a 3 cm longitudinal incision approximately 5 cm lateral to the midline. The two ECG leads were placed subdermally and electrodes were sutured near each end of the sternum. Animals were allowed to recover for one week before experimentation.

The biotelemetry system used in this study is similar to that described by Smith and Salb (1975) and is available from Biotelemetry Systems, Inc. (P.O. Box 10, Rush, N.Y. 14543, U.S.A.). The transmitter employs a magnetic reed switch (Smith and Crowder, 1974) and a CMOS integrated circuit memory so the transmitter may be switched on and off (Smith, 1980). This extends battery life to several months of intermittent use. The 150 MHz subcarrier signal was received with a Lafayett VHF receiver (Stock No. 99-3431L) and portable VHF antenna (Biotelemetry Systems, Inc.). The subcarrier tone ($\simeq$ 1 KHz) was recorded on a cassette recorder (Sony TC-45) and demodulated using a two-channel temperature and ECG demodulator (Biotelemetry Systems, Inc., Model D2-1) and physiograph (Narco-Biosystems, Inc., Model 4B). Heart rate was counted as QRS intervals on the physiograph record.

For each experiment, the rabbit was moved to a 50 gallon aquarium and its transmitter turned on with a magnet. The animal was then left undisturbed for 1 h. Then heart rate was recorded continuously for 25 min. During the first 10 min, the animal remained undisturbed, then the investigator entered the room and approached the rabbit for 5 min. Following disturbance, 10 min of recovery heart rate was recorded. Several replicate experiments were obtained for each radio tagged rabbit.

Atropine was administered in a second series of experiments. In most cases control and atropine experiments were done on the same day. Preliminary experiments demonstrated 4.3 mg kg^{-1} atropine had a maximal heart rate effect within 10 min of IM injection and the drug lasted for at least one hour. Additional atropine did not further increase heart rate. A dosage of 4.3 mg kg^{-1} ($\bar{x} \pm$ SE) atropine sulfate (Lilly 0.4 mg ml^{-1}) was injected and heart rate measurements began 10 min later. As in the control experiments 10 min of prestimulus data were obtained, then 5 min of disturbance and 10 min of recovery data were recorded. Heart rate was counted for 30 s intervals throughout the experiments. The results were plotted and compared with a *t*-test.

RESULTS

Each attempt to frighten the caged rabbit resulted in bradycardia ($P < 0.01$). Upon sensing the presence of the investigator, the rabbit would usually drop to a crouching position and lower its ears. It remained in this position even if the investigator approached to within 1 m. Heart rate remained low as long as the rabbit was motionless. If the investigator startled the rabbit so that it moved, tachycardia resulted and the data were not used. Only occasionally did a rabbit actually jump and attempt to escape.

Atropine injection increased heart rate above control values by 21.5, 29.5 and 20.2 percent during prestimulus, stimulus and poststimulus, respectively. Fear bradycardia, although reduced, was still present in each rabbit tested. Figure 1 shows the heart rate response of an untreated and atropine treated rabbit. The onset of bradycardia was sudden with recovery requiring longer and recovery tachycardia was not observed. Figure 2 shows typical ECG records at the beginning of disturbance for control and atropine treated rabbit. Control animals often exhibited a marked fear bradycardia arrhythmia that was abolished with atropine treatment. Stimuli calculated to cause fear (Table 1) reduced heart rate by 25.4 percent which subsequently recovered to within 5.3 percent of prestimulus heart rate. Table 2 contains results of atropine experiments. Fear bradycardia was 20.5 percent with recovery to 6.3 percent of prestimulus values. Figure 3 shows the combined responses of untreated and atropine treated rabbits.

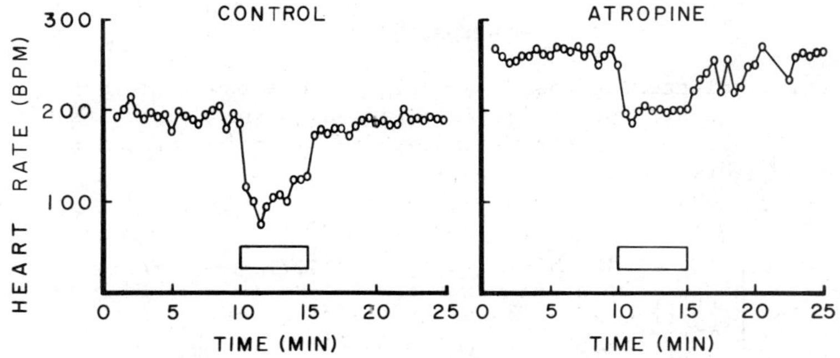

Fig. 1. Heart rate response to fear of a 995 g female cottontail rabbit, *Sylvilagus floridanus*. Bar indicates 5 min period of disturbance.

DISCUSSION

The heart rate response to fear of captive eastern cottontail rabbits was similar to that described for free ranging woodchucks (Smith, 1978) and alligator (Smith *et al.*, 1974). In each case when the animal was approached, heart rate declined suddenly and remained low as long as the disturbance lasted. The end of fear bradycardia in both the woodchuck and rabbit was often marked by movement.

Other results of this study are in contrast with those previously reported. Attempts to frighten restrained woodchucks or alligators resulted in aggressive behavior and tachycardia. Captive rabbits responded to disturbance with behavioral 'freezing' and bradycardia. Woodchucks showed a consistent recovery tachycardia and marked sinus respiratory arrhythmia (Smith and Causby, 1980, this volume) that was absent in the rabbits.

Physiologically, the fear bradycardia in the alligator is probably similar to classic diving bradycardia. Blood is shunted away from areas that can tolerate hypoxia, run off is reduced and heart rate drops while still maintaining systemic blood pressure (Andersen, 1966). Perhaps a similar reduction in oxygen consumption occurs in

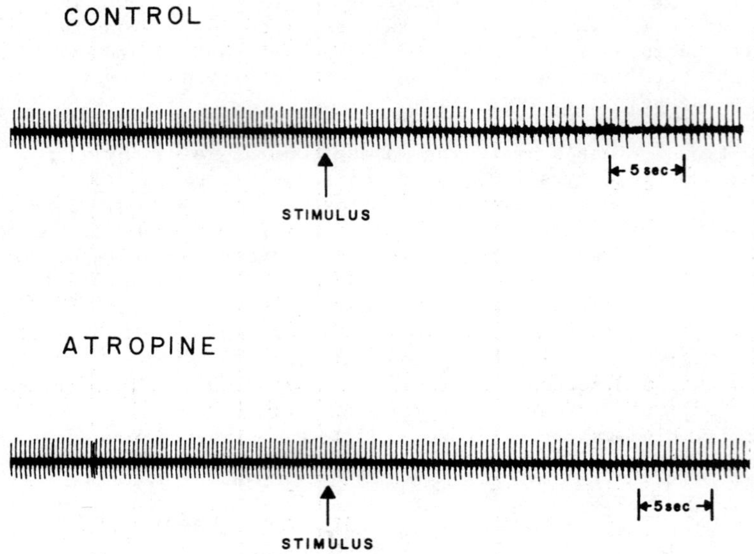

Fig. 2. Electrocardiogram showing the onset of fear bradycardia for an untreated and atropine treated rabbit. Stimulus arrow indicates the beginning of 5 min of disturbance.

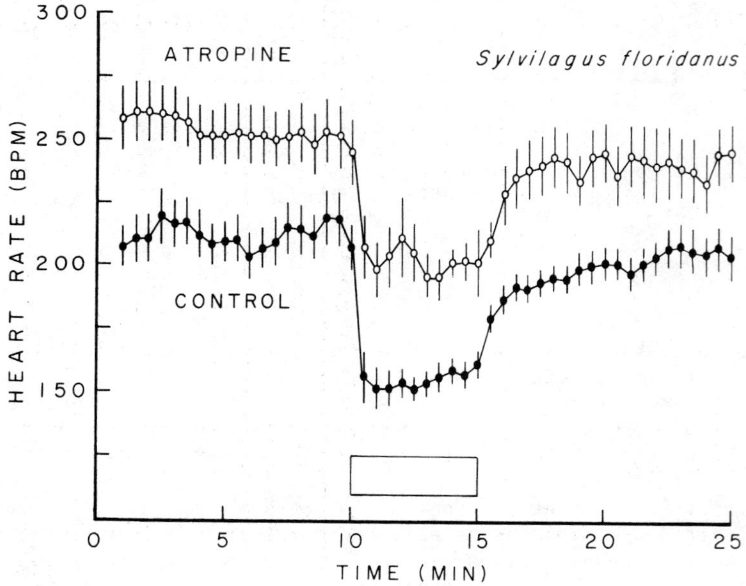

Fig. 3. Combined heart rate response of eastern cottontail rabbits to fear. Open circles indicate atropine treatment and solid circles show control means. Vertical lines indicate 2 standard errors. Open bar indicates 5 min period of disturbance.

Table 1.
Results of fear bradycardia experiments for untreated eastern cottontail rabbits

Weight (g)	No. of experiments	Prestimulus ($\bar{x}$ ± SE bpm)	No. of measures	Stimulus ($\bar{x}$ ± SE bpm)	No. of measures	Recovery ($\bar{x}$ ± SE bpm)	No. of measures
995	6	215 ± 2.0	119	115 ± 3.2	60	201 ± 1.7	113
888	2	250 ± 4.3	40	196 ± 4.3	20	250 ± 4.5	40
793	5	218 ± 3.5	100	160 ± 3.3	50	196 ± 2.2	60
1039	4	176 ± 1.9	99	142 ± 1.9	50	175 ± 0.8	98
Grand Mean		209 ± 1.2		156 ± 1.4		198 ± 0.7	
Δ% of Prestimulus				-25.4		-5.3	

Table 2.
Results of fear bradycardia experiments for atropine treated eastern cottontail rabbits

Weight (g)	No. of experiments	Prestimulus ($\bar{x}$ ± SE bpm)	No. of measures	Stimulus ($\bar{x}$ ± SE bpm)	No. of measures	Recovery ($\bar{x}$ ± SE bpm)	No. of measures
995	5	258 ± 2.2	85	202 ± 2.6	44	265 ± 2.2	70
1046	1	280 ± 2.6	20	225 ± 3.1	10	—	—
888	2	359 ± 3.8	27	298 ± 7.5	15	320 ± 5.8	33
793	5	251 ± 3.2	98	192 ± 3.7	50	229 ± 3.0	100
1039	4	210 ± 1.3	75	174 ± 2.3	40	193 ± 1.3	80
Grand Mean		254 ± 0.8		202 ± 1.4		238 ± 1.0	
Δ% of Prestimulus				-20.5		-6.3	

disturbed woodchucks. They retreat to burrows where oxygen levels may drop to a few percent (Kenerly, 1964).

Rabbits are not stressed with hypoxia. Perhaps fear bradycardia helps the animal remain invisible to a predator. A classic 'fight or flight' response with an increase in heart rate, respiratory rate, and overall 'nervousness' might give away the location of a hiding rabbit. It is significant that, unlike many diving animals and the woodchuck displaying fear bradycardia, rabbits showed no recovery tachycardia. This suggests an oxygen debt was not incurred. Apparently metabolism was simply reduced during the disturbance period.

The fact that atropine did not abolish fear bradycardia was unexpected. Diving bradycardia can be largely abolished by atropine injection in a variety of vertebrates (Andersen, 1966). Atropine treatment reduced the magnitude of fear bradycardia by only about 20 percent. This suggests the parasympathetic portion of the autonomic system plays a role in fear bradycardia but is not solely responsible.

Mammalian heart rate can be reduced by two separate mechanisms. First, an increase in acetylcholine released by increased parasympathetic firing of fibers in the heart causes bradycardia. Secondly, reduction in the release of norepinephrine from sympathetic fibers in the heart and/or a reduction in circulating norepinephrine in the blood can cause bradycardia. Data from the present study suggests that removal of sympathetic tone plays the dominant role in fear bradycardia in rabbits. The sudden onset suggests neural instead of chemoreceptor involvement.

In summary, caged eastern cottontail rabbits respond to threat with a marked reduction in heart rate. Onset is sudden and there is no recovery tachycardia. Atropine reduced, but did not abolish the response, suggesting sympathetic as well as parasympathetic involvement. The adaptive significance of fear bradycardia in rabbits is unclear but may be related to aiding the animal in remaining invisible to a possible predator.

Acknowledgements — This research was made possible by support from NIH Grant Minority Biomedical Support Program No. 5 S06 RR08123 and MARC Grant No. 5 R32 GM07694. The Caesar Cleberg Foundation generously supported portions of this research. Several students helped with trapping, surgery and data acquisition. Especially helpful were: Steve Adams, Lisa Causby, David Cornsilk and Patty Harjo. Everett Grigsby generously provided release time and laboratory space. Larry Peters, Dennis Bird and Lou Ann Roberts offered valuable suggestions relating to surgery, anesthesia, and use of atropine. Travel funds (for ENS) were supplied by the College of Arts and Sciences and a faculty grant from Northeastern State University. We are indebted to Charles Amlaner, Jr., Oxford University, for organizing this conference and for the opportunity to participate. Debbie Boatman typed several editions of the manuscript.

REFERENCES

Andersen H. T. (1966) Physiological adaptations in diving vertebrates. *Physiol. Rev.* 46, 212-243.

Belkin D. A. (1968) Bradycardia in response to threat. *Am. Zool.* 8, 775.

Essler W. O. and Folk G. E. (1961) Determination of physiological rhythms of unrestrained animals by radio telemetry. *Nature* 190, 90-91.

Kenerly T. E., Jr. (1964) Microenvironmental conditions of pocket gopher burrows. *Texas J. Sci.* 16, 395-441.

Ruff R. L. (1971) Telemetered heart rates of free-ranging Uinta ground squirrels in response to social interactions. Ph.D. dissertation. Utah State University, Logan, Utah.

Smith E. N., Allison R. D. and Crowder W. E. (1974) Bradycardia in a free ranging American alligator. *Copeia* 1974, 770-772.

Smith E. N. and Crowder W. E. (1974) Implantable ECG transmitter employing a magnetic switch. *J. Appl. Physiol.* 36, 634-635.
Smith E. N. and Salb T. J. (1975) Multichannel subcarrier ECG, respiration and temperature biotelemetry system. *J. Appl. Physiol.* 39, 331-334.
Smith E. N. (1978) Fear-evoked bradycardia of free ranging woodchuck. *The Physiologist* 21, abstract.
Smith E. N. (1980) Physiological radio telemetry of vertebrates. In C. J. Amlaner, Jr. and D. W. Macdonald (eds.), *A Handbook on Biotelemetry and Radio Tracking*, pp. 45-55. Oxford: Pergamon Press.
Smith E. N. and Causby L. A. (1980) Respiratory sinus arrhythmia in free ranging woodchucks, *Marmota monax*. In C. J. Amlaner, Jr. and D. W. Macdonald (eds.), *A Handbook on Biotelemetry and Radio Tracking*, pp. 557-562. Oxford: Pergamon Press.

Respiratory Sinus Arrhythmia in Free Ranging Woodchucks, *Marmota monax*

E. N. SMITH and L. A. CAUSBY

Division of Natural Science and Mathematics, Northeastern State University, Tahlequah, Oklahoma 74464, U.S.A.

Abstract — Multichannel temperature and ECG biotelemetry transmitters were surgically placed in the abdominal cavity of two mature woodchucks, *Marmota monax*. Following recovery they were released on a 7.6 ha wooded island on Lake Tenkiller in eastern Oklahoma (Cherokee County). Telemetry data were recorded in the field by a magnetic recorder; the record was demodulated in the laboratory and heart rate was counted. Respiration appeared as amplitude modulation of the QRS wave. Of 343 two minute records, 32 showed a marked cardio-respiratory synchrony with heart rate dropping about 50 percent between breaths and accelerating with each inhalation. Heart rate values for a 3.1 kg male woodchuck went from 71 beats per minute (bpm) between breaths to 135 bpm after inhalation and from 87 bpm to 160 bpm for a 2.2 kg female. This is the first report of cardio-respiratory synchrony for a free ranging mammal. The adaptive significance remains to be elucidated but may be related to obligate exposure to hypoxia of these fossorial mammals.

INTRODUCTION

Rhythmic alteration in heart rate with respiration, called respiratory sinus arrhythmia (RSA), has been well studied in dogs (Anrep, Pascual and Rossler, 1936; Hainsworth, 1974; Haymet and McCloskey, 1975; Hamlin, Smith and Smetzer, 1966), cats (Black and Torrance, 1971; Eldridge, 1972) and man (Bainbridge, 1920; Melcher, 1976; Freyschuss and Melcher, 1975). It is generally agreed that RSA depends on autonomic activity, especially the rhythmic alteration of vagal impulses to the sinoatrial node (Wheeler and Watkins, 1973). The response has also been described for fish (Shelton and Randall, 1962; Capra, 1976), and crocodilians (Huggins, Hoff and Pena, 1969). Although much has been written describing the response, its mediation, and its possible origin, little is known about its possible adaptive significance. Other studies have investigated the response of anesthetized or restrained animals. This is the first report of RSA from unrestrained mammals.

Woodchucks are large fossorial rodents usually living in extensive burrows. Several comprehensive papers are available about woodchuck natural history (Grizzell, 1955; Bronson, 1962; Merriam, 1966). Woodchuck hibernation has been thoroughly documented (Bailey and Davis, 1965; Snyder, Davis and Christian, 1961). Except for telemetrically measured body temperature (Hayes, 1976) and fear bradycardia (Smith, 1978), little is known about the physiology of free ranging woodchucks. The purpose of this paper is to report respiratory sinus arrhythmia from free ranging woodchucks, *Marmota monax*. We will also discuss the possible adaptive significance of RSA for the woodchuck.

METHODS

One mature woodchuck of each sex was captured in eastern Oklahoma (Cherokee County) by box traps (Havahart, No. 2). Woodchucks were anesthetized with Ketamine Hydrochloride (Parke-Davis, Vetalar 100 mg ml^{-1}) and the multi-channel ECG and temperature transmitters were inserted into the abdominal cavity through a 3 cm longitudinal incision approximately 5 cm lateral to the midline. ECG leads were placed subdermally. Electrodes were sutured near each end of the sternum. After a one week recovery period the woodchucks were released on 7.6 ha Buzzard Island in Lake Tenkiller (Cherokee County, Oklahoma). The island has an elevation of approximately 20 m above mean lake level and is wooded with mixed hardwood (predominantly Oak and Hickory). The southeast end of the island contains limestone outcroppings and the woodchucks dug extensive burrows under the large rocks. Natural food and water were available.

The biotelemetry system used in this study is similar to that described by Smith and Salb (1975) and is available from Biotelemetry Systems, Inc., P.O. Box 10, Rush, N.Y. 14543, U.S.A. A magnetic switch and CMOS integrated circuit memory provided a way to turn the implanted transmitter on and off (Smith, 1980). The 150 MHz signal was received with a Lafayett VHF Receiver (Stock No. 99-3531L) and portable directional VHF antenna (Biotelemetry Systems, Inc.). The subcarrier tone was recorded on a cassette recorder (Sony, TC-45) and demodulated in the laboratory using a two channel temperature and ECG demodulator (Biotelemetry Systems, Inc. — Model D2-1) and physiograph (Narco-Biosystems, Inc., Model 4B). Instantaneous heart rate was obtained by measuring the QRS to QRS interval. Respiration appeared as a regular amplitude modulation of the QRS waves. Preliminary observations in the laboratory confirmed the relationship between ventilatory movements and QRS amplitude. Inspiration is indicated by increasing QRS amplitude and expiration by decreasing QRS height.

Field measurements were made for two minutes at 10, 30, or 60 min intervals. Field observations were obtained for 2-3 weeks, on each woodchuck. Although the animals were not monitored continuously, several continuous 24 h measurements were made.

RESULTS

There are two distinct types of cardio-respiratory patterns in free ranging woodchuck. Heart rate changed only about 10 percent with respiration most of the time. On other occasions, especially when the animals were inactive, heart rate varied profoundly with each respiratory cycle. Of 343 two minute records, 32 showed a marked cardio-respiratory synchrony or respiratory sinus arrhythmia (RSA).

Figures 1 and 2 show typical records for each cardio-respiratory pattern. Figure 3 shows mean values of instantaneous heart rate for the two patterns of cardio-respiratory synchrony. In the records displaying marked RSA the end of expiration is marked by a profound bradycardia. Inspiration results in tachycardia. During marked RSA mean heart rate values for the male increased from 71.1 ± 1.06 ($\bar{x}$ ± SE) bpm between breaths to 135.3 ± 2.92 bpm during expiration. For the female mean instantaneous heart rate increased from 87.3 ± 1.1 bpm between breaths to 160.2 ± 3.8 bpm during expiration. Mean heart rate during low RSA increased from 186.4 to only 204.7 during expiration.

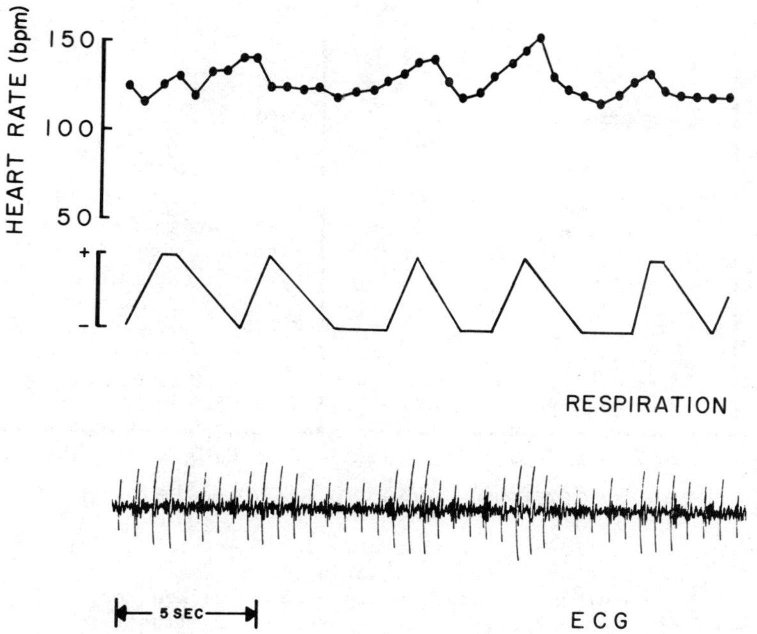

Fig. 1. Typical cardio-respiratory pattern with a low level of respiratory sinus arrythmia. Upper line indicates instantaneous heart rate determined by measurement of each QRS interval. Respiration determined from QRS height. Plus indicates inhalation, minus indicates exhalation. Lower line represents telemetry obtained electrocardiogram.

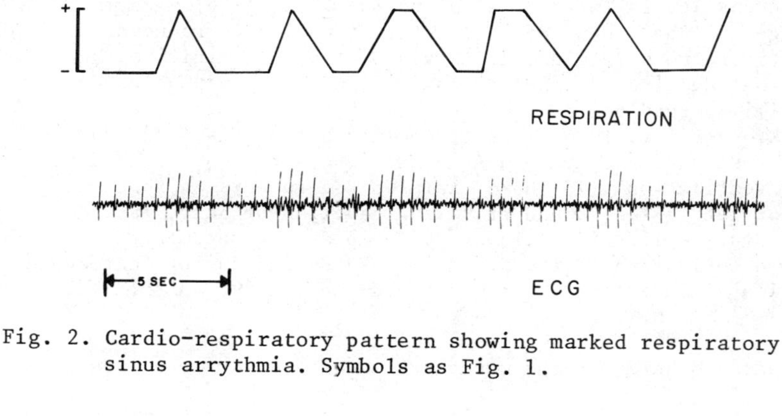

Fig. 2. Cardio-respiratory pattern showing marked respiratory sinus arrythmia. Symbols as Fig. 1.

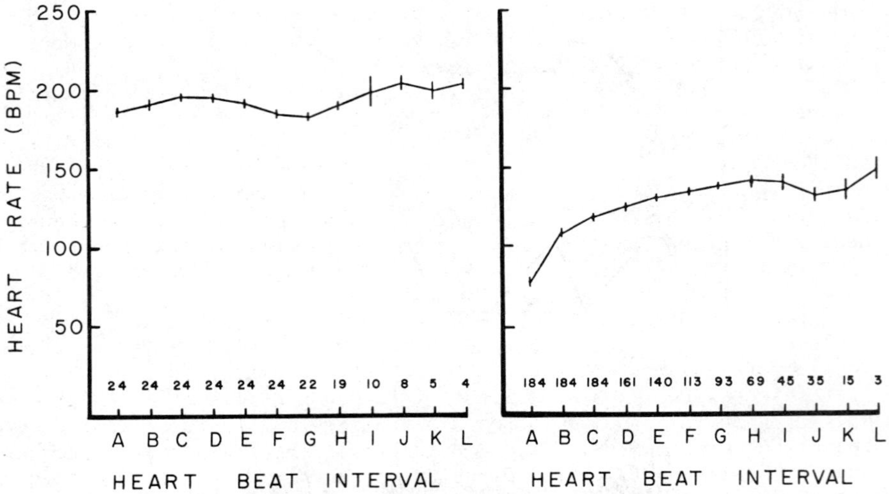

Fig. 3. Relationship of instantaneous heart rate to individual heart beat intervals during each breath. Left hand panel shows relationship without marked respiratory sinus arrythmia. Right hand panel illustrates marked respiratory sinus arrythmia. Vertical lines indicate ± 1 SE.

DISCUSSION

Several mechanisms are thought to play a part in evoking respiratory changes in heart rate (see Melcher, 1976, for a recent review of the literature). There is evidence indicating RSA is caused by a pulmonary reflex with the afferent portion of the reflex in the vagus nerve. Bainsbridge (1920) suggested RSA could be explained by increased diastolic filling of the heart with each inspiration. Recent evidence confirms the Bainsbridge effect (Horwitz and Bishop, 1972; Vather and Braunwald, 1975). A systemic arterial baroreflex has been postulated (Matthes and Ebeling, 1948) in creating RSA due to the respiratory effects on arterial blood pressure. Mechelke (1953) suggested an interplay between arterial baroreflex and the Bainbridge reflex. There is also thought to be a central nervous reflex (Hamlin, Smith and Smetzer, 1966) because RSA often persists in animals when respiratory movements are paralyzed or in open-chest preparations. Interestingly, RSA is positively correlated with Pco_2 levels in arterial blood (Melcher, 1976). Haymet and McCloskey (1975) suggested that when the circulatory time lag is reduced, as in exercise, the chemoreceptor response would enhance ventilation and minimize the reflex bradycardia. This would be adaptive in assuring maximum pulmonary perfusion during maximum ventilation. Similar results were obtained for the Port Jackson shark, *Heterodontus portus jackson* (Capra, 1976), and crocodilians (Huggins, Hoff and Pena, 1969). Bradycardia occurred during apnea and a marked tachycardia occurred during infrequent periods of ventilation. In these examples RSA appears to function in reducing heart rate when oxygen is less abundant. The reduction of heart rate during periods of low oxygen availability is well developed in diving vertebrates (Andersen, 1966), during human childbirth (Weisbrot *et al.*, 1958) and during fear-evoked bradycardia of crocodilians (Andersen, 1961; Gaunt and Gans, 1969; Smith, Allison and Crowder, 1974).

The RSA described here for the woodchuck is of greater magnitude than previously described for any mammal. Woodchucks are fossorial, spending a significant amount of time underground. Underground living offers relative freedom from many predators and provides safe quarters for bearing young, sleeping and hibernating. It also exposes the individual to a low oxygen and high carbon dioxide environment. Kenerly (1964) found pocket gopher burrow O_2 as low as 6 percent and CO_2 exceeding 2 percent. A thorough study by Burlington *et al.* (1971) demonstrated a significant difference in response to hypoxia of woodchuck. In the rabbit, dog or man, hypoxia results in increased cardiac output. Radio tagged eastern cottontail rabbits did not show a marked RSA (Smith and Worth, 1980, this volume). The woodchuck exhibited decreased cardiac output with maintenance of mean arterial pressure. Blood flow was reduced to skin, skeletal muscle, white fat and brown fat with no observable change in blood flow to heart, diaphragm, kidney, liver, stomach and intestines. Woodchuck hemoglobin has a higher affinity for oxygen than do other rodents living above ground (Hall, 1965).

The available evidence indicates woodchucks are physiologically well equipped to endure hypoxia. Perhaps the marked RSA is but another facet of their resistance to hypoxia. It is energetically expensive to maintain a high rate of cardiac output continuously. An animal exposed to hypoxia can reduce circulatory costs by reducing cardiac output between breaths. Diving vertebrates shunt blood away from areas that can withstand hypoxia and maintain perfusion of heart and brain. The oxygen debt incurred is repaid upon surfacing. It would be maladaptive for resting woodchucks to incur an oxygen debt (for hours at a time). The marked RSA described here might enable them to reduce oxygen consumption which along with other physiological adaptations enable them to tolerate obligate hypoxia inherent in fossorial existance.

Acknowledgements — Portions of this research was supported by NIH (MBS) Grant No. 5 S06 RR08123 and NIH National Institutes of General Medical Science, MARC Grant No. 5 R32 GM07694. Faculty grant from Northeastern Oklahoma State University to ENS is gratefully acknowledged. Several people helped with trapping, surgery, and the collecting and analysis of data. Especially helpful were Steve Adams, Patricia Harjo, Lisa Walker, Patricia Williams and Robert Woodruff. Edgar Folk, University of Iowa Medical School, provided early stimulation of the study, offered valuable suggestions regarding the use of radio telemetry and critically reviewed an early version of the paper. Debbie Boatman patiently typed and re-typed the manuscript. Travel funds (for ENS) were supplied by the College of Arts and Sciences and a faculty grant from Northeastern State University. We are indebted to Charles Amlaner, Jr., Oxford University for organizing this conference and for the opportunity to participate.

REFERENCES

Andersen H. T. (1961) Physiological adjustments to prolonged diving in the american alligator, *Alligator mississippiensis*. *Acta. physiol. scand.* 53, 23-45.

Andersen H. T. (1966) Physiological adaptations in diving vertebrates. *Physiol. Rev.* 46, 212-243.

Anrep G. V., Pascual W. and Rossler R. (1936) Respiratory variations of the heart rate. 1. The reflex mechanism of respiratory arrhythmia and the interrelations between the central and the reflex mechanism. *Proc. R. Soc. Lond.* Ser. B. 191-217.

Bailey E. D. and Davis D. E. (1965) The utilization of body fat during hibernation in woodchucks. *Can. J. Zool.* 43, 701-707.

Bainbridge F. A. (1920) The relation between respiratory and the pulse rate. *J. Physiol.* 54, 192-202.

Black A. M. S. and Torrance R. W. (1971) Respiratory oscillations in chemoreceptor discharge in the control of breathing. *Resp. Physiol.* 13, 221-237.

Bronson F. H. (1962) Daily and seasonal activity patterns in woodchucks. *J. Mammal.* 43, 425-427.

Burlington R. F., Vogel J. A., Burton M. and Salkovitz I. A. (1971) Cardiac output and regional blood flow in hypoxia woodchuck. *Am. J. Physiol.* 220, 1565-1568.

Capra M. F. (1976) Cardio-respiratory relationships during 'euponea' and respiratory arrhythmias in the Port Jackson shark, *Heterondontus portus jacksoni*. *Comp. Biochem. Physiol.* 53A, 259-262.

Eldridge F. L. (1972) The importance of timing on the respiratory effects of intermittent carotid body chemoreceptor stimulation. *J. Physiol.* 222, 310-333.

Freyschuss U. and Melcher A. (1975) Respiratory sinus arrhythmia in man: relation to cardiovascular pressures. *Scand. J. Clin. Lab. Invest.* 36, 221-229.

Gaunt S. S. and Gans C. (1969) Diving bradycardia and withdrawal bradycardia in *Caiman crocodilus*. *Nature* 223, 207-208.

Grizzell R. A. (1955) A study of the southern woodchuck, *Marmota monax*. *Am. Midland Nat.* 53, 257-293.

Hainsworth R. (1974) Circulatory responses from lung inflation in anesthetized dogs. *Am. J. Physiol.* 226, 247-255.

Hall F. D. (1965) Hemoglobin and oxygen: affinities in seven species of sciuridae. *Science* 148, 1350-1351.

Hamlin R. L., Smith C. R. and Smetzer D. L. (1966) Sinus arrhythmia in the dog. *Am. J. Physiol.* 210, 321-328.

Hayes S. R. (1976) Daily activity and body temperature of the southern woodchuck, *Marmota monax*, in northwestern Arkansas. *J. Mammal.* 57, 291-298.

Haymet B. T. and McCloskey D. I. (1975) Baroreceptor and chemoreceptor influences on heart rate during the respiratory cycle in the dog. *J. Physiol.* 245, 699-712.

Horwitz L. D. and Bishop V. S. (1972) Effect of acute volume loading on heart rate in the conscious dog. *Circ. Resp.* 30, 316-321.

Huggins S. E., Hoff H. E. and Pena R. V. (1969) Heart and respiratory rates in crocodilian reptiles under conditions of minimal stimulation. *Physiol. Zool.* 42, 320-333.

Kenerly T. E., Jr. (1964) Microenvironmental conditions of pocket gopher burrows. *Texas J. Sci.* 16, 395-441.

Matthes K. and Ebeling J. (1948) Untersuchungen uber die atemschwankungen des blutcrucks and der pulsfrequenz beim menshen. *Pflugers Arch. ges. Physiol.* 250, 747-768.

Mechelke K. (1953) Uber die atemschwankungen des blutdrucks and der pulsfrequenz biem menschen. *Arch Kreisaufforsch.* 19, 204-220.

Melcher A. (1976) Respiratory sinus arrhythmia in man, a study in heart rate regulating mechanisms. *Acta physiol. scand.* (S 435), 3-31.

Merriam H. G. (1966) Temporal distribution of woodchuck interburrow movements. *J. Mammal.* 47, 103-110.

Shelton G. and Randall D. J. (1962) The relationship between heart beat and respiration in a teleost fish. *Comp. Biochem. Physiol.* 7, 237-250.

Smith E. N., Allison R. D. and Crowder W. E. (1974) Bradycardia in a free ranging american alligator. *Copeia* 1974, 770-772.

Smith E. N. and Salb T. J., Jr. (1975) Multichannel subcarrier ECG, respiration and temperature biotelemetry system. *J. Appl. Physiol.* 39, 331-334.

Smith E. N. (1978) Fear-evoked bradycardia of free ranging woodchuck. *The Physiologist* 21(4), abstract.

Smith E. N. (1980) Physiological radio telemetry of vertebrates. In C. J. Amlaner, Jr. and D. W. Macdonald (eds.), *A Handbook on Biotelemetry and Radio Tracking*, pp. 45-55. Oxford: Pergamon Press.

Smith E. N. and Worth D. J. (1980) Atropine effect on fear bradycardia of the eastern cottontail rabbit *Sylvilagus floridanus*. In C. J. Amlaner, Jr. and D. W. Macdonald (eds.), *A Handbook on Biotelemetry and Radio Tracking*, pp. 549-555. Oxford: Pergamon Press.

Snyder R. L., Davis D. E. and Christian J. J. (1961) Seasonal changes in the weights of woodchuck. *J. Mammal.* 42, 297-312.

Vather S. F. and Braunwald E. (1975) Cardiovascular control mechanisms in the conscious state. *New Engl. J. Med.* 293, 970-976.

Weisbrot I. W., James L. S., Prince E. E., Holaday D. A. and Apgar V. (1958) Acid-base homeostasis of the newborn infant during the first 24 hours of life. *J. Pediat.* 52, 395-403.

Wheeler A. T. and Watkins P. J. (1973) Cardiac denervation in diabetes. *Br. Med. J.* iv, 584-586.

Radio Telemetry of Heart Rate and Temperature in Fallow Deer

E. JOHNSON, M. MILLER and R. H. SMITH

Zoology Department, The University, Whiteknights, Reading, U.K.

Abstract — A transmitter is described which is capable of switching to a number of channels of information in sequence. The unit uses a timed switching circuit to link either ECG electrodes or one of two thermistor circuits to a transmitter. The frequency output (clicks per second) of a thermistor circuit depends on thermistor temperature and might conveniently be made approximately equal to the heart rate.

The unit is housed in a box 8 cm^2 × 2 cm which needs to be mounted externally; this has proved impractical for use with wild species such as deer but could be used successfully with domesticated species. We are currently developing a totally implantable unit which will require a very small long life power unit and/or a disconnecting switch between transmissions.

AIMS OF STUDY

On behalf of the Forestry Commission (U.K.) we are studying a herd of fallow deer (*Dama dama*) living feral in the West Midlands of England, which contains a proportion with long coats (Springthorpe, 1969). In order to assess whether the long silky coat is a physiological advantage or disadvantage when compared with the normal short, coarse coat we wish to monitor various physiological parameters from the animals under different environmental conditions. Measurements of skin and body temperature, respiration and heart rate can provide useful indicators of the adaptability of mammals to environmental conditions, and it is possible to predict metabolic rate from a measure of heart rate and temperature when calorimetric studies are impractical (Holter *et al.*, 1976). The handling of such timid, wild species as deer can interfere with their physiological responses and we therefore wished to monitor the physiological parameters by telemetry.

PRELIMINARY DEVELOPMENT OF A TRANSMITTER

PARTICULAR PROBLEMS OF TRANSMISSION IN BRITAIN

Implantable transmitters for telemetry of heart rate and temperature have been developed in the U.S.A. (Skutt *et al.*, 1973) which are not crystal controlled.

However, there are stringent regulations in Britain governing the transmission and control of RF oscillation frequencies. A licence for medical and biological telemetry allows the operation of equipment in the band 104.675 MHz ± 2.0 KHz (Home Office leaflet HO 265). With such a narrow wave band available it seemed desirable to develop a unit capable of switching to a number of channels of information in sequence, so that various parameters could be monitored on the same RF frequency.

CIRCUIT

The transmitter (Fig. 1) is built from four basic circuits. The first is a timed switching circuit (Fig. 2a) controlling the sequence and length of transmission of

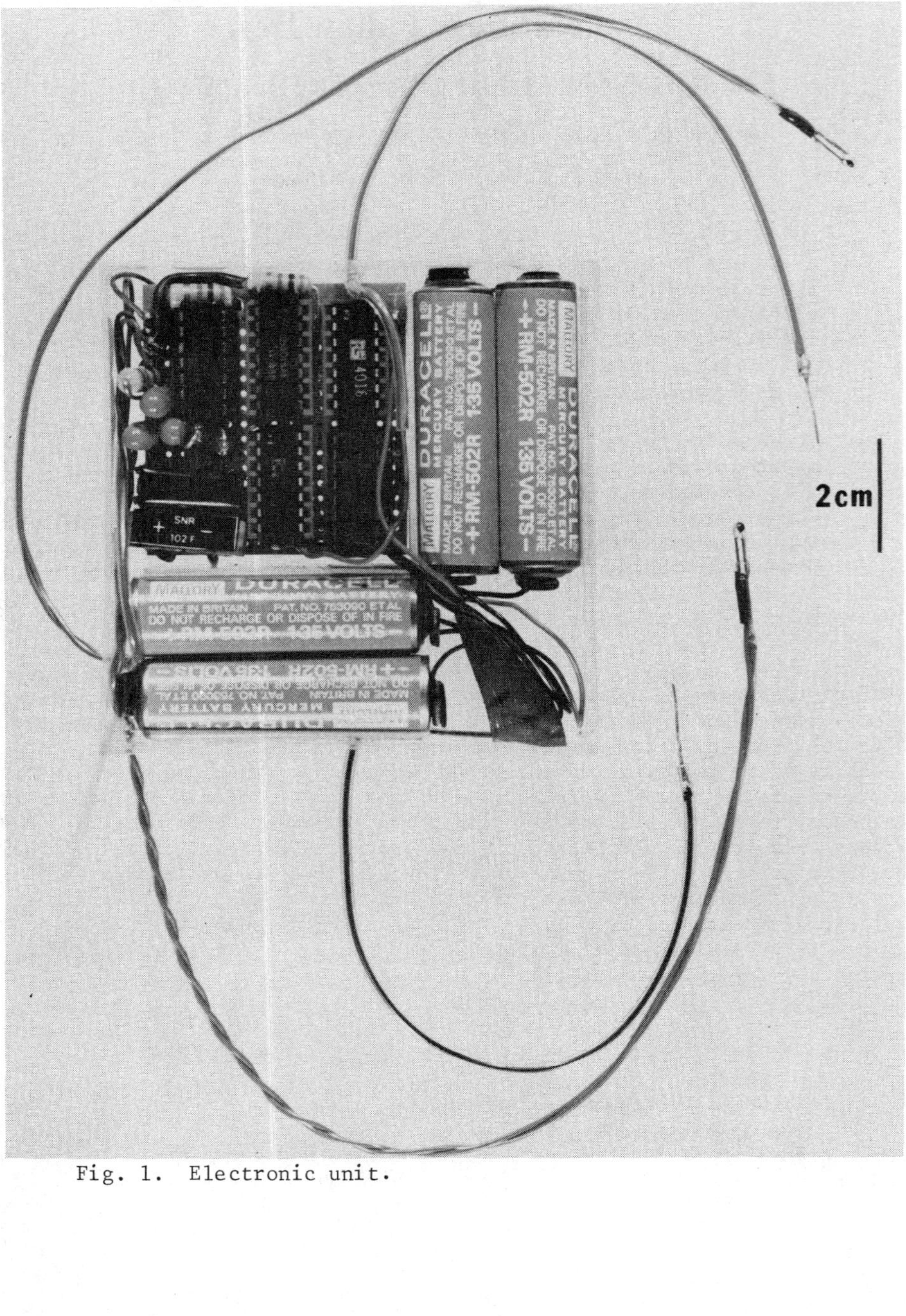

Fig. 1. Electronic unit.

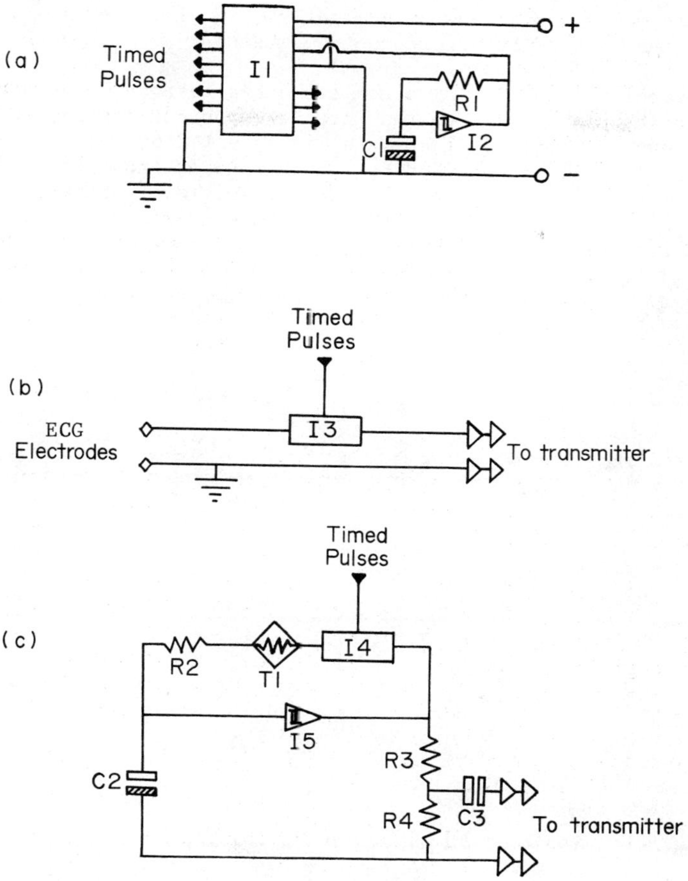

Fig. 2. Electronic unit: circuits. (a) Switching circuit; (b) ECG; (c) thermistor; R1 – 130 kΩ; R2 – 10 kΩ variable resistor, select for appropriate 'click' rate; R3 – 10 kΩ; R4 – 33 Ω; C1, C2 – 22 μF; C3 – 0.002 μF; T1 – 25 kΩ at 20°C thermistor; I1 – CD4017 decade counter; I2, I5 – 74C14 hex schmitt trigger integrated circuit; I3, I4 – CD4016 quad bilateral solid state switch.

information from the physiological sensors (thermistors and ECG electrodes). The switching circuit enables consecutive transmission of up to ten channels of information. The clock generator (I2) produces timed pulses which cause the decade counter (I1) to switch from one output channel to the next every 12 s. As our unit has only three sensors, several outputs from the decade counter are tied together and one is left blank as a marker giving, for example, 60 s of ECG – 24 s of output from thermistor 1 – 12 s blank marker – 24 s of output from thermistor 2 – 60 s of ECG etc.

The ECG circuit (Fig. 2b) simply consists of a solid state switch (I3) controlled by output pulses from the switching circuit so that ECG potential is either switched through to the transmitter or not. There is no preamplification of the ECG signal. The electrodes comprise 15 mm of flattened gold wire with the leads threaded and sealed in biologically inert polyvinyl tubing. These leads are placed under the skin with the electrodes implanted intradermally (Fig. 1). In this position a clear ECG can be monitored free of interference from muscular activity (Fig. 3a).

The thermistor circuits (Fig. 2c) also use a clock generator integrated circuit (I5) and incorporate a thermistor (T1) so that frequency output depends on thermistor temperature. Frequency range ('click' rate) is selected by varying the series resistor (R2). The signal is switched through by I4, attenuated 1:303 and buffered by a small capacitor (C3) before reaching the transmitter. The present thermistor circuits are calibrated to produce one click per second at 40°C (Fig. 3b). This rate has proved to be inaccurate in determining the temperature over the 24 s available and the rate should be increased. As a rule of thumb, we suggest a click rate approximately equal to the heart rate so that both can be measured using the same equipment.

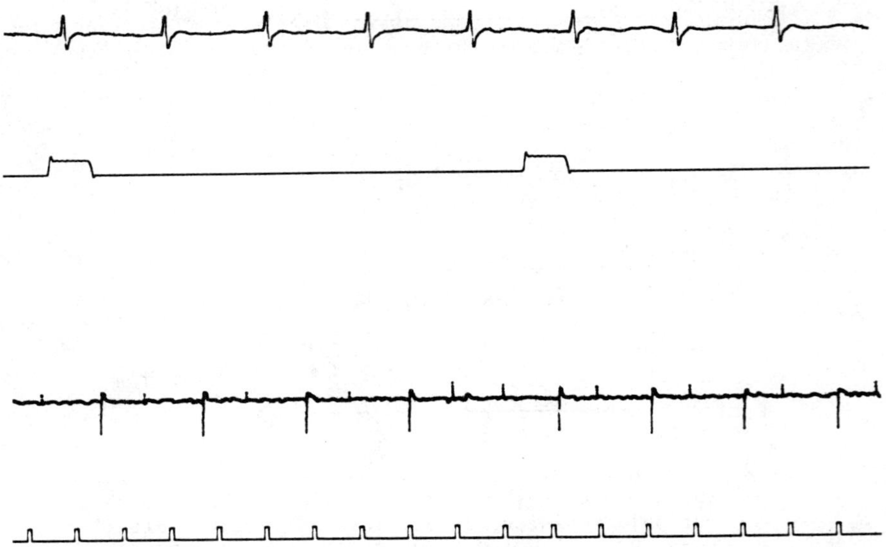

Fig. 3. (a) Heart trace (ECG). (b) Temperature trace (time trace for a and b; 1 s).

The transmitter is an unmodified FM biotelemetry transmitter supplied by Dynamic Electronics (SNR 102F) with output power -90 dB at 15 m and a current drain of 3 mA. The current drain of the whole electronic unit is approximately 8.3 mA. Power is provided by Mallory Duracell 1.35 V mercury batteries (RM 502R) and two parallel sets of two in series allow four weeks continuous transmission.

The signal is received using a Dynamic Electronics SNR 112R receiver with automatic frequency control. A permanent record can be made by feeding the signal from the receiver into a chart recorder or a tape recorder.

PROBLEMS AND PROSPECTS

A variety of harnesses and saddles tc mount and protect the unit on deer have been investigated without success; fallow ceer seem capable of removing attachments from any part of the body. However, the unit as described could be used on domesticated species such as cattle and sheep. We are currently working towards the development of a totally implantable unit. Problems still to be overcome are the requirement for small but long life batteries and/or the inclusion of a switch capable of disconnecting the unit between transmissions.

Acknowledgements — We wish to thank Paul Howey and Robert Forrester for helpful discussions; Janet Hornby, Kathleen Llewellyn and Mr. E. Snowden for practical assistance.

REFERENCES

Holter J. B., Urban W. E., Hayes H. H. and Silver H. (1976) Predicting metabolic rate from telemetered heart rate in white-tailed deer. *J. Wildl. Mngmt.* 40, 626-629.

Skutt H. R., Bock F. M., Haugsted P., Holter J. B., Hayes H. H. and Silver H. (1973) Low-power implantable transmitters for telemetry of heart rate and temperature from white-tailed deer. *J. Wildl. Mngmt.* 37, 413-417.

Springthorpe G. (1969) Long-haired fallow deer at Mortimer Forest. *J. Zool., Lond.* 159, 537.

The Use of Radio Telemetry in the Studies of Diving and Flying of Birds

P. J. BUTLER

Department of Zoology and Comparative Physiology, The University of Birmingham, Birmingham, U.K.

Abstract — When the head of a restrained duck is placed under water, there is a progressive reduction in heart rate. By using radio telemetry, it has been demonstrated that during natural submersion there is no maintained reduction in heart rate below that seen when the duck is swimming. The classical idea that during diving in ducks there is selective vasoconstriction and anaerobiosis is thus open to question. The measurements of respiratory and cardiovascular variables in free flying birds have previously been restricted to flights of relatively short duration (< 20 s). Barnacle geese have been trained to fly behind a truck by imprinting them on a human. Heart rate and respiratory frequency have been measured by radio telemetry during flights averaging 14.4 min and at a mean air velocity of 18.7 m s^{-1}.

DIVING

The ability of some birds and mammals to spend prolonged periods of time under water has fascinated physiologists for over a century (for detailed review of early literature see Andersen, 1966). The responses which are widely accepted as being responsible for this ability were proposed by Irving in 1934. He suggested that a large part of the systemic circulation was reduced and that the oxygen in the body was used preferentially by the heart and brain, which are particularly sensitive to oxygen lack, while the muscles and other tissues underwent anaerobiosis.

The studies of Scholander (1940) added much weight to Irving's theory. He reported that, when restrained domestic ducks and penguins (*Pygoscelis papua*) were placed under water, there was a progressive reduction in heart rate. The penguins could withstand forced submersion for up to 6-7 min whereas the domestic duck could survive for twice this length of time. In ducks, oxygen usage decreased during submersion, and there was a relatively slow increase in blood lactate during the period of submersion, but a much quicker and greater increase after surfacing. When small pieces of flesh were removed from the pectoral muscles during a period of submersion there was virtually no bleeding. This was in distinct contrast to the situation when the birds were breathing, particularly just after a period of submersion. By using a radioactive technique to determine the regional distribution of blood in ducks during head submersion, Johansen (1964) found that flow to the muscles declined whereas flow to the left ventricle and to the diencephalon increased.

The cardiovascular changes that occur in restrained domestic ducks during a 2 min period of head submersion are summarized in Fig. 1. If either bradycardia or vasoconstriction are prevented from occurring during head submersion by specific blocking drugs, then arterial oxygen tension declines more rapidly than in untreated ducks (Butler and Jones, 1971). All of these cardiovascular adjustments can be explained in terms of the stimulation of peripheral receptors, particularly the arterial chemoreceptors, as the blood becomes progressively more hypoxic and hypercapnic (Jones and Purves, 1970; Butler and Taylor, 1973). However, in order to conserve effectively the oxygen in the body for use by the CNS and heart, these

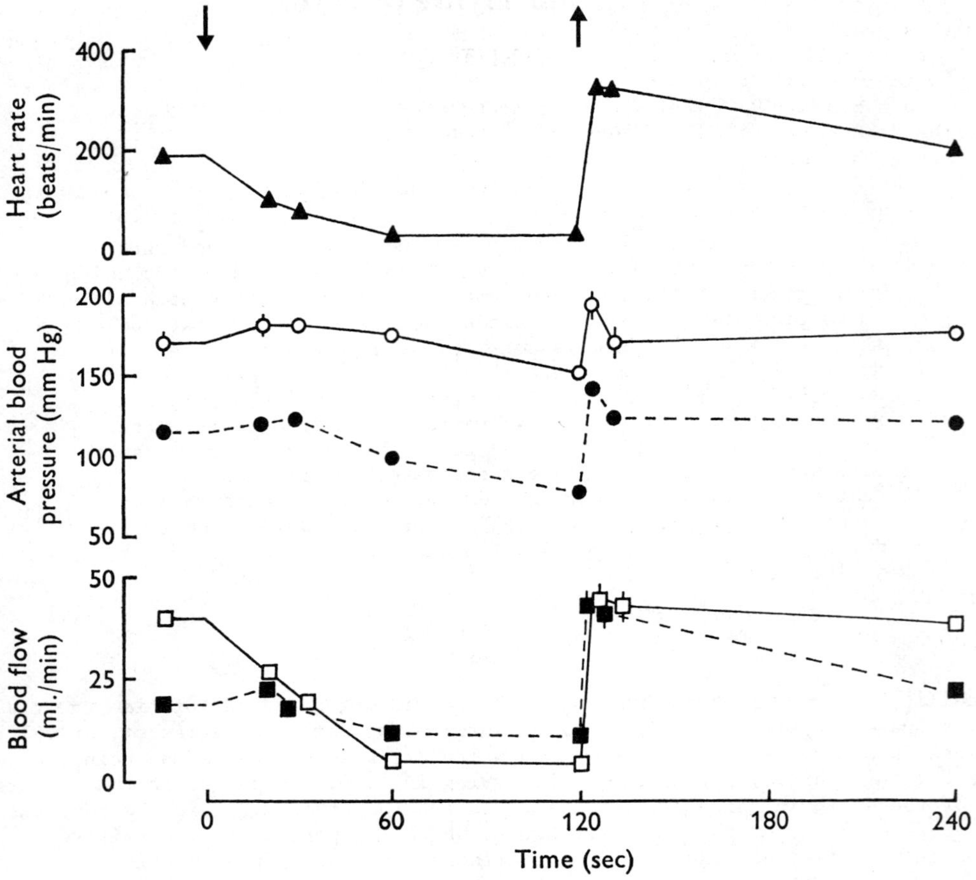

Fig. 1. Effect of enforced head submersion on heart rate and regional blood flow in restrained domestic ducks (*Anas platyrhynchos*). Downward pointing arrow indicates submersion, upward pointing arrow indicates emersion. From above downwards: ▲, heart rate; O, systolic pressure; ●, diastolic pressure; □, sciatic artery blood flow; ■, common carotid artery blood flow. Mean results from 36 'dives' on 13 ducks. Vertical lines through each symbol indicate ± S.E. of mean. Where vertical lines are absent, the S.E. of the mean is within the limits of the symbol. From Butler and Jones, 1971, by permission.

adjustments would have to occur 'at the very onset of apnea' (Irving, 1966). Whether or not this does occur during natural submersion of birds was not known, and there was a clear need for physiological data from free diving birds.

Millard, Johansen and Milsom (1973) used radio telemetry to record pulsatile blood flow through the femoral artery in free-diving penguins (*Pygoscelis papua* and *P. adeliae*). A Doppler shift ultrasonic blood flowmeter (Franklin *et al.*, 1966) modulated the frequency of a miniature FM transmitter which was attached to a float. Butler and Woakes (1979) used an implantable 2 channel FM transmitter (Woakes and Butler, 1975) which allowed the measurement of respiratory frequency and heart rate in naturally diving pochords and tufted ducks (*Aythya ferina* and *A. fuligula* respectively). In both studies, it was found that the birds performed a number of dives in quick succession and that heart rate did decline rapidly at the time of submersion. In the penguins, heart rate during the period of submersion was maintained at approximately 33 percent of the rate measured when the birds were swimming at the surface, and blood flow through the femoral artery was 25 percent of the value when the birds were swimming on the surface. These reductions in heart rate and blood flow through the femoral artery were not as great as those during forced dives but as the authors point out, during the natural dives they were getting 'a composite of diving and exercise responses'.

In the ducks, there were distinct increases in respiratory frequency and heart rate before the first dive of a series (Fig. 2). Following the immediate reduction in

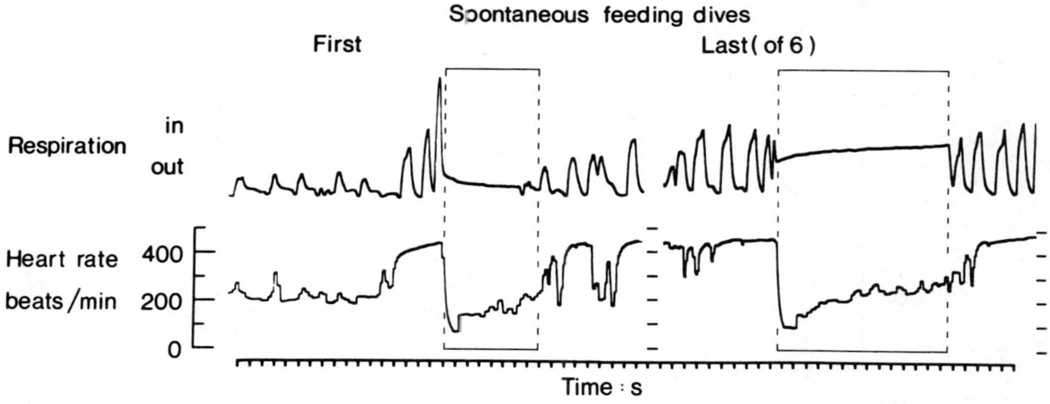

Fig. 2. Traces showing heart rate and respiratory frequency associated with the first and last spontaneous dives of a series performed by a ♂ tufted duck (*Aythya fuligula*) on a pond 0.65 m deep. In each case the period of submersion is indicated by the dashed vertical line joined by horizontal bars. After Butler and Woakes, 1979, by permission.

heart rate upon submersion, there was a progressive increase in cardiac frequency during the remainder of the dive, so that after 6-8 s of submersion, it was close to the rate recorded when the bird was swimming on the surface. This rate was maintained throughout the remainder of the dive. In this series of observations (Butler and Woakes, 1979) the water was only 0.65 m deep, and the average duration of the dives was 10 s for the tufted ducks. More recent data have been obtained from tufted ducks on a pond 2.0-2.2 m deep where dives in excess of 35 s (which is close to the maximum dive duration for these birds — Dewar, 1924) have been recorded. The initial changes in heart rate were similar to those reported in the earlier study. Heart rate continued to increase slightly after the first 10 s of submersion, but 1-2 s before the birds surfaced there was a distinct increase in heart rate (Fig. 3). Similar increases in heart rate just before surfacing were also seen in penguins by Millard *et al.* (1973). These anticipatory changes in heart rate, together with those just before the first dive of a series, are likely to be of central nervous system origin (Millard *et al.*, 1973; Butler and Woakes, 1979).

In all birds studied, a large proportion of the natural dives are of less than 1 min duration (Scholander, 1940; Eliassen, 1960; Kooyman, 1975) and, with the exception of the Emperor penguin, *Aptenodytes forsteri*, (Kooyman, 1975), it is doubtful whether any bird dives spontaneously for longer than 2.5 min. Eliassen (1960) calculated that the guillemot (*Uria aalge*), with an average dive duration of 20 s, would be able to metabolize aerobically while swimming under water at a velocity of 2 m s^{-1} for the whole of this period by using the oxygen stored in the body. There would, therefore, be no need for vasoconstriction in the muscles. The respiratory system contains most of the stored O_2 and the lungs could be supplied by air from the air-sacs by the action of the limbs during under water swimming. Similar calculations have led to a similar conclusion for the pochard duck (Woakes, 1978). The O_2 stores may be enhanced in ducks before the first dive of a series by the increased ventilation and heart rate just before submersion (Fig. 2), and they could be replaced by the few deep breaths between dives.

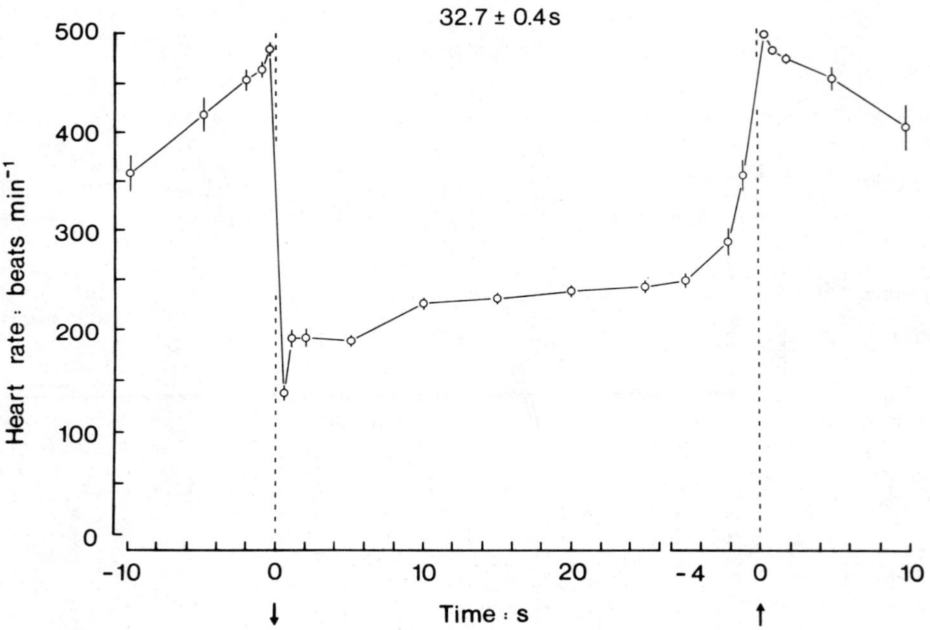

Fig. 3. Graph showing heart rate in tufted ducks (*Aythya fuligula*) before, during and after spontaneous dives (not necessarily the first in a series) in excess of 30 s duration. The birds were on a pond 2 m deep. Mean results from 25 dives by 3 ducks. Vertical lines through each symbol indicate ± S.E. of mean. Where vertical lines are absent, the S.E. of mean is within the limits of the symbol. The mean duration of the dives ± S.E. of mean is given between the vertical dashed lines which delimit the period of submersion.

Eliassen's original work and arguments have been criticized for faults in methodology and logic (Johansen, 1964; Andersen, 1966; Irving, 1966). However, if bradycardia is an integral part of the total response to diving as proposed by Irving, then the absence of such a bradycardia in two species of ducks during natural diving gives some credence to Eliassen's conclusion. Radio telemetry has allowed physiological data to be obtained from birds diving naturally. Some of the results have raised a question mark against a theory which has gained much of its support from so-called dives of abnormally long duration on restrained, inactive birds. It should now be possible to design experiments on free-diving birds to make more meaningful tests of Irving's theory and to examine the influence of peripheral receptors, such as arterial chemoreceptors on heart rate during natural dives.

FLYING

"If one excludes the ratites and penguins, everything birds do is in one way or another linked to flight." Although Bartholomew's (1974) statement may be a slight exaggeration, it is not far from the truth, and some of the achievements of birds during flight are remarkable. Lesser snow geese (*Anser c. caerulescens*) fly non-stop for 2,700 km (Ogilvie, 1978) while bar headed geese (*A. indicus*) have been observed flying at altitudes up to 9,000 m (where the oxygen tension is 45-50 mmHg) during their migration across the Himalayas (Swan, 1961). These extreme examples prompt a number of questions. How much energy do birds use when flying? How is body temperature regulated and how much water is lost in evaporative cooling? Are there special features of the respiratory and circulatory systems that are important during flight?

There are obvious technical difficulties in obtaining physiological data from flying birds. LeFebvre (1964) used a radio isotopic method to get indirect measurements of metabolism during long flights of pigeons (*Columbia livia*) whereas Tucker (1966, 1968) and Pennycuick (1968) trained birds to fly in a wind tunnel. Pennycuick made a number of measurements from ciné films of the flying birds and used a method based on helicopter theory for calculating the power required to fly at a given velocity. Tucker measured oxygen uptake directly either by having the birds fly in a sealed wind tunnel (1966), or by placing a loose-fitting mask over the face of the bird and drawing air through the mask at a known rate (1968). The advantage of Tucker's methods is that any theory can be tested against actual measurements. The objections to his techniques are that it is difficult to estimate the effect of the mask and trailing tubes on the oxygen uptake of the birds and that the wind tunnel may impose unnatural restrictions on the flight patterns of the bird. By using a commercially-available FM transmitter (Narco Bio Systems) attached to the back of the birds, it has been shown that the pattern of wing beating of free flying pigeons is different from that seen in pigeons flying in a wind tunnel (Butler, West and Jones, 1977).

Radio telemetry would seem to be an obvious technique to use to study certain aspects of flight in birds. When using it during studies of orientation of the mallard duck (*Anas platyrhynchos*), Lord, Bellrose and Cochran (1962) accidentally found that they were able to record respiratory activity of the bird. The transmitter which was strapped to the back of the animal, weighed 38 g (including the batteries) and had a range in excess of 150 m. At rest the duck had a respiratory frequency of 14 breaths min^{-1} and after it had been flying for more than 5 min and at a ground velocity of 18 m s^{-1}, the respiratory frequency was 96 breaths min^{-1}. Eliassen (1963 a and b) mounted a transmitter weighing 40 g on the back of mallards or great black-backed gulls (*Larus marinus*) and found that in the gull there was an increase in arterial pulse pressure but little change in heart rate during a flight of 100 m distance. Roy and Hart (1963, 1966) built transmitters for use with free flying birds. Again the transmitters were mounted externally and weighed 36 g and 30 g respectively. In order to obtain ventilation volume and oxygen uptake, a mask was placed over the face of the bird. The results obtained have been published in five major papers (Hart and Roy, 1966, 1967; Berger, Hart and Roy, 1970, 1971; Berger, Roy and Hart, 1970). In all except one paper on temperature regulation (Hart and Roy, 1967) the flights were of short (< 20 s) duration. In all birds studied by these workers (pigeon; evening grosbeak, *Hesperiphona vespertina*; ring billed gull, *Larus delawarensis*; and black duck, *Anas rubripes*), heart rate increased by 2-4× its resting value during flight. Oxygen uptake increased by 12-14× resting and, with the exception of the pigeon, ventilation volume increased by a similar proportion. Respiratory heat loss was approximately 20 percent of total heat production for pigeon and black ducks and except at very low environmental temperatures, water loss exceeded metabolic water production. Core temperature of the pigeon increased during flight and local heat flow over the breast muscles during flight was 6-7× the resting value. It was suggested that pigeons can control the rate of heat loss during flight by vasomotor regulation and by regulating the passage of air through the feathers.

Torre-Bueno (1976) was rather critical of all the previous experiments in which radio telemetry had been used with flying birds. The transmitters were bulky, mounted externally and would most likely affect the flying performance of the birds. In most instances the flights were of less than 20 s duration. He implanted a small transmitter (Southwick, 1973) into starlings (*Sturnus vulgaris*) which had been trained to fly in a wind tunnel and was able to record their core and skin temperatures during flights of 0.5-2 h duration. As in pigeons (Hart and Roy, 1967), core temperature was independent of ambient temperature and Torre-Bueno suggests that insulation is adjusted to maintain a high body temperature which leads to an increase in the efficiency of the muscles.

In the absence of a wind tunnel, radio telemetry has been of limited use with flying birds. With a short-range transmitter, reception (and maybe the transmitter itself) is lost very quickly unless the bird is enclosed in a large room or restrained on a line. The flights are then of short duration. Increasing the power output of the transmitter and maintaining a reasonably long battery life increases the bulk of the system. Ideally a small implantable system is required which can be used for flights of relatively long duration (> 30 min). This can only be achieved by keeping the receiver close to the flying bird.

Barnacle geese (*Branta leucopsis*) and Canada geese (*B. canadensis*) have been raised from eggs and imprinted upon a human. Eventually they will follow this human, even when he is in the back of a moving truck. Using a two channel implantable transmitter (Woakes and Butler, 1975), measurements of heart rate and respiratory frequency have been obtained from barnacle geese, during flights of an average duration of 14.4 min and at a mean air velocity of 18.7 m s^{-1} (Fig. 4). Heart rate increased substantially just before take-off and rose further as flight began. This rate was maintained throughout the flight and was independent of air velocity between 15 and 26 m s^{-1}. On average, resting heart rate was 72 beats min^{-1}, and just before landing it was 515 beats min^{-1}. Respiratory frequency was at its highest value during flight at take-off. It then declined and settled at an average rate of 100 breaths min^{-1} (*cf.* value for mallard duck obtained by Lord *et al.*, 1962). At rest, respiratory frequency was, on average, 8.5 breaths min^{-1}. Heart rate returned to its preflight value within 5 min of landing but respiratory frequency increased to over 250 breaths min^{-1} as the birds panted after landing.

A multichannel transmitter is being developed so that EMGs from the pectoral muscles and temperature from various parts of the body can be monitored in imprinted Canada geese. The development which is potentially most exciting is that of an implantable transducer which will enable ventilation volume and oxygen uptake to be measured on trained Canada geese or swans during long flights (Woakes and Butler, 1980, this volume). It may then be possible to answer some of the queries posed at the beginning of this section and to test the various theories of flight on free-flying birds that are unhindered by back packs, trailing leads or masks.

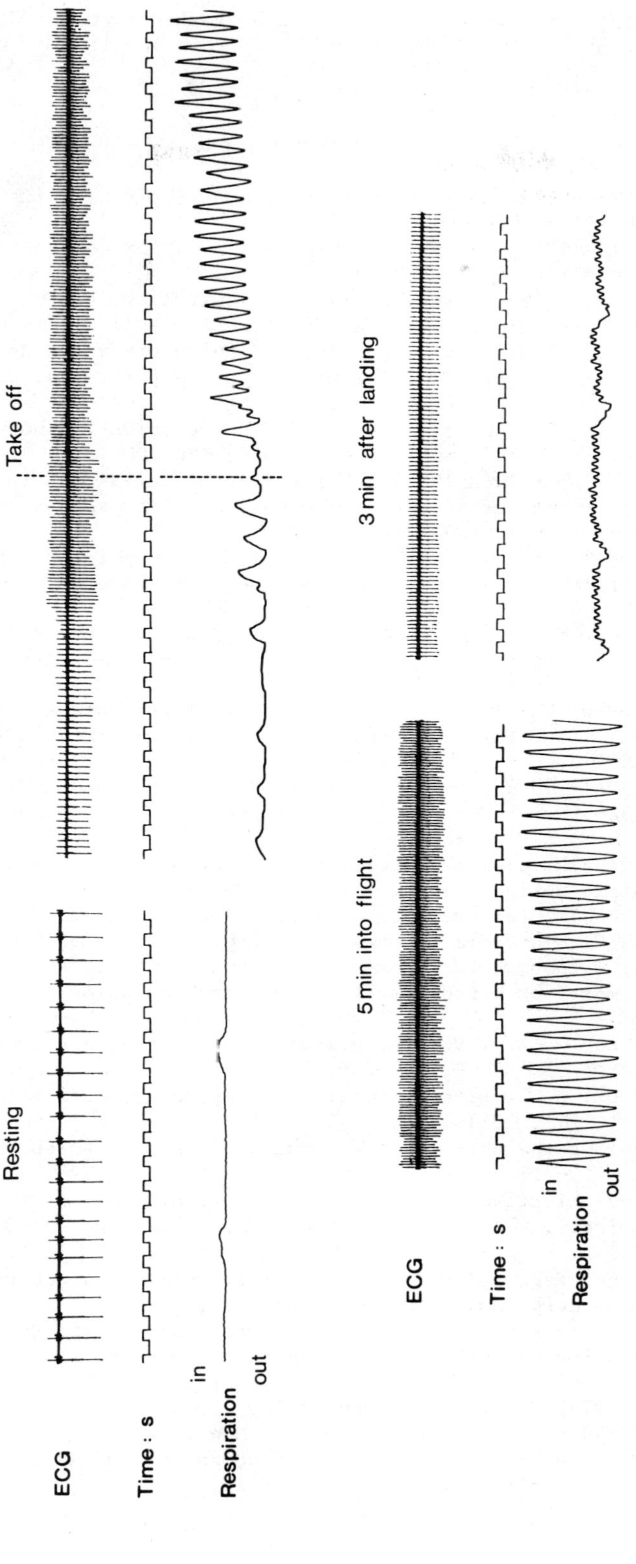

Fig. 4. Traces showing heart rate and respiratory frequency before, during and after the flight of a ♀ barnacle goose (*Branta leucopsis*). The bird flew behind a truck at an average ground velocity of 18.1 m s^{-1} for a duration of 13 min 40 s.

Acknowledgements — This work was supported by grants from the Science Research Council. The goose eggs and the diving ducks were supplied by the Wildfowl Trust, Slimbridge.

REFERENCES

Andersen H. T. (1966) Physiological adaptations in diving vertebrates. *Physiol Rev.* 46, 212-243.

Bartholomew G. A. (1974) In R. A. Rayner (ed.), *Avian Energetics*, p. 330. Cambridge Mass: Nuttal Ornithological Club.

Berger M., Hart J. S. and Roy O. Z. (1970) Respiration, oxygen consumption and heart rate in some birds during rest and flight. *Zt. vergl. Physiol.* 66, 201-214.

Berger M., Hart J. S. and Roy O. Z. (1971) Respiratory water and heat loss of the black duck during flight at different ambient temperatures. *Can. J. Zool.* 49, 767-774.

Berger M., Roy O. Z. and Hart J. S. (1970) The coordination between respiration and wing beats in birds. *Zt. vergl. Physiol.* 66, 190-200.

Butler P. J. and Jones D. R. (1971) The effect of variations in heart rate and regional distribution of blood flow on the normal pressor response to diving in ducks. *J. Physiol.* 214, 457-479.

Butler P. J. and Taylor E. W. (1973) The effect of hypercapnic hypoxia accompanied by different levels of lung ventilation, on heart rate in the duck. *Respir. Physiol.* 19, 176-187.

Butler P. J., West N. H. and Jones D. R. (1977) Respiratory and cardiovascular responses of the pigeon to sustained, level flight in a wind-tunnel. *J. exp. Biol.* 71, 7-26.

Butler P. J. and Woakes A. J. (1979) Changes in heart rate and respiratory frequency during natural behaviour of ducks, with particular reference to diving. *J. exp. Biol.* 79, 283-300.

Dewar J. M. (1924) *The Bird as a Diver*. London: H. F. and G. Witherby.

Eliassen E. (1960) Cardiovascular responses to submersion asphyxia in avian divers. *Arbok. Univ. Bergen* 2, 1-100.

Eliassen E. (1963a) Preliminary results from new methods of investigating the physiology of birds during flight. *Ibis* 105, 234-237.

Eliassen E. (1963b) Telemetric registering of physiological data in birds in normal flight. In L. E. Slater (ed.), *Bio-telemetry*, pp. 257-265. Oxford: Pergamon Press.

Franklin D. L., Watson N. W., Pierson K. E. and Van Citters R. L. (1966) A technique for radiotelemetry of blood flow from unrestrained animals. *Am. J. Med Electron.* 5, 24-28.

Hart J. S. and Roy O. Z. (1966) Respiratory and cardiac responses to flight in pigeons. *Physiol. Zool.* 39, 291-306.

Hart J. S. and Roy O. Z. (1967) Temperature regulation during flight in pigeons. *Am. J. Physiol.* 213, 1311-1316.

Irving L. (1934) On the ability of warm-blooded animals to survive without breathing. *Sci. Mon.* 38, 422-428.

Irving L. (1966) Elective regulation of the circulation in diving animals. In K. S. Norris (ed.), *Whales, Dolphins and Porpoises*, pp. 381-395. London: Cambridge University Press.

Johansen K. (1964) Regional distribution of circulating blood during submersion asphyxia in the duck. *Acta physiol. scand.* 62, 1-9.

Jones D. R. and Purves M. J. (1970) The carotid body in the duck and the consequence of its denervation upon the cardiac responses to submersion. *J. Physiol.* 211, 279-294.

Kooyman G. L. (1975) Behaviour and physiology of diving. In B. Stonehouse (ed.), *Biology of Penguins*, pp. 115-137. London: Macmillan.

LeFebvre E. A. (1964) The use of D_2O^{18} for measuring energy metabolism in *Columbia livia* at rest and in flight. *Auk* 81, 403-416.

Lord R. D., Bellrose F. C. and Cochran W. W. (1962) Radio telemetry of the respiration of a flying duck. *Science, N.Y.* 137, 39-40.

Millard R. W., Johansen K. and Milsom W. K. (1973) Radiotelemetry of cardiovascular responses to exercise and diving in penguins. *Comp. Biochem. Physiol.* 46A, 227-240.

Ogilvie M. A. (1978) *Wild Geese*, p. 295. Berkhamspted, Herts: T. and A. D. Poyser.

Pennycuick C. J. (1968) Power requirements for horizontal flight in the pigeon *Columbia livia*. *J. exp. Biol.* 49, 527-555.

Roy O. Z. and Hart J. S. (1963) Transmitter for telemetry of biological data from birds in flight. Bio-Med Electronics BME, *IEEE Trans.* 10, 114-116.

Roy O. Z. and Hart J. S. (1966) A multichannel transmitter for the physiological study of birds in flight. *Med Biol. Engng* 4, 457-466.

Scholander P. F. (1940) Experimental investigation on the respiratory function in diving mammals and birds. *Hvalrad-Skrift.* 22, 1-131.

Southwick E. E. (1973) Remote sensing of body temperature in a captive 25g bird. *Condor* 75, 464-466.

Swan L. W. (1961) The ecology of the high Himalayas. *Sci. Am.* 205, 68-78.

Torre-Bueno J. R. (1976) Temperature regulation and heat dissipation during flight in birds. *J. exp. Biol.* 65, 471-482.

Tucker V. A. (1966) Oxygen consumption of a flying bird. *Science, N.Y.* 154, 150-151.

Tucker V. A. (1968) Respiratory exchange and evaporative water loss in the flying budgerigar. *J. exp. Biol.* 48, 67-87.

Woakes A. J. (1978) Biotelemetry and its application to the study of avian physiology. M.Sc. Thesis University of Birmingham.

Woakes A. J. and Butler P. J. (1975) An implantable transmitter for monitoring heart rate and respiratory frequency in diving ducks. *Biotelemetry* 2, 153-160.

Woakes A. J. and Butler P. J. (1980) An implantable transducer for the measurement of respiratory air flow. In C. J. Amlaner and D. W. Macdonald (eds), *Handbook on Biotelemetry and Radio Tracking*, pp. 287-292. Oxford: Pergamon Press.

The Use of Radio Telemetry Devices to Measure Temperature and Heart Rate in Domestic Fowl

I. J. H. DUNCAN and J. H. FILSHIE

Agricultural Research Council's Poultry Research Centre, King's Buildings, West Mains Road, Edinburgh EH9 3JS, U.K.

Abstract — Since modern, intensive, husbandry systems may affect adversely both the productivity and welfare of domestic fowl, an understanding of the effects of short term stress is urgently required. Work has already shown what type of situations are liable to frustrate or frighten chickens and what the behavioral responses are likely to be.

In the experiments described, these behavioral changes were correlated with physiological changes using radio telemetry techniques, which were particularly suitable since they resulted in minimal interference with the birds. Thin-film hybrid radio telemetry devices, designed to transmit temperature and electrocardiogram (ECG) signals, were attached to, or implanted in, domestic hens. The signals were picked up on a commercial receiver and recorded on magnetic tape. The ECG signals were analyzed on a small computer using an autocorrelation technique to give heart rate.

Records were taken from undisturbed birds to get some idea of the normal range of temperature and heart rate during maintenance activities. Birds were then exposed to various visual and auditory frightening stimuli. Shank temperature was found to be a good indicator of the Alarm Reaction. Interesting differences in fear reactions were found between strains of fowl. 'Flighty' strains of hen showed far more avoidance and panic to visual stimuli than 'placid' strains. However, the heart rate of the so-called 'placid' birds rose almost as much and took longer to recover than that of the so-called 'flighty' birds.

INTRODUCTION

In recent years, much research effort has gone into studying stress in farm animals. There are two main reasons for this. First, the production or productivity of animals which are stressed may be adversely affected. Second, there has been great public interest in the welfare of animals kept under modern, intensive husbandry systems and it is inevitable that any assessment of welfare must consider the concept of stress.

There are two major ways in which domestic animals may be stressed by modern husbandry methods. They may be crowded together by the intensive nature of the system and they may be frustrated, frightened, bored or over stimulated by the artificial nature of the environment. Among the most intensive and most artificial environments are those created by the modern poultry industry, with the battery cage for laying hens a typical example. Work at the Poultry Research Centre, Scotland, has already shown what types of situations are liable to frustrate or frighten chickens and what their behavioral responses are likely to be (Duncan and Wood-Gush, 1971, 1972, 1972a; Murphy, 1977, 1978). But are birds which are showing these behavioral symptoms necessarily stressed physiologically? In order to answer that question it is necessary to measure the physiological reactions of chickens while they are being frustrated or frightened. The physiological responses of animals to stressful stimuli are usually divided into the Alarm Reaction (also known as the Defence Reaction and Fight or Flight Syndrome) (Draper and Lake, 1968) and the longterm response commonly known as the General Adaptation Syndrome (Selye, 1950). There is evidence that the Alarm Reaction may be relatively more important than the General Adaptation Syndrome in birds compared to mammals (Draper and Lake, 1968). In the Alarm Reaction there is arousal of the sympathetic nervous system and release of adrenaline from the medullary tissue of the adrenal glands with, of course, many secondary effects such as increased heart rate, and redistribution of the blood in the tissues. Evidence for the occurrence of the Alarm Reaction should ideally be based on measurements of the primary physiological responses, that is increased sympathico-adrenal activity. However, these measurements are very difficult to make so the evidence is usually based on some of the secondary changes, and even they are not easy to measure. Conventional techniques may require restraint or the attachment of leads or probes, procedures which are themselves likely to be stressful. For this reason it was decided to use radio telemetry techniques to measure physiological changes associated with the Alarm Reaction in domestic fowl. In a study of stress of this type, radio telemetry techniques are particularly suitable since they result in minimal interference to the experimental subject.

RADIO TELEMETRY TRANSMITTER

The transmitter is manufactured as a thin-film hybrid microcircuit on a 20 × 10 mm substrate. It is capable of transmitting both temperature and biopotentials such as the electrocardiogram (ECG) by summing the outputs from the separate front ends into a frequency modulated Colpitts oscillator operating at 104.5 MHz. The oscillator has a modulation sensitivity of approximately 15 KHz mV^{-1} and can be temperature compensated to better than 1 KHz $^{o}C^{-1}$. The circuit diagram of the transmitter is shown in Fig. 1. The temperature circuit is an astable multivibrator, the mark/space ratio of which is modulated by a thermistor (Duncan, Filshie and McGee, 1975). The ECG amplifier is a low current two-stage amplifier with an overall gain capability of 200 and a frequency response of 5 Hz-7 KHz. For the avian ECG the gain is reduced to 20 thus producing an oscillator frequency deviation of 30 KHz for the 100 μV QRS complex. The transmitter front end consumes approximately 50 μA and the oscillator current consumption can be set from 50 μA to 1 mA, which allows the range to be varied from 1 to 50 m.

EXPERIMENTS WITH BIRDS

SKIN TEMPERATURE STUDIES

It was stated earlier that during the Alarm Reaction there is a redistribution of blood in the tissues. Blood is shunted away from the viscera and skin into the main muscle masses to prepare the animal for fight or flight. The shunting of blood away from the viscera may be particularly important in the case of laying hens, which are forming eggs on three out of four days throughout the year. The egg formation

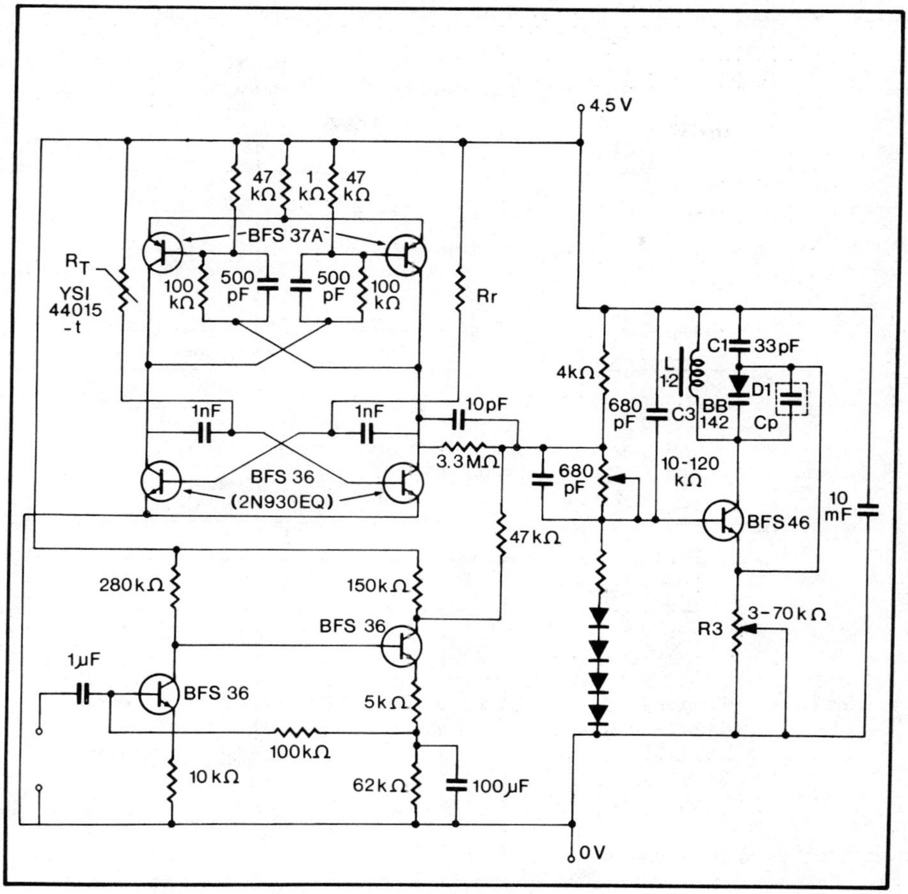

Fig. 1. Circuit diagram of two-channel biotelemetry transmitter.

process is almost continuous with the yolk spending about 6 h in the oviduct receiving albumen and about 18 h in the uterus receiving membranes and shell. To cope with this massive formation process, the reproductive tract of the hen receives over 20 percent of the cardiac output (Draper and Lake, 1968). Obviously anything which interferes with the blood flow to this region carries with it the risk of disturbing egg formation. Of course, it is not easy to demonstrate directly that the flow of blood to the reproductive tract is disturbed when the hen gets a fright. However, we were able to show that blood is shunted away from the feet and lower legs and this is accompanied by a drop in temperature of these regions (Duncan, Filshie and McGee, 1975). The likelihood is that there is a simultaneous shunting of blood away from the viscera, including the reproductive tract.

In these experiments on skin temperature, the radio transmitter was attached externally to a harness made of nylon strapping worn by the bird. The transmitter and battery pack were situated on the bird's back and the thermistor lead was taken down behind the wing and taped to the inside of the tarso-metatarsal part of the leg. The hens were tested in a Perspex cage measuring 60 × 60 × 60 cm. The signal was picked up on a commercial VHF receiver through a dipole antenna outside the cage and passed via a decoder to a Bausch and Lomb VOM5 pen recorder. Because the thermistors were interchangeable, it was possible to calibrate the system by replacing the thermistor with resistors corresponding to the maximum and minimum temperatures of the required working range.

Figure 2 shows the drop in temperature which occurred when the birds were given a fright by waving a black cloth over the cage to simulate an aerial predator.

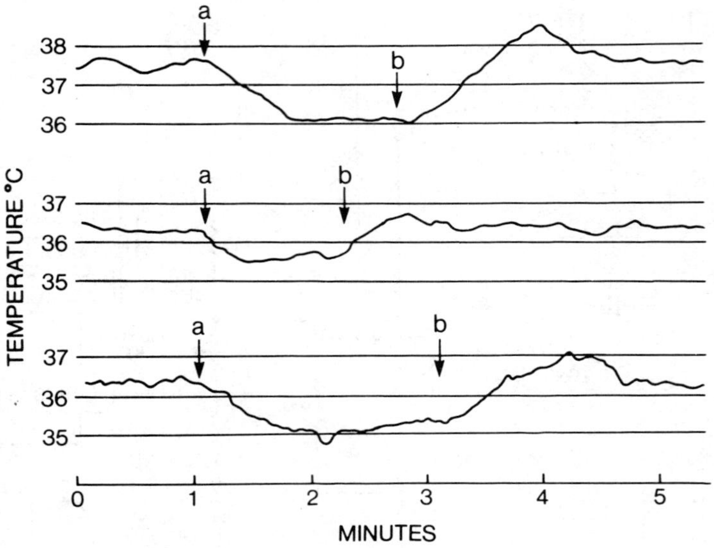

Fig. 2. Temperature traces from the lower leg of three domestic hens. A black cloth was waved over the feeding birds at (a). The birds stopped alarm-calling and resumed feeding at (b).

HEART RATE STUDIES

One of the disadvantages of having radio telemetry devices attached externally, particularly in the case of birds, is that they do pay attention to them, even when they have been fitted with dummy devices for a week before testing. It was therefore decided that the transmitters for heart rate studies should be implanted subcutaneously. A 40 mm incision was made in the skin covering the right pectoral muscle with the bird under pentobarbitone sodium anesthesia. The device, coated in silicon rubber, was slipped under the skin with the battery pack lying over the left, and the transmitter over the right, pectoral muscle and both held in place by a suture round the connecting lead and through the connective tissue anterior to the keel bone. Silver ring electrodes were sutured to the connective tissue covering the keel bone, one at the anterior end and the other 60 mm posterior to it. These electrode placements gave a large, distinct QRS complex as shown in Fig. 3, which was suitable for analysis. The birds were allowed two days to recover and testing began on the third day after surgery. The testing cage, antenna and receiver were the same as those used for the skin temperature studies. The ECG signal was stored on magnetic tape for later analysis by a small computer. Initially the QRS complex was used to trip a Schmitt Trigger and the elapsed time was measured until the next QRS wave. Latterly, an autocorrelation technique has been used which is more reliable, particularly when muscle noise and electrical interference are present (Clark and Filshie, 1977).

The normal heart rate of the undisturbed domestic fowl is much lower than was previously thought. Conventional methods involving trailing leads have yielded 'normal'

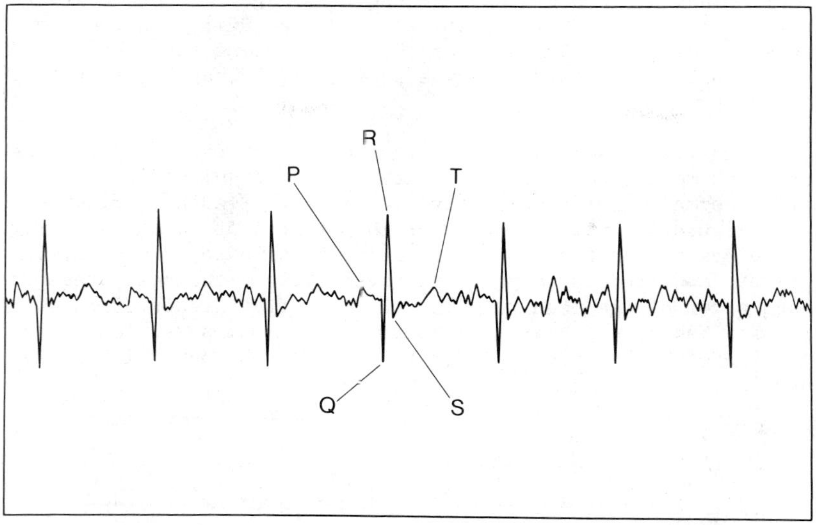

Fig. 3. A typical telemetered electrocardiogram of a hen with electrodes 60 mm apart on the keel.

rates of 350-450 beats min^{-1} (Ringer, Weiss and Sturkie, 1957). In the present experiments, heart rate varied between 280 and 320 beats min^{-1} when the birds were active but undisturbed during the day and fell to 230 beats min^{-1} at night.

We were particularly interested in differences that might exist between two commonly used commercial strains of laying fowl. One was a light weight hybrid derived from the White Leghorn breed and usually described as 'flighty' in nature. The other strain was a medium-weight hybrid derived from the Light Sussex and Rhode Island Red breeds and commonly described as 'placid'. For comparison we also included a third strain of broiler chicken which was a heavy-hybrid kept for meat production and usually described as 'placid'. Everyone who has dealt with these strains would agree with this description of 'flighty' or 'placid'. However, it should be pointed out that this classification is based entirely upon the birds' reactions to human beings. When approached, the flighty birds show extreme avoidance and violent escape behavior whereas the placid birds show only mild avoidance. Recently however, Murphy (1977, 1978) has shown that these strains behave differently when tested with other fear-inducing stimuli.

Apart from human beings, the most potent stimulus for eliciting avoidance and escape in domestic fowl is a balloon being inflated quickly close to the bird. There seems to be something about the looming properties of an inflating balloon which birds find very alarming. This was used as one of the frightening stimuli. A red balloon attached to the center of one of the walls of the test cage on the inside was inflated for about 4 s, held inflated for 4 s and then allowed to deflate in 2 s. The test cage was situated behind a one-way screen and the inflation procedure was controlled out of sight of the bird. The stimulus was presented five times with an interval of 10 min between each presentation.

For comparison another, more natural, stimulus was also used. This was a 10 s sound recording of other birds giving high intensity, ground predator, alarm calls (Wood-Gush, 1971). This stimulus was also presented five times with an interval of 10 min between each presentation.

Six birds of each of the three strains aged between 12 and 15 months were used in the experiment. They were each placed in the test cage for three days to settle down. They then received the visual stimuli and the auditory stimuli on successive days in a random order. The behavior of the birds during the tests was recorded as a commentary on another channel of the same magnetic tape which was recording the ECG.

In response to the visual stimulus, the light-hybrid birds showed extreme panic, squawked and tried to escape. However, they recovered quickly and most of them were performing normal maintenance activities such as feeding, drinking and preening within 5 min. Figure 4 gives a summary of the effects on heart rate. It can be seen that there was a huge increase in heart rate following presentation of the stimulus but this had dropped to pre-test levels before the next presentation. The small standard error of the mean rate immediately after presentation suggests that perhaps this rate (about 450 beats min^{-1}) is the maximum possible physiologically. There was no sign of habituation, either behaviorally or physiologically, after five presentations.

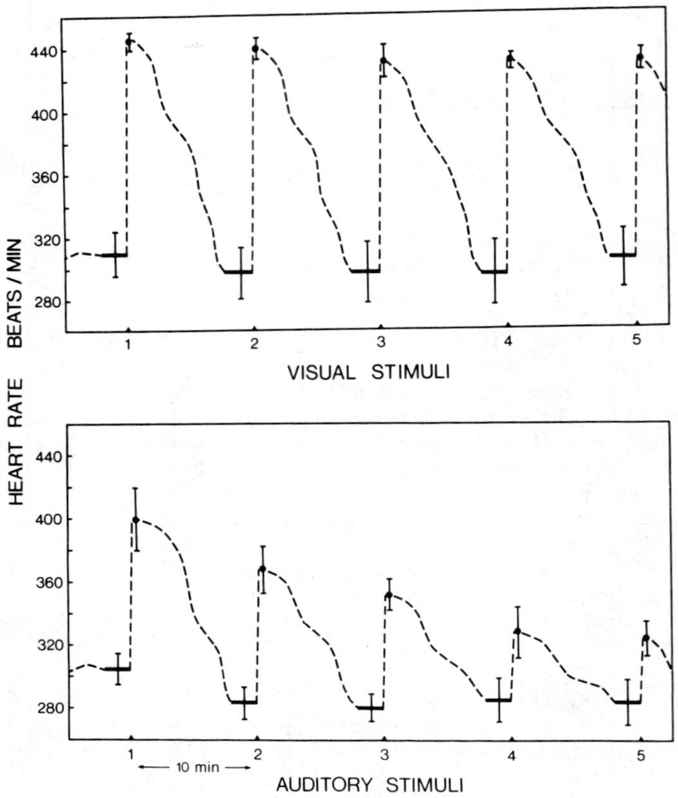

Fig. 4. The heart rate of light-hybrid hens (*n*=6) presented five times with a visual stimulus (above) and an auditory stimulus (below). The bars show the average heart rate (±S.E.) for 2 min before, and the dots show the average of the highest heart rate reached (±S.E.) in 20 s after, stimulus presentation.

The medium-hybrids and broilers also panicked when the balloon was inflated. However, although they showed a lot of alarm and escape, the reaction did not seem as violent as that of the light-hybrids. An interesting feature was that the escape behavior tended to give way to an alert posture with the bird standing motionless fixating visually some aspect of the environment. The posture was not rigid, in fact the birds looked quite relaxed, but once adopted, it tended to be maintained for a long period of time and often the birds had not resumed maintenance activities when the next stimulus was presented. Figures 5 and 6 summarize the effect on heart rate. It can be seen that the increase in rate tended to be less and more variable than that shown by the light-hybrids but there was a tendency for the rate not to drop back to the original pre-test level. Again, there was no sign of habituation after five presentations.

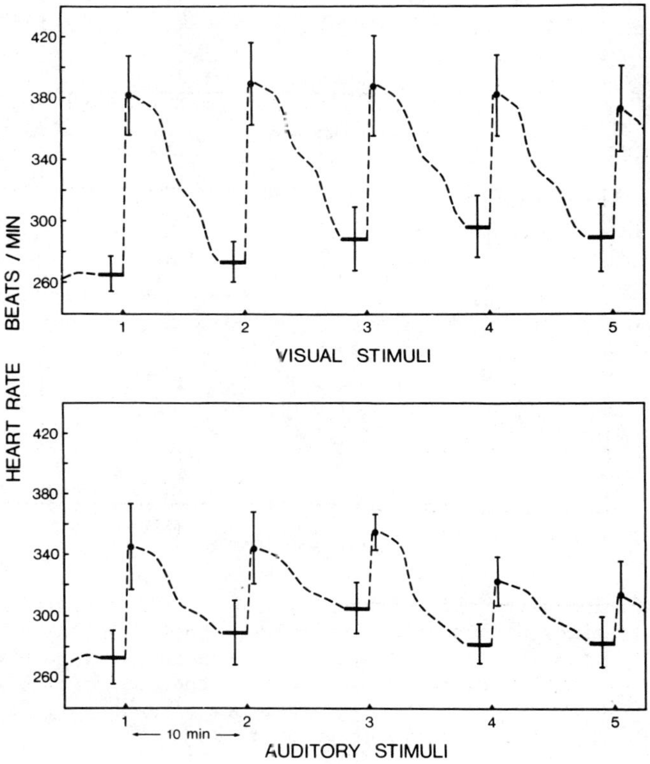

Fig. 5. The heart rate of medium-hybrid hens (n=6) presented five times with a visual stimulus (above) and an auditory stimulus (below). The bars show the average heart rate (±S.E.) for 2 min before, and the dots show the average of the highest heart rate reached (±S.E.) in 20 s after, stimulus presentation.

All the strains behaved similarly to the auditory stimulus. At the first presentation the birds showed a large startle response followed by escape but not the panic which characterized the balloon presentation. There was also obvious habituation and by the fifth stimulus the birds were paying little attention. The heart rate response correlated well with this behavior and a summary is shown in Figs 4-6.

Therefore the biggest difference between the strains was that the so-called 'flighty' birds showed extreme panic to the balloon and a huge increase in heart rate but recovered fairly quickly whereas the so-called 'placid' birds showed slightly less panic and less increase in heart rate and adopted an alert motionless posture for a long time afterwards. Inspection of individual heart rate records revealed that

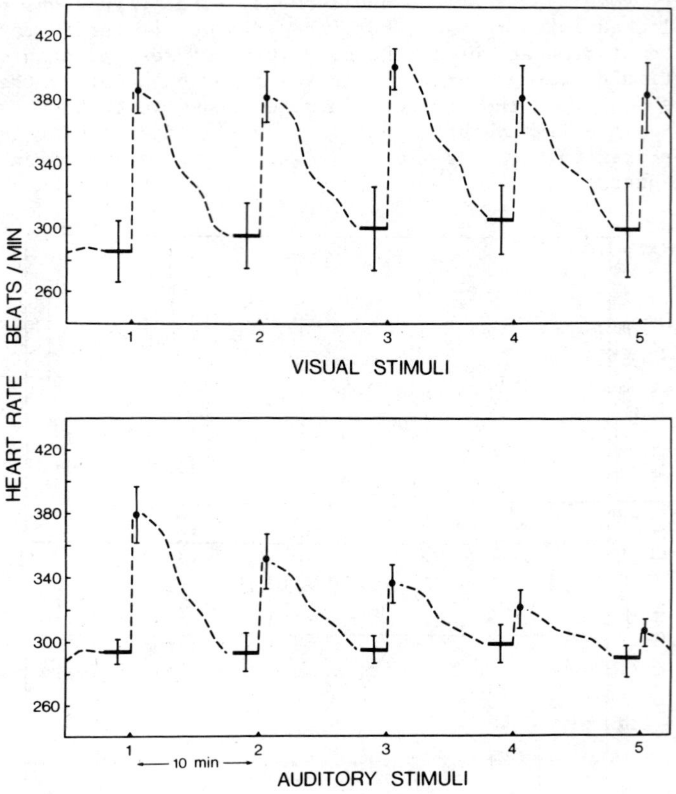

Fig. 6. The heart rate of broiler hens (n=6) presented five times with a visual stimulus (above) and an auditory stimulus (below). The bars show the average heart rate (±S.E.) for 2 min before, and the dots show the average of the highest heart rate reached (±S.E.) in 20 s after, stimulus presentation.

heart rate remained high as long as this motionless posture was maintained. Examples of records from the light-hybrid and a medium-hybrid are shown in Figs. 7 and 8.

Radio telemetry techniques have thus proved invaluable in this study of short term stress. Changes in skin temperature and heart rate have been found in response to stimuli which frighten the birds behaviorally. Some of the physiological changes correlate well with obvious behavioral changes whereas others correspond to more subtle changes in behavior. The differences that have been found between strains of domestic fowl in response to a frightening visual stimulus suggests that some reappraisal is required of the classification of birds as 'flighty' or 'placid', since docile birds may be as frightened as flighty birds in physiological terms.

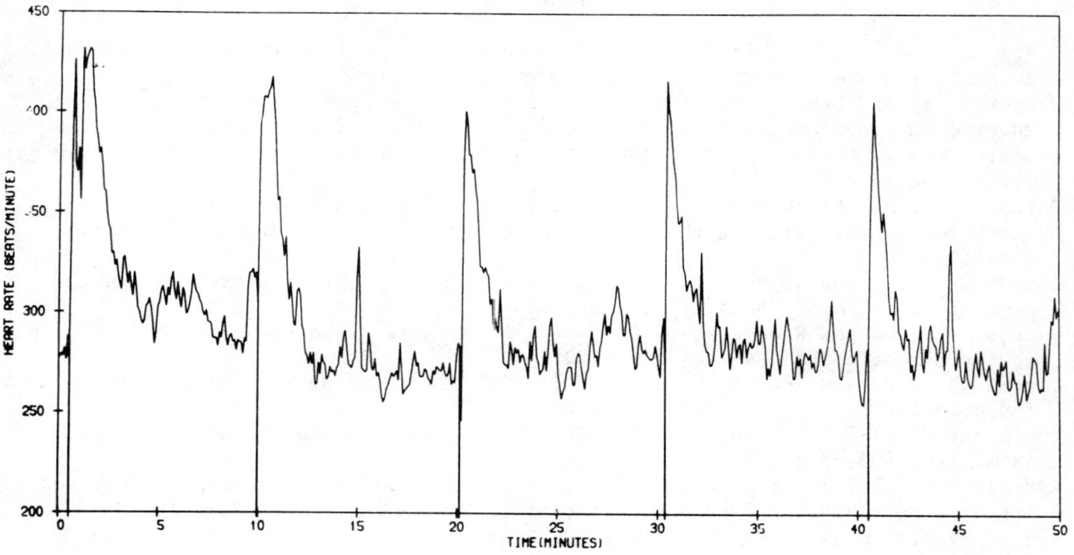

Fig. 7. Heart rate record of a light-hybrid hen presented five times with a visual stimulus. The heart rate line comprises 3 s averages at 5 s intervals. The stimulus presentations are denoted by vertical cursors below the heart rate line.

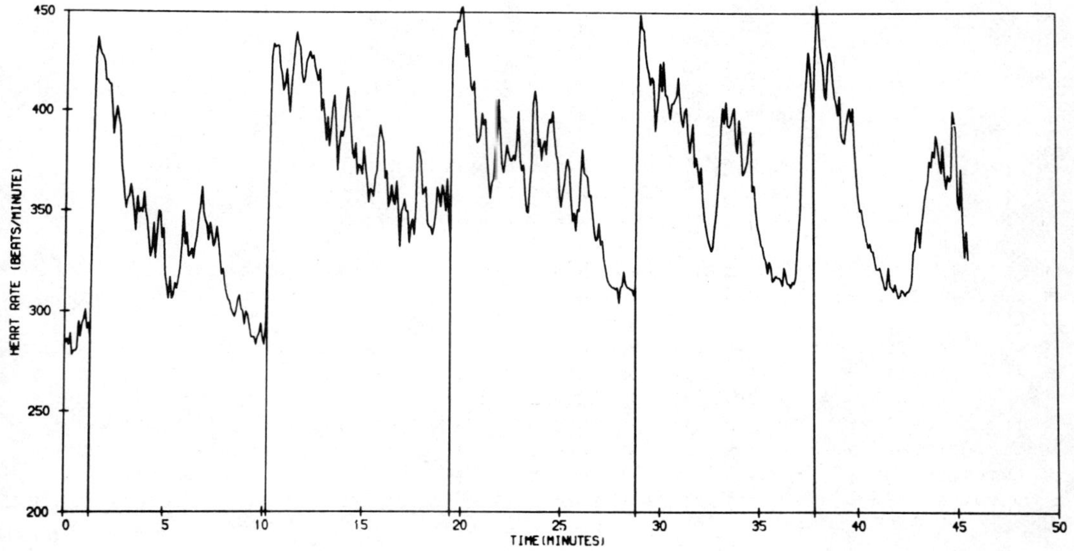

Fig. 8. Heart rate record of a medium-hybrid hen presented five times with a visual stimulus. The heart rate line comprises 3 s averages at 5 s intervals. The stimulus presentations are denoted by vertical cursors below the heart rate line.

REFERENCES

Clark J. S. B. and Filshie J. H. (1977) Autocorrelation techniques for measuring avian heart rates. *Med. biol. Engng Comput.* 15, 656-665.

Draper M. H. and Lake P. E. (1968) Physiological reactions of the laying fowl to adverse environments, with special reference to the defence reaction. In T. C. Carter (ed.), *Environmental Control in Poultry Production*, pp. 87-100. Edinburgh: Oliver and Boyd.

Duncan I. J. H., Filshie J. H. and McGee I. J. (1975) Radiotelemetry of avian shank temperature using a thin-film hybrid microcircuit. *Med. biol. Engng* 13, 544-550.

Duncan I. J. H. and Wood-Gush D. G. M. (1971) Frustration and aggression in the domestic fowl. *Anim. Behav.* 19, 500-504.

Duncan I. J. H. and Wood-Gush D. G. M. (1972) An analysis of displacement preening in the domestic fowl. *Anim. Behav.* 20, 68-71.

Duncan I. J. H. and Wood-Gush D. G. M. (1972a) Thwarting of feeding behaviour in the domestic fowl. *Anim. Behav.* 20, 444-451.

Murphy L. B. (1977) Responses of domestic fowl to novel food and objects. *Appl. Anim. Ethol.* 3, 335-349.

Murphy L. B. (1978) The practical problems of recognizing and measuring fear and exploration behaviour in the domestic fowl. *Anim. Behav.* 26, 422-431.

Ringer R. K., Weiss H. S. and Sturkie P. D. (1957) Heart rate of chickens as influenced by age and gonadal hormones. *Am. J. Physiol.* 191, 145.

Selye H. (1950) *Stress: The Physiology and Pathology of Exposure to Stress*. Montreal: Acta Inc.

Wood-Gush D. G. M. (1971) *The Behaviour of the Domestic Fowl*, pp. 39-40. London: Heinemann.

Changing Heart Rates of Herring Gulls when Approached by Humans

N. J. BALL and C. J. AMLANER, JR.

Animal Behaviour Research Group, Department of Zoology, University of Oxford, South Parks Road, Oxford OX1 3PS, U.K.

Abstract — Herring gulls (*Larus argentatus*) respond to an approaching human by an 'emotional' tachycardia. This response, which did not habituate, had a form dependent on the distance between the gull and an approaching human. It could also be modified by changing the speed of approach. Because there are reasons for believing this to be of adaptive significance, it is suggested that the response can be used in the study of animal behavior.

INTRODUCTION

Considerable changes in the heart rate of an animal may result from the approach of a potential predator (Belkin, 1968). A decrease in heart rate (bradycardia) has been recorded, under various circumstances, from ground squirrels, *Citellus armatus* (Ruff, 1971), willow grouse, *Lagopus lagopus* (Gabrielsen *et al.*, 1977) and woodchucks, *Marmota monax* (Smith, 1978), while increases (tachycardia) occur in herring gulls, *Larus argentatus* (Kanwisher *et al.*, 1978) and diving ducks, *Aythya ferina* and *A. fuligula* (Butler and Woakes, 1979).

There appears to be little evidence as to the function of these responses. Gabrielsen *et al.* (1977) comment that: (1) heart rates in gulls may double without a visible change in behavior (Kanwisher *et al.*, 1978); (2) human heart rates prior to a parachute jump may double without a change in metabolism (Ursin, Baade and Levine, 1977); (3) there is individual variation: flight may occur from a state of either bradycardia or tachycardia in willow grouse when they are approached. For reasons such as these, Gabrielsen *et al.* (1977) call the decreased heart rate of these grouse 'emotional' bradycardia.

The 'emotional' nature of these changes suggests interesting possibilities for the study of animal behavior. It may be that the response is determined by characteristics of the predator (size, potential danger, speed etc.) and of the subject's behavior (how important it is; for example, protecting young may be more important than resting). Standardizing some of these factors, modifying others and correctly interpreting the heart rate changes could therefore give us an index of the animal's 'assessment' of the value of its behavior. This pilot study was designed to investigate these possibilities by evaluating the shape of the reaction curves for captive herring gulls and to see if the curves could be experimentally modified.

METHOD

A captive herring gull (*Larus argentatus*) was placed in the experimental chamber (Fig. 1) 0.5 h before the start of a session, each of which comprised about 10 trials. Only one session was conducted each day. A normal trial (the 'standard approach') consisted of the experimenter walking right up to the gull starting from a spot 50 m away and then turning, and walking back. The first trial of any session was always discarded as we were interested only in steady-state responses. Subsequent trials were at 5 min intervals.

Fig. 1. A herring gull in the experimental chamber. Plexiglas on the front helps to reduce noise. The gull can stand and walk but not fly in the chamber.

Two experiments differed from this pattern:

1. The 'interrupted approach' in which the experimenter stopped for 40 s at every 5 m during the walk. By comparing heart rates while the experimenter was moving and while stationary, with those obtained from the standard approach, the influences of time, movement and distance were investigated.

2. Different speeds of approach. Fast (1.6 ms^{-1}), medium (0.7 ms^{-1}) and slow (0.3 ms^{-1}) speeds were compared. A 3 × 3 Latin Square was used to determine the order of 9 trials.

An AM-FM transmitter was modified from Morrow and Taylor (1976) to evaluate heart rates of free ranging animals. The transmitter (Fig. 2) consists of a two stage Darlington amplifier, subcarrier oscillator (1-2 KHz) and a crystal controlled RF

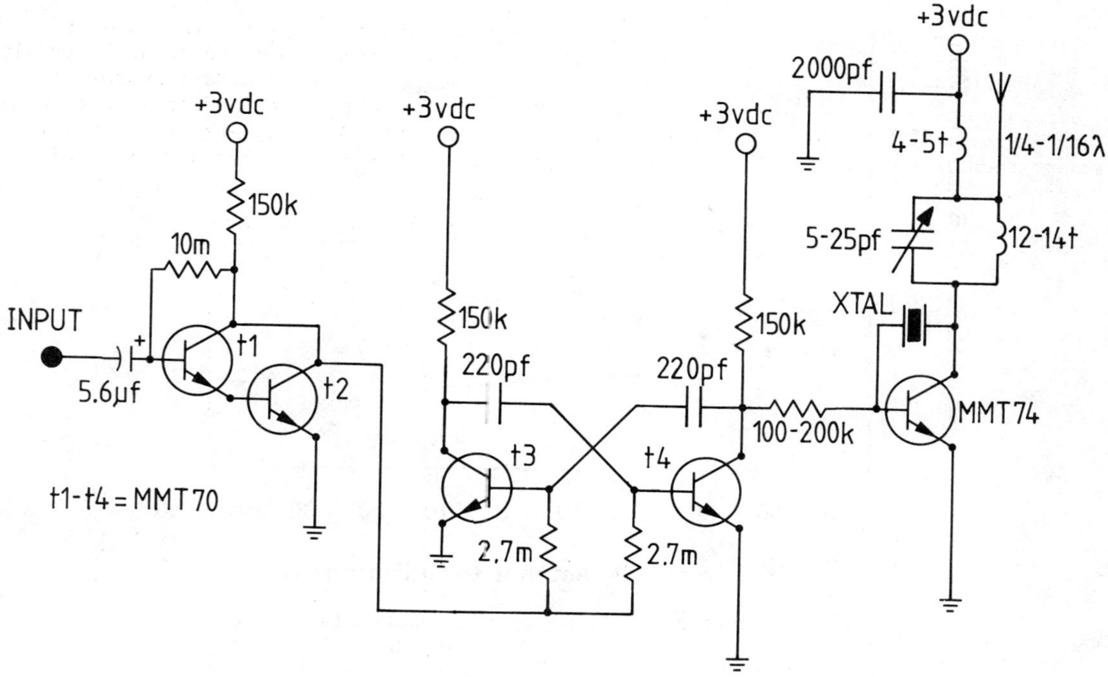

Fig. 2. An AM-FM bipotential transmitter (after Morrow and Taylor, 1976) for the 150-180 MHz band. The crystal (Xtal) = 0.33x frequency and the coils are wound on a 3.18 mm form.

carrier (173 MHz). Input voltages of 0.2-2.0 mV caused a shift in subcarrier frequency which was detected by a Tandy Corporation VHF receiver. The signal was recorded directly on magnetic tape and also demodulated by a phase locked loop for graphical display and analysis.

For most trials, heart rates were calculated by counting the number of beats every 2 s. The mean of 2 consecutive inter-heart beat intervals was used in the experiments on habituation and on the effect of varying speed, where 2 s periods were too long. Wherever relevant, standard errors are shown.

RESULTS

The relationship between heart rate and distance is shown in Fig. 3. There are two periods of slow increase in heart rate (50-10 m, 5-0 m) 'sandwiching' a period of much faster increase. It is composed of the mean of 10 consecutive trials within a session, during which time there was no habituation (Fig. 4). Figure 5 is a comparison between an interrupted approach and a standard approach. At most distances, heart rates for the standard approach are more similar to the values when the experimenter is moving during an interrupted approach than when stationary. In Fig. 6 the influence of the experimenter's speed of movement on the shape of the heart rate-distance curve is shown. In general, faster speed of movement gives an earlier, more drawn out and, ultimately, perhaps a more extreme reaction.

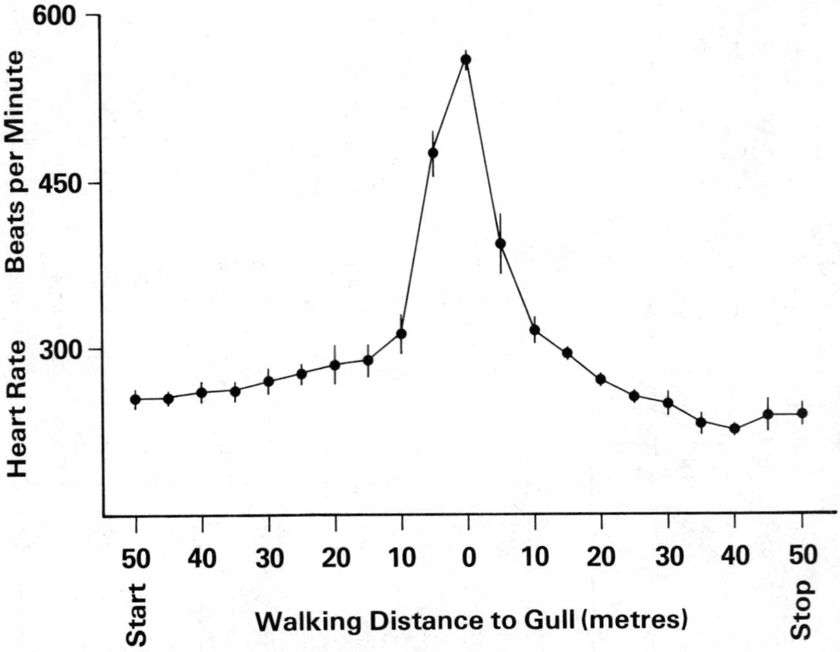

Fig. 3. Mean heart rates (± S.E.) for ten trials on the same gull.

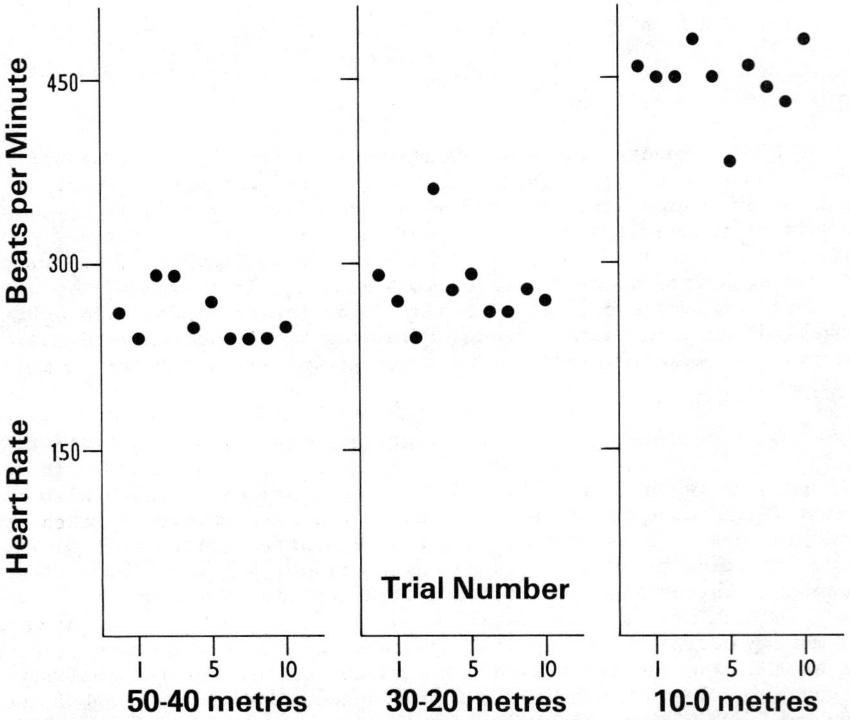

Fig. 4. Each point is the mean heart rate for the 3 distances (at 5 m apart) within the range indicated, i.e. 50–40 m.

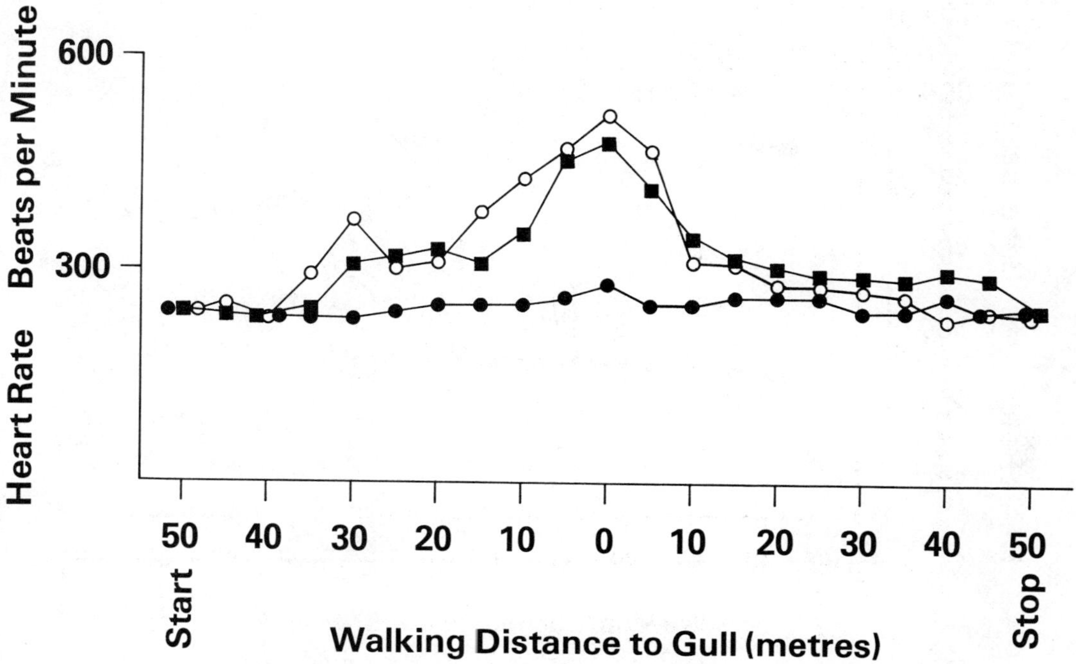

Fig. 5. Heart rates of the gull at specific distances from the experimenter. Two different trials are given: (i) The standard approach but the experimenter stops every 5 m for 40 s. ● Heart rates while stationary. o Heart rates while moving. (ii)■ Heart rates during standard approach.

DISCUSSION

Tachycardia (an increase in heart rate from about 240 to nearly 600 beats min^{-1}) was found in captive herring gulls when approached by a human. The form of the heart rate-distance curve (Fig. 3) may reflect the gull's evaluation of the danger represented by the approaching human, implying that danger increases slowly when the predator is far away and at a 'critical distance' it increases suddenly. The smaller changes in heart rate at very short distances may either indicate little difference in danger whether the human is 5 or 0 m away, or alternatively that a maximum heart rate has been reached.

It is important to demonstrate that approach distance influences heart rate and not some factor correlated with it. For example, if heart rate varied with time after physiological stimulation, Fig. 3 would imply a false relationship with distance. An interrupted approach was therefore compared with a standard approach (Fig. 4). The 40 s interruption in walking made little difference to the form of the curve (except when the experimenter was stationary), suggesting that heart rates are indeed distance specific under these conditions.

Is the heart rate-distance curve (Fig. 3) fixed, or is it flexible and capable of being modified? When the experimenter was stationary during the interrupted approach (Fig. 4), the gull's heart rate reduced to the basal level. Motion of the predator can therefore influence the shape of the curve. An experimental manipulation of the approach speed (Fig. 6) also resulted in changes to the response. A faster speed led to an earlier, more drawn out and perhaps a more extreme reaction. Both of these

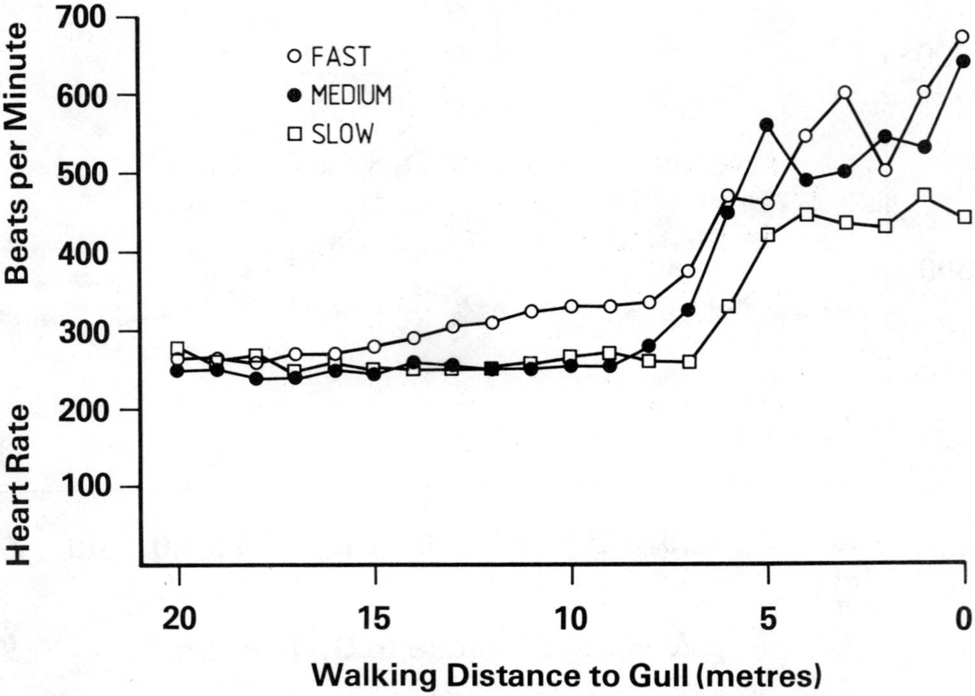

Fig. 6. Approaches were made at three speeds. o Fast (1.6 ms^{-1}), ● Medium (0.7 ms^{-1}), and □ Slow (0.3 ms^{-1}).

manipulations are assumed to represent changes in the amount of danger presented to the predator, and the changes in the response are more extreme for greater amounts of danger. It therefore seems that these 'emotional' heart rate responses are adaptively significant and that the form of the tachycardia may give information about the gull's assessment of the danger it is exposed to during an approach. This possibility is currently being investigated in field experiments (Ball and Amlaner, in preparation).

REFERENCES

Belkin D. A. (1968) Bradycardia in response to threat. *Am. Zool.* 8, 775.

Butler P. J. and Woakes A. (1979) Changes in heart rate and respiratory frequency during natural behaviour of ducks, with particular reference to diving. *J. exp. Biol.* 79, in press.

Gabrielsen G., Kanwisher J. and Steen J. B. (1977) 'Emotional' bradycardia: a telemetry study on incubating Willow grouse (*Lagopus lagopus*). *Acta physiol. scand.* 100, 255-257.

Kanwisher J. W., Williams T. C., Teal J. M. and Lawson K. O., Jr. (1978) Radiotelemetry of heart rates from free ranging gulls. *Auk* 95, 288-293.

Morrow J. L. and Taylor D. H. (1976) A telemetry system for recording heart rates of unrestrained game birds. *J. Wildl. Mngmt.* 40, 359-360.

Ruff R. L. (1971) Telemetered heart rates of free living Uinta ground squirrels in response to social interaction. Ph.D. dissertation. Utah State Un., Logan, Utah.

Smith E. N. (1978) Fear-evoked bradycardia of free ranging woodchuck. *The Physiologist* 21, abstract.

Ursin H., Baade E. and Levine S. (1977) *Coping Men — a study of human psychobiology*. London: Academic Press.

Telemetry of Heart Rate as a Possible Method of Estimating Energy Expenditure in the Redshank *Tringa totanus* (L.)

P. N. FERNS,* I. H. MACALPINE-LENY* and
J. D. GOSS-CUSTARD**

**Zoology Department, University College, Cathays Park, Cardiff CF1 1XL, U.K.*
***Institute of Terrestrial Ecology, Furzebrook Research Station, Wareham, Dorset, U.K.*

Abstract — In order to determine the optimal foraging strategies of free living birds, some measure is required of the energetic costs of different types of foraging activities. Conventional calorimetry techniques cannot be used for this purpose. If oxygen consumption can be predicted from heart rate, the latter can be monitored by radio telemetry and used as a method of estimating energy expenditure of birds in the field.

A VHF telemetry system was developed to monitor heart rate in laboratory reared Redshanks. The relationship between heart rate and oxygen consumption was then measured in a small open circuit respirometer, using a paramagnetic oxygen analyzer. Under these circumstances, there was a good correlation between heart rate and oxygen consumption, although the relationship varied in the same individual bird on different days and between different individuals.

The relationship between heart rate and foraging activity was measured on a small laboratory mud flat, using a manually operated activity recording system. There was a significant correlation between heart rate and activity but great care needs to be taken when attempting to apply these results to the field situation. The technique does, however, offer considerable promise for the future.

INTRODUCTION

The concept of optimal foraging has aroused considerable interest among behavioral ecologists during recent years (MacArthur and Pianka, 1966; Charnov, 1976; Krebs, 1978). The central question is whether or not animals maximize their net rate of intake of energy when foraging for food under natural conditions. To answer this, it is necessary to know both the rate of intake of various types of food and the amounts of energy expended in locating, capturing and consuming it.

The Redshank, *Tringa totanus* (L.) is a widely distributed wading bird which feeds predominantly on mud and sand dwelling intertidal invertebrates. Considerable information already exists on the rates of foraging and food intake in this species (Goss-Custard, 1970, 1977a). Furthermore, rates of food intake can readily be converted into energy units, and comparisons made between the energy intake when feeding on different types and sizes of prey (Goss-Custard, 1977b). However, no established methods are available for determining the energy expended during various types of foraging activities. This paper presents an interim report of an investigation into the possibility of using heart rate, monitored by telemetry, as a means of measuring the metabolism of foraging Redshanks. The relationship between heart rate and metabolism was first established by measuring oxygen consumption in a respirometer. Heart rate was then monitored in birds feeding on a small laboratory mud flat.

METHODS

Six Redshanks were used in this study. They were reared from eggs collected under licence in Avon and were maintained on a diet of chopped liver, maggots, fish and turkey starter (Goss-Custard, Wilkins and Kear, 1971).

Full details of the telemetry and recording system are given by Macalpine-Leny (1978). The transmitter consisted of a 102 MHz Hartley oscillator with capacitive feedback to the emitter, which was frequency modulated by changes in the base current (Fig. 1). The circuit was constructed from discrete miniature and subminiature components and encapsulated in Araldite. The resultant package was mounted on top of a Perspex battery holder housing two RM312 Mallory cells. A current drain of 2.5 mA gave a battery life of about 12 h and a range of several meters. Each transmitter measured 22 × 14 × 13 mm, weighed about 5.6 g, and had an input sensitivity of 4 kHz mV^{-1}. ECG was picked up by means of small safety pin electrodes, pinned to a fold of skin. The positive electrode was located above the anterior tip of the sternum and the indifferent electrode dorsally at the base of the tail. This typically produced a positive R wave of 1 mV at the input to the transmitter.

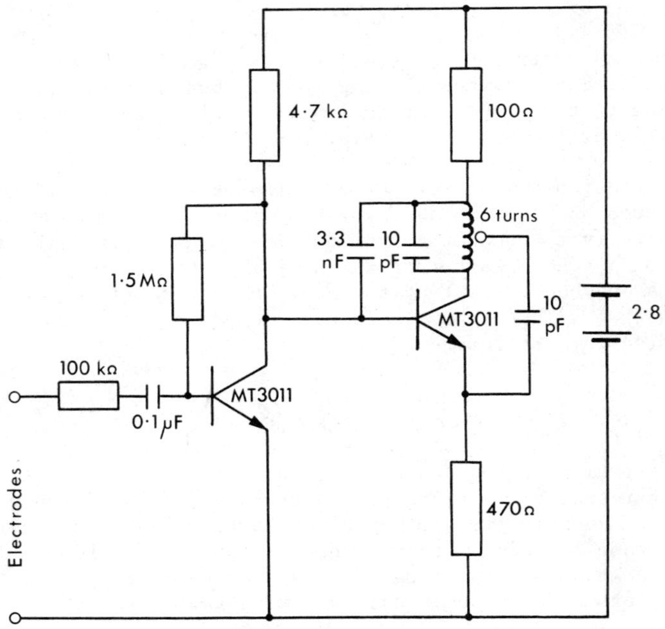

Fig. 1. Transmitter circuit.

The receiving system consisted of an indoor FM antenna and a Leak Stereofetic FM tuner. The output from the tuner was fed via a pair of Sallen and Key active high pass and low pass switchable filters to one channel of a Sony TC252 stereo tape recorder. The amplified ECG was then passed through a current-operated R wave detector circuit which compensated for changes in R wave magnitude (Evans, 1975). The output, consisting of pulses of constant magnitude and duration, was used to trigger a Devices Instantaneous Ratemeter (Type 2751).

Oxygen consumption was measured in an open circuit system. Individual Redshanks were sealed in a cylindrical Perspex respirometer, 190 mm high and 140 mm in diameter, immersed in a water bath at -7.5°C. Air was passed through the respirometer at a known rate (about 500 cm^3 min^{-1}) and the oxygen content of the effluent, after water vapor and CO_2 had been removed, was measured using a Taylor Servomex OA272 paramagnetic oxygen analyzer. When the system had reached equilibrium (20 min to one hour), two or three measurements of oxygen content and heart rate were recorded. Measurements were made at different heart rates by progressively raising the temperature of the water bath to 15°C in six steps, allowing adequate time for equilibrium to be reached at each temperature.

Foraging activity was studied on a small laboratory mud flat, one meter square, arranged in such a way that the observer could not be seen by the experimental bird. Each Redshank was introduced to the mud flat for at least two days prior to the start of a trial. The transmitter was fitted at the beginning of the second day and the bird was starved for 24 h. At the start of each trial, the Redshank was removed and 30, 60 or 100 maggots were scattered at random on the surface of the mud. The bird was then allowed to reenter from a side chamber and its behavior was recorded using a simple push button activity box. This converted each of four continuous subdivisions of activity — standing, searching, feeding and preening — into a different analog voltage, upon which were superimposed individual paces, pecks and prey captures. This information was displayed alongside instantaneous heart rate on the second channel of a Devices M2 chart recorder. Each trial continued until the Redshank ceased moving actively.

RESULTS

158 measurements of oxygen consumption and heart rate were made on four birds in a total of 11 trials. There was a highly significant correlation between oxygen consumption and heart rate in every trial, with an average of 94.1 percent of the variation in oxygen consumption being accounted for. Analysis of covariance showed that the regressions for bird RS1 differed significantly in the three trials ($F_{2,33}$ = 46.8, $P < 0.01$), as it also did in RS2 (3 trials, $F_{2,44}$ = 56.9, $P < 0.01$) and RS6 (2 trials, $F_{1,26}$ = 141.3, $P < 0.01$). There was no significant difference between the regressions for RS3 (3 trials, $F_{2,40}$ = 0.2, $P > 0.10$). In the case of RS6, there was no significant difference in the slopes of the regression lines between the two trials (t = 1.57, $P > 0.10$) and therefore they must have differed in elevation. Comparisons between the closest pairs of regression coefficients in RS1 and RS2 suggest that all three trials differed in slope in RS1 (trial 1 × trial 3, t = -7.16, $P < 0.001$; trial 3 × trial 2, t = -8.06, $P < 0.001$) and that the third trial differed from the other two in RS2 (trial 2 × trial 1, t = -0.72, $P > 0.10$; trial 1 × trial 3, t = -7.90, $P < 0.001$).

These differences show that the relationship between oxygen consumption and heart rate is not consistent. However, the absolute magnitude of the differences is small, and they are statistically significant only because the correlations are almost perfect. Examination of the results obtained for every bird (Fig. 2), showed a similar range of variation in each individual. There were no statistically significant differences between birds. We feel justified therefore in combining all the results to obtain the most realistic estimate of the relationship and its range

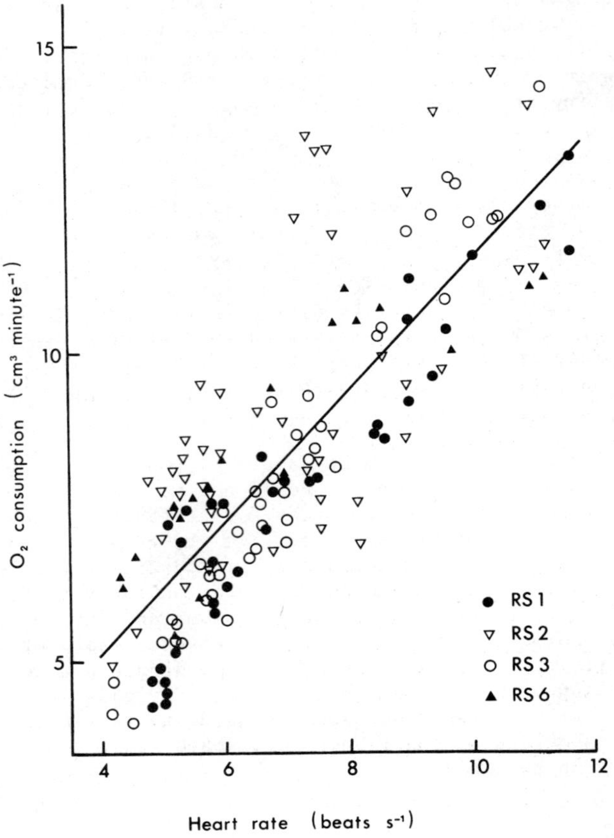

Fig. 2. Relationship between oxygen consumption and heart rate in the respirometer (all trials combined for each bird).

of variation. Differences in weight between individual Redshanks (95-115 g) did not account for a significant amount of the variation in oxygen consumption. The overall equation, based on the results in Fig. 2, is as follows

$$VO_2 = 1.062\ HR + 0.866$$

(where, VO_2 = oxygen consumption in $cm^3\ min^{-1}$, and HR = heart rate in beats s^{-1}). This regression is highly significant ($r = 0.83$, $P < 0.001$) and accounts for 69.3 percent of the observed variation in oxygen consumption. The 95 percent confidence limits for predicting mean values are given by $\pm\ 1.980\ \sqrt{0.001134 + 0.003202\ HR^2}$. For example, the predicted oxygen consumption at a heart rate of 10 beats s^{-1} is $11.49 \pm 1.12\ cm^3\ min^{-1}$.

Observations of foraging activity included 17 trials on six Redshanks, and involved five hours of active feeding. Heart rate was clearly affected by factors other than the level of activity. For example, the heart rate of a bird asleep with its head under its wing typically rose from about 5 beats s^{-1} to about 10 beats s^{-1} when someone entered the laboratory or when another Redshank gave an alarm call. This was accompanied by little outward change in the bird, other than the opening of its eyes. The average heart rate of alert Redshanks (±SD) in the side chamber prior to the commencement of each trial, was 6.8 ± 1.0 beats s^{-1}, and the average rate in the first minute after they had walked onto the mud flat and commenced feeding was 10.7 ± 1.3 beats s^{-1} (Fig. 3). Active foraging thus caused a rise in heart rate of about 3.9 beats s^{-1}, equivalent to a rise in oxygen consumption of 4 cm^3 min^{-1}. This is an increase of 51 percent over standing alert levels. There was usually a delay between the beginning of a period of intense activity, such as rapid pacing, and the consequent rise in heart rate. This delay amounted on average to 12.7 s, but the range of values recorded was large (-5 to +23 s), and occasionally a rise in heart rate actually preceded the onset of activity. There was a steady decline in heart rate throughout the course of each trial as the amount of activity decreased. Maximum heart rates of about 14 beats s^{-1} were recorded after bouts of wing fluttering and short flights. It was not possible to monitor heart rate in flight using the safety pin electrodes because of interference from EMG. There was no significant correlation between heart rate, either before or after commencement of a trial, and the temperature of the mud flat cage, though the latter only varied between 18 and 22°C. More detailed analysis of these results is in progress, with the specific objective of assessing the average heart rate increments associated with different amounts of pacing and pecking.

DISCUSSION

The difficulties associated with the use of heart rate as a means of measuring metabolism have been described by Johnson and Gessamen (1973). They concluded that heart rate can be a fair to good index of the hourly metabolism of resting or moderately exercising individuals. The main problem is that oxygen consumption can be affected by both the cardiac stroke volume and the difference in oxygen content of the arterial and venous blood (a-v O_2), as well as by the heart rate itself (as predicted by the Fick equation). Changes in stroke volume are probably of minor importance in birds, but changes in a-v O_2 are significant (e.g. Butler, West and Jones, 1977). It is thus possible that the tachycardia induced in inactive Redshanks by conspecific alarm calls could be accompanied by a-v O_2 changes.

The strong correlation between oxygen consumption and heart rate which was observed in Redshanks, and the fact that the regressions varied between different birds on the same day and in the same bird on different days, are markedly similar findings to those of Morhardt and Morhardt (1971) for several species of rodents.

Three factors need to be carefully considered before any attempt can be made to extrapolate these results to the field. Firstly, there is a possibility that, because the temperature necessary in the respirometer to produce heart rates as high as those of foraging Redshanks were many degrees colder than that of the mud flat trials themselves, there might have been differences in a-v O_2 levels in these two situations. Secondly, the mud flat was of small size and it is therefore possible that the average length or speed of pacing may have been restricted. Thirdly, all the prey items were scattered on the surface of the mud and therefore no estimate can be made of the energy expenditure involved in probing. Despite these reservations, the technique of predicting the metabolism of foraging birds from their heart rates offers considerable promise.

Acknowledgements — We are indebted to the following for their advice, encouragement or assistance — Professor D. Bellamy, Dr. B. W. Watson, Dr. B. T. Evans, Dr. P. J. Butler and P. Wilkins.

Fig. 3. Redshank with telemeter attached, foraging on the laboratory mud flat.

REFERENCES

Butler P. J., West N. H. and Jones D. R. (1977) Respiratory and cardiovascular responses of the pigeon to sustained, level flight in a wind-tunnel. *J. exp. Biol.* 71, 7-26.

Charnov E. L. (1976) Optimal foraging: attack strategy of a mantid. *Am. Nat.* 110, 141-151.

Evans B. T. (1975) The long term monitoring and analysis of the instantaneous heart rate of rats using implanted VHF telemetry capsules. Unpublished Ph.D. Thesis, Univ. of London.

Goss-Custard J. D. (1970) The responses of redshank (*Tringa totanus* (L.)) to spatial variations in the density of their prey. *J. Anim. Ecol.* 39, 91-113.

Goss-Custard J. D. (1977a) The energetics of prey selection by redshank, *Tringa totanus* (L.), in relation to prey density. *J. Anim. Ecol.* 46, 1-19.

Goss-Custard J. D. (1977b) Optimal foraging and the size selection of worms by redshank, *Tringa totanus*, in the field. *Anim. Behav.* 25, 10-29.

Goss-Custard J. D., Wilkins P. and Kear J. (1971) Rearing wading birds in captivity. *Avicult. Mag.* 77, 16-19.

Johnson S. F. and Gessamen J. A. (1973) An evaluation of heart rate as an indirect monitor of free-living energy metabolism. In J. A. Gessamen (ed.) *Ecological Energetics of Homeotherms*. pp. 44-54. Utah State University Press.

Krebs J. R. (1978) Optimal foraging: decision rules for predators. In J. R. Krebs and N. B. Davies (eds.) *Behavioural Ecology*. pp. 23-63. Oxford: Blackwell Scientific Publications.

Macalpine-Leny I. H. (1978) The develcpment of a VHF telemetry system to investigate the relationship between oxygen consumption and heart rate in the Redshank, *Tringa totanus*. Unpublished M.Sc. Thesis, Univ. of Wales.

MacArthur R. H. and Pianka E. R. (1966) On the optimal use of a patchy environment. *Am. Nat.* 100, 603-609.

Morhardt J. E. and Morhardt S. S. (1971) Correlations between heart rate and oxygen consumption in rodents. *Am. J. Physiol.* 221, 1580-1586.

Telemetry of Heart Rate as a Measure of the Effectiveness of Dispersal Inducing Stimuli in Seagulls

J. F. STOUT and E. R. SCHWAB

Biology Department, Andrews University, Berrien Springs, Michigan 40104, U.S.A.

Abstract — Seagulls on and around airport runways represent a significant hazard. Several procedures have been developed to disperse the aggregated gulls. Research supported by the U.S. Air Force Bird Air Strike Hazard (BASH) program has been conducted to develop new dispersal procedures and to evaluate the relative effectiveness of alternative techniques. Ten ring-billed gulls (*Larus delawarensis*) have been maintained in the laboratory. An FM transmitter has been developed that is crystal controlled and transmits heart rate over at least a 0.4 km range using a low cost high fidelity tuner as the receiver. The unpotted weight of the transmitter, minus batteries is 12 g and 25-30 g when ready for operation. Recordings of the FM modulated ECG were made on magnetic tape. The demodulated ECG was evaluated using a PDP 8E computer. Prerecorded distress calls, pyrotechnics and models of dead seagulls were compared with respect to their effectiveness on elevating the heart rate of a gull and keeping it elevated. The models proved most effective in maintaining heart rates over relatively long periods of time. The effect of the models on a gull's heart rate correlates well with simultaneous observations of the gull's behavior, and with field experiments and observations on the effectiveness of models on dispersal, as compared with distress calls and/or pyrotechnics.

INTRODUCTION

Seagulls on and around airport runways represent a significant hazard for aircraft. Seagulls are involved in more than 50 percent of the bird strikes with commercial and military aircraft (Thorpe, 1975). Of these seagull strikes, more than half occur during landing, takeoff or taxiing (Nemergut, 1973).

Steps to control gulls, and other problem bird species involve environmental modifications, making the airport less attractive to birds, and dispersal inducing procedures (Blokpoel, 1976). The most common methods used to disperse gulls, or other problem species, on an airport runway include driving a vehicle toward the group of gulls, while playing their distress call loudly. Pyrotechnics are also frequently discharged, causing loud noises. Although these, and other forms of harassment will usually cause loafing gulls to disperse, they often return rather quickly. Habituation to the frightening stimuli is also commonly observed whether they be pyrotechnics or distress calls (Blokpoel, 1976; Stout *et al.*, 1975).

Research, supported by the U.S. Air Force Bird Airstrike Hazard (BASH) program has been conducted to develop new dispersal procedures and to evaluate the relative effectiveness of alternative dispersal procedures. Prolonged and intensive studies of the communication of gulls (Stout and Brass, 1969; Stout, 1975; Galusha and Stout, 1977; Amlaner and Stout, 1978) using feathered models (dummies) of gulls, mounted in different display postures, indicated that body position was a very important informational element in gull communication. As a consequence we have developed models of seagulls which are effective in dispersing conspecifics from the sites where the models are deployed (Stout *et al.*, 1975). Gulls return to sites where models are left in position much less frequently than they do the same areas after the models are removed. The models have remained effective during continuous deployment of three to six months.

The work presented in this paper is a first step in evaluating the relative effectiveness of dispersal inducing stimuli, using radio transmitter ECG as a measure of the gulls response to each stimulus. The laboratory studies reported here will be continued and extended to include relevant field experiments.

MATERIALS AND METHODS

Ten ring-billed gulls (*Larus delawarensis*) were captured and maintained in the laboratory. Experimental birds were transferred individually to the room where all trials were conducted. This experimental room was 4 × 2.5 × 2.75 m, and had two perches, consistently used by the gulls.

A transmitter was designed and constructed (similar to Smith, 1974, except that it was crystal controlled) that operated on the commercial FM band. A standard high fidelity tuner was used as the receiver. The frequency modulated signal was recorded on magnetic tape. Later, the tape recorded signal was demodulated using a phase locked loop circuit to reproduce the ECG signal. The demodulated ECG was fed into a Schmitt trigger input of a PDP8e laboratory computer. The computer was triggered on the QRS component of the ECG, and programmed to calculate the heart rate, in beats min^{-1}, from the average of five successive QRS intervals. The results were plotted on a graphics terminal and photographed.

The transmitters were potted in urethane foam and weighed 25 to 30 g, including lithium batteries. A transmitter was mounted on the gull's back, just behind the neck, using a chest harness. Safety pin electrodes were attached through the skin just below the neck, and in the middle of the back. After an initial attempt to reject the transmitter, it was accepted within a few minutes to several hours, and the birds flew and appeared to behave normally in the laboratory.

After mounting the transmitter on a bird, it was placed in the experimental room until the end of experimentation with that bird. Soon after acceptance (cessation of struggling), heart rates decreased to their minimal, base line values (160 to 250 beats min^{-1}), and remained relatively stable. The birds were left in the experimental room for at least one night before experiments began.

Trials involved the application of one of three procedures shown to be effective in causing dispersal of gulls in the field. The stimuli included: (a) Playback of the pre-recorded distress call (at 90 dB, 1 m from the speaker). (b) Placing a fiberglass model of a dead ring-billed gull at the site most frequently used by the gull for sitting, resting, preening, etc. (c) Exploding a firecracker in the room with the gull.

Heart rates were recorded before and during the application of a stimulus, and recordings were usually continued after the stimulus had been applied until the heart rate had stabilized. The behavior of the experimental gull was monitored using closed circuit TV. A description of the bird's behavior was recorded simultaneously with heart rate on the other track of a stereo tape.

RESULTS

Figure 1 shows the effects of various disturbances on the heart rate of a ring-billed gull. In fig. 1 the disturbances and flight behavior were all followed by a rapid return of the heart rate to its baseline rate.

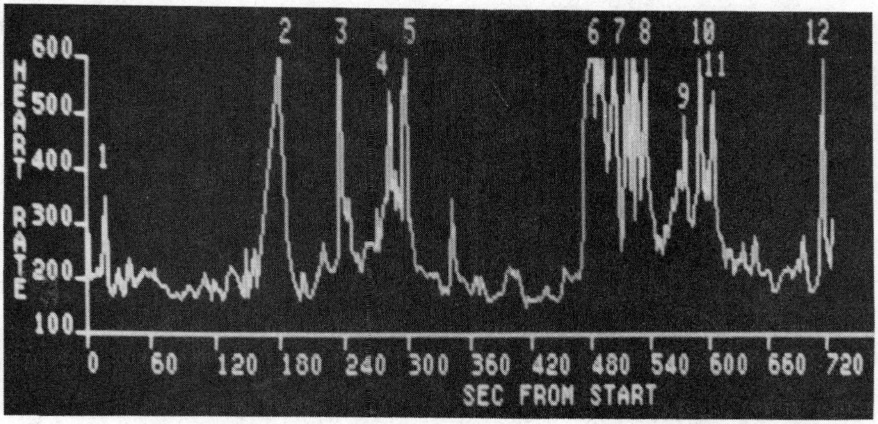

Fig. 1. Changes in the heart rate of a ring-billed gull in response to various human disturbances. (1) Noise outside room, (2) hand waved in front of window, (3-5) gull flew from one end of the room to the other and landed, (6) experimenter walked into the room and left immediately, (7-11) gull flew around room.

Figure 2A shows the response of a gull at 1 to the experimenter's entry into the room, followed by an immediate exit. The gull flew very briefly in response and then settled down. Prior to the experimenter's walking into the room the heart rate had remained at about 300 beats min^{-1}. It returned to 300 beats min^{-1} within 90 s of the experimenter's exit. However, the heart rate continued to decrease below the preceding baseline. Starting with 2 and ending with 3 in Fig. 2A, a distress call was played back at 90 dB for one min. The heart rate accelerated from about 250 beats min^{-1} to somewhat more than 350 beats min^{-1} with transient increases to as high as 450 beats min^{-1}. The bird did not fly during or after the playback of the distress call. Very shortly after the distress call was turned off at 3, the heart rate returned to its prestimulus level, and then continued to decrease to levels varying between 200 and 250 beats min^{-1}. The initial response of the fifth distress call (Fig. 2C) was walking and then sitting, with a moderate increase in heart rate until the stimulus was turned off. Following stimulus cessation, the heart rate rapidly decreased to a relatively constant, prestimulus rate. In general the acceleration in heart rate was smaller for the fifth playback, than the preceding responses. Each of the five playbacks resulted in progressively declining heart rates, thus showing habituation to playback of the distress call.

The effect of pyrotechnics on heart rate is illustrated in Fig. 3, which begins with the gull flying to the floor at 1. During 2 it is sitting at rest on the floor.

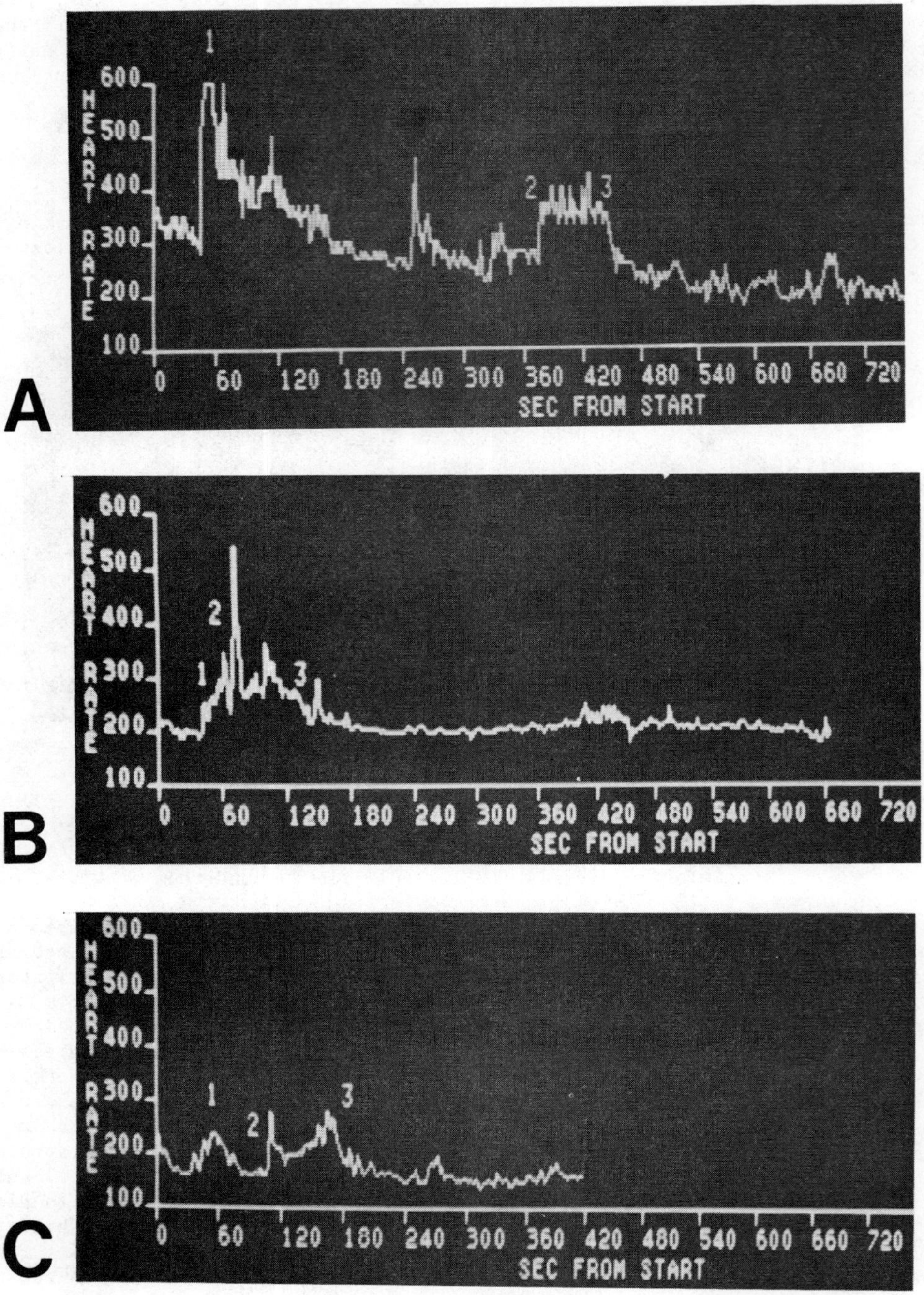

Fig. 2. The effects of the (A) first, (B) second, and (C) fifth playbacks of ring-billed gull distress calls on the heart rate of a ring-billed gull. The distress call is turned on at 2 and off at 3 in each figure.

At 3, a firecracker was slipped under the door of its room but did not go off. The bird flew up to a perch and its heart rate rapidly returned to normal. A second firecracker was slipped under the door at 4 and the heart rate increased briefly and began to decrease when the firecracker exploded at 5.

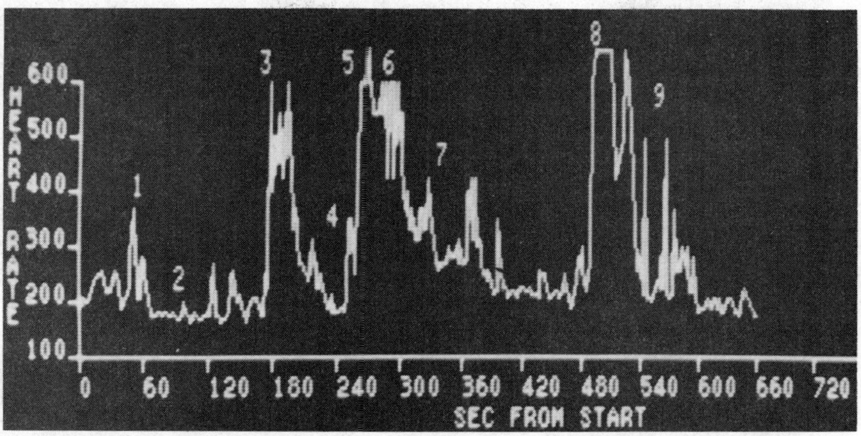

Fig. 3. The effect of the explosion of a firecracker on the heart rate of a ring-billed gull. The firecracker exploded at 5.

The effect of a fiberglass model (dummy) of a dead ring-billed gull on the heart rate of the experimental gull is shown in Fig. 4A. The experimenter entered the gull's room at 4 and placed a fiberglass model on its favored perch. The gull flew back and forth between perches (5-9). 10 represents a 'run' in heart rate unaccompanied by flight and it then sits on the other perch, while the heart rate remained elevated above the basal rate for the remainder of the recording, with 'runs' in heart rate unaccompanied by flight or walking at 12 and 13. As neither entry into the experimental room, nor leaving other novel objects caused such prolonged increases in heart rate, it seems apparent that the prolonged period of time until the heart rate returned to a resting level was a response to the model of a dead gull. A similar slow return to a basal rate for a different gull is seen in Figs. 4B and C. Each approach toward the model (2, 3, 4 and 5) was accompanied by short term increases in heart rate, while it generally remained higher than the usual basal rate for this gull (200 beats min^{-1}).

DISCUSSION

Increases in the heart rate of gulls, in response to threatening stimuli have been noted by Kanwisher *et al.* (1978) in response to human sounds and movements, and in response to the approach of humans by Ball and Amlaner (1980, this volume), Kanwisher *et al.* (1978) also described an increase in heart rate that occurred just before the gull flew. The ptarmigan (Gabrielson, 1979) responds to disturbance by remaining motionless, depending on its cryptic coloration for protection, and experiences a heart rate decrease when disturbed. Thus, it appears that heart rate changes that occur in threatening situations are correlated with the animal's behavior most likely to follow. Interpreted in this light, a more prolonged increase in heart rate for gulls is indication of an increased readiness to escape or flee.

We observed increases in heart rate in the ring-billed gull prior to flight and in response to disturbances and threatening stimuli as did Kanwisher *et al.* (1978) and

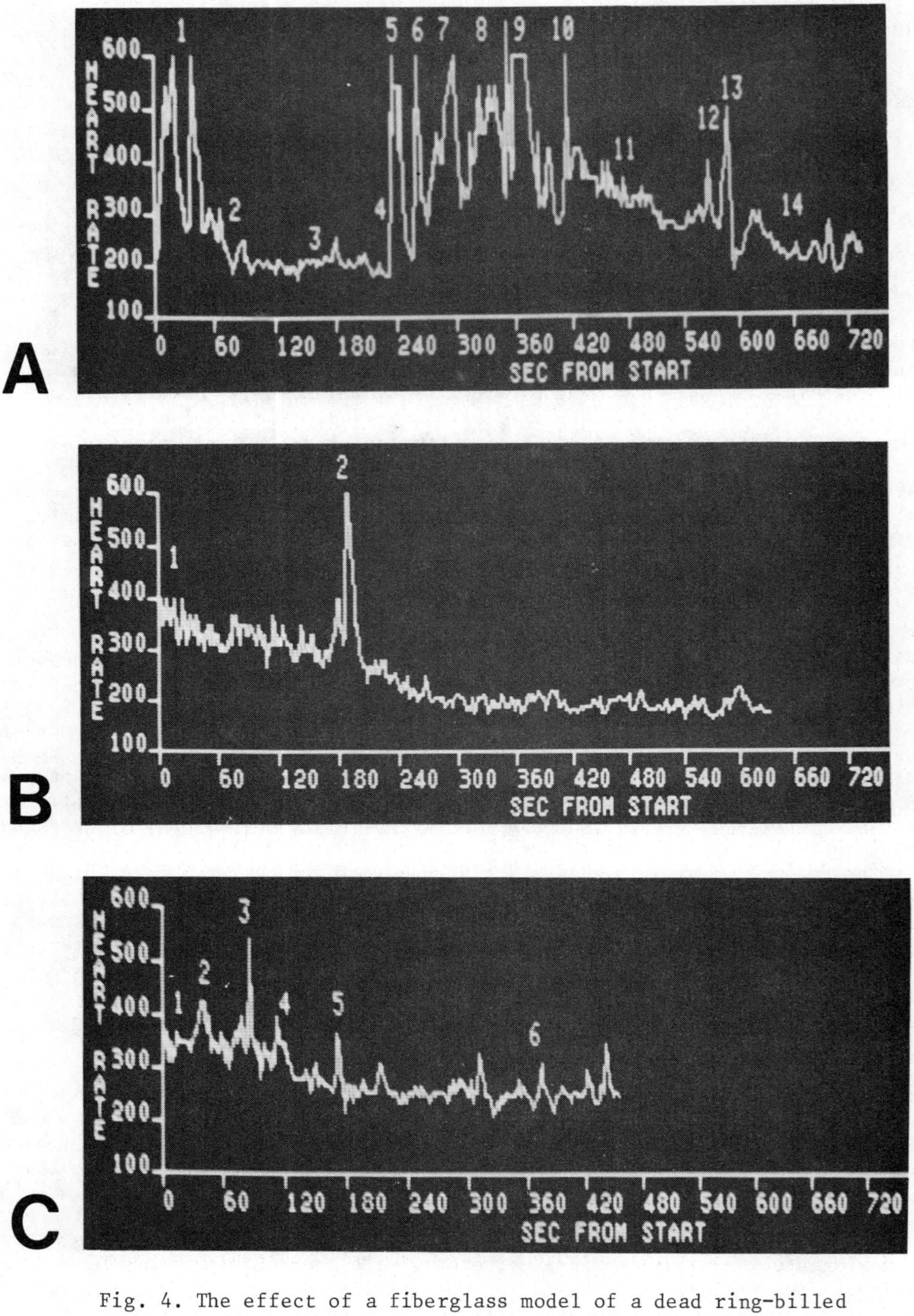

Fig. 4. The effect of a fiberglass model of a dead ring-billed gull on the heart rate of a ring-billed gull.

Ball and Amlaner (1980, this volume) for the herring gull. For ring-billed gulls, the heart rate accelerates to 600 beats min^{-1} just before takeoff in the laboratory. Upon landing, the rate rapidly returns to baseline levels. Changes in response to other disturbances can result in acceleration to as much as 600 beats min^{-1} without subsequent flight. However, the heart rate returns to basal rates, often within 30 s of the cessation of the disturbance. Thus, it is likely that stimuli which result in a prolonged elevation of heart rate, keep the gull more physiologically ready for escape.

The effects of stimuli normally used for dispersing gulls from airports may be compared in their tendency to cause prolonged elevations in heart rate. From the data presented in this paper, it is apparent that the distress call of ring-billed gulls raised the heart rate of this species during playback (Fig. 2). However, the heart rate returned to prestimulation levels shortly after the call was turned off. Habituation (Fig. 2) was also apparent in the responses of heart rate to sequential playbacks of the distress call. In natural situations, gulls usually dispersed from the site of distress call playback, but often returned to this site within one or two min (Stout *et al.*, 1975). The limited, longer term dispersal effectiveness of the distress call correlates well with its very short term elevation of heart rates.

Pyrotechnics also caused elevation in heart rate (Fig. 3). However this increase was rather short lived, as the heart rate returned to baseline relatively quickly. Loud noises in the field are often ignored by gulls, especially if they are repeated very frequently. There is also no longterm effect of the pyrotechnic devices on the dispersal of gulls, or keeping them from returning to the dispersal site.

The models of dead gulls developed for use in dispersal of gulls from airports were the most effective of three stimuli used, in maintaining prolonged elevations in the heart rate (Fig. 4). Laboratory behavioral responses to the models included avoiding the area where the model was placed, alarm calling, and continuous flying from perch to perch. These responses were not observed as a response to the other two stimuli, or during any other part of the experiments other than in the presence of the model. The acceleration of the heart rate as the gull approached a model (Fig. 4) further demonstrated the 'threatening' effects of the model. The prolonged elevation of heart rate correlates very well with the observed longterm dispersal effectiveness of the models.

This is a preliminary report of the effects of dispersal inducing stimuli on gulls. We suggest that the effectiveness of these stimuli can be meaningfully compared by evaluating their respective impacts on heart rate. Using these techniques, it is our plan to develop more effective auditory and visual stimuli, for use in gull dispersal. The work will be continued and expanded by comparing the effects of these dispersal inducing stimuli on heart rates of gulls in field studies.

REFERENCES

Amlaner C. J. and Stout J. F. (1978) Aggressive communication by *Larus glaucescens*. Part VI: Interactions with a remotely controlled, locomotory model. *Behaviour* 66, 223-251.

Ball N. J. and Amlaner C. J., Jr. (1980) Reactions of herring gulls (*Larus argentatus*) to approaching humans. In C. J. Amlaner, Jr. and D. W. Macdonald (eds.), *A Handbook on Biotelemetry and Radio Tracking*, pp. 589-594. Oxford: Pergamon Press.

Blokpoel H. (1976) *Bird Hazards to Aircraft*, chapters 2 and 5. New York: Books Canada Inc.

Gabrielson G. W. (1979) Tachycardia during egg-hypothermia in incubating ptarmigan In C. J. Amlaner, Jr. and D. W. Macdonald (eds.), *Abstract Book* for International Conference on Telemetry and Radio Tracking in Biology and Medicine, p. 17. London: Biological Engineering Society.

Galusha J. G. and Stout J. F. (1977) Aggressive communication by *Larus glaucescens*. Part IV. Experiments on visual communication. *Behaviour* 62, 222-235.
Kanwisher J. W., Williams T. L., Teal J. W. and Lawson K. O. (1978) Radiotelemetry of heart rates from free ranging gulls. *Auk* 95, 288-293.
Nemergut J. P. (1973) A review of the USAF bird strike problem. Proceedings of the Conference on Transparent Aircraft Enclosures. Air Force Materials Laboratory, Wright-Patterson AFB, Ohio. Tech. Report AFML-TR-73-126.
Smith E. N. (1974) Multichannel temperature and heart rate radio telemetry transmitter. *J. Appl. Physiol.* 36, 252-255.
Stout J. F. (1975) Aggressive communication by *Larus glaucescens*. Part III. Description of the displays related to territorial protection. *Behaviour* 55, 181-208.
Stout J. F., Gillett W. H., Hayward J. L., Jr. and Amlaner C. J., Jr. (1975) Dispersal of seagulls in an airdrome environment. Air Force Office of Scientific Research, Arlington, Va. Tech. Report AFWL-TR-74-324.
Stout J. F. and Brass M. E. (1969) Aggressive communication by *Larus glaucescens*. Part II. Visual communication. *Behaviour* 34, 42-54.
Thorpe J. (1975) Bird strikes during 1973 to European registered civil aircraft. Proceedings of the tenth meeting of the bird strike committee Europe.

The Conflict between Feeding and Overheating in the Aldabran Giant Tortoise

I. R. SWINGLAND* and J. G. FRAZIER**

**Animal Ecology Research Group, Department of Zoology, South Parks Road, Oxford OX1 3PS, U.K.*
***c/o Smithsonian Institution, Washington, D.C., U.S.A.*

Abstract — Some giant tortoises (*Geochelone gigantea*) migrate to the relatively shadeless coast during the rainy season to obtain the seasonal flush of food which results in migrant females laying larger clutches than those tortoises that stay inland. The migrants face the conflict of maximizing food intake and minimizing the risk of death from overheating. Internal body temperature was measured using radio pills; other observations were made of ambient temperature, critical thermal maximum, sward height, tortoise density and body size distribution of grazing tortoises in relation to distance from the nearest shade center. The results are: (i) on cool days tortoises are active all day but on hot days they shade at midday, (ii) CTMax is 36-38°C, (iii) larger tortoises spend longer in the shade, (iv) the need to shade imposes limitations on feeding range, (v) different size tortoises do not have different size feeding ranges.

INTRODUCTION

The giant tortoises *Geochelone gigantea* of Aldabra Atoll are divided into three isolated populations. Two of these are relatively sedentary, but a proportion of the third, which inhabits the island of Grande Terre, migrate to the coast each rainy season (Swingland and Lessells, 1979). The migrant tortoises benefit from their activity through access to the seasonal flush of coastal vegetation which responds more quickly to the start of the rains than the vegetation inland. As a result, migrant females are able to lay larger clutches than their conspecifics who have remained inland.

However there are disadvantages for migrants on the coast. The rainy season is the hottest time of year and in the coastal area there are only isolated shade centers, often single trees, to which the tortoises must retire during the heat of the day. Risk of overheating not only limits the maximum distance from each center at which the tortoises can feed, but there is also evidence that competition for shade results in the death of the smaller tortoises, particularly the females (Swingland and Lessells, 1979). Indeed, death from overexposure is the major apparent cause of death.

As the rainy season progresses and the coastal vegetation is depleted the tortoises return to feed on the inland vegetation where shade and nest sites are more abundant.

While an individual is on the coast it faces the conflict between maximizing food intake and minimizing the risk of death from overheating. This paper is concerned with this conflict and the influence of internal body temperature on tortoise behavior. It forms part of a larger study concerning the ecology of the giant tortoise (Swingland, 1977, 1978; Bourn and Coe, 1978; Swingland and Coe, 1978, 1979; Coe, Bourn and Swingland, 1979; Swingland and Lessells, 1979; Gould and Swingland, in press; Coe and Swingland, in press).

METHODS

Biotelemetry measurements of internal body temperature were made by Frazier (1971). Seven tortoises were fed radio pills (Mackay, 1968) which transmitted a signal whose frequency varied linearly with internal body temperature. The radio signal was detected by an antenna attached to the animal's carapace and converted by a receiver to a galvanometer deflection calibrated in degrees Celsius. A thermistor probe was also fixed to the top of the carapace and protected with a small sun-shield. This measured the air temperature immediately around the animal. The seven individuals were followed from three to fourteen days and their internal and external body temperatures determined at regular intervals.

Observations were also made of the Critical Thermal Maximum (CTMax); the temperature at which an individual becomes incapable of escaping the conditions contributing to its death (Heatwole, 1976), and of sward height, tortoise density, and the body size distribution of grazing tortoises in relation to distance from the nearest shade center.

RESULTS

The internal body temperature of tortoises follows a regular daily cycle (Fig. 1) with maximum temperatures being reached in the afternoon and minimum temperatures in the early morning. The details of the daily cycle depend on both tortoise body weight and ambient temperature. The temperature records clearly show that, as might be expected, the smaller tortoises have a lower thermal capacity than their larger conspecifics, and heat up and cool down faster.

Measurements of three dying tortoises suggest that the CTMax of giant tortoises is 36-38°C (cf. CTMAx of 39.8-43.3°C for *Chelodina longicollis* (Webb and Witten, 1973)). Tortoises sometimes die of overheating several hours after entering shade; the temperature record for the largest tortoise suggests that this is the result of a continuing rise in body temperature at this time. Thus tortoises must cease feeding and seek shade at such a time that their body temperature does not *ultimately* exceed some critical level.

On warm (cloudless) days (upper trace in Fig. 1), the tortoises must seek shade during the hottest part of the day. Due to their lower thermal capacities, small tortoises cool down more rapidly in the afternoon and have a longer active feeding period than large tortoises at this time of day. However, large tortoises do not have a corresponding gain in the morning; although their body temperatures rise more slowly, their temperatures continue to rise after they have sought shade, and they must therefore seek shade at the same time as (and therefore lower internal temperature than) smaller tortoises.

On cool (cloudy) days (lower trace in Fig. 1) there is no need for tortoises to seek shade; the maximum ambient temperature is less than their CTMax, and therefore

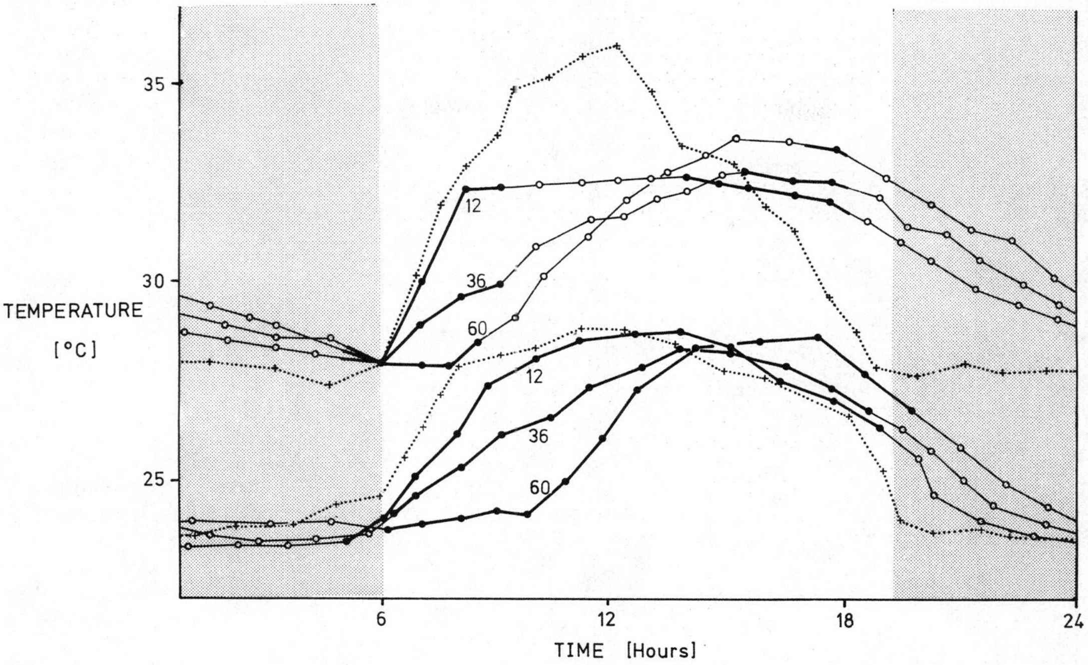

Fig. 1. The internal body temperatures of giant tortoises in relation to ambient temperature (after Frazier, 1971). Upper traces: warm (cloudless) day; lower traces: cool (cloudy) day, internal body temperature of tortoises when active ●— and inactive —o—; the weights of individuals are given in kilograms next to each record; ambient temperature ···×··· .

individuals do not risk death from overheating even when their body temperature reaches the maximum ambient temperature. Tortoises of different sizes differ only in the time at which they reach their maximum internal temperature, the time lag between ambient and internal being longer in larger individuals. In the early rainy season there are cloudy periods of three to four days when the tortoises do not stop feeding during the middle of the day. Individuals may move further and further from shade and on these occasions up to 50 percent of the tortoises may leave a shade center and be replaced by others (Swingland and Lessells, 1979).

The height of the single species *Sporobolus virginicus* grass sward increases only slightly at distances up to 450 m from shade. Beyond this distance the height of the sward increases more abruptly (Fig. 2). Concomitantly, the grazing density of tortoises decreases away from the shade center (Fig. 2). There is no discernible trend in body sizes of the tortoises feeding on the grass sward, i.e. small tortoises do not feed further from the shade than larger individuals or *vice versa*. Equally, there is no sex difference in this respect. Moreover, individual tortoises do not tend to feed at a similar distance from shade on successive days.

DISCUSSION

The results of the biotelemetry measurements of internal body temperature show that whereas tortoises on cloudy days are not at risk from overheating, tortoises on

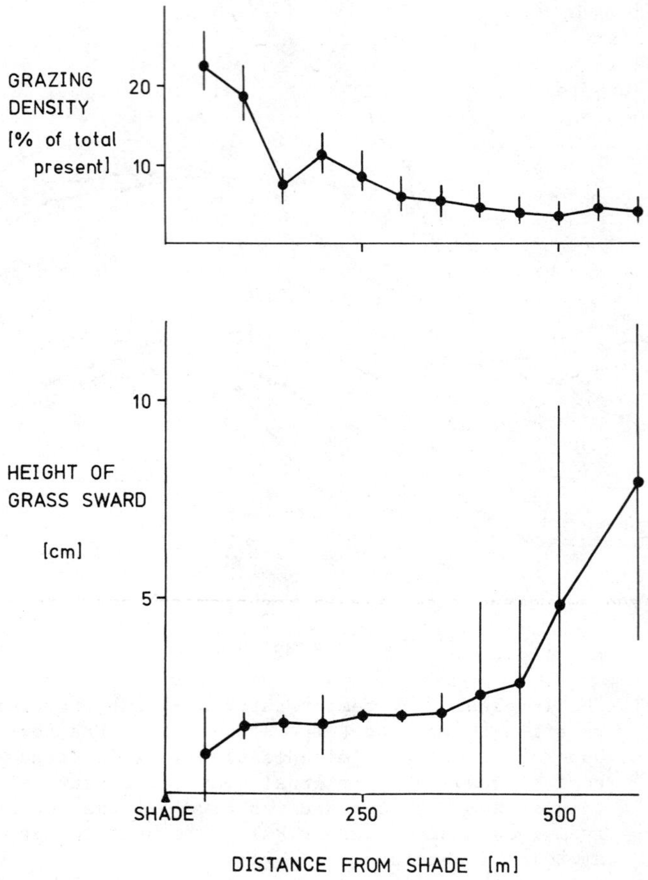

Fig. 2. Grazing density and height of the grass sward in relation to distance from shade. Grazing density observations were made on five separate mornings and five samples of sward height were taken at each distance from shade. ● mean. Vertical lines indicate 95 percent confidence limits for the mean.

cloudless days must seek shade in the middle of the day in order to prevent their internal body temperature becoming dangerously close to their CTMax. In addition large tortoises appear to spend longer in the shade each day than their smaller conspecifics. The distribution of grazing tortoises and the height of vegetation in relation to the distance from shade clearly reflect the limitation imposed on the tortoises by this need to return to shade. Moreover, since the height of vegetation increases away from the shade, each tortoise faces a decision in allocating its time between travelling and feeding in order to maximize its food intake; the more time that an individual spends travelling, the taller the vegetation it may feed on, but the shorter the time available for feeding. In view of the apparent differences in the total time available for travelling and feeding and the likely differences in walking rate and food intake in relation to vegetation height between the different size tortoises, it is surprising that there are no differences in the distance that different size tortoises feed from shade.

A question that remains unanswered is why the tortoises do not feed at night, for the tortoises do not return to the shade center at this time, but remain inactive (asleep?) wherever they happen to be on the grass sward at dusk. It is unlikely that the tortoises are sufficiently discriminating grazers that they cannot feed efficiently at low light intensities, nor do they have any enemies that would make feeding at night dangerous. The biotelemetry records show that, even if digestion is temperature dependent, the tortoises cannot be prevented from feeding by low rates of digestion on warm nights because internal body temperatures are as high as when feeding during cool days (see Fig. 1). The possibility remains that their rate of intake is limited in some way by a 'digestive bottleneck' (Kenward and Sibly, 1977) unrelated to internal temperature. Without further information on their rate of digestion in relation to body temperature, and on the pattern of food intake on cool days when they are free to feed for up to twelve hours continuously, it is impossible to make an adequate assessment of this suggestion.

Acknowledgements — We would like to thank the Royal Society, the Royal Society Aldabra Research Station, the Natural Environment Research Council, Malcolm Coe, Kate Lessells, Geoffrey Taylor and Rob Wilderspin for their help.

REFERENCES

Bourn D. and Coe M. (1978) The size, structure and distribution of the giant tortoise population of Aldabra. *Phil. Trans. R. Soc., Lond.* B 282, 139-175.

Coe M., Bourn D. and Swingland I. R. (1979) The biomass, production and carrying capacity of giant tortoises on Aldabra. *Phil. Trans. R. Soc., Lond.* B 286, 163-176.

Coe M. and Swingland I. R. Giant tortoises. In D. Stoddart (ed.), *Biogeography and Ecology of the Seychelle Islands.* Netherlands: Junk, in press.

Frazier J. G. (1971) Behavioural and ecological observations on giant tortoises on Aldabra Atoll. D.Phil. thesis, University of Oxford.

Gould M. and Swingland I. R. The tortoise and the goat: interactions on Aldabra Island. *Biol. Cons.*, in press.

Heatwole H. (1976) *Reptile Ecology.* University of Queensland Press.

Kenward R. E. and Sibly R. M. (1977) A woodpigeon (*Columba palumbus*) feeding preference explained by a digestive bottle-neck. *J. appl. Ecol.* 14, 815-826.

Mackay R. S. (1968) *Bio-medical Telemetry.* London: John Wiley.

Swingland I. R. (1977) Reproductive effort and life history strategy of the Aldabran giant tortoise. *Nature, Lond.* 269, 402-404.

Swingland I. R. (1978) Marking reptiles. In B. Stonehouse (ed.), *Animal Marking: Recognition Marking of Animals in Research.* London: Macmillan.

Swingland I. R. and Coe M. (1978) The natural regulation of giant tortoise populations on Aldabra Atoll. Reproduction. *J. Zool. Lond.* 186, 285-309.

Swingland I. R. and Coe M. (1979) The natural regulation of giant tortoise populations on Aldabra Atoll. Recruitment. *Phil. Trans. R. Soc., Lond.* B 286, 177-188.

Swingland I. R. and Lessells C. M. (1979) The natural regulation of giant tortoise populations on Aldabra Atoll. Movement polymorphism, reproductive success and mortality. *J. Anim. Ecol.* 48, 639-654.

Webb G. J. W. and Witten G. J. (1973) Critical thermal maxima of turtles: validity of body temperature. *Comp. Biochem. Physiol.* 45A, 829-832.

Salinity Telemetry from Estuarine Fish

R. MANI and I. G. PRIEDE

Department of Zoology, University of Aberdeen, Aberdeen, U.K.

Abstract – Salmon migrating up an estuary undergo profound physiological changes in transferring from seawater to freshwater. Renal function for example, changes from a water conservation condition to extreme diuresis. It is not clear in the wild, whether salmon chose to move through salinity gradients slowly or rapidly. In order to examine these changes a two channel ultrasonic transmitter is being developed.

Salinity and temperature changes will be continuously monitored using an ion selective membrane transducer and small thermistors respectively. The output of these transducers will be dc voltage levels which will be converted to frequency and impressed upon an oscillator with a center frequency of 77 KHz. This paper describes the development of this device.

INTRODUCTION

Ultrasonic transmissions have been used to telemeter environmental data (Stasko and Pincock, 1977) and physiological parameters from fish in the wild (Ireland and Kanwisher, 1978; Priede, 1978). They have also been widely used with great success in tracking (Hawkins *et al.*, 1974; Young *et al.*, 1976; Stasko, 1975). The only environmental parameters monitored so far using telemetry are pressure and temperature (Ireland and Kanwisher, 1978). This paper describes a method of monitoring salinity and temperature from fish in estuaries using ultrasonic telemetry.

PHYSIOLOGY

For animals to survive in an aquatic environment, their physiology must be capable of osmotic and ionic regulation in relation to their surroundings which may be freshwater, seawater or an intermediate brackish medium. The Atlantic salmon (*Salmo salar* L.) is a complex animal that survives in both freshwater and saline environments. In seawater, the salmon drinks in large volumes. It keeps fluid losses low by excreting small quantities of concentrated urine which contain excess magnesium sulphate ions dissolved in it. Excess sodium and chloride ions are lost by diffusion through the gill surfaces. In freshwater, the salmon drinks less and obtains solutes from food and through increased diffusion through gill surfaces. Excess water is excreted as dilute urine. The transition from the marine to the freshwater environment therefore entails profound physiological adjustments and it is by means of telemetry that we hope to determine the time course of these adjustments.

SYSTEM

Figure 1 is the system diagram which shows how salinity and temperature are converted to frequency pulses. The present design converts salinity data to frequencies between 1-10 KHz and temperature data to frequencies between 300 and 800 Hz. These frequencies are alternately connected by the switch to the carrier oscillator. The tracker will therefore hear two distinctly different tones from the tracking receiver. The demodulated output of the receiver will be filtered and converted to voltage levels for data processing. Figure 4 is a circuit diagram of the transmitter.

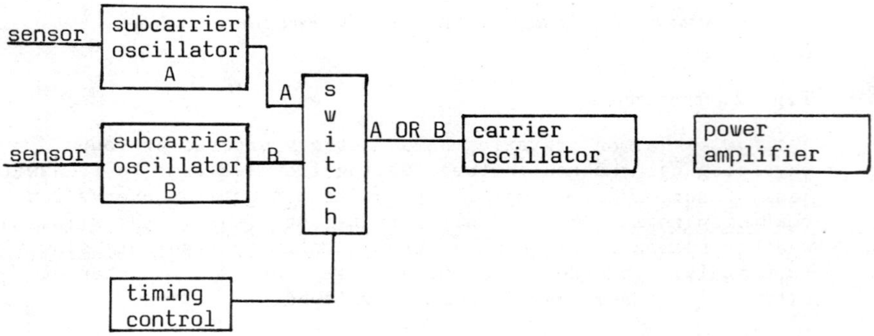

Fig. 1. Schematic of two channel ultrasonic transmitter.

TRANSDUCER

Temperature is converted using a thermistor bead (1.5 mm diameter and 5 mm long) mounted on the housing of the transmitting tag. The resistance of the thermistor varies linearly over a wide range of temperatures in this circuit. The range of interest is 0 to 15°C.

Salinity is monitored by a Perspex cell shown in Fig. 2. This cell, based on the design of the flow cell salinometer by Wilson (Wilson, 1971), has two chambers which are filled with a seawater solution of known concentration. The top ends of the cells are fitted with membranes of which one is selectively permeable to anions and the other to cations. Silver-Silver Chloride electrodes are glued into the chambers. A faceplate with holes of the same diameter as the chambers is screwed onto the cell. Rubber 'O' rings between the membranes and the front plate seal the chambers. Figure 2B shows the faceplate.

The output of the transducer in millivolts is given by:

$$V = K \cdot \frac{2RT}{F} \log \frac{S_1}{S_2} \tag{1}$$

where K = Empirical constant
R = Gas constant
T = Temperature in degrees Kelvin °K
F = Faraday's constant
S_1 = Concentration of solution sealed in chambers
S_2 = Concentration of solution under test.

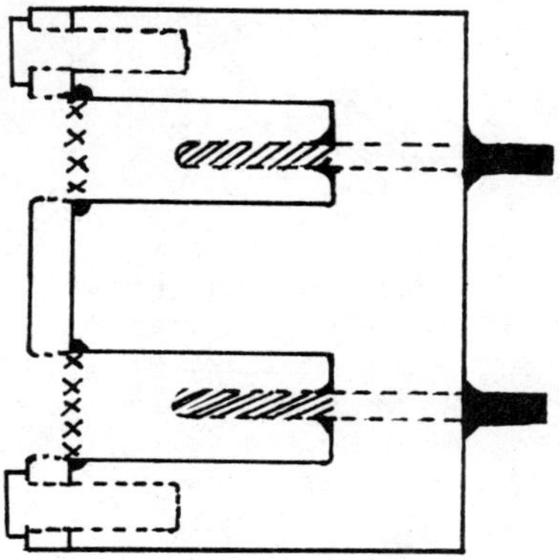

(A)

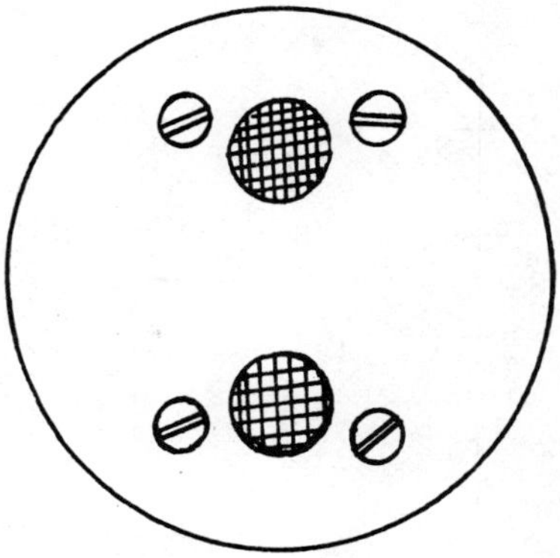

(B)

Fig. 2. Transducer.

Results obtained from a typical transducer are given in Table 1. The results were obtained with the transducer placed in test solutions of different salinity which were also measured using a Goldberg pocket refractometer. The test solutions were continuously stirred and the output of the transducer was measured with a millivoltmeter in series with a 18 MΩ resistor. Currents drawn across the membranes were therefore of nanoampere levels.

Table 1.
Concentration of solution in chambers — 14°/oo

Salinity of medium °/oo	Mean output of transducer mV
34.5	-9.2
25	-5.26
23	-4.13
17	-0.96
14	2.13
8	6.8
7	9.6
4	13.93
3	15.53
2	24.9

Figure 3 is a plot of the transducer output against the logarithm of the concentration. The slope of this line is -25.35 Linear Regression of the data yields a correlation coefficient of 0.99.

SUB CARRIER OSCILLATOR (SCO)

The analog outputs of each transducer are amplified and used to control the voltage

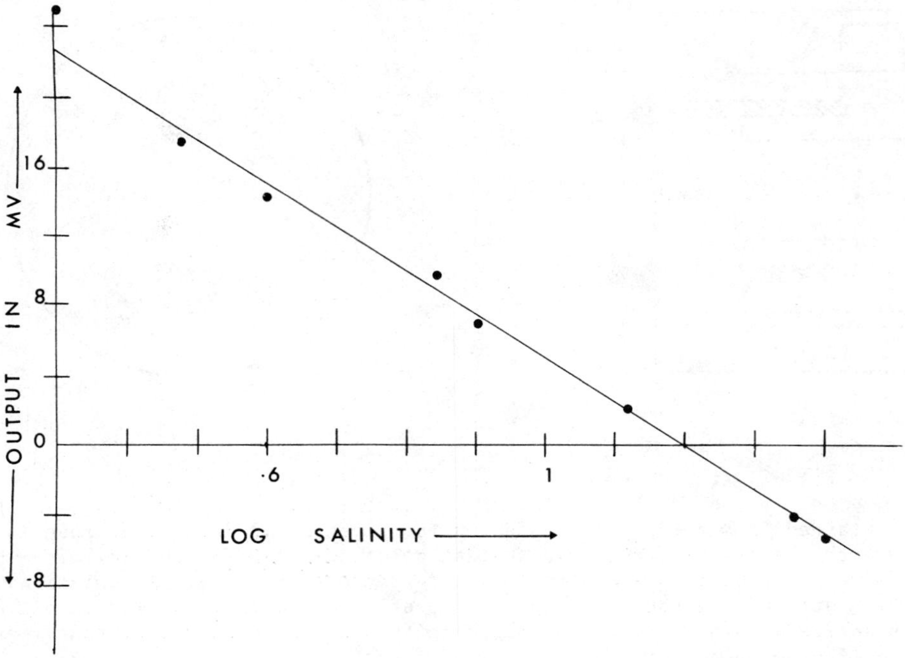

Fig. 3. Characteristic curve of salinity transducer.

controlled oscillator (VCO). The present system uses CD 4046 as VCOs. These chips were chosen for their stability and linearity over a wide range of control voltages. With the VCOs controlled by the rail voltage i.e. 2.8 V, the SCO outputs are 3.3 KHz and 300 Hz for the salinity and temperature channels respectively.

GATING ASSEMBLY

The SCO outputs are fed into a switch which is controlled by a local oscillator that puts out half second pulses. The switch connects each channel alternately to the carrier oscillator.

CARRIER OSCILLATOR AND OUTPUT STAGE

The carrier oscillator is a transistor stage with the collector circuit inductively coupled into the base of T1 (Fig. 4). The collector circuit is tuned by a slug in the coil. The output of this oscillator is a sine wave of amplitude twice the voltage between the collector and emitter at 77 KHz. Output is fed into a Class B amplifier which amplifies the input and inductively feeds the ultrasonic crystal which is a piezoelectric ceramic cylinder 3 mm long, 12.75 mm diameter and 2.45 mm thick. The crystal resonates at 77 KHz with an output of 15 V peak to peak. The system when miniaturized, will be driven by a 2.8 V lithium battery.

RECEIVER

A battery operated receiver and data acquisition module are under development. We hope to decode the frequencies by converting to voltages which, when plotted on the appropriate calibration curve, will provide salinity and temperature data.

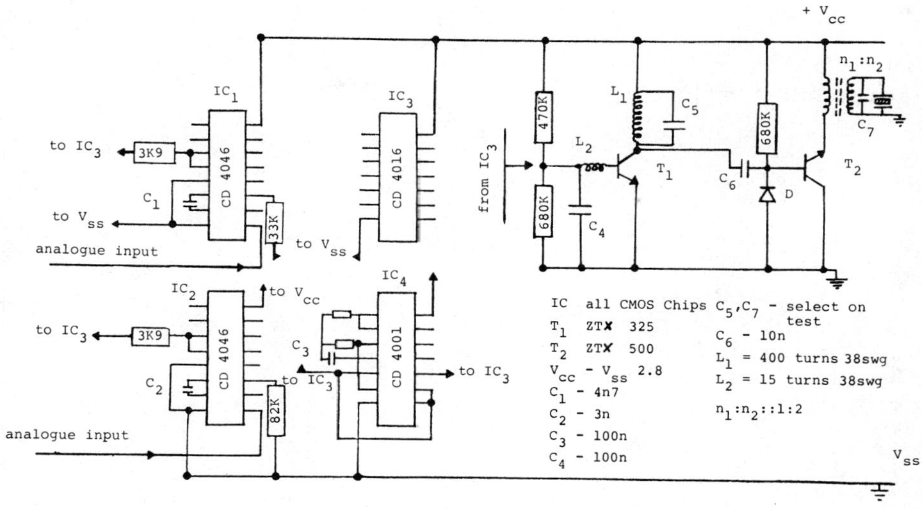

Fig. 4. Circuit diagram of two channel ultrasonic transmitter.

Freshwater evaluation and trials on the Lune estuary (U.K.) are planned for late spring and summer of 1979. We hope to have quantitative assessment of the transmitting tag in terms of range, life expectancy and reliability together with a clearer picture of salmon migration.

Acknowledgements — This pilot study is supported by a NERC grant to Dr. I. G. Priede. We would like to express our gratitude to Mr. Roland Smith for making the prototype transducer, Mr. A. Lucas and Mr. B. Duthie for their help with the visual presentation and Mrs. L. Emslie for typing the mansucript.

REFERENCES

Hawkins A. D., Maclennan D. N., Urquhart G. G. and Robb C. (1974) Tracking cod *Gadus morhus* L. in a Scottish sea loch. *J. Fish Biol.* 6, 225-236.

Ireland L. C. and Kanwisher J. W. (1978) *The Behavior of Fish and Other Aquatic Animals*, Vol. 9, pp. 341-379. London: Academic Press.

Priede I. G. (1978) *Rhythmic Activity of Fishes*, pp. 153-167. London: Academic Press.

Stasko A. B. (1975) Progress of migrating Atlantic salmon tracking. *J. Fish Biol.* 7, 329-338.

Stasko A. B. and Pincock D. G. (1977) Review of underwater biotelemetry, with emphasis on ultrasonic techniques. *J. Fish. Res. Bd Can.* 34, 1261-1285.

Wilson T. R. S. (1971) A portable flow cell membrane salinometer. *Limnol. Oceanogr.* 16, 581-586.

Young A. H., Tyler P. and Holliday F. G. T. (1976) New developments in ultrasonic fish tracking. *Proc. R. Soc. Edin.* (*B*) 75, 145-155.

V. Applications in Radio Tracking

The Serengeti Radio Tracking Program, 1971-73

B. BERTRAM

Research Centre, King's College, Cambridge, U.K.

Abstract — The wildlife radio tracking program of the Serengeti National Park in Tanzania during 1971 to 1973 is described. Radio transmitters were used on 14 species — lion, leopard, wildebeest, buffalo, topi, giraffe, elephant, two hyrax spp., hyaena, jackal, vulture, and two tortoise spp. — and the contribution of radio tracking to the studies of these species is summarized. Observation of the animals played a large part in these studies. Many of the transmitters were recovered and reused. A few recommendations concerning radio tracking in wildlife research are made.

INTRODUCTION

The Serengeti National Park in Tanzania is famous for its huge herds of migratory wildebeest, zebra and gazelles, for its wide open grass plains over which these herds wander in the wet season, and for its variety of large predator species. It is considered by many people to be the finest wildlife area in the world. Among those who hold this view are the biologists who have worked at the Serengeti Research Institute (S.R.I.), an organization set up by Tanzania National Parks in 1966, with two main objectives of discovering and describing the ecology of the Serengeti region, and providing information on which scientific management of the National Park can be based. At any one time, about 15 biologists from several different parts of the world, including our host country Tanzania, were engaged in this work, on different projects which fitted in with S.R.I.'s overall research objectives.

Our task was made challenging as well as stimulating by the scale. How does one find one wildebeest in a herd of a hundred thousand? How does one find an elephant which may be almost anywhere in the thirteen thousand km^2 of the National Park, or outside it? How does one find out what an animal is up to at night? There were many aspects of our various projects where we needed to be able to do such things, and as a result of a grant for radio tracking in 1970 we were able to do them. The donors, Mr. Royal Little and the New York Zoological Society, generously provided funds sufficient for a large scale radio tracking program, aspects of which are summarized below.

Our radio tracking program differed from most in that observation was also important. We did not simply put large numbers of transmitters on animals and obtain large numbers of fixes and lines on maps; in general we fitted a smaller number of transmitters and used them to find and observe those animals. We considered that fewer fixes but with data on behavior and reproduction, on feeding and habitat, were likely to be of more value. In most cases, radio tracking supplied extra information to answer specific questions arising in the course of studies which were already under way.

In this paper I shall describe briefly our equipment needs and how they were met, outline the variety of projects and species on which we made use of radio tracking, and make some generalizations based on radio tracking wildlife in the Serengeti.

EQUIPMENT

The two prime requirements of our radio tracking equipment were reliability through thorough field testing, and interchangeability. All field workers want reliability in their equipment; but our need was probably greater than average because of the degree of isolation from technical expertize. To send faulty equipment from the Serengeti for repairs in Europe or North America would have caused huge delays, wasted days, administrative problems, and probably ulcers. To have transmitters fail in the field would have caused the same problems of lost time and lost data as anywhere else.

We also wanted our equipment to be interchangeable, so that if one scientist's receiver failed he could use another's, and so that a transmitter could be used for any species or project in need of one. We therefore settled for a standard frequency (148.5 kHz) transmitter and receiver for almost all purposes. We chose the SM2 (or for small species, the SM1) transmitter and receiver manufactured by AVM Instrument Co., Champaign, Illinois, with Mallory ZM12 (or sometimes larger) mercury batteries. Battery life was on the whole poor, probably because of shipment and storage problems, but the transmitter performance was generally high.

This standardization naturally resulted in some loss of efficiency. For example, a lower frequency would probably have given longer range in rough or thickly vegetated terrain. Again, because a mammal as large as an elephant could carry enough batteries to power an SM2 transmitter for a century (or longer than the lifetime of the investigator or his subject), it would have been in some ways more efficient to have used a higher-power, longer-range transmitter. But then it could not have been used on a jackal instead, if that was the greatest need by the time the specially ordered transmitter had arrived. At that time there was no standard, reliable, field-tested transmitter for large mammals, so we opted for interchangeability and reliability, rather than maximum range. Most tracking was by means of a 3-element yagi antenna, either held by hand above a vehicle, or strapped to an aircraft wing strut.

PROJECTS EMPLOYING RADIO TRACKING

LIONS AND LEOPARDS

My own research on lions (*Panthera leo*) and leopards (*Panthera pardus*) probably benefitted most from radio tracking. In investigating their social organization and ecology, simply to find the predators and their kills was the major problem. Radio tracking resulted in a three- or four-fold increase in the number of lions seen per day, while the rate of leopard sightings rose from around one per three weeks to one or two per day. In addition, I could find and follow them at night. Other advantages resulted from being able to find the same individuals reliably and regularly, for example being able to relate their daily movements, hunger, hunting success and reproductive performance (Bygott, Hanby and Bertram, in preparation). A further advantage, particularly striking in the case of leopards, was that the animals rapidly became much better habituated to vehicles through repeated peaceful contacts with mine. Not only could I approach and observe a leopard much closer because it would let me, but also I could observe its hunting attempts from much further away, secure in the knowledge that I could always find it again if it disappeared.

Because I needed to find and observe each animal, 8 lions and 2 leopards radio collared was the maximum I could manage. In fact I rarely had as many, because of problems mainly with battery failure and, at first, with animals shedding their collars. A total of 34 collars were put on lions and leopards over the 3 year period, and 31 of them taken off again, 18 by me and 13 by the lion or leopard. I chipped open all recovered collars to replace batteries; in 20 out of 22 cases the transmitter could be used again satisfactorily. Improved collar design, from a rigid acrylic ring to flexible machine belting, resulted in fewer collar breakages (Bertram, 1976a).

WILDEBEEST

The Serengeti wildebeest (*Connochaetes taurinus*) population, over a million strong, are migratory, moving to the open grass plains in the wet season and through the woodlands in the dry. Two projects involving wildebeest investigated the details of their movements.

Inglis (1976) placed radio collars on 16 animals and tracked them from the air over an eight month period. He showed that their average fix-to-fix daily rate of travel was about 3.75 km day^{-1}. They went faster when approaching or leaving the plains, and slower when on the plains where they oscillated back and forth within individually distinct ranges, dependent on finding water and green forage. Their huge herds appeared to be fluid entities, with no fixed subgroupings within them (Inglis, 1976).

Kreulen (1975) followed some of these radio collared individuals on the ground as part of his project of examining the habitat requirements of the species. The radio provided a label, without which it would have been impossible to tell that he was observing the same part of the moving mass. Wildebeest were followed for continuous periods of a week, and data collected on a wealth of environmental features, on food and feeding behavior, on activity patterns, and on interactions with other species (Kreulen, personal communication).

OTHER UNGULATES

In the course of his study of buffalo (*Syncerus caffer*), Sinclair (1977) needed to investigate the home ranges of the big breeding herds, to discover how distinct these ranges were from those of neighboring herds, and how much energy the animals used in locomotion about their range. By radio tracking two female buffalo in neighboring herds for almost a year, he showed that there was almost no overlap in the two ranges; that the herds moved about 2.7 km day^{-1}; and that in doing so they consumed a quantity of energy approximately three times that of the basal metabolic energy of inactive animals.

Topi (*Damaliscus korrigum*) in the Serengeti, are mainly sedentary, unlike their close relatives the wildebeest. Duncan (1975) radio collared two female topi to determine, on the rare occasions that they left their territories, what made them go and where they went to. Invasion of the area by dense masses of grazing, migrating wildebeest was the main cause of the usually temporary departures of the topi, who moved up to 20 km away.

Mejia was carrying out a study of giraffe (*Giraffa camelopardalis*) social behavior in the Serengeti, as part of which he wanted to follow changes in social interactions between mothers and their calves at intervals as the calves grew up. Because giraffes move over large areas and because they provide high transmitting stations, radio tracking was both necessary and productive in enabling Mejia to locate his mother-calf pair as required (C. Mejia, personal communication).

ELEPHANTS AND HYRAXES

Elephant (*Loxodonta africana*) bulls are usually found apart from the far-ranging breeding herds, the young males separating from them at 10-12 years old (Hendrichs and Hendrichs, 1971). Croze showed that the process of separation was not the sudden and irreversible one it had previously been thought to be. When the youngest member of a bull group was fitted with a radio collar, he immediately left his companions and headed back to join up again as an integrated member of a family unit. He had clearly run back to mother and her relatives, and he remained with them for a month before leaving to join a bull group again (H. Croze, personal communication).

The behavior and ecology of hyraxes (*Heterohyrax brucei* and *Procavia johnstoni*), rodent-like members of the elephant family, were studied by Hoeck (1975). They inhabit the granite outcrops (kopjes) dotted like islands throughout the Serengeti. By trapping and radio collaring several hyraxes, Hoeck was able to show that they travel at intervals between the members of closely adjacent kopje groups, but that there is almost no exchange between the hyrax populations of kopjes more distant than a few hundred meters (H. Hoeck, personal communication). They are probably deterred both by the risk of predation en route, and by the territorial behavior of the hyrax colony at their destination.

HYAENAS AND JACKALS

Striped hyaenas (*Hyaena vulgaris*) are scarce, nocturnal and elusive in the Serengeti, compared with the spotted hyaena (*Crocuta crocuta*) which Kruuk (1972) showed to be a clan-living pack-hunting species. Information could only be obtained on striped hyaenas by radio collaring them, and tracking them during their nocturnal wanderings. Kruuk (1976) found that they were omnivorous, mainly scavenging but also eating insects, fruits and small vertebrate prey, and that they travelled an average of 19 km per night in the course of their foraging, within home ranges of 40-70 km^2.

Lamprecht found that the black-backed jackal (*Canis mesomelas*) apparently occupies a rather similar niche. By radio tracking them he showed that they are also omnivorous, but that they do their foraging within much smaller territories than those of hyaenas. Most adult jackals remained throughout the year in their territories, breeding there during July to September (J. Lamprecht, personal communication).

VULTURES

Rüppell's griffon vulture (*Gyps ruppelli*) is one of the six vulture species in the Serengeti. It breeds on the Gol cliffs outside the park to the east, but feeds mainly on caracases left in the wake of the migrating herds of wildebeest and zebra.. By fitting radio transmitters to two vultures, C. Pennycuick, (personal communication) was able to show that the birds can commute great distances (up to 135 km from the breeding cliffs) and at considerable speeds (80 km in less than 9 h). They do so entirely by gliding, an efficient form of travel which gives them a competitive edge over the terrestrial scavengers (Houston, 1979; Pennycuick, 1979).

TORTOISES

As an interesting sideline, I fitted radio transmitters to two tortoises, animals which are almost invisible among long vegetation. One, a hingeback tortoise (*Kinixys belliana*) remained for 2 years within the 190 m area where she was first found. The tortoise aestivated in a favorite hole under a rock in the dry seasons, from whence she emerged to potter about her range in the wet seasons, at a maximum sustained speed of 12 m h^{-1}. It was obvious that this tortoise knew well the geography of her range (Bertram, in press, a). The second tortoise, a large leopard tortoise (*Geochelone pardalis*) was found 7.5 km east of S.R.I. After fitting her with a transmitter she was released at S.R.I. She remained nearby for 3½ months, then started travelling when the rains commenced, and in 53 days went back 7.5 km to within 50 m of her capture point. She overshot, but returned to stay within 2 km of this point for the next 4½ months (Bertram, in press, b). Thus tortoises seem to know their home and to be able to find their way home.

CONCLUSIONS

Radio tracking is a technique, not an end in itself. Obtaining fixes on animals' locations is only one possible product of the technique. What is important is what the scientist does with the information on locations, and he can usually do a great deal more with it if he can observe the animal and so collect further information on its immediate environment, its companions, and its (and their) feeding and social behavior. Knowing where an animal went becomes a great deal more interesting and useful if one also knows why it went there. In many cases, using radio tracking can help in providing data on both points.

Our radio tracking program illustrates some of the variety of ways in which radio tracking can benefit wildlife studies. One type of advantage is in increasing the efficiency of finding any animal of the study species to observe, as shown by the lion, leopard and hyaena projects. A second advantage is in enabling the scientist to find particular individuals for observation at regular intervals, as in the lion, hyrax, topi, giraffe, buffalo and jackal studies. In these cases, searching would probably reveal the animal eventually, but would waste an enormous amount of time. In some cases, however, the animal would not be found at all without radio tracking. This can be because the animal is completely hidden by its environment (tortoise and leopard projects); or because the animal is hidden among a multitude of its companions (wildebeest); or because the animal has moved too far away (vultures, *par excellence*, and particularly wildebeest and elephant studies). A failure to find an animal usually tells one nothing, often not even the reason for the failure.

Range or territory sizes compiled without a fairly high degree of certainty of discovering the animal wherever it is are inevitably rather tenuous. My best known lion pride provided an example. After starting to radio track I could find them every day (whereas previously I might have managed, with infinitely more effort, on perhaps one day in three, had their range been known reliably). On only one occasion, out of several hundred, I tracked the radio collared lioness of this pride far outside her pride's area, and well outside my study area, to find her eventually 10 km beyond her normal territory. Without radio tracking I would not even have suspected such excursions.

I have touched on two other less obvious advantages of radio tracking. One is in being able to watch animals from much further away because there is no risk of losing them permanently (lion, leopard, and hyaena studies). The other is in habituating the animal so that it can be observed from much closer (leopard project particularly).

RECOMMENDATIONS

In connection with wildlife studies, I should like to make first a plea and then four practical recommendations. The plea is that a simple small standby transmitter system be devised to be incorporated with the main transmitter system. It need not give long range, but needs to be reliable enough so that the animal or its collar can be found by an intensive search following the loss of the main signal. A few months' valuable data on an individual with which contact is suddenly lost becomes less useful, and not knowing the cause of the loss of signal hinders making improvements. A spare small transmitter, perhaps switched on by the failure of the main one, would overcome this problem. It would proably not add to the expense of the project either, because more collars would be retrieved to be reused. Our busiest transmitter in the Serengeti functioned consecutively on a hyaena and three lions, and was still reusable again after its 2½ years of intermittent use.

For the wildlife field worker engaging in radio tracking work, I would make the following recommendations, on the basis of our Serengeti experience, to complement the usual technical advice.

1. It is well worth carrying at least one spare radio collar with you always, for two reasons. One is that brief opportunities may arise to replace a dying collar or fit a new one. The second is that you always have in hand the opportunity to check the performance of your equipment at once, and so detect and often remedy defects in it without delay.

2. It is always worthwhile to time routinely the signal rate of each transmitting animal, thereby getting advance warning of impending battery failure.

3. It is worth putting radio collars onto more than one animal of a social group, if possible, so that the group can still be found despite the loss of signals from one of them.

4. It is well worth dealing with all animals as gently and quietly as possible, so that they can be observed closely and without disturbance, and so that collars can be fitted or removed easily.

I shall conclude by citing an example from my work which illustrates these recommendations. I noted one morning that the signal rate from one of my radio collared leopards was slowing down, denoting battery failure (Recommendation 2). It had taken me 6 months originally to find her and to get close enough to dart and collar her, and I needed to get more data from her. I did not have a spare collar with me (Recommendation 1) because earlier I had seized an opportunity to put it on another animal. So I tracked a lioness with a functioning collar and immobilized her (Bertram, 1976b) with a low dose. She was a good approachable animal (Recommendation 4) so the low dose was sufficient and she therefore came round quickly (Bertram and King, 1976). Another lioness in her pride also wore a radio collar (Recommendation 3) so I knew that I would be able to find the pride again. I returned to the leopard, who fearlessly (Recommendation 4) allowed me to approach and dart her, and fit the lioness' collar. As a result no data were lost, only a few hours and a few milligrams of anesthetic. The whole procedure was only possible because these recommendations were followed, except to the extent that my spare transmitter was not in my car but round the neck of a spare lioness.

Acknowledgements — I am grateful to the Directors and staff of Tanzania National Parks and the Serengeti Research Institute for permission, assistance and facilities for my research in the Serengeti. This is S.R.I. Publication No. 275. I am indebted to my co-workers for accounts of their radio tracking work, in some cases unpublished, and I apologise for the cases where no doubt I have not done full justice to it. I am grateful for financial support from the Natural Environment Research Council, the Royal Society, and the African Wildlife Leadership Foundation.

REFERENCES

Bertram B. C. R. (1976a) *Studying Predators*. Nairobi: African Wildlife Leadership Foundation.

Bertram B. C. R. (1976b) Lion immobilization using phencyclidine (Sernylan). *E. Afr. Wildl. J.* 14, 233-235.

Bertram B. C. R. (1979a) Homing by a leopard tortoise in the Serengeti. *Afr. J. Ecol.* 17, (in press).

Bertram B. C. R. (1979b) Home range of a hingeback tortoise in the Serengeti. *Afr. J. Ecol.* 17, (in press).

Bertram B. C. R. and King J. M. (1976) Lion and leopard immobilization using Ci-744. *E. Afr. Wildl. J.* 14, 237-239.

Duncan P. (1975) Topi and their food supply. Unpublished Ph.D. thesis, Nairobi University.

Hendrichs H. and Hendrichs U. (1971) *Dikdik und Elephanten*. Munich: Piper.

Hoeck H. N. (1975) Differential feeding behaviour of the sympatric hyrax *Procavia johnstoni* and *Heterohyrax brucei*. *Oecologia* 22, 15-47.

Houston D. C. (1979) The adaptations of scavengers. In A. R. E. Sinclair and M. Norton-Griffiths (eds.) *Serengeti: the Dynamics of an Ecosystem*, (in press). Chicago: Chicago University Press.

Inglis J. M. (1976) Wet season movements of individual wildebeests of the Serengeti migratory herd. *E. Afr. Wildl. J.* 14, 17-34.

Kreulen D. (1975) Wildebeest habitat selection on the Serengeti plains, Tanzania, in relation to calcium and lactation: a preliminary report. *E. Afr. Wildl. J.* 13, 297-304.

Kruuk H. (1972) *The Spotted Hyena*. Chicago: Chicago University Press.

Kruuk H. (1976) Feeding and social behaviour of the striped hyaena (*Hyaena vulgaris* Desmarest). *E. AFr. Wildl. J.* 14, 91-111.

Pennycuick C. J. (1979) Energy costs of locomotion, and the concept of 'foraging radius'. In A. R. E. Sinclair and M. Norton-Griffiths (eds.) *Serengeti: the Dynamics of an Ecosystem*, (in press). Chicago: Chicago University Press.

Sinclair A. R. E. (1977) *The African Buffalo*. Chicago: Chicago University Press.

The Social Organization of a Nocturnal Primate Revealed by Radio Tracking

S. K. BEARDER[1] and R. D. MARTIN[2]

Wellcome Laboratories of Comparative Physiology, Institute of Zoology, The Zoological Society of London, U.K.

Abstract — During a two year field study in South Africa, all members of a community of lesser bushbabies (*Galago senegalensis moholi*) within a 1 × 1 km study area were captured, individually marked, and recaptured at regular intervals. Daily sightings of radio tagged animals at their sleeping sites offered a reliable basis for mapping individual home ranges. An analysis of social contact and dispersal during this relatively gregarious phase of the daily activity cycle is supplemented by quantitative data from direct observation at night with the aid of red light, including 57 nights on which single bushbabies were followed throughout the active period. A basic system of male and female territoriality between well established individuals of the same sex is obscured by a complex pattern of ranging and social contact exhibited by lower ranking animals. Periodic 'mating' and 'removal' migrations are described briefly. Despite the predominantly solitary habits of this species, social interrelationships are similar to those found within 'age-graded male' groups of more gregarious primates.

INTRODUCTION

The 'nocturnal solitary prosimians' (bushbabies, lorises, tarsiers, and some lemurs) share a number of basic social characteristics, yet four distinct types of organization are apparent and clear interspecific (and probably intraspecific) differences exist within each type (Charles-Dominique, 1978). With radio tracking the field worker can now collect systematic records and so can compare social systems quantitatively, even with secretive species in dense vegetation and on difficult terrain. The nocturnal prosimians are particularly suitable subjects for radio tracking and it has therefore been possible to examine their social organization in depth (Charles-Dominique, 1977a, 1977b).

Basic information on the behavior and ecology of lesser bushbabies, *Galago senegalensis moholi*, has been obtained in a previous study (Bearder, 1969; Bearder and Doyle, 1974a) and a series of dyadic introductions in the laboratory has illustrated the

[1]Department of Social Studies, Oxford Polytechnic, Headington, Oxford, U.K.
[2]Department of Anthropology, University College London, Gower Street, London, U.K.

nature of recognition and social status between individuals (Bearder and Doyle, 1974b). This article examines the social behavior of wild *Galago senegalensis* in the northern Transvaal with particular reference to five measures of social contact: (i) social interactions at night; (ii) migration patterns; (iii) organization of home ranges; (iv) daytime sleeping associations; and (v) mating behavior.

STUDY AREA AND METHODS

A flat, easily accessible area of open thornveld was chosen on the farm 'Mosdene', the site of a previous study, in the northern Transvaal, South Africa (28°47'E: 24°35'S). The broken vegetation, with a canopy approximately 3-5 m high, was sufficiently homogeneous to allow ample opportunities for bushbaby immigration and emigration and to minimize the effects of vegetation type upon ranging behavior. The predominant trees were *Acacia karroo, A. tortilis* subsp. *heteracantha. A. nilotica* subsp. *kraussiana* and *Euclea undulata.* Saplings of the same species together with *Diospyros lyciodes* and *Maytenus heterophylla* dominated the sparse understorey. The climate was characterized by wet summers (October-March) and predominantly dry winters (May-August) with an average annual rainfall of 610 mm (1914-1977) (E. A. Galpin, personal communication). The only other primate species on the farm was the vervet monkey, *Cercopithecus aethiops*.

Bushbabies were captured with modified Chardonneret bird traps after a procedure described by Charles-Dominique and Bearder (1979). Up to 15 traps were distributed throughout a 1 × 1 km main 'study area'. All bushbabies taken in these traps were considered as the main 'study animals', even if they habitually moved beyond the artificial limits of the study area. Full time monitoring of behavior in relation to selected environmental variables was conducted between August, 1975 and August, 1977.

Trapping was continued until no unmarked individuals could be found within the study area. Weight, reproductive condition and standard measurements were noted for each animal. Longterm marking was achieved with coded ear-notches and by dye-marking or fur-clipping the tail to ensure identification by day or night even when no transmitter was fitted. Repeated trapping programs provided a useful means of monitoring physical changes within the population and of marking recruits. Trapping was gradually extended into other regions by way of a control.

The radio equipment consisted of 20 AVM SM1 transmitters of different frequencies powered by Mallory MP 675 HT2 batteries. A portable LA 12 receiver (frequency range 148.000-148.650 MHz) was used with a 3 element Yagi antenna. The transmitter and battery were mounted with an acrylic casing onto an inconspicuous reversed leather belt lined with smooth Polyvinyl chloride (PVC) which covered the antenna wire. The belt was fitted loosely around the animal's waist and was held together by an easily applied and suitably camouflaged brass rivet. The total weight of the radio package was eventually reduced to 8.0 g and its use was restricted to bushbabies in excess of 170 g. The effective reception distance was 200-300 m and the average life of the transmitter battery was 2 months. It proved possible to keep working transmitters on at least some of the bushbabies throughout the study owing to the ease with which they could be trapped, the reliability of the transmitters and the facility with which it was possible to remove and replace an old battery. Not one transmitter failed during a combined total transmission period of 9.5 years.

Up to 20 radio tagged individuals and their companions were identified at their sleeping places, with the aid of binoculars, in each working day for 12 months between October, 1975 and September, 1976. Less regular spot locations were made at night using a red flashlight which produced a brilliant tapetal reflection. At this time continuous all-night observations of selected individuals were conducted at approximately 10 day intervals, when behavior, position and ambient temperature was sampled every 10 min. In the second half of the study the number of transmitters in use at any one time was reduced, with a consequent reduction in the number of animals identified during the day. With fewer animals to follow it became possible to stabilize all-night observations to weekly intervals and to examine in more detail the activity and ranging habits of particular individuals. Loss of data on the study population was avoided by a continual shift in emphasis from one part of the study area to another by fitting transmitters in a new sector and removing old ones every 2 months.

The positions of numbered sleeping trees and those marked at 10 min intervals during continuous observations were measured by pacing the ground in relation to the nearest intersections of pathways. Each tree was subsequently plotted on a large scale aerial photograph (1 cm = 7.8 m) and its location converted into a 6 figure grid reference using an overlay grid to a scale of 25 × 25 m. All other positional plots were recorded to within the nearest 25 × 25 m square at the time of sighting by referring to a small aerial photograph having the appropriate grid. Travel distances could be measured accurately from the large scale map and areas were calculated using a planimeter.

RESULTS

Forty-eight bushbabies were seen within the study area and 2 adult males were introduced experimentally during the 2 year period. Seventy-three transmitter belts, having an average working life of 48 days, were fitted to 37 of these animals to yield over 9000 positional records. The majority of these sightings involved sleeping site records (n = 4202) and 10 min plots obtained during 57 all-night observations (n = 3456). The remainder are accounted for by independent spot-locations at night (n = 1172) and by captures (n = 260).

Inevitably some bushbabies were studied more intensively than others and fluctuations in the population structure ensured that the majority were followed for different lengths of time. For example, the adult/subadult population of 31 animals at the start of the study gained 10 but lost 21 individuals before the end; the remaining 20 included only 13 survivors from the original community. Such recruitment and loss causes frequent social change and it is therefore desirable to select a study period long enough to allow collection of a sufficiently large sample of observations yet short enough to ensure a reasonable degree of social stability. Thus, all sightings of the best studied bushbabies are used to determine a reasonable measure of home range and to demonstrate migration, whereas the nature of home range relationship is shown for only the first spring and summer seasons when climatic and population fluctuations were low and the possible adverse effects of observer interference were at a minimum. Details of social change through time are not considered here.

SOCIAL STATUS

Laboratory introductions show that 'strangers' of the same age/sex class are generally incompatible whereas companions of similar status, having matured together, may show little mutual antagonism. Adult females are fiercely aggressive towards strange adult females, sometimes fighting to the death. Adult males introduced to one another by removal of a partition between their home cages show mutual attack, often inflicting serious injuries without resort to vocalizations characteristic of submission. Maturing males eventually dominate all females after a period of ambiguity when they are subadult. It seems therefore that status is primarily determined by age and sex but it is modified by previous associations, familiarity with the environment and individual differences in aggressiveness (Bearder and Doyle, 1974b).

Females in the wild displayed to one another near 'border zones' in some cases (vocal exchanges — 'barks') but associated amicably in others. No obvious behavioral distinction could be made between adults and subadults, despite signs of differences in rank. Wild males which appeared to display a continuous gradation in physical characteristics could be grouped into distinct categories reflecting differences in their social behavior and age. The descriptive terms of Charles-Dominique (1972) for social classes in *Galago demidovii* have been employed although they are not strictly comparable. In *G. senegalensis* 3 categories were distinguished on the basis of 218 encounters involving known males:

'Central A' — males which dominate or displace all other bushbabies they meet at night while showing territorial behavior towards neighboring high ranking males.

A dominant bushbaby can be defined as one which initiates social contact or agonism. A submissive animal generally descends, uttering characteristic vocalization, and if not chased may sit quietly on the ground until the other moves on.

'A' males fall at the upper end of the size/weight range; they have well developed glandular areas on the throat and scrotum and a characteristic smell of the urine and pelage. They exhibit bold assertive movements during range 'patrols' which involve vocal exchanges (barks) with adjacent A males and they gain prior access to females and sleeping sites.

'Central B' — non-territorial males which are mutually intolerant during social encounters, displacing one another within a linear dominance hierarchy.

'B' males occupy the middle to upper end of the size/weight range. They show various degrees of glandular development and do not have a distinctive smell. They are submissive towards A males at all times, rarely emitting barks which are never given in the context of border exchanges.

Subadults/juveniles — males which exhibit amicable social contact with all other social classes.

A MEASURE OF HOME RANGE

As a measure of *relative* home range size and position, the 'minimum sleeping area' is defined as the area of a convex polygon enclosing the outermost sleeping sites seen to be used by an individual during a period of social stability. Figures 1 and 2 show 9 examples of the minimum sleeping area (heavy lines) for well-studied bushbabies in relation to the total sightings of each animal *outside the mating season*. Details of the intensity of sampling for the four social classes are presented in Table 1.

Figure 1 depicts the movements and spot-locations of four bushbabies of different age, sex and origin which remained more or less sedentary during the time they were known. With one notable exception (male M; not shown in Fig. 1) there was considerable similarity between the minimum sleeping area of sedentary animals (a_1) and the area which would include *all* points of resighting, the 'minimum home range' (a_2) (Mohr, 1947; Layne, 1954). The average degree of overlap (a_1/a_2 percent), excluding male M, was 83 percent (N = 12; S.E. = 3.37; range 58-96). In spite of the different sample sizes of sleeping site records (51-265) the sleeping areas of A and B males are clearly larger than those of females and juveniles.

The sleeping area of male M was unusual, covering only 42 percent of his minimum home range even though it was based upon the greatest number of daily observations (265). This male was particularly slow to adopt 'new' sleeping trees (one discovered every 5.3 days of tracking compared to a range of one every 1.4 to 2.9 days for other B males).

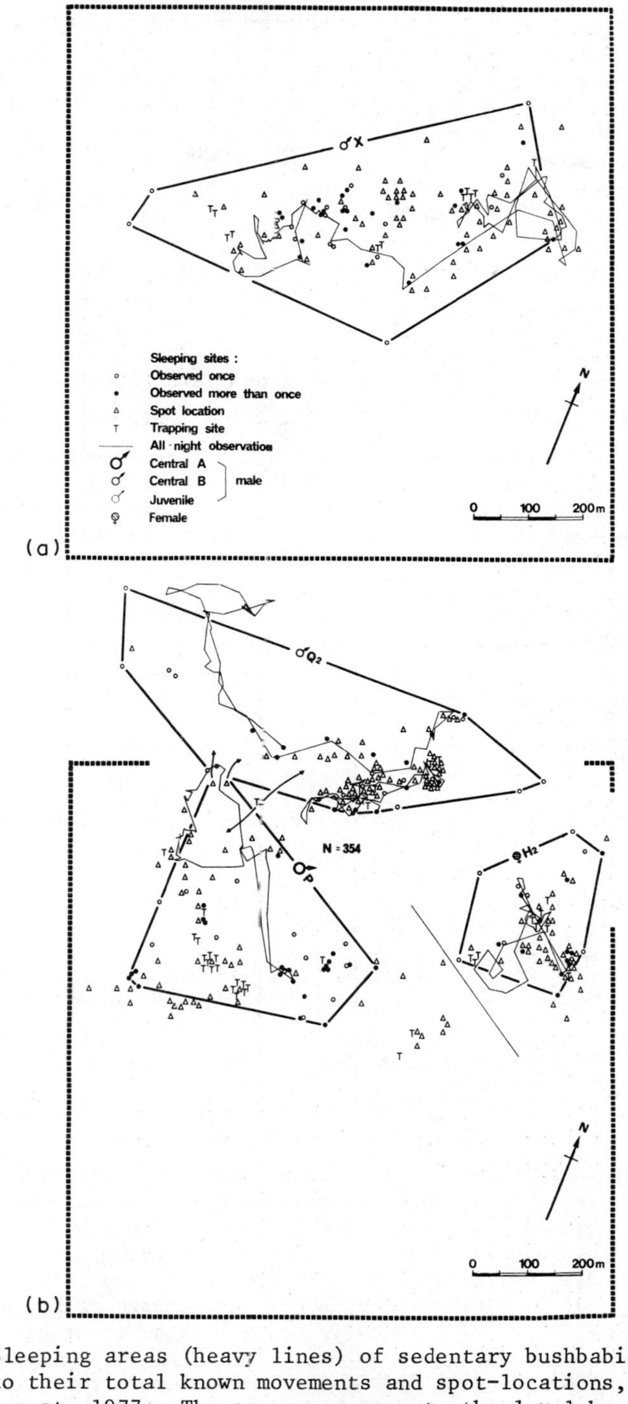

Fig. 1. Sleeping areas (heavy lines) of sedentary bushbabies in relation to their total known movements and spot-locations, October, 1975–August, 1977. The square represents the 1 × 1 km study area. Clumping of symbols indicates the intensity of range use. (a) B male, X (N = 349). (b) A male, P (N = 354); B male Q_2 (N = 571); Female, H_2 (N = 197). Note the relative size of the sleeping areas of each social class.

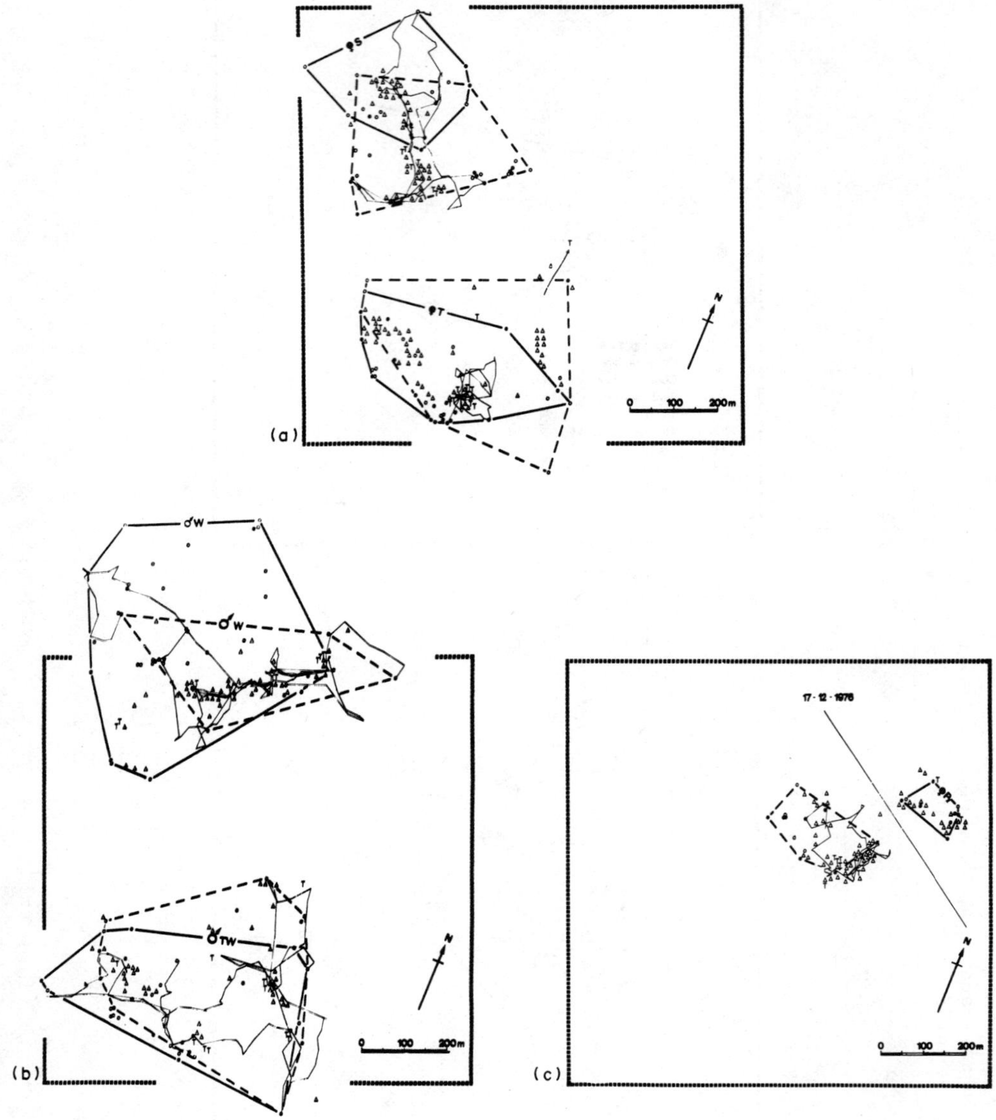

Fig. 2. Examples of sleeping area shifts by lesser bushbabies in response to major changes in the social environment. New range = heavy broken lines. (a) Disappearance of neighboring females: Female S (sleeping locations: $n_1 = 63$, $n_2 = 106$); Female T ($n_1 = 140$, $n_2 = 22$). (b) Disappearance of A male and change of status: Male W ($n_1 = 74$, $n_2 = 84$). Death of a territorial neighbor: A male, TW ($n_1 = 93$, $n_2 = 71$). (c) Disappearance of a neighboring female and independence from mother: Female P_2 ($n_1 = 100$, $n_2 = 47$).

Table 1.
Mean sample size and range of sightings used to establish a measure of home range in *G. senegalensis* (October, 1975-August, 1977)

	Time known (months)	Sleeping site plots	10-min plots	Miscellaneous sightings	Capture/recapture	Total sightings
Central A males	21	175	145	46	13	379
(N = 3)	20-22	158-202	75-191	49-56	7-21	354-397
Central B males	16	151	157	51	9	368
(N = 9)	9-23	51-265	42-393	21-100	4-14	158-571
Females	18	155	87	46	9	297
(N = 5)	11-22	112-198	147	25-59	4-11	197-367
Juvenile/subadult	11	140	118	33	3	294
(N = 1 male)						

Potential distortions in sampling are illustrated in Fig. 2, which shows range shifts in response to major alterations in population structure (unpublished data). These include (i) disappearance of neighboring rivals (S, T, TW); (ii) independence of a young female from her mother (P_2); and (iii) a change in status from B male to A male (W). Once again, sleeping areas measured before or after the shift provided a realistic estimate of the area actually traversed (n_1/n_2 percent: 84; S.E. = 4.38; range 61-100).

The great advantage of the minimum sleeping area method for this species is that it reduces the bias of unequal sampling by using only the most easily observed feature of range use (sleeping location) and one that is least likely to show short term variability.

MIGRATION PATTERNS

The pronounced effects of social change upon the apparent size of home ranges are further illustrated through consideration of migration. The displacement of female P_2 can be considered as a minor migration which was the most extensive observed in the case of a female. Males, however, periodically showed more extensive migration of two kinds which are defined following the terminology of Baker (1978).

Mating Season Migrations

Bushbabies in the study area exhibited estrus synchrony, showing two brief estrus peaks each year spaced in such a way that a female could potentially rear two sets of twins within the spring and summer prior to the onset of drought and cold. The start of estrus was usually clearly indicated by vaginal swelling and rupture of the membranes which sealed the opening during anestrus. Post-partum estrus also occurred. Estrus, mating and associated behavior was observed each year between May 5 and June 13 and between September 14 and October 5. Within these periods males were found to make wide ranging movements into areas where they had not previously been seen, where they associated with females which they did not usually contact (Fig. 3).

Male A provided several examples of mating season migrations, some of which were followed in detail on two consecutive nights of continuous observation (Fig. 3a). This male visited at least 3 females up to 1.2 km apart within 10 days. He moved and slept as much as 700 m away from his normal sleeping area, to which he returned to re-establish sedentary activity after the estrus period had ended. These greatly increased movements probably resulted from olfactory signals released by the female. They can be defined as 'regular, calculated exploratory and return migrations' of short duration which serve to increase a male's potential reproductive success. Further examples of this type of migration, including that of a A male, are shown in Fig. 3b (see Mating Behaviour).

Pre-reproductive Migrations

Six bushbabies (5 males 1 unsexed) moved into the study area as adults or subadults. Four of these became completely integrated into the social system (including Q_2; Fig. 1) while two remained for a few days and were not seen subsequently. Four juvenile/subadult males (R, B_2, M_2, N_2), present at the start of the study, disappeared shortly before reaching adulthood, suggesting that they may have moved as part of a pre-reproductive migration. On the other hand, a further two males (A, G_2) matured within the range of their birth.

ORGANIZATION OF RANGES

The relative positions of home ranges based on minimum sleeping areas during October, 1975-April, 1976 are shown separately for each social class in Fig. 4.

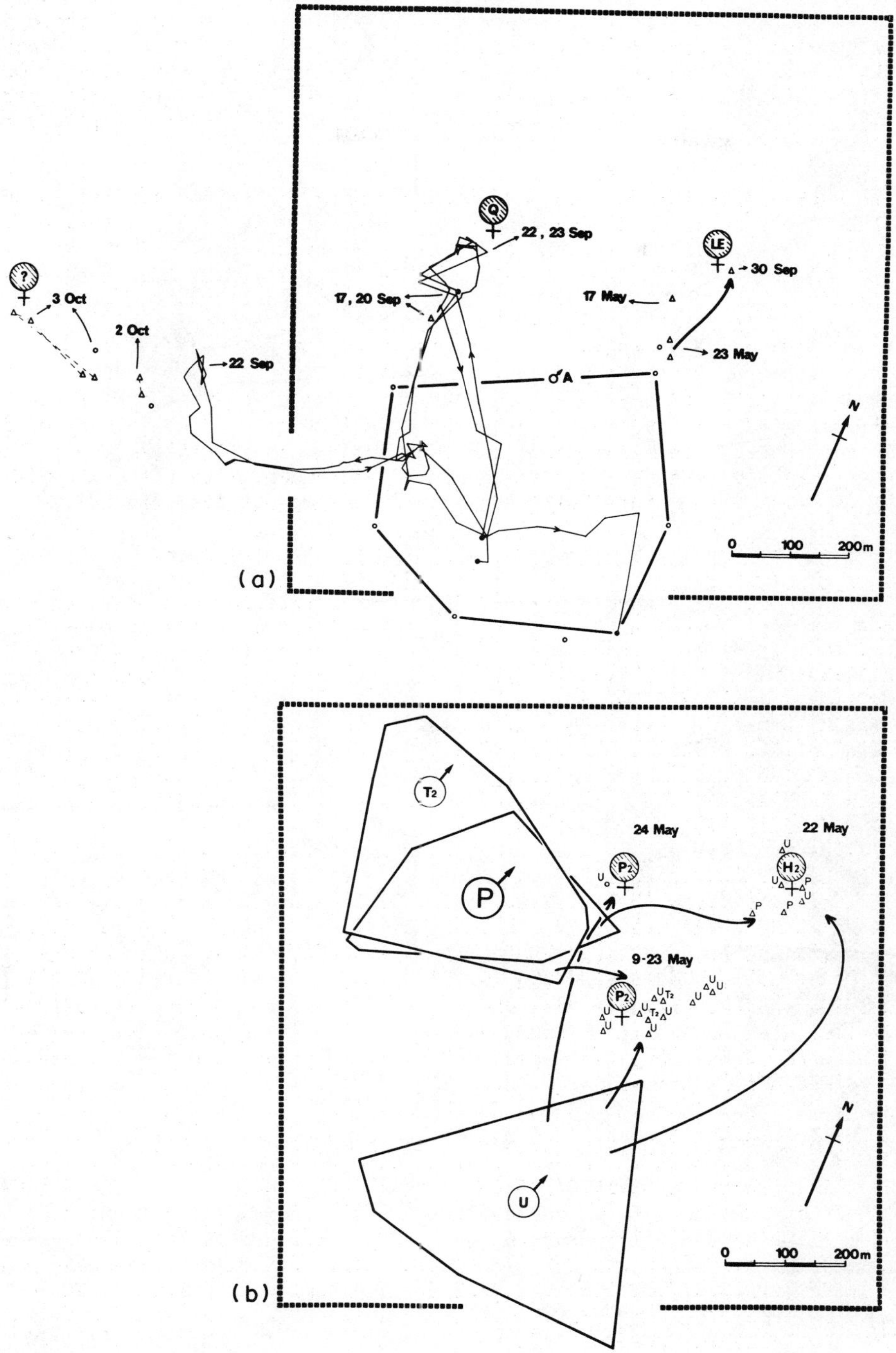

Fig. 3. Examples of migration by male bushbabies away from sleeping areas measured in the months immediately preceding and following mating. Note the unusual, short term, excursions to visit estrous females well outside the normal range.

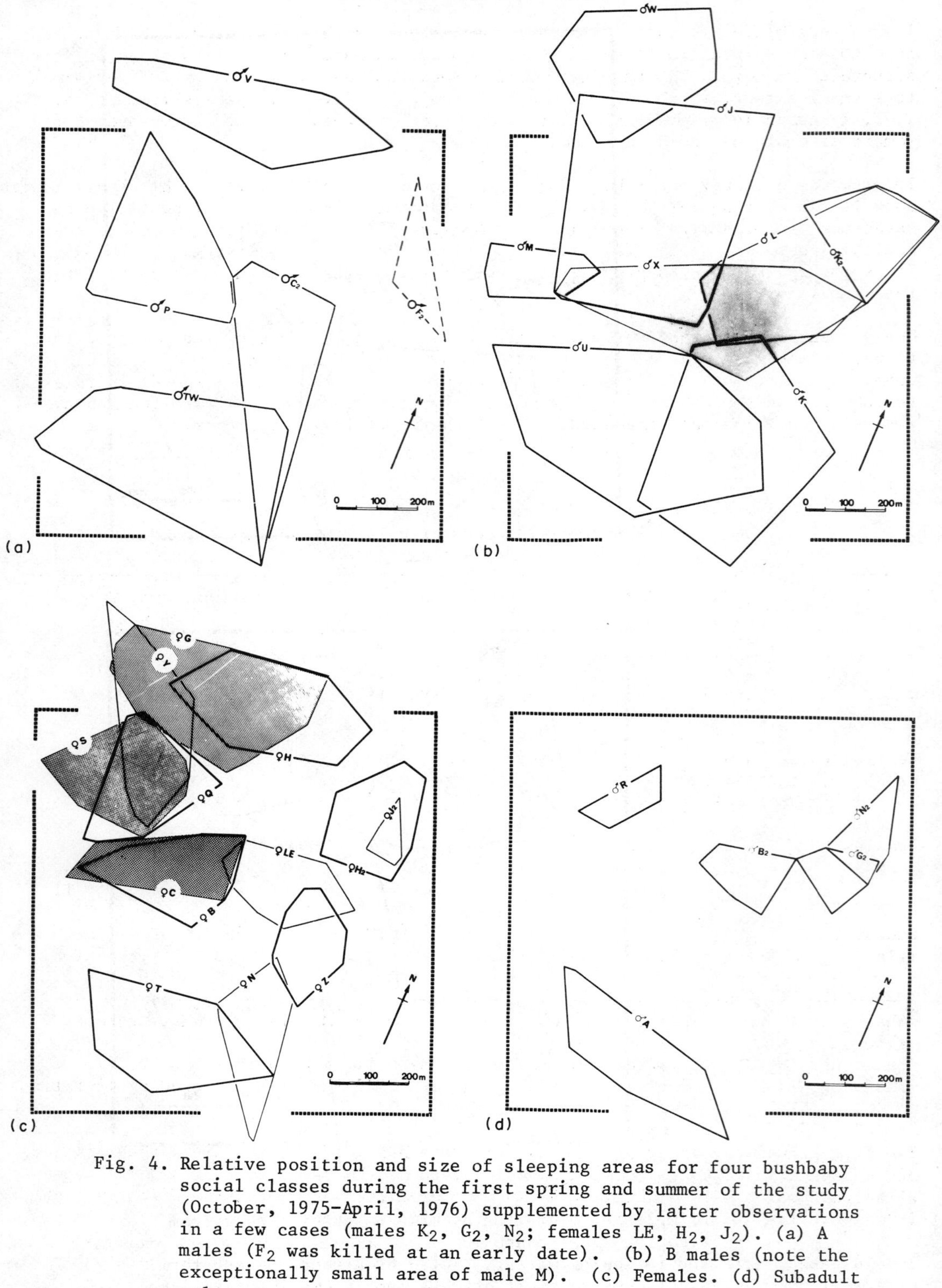

Fig. 4. Relative position and size of sleeping areas for four bushbaby social classes during the first spring and summer of the study (October, 1975–April, 1976) supplemented by latter observations in a few cases (males K_2, G_2, N_2; females LE, H_2, J_2). (a) A males (F_2 was killed at an early date). (b) B males (note the exceptionally small area of male M). (c) Females. (d) Subadult males.

The pattern of female ranges varies between those which are more or less separate, and those which are almost entirely congruous. In fact this appears to be an age-structured pattern. The ranges of five parous females (C, H, N, Q, LE) overlapped to a small extent, if at all, whereas the ranges of five nulliparous females (G, S, T, Y, Z) extensively overlapped those of 'older' females. The relative age of the remainder could not be determined.

In view of the territorial behavior of A males it is not surprising that their ranges show little overlap, although some is to be expected in order to allow reciprocal assessment through olfaction (Charles-Dominique, 1977, 1978). The ranges of sub-adult males appear to be distributed at random, sometimes showing extensive overlap. It is the complex overlapping arrangement of the ranges of B males which is perhaps the most interesting.

Measurement of home range area was only attempted in the case of radio tagged individuals which were seen on more than 30 days in the spring and summer. Table 2 compares home range size with a second measure of ranging (travel speed) in relation to the size and weight of each class. No significant variations were found between parous and nulliparous females, which are therefore lumped together.

Table 2.
Physical characteristics and range size of *G. senegalensis* (October, 1975-April, 1976) compared with mean travel speed for each social class

	Head length cm	Body length cm	Weight[1] g	Home range[2] ha	Mean travel speed[3] $m\ h^{-1}$ moon	no moon
Central A males	4.6 4.5-4.7 N = 5	11.9 11-13 N = 5	226 208-242 N = 5	11.0 9.5-15.6 N = 4	157 65-211 N = 3	70 63-85 N = 3
Central B males	4.6 4.5-4.7 N = 8	11.8 11-12 N = 8	211 202-224 N = 8	15.9 9.9-22.9 N = 7	159 75-240 N = 9	84 47-168 N = 7
Females	4.4 4.1-4.7 N = 13	11.8 11-13 N = 13	200 177-221 N = 13	6.7 4.4-11.7 N = 11	109 45-180 N = 8	78 27-140 N = 11
Subadult males	4.5 4.4-4.6 N = 5	11.6 11-12 N = 5	189 183-195 N = 5	6.9 N = 1	83 74-98 N = 3	95 67-108 N = 3

[1]When more than one measurement was taken for each animal the average figure is given.
[2]Based on 34-142 daily sightings per animal (Mean 84, S.E. 6.14, S.D. 29.47).
[3]Based on 413 h of direct observation, October, 1975-May, 1977 (excluding migrations).

Two size groups are apparent: A and B males were significantly larger than females and subadult males on the basis of head length ($F = 7.54$ df 1, 29 $P < 0.05$). However, although B males were equivalent in size to A males, they were significantly lighter in weight ($F = 5.35$ df 1, 11 $P < 0.05$), which is in accord with their social status.

Female ranges were similar in size to that of a single subadult male whose movements were well documented. These ranges were considerably smaller than those of A or B males (Mann-Whitney U test: $n_1 = 10$, $n_2 = 12$; $U = 4$; $P < 0.001$), the ranges of B males tending to be particularly large. The validity of this difference is strengthened through measurement of path length during direct observation. The speed of travel of all classes was much the same when there was no moon visible. Once the moon rose, females (and subadult males) showed little change in the extent of their movement but older males travelled significantly further each hour, both in relation to females (8 females, 11 males: $U = 18$, $P < 0.025$) and to their movements during periods without a moon ($n_1 = 10$, $n_2 = 11$; $U = 14$; $P < 0.01$).

A MEASURE OF ASSOCIATION

Interrelationships between ranges become clear when they are viewed in conjunction with knowledge of the degree of social contact between individuals. Bushbabies fluctuate between a solitary active phase and a gregarious resting phase during each 24 h period and therefore a particularly suitable index of association is provided by the formula $a = 2N/n_1 + n_2$ where $n_1 + n_2$ represent the number of times that a pair of bushbabies were seen in the daytime and N the number of times they were sleeping together (see Schaller, 1972, p. 36). The resulting pattern of association is illustrated in Fig. 5.

Females with overlapping home ranges are those which associate either regularly or occasionally (females T and N were seen together once at a later date). Later observations confirmed that female/female associations were in fact kin groups showing various degrees of stability. Thus the tolerance noted between female companions in the laboratory and the intolerance of strangers is clearly related to the system of social spacing operating in the wild.

All males associate with females on a regular basis (shown by the many connecting lines between males and females) but apart from associations involving juveniles or subadults, males never sleep together. It should be noted that central B males maintain social links with females which may be more extensive than those of central A males (male J for example was seen with 6 females whereas P slept with 5).

The picture to emerge when the home ranges of each social class are superimposed is shown in Fig. 6. Only animals which come into extensive contact with the range of a single A male are included for the sake of clarity. In this case the A male 'controls' 5 females and tolerates 3 B males whose ranges extend in all directions, bringing them into contact with adjacent 'groups'.

MATING BEHAVIOR

Considering that certain large males 'defend' a territory (Pitelka, 1959) against encroachment by dominant neighbors while allowing free access to subordinate males of a similar size, it is safe to assume that the defended resource is not, for example, food but probably females. In that case it might be predicted that B males would be excluded from mating in some other way. At first sight the mating season migration noted above and the sexual development of B males seems to contradict this conclusion. No correlations could be found, for example, between social position, and the size of the scrotum. An answer is provided by observations of mating and associated behavior which are described briefly.

Visits by males to estrous females become increasingly frequent until, at the peak of estrus, a cluster of males may form around each female. Despite the confusion of almost continuous chasing and repeated vocalizations, it was once possible to discern a distinct organization within a cluster of marked animals. The female and the resident A male remained together in the center surrounded by 3 resident and 2 immigrant males, including an A male from an adjacent territory. The group disperse

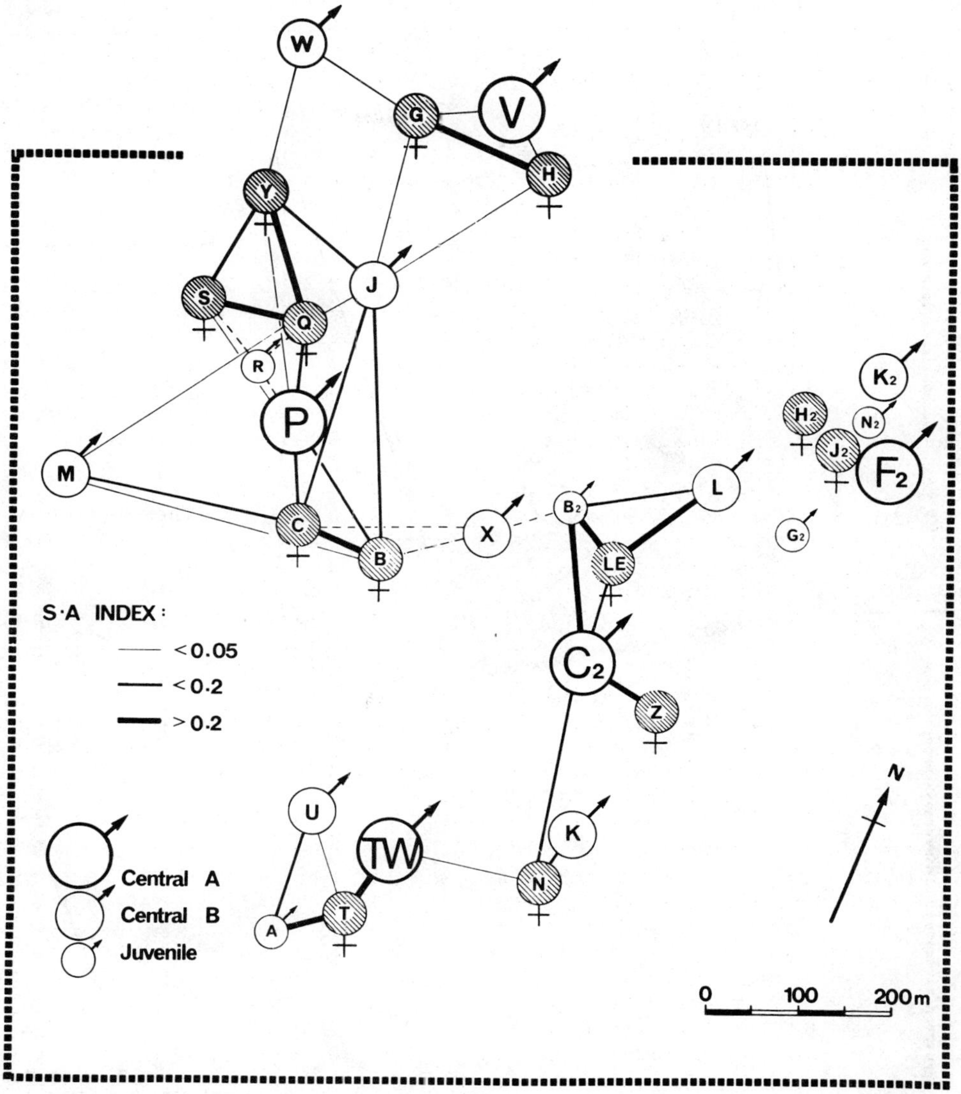

Fig. 5. Social contact between bushbabies in the study area, October, 1975–April, 1976. Each animal is represented at the center of its sleeping area by its initial and a symbol appropriate to its social class. Line thickness is proportional to the degree of association between individuals at sleeping sites (Index calculated for bushbabies seen on > 25 days). Five 'groups' can be distinguished, each centered upon a single A male, with B males making contacts between groups.

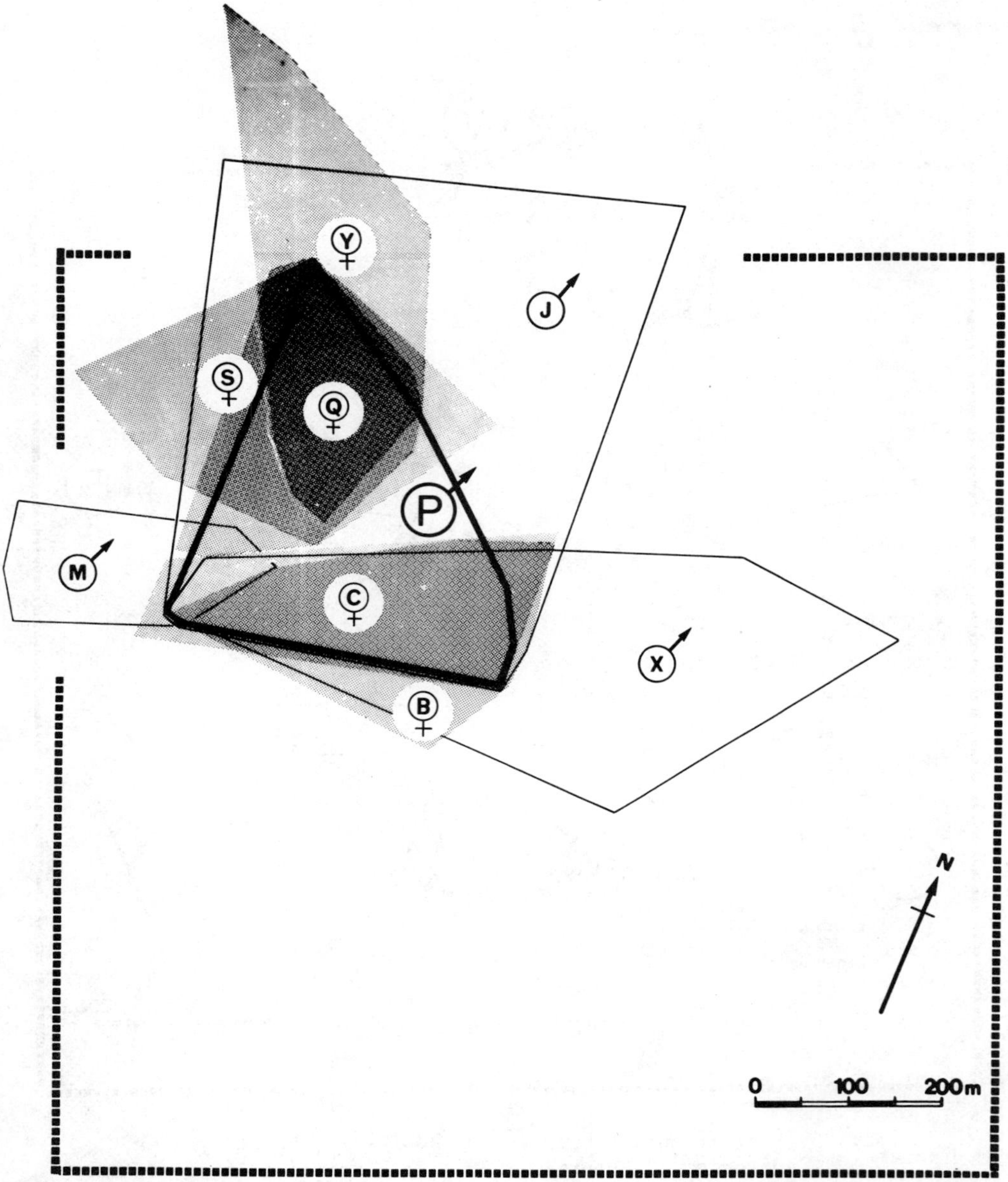

Fig. 6. The interrelationship of sleeping areas for a single group of bushbabies centered around an A male, October, 1975-April, 1976. Male P controls five females from two separate associations and tolerates three B males which are free to visit females falling under the control of neighboring A males.

at dawn and no further interest was shown towards the female on the following evening. It seems likely that B males will only obtain a chance to mate if the resident 'control' male is prevented from being present. The observation that even an A male will encroach upon a neighboring territory at the peak of estrus may suggest that this possibility is sometimes rewarded.

DISCUSSION

Sightings of all individuals within the study area can be explained in terms of slowly evolving home range movements and two types of migration. It appears that all bushbabies tend to be sedentary for long periods within an established social system. The pattern of female territoriality ensures that females are spaced out fairly evenly in suitable habitats, no doubt showing range expansion and contraction within certain limits depending on population density. Displacements of females were limited to short range shifts into adjacent areas that had been recently vacated. The movements of A males are similarly influenced by the presence of aggressive neighbors and marked changes in their territories were also related to the death or disappearance of rivals. B males on the other hand, through domination of females and active avoidance of direct conflict with A males, are able to exploit large ranges which are not restricted by social pressures. Maturing males are free to extend their ranges in any direction and to establish regular contact with females. It is not therefore surprising that these bushbabies were the ones to migrate.

The selective pressures acting upon females, which have encouraged territorial defence, are probably concerned with access to environmental resources which contribute towards maximum fitness for themselves and their offspring. The ranging patterns of males are more likely to be mainly determined by sexual selection. Each female within the study area associated with a single A male (primary consort male) who was probably able to inseminate his 'controlled' female or females on most occasions. The remaining males were faced with three options: they could attempt to find an estrous female that was temporarily unguarded; they could wait their turn to reach the top of the local social hierarchy, or; they could migrate, with a chance of finding an area where the opportunities for successful reproduction would be increased. There is evidence that *all* three theoretical strategies were in fact adopted by B males in the study area (secondary consort males).

The use of new sets of sleeping sites by an individual, which signified a range shift, often occurred abruptly and invariably coincided with an alteration in the social environment. The mechanisms underlying the rapid appreciation of a vacancy are poorly understood but the elegant radio tracking study of Charles-Dominique (1978) in the case of *Galago alleni* demonstrates that urine washing is important and vocalizations, particularly 'vocal advertisements' or 'spacing calls', are also involved. In this context it is interesting that the most obvious bouts of urine washing and olfactory investigation seen in *G. senegalensis* took place in sleeping-trees, which were sometimes visited and checked systematically during the course of a night (personal observation). It is quite possible that this provides information concerning the identity of bushbabies which slept there and the recency of their last visit (Schilling, 1979). There are several possible reasons for the use of peripheral sleeping sites but communication near territorial boundaries may certainly be involved.

Field and laboratory data on the social status of known-age bushbabies provide evidence that position within the male hierarchy is largely dependent on relative age and it follows that A males are the oldest males in a community. It is tempting to compare this system of ordered reproduction in a solitary foraging species with the 'age-graded male groups' of more gregarious primates (Eisenberg, Muckenhirn and Rudran, 1972). Lesser bushbabies are effectively divided into 'groups' consisting of several adult females and their offspring which have regular contact with a similar number of adult males organized by linear dominance based on age. This social organization shows remarkable convergence with that of the brush-tail possum

(*Trichosurus vulpecula*) in Australia (Winter, 1975). It also shares important characteristics with the societies of other nocturnal prosimians, yet each species so far examined shows distinctive variations (see Charles-Dominique, 1977; Martin, 1972). Identification of the reasons for similarities and differences in social behavior depends upon detailed studies, which, in the case of mammals, are often only possible with the aid of radio tracking.

Acknowledgements — Our sincere thanks go to Mr. and Mrs. E. A. Galpin and Mr. and Mrs. R. G. Galpin for their continuous support and hospitality. Invaluable field assistance was provided by C. Bearder, R. Brooks, J. Cunningham, P. Fleischack, D. Freeman, S. Galpin, C. Harcourt, Z. Katzir, I. Lemon, K. McDonald, T. Olson, N. Passmore, P. Small, W. Tarboton, I. Temby and R. Toms. Mr. T. Dennett provided expert help with all aspects of photography throughout the study and Ms. C. Nutkins kindly typed the manuscript. The field work was financed by a grant-in-aid from the Royal Society (London) and supplementary allowances from the Boise Fund (Oxford). Subsequent analysis was generously supported by the Science Research Council and the Zoological Society of London.

REFERENCES

Baker R. R. (1978) *The Evolutionary Ecology of Animal Migration*. London: Hodder and Stoughton.

Bearder S. K. (1969) Territorial and intergroup behaviour of the lesser bushbaby, *Galago senegalensis moholi* (A. Smith), in semi-natural conditions and in the field. M.Sc. dissertation, Witwatersrand University.

Bearder S. K. and Doyle G. A. (1974a) Ecology of bushbabies, *Galago senegalensis* and *Galago crassicaudatus*, with some notes on their behaviour in the field. In R. D. Martin, G. A. Doyle and A. C. Walker (eds.), *Prosimian Biology*, pp. 109-130. London: Duckworth.

Bearder S. K. and Doyle G. A. (1974b) Field and laboratory studies of social organization in bushbabies (*Galago senegalensis*). *J. Hum. Evol.* 3, 37-50.

Charles-Dominique P. (1972) Ecologie et vie sociale de *Galago demidovii* (Fischer 1808; Prosimii). *Z. Tierpsychol., Bieheft* 9, 7-41.

Charles-Dominique P. (1977a) *Ecology and Behaviour of Nocturnal Primates: Prosimians of Equatorial West Africa*. London: Duckworth.

Charles-Dominique P. (1977b) Urine marking and territoriality in *Galago alleni* (Waterhouse, 1837 — Lorisoidea, Primates) — a field study by radio-telemetry. *Z. Tierpsychol.* 43, 113-138.

Charles-Dominique P. (1978) Solitary and gregarious prosimians: evolution of social structure in primates. In D. J. Chivers and K. Joysey (eds.), *Recent Advances in Primatology*, Vol. 3: *Evolution*, pp. 139-149. London: Academic Press.

Charles-Dominique P. and Bearder S. K. (1979) Field studies of lorisid behaviour: methodological aspects. In G. A. Doyle and R. D. Martin (eds.), *The Study of Prosimian Behaviour*, pp. 567-629. London: Academic Press.

Eisenberg J. F., Muckenhirn N. A. and Rudran R. (1972) The relation between ecology and social structure in primates. *Science* 176, 863-874.

Layne J. N. (1954) The biology of the red squirrel, *Tamiasciurus hudsonicus loquax* (Bangs), in central New York. *Ecol. Monogr.* 24, 227-267.

Martin R. D. (1972) A preliminary field study of the lesser mouse lemur (*Microcobus murinus*, J. F. Miller 1777). *Z. Tierpsychol., Bieheft* 9, 43-89.

Mohr C. O. (1947) Table of equivalent populations of North American small mammals. *Am. Midl. Nat.* 37, 223-249.

Pitelka F. A. (1959) Numbers, breeding schedule and territoriality in pectoral sandpipers of northern Alaska. *Condor* 61, 233-264.

Schaller G. B. (1972) *The Serengeti Lion: A Study of Predator-Prey Relations*. Chicago: The University of Chicago Press.

Schilling A. (1979) Olfactory communication in prosimians. In G. A. Doyle and R. D. Martin (eds.), *The Study of Prosimian Behaviour*, pp. 461-542. London: Academic Press.

Winter J. W. (1975) The behaviour and social organization of the brushtail possum (*Trichosurus vulpecula:* Kerr). Ph.D. thesis, University of Queensland.

Radio Tracking in the Study of Bovine Tuberculosis in Badgers

C. L. CHEESEMAN and P. J. MALLINSON

Ministry of Agriculture, Fisheries and Food, Agricultural Science Service, Worplesdon Laboratory, Tangley Place, Worplesdon, Guildford, Surrey GU3 3LQ, U.K.

Abstract — Radio tracking techniques have been employed as part of the study to investigate the role of badgers in the transmission of bovine tuberculosis. The advantages and disadvantages of using traps, snares or nets for catching badgers are discussed. Ketamine hydrochloride administered intramuscularly has proved to be safe and effective for immobilization. Crystal controlled radio transmitters incorporating internal iron dust core antennas have produced satisfactory results. Various methods of encapsulation and attachment of transmitters are described. Hand portable equipment with an H-Adcock antenna is used for radio tracking. Some aspects of field strategy and sources of error are described. The value of radio tracking for investigating home range and movements of badgers is illustrated by comparing information obtained from radio tagged badgers with that obtained by bait marking.

INTRODUCTION

When it became known in 1971 that badgers (*Meles meles* L.) might be a reservoir of bovine tuberculosis, the Ministry of Agriculture was faced with several questions concerning the ecology and behavior of badgers. Why, for instance, is tuberculosis (TB) mainly restricted to the southwest of England? How is it likely to spread both among badgers and from badgers to cattle? Also, with the need to implement control measures against badgers, how extensive should such control operations be around farms where TB breakdowns were attributed to badgers? Some aspects of these questions could only be answered by field investigations of the behavior, range and movements of badgers.

This study began in 1975 shortly after the commencement of the Ministry's current gassing program. The main objective was to reveal aspects of badger ecology and behavior related to their role in the transmission of TB. A longterm study area was selected in the Cotswolds where bovine TB was a persistent and serious problem. Radio tracking would clearly be the most important research technique. We had the advantage that some years earlier the MAFF had set up a fox study in Wales and radio tracking equipment had been developed for this work. A complete system of receiver, directional antenna and transmitters had been designed, developed and built by the late Gordon Ashwell (Taylor and Lloyd, 1978). He devised a transmitter with the advantage of an iron-dust core loop antenna that could be encapsulated with the rest of the components. This avoided the problem of water penetration through connecting leads and made transmitter attachment easier. This design had been well proven in work on foxes and in a modified version, for rats, so there was no difficulty in adapting it for use on badgers.

CATCHING AND HANDLING BADGERS

The methods of capture included trapping, snaring, and netting. Whichever method is used the intention should be to catch, handle, and release the animal with the least possible stress. Cage traps produce best results in the summer months when food for badgers is in short supply. Traps are placed near the holes of a sett, which showed signs of badger activity, with the entrance facing the hole (Fig. 1a). The floor of the trap should be firmly bedded down in the soil to make it steady. A trip cord is used as a trigger mechanism and peanuts are used for bait. With pre-baiting, the overall success rate of cage traps in the study area is one capture per six trap-nights, but in mid summer success rates better than one capture per two trap-nights have been achieved.

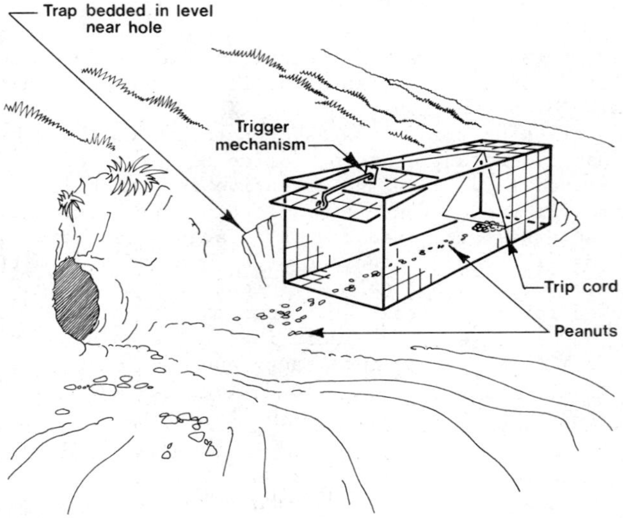

Fig. 1a. Cage trap set for badgers.

Snares are more efficient and easier to handle then cage traps and in our opinion cause less stress to captive animals when used correctly. An average of one capture per three snare-nights has been achieved and snares work well at all times of the year. A large diameter loop set low on a run will usually catch a badger around the body (Fig. 1b). This is more successful than a smaller loop set high with the intention of catching round the neck. An angle-iron stake is used to anchor the snare which should be placed where a captured animal cannot become entangled in branches, saplings or other obstructions. Snares should be visited as frequently as possible or ideally kept under continuous observation. In more than 50 captures using snares there have not been any injuries, whereas teeth and claws are sometimes damaged in traps.

Thirty-one captures have been made using a hand net. A large 1.25 m gape, fisherman's landing net with a strengthened frame is used. This is perhaps the best method in terms of inducing least stress on the animal. An advantage with this method is that selected individuals may be caught, (e.g. to recapture a radio tagged individual for transmitter replacement before the batteries expire). This method is only

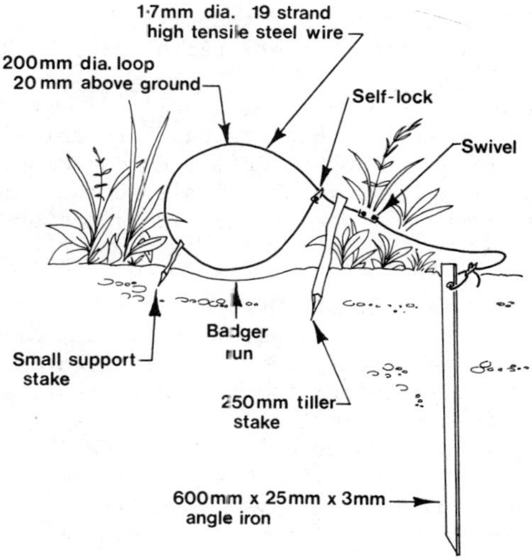

Fig. 1b. Snare set for badgers.

suitable for use on open ground. Depending on the uneven nature of the terrain, approach may be made either on foot by two people, one with the net and the other with a powerful handlamp, or alternatively, the initial approach may be made using a vehicle.

It is necessary to restrain a badger in order to administer an anesthetic. A forked stick may be held over the shoulders of a snared or netted badger, and a number of wooden rods passed through the mesh may be used to restrict the movements of a trapped badger. Ketamine hydrochloride ('Vetalar') is widely used for anesthesia of domestic animals and has proved very suitable for badgers. Hunt (1976) and Mackintosh *et al.* (1976) successfully used 15-20 mg kg^{-1} intramuscularly to immobilize badgers, the latter authors recommending injection into the thigh for best results. In a sample of 194 badgers immobilized with ketamine during our studies there were no mortalities or apparent ill effects. A mean dose of 30.8 mg kg^{-1} was used to produce a mean time to full relaxation of 1 min 45 s with a mean time to recovery of 44 min. The weight of captive badgers has to be estimated but ketamine appears to have a wide safety margin. Thus, a dose of approximately 30 mg kg^{-1} produces full relaxation for a period of approximately 45 min. This is sufficient time to fit a transmitter, mark with ear tags and a tattoo, as well as take blood and other samples.

TRANSMITTER CHARACTERISTICS AND ENCAPSULATION

A crystal controlled transmitter emitting a pulsed signal is used. A complete description of this transmitter is given by Taylor and Lloyd (1978). The radio frequency was in the range 102.285-102.315 MHz, though we have now changed to the new 104.6-105 MHz band allocated by the Home Office in Britain. Individual transmitters can be identified by their radio frequency, and by different pulse repetition rates. Confusion in transmitter identity can arise when two transmitters of similar frequency are used in the same area. With the type of receiver used this can be overcome by tuning to the zero beat frequency of one transmitter while still listening to the audible signal of the other. An explanation of this is given in Taylor and Lloyd (1978). A 2000 mAh mercury battery gives continuous transmission for about 400 days, but this will depend on a number of variables such as pulse length and repetition rate. In our experience the working life never matches the calculated theoretical life, due partly to the unknown quality and limited shelf life of mercury cells, so it is advisable to recover transmitters in good time. Harding *et al.* (1976) describe a means of checking battery reliability by X-ray examination.

In hilly terrain, our transmitters give an effective range of up to 1 km. This is sufficient for radio tracking badgers in this region where animal movements are normally contained within an area of about 50 ha. Limited transmission range reduces overlap where it is intended to work with several animals in the same study area. Signal penetration through the ground is good, transmitters having been located and recovered from depths of up to 2 m.

The transmitter incorporates a switching circuit which enables pulse length and repetition rate to be predetermined, so allowing economy of battery life and assistance in individual transmitter identification through different repetition rates. Changes in pitch caused by movement of the transmitter package against the body of a badger can convey additional information to the tracker. With experience it is possible to distinguish signals characteristics of a walking or running badger, and to check that a badger below ground still has its transmitter attached (or is still alive) by stamping on the ground above the sett. The disturbance causes a badger to stir with a consequent change in the pitch of the signal.

Various encapsulating media have been tried. If a betalight is to be incorporated with the transmitter then a clear medium is required. An advantage of clear media is that old batteries may be more easily cut out and the transmitter re-encapsulated with new ones. With an internal coil antenna, the type of encapsulant used can markedly affect the properties of the radio signal generated and transmitter tuning. The type of encapsulant used is therefore important in minimizing capacitive loading on the coil after encapsulation.

METHODS OF ATTACHMENT

As the diameter of a badger's neck is almost the same as that of its head it was considered that a harness would be the best method of transmitter attachment (see Kruuk, 1978). Several different designs and materials were tried on captive and wild badgers. The most successful was constructed of split rawhide with noncorrosive fittings. This was tailor made to fit each individual badger (Fig. 2a). However, the problems of loss and abrasion could not be completely overcome.

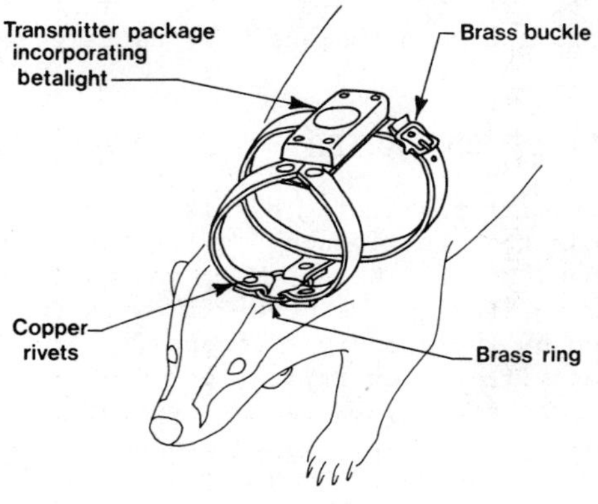

Fig. 2a. Transmitter and betalight attachment by harness.

Implantation was considered as an alternative but it was decided to try a simple collar before taking this step. The first few collars used proved very successful, and of the 17 fitted to wild badgers none has been lost and abrasion does not occur. Certain modifications had to be made to transmitter encapsulation for collar attachment and it was also necessary to encapsulate the betalight separately and mount it on top of the collar (Fig. 2b). The heavier transmitter package holds the betalight in this position. Holes are drilled through the transmitter package which is fixed to the collar with brass bolts. The head of each bolt must be filed down to make a flush fitting on the inside of the collar when the nut on the outside is tightened.

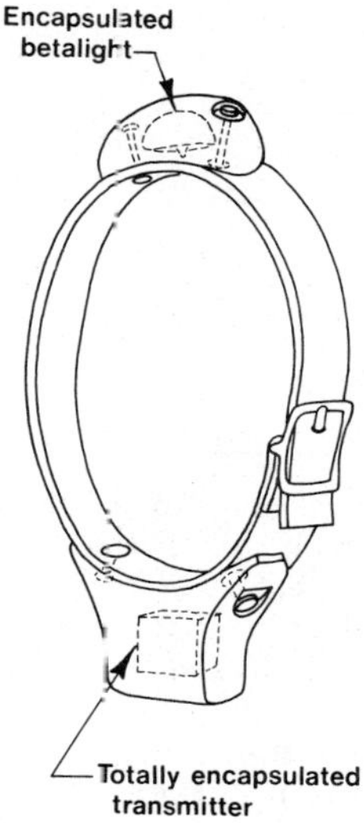

Fig. 2b. Transmitter and betalight attachment by collar.

The collar is made from a strip of split rawhide 300 mm long, 22 mm wide and 4 mm thick. Copper rivets are used to attach a square recessed brass buckle. A hole may be punched when the collar is fitted. As a rough guide the collar should be tightened until it is just possible to slip two fingers between the collar and neck of the badger. Inner edges of the collar are filed smooth to remove sharp corners and the leather treated with either neatsfoot oil or Hydrophane (made by Hydrophane Labs. Ltd.) to soften and preserve it. Collars recovered after periods of up to seven months showed very little sign of wear. Although the leather becomes hard after a time it has by then taken the shape of the neck and does not cause any irritation or discomfort to the animal. Split rawhide has the advantage of being reasonably supple, while the layer of untanned leather through the middle prevents stretching.

We believe that badgers quickly become accustomed to wearing a collar and no overtly abnormal behavior has been observed as a result of fitting a transmitter in this way. The collar, complete with all attachments weighs 150 g. This is approximately one and half percent of the weight of an adult badger and well within the five percent limit suggested as a maximum by Macdonald (1978).

TRACKING

In our work radio tracking is essentially used as an aid in visually locating badgers so that direct observations may be made. This has been termed 'predictive' tracking (Macdonald, 1978). A portable set of equipment is used consisting of a six-channel receiver plus a hand held H-Adcock antenna. At the 102.3 MHz wavelength the H-Adcock antenna is possibly a little cumbersome but it has good gain and directional properties. One disadvantage is that it has two nulls and it is not immediately possible to decide whether the source is in front of or behind the bearing of each null. The problem of reciprocal nulls can be overcome by moving to a new position and taking a new bearing. Telescopic elements are used to make the antenna collapsible for easier transportation. Full specifications and details of construction of the H-Adcock antenna are given in Taylor and Lloyd (1978).

Once a bearing has been obtained, the observer walks along the line of the bearing until signal strength indicates that the badger is near. A betalight is a very useful aid in visual location. Hemispherical betalights 25 mm in diameter and 800 μL in brightness are used for badgers. Protection and a means of mounting on the collar is provided by encapsulating the betalight in clear casting resin. Betalights can be seen at up to 100 m with the naked eye and up to 300 m with 7 × 50 binoculars. Behavioral observations are made using infra red binoculars from a distance of 50-150 m.

A considerable degree of observer stealth is required and it is important to wear soft clothing so that movements are silent. Approach should always be made up wind and if possible the movements of a badger should be anticipated in order to avoid leaving one's scent across its path.

Woodland, particularly coniferous plantations, tends to polarize radio waves vertically which are normally horizontally polarized thus the H-Adcock antenna must be held vertically. If reception is difficult and vertical polarization is suspected the antenna may be tilted to the horizontal mode to obtain a stronger signal. Under these conditions it is difficult to obtain an accurate bearing but having established the cause of the problem it may be possible to move to a new location where conditions are better for normal reception. A detailed account of other sources of error in bearing accuracy is given in Tester (1971). Although we have tried to convey a few of the more fundamental aspects of badger radio tracking strategy, full coverage of field application and technique is beyond the scope of this paper. However, we would emphasize to newcomers that radio tracking always involves a multitude of unexpected difficulties which can only be overcome with patience and experience. Further practical guidance is given in Taylor and Lloyd (1978).

RESULTS

A short account of some of the results of this work is given in order to illustrate the value of radio tracking as a research tool in the study of badgers. Badgers are social carnivores living together in groups of up to twelve individuals. Each social group occupies a territory with a main sett as its headquarters. (The term 'territory' is used here to describe the area defended by a social group of badgers, while 'home range' is used to describe the area used by an individual badger.) Territorial boundaries are marked by a series of latrines which are also used by neighboring groups and provide a mutually respected boundary marker. By bait marking it is possible to delineate the territory of each social group (Kruuk, 1978). A bait of peanuts and honey (very attractive to badgers) is mixed with an indigestible colored plastic marker. Different main setts in an area are fed bait containing different coloured markers. Subsequent recoveries of markers in the feces show which latrines badgers from each sett have visited. More conclusive results are obtained if this is done in the spring when, at the time of breeding, territory marking is at a peak. The territory of each social group will become apparent when the results are plotted on a map (Fig. 3). Territories are not, however, entirely exclusive, since the results show that badgers will transgress territorial boundaries. This can be seen where bait markers from group B were found in one latrine on the opposite boundary of group C. Also, droppings are somtimes found containing two different markers, indicating that a badger has visited two main setts.

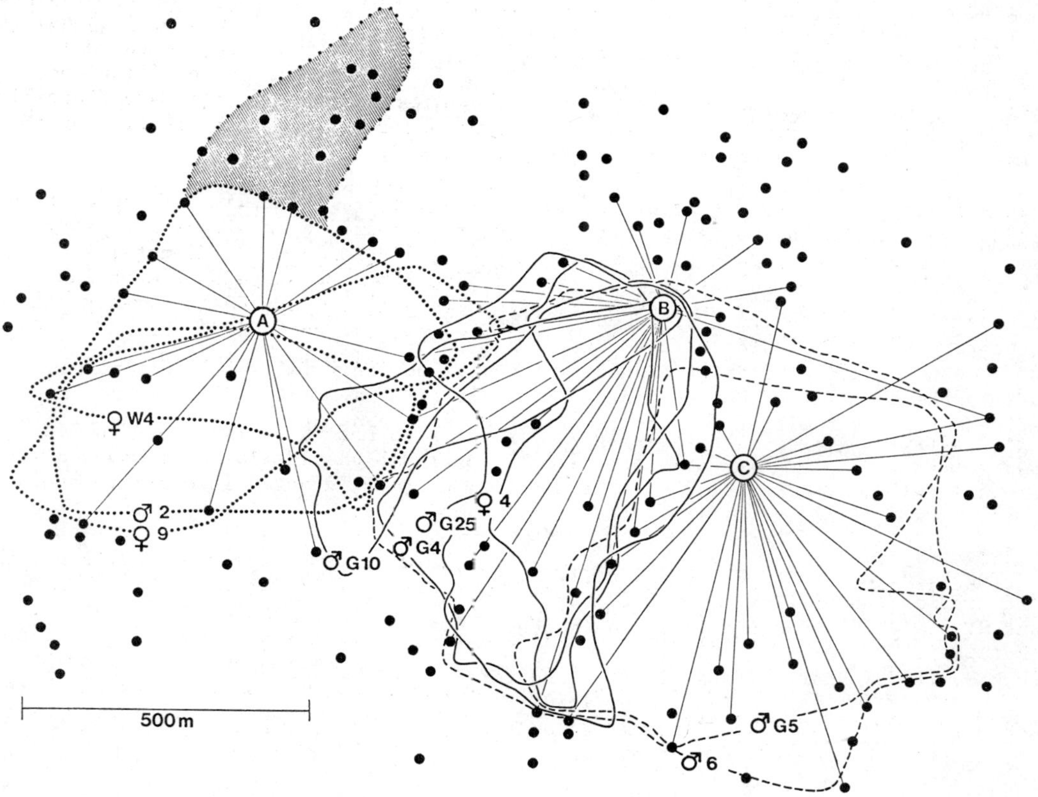

Fig. 3. Territories of badger social groups and home ranges of individual badgers revealed by bait marking and radio tracking respectively. Lettered circles represent main setts; large dots are latrines; straight lines from the main setts indicate those latrines where bait markers were recovered; home ranges of individual badgers from setts A, B and C are shown by dotted, continuous and broken lines respectively.

Figure 3 also shows the home ranges of some individual badgers from each of these three social groups as revealed by radio tracking. Each badger was observed over a period of not less than 60 days, which was demonstrated to be sufficient, if frequent observations were made, to reveal the main areas of an individual's home range at any season of the year. From radio tracking results it is evident that male badgers often have larger home ranges than females and utilize most if not all of the social group territory revealed by bait marking. This is in contrast to Kruuk's (1978) findings in Wytham Wood, Oxfordshire, U.K., where all male and female members of a social group covered more or less the same home range. In the breeding season males may make sallies outside their group territory, probably to mate with females in neighboring social groups. Such a movement is seen in the shaded area of the home range of male 2 in Fig. 3. Kruuk (1978) suggested that males sometimes occupy two otherwise separate social group territories. An example of this is seen in the case of male G5 whose home range covered the territories of groups B and C. Male G10 represents an unusual example of different movements. When first caught this individual was resident in the main sett with other members of social group B.

After a short period of time, unfortunately not long enough to find out its normal home range, this badger took up residence in an outlying hole in an area of the boundary between Group B and Group A. For the following few weeks he lived a solitary existence in a relatively small area, mainly around farm buildings, before becoming very ill and eventually dying. Post mortem examination revealed that this badger had been suffering from bovine tuberculosis, which might explain the unusual behavior observed.

These results serve to illustrate that much more detailed information is obtainable using radio tracking in conjunction with visual observation. The bait marking method alone gives only an impression of the way badger social groups are spatially organized. It is necessary to use radio tracking in order to understand the complex way in which members of each group utilize the available habitat. However, the success of radio tracking will depend ultimately on the equipment used which needs to be robust and reliable. Transmitters must continue to operate for long periods, and a means of transmitter attachment which will not only last but cause no discomfort to the animal must be developed. There is still a great deal of scope for refinement of transmitter design, construction and means of attachment, as well as the practical application of radio tracking in the field.

REFERENCES

Harding P. J. R., Chute F. S. and Doell A. C. (1975) Increasing battery reliability for radio transmitters. *J. Wildl. Mngmt.* 40, 357-358.

Hunt H. S. (1976) Anaesthesia of the European badger using Ketamine hydrochloride. *Vet. Rec.* 98, 94.

Kruuk H. (1978) Spatial organisation and territorial behaviour of the European badger *Meles meles*. *J. Zool.* 184, 1-19.

Macdonald D. W. (1978) Radio-tracking: some applications and limitations. In B. Stonehouse (ed.), *Animal Marking: Recognition Marking of Animals in Research*, pp. 192-204. London: Macmillan.

Mackintosh C. G., MacArthur J. A., Little T. W. A. and Stuart P. (1976) The immobilisation of the badger (*Meles meles*). *Br. vet. J.* 132, 609-614.

Taylor K. D. and Lloyd H. G. (1978) The design, construction and use of a radio-tracking system for some British mammals. *Mammal Rev.* 8, 117-141.

Tester J. R. (1971) Interpretation of ecological and behavioural data on wild animals obtained by telemetry with special reference to errors and uncertainty. *Proc. Symp. Biotelemetry, Pretoria,* (1971). Council for Scientific and Educational Research, Pretoria, pp. 385-407.

Radio Tracking of *Rattus norvegicus* on Farms

A. R. HARDY and K. D. TAYLOR

Rodent Pests Department, Ministry of Agriculture, Fisheries and Food, Tolworth Laboratory, Hook Rise South, Tolworth, Surrey, U.K.

Abstract — 36 radio tagged rats were tracked in three separate farmland areas and their range of movements related to the spatial and temporal distribution of food sources. In arable farmland where rats lived in hedgerows, movements in winter were initially restricted and near to established bait points. Removal of the bait caused an increase in the range of movements with a maximum recorded journey of 3.3 km in a single night. Rats associated with food stores in farm buildings in winter also showed very restricted movements. Several individuals were found to visit farm buildings regularly in winter from up to 700 m distant. Juvenile rats in summer were shown daily to move small distances from buildings to feed in an adjacent field of growing grain. The implications of this study for control are discussed.

INTRODUCTION

Found in a wide range of terrestrial habitats, the brown rat (*Rattus norvegicus* Berkenhout) is a frequent pest of both rural and urban situations. Although established rat populations in agricultural premises can be successfully controlled, invasion may occur again within a few months. Lack of information on the ranges of individual rats together with the inadequacies of trap-based studies of this species, which is difficult to recapture, prompted investigation using radio tracking techniques. Work reported here was conducted on three arable farms in southeastern England to examine the range of movements of rats in farmland and to relate them to the use of identified food sources.

METHODS

Radio equipment was specially designed by G. E. Ashwell and is fully described elsewhere (Taylor and Lloyd, 1978). Rats were captured in wire 'Bledorberry' cage traps, or smoked, ferreted or dug out of their burrows. While immobilized in a Perspex holding box, each rat was anesthetized with methoxyfluorane ('Penthrane') before a transmitter, attached to a plastic or nylon collar, was fitted round its neck. Completed collars weighed between 8 and 9 g and only rats weighing upward of 200 g (10 weeks old) were radio tagged. Collars on large rats also carried 'betalights' for visual tracking at night. When fully recovered (15-30 min) rats were released at their point of capture.

Tracking was done on foot using a handheld directional H-adcock antenna (Taylor and Lloyd, 1978). Locations of moving rats were accurate to within 15 m; positional fixes on stationary animals and visual fixes were more accurate. Maximum transmitter range was 400 m but tracking was normally undertaken at distances of less than 60 m without disturbance to the rats (Taylor, 1978; Taylor and Quy, 1978). Transmitter life varied up to 7 months. Successive radio fixes were taken at minimum intervals of 10 or 15 minutes. Pitch variation was used to identify activity by transmitter bearing rats (Taylor and Lloyd, 1978).

Twenty-one rats in open farmland were studied on a large arable and dairy farm in Hampshire. Plain wheat was provided at covered bait points in September to provide a stable food supply to rats living in hedgerows. Bait was removed in January and radio tracking was performed between November and April (Taylor, 1978). At a small arable and dairy farm in Surrey, 10 rats were studied around farm buildings. Additional food was not provided and animals were tracked from October to February. Radio tagged rats were observed directly through infra-red binoculars (De Oude Delft IRH6ML). A limited study was made of juvenile rats on a small mixed farm in Surrey from June to August. Miniature transmitters weighing 3.5 g were used to tag 5 rats weighing about 100 g, approximately 6 weeks old (Taylor and Lloyd, 1978).

RESULTS

ARABLE FARMLAND

Rats in arable farmland showed linear ranges primarily oriented along hedgerows in which their homesite burrows were sited (Taylor, 1978). Males ranged more widely than females as seen clearly in Figs. 1 and 2; mean range length (Stickel, 1954) for 7 males was 660 m, and for 10 females, 340 m. Homesites were changed frequently and for reasons generally not known except in response to other individuals. Although most movements occurred in cover within or beside hedgerows, several rats regularly crossed up to 500 m of open ground away from cover (male 5B in Fig. 1).

Many observed movements could be related to the availability of food. Initially when surplus wheat was provided in the bait boxes, rats showed restricted ranges and were rarely recorded more than 100 m from their homesites. However when bait was removed, ranges were extended with a maximum recorded journey of 3.3 km in a single night (Taylor and Quy, 1978). Two males regularly visited farm buildings from homesites 500 m distant following bait removal, but failed to become established. Four rats (2 male, 2 female) spent many hours, presumably feeding, in a kale field where they took refuge in single entrance 'resting' burrows. Another female fed nightly on sown barley and earthworms in an open cultivated field, 200 m distant from her homesite. Additional food sources identified were a slurry pit (3 rats), cattle feed in buildings, household compost, spoiled grain and rotting maize.

Most rats spent daylight hours sleeping in their burrows and did not become active until sunset or just after (Taylor, 1978). The next one to two hours were generally spent in the vicinity of the homesite before moving away and a peak of activity was observed 4 to 5 h after sunset (Fig. 3). Return to homesites often occurred in the succeeding 4 h. Diurnal activity was shown by three rats (two occasionally, one predominantly) which were never recorded more than two meters from cover.

FARM BUILDINGS

In contrast to those in open farmland, rats associated with farm buildings occupied

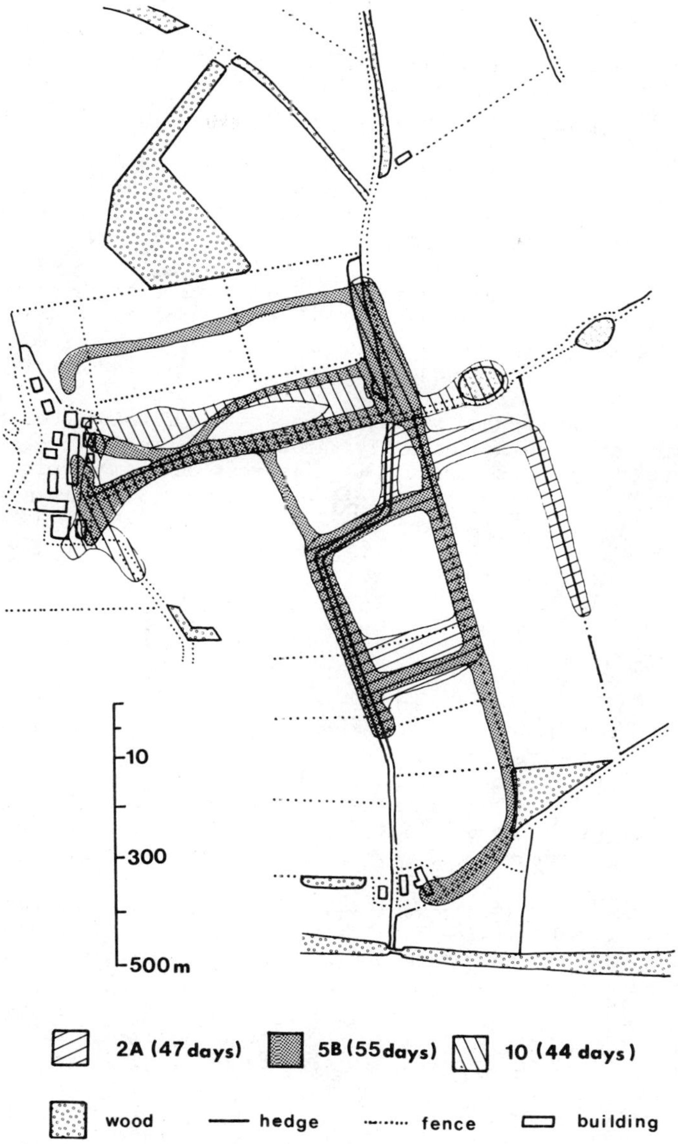

Fig. 1. Part of the arable study area showing the observed ranges of three male rats (2A, 5B, 10).

small home ranges (mean range length 65 m for 5 males and one female). Around the buildings the longest range, 110 m, represents only 8.7 percent of the maximum recorded rangelength in open farmland (Taylor, 1978). Ranges of 5 rats are shown in Fig. 4 together with the locations of major food sources on the farm. A high degree of overlap was recorded with 5 males and two females using the same area of meal-bin sheds and an adjacent hay barn (A and D, Fig. 4). Homesites, normally adjacent to or under food stores, were changed less frequently than in hedgerows. This was usually in response to disturbance within the buildings, e.g. following the clearance of animal pens in which it lived, male 6 moved out of the dairy and established a new homesite in a hay shed 60 m away (Fig. 4).

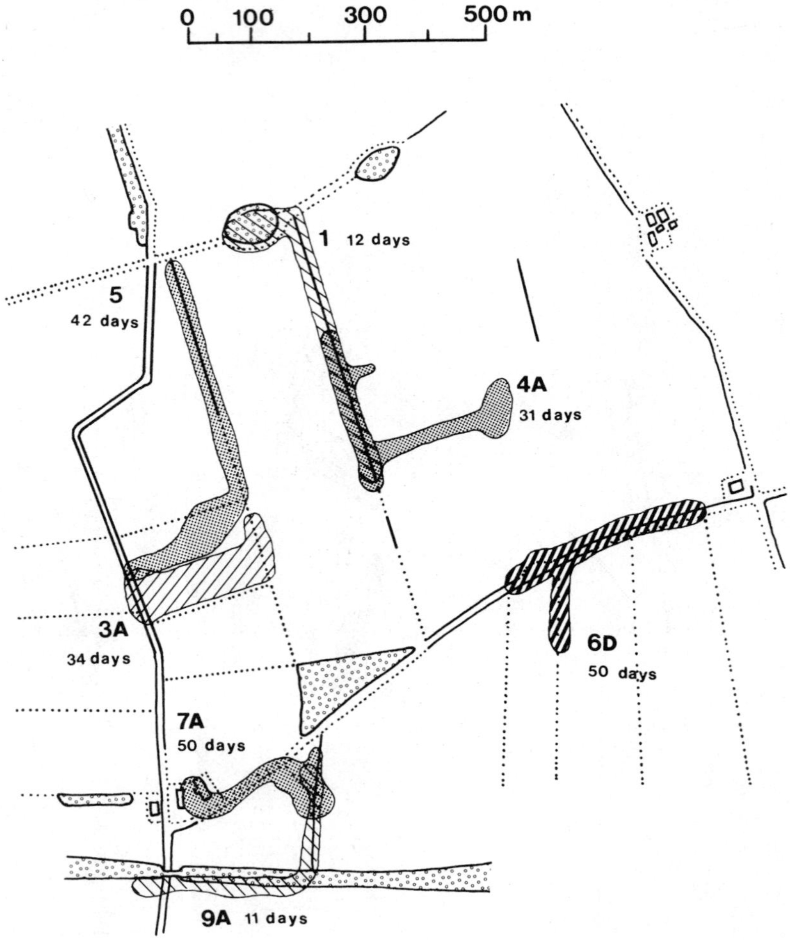

Fig. 2. Part of the arable study area showing the observed ranges of 7 female rats (1, 3A, 4A, 5, 6D, 7A, 9A). Not all rats were tracked concurrently. The same conventions are used as in Fig. 1.

[All rats were nocturnal and while active showed little movement away from food stores.] Direct observations (through infra red viewing equipment) revealed that [animals rested for long periods]. [Rats spent most time in or around meal-bin sheds where they fed on crushed grain and spillage (Fig. 5). Bales of hay and straw provided both homesites by day and food by night]. Only two of 9 tagged rats used a granary shed in which was stored bulk wheat. Some individuals were occasionally observed eating grass beside the farm buildings. [Little time was spent in open habitats away from buildings] (Fig. 5). Despite considerable overlap of ranges (38 rats were trapped in a small area around the buildings during an 18 week period) [little interaction was observed between rats] and few agonistic shrieks were heard at night.

Two peripheral rats were trapped beside a stream, 70 m from the main farm buildings (Fig. 4). Male 15B subsequently had its homesite in hay bales at a stable 375 m

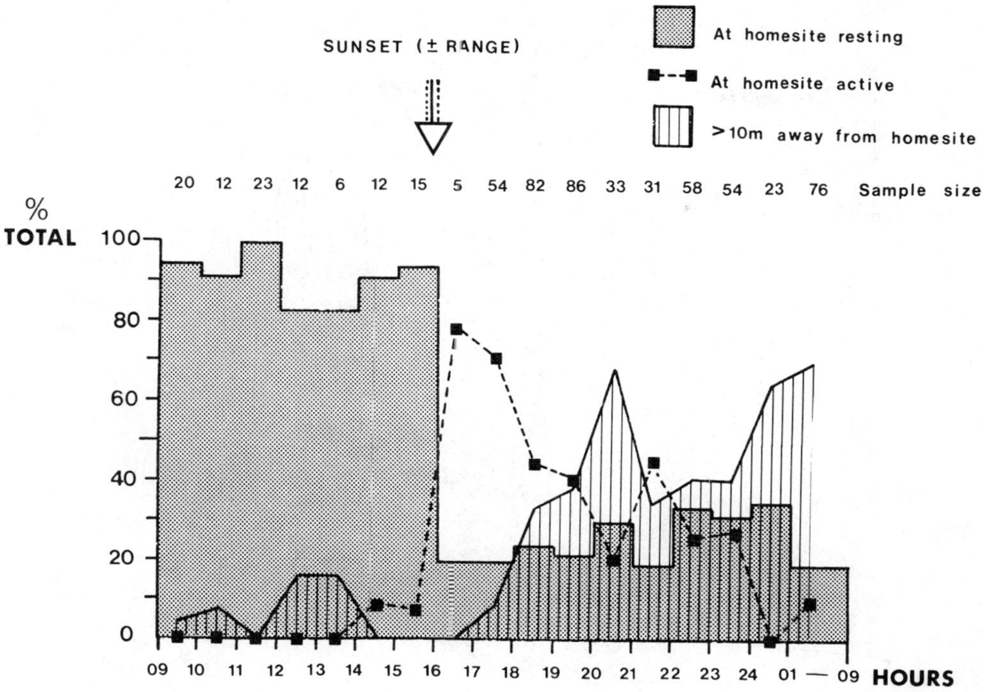

Fig. 3. Activity of three rats (males 2 and 6, female 5) in arable farmland in December. Pooled data for 602 radio fixes taken at 15 min intervals.

distant from and overlooking the farm on which it was trapped. It was never observed again at the original farm. Male 11, a large rat weighing 560 g, lived in an isolated straw barn, surrounded by pasture, 750 m down the valley from its site of capture. However several return visits were made to the original farm. This rat was observed to leave the vicinity of its homesite barn two to three hours after dusk and to cross open pasture to a stream bank. Running beside the stream, the male covered 750 m in 35 min to reach the outskirts of the original farm where it fed in woodland from a feeding station for pheasants (*Phasianus colchicus* L.). The succeeding day was usually spent in a homesite adjacent to its capture point before retracing its path to the straw barn the next night. In December, this dominant rat established a new homesite on the original farm in hay bales displacing the smaller male 2, 310 g, which moved to a new homesite 60 m away from the buildings (Fig. 4).

Five young rats (95-130 g), trapped at various places near buildings on a third study farm, were fitted with miniature transmitters. They all used the same homesite and were assumed to be siblings. Initially, rats moved short distances around the farm buildings where there was limited food in last year's hay and straw bales. Soon all started foraging regularly in a field of growing barley up to 200 m from their homesite. Rats always moved singly and access was by a well marked run that was also used by adult rats (Fig. 6). Foraging was initially at any time of day or night. Rats frequently went into the barley in the midafternoon and returned after dark. Later, their habits became more nocturnal. Apart from their homesite, juvenile rats never visited other farm buildings where stock were fed and food was apparently plentiful. These buildings were occupied by unmarked rats. Later on, the female moved to a burrow under a tree stump beside the barley field. She never moved more than 15 m away and probably had a litter.

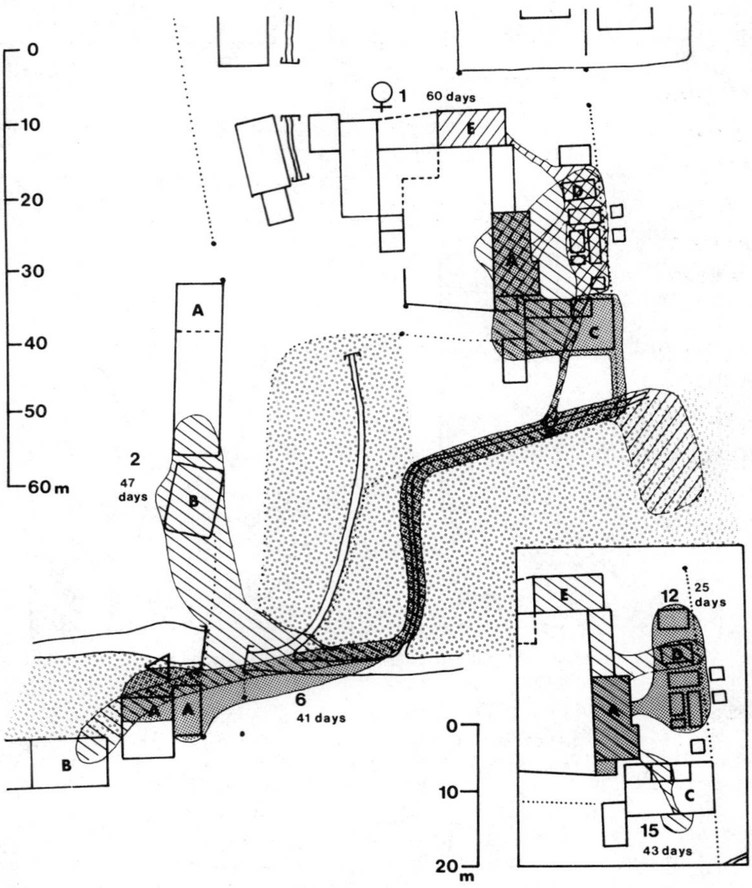

Fig. 4. Map showing the observed ranges of female 1 and 4 male (2, 6, 12, 15) rats around farm buildings in the second study area. Rats were tracked concurrently. Food stores, labelled A-E, are identified in Fig. 5. Two peripheral rats (males 11 and 15B) were trapped at (∇).

In general, adverse weather conditions appeared to have little effect on the movements of rats in arable farmland. However, around farm buildings rats were clearly less active on bright moonlight nights. Most animals were observed sheltering under sheds and associated harborage. Snow, lying to a depth of 15 cm, also appeared to reduce activity outside farm buildings. No tracks were seen and regular runs were unused.

DISCUSSION

Most rats that were radio tracked had identifiable food sources that were related to farming and many of the observed movements were from homesite to food source. This study shows that rats near established food sources have small ranges while animals in open farmland range widely in response to lower food availability. Range

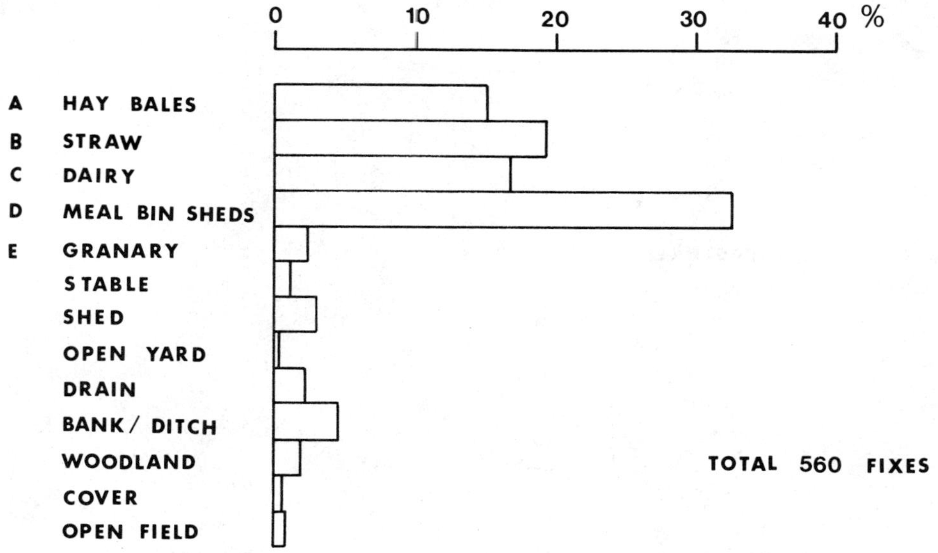

Fig. 5. Recorded habitat locations of 8 rats (7 males, 1 female) around the farm at minimum intervals of 10 min. All rats were active and away from their homesites. Food stores, A–E are indicated in Fig. 4.

size was also shown to vary indirectly with the availability of food for *Rattus exulans* Peale radio tracked in sugarcane (Lindsey *et al.*, 1973; Nass, 1977). In addition, males ranged more widely than females as was found in the present study.

Although some rats in the field populations regularly visited farm buildings, only a large dominant animal, male 11 in the second study area, became established. It is possible that social interactions by resident rats prevented peripheral individuals moving in (Calhoun, 1962). Night vision equipment is clearly important to supplement data from radio tracking which pinpoints an animal in both space and time but does not reveal what it is doing. However, difficulty was experienced around and within buildings where individual rats were seen infrequently.

Predation may be very important to rat populations, both around buildings and in more open habitats. Recorded movements were largely confined to cover but wide ranging individuals may be more at risk. Five out of 21 rats tracked in hedgerows were killed by predators (2 stoat *Mustela erminea* L., 2 fox *Vulpes vulpes* L. and 1 cat *Felis catus* L.) and it is notable that both peripheral males in the second study area were killed by foxes.

Lower recorded activity under bright moonlight probably reduces the risk of predation. Cats and tawny owls (*Strix aluco* L.) were regularly observed hunting round the farm buildings. Sanderson and Sanderson (1964) found that *Rattus mülleri* Jentink confined movements in Malayan forest to heavy cover under bright moonlight. Nass, Hood and Lindsey (1971) report that the removal of harborage when harvesting sugarcane was followed by rapid predation of field dwelling *Rattus exulans*.

The studies reported here have clear implications for the control of infesting rat populations. Individuals show a much greater range of movement than previously

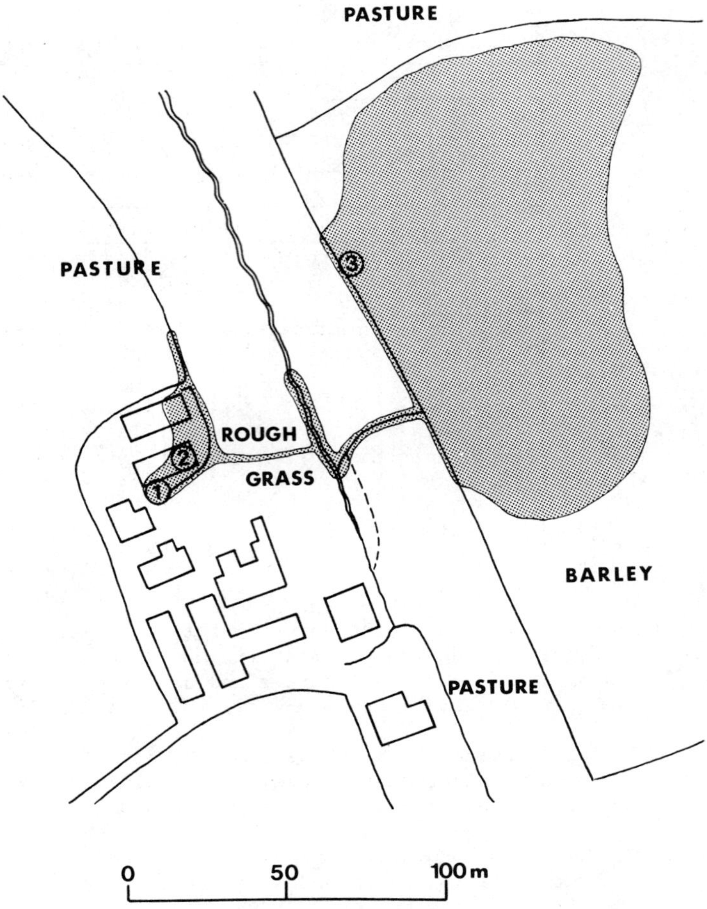

Fig. 6. The range covered by 5 sibling rats (1 female, 4 males) marked while young and followed in the third study area for around two months (pooled data from over 500 positional fixes). (1) Homesite under elm stump 15-24 June; (2) homesite in hay bales 25 June-3 August; (3) homesite of female, 20 July-3 August.

recognized. Buildings may be regularly visited by field living rats up to 500 m distant. If rat populations are controlled in the buildings alone potential immigrants are already in contact and rapid reinfestation may take place. This may explain why protracted treatments are sometimes necessary. However field populations at low densities, particularly males may leave very few rat signs (Taylor, 1978; Greaves, personal communication) which are not easily detected and thus treatment of such populations is difficult. Investigations of peripheral rat movements following the removal of adjacent populations continue with particular emphasis on the dispersal of juvenile rats.

Acknowledgements — We are particularly indebted to the late G. E. Ashwell who designed and built our original radio equipment. We thank P. Fox and R. Quy for invaluable help in the field and R. Stockall for making many of the transmitters.

REFERENCES

Calhoun J. B. (1962) *The Ecology and Sociology of the Norway Rat*. Public Health Service Publication no. 1008. Maryland. U.S. Dept. of Health, Education and Welfare.

Lindsey G. D., Nass R. D., Hood G. A. and Hirata D. N. (1973) Movement patterns of Polynesian rats (*Rattus exulans*) in sugarcane. *Pacif. Sci.* 27, 239-246.

Nass R. D. (1977) Movements and home ranges of Polynesian rats in Hawaiian sugarcane. *Pacif. Sci.* 31, 135-142.

Nass R. D., Hood G. A. and Lindsey G. D. (1971) Fate of Polynesian rats in Hawaiian sugarcane fields during harvest. *J. Wildl. Mngmt.* 35, 353-356.

Sanderson G. C. and Sanderson B. C. (1964) Radio-tracking rats in Malaya — a preliminary study. *J. Wildl. Mngmt.* 28, 752-768.

Stickel L. F. (1954) Home ranges in small mammals. *J. Mammal.* 35, 1-15.

Taylor K. D. (1978) Range of movement and activity of common rats (*Rattus norvegicus*) on agricultural land. *J. appl. Ecol.* 15, 663-677.

Taylor K. D. and Lloyd H. G. (1978) The design, construction and use of a radio-tracking system for some British mammals. *Mammal Rev.* 8, 117-141.

Taylor K. D. and Quy R. J. (1978) Long distance movements of a common rat (*Rattus norvegicus*) revealed by radio-tracking. *Mammalia* 42, 63-71.

The Determination of Free Ranging Rodent Activity by Telemetry

L. J. SHIELDS

Department of Biology, University of California at Los Angeles, Los Angeles, California 90024, U.S.A.

Abstract — A method of longterm, continuous determination of free ranging rodent activity based upon detecting gross postural changes is described. Gross postural changes produce frequency modulation in a small, inexpensive transmitter that is also suitable for radio tracking. The sensitivity of this method allows the fine structure of activity patterns to be resolved to an extent not previously possible. A full sized template of a transmitter is included.

INTRODUCTION

The many vigorous behaviors, i.e., locomotion, climbing, digging, grazing, etc. performed by rodents during their active periods necessarily involve changes in posture. Relative to the postural changes that occur during rest and sleep, those that occur during active periods are frequent and gross (Knowlton *et al.*, 1968). By detecting and recording continuously these gross postural changes, a detailed record of virtually all vigorous activity can be obtained. A telemetry system employing this principle has been used to obtain longterm, continuous records of the diel activity patterns of several species of free ranging rodents, primarily, *Neotoma fuscipes*, *Peromyscus californicus* and *Microtus californicus* (Kawai, 1973). Figure 1 shows the patterns for 10 consecutive days from individuals of each of these species. Such precise data allow the formulation and testing of detailed hypotheses concerning the effects of biotic and abiotic variables on activity. Furthermore, because vigorous activity *per se* is detected, rather than only that set of activities involving locomotion — which differs in composition and importance from species to species — such data may be especially useful in comparative studies.

THE TRANSMITTER

The transmitter circuit (Fig. 2) is a blocking oscillator with a band width of about 3 MHz and a blocking frequency of from 10 to 20 Hz (see Mackay, 1968). These transmitters are designed to operate in the FM entertainment band, 88-108 MHz with a typical duty cycle of 5 percent. The transmitters are worn as collars. A single wire loop functions as the tuned circuit inductor, loop antenna and means of attachment. The collar is attached by breaking the loop, fitting it around the animal's neck and resoldering, similar to the method described by Mech *et al.* (1965).

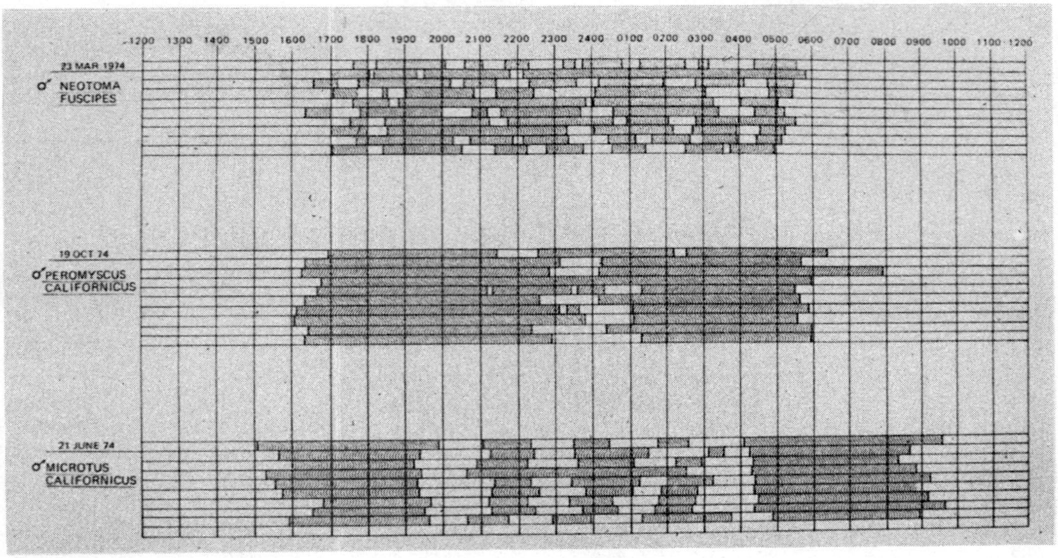

Fig. 1. Representative activity patterns of *Neotoma fuscipes*, *Peromyscus californicus* and *Microtus californicus*.

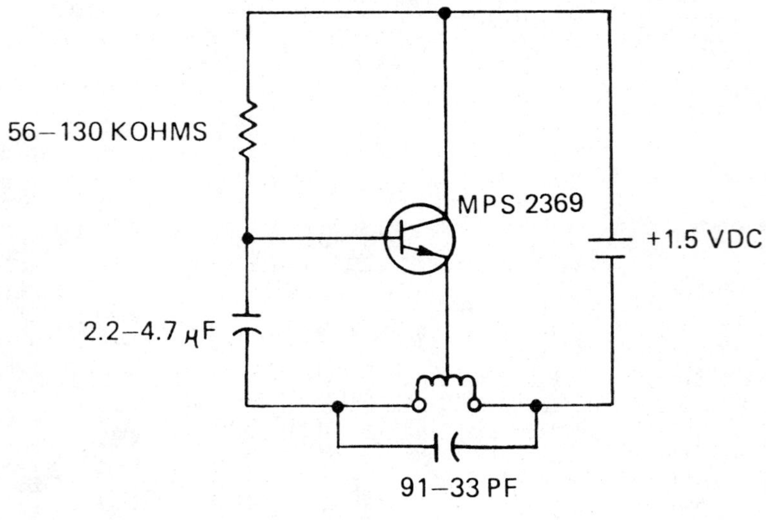

Fig. 2. Schematic diagram of transmitter circuit.

Transmitters are constructed over a wide size range. Their weight and longevity are determined primarily by the size of the battery used. Transmitters suitable for mouse-sized animals, 15-90 g, weight 1.1-2.0 g and transmit for approximately 12 days. Those for rat-sized animals, 90-500 g, weight 3.5-7 g and transmit approximately 55 days. Table 1, is a list of parameters of transmitters that have performed successfully in the field. The components for these transmitters cost less than US $4.00 per unit.

When used in conjunction with a hand held, FM band transistor radio equipped with a

Table 1.
Transmitter component values and parameters for variously sized rodents. Transmitters to operate at 90 MHz. Life span (days) for transmitters using batteries with soldering tabs

Loop diameter	Tank capacitor	Battery	Transmitter weight	Life span	Representative rodents
12 mm	82 pF	S 13	1.1 g	12	*Onychomys torridus* (17 g)
15 mm	75 pF	S 13	1.3 g	12	*Peromyscus californicus* (35 g)
20 mm	56 pF	S 13	1.7 g	12	*Dipodomys agilis* (50 g)
25 mm	45 pF	S 76	3.5 g	55	*Cittelus mohavensis* (100 g)
33 mm	39 pF	S 76	4.0 g	55	*Neotoma fuscipes* (300 g)
38 mm	33 pF	S 76	5.5 g	55	*Otospermophilus beecheyi* (500 g)

1/4 wave whip antenna, the reception range is from 30 to 100 m. This performance has been consistent in thick brush at the bottom of a deep canyon where most field experiments have been conducted. Performance in the desert is generally about 30 percent better. Transmission from underground depends entirely upon local soil type and water content. Radio location is readily accomplished by means of a FM receiver equipped with a telescopic whip antenna. Initially, the fully extended antenna is held vertically and suspected locations of the animal are investigated until a strong signal is obtained. The antenna is then swept in the horizontal plane so that a null can be detected. If no null is found, i.e., if the signal is so strong that the audio output is saturated with the antenna in any position, then the length of the antenna is decreased and the procedure is repeated. Decreasing the length of the antenna has the effect of decreasing the radio frequency gain of the receiver. At close range, less than 10 m, only a few cm of antenna may suffice. Once a null is detected, the animal is located by standard triangulation procedure to an accuracy of about 10 degrees.

DETERMINATION OF ACTIVITY

The 'click' rate of the transmitter is to a large extent determined by the time constant of RC2 (Fig. 2). However, 'stray' capacitance between the animal and the transmitter also affects the 'click' rate such that an increase in the animal to transmitter capacitance decreases the blocking frequency. Pulse spacing (the interval between the onset of two consecutive clicks) (Graf, 1970) is about 10 percent shorter when the transmitter is in intimate contact with the animal in the 'extreme curled' posture than in the 'extreme extended' postures (Fig. 3), when the bulk of the transmitter hangs freely from the animal's neck. It is therefore possible to establish a pulse spacing threshold. The crossing of which represents a 'head down' to a 'head up' movement. This postural change is identified and is used as the criterion for activity. By means of appropriate electronics (Shields, 1976) (Fig. 4) it is possible to measure automatically the duration of each pulse spacing and on that basis, to determine continuously whether the animal is in a 'curled' or 'extended' posture. Since the pulse spacing is about 75 ms, any movement that causes the animal to be in an 'extended' posture for more than 75 ms will be detected, i.e., the sampling rate for postural change, and hence, activity, is the same as the blocking rate of the transmitter, about 14 Hz. Because the pulse spacing changes instantaneously with changes in posture, a continuous, longterm record of split second resolution can be obtained. The discriminating system is capable of analyzing any signal that lasts longer than seven consecutive pulse spacings — about 1/2 s. Depending upon the pulse spacings, one or more outputs of the discriminating circuits are activated, indicating that the animal is: (1) present but not active, (2) active, or (3) noise is being

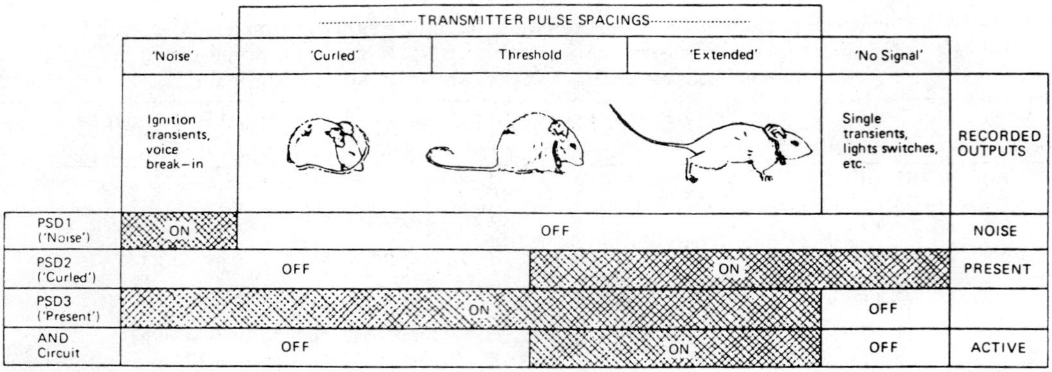

Fig. 3. Postures associated with transmitter pulse spacings.

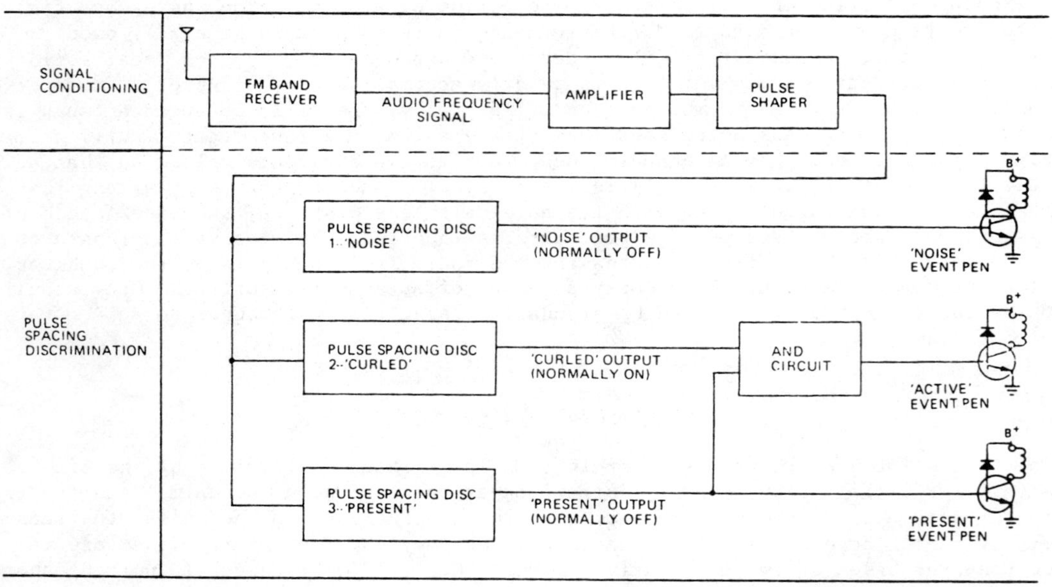

Fig. 4. Block diagram of pulse spacing discriminator circuit.

received. The 'present', 'active' and 'noise' outputs operate pens on a multichannel event recorder operating at a chart speed of 3" (7.6 cm) h^{-1}. An accuracy of 1 min is readily obtained.

MONITORING ANIMALS IN THE FIELD

In the field animals are monitored by receivers located within their home ranges. Power is supplied to and audio frequency signals obtained from, these receivers by means of three-wire shielded cables. Because 'transistor' radios are designed to operate over a wide voltage range and the audio frequency output has a very low impedance (8 Ω), voltage drop and electronic noise on long cables is not a problem. Cables as long as 400 m have been employed successfully. Large coffee cans make excellent containers for receivers in the field. They can be made completely water tight and being good 'litter mimics', attract little attention from potential vandals.

The type of antenna used depends upon the size of the area to be monitored. Small loop antennas of about 1/4 wave length in diameter are used to monitor specific locations, such as nest sites. These antennas are horizontal and almost always very close to the ground. Within the loop reception is continuous. Maximum area coverage is obtained by using an omni-directional vertical 1/2 wave length or greater whip antenna. A single receiver of average quality can give good coverage of a circle about 30 m in radius. Within this distance signal dropout from an active animal rarely lasts longer than one minute. If more than one receiver is used, coverage is considerably improved. Three receivers may be arranged so that signal drop out rarely occurs.

However, in general the most effective use of time and materials is to use two receivers, one 'site' receiver to monitor activity in and around the animal's main nest and one 'area' receiver to obtain the widest possible coverage. This arrangement gives precise data concerning the manner in which the animal uses its nest, including the onset and cessation of activity, and also allows one to eliminate the possibility that the animal is inactive at some other location when away from the nest being monitored by the 'site' receiver.

RELIABILITY OF THE METHOD

For 12 years — until it was destroyed by a brush fire in 1978, I lived at my study site. This enabled me to keep the discriminating electronics and recording equipment in my living room, while monitoring animals 30 to 400 m from my residence. This allowed me to make hundreds of comparisons between the audio signal from one of the field receivers (as displayed on an oscilloscope) and the output of the discriminating system and so to immediately verify the actual behavior of the animal by means of radio tracking.

The only occurrence of ambiguous data was immediately after heavy rains when *M. californicus* active in very wet grass, apparently became soaked, causing drastic changes in the transmitter pulse spacings. Similar events have not occurred with *P. californicus* or *N. fuscipes*. Both of these animals stay in their nests during heavy rains and apparently move carefully enough to avoid getting wet when foraging immediately after rains.

IN THE FUTURE

The methods described in this paper are only a first step in using electronics to monitor free ranging rodents. There is little doubt that discrete behaviors, e.g., locomotion, feeding, etc., cause unique patterns of transmitter modulation, and that by using computer analysis similar to that used to recognize EEG patterns, these patterns could be automatically recognized and recorded. Data of this type would add immeasurably to our understanding of over-all bio-energetic strategies and the way in which different species partition their energy budgets.

It may well be that the most efficient approach to obtaining even more detailed data than can be obtained by the methods described in this paper is to concentrate, not on the design and construction of more sophisticated transmitters, but on more sophisticated methods of analyzing the wealth of information that is contained in the signals from the reliable, simple, and inexpensive transmitters used in this system.

GETTING STARTED

Many investigators dismiss small mammal telemetry because they believe it would be too expensive or complex. Also, obtaining specialized components may present a problem. Hopefully, the transmitter shown in Fig. 5 will alleviate many of these problems and encourage persons with only a slight knowledge of electronics to attempt radio tracking. Figure 5 is a full-sized template of a circuit used to test the performance of individual transistors. Constructed as shown the circuit will operate at about 90 MHz. Do not be concerned if your components are not the exact physical size of the ones in the template. Virtually any combination of good quality components will work in this sized transmitter. The best transistor for this circuit seems to be a Motorola MPS 2369, an economy plastic package transistor that costs about US $0.50. However, a high percentage of many types of radio frequency amplifiers or high speed switching transistors will work. Silver-oxide batteries can be obtained in hearing aid or photography stores. With care, it is possible to solder directly to these batteries. One must use a good flux and polish the area with sanding paper just prior to soldering. Alternatively, batteries with soldering tabs may be specially ordered.

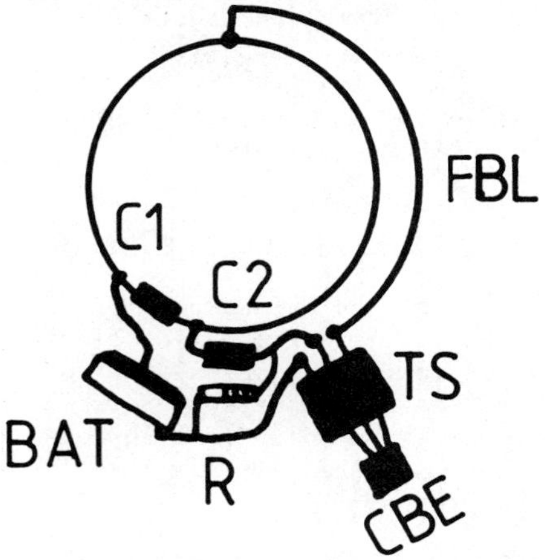

Fig. 5. Transistor test circuit. BAT = battery (S-76), TS = transistor socket, FBL = feedback loop, C1 = tank capacitor (39 pF), C2 = timing capacitor (3.3 mF), R = resistor (100 KΩ), CBE = collector, base and emitter of transistor. Diagram is full-sized and may be used as a template.

The circuit can be converted to a transmitter suitable for rat-sized animals by leaving out the transistor socket and 'condensing' the rest of the circuit. Do not be afraid to rearrange the geometry of the components to obtain optimum packaging. Construction is greatly simplified if the components (except the battery) are glued together with a cyanacrylate glue before soldering commences. After soldering is complete, the entire transmitter must be potted with several thin layers of 'Q-dope' (polystyrene dissolved in acetone) and then finally with 5 min epoxy used sparingly. Never attempt to build a transmitter with a transistor that has not been tested!

REFERENCES

Graf R. F. (1970) *A Modern Dictionary of Electronics*. New York: Howard Sams and Co.

Kawai M. and Mito Y. (1973) Quantitative study of activity patterns and postures of Formosan monkeys by the radio-telemetric technique. *Primates* 14, 179-194.

Knowlton F. F., Martin P. E. and Haug J. C. (1968) A telemetric monitor for determining animal activity. *J. Wildl. Mngmt.* 32, 943-948.

Mackay R. S. (1968) *Biomedical Telemetry: Sensing and Transmitting Biological Information from Animals and Man*. New York: John Wiley.

Mech L. D., Kuechle V. B., Warner D. W. and Tester J. R. (1965) A collar for attaching radio transmitters to rabbits, hares and raccoons. *J. Wildl. Mngmt.* 29, 946-954.

Shields L. J. (1976) Telemetric determination of free-ranging rodent activity: The fine-structure of *Microtus californicus* activity patterns. Unpublished Ph.D. Thesis. Los Angeles: University of California.

A Radio Tracking Study of Young Translocated Elephants

N. FAIRALL

Mammal Research Institute, University of Pretoria, Pretoria 0002, South Africa

Abstract — Because of the large size of the African elephant, translocations have to be done using juvenile animals. The social structure of the elephant herd raises doubt about the success of such a venture. The present study concerns a translocation of juvenile elephants and indicates that they had reasonably small home ranges in the short term, but that they are affected by human disturbance. The elephants developed a functional group structure indicating that such a translocation can be successful.

INTRODUCTION

It has become common practice in Africa to relocate or reintroduce large mammals to areas from which they have been previously exterminated. However, there are no reports on relocating elephants, probably due to the difficulties imposed by their large size. For this reason any attempt at such relocation must involve subadults or juveniles. However, elephants have evolved a well developed social system in which juveniles are protected and remain in well defined family groups within a herd (Buss and Smith, 1966). The effect of removing the adults from a herd on the behavior and survival of lone groups of juveniles has never been documented. The present study was aimed at establishing the effects on behavior of relocating juvenile groups of elephants during the reestablishment of this species in a private game reserve in the Transvaal.

MATERIAL AND METHODS

STUDY AREA

The Sabi Sand Game Reserve consists of a group of unfenced privately owned farms operated as a conservation unit of approximately 300 km^2. The area adjoins the Kruger National Park and has the full complement of game occurring in the Transvaal Lowveld. Elephant occurred throughout the area in the past and lone bulls still occur. Naturally breeding herds have, however, not been resident in the area for at least 50 years. The Reserve centers on about 24°S, 31°E. It has a mean elevation of 1200 m, varying from about 100 to 1300 m. Three high points within 3 km of each other rise to 66 and 88 m above the surrounding terrain which is relatively level and cut by a number of drainage lines flowing to the Sand River, which is the most important feature of the topography (Fig. 1). The area is transversed by roads and tracks but as most of these were made since the last aerial survey, fixed points on the map are scarce. The combination of flat terrain and lack of fixed points therefore, makes triangulation from radio signals difficult.

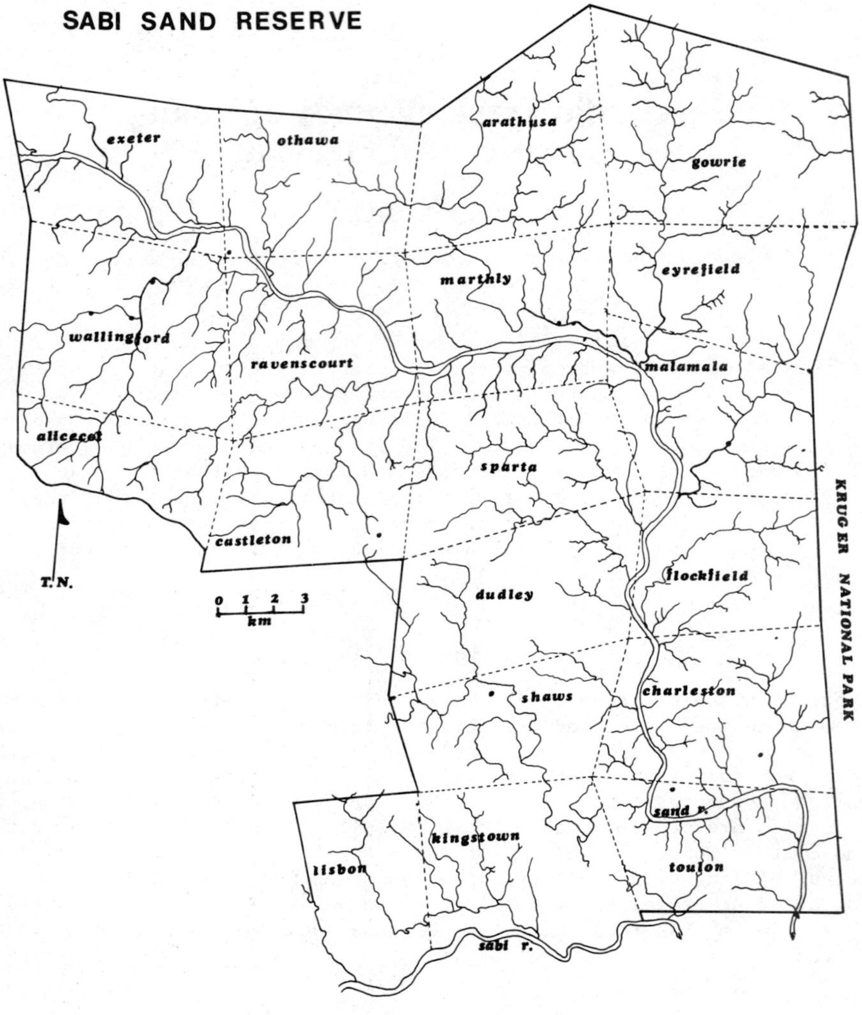

Fig. 1. Map of the Sabi Sand reserve showing the main drainage lines.

The vegetation of the area is diverse (Fig. 2) consisting of reasonably dense woodland dominated by *Acacia* in the drainage lines, and mixed *Combretum* on the ridges in between. Grass cover ranges from extremely dense on the dolorite soils, to sparse on the granite derived soils on the ridges.

ANIMALS

Two groups of elephants are at present being studied. Group 1, consisting of 2♂ and 5♀, was introduced in 1977, collar marked and released, but only sporadic

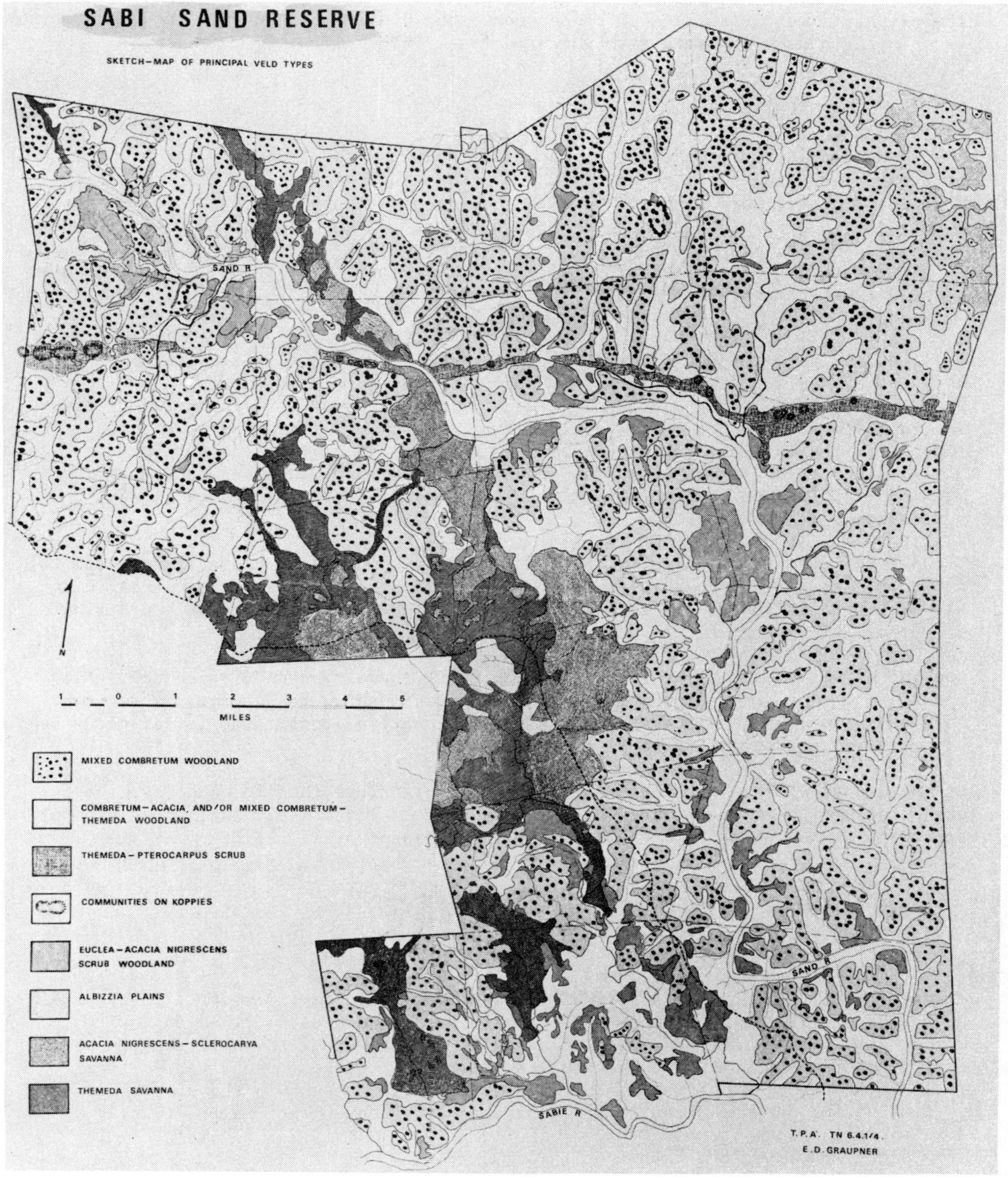

Fig. 2. Sketch map of the main vegetation types of the Sabi Sand reserve.

observations were made until February, 1979 when one animal was caught and fitted with a radio transmitter. Group 2 (2 ♂ and 7 ♀) was introduced in April, 1978. After a quarantine period of one month three of the elephants were immobilized, radio transmitters fitted and the elephants were then immediately released. All the elephants were estimated to be between 5 and 9 years of age when they were introduced but did not originate from the same family group or herd.

TRACKING

The 84 Mhz radio tracking system used was designed by the South African Council for Scientific and Industrial Research (CSIR). Tracking was done on three consecutive days every month, the elephants' position being determined every hour for eight consecutive hours. Elephants were located by driving through the area until a signal was received, and bearings were then taken from fixed points on a 1:50,000 map of the area using a hand held yagi antenna and a compass. The position of the elephant was determined by triangulation.

RESULTS

The area occupied by the elephants at each tracking period is illustrated in Fig. 3. The range covered in October is derived from a few radio fixes and some visual sightings. It is clear that the elephants have used relatively small areas (mean = 3.9 km^2) during any specific month. Mean daily distances covered was 4.7 km and mean hourly distance covered 2.3 km with the longest distance measured in one 8 h period being 9 km.

Group 2 has remained mainly in *Combretum* woodland but this is possibly coincidental as Group 1 has not shown this tendency. Both groups have always stayed near water although this factor is not important at present, with water being well distributed and plentiful during the study period.

One factor not shown by the figures is important, namely human disturbance. All the areas in which the elephants of Group 2 were found up to December, were prone to disturbance, whereas the area occupied by Group 1 was not. The latter group had remained in their area for about 2 years.

It becomes clear from tracking data and visual sightings that in each case the groups stayed together and developed family bonds. Three individuals have disappeared from Group 2 and another is known to have been caught by a lion.

DISCUSSION

EQUIPMENT

The equipment has functioned satisfactorily and the transmitters still in use are operating well after 11 months. Failures to date have been ascribed to faulty mounting. Accuracy of the system is adequate for the study as positions can be determined to within 1 km^2 on beacon transmitters, and their range varies from 2 to 10 km depending on terrain. The main drawback at present is the size of the yagi receiving antenna which causes considerable inconvenience in the thickly wooded area where the work is being done.

ANIMALS

Very few detailed studies of habitat requirements for elephants have been determined. Douglas-Hamilton (1971) reported on a radio tracking study of East African elephant

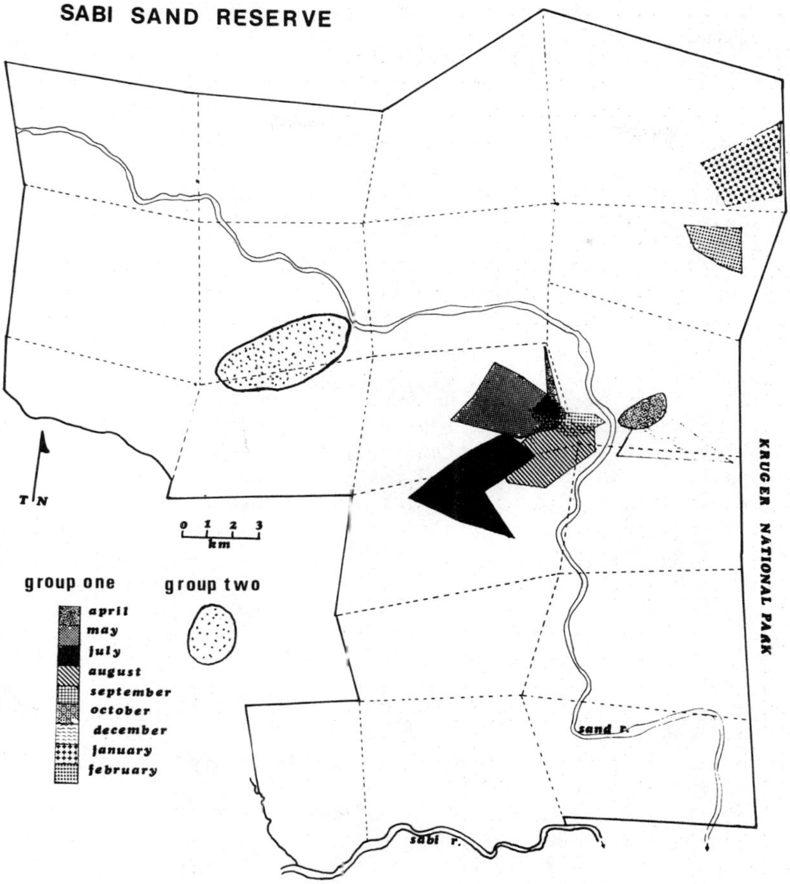

Fig. 3. Ranges of the elephants in each month of the study (October is an approximate range from visual records).

and showed that home ranges varied between 23 km^2 and 330 km^2, with daily changes of position being between 2.9 and 21.4 km. In the same area different family groups had differing movement patterns and he interpreted these in terms of the nervousness of the groups and showed that some groups were affected by human disturbance.

In the most detailed study to date, Leuthold (1977) gave mean home ranges of 303 and 281 km^2 for males and females respectively, with the smallest ranges being 41 and 50 km^2 for the two sexes, but his data did not allow the determination of daily movements.

The data in Fig. 3 indicate that each month's home range is discrete, although possibly segregating into two areas. The whole range is not considered a home range as the animals have not tended to return to their old haunts. The mean monthly home range is much smaller than that reported by Douglas-Hamilton (1971) but the daily movements (8 h period) are somewhat larger than those he found. Using his reasoning this suggests that the habitat is most suitable.

Although the elephants were most frequently found in *Combretum* veld it is thought that this is fortuitous as most of the Reserve provides adequate habitat as described in the literature (Van Wyk and Fairall, 1966; Laws and Parker, 1968; Wing and Buss, 1970). Water, although important for elephants, has not been shown to affect habitat choice in the present study being freely available and well distributed in the Reserve.

The influence of human disturbance was possibly the major factor influencing the movements of Group 2, as this group gradually moved out of an area where there was continuous traffic and hunting while Group 1 has evidently been much more stable in the relatively quiet area they have occupied since being released.

The data obtained to date indicate that juvenile elephants develop a functional group structure when relocated, and indicate that they can be successfully reestablished if suitable sanctuary areas are available.

Acknowledgements — Thanks are due to the owners of the Sabi Sand Reserve for allowing the study to be undertaken on their property, especially to John and Dave Varty who were the moving force behind the translocation and who provided accommodation and great hospitality. The CSIR provided the radio equipment and many colleagues and personnel have helped with the fieldwork while Professor John Skinner, the Director of the Mammal Research Institute has provided much moral support, especially when things did not go according to plan. The study is financed by the owners of Sabi Sand Game Reserve, the University of Pretoria and the SCIR.

REFERENCES

Buss I. O. and Smith N. S. (1966) Observations on reproduction and breeding behaviour of the African elephant. *J. Wildl. Mngmt* 30, 375-388.

Douglas-Hamilton I. (1971) Radio-tracking of elephants. *Proc. Symp. on Biotelemetry*. CSIR, Pretoria, South Africa.

Laws R. M. and Parker I. S. C. (1968) Recent studies on elephant populations in East Africa. *Symp. Zool. Soc. Lond.* 21, 319-359.

Leuthold W. (1977) Spatial organization and strategy of habitat utilization of elephants in Tsavo National Park, Kenya. *Z. Saugetierkunde*. 42, 358-379.

Van Wyk P. and Fairall N. (1969) The influence of the African elephant on the vegetation of the Kruger National Park. *Koedoe* 12, 57-89.

Wing L. D. and Buss I. O. (1970) Elephants and forests. *Wildl. Monogr.* 19.

Territorial Behavior of the Roe Buck as Determined by Radio Tracking: Qualitative and Quantitative Analysis of Territorial Movements

A. SEMPERE

Centre d'Etudes Biologiques des Animaux Sauvages, 79360 Beauvoir-sur-Niort, France

Abstract — The roe deer (*Capreolus capreolus* L.) exhibits temporary territorial behavior induced by sexual activity and defends his territory for only part of the year. Using radio tracking we have studied the effect of the forest structures upon territorial behavior and relationships between bucks. Qualitative and quantitative analyses of movements have been made according to the square method of Schneider *et al.* (1971) and by speed analysis. The results show the development of territorial behavior by stages, faster in early spring than in winter (1.58 ± 0.05 km h^{-1} versus 0.57 ± 0.05 km h^{-1} SEM) which results in an increase of spring home range which is more important when home ranges are located in thick wood than in clear forest. Structure does not influence the mean speed locomotion (1.402 ± 0.109 km h^{-1} in thick wood versus 1.684 ± 0.085 km h^{-1} Standard Error of the Mean (SEM) in forest). The relations between neighboring bucks are more difficult in thick wood than in clear wood so there is a large control area to assure his integrate territory.

INTRODUCTION

For a wild mammal, locomotion plays an essential part in its adaptation to the environment. Its survival depends on many factors, e.g. feeding, breeding, predators. Qualitative variation of temporal or spatial features of the environment or the nature of inter or intraspecific relationships may also exert an effect on the animal's behavior. The notion of territoriality from Burt (1943) that we have considered in this study is very well correlated with the cycle of genital activity and in the roe deer (*Capreolus capreolus* L.), involves ecological and behavioral adaptations which are expressed as increasing aggression between bucks.

According to St-Girons and St-Girons (1959), we can consider two categories of territorial behavior. (1) Continuous territorial behavior related to primary functions (feeding, drinking). (2) Temporary territorial behavior induced by sexual activity. The latter behavior results in rapid development of intraspecific relationships between mammals.

The roe buck belongs to the second category and defends his territory for only part of the year. This increase of aggresivity corresponds to a significant increase of androgen secretion and has a profound effect on population density and structure.

In this study, we attempted to show the effect of the structure of the forest upon the territorial behavior and relationships between bucks. In the Chizé Forest we have an automatic continuous radio tracking system based upon triangulation (Deat *et al.*, 1980, this volume). The forest is an European temperate deciduous biome. The major tree species are oak (*Quercus* sp.), beech (*Fagus sylvatica*) and different conifers (*Pinus maritina, P. sylvestris*). In this study, we have taken into account only four different types of forest structure: (1) cutting back, with many stubs; (2) brushwood which is the natural development of cutting back with some glades and where animals find food and lairs; (3) copsewood; and (4) forest.

The regularity of the bearings (one point every 4 min), and a large sample size (12,000 per month) permitted the application of a correction factor which determined the rate of utilization and frequency of use of the home range by each buck. This method is similar to that applied by Schneider, Mech and Tester (1971) and takes into account the precision of the system.

RESULTS

The winter home range has been studied in relation to the use of different structures of the biotope by the bucks. In winter, the size of the range varied from 26 to 39 ha (Fig. 1). The mean size of winter range is 31.70 ± 1.55 ha Standard Error of the Mean (SEM). The mean size of the home ranges established in copsewood and forest is larger than for home ranges located in thick wood, 34.6 ± 2 versus 28.8 ± 1.3 ha (SEM).

In early spring we recorded a significant increase in the size of home range for one buck (no. 258) who lived in thick wood (from 31.51 to 65.24 ha) and a smaller increase on another buck (no. 285) in copsewood (from 24.49 to 35.95 ha). At the same time males began to be aggressive and this behavior plays a decisive part in the establishment of a territory. One result was an increase in locomotory activity. The pattern of locomotion became more linear (Fig. 2). The bucks walked quickly round their home ranges with a rapid increase of their mean speed. The mean speed of feeding movements was in the same order in winter as in early spring, but territorial movements were faster (0.57 ± 0.05 km h^{-1} during feeding movements versus 1.58 ± 0.05 km h^{-1} during territorial movements).

A qualitative analysis of the home range use movements allows the dissociation of feeding, bedding and territorial movements. The quantitative analysis of the utilization of home range during feeding and resting shows that there is a tendency to concentrate around the center of the home range whereas territorial behavior predominated at the periphery (Fig. 3). Structure of the biotope had no effect on the locomotion speed (Table 1). Territorial travels were as fast in dense undergrowth (such as thick wood) as in clear forest (1.402 ± 1.109 versus 1.684 ± 0.098 km h^{-1} SEM). The mean locomotion speed recorded on roads and paths for two bucks (1.882 ± 0.166 for buck no. 185 and 1.739 ± 0.085 for buck no. 258) are not significantly different from those recorded in any of the forest types.

DISCUSSION

An increase in aggression corresponds to a change in the ethological relationships between bucks. These relations vary with the ecological characteristics of the biotope. In clear structures (copsewood, forest, cutting back, meadow) bucks are easily able to locate other territorial bucks because they can use all their senses (vision, smell and hearing) (Mauget and Sempéré, 1978).

The development of home range size in spring is a consequence of territorial behavior. This new area corresponds to a zone which is frequented only during this new

n° buck	dates	forest structure	size of home range (ha)
131	9/8/74-9/9/74		30.75
185	27/2/75-31/3/75		29.49
186	24/10/75-4/1/76		38.95
258	29/1/76-26/2/76		31.51
280	10/9/77-23/11/77		35.70
481	20/12/78-20/1/79		26.60
487	20/12/78-5/1/79		26.40
635	29/11/78-20/1/79		34.10

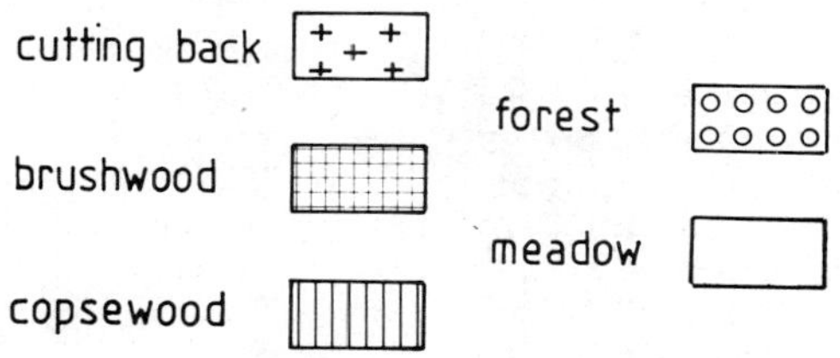

Fig. 1. Winter home range size of bucks living in different forest structures.

locomotory activity. A qualitative analysis of home range utilization (Sempéré, 1979) has shown that all lairs (day lairs and night lairs) are located in the center of the home range (which corresponds to the winter home range).

The relations between neighboring bucks and the boundaries of their territories are well defined. This territorial behavior results in a well defined and relatively narrow peripheral area. In contrast if a buck lives in thick wood the relations with his neighbor are hampered by dense forest structure and the animals can use only two senses (smell and hearing); then, the localization of neighboring bucks is very difficult. In this case it is necessary for bucks to establish a large control area on the periphery of their home range in order to assure their integral territory. This control may be maintained through a network of trails which enable bucks to progress as quickly in thick wood as in clear forest structures.

The manifestations of territorial behavior follow an increase in testosterone secretion (Short and Mann, 1966; Bramley, 1970; Barth *et al.*, 1975; Sempéré, 1978). This demonstrates the close relationship between hormonal regulation, behavior and biotope that we can summarize as the Eco-Etho-Physiology of the roe buck.

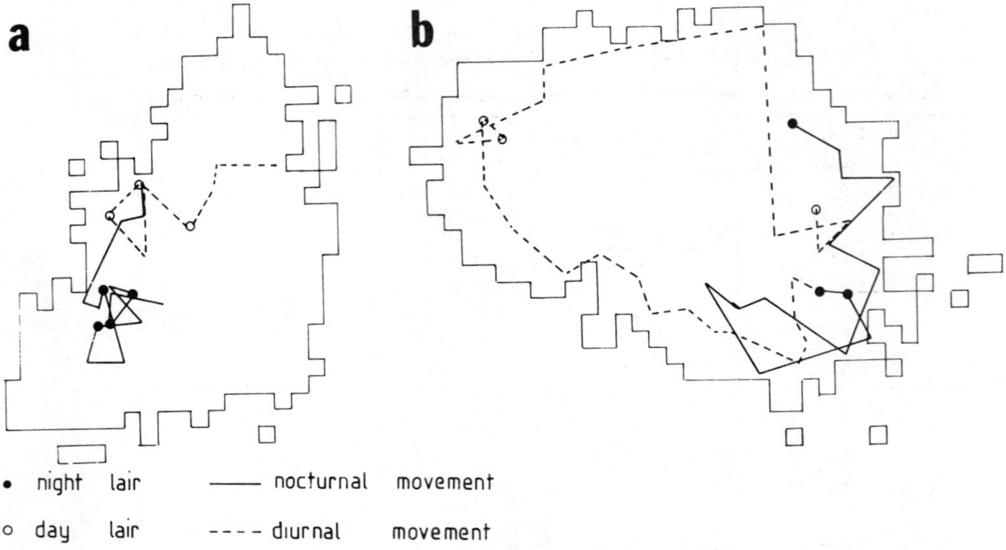

Fig. 2. Locomotory activity of buck No. 258. (a) Feeding movements in winter. (b) Territorial movements in early spring.

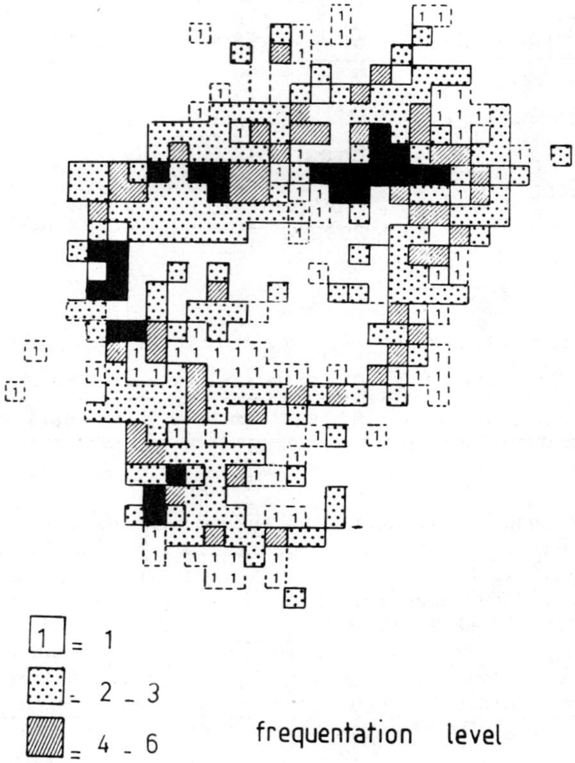

Fig. 3. Peripheral pattern of territorial movements around the spring home range.

Table 1A.
Buck 258 (20-2 — 28-2) winter's speeding moves

Distance (km)	Time (mn)	Speed (km h^{-1})	Mean speed (km h^{-1})
0.180	20	0.540	
0.450	49	0,551	
0.360	45	0,480	
0.075	8	0.562	
0.180	12	0.900	
0.070	8	0.525	
0.180	16	0.675	± 0.570
0.150	16	0,562	± 0.05
0.500	53	0.566	
0.400	80	0.300	

Table 1B.
Buck 258 early spring territorial moves in cutting back

Distance (km)	Time (mn)	Speed (km h^{-1})	Mean speed (km h^{-1})
0.180	8	1.350	
0.270	16	1.012	
0.510	20	1.530	
0.300	16	1.125	
0.390	12	1.950	
0.450	16	1.687	
0.300	20	0.900	1.442 ± 0.136
0.150	8	1.125	
0.450	12	2.250	
0.300	12	1.500	

Table 1C.
Buck 258 early spring territorial moves in thick wood

Distance (km)	Time (mn)	Speed (km h^{-1})	Mean speed (km h^{-1})
0.300	16	1.125	
0.230	12	1.150	
0.150	8	1.125	
0.360	12	1.800	
0.320	16	1.200	1.402 ± 0.109
0.480	24	1.200	
0.300	12	1.500	
0.330	12	1.650	
0.390	20	1.170	
0.420	12	2.100	

Table 1D.
Buck 258 early spring territorial moves in copsewood

Distance (km)	Time (mn)	Speed (km h^{-1})	Mean speed (km h^{-1})
0.270	12	1.350	
0.360	12	1.800	
0.210	8	1.575	
0.300	16	1.125	
0.290	16	1.087	
0.300	16	1.125	1.243 ± 0.073
0.180	8	1.350	
0.450	20	1.350	
0.300	12	1.500	
0.390	20	1.170	

Table 1E.
Buck 258 early spring territorial moves in forest

Distance (km)	Time (mn)	Speed (km h^{-1})	Mean speed (km h^{-1})
0.180	8	1.350	
0.270	8	2.025	
0.330	16	1.237	
0.300	12	1.500	
0.240	8	1.800	
0.210	8	1.575	1.684 ± 0.098
0.150	4	2.250	
0.360	12	1.800	
0.240	8	1.800	
0.300	12	1.500	

REFERENCES

Barth D., Gimenez T., Hoffman B. und Karg H. (1976) Testosteron konzentrationen im peripheren blut beim rehbock (*Capreolus capreolus*). Saisonal und rhythmishe veraendereungen sowie bezichungen zueinizen verhalten sweeden und anderen saisonabhaneigien parametern. *Z. Jagdwissench (Dish)* 22, 134-148.

Bramley P. S. (1970) Territoriality and reproductive behaviour of roe deer. *J. Reprod. Fert.* 11, 43-70.

Burt W. H. (1943) Territoriality and home range concepts as applied to mammals. *J. Mammal.* 24, 346-352.

Deat A., Mauget C., Mauget R., Maurel D. and Sempéré A. (1980) The automatic, continuous and fixed radio tracking system of the Chizé Forest: Theoretical and practical analysis. In C. J. Amlaner, Jr. and D. W. Macdonald (eds.), *A Handbook on Biotelemetry and Radio Tracking*, pp. 439-451. Oxford: Pergamon Press.

Mauget R. et Sempéré A. (1978) Comportement locomoteur déterminé par radiotracking chez deux ongulés sauvages en liberté: le chevreuil (*Capreolus capreolus* L.) et le sanglier (*Sus scrofa* L.). *Biol. Behav.* 3, 331-340.

Saint-Girons H. et Saint-Girons Mc. (1959) Espèce vital, domaine et territoire chez les vertébrés terrestres. *Mammalia* 23, 448-476.

Schneider D. G., Mech L. D. and Tester J. R. (1971) Movements of female racoons and their young as determined by radiotracking. *Anim. Behav. monog.* 4 part 1, 42 p.

Sempéré A. (1978) The animal cycle of plasma testosterone and territorial behaviour in the roe deer. *Environmental Endocrinology*. Berlin: Springer Verlag.

Sempéré A. (1979) Utilisation et évolution du domaine vital chez le chevreuil mâle européen determinées par radiotracking. *Biol. Behav.*, in press.

Short R. V. and Mann T. (1966) The sexual cycle of seasonally breeding mammal. The roe-buck (*Capreolus capreolus*). *J. Reprod. Fert.* 12, 337-351.

Home Ranges and Patterns of Distribution of Foxes (*Vulpes vulpes*) in an Urban Area, as Revealed by Radio Tracking

S. HARRIS

Department of Zoology, University of Bristol, Woodland Road, Bristol BS8 1UG, U.K.

Abstract — The red fox has been the subject of a number of radio tracking studies. These have produced a variety of results, but most studies have suggested that the species is territorial. This paper presents some preliminary results of a study in an urban area of Bristol, U.K. The patterns of distribution and home ranges of breeding vixens are shown. Here the breeding females have overlapping home ranges, and exhibit little, if any, territorial behavior. The results are compared with those from other studies and some tentative ideas to explain this pattern of distribution discussed. Current work is designed to investigate the pattern of resource utilization by the urban fox.

INTRODUCTION

Foxes are a common element of the fauna of many towns and cities in Britain. Harris (1977) described the distribution and habitat preferences of foxes in London, and continued the accounts of Teagle (1967) and Beames (1972) in describing the spread of the fox population in London. Harris also reviewed the literature on the occurrence of larger mammals in North American and continental European cities. He concluded that the urban fox was a British phenomenon. However, this is not entirely true, since foxes are found in some European cities, such as Stockholm (R. J. Berry, personal communication) and Copenhagen (J. Cotton, personal communication), while Brosset (1975) described the food habits of foxes living in the Senart forest on the outskirts of Paris. The European urban fox populations at present do not appear to be as widespread as those in Britain.

Harris (1977) also described the sites selected for natal dens in London; he concluded that a quiet suburban back garden was the most favored site, the cubs being born under raised garden sheds, summerhouses, or in earths dug in banks in gardens. Other natal sites were found under portable huts, such as those used on building sites or as school room extensions, where disturbance was limited to the working day.

Since September, 1977 the author has been studying foxes in the urban areas of Bristol. In the suburbs built in the last fifty years, which consist mainly of semi-detached houses, the selection of natal sites conforms with the data from London. In the older, Victorian parts of Bristol, many litters of cubs are born under the floor boards of occupied and unoccupied houses, access being gained through broken air bricks. In one instance, cubs were born under the floor boards of the kitchen of an occupied house; there was a broken floor board in the kitchen and the foxes entered and left the kitchen via the cat-flap. These foxes were tolerated by the residents; including a cat and dog also in the house.

With the urban fox population living in such close proximity to man and his pets, there is a potentially serious situation in the event of an epizooty of rabies. The present study is designed to examine some of the factors that might affect the spread of rabies in an urban fox population, in particular fox movement patterns and resource utilization, and adaptations of fox social organization to such a diversified habitat. Bristol is an ideal study area since it contains a variety of urban habitats within a relatively small area. The present paper describes the preliminary results of a study of the distribution of breeding females and their natal earths on one study area in a northwest suburb of Bristol, U.K. This area consists predominantly of semidetached houses, interspersed with playing fields, allotments, a cemetery and pieces of waste ground.

METHODS

Adult foxes were caught in 'Hunter Humane' fox traps (W. Flanagan & Son, Ltd., Liverpool, U.K.), modified by removing the treadle, and by snaring, by driving them out of cover into nets, or by poking them out from under sheds with drain rods and catching them in purse nets. They were immobilized with an intramuscular injection of 1 ml of 100 mg ml^{-1} ketamine hydrochloride ('Vetalar', Parke, Davis & Co., Gwent, U.K.). The animals were tagged with numbered white plastic 'rototags' (Dalton Supplies Ltd., Henley-on-Thames, Oxfordshire, U.K.), and fitted with radio collars on 173 MHz. These contained two Mallory 'Duracell' 1.35 V mercury batteries (RM12R). The collar consisted of a piece of car battery earthing strap, which also functioned as the antenna (as described in Macdonald, 1978); the complete transmitter unit and collar weighed approximately 200 g. Transmitter life averaged four months.

Standard body parameters were recorded, the animal aged by incisor wear (Harris, 1978a), and the condition of the animal, presence of old bone fractures (Harris, 1978b) and any distinguishing marks noted. Prior to release the animal was returned to its earth or left in the trap to recover fully from the anesthetic.

The foxes were tracked on foot using an LA12 receiver (AVM Instrument Co., Champaign, Illinois 61820, U.S.A.), with a three-element hand held Yagi antenna. If the animal moved out of the city, reception range exceeded 1 km in open farmland; amongst semidetached and detached housing areas the range was reduced to 200 to 300 m.

Radio tagged foxes spent much of their time in the back gardens of houses. Tracking was done from the road outside the house, usually at a range of less than 50 m. If the animal's exact position was in doubt, cross-bearings were taken from other roads around that block of houses, or by going into a nearby back garden. When an animal was moving around in a block of houses, it was possible to determine from signal strength, noises from the animal, and by taking a number of bearings around the block of houses, not only which garden the animal was in, but also its approximate position in the garden. When possible, animals were observed, using Old Delft PB4DS Passive Binoculars or IRH-6ML Infra red Binoculars (Old Delft England Ltd., Surrey, U.K.). Field data were recorded on a pocket dictaphone.

RESULTS

The home ranges of seven adult breeding females are shown in Fig. 1. Home range size varied from 25.7 to 78.2 ha (mean 45.4 ± 6.9 ha). Adult females would also occasionally move up to 350 m outside their established range; these movements were

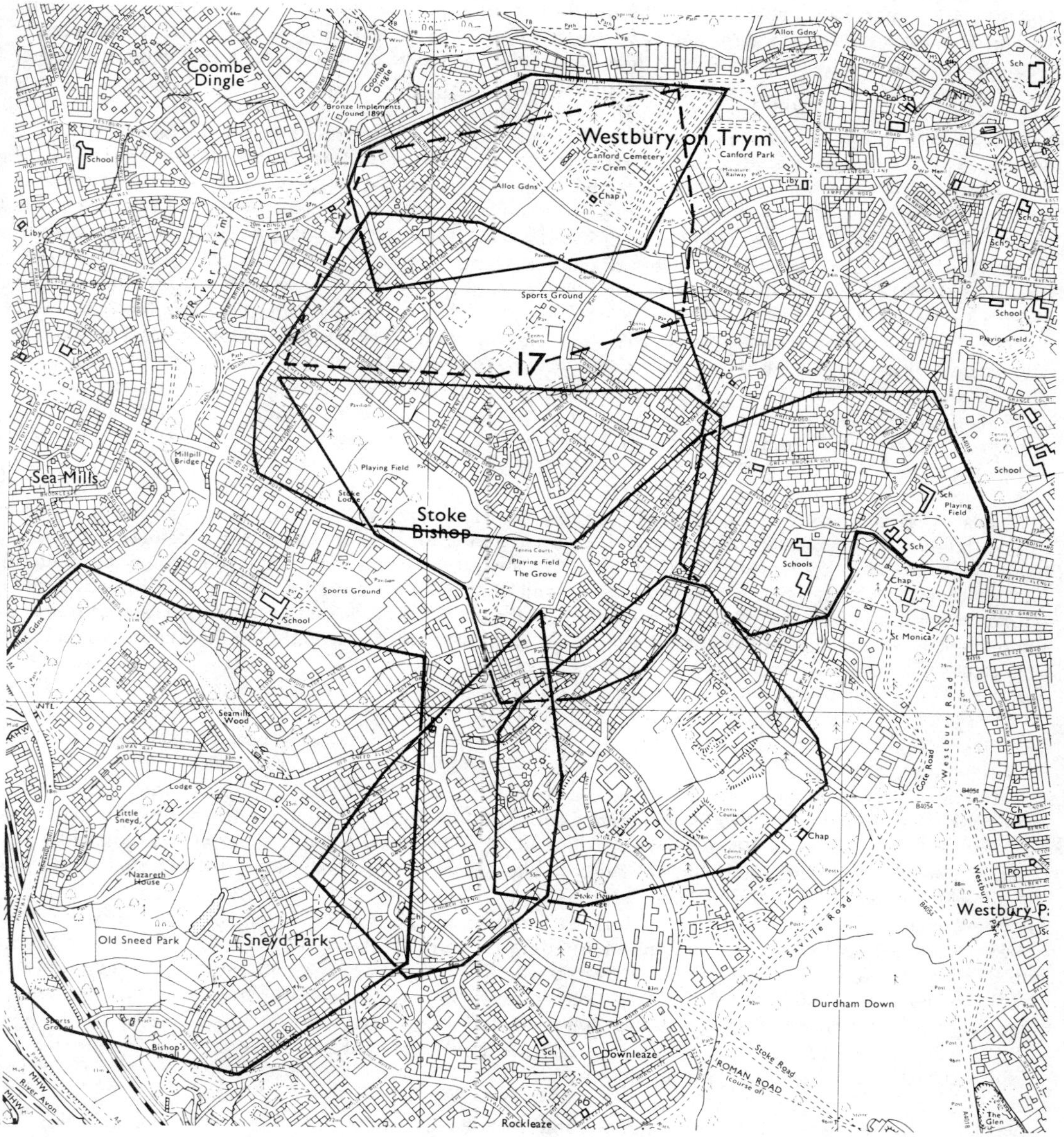

Fig. 1. Home ranges of seven breeding vixens (solid lines) in an urban area of Bristol. The four northerly animals were studied between April and September, 1978, the three southerly animals between October, 1978 and January, 1979. The range of one barren vixen is shown with a broken line. A number of other breeding vixens in this area were not studied. The grid lines are one kilometer apart.

either made quickly to a specific point, presumably a known food source, and were followed by a direct return to the normal range, or were slow movements out from the usual range, feeding en route. Relatively few adult dog foxes have been studied, but the limited data available suggest that the range of the dog fox conforms closely to that of the breeding female. Within each breeding range additional barren females may have been present; a maximum of two on each range was recorded. These barren females were usually yearlings. In suburban London, Harris (1979) found that up to 52 percent of yearling vixens were barren. Most of the barren yearlings studied, occupied only part of the range of the breeding female.

There were exceptions to this general pattern. Female 17, shown in Fig. 1, was a four year old barren animal. She occupied part of the ranges of two breeding females, one a year old and one two years old. Female 17 spent much of her time with a three year old dog fox that was the mate of the yearling vixen. She often moved around with the dog fox or lay up with him and was only rarely encountered with the yearling vixen, but would utilize the same lying-up places as the two year old vixen, though not necessarily at the same time. I believe that the two breeding vixens were the daughters of No. 17 of previous years, and the dog fox, then the mace of the yearling vixen, had been the mate of No. 17 in previous years. Female 17 ranged over 48.5 ha.

Animals from adjacent groups would share feeding sites, particularly gardens in which the householder would supply food in abundance. Up to three breeding females have been caught at a single trap point. Breeding females would range up to and past the natal sites of other females, and natal dens could be very close; in London there was one instance of two separate litters of cubs being reared in adjacent gardens (S. Harris, unpublished data). In parts of Bristol five litters of cubs per square kilometer have been recorded.

DISCUSSION

There have been a number of studies on red fox home ranges, and these have produced a variety of results. The most conspicuous feature of the present study is the overlapping pattern of the home ranges. Lloyd (1977) also reported that ranges were mostly overlapping and non-exclusive, although small areas, such as in the vicinity of cubbing dens, could have been exclusive. Browne (1978) suggested that vixens guarded the den with young cubs, but otherwise were not territorial, and would travel freely through territories occupied by other pairs. Most other studies found that adult vixens were territorial, at least in the breeding season (Scott, 1943, 1947; Ables, 1969, 1975; Sargeant, 1972; Montgomery, 1974; Storm and Montgomery, 1975; Niewold, 1976; Storm *et al.*, 1976; Johnson and Sargeant, 1977; Macdonald, 1977, 1980). Phillips (1971) even found evidence for territories being established by a number of foxes in a 5.3 ha enclosure.

Several authors have described occasional movements outside the established home range; Ables (1969) noted such unexplained trips and Niewold (1974) found that twelve out of twenty radio tracked adult foxes every now and then made excursions. Niewold listed four types of movement outside the home range; one of these consisted of irregular and quickly performed trips, often directed towards specific food sources, especially by lactating vixens. Similar movements were observed in the present study. Niewold found that the fox would remain at the food source for a number of hours. Many authors have described numbers of adult foxes exploiting a single food supply e.g. Alexander *et al.* (1967) observed up to five foxes at one time in a lambing paddock. Browne (personal communication) suggested that when a vixen with cubs crossed the home range of a neighboring breeding vixen, it was usually to hunt at a rabbit colony, such a food source being absent on her own range.

In the urban area, foxes eat a diversity of food types. The major items consist of small birds and food put out by the residents, either for the foxes, or for badgers, birds, hedgehogs and squirrels; this is supplemented by worms, insects, fruit and domestic pets (S. Harris, unpublished data). Such food sources are both readily available and widely dispersed throughout the urban area, but they are variable locally in their availability. Food put out for the foxes is usually dependent upon that particular house having some suitable scraps, and even when available, household scraps are of very variable calorific value. Also the amount of food put out for the birds depends on the severity of the weather on a particular day, and the intensity of the publicity from bird protection societies. Small birds are caught and eaten predominantly during the nesting and moulting periods. Thus even though food of some sort can be found in the majority of the gardens in the urban area on most nights of the year, the quantity and calorific value of the food available from each particular garden is variable and difficult to predict. The diverse pattern of food availability in the urban area would be difficult to defend, and this may explain why the foxes in Bristol are not maintaining a rigid territorial system. In an analogous situation, Kruuk and Hewson (1978) found that otters (*Lutra lutra*) living on the Scottish coast had over-lapping home ranges, and they suggested that such an arrangement was a function of resource defendability. The pattern of resource exploitation by urban foxes is currently being examined in an attempt to explain this pattern of over-lapping home ranges.

Acknowledgements — This work was carried out while in receipt of a grant from the Nature Conservancy Council, which is gratefully acknowledged.

REFERENCES

Ables E. D. (1969) Home-range studies of red foxes (*Vulpes vulpes*). *J. Mammal.* 50, 108-120.

Ables E. D. (1975) Ecology of the red fox in North America. In M. W. Fox (ed.), *The Wild Canids — Their Systematics, Behavioral Ecology and Evolution*, pp. 216-236. Amsterdam: Van Nostrand Reinhold.

Alexander G., Mann T., Mulhearn C. J., Rowley I. C. R., Williams D. and Winn D. (1967) Activities of foxes and crows in a flock of lambing ewes. *Aust. J. exp. Agric. Anim. Husb.* 7, 329-336.

Beames I. (1972) The spread of the fox in the London area. *Ecologist* 2, 25-26.

Brosset A. (1975) Regime alimentaire d'une population suburbaine de renards au cours d'un cycle annuel. *Terre et la Vie* 29, 20-30.

Browne M. (1978) Hunted fox: a study of fox hunting and its effects on fox populations. In N. Dunstone (ed.), *Carnivore Biology*, pp. 28-29. Mammal Society.

Harris S. (1977) Distribution, habitat utilization and age structure of a suburban fox (*Vulpes vulpes*) population. *Mammal Rev.* 7, 25-39.

Harris S. (1978a) Age determination in the red fox (*Vulpes vulpes*) — an evaluation of technique efficiency as applied to a sample of suburban foxes. *J. Zool., Lond.* 184, 91-117.

Harris S. (1978b) Injuries to foxes (*Vulpes vulpes*) living in suburban London. *J. Zool., Lond.* 186, 567-572.

Harris S. (1979) Age-related fertility and productivity in red foxes, *Vulpes vulpes*, in suburban London. *J. Zool. Lond.* 187, 195-199.

Johnson D. H. and Sargeant A. B. (1977) Impact of red fox predation on the sex ratio of prairie mallards. *U.S. Dept. Int. Wildl. Res. Rept.* 6, 1-56.

Kruuk H. and Hewson R. (1978) Spacing and foraging of otters (*Lutra lutra*) in a marine habitat. *J. Zool., Lond.* 185, 205-212.

Lloyd H. G. (1977) Fox *Vulpes vulpes*. In G. B. Corbet and H. N. Southern (eds.), *The Handbook of British Mammals*, pp. 311-320. Oxford: Blackwell Scientific Publications.

Macdonald D. W. (1977) The behavioural ecology of the red fox. In C. Kaplan (ed.), *Rabies — The Facts*, pp. 70-90. Oxford: Oxford University Press.

Macdonald D. W. (1978) Radio-tracking: some applications and limitations. In B. Stonehouse (ed.), *Animal Marking: Recognition Marking of Animals in Research*, pp. 192-204. London: Macmillan.
Macdonald D. W. (1980) *Rabies and Wildlife: a Biologist's Perspective*. Oxford: Oxford University Press. (in press).
Mongtomery G. G. (1974) Communication in red fox dyads: a computer simulation study. *Smithson. Contr. Zool.* 187, 1-30.
Niewold F. J. J. (1974) Irregular movements of the red fox (*Vulpes vulpes*) determined by radio tracking. *Trans. Int. Congr. Game Biol.* 11, 331-337.
Niewold F. J. J. (1976) Aspecten van het sociale leven van de vos. *Natura* 73, 1-8.
Phillips R. L. (1971) Notes on the behavior of red foxes in a large enclosure. *Proc. Iowa Acad. Sci.* 78, 36-37.
Sargeant A. B. (1972) Red fox spatial characteristics in relation to waterfowl predation. *J. Wildl. Mngmt.* 36, 225-236.
Scott T. G. (1943) Some food coactions of the northern plains red fox. *Ecol. Monogr.* 13, 427-479.
Scott T. G. (1947) Comparative analysis of red fox feeding trends on two central Iowa areas. *Iowa Agr. Exp. Sta. Res. Bul.* 353, 426-487.
Storm G. L. and Montgomery G. G. (1975) Dispersal and social contact among red foxes: results from telemetry and computer simulation. In M. W. Fox (ed.), *The Wild Canids — Their Systematics, Behavioral Ecology and Evolution*, pp. 237-246. Amsterdam: Van Nostrand Reinhold.
Storm G. L., Andrews R. D., Phillips R. L., Bishop R. A., Siniff D. B. and Tester J. R. (1976) Morphology, reproduction, dispersal, and mortality of midwestern red fox populations. *Wildl. Monogr.* 49, 1-82.
Teagle W. G. (1967) The fox in the London suburbs. *Lond. Nat.* 46, 44-68.

The Ranging Behavior of a Maturing Female Red Fox, *Vulpes vulpes*

N. G. HOUGH

Animal Behaviour Research Group, Department of Zoology, University of Oxford, South Parks Road, Oxford OX1 3PS, U.K.

Abstract — This paper attempts to answer the questions: what happens to the young foxes which leave their parental group and why do they disperse? Detailed radio tracking data, recorded over four months are described and these show how a maturing female red fox established and abandoned several different home ranges. Several hypotheses for the reasons why this dispersal occurred are outlined and discussed.

It has recently been shown that, in some habitats, foxes are territorial and live in groups of one male and several females (Macdonald, 1977, 1980). Group sizes of more than six adults have never been found and most of them are smaller; and although not all vixens breed, more cubs are produced in many years than are incorporated into the group and, of these, most males and some females are compelled to disperse. One of many interesting questions is what happens to the young foxes which are not absorbed into their parental group? So far, this question has not been answered at a behavioral level, although several population studies have shown trends in dispersal (e.g. Storm *et al.*, 1976).

This paper describes, in detail, the movements of a female red fox of known origin. One of six cubs born in an urban garden in Oxford, U.K., was observed in this garden prior to being trapped in mid September, 1978, when about six months old. The fox (number V18) was equipped with a transmitter incorporating a mercury switch, indicating movement by changes in pulse rate, and this allowed determination of both location and activity. (Transmitter was produced by Telemetry Systems Inc., Model No. LT43-2TM-LD-M-MS.) Over a period of four months the vixen was tracked on more than 60 nights, both by car (using a roof mounted Yagi antenna) and on foot (using a hand-held dipole antenna). Direct observations were frequently obtained using binoculars and image intensifying equipment. Location and activity were monitored continuously throughout the night and radio fixes were recorded on a large scale map. Figure 1 shows the radio locations recorded, and temporary home range borders drawn using the 'Convex Hull' method. (The Convex Hull is the smallest convex polygon containing all the radio fixes — Southwood, 1966.) Although this method can be criticized (see Macdonald, Ball and Hough, 1980, this volume), its use is appropriate in this case as these borders were confirmed by direct observations. Initially the fox moved about within its natal range (Fig. 2A), establishing very distinct borders, outside which it did not venture during the first 5 weeks of tracking.

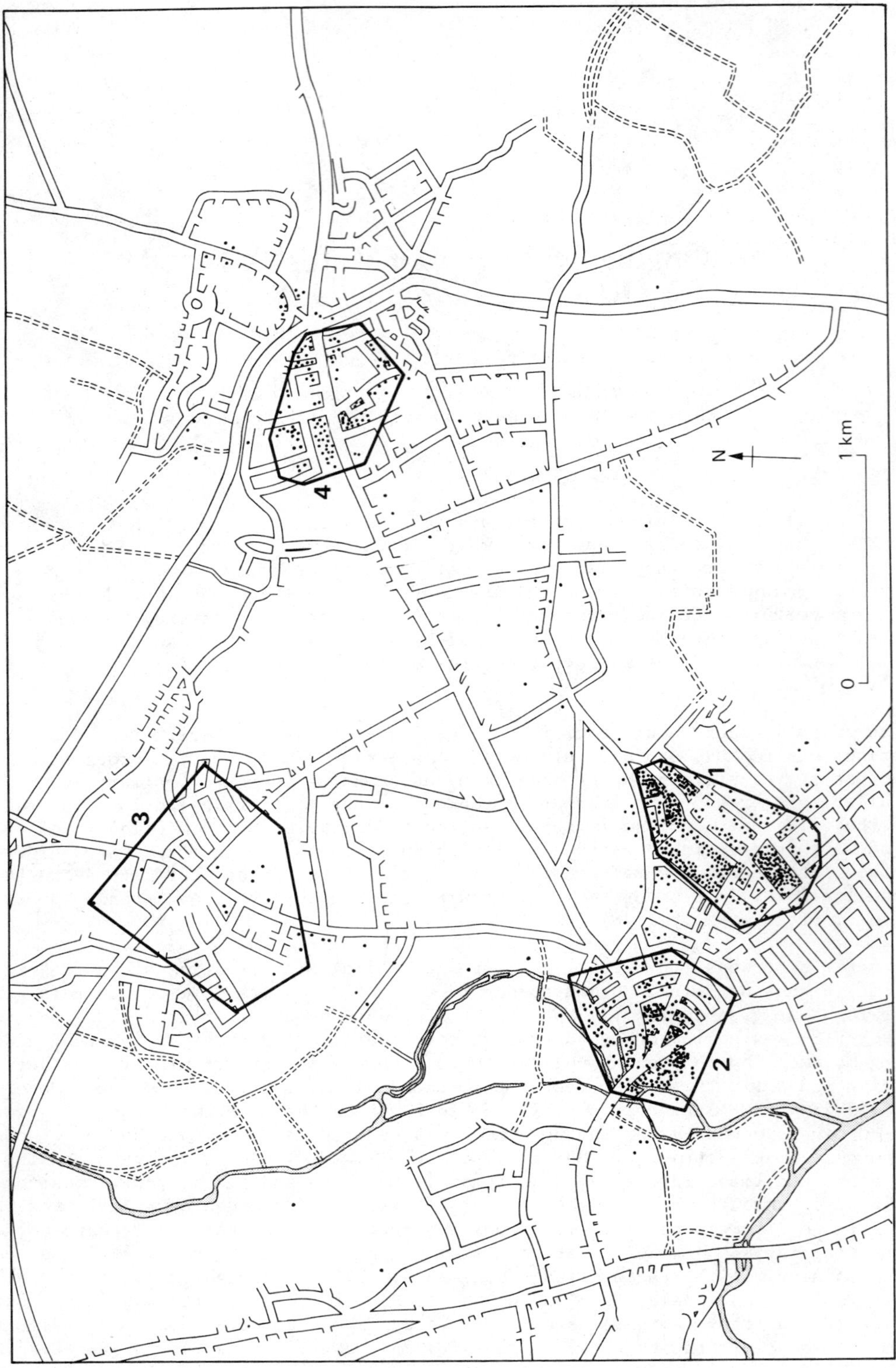

Fig. 1. All radio locations recorded between 18 September, 1978 and 4 January, 1979.

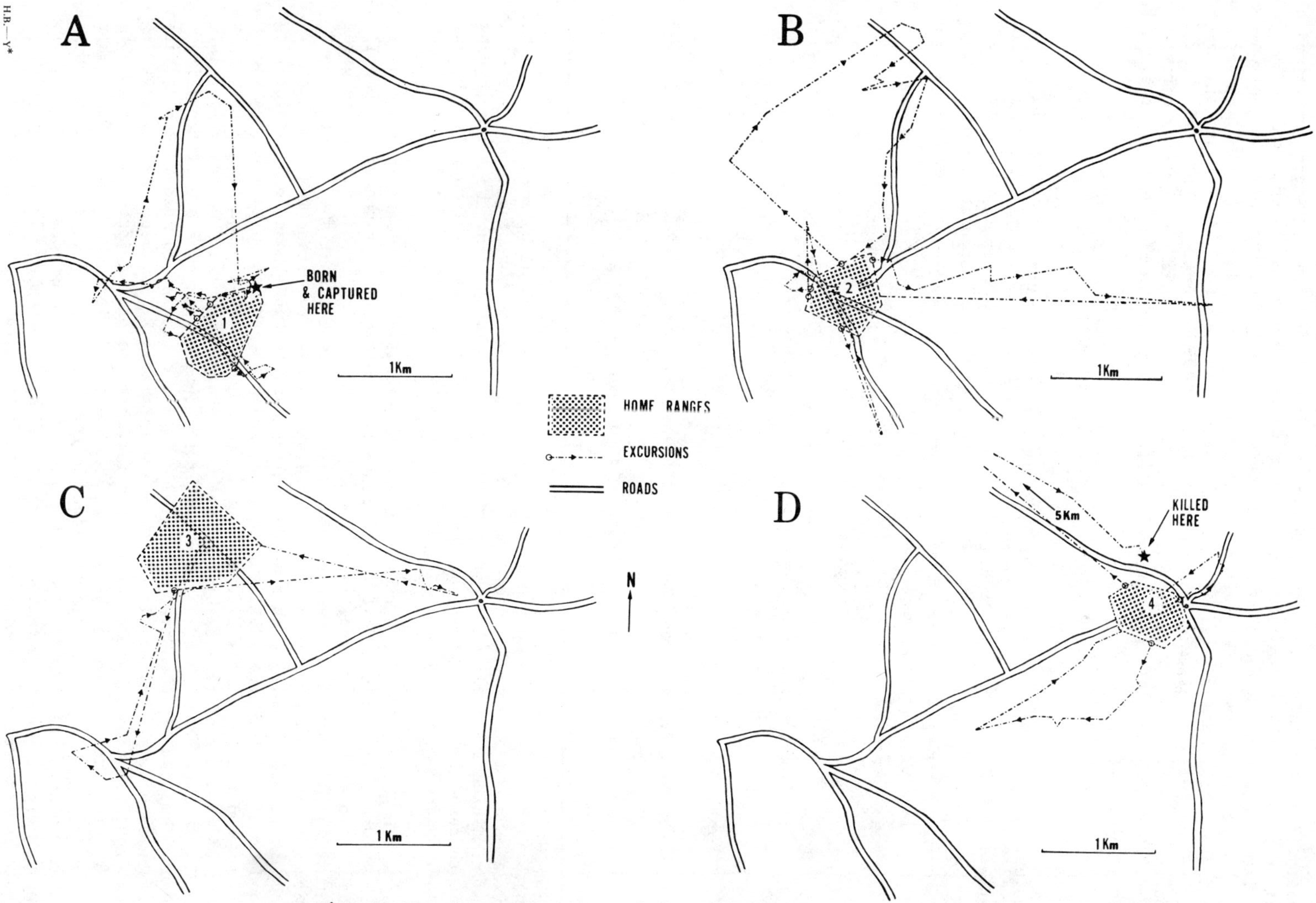

Fig. 2. The four ranges occupied and excursions from these ranges.

Observations were often made of interactions between this radio tagged female and a male cub (number D17) from the same litter, which was also equipped with a radio collar. This male shared a very similar home range (M. Newdick, personal communication); both these foxes were often seen feeding from fish and chip wrappers discarded in the streets. No aggressive encounters were seen, which suggests that this male did not directly precipitate the ensuing changes in the female's behavior. On 8 and 26 October, 1978 the vixen left home range 1, making short excursions before returning; these and all other excursions are illustrated in Fig. 2, while their dates are shown in Table 1.

Table 1.
The dates of occupation, excursions from and areas of home ranges

Range	Dates of Occupation	Dates of Excursions	No. days occupied	Area (hectares)
1	18 Sept. 78 to 15 Nov. 78 (captured)	8 Oct. 78 26 Oct. 78 9 Nov. 78 13 Nov. 78 14 Nov. 78	58	31.8
2	15 Nov. 78 to 30 Nov. 78	15 Nov. 78 20 Nov. 78 26 Nov. 78 28 Nov. 78 30 Nov. 78	15	32.4
3	30 Nov. 78 to 5 Dec. 78	3 Dec. 78	5	64
4	5 Dec. 78 to 4 Jan. 79 (killed)	11 Dec. 78 12 Dec. 78 4 Jan. 79	30	28

Another short excursion occurred on 9 November, 1978, and on 13 November, 1978 the vixen embarked on a much longer one, travelling to areas more than 4 km from range 1, and remaining approximately 1 km outside her natal range during the following day before returning very quickly at dusk. This return to her natal range was brief, as she passed along its northern edge and down to area 2 (Fig. 2B).

The fox never returned to her natal range but established a clearly defined area 0.5 km west, which was almost identical in area to the first range (see Table 1). What provoked this dispersal is unknown, but some possibilities will be discussed later. For 10 days this vixen then appeared to become established within area 2. During a night when there were frequent disturbances, two other foxes were observed; these were seen chasing each other in an aggressive manner, and both were barking loudly. They also individually chased the radio tagged fox on several occasions, which resulted in all three foxes running along main roads. The female then spent most of the rest of this night inactive in a drain which she had used daily since moving to this area. The following night the female left this area (travelling about 4 km) and explored area 3 (Fig. 2C) before returning to area 2 quickly, just before dawn. Three nights later she again left area 2 and spent the day sleeping just south of area 3, and after briefly returning to area 2 (from 01.25 to 02.00) she returned once again to the area in which she had slept the previous day. She was seen, briefly, while being followed by another fox, and subsequently moved very quickly west, exploring area 4 (Fig. 2D), before returning to area 3. After a few days in area 3, the vixen moved to area 4 and by 10 December, 1978 it was clear that the home range shown in Fig. 2D was becoming established. Apart from one excursion back in the direction of area 1, and her subsequent return to area 4, the vixen's range seemed to have stabilized once again. However, on 4 January, 1979, the fox was located about 5 km northwest of area 4. By this time it was clear that she was already returning from an excursion. Within 20 min she was within 200 m of the edge of area 4, but was trapped and killed at this point by a gamekeeper.

DISCUSSION

Initially, the fox moved within its natal range, but subsequently left this area, establishing and abandoning several other ranges. The movements which led to the establishment of these ranges, were in contrast to the feeding excursions noted by Niewold (1974).

There are many possible reasons why a fox may leave its home range:

1. Expulsion by a dominant fox: It is possible that another fox, either from within its own group or from outside, forced the radio collared fox to leave her natal range. However, no aggressive encounters were observed in range 1, and although it is possible she was forced to leave, the observations suggest this did not happen. One thing is clear: excursions tended to occur in the last few days of occupancy of a home range, and these excursions tended to be to areas to which the fox subsequently moved. It is therefore possible that aggressive encounters did occur in range 1, but were not seen; in area 2 the female was seen being chased, and subsequently left this area.

2. Left in search of a mate: Although it is unknown if the vixen was sexually mature at the time of dispersal, at six months it is likely that she was (Layne and McKeon, 1956). In range 2 she was seen on several occasions with a large male fox. All interactions seen were amicable, and the female took great interest in the urination sites of the male; several times she investigated the male's marks and also marked over them. This double marking has been described by Rothman and Mech (1979) in wolves as a common event during pair formation.

The male was also seen following, but not chasing, the female; this often occurs prior to mating in the red fox (Macdonald, 1977), and so at this time it seemed likely that the female had paired with this male. However, it became obvious this was not the case when she later left this area.

3. A shortage of safe daytime resting places: This may be an important factor in determining the usefulness of a range to a fox. It is possible that the loss of all safe daytime resting places (for example, to dominant animals within the group) could force a fox to leave a range. After approximately half of the excursions witnessed, the fox still used the same daytime sleeping locations upon returning, which suggests it was not the loss of daytime resting sites that caused this fox to disperse.

4. Dispersed in search of a more profitable area in terms of food supply: No matter how 'good' a home range may be, it is always possible that something better exists elsewhere, and so you would expect that a fox would occasionally explore outside its range; the data for this vixen do not support or disprove this hypothesis.

5. The direct effects of weather extremes: Maximum and minimum temperatures were examined for the days on which excursions occurred. Although no clear picture emerged, they suggest that extremes of temperature may be a contributing factor.

These data illustrate the complexity of the social behavior of red foxes, and show how little we currently understand about their social system. The failure of this fox's behavior to fit any single hypothesis not only illustrates the expansive gaps in our understanding, but also the necessity for further detailed study of the behavior of individual foxes.

Acknowledgements — I am grateful to the following for their help and advice: C. J. Amlaner, Jr., N. J. Ball, D. W. Macdonald, M. T. Newdick, S. A. Rolls, Mr. and Mrs. Dubber and Mr. and Mrs. Ward. Figure 1 was kindly drawn by Oxford University Press. This work was funded in part by the Medical Research Council.

REFERENCES

Layne J. N. and McKeon W. H. (1956) Some aspects of red fox and grey fox reproduction in New York. *N.Y. Fish Game J.* 3, 44-74.

Macdonald D. W. (1977) The behavioural ecology of the red fox, *Vulpes vulpes*: a study of social organisation and resource exploitation. D.Phil. thesis, Oxford.

Macdonald D. W. (1980) Social factors affecting reproductive success in the red fox, *Vulpes vulpes*. In E. Zimen (ed.), *The Red Fox: Behaviour and Ecology*. Proc. of the Symposium on Fox Biology, Saarbrücken, 1979. The Hague: W. Junk.

Macdonald D. W., Ball F. and Hough N. G. (1980) The evaluation of home range size and configuration from radio tracking data. In C. J. Amlaner, Jr. and D. W. Macdonald (eds.), *A Handbook of Biotelemetry and Radio Tracking*, pp. 405-424. Oxford: Pergamon Press.

Niewold F. J. J. (1974) Irregular movements of the red fox (*Vulpes vulpes*) determined by radio tracking. *XI Int. Congress of Game Biology*, pp. 331-337, Stockholm.

Rothman R. J. and Mech L. D. (1979) Scent-marking in lone wolves and newly formed pairs. *Anim. Behav.*, in press.

Southwood T. R. E. (1966) *Ecological Methods, with Particular Reference to the Study of Insect Populations*. 391 pp. London: Methuen.

Storm G. L., Andrews R. D., Phillips R. L., Bishop R. A., Siniff D. B. and Tester J. R. (1976) Morphology, reproduction, dispersal and mortality of mid-western red fox populations. *Wildl. Monog.* 49, 1-82.

Home Range and Activity Rhythm of Adult Male Foxes during the Breeding Season

D. MAUREL

Centre d'Etudes Biologiques des Animaux Sauvages, 79360 Beauvoir-sur-Niort, France

Abstract — The activity rhythm of the fox, *Vulpes vulpes*, is characterized by synchrony with sunset and sunrise. This rhythm is polyphasic with a slight diurnal activity period correlated with the return to the main burrow area. The beginning of nocturnal activity is closely correlated with sunset and in 50 percent of the observations cessation occurs in the hour before sunrise, but the fox can exhibit diurnal activity two hours after sunrise. The duration of this diurnal activity varies depending on the time the fox previously had spent in captivity. The home range of the adult male fox determined in the Chizé Forest during the breeding season is about 500 ha. Variations observed in the 3 animals during the study period can be related to the age and the duration of the animal's captivity. In each case the area was not used equally by the fox but the range consisted of a series of pathways which connect highly used areas.

INTRODUCTION

In the red fox, the breeding season occurs during the winter time: in the male, the plasma testosterone level starts to increase at the end of autumn, the maximum level is reached in February. From April to November in the Chizé Forest (west France) there is a period of sexual inactivity. During the breeding period we have studied the activity rhythm and the home range of 3 adult male foxes equipped with radio collars and automatically radio tracked during two months. The radio tracking system of the Chizé Forest has been described by (Deat *et al.*, 1980, this volume). The large number of radio locations (maximum: 360 24 h^{-1}) favors the use of a continuous recording system. The method used to determine the home range consisted of connecting the peripheral locations obtained on a tracking map summarizing several daily tracking maps. Thus, the home range surface is determined during the whole period of the animal's activity. The animal's home range and its utilization can be represented by the square method (e.g. Sempéré, 1979) on the roe deer. The activity rhythm determination has been carried out by recording the number of minutes that the fox was motionless or moving, for every hour of the day. Thus we obtained for each fox, every day and every hour, the amount of time spent at rest and moving. In order to correct possible mistakes owing to signal loss the actual activity has been expressed as percentage of time active in relation to the time at rest. Signal loss may result either from movements too quick to be recorded or from movements made in forest zones where the radio reception was poor. Data from particular times of day have been summed to give an overall activity rhythm for the whole experiment.

HOME RANGE

The size of fox home ranges varies considerably, as revealed by the radio tracking. Sargeant (1972) delimits 3 home ranges of respectively 645 ha, 796 ha, and 858 ha; the greatest area was found by Storm *et al.* (1976) with 960 ha and the smallest by Ables (1969a): 57.5 ha. The latter explains that the range variability may be correlated to food availability (larger home ranges in poorer areas). Our results (Fig. 1), 488 ha (fox 300), 464 ha (fox 621) and 600 ha (fox 726) are within the limits quoted in the literature. Of the three foxes, two animals had a similar size home range, the third (fox 726) had a larger home range; we can explain this difference by the behavior of the animal observed on the daily tracking map. This fox was a subadult without a previously secured home range, caught as a cub in the forest and released

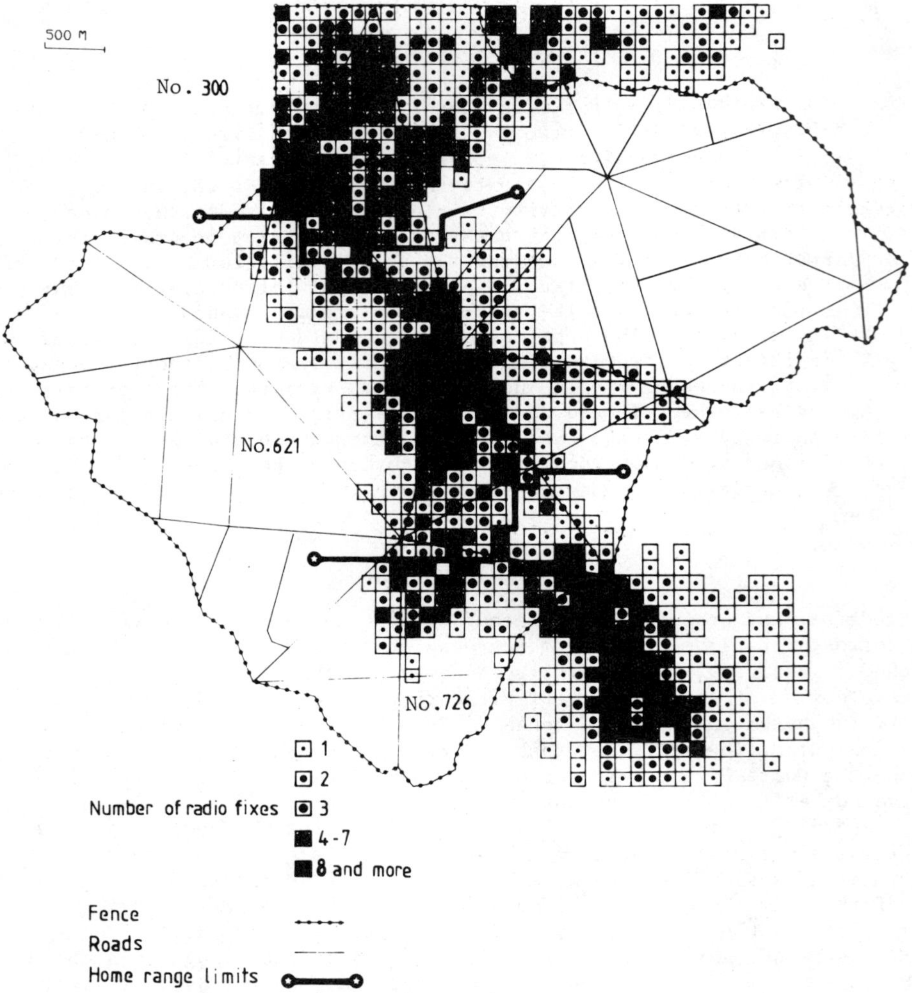

Fig. 1. Home range of the three adult male foxes No. 300 (488 ha), No. 621 (464 ha) and No. 726 (394 ha) during two months of radio tracking in winter.

when adult (after five months of captivity). This animal exhibited exploratory activity with long treks along the roads and paths of the forest and around the fence; after 15 days of exploratory activity, he settled in an area of the forest, 394 ha in size. Foxes 300 and 621 were caught as adults, their great stability in the forest suggest that they may have returned to their previous home ranges. For fox 300 (one day of captivity) this reoccupation is immediate, but for fox 621 (one year of captivity) this reoccupation occurs after several days. The home ranges of the foxes were not occupied homogeneously. The home ranges consisted of a series of pathways which connect highly used areas. These three home ranges were contiguous and never overlapped; we can suppose that, in this period of the reproductive cycle there is territorial behavior and the area used is a territory. Thus, the total area necessary during the reproduction period is about 500 ha; if we consider a social structure as described by Macdonald (1980) for some habitats in Great Britain (1 male, 1 dominant vixen and 2 or 3 subordinate vixen) we can estimate that the fox population density in the Chizé Forest is about one fox for 100 ha. However, Macdonald stresses the intraspecific variation in fox behavior, so these figures may not apply to the Chizé Forest.

ACTIVITY RHYTHM

The three foxes exhibited diurnal polyphasic locomotor activity rhythms (Fig. 2). The activity occurs generally during the night and exhibits two maxima: at sunset and sunrise. These observations are comparable to those of Ables (1969b) and Himler and Tembrock (1972). However, we have observed that the activity during the diurnal phase is more important in the two animals which were held in captivity for longer times. Fox 621 (one year of captivity) was active more often during the light phase than fox 726 (five months of captivity) while fox 300, released immediately after his capture, exhibited negligible light phase activity. There was a close correlation between the beginning and the end of nocturnal activity and sunset and sunrise (Fig. 3). Nocturnal activity begins during the hour following sunset for 60 percent of the observations, and ends in the hour preceding sunrise for 50 percent of the observations but the fox may exhibit diurnal activity two hours after sunrise. This diurnal activity consists of successive burrow changes occurring in the morning. Differences in activity rhythm can be related to individual variability and we observed that, although all three foxes exhibited activity rhythms synchronized with the day-night cycle, no two animals behaved the same.

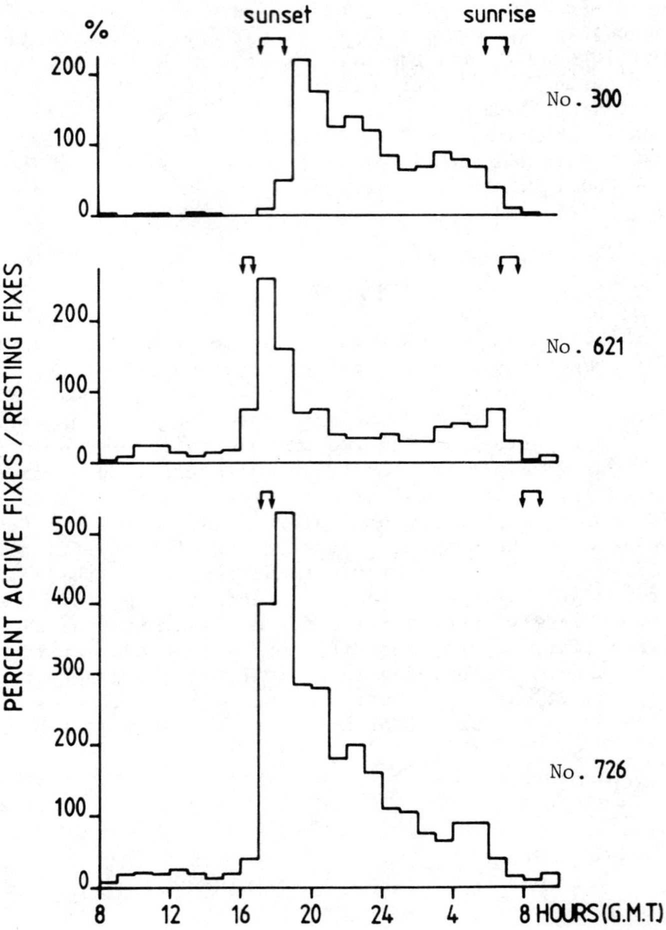

Fig. 2. Activity rhythm of the adult male foxes Nos. 300, 621 and 726 during the breeding season, expressed as percentage of active fixes/resting fixes.

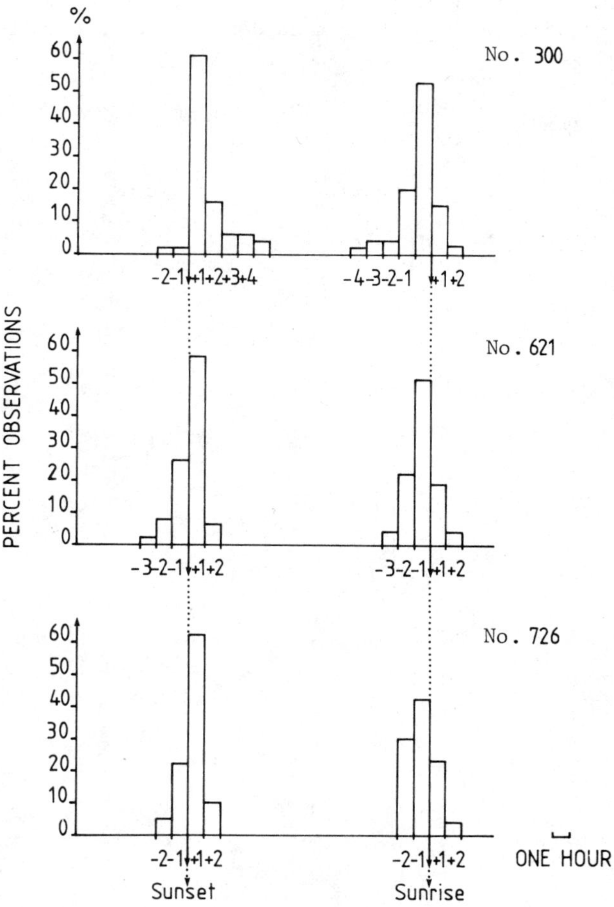

Fig. 3. Beginning and end of nocturnal activity for the three adult male foxes during the breeding season in relation to sunset and sunrise expressed as a percentage of the observations in the first, second or third hour before or after sunset or sunrise.

REFERENCES

Ables E. D. (1969a) Home range studies of red foxes (*Vulpes vulpes*). *J. Mammal.* 50, 108-120.

Ables E. D. (1969b) Activity studies of red foxes in Southern Wisconsin. *J. Wildl. Mngmt.* 33, 145-153.

Deat A., Mauget C., Mauget R., Maurel D. and Sempéré A. (1980) The automatic, continuous and fixed radio tracking system of the Chizé Forest: theoretical and practical analysis. In C. J. Amlaner, Jr. and D. W. Macdonald (eds.), *A Handbook on Biotelemetry and Radio Tracking*, pp. 439 -451. Oxford: Pergamon Press.

Hilmer H. von and Tembrock G. (1972) Zur aktivitätsperiodik bei *Vulpes und Alopex*. *Zool. Jahrb.* (*Phys.*) 68, 297-324.

Macdonald D. W. (1980) *Rabies and Wildlife: A Biologist's Perspective*. Oxford: Oxford University Press.

Sargeant A. B. (1972) Red fox spatial characteristics in relation to waterfowl predation. *J. Wildl. Mngmt.* 36, 225-236.
Sempéré A. (1979) Utilisation et évolution du domaine vital chez le chevreuil mâle européen déterminé par radiotracking. *Biol. Behav.*, in press.
Storm G. L., Andrews R. D., Phillips R. L., Bishop R. A., Siniff D. B. and Tester J. R. (1976) Morphology, reproduction, dispersal and mortality of midwestern red fox populations. *Wildl. Monogr.* 49, 1-82.

Movements and Daily Activity Pattern of Radio Tracked Male Stoats, *Mustela erminea*

S. ERLINGE

Ecology Building, 223 62 Lund, Sweden

Abstract — Movements and daily activity patterns of the stoat, *Mustela erminea*, were examined by radio tracking. Four males were selected to exemplify movement patterns of territorial and non-territorial individuals and also seasonal variations in daily activity patterns. Shorter or longer spells of activity, generally lasting for 10 to 45 min, alternated with periods of rest. There was a seasonal change in daily activity; the stoats tended to be nocturnal in winter and diurnal in summer. The increased diurnal activity in summer was associated with increased movements, whereas the duration of hunting activity during daytime and nighttime was less different.

INTRODUCTION

The ecology of a stoat population in southern Sweden is being examined. Radio tracking is used when examining the spatial distribution of the stoats and their daily activity pattern. This paper reports on the movements and daily activity of four male stoats selected to illustrate seasonal variations and differences in movements of territorial and non-territorial individuals.

STUDY AREA AND METHODS

The investigation was carried out in the Revinge area of southern Sweden. This area of about 4000 ha, which is used for military training and cattle grazing, contains extensive marshes. These are the favored habitats of the stoats. So far 18 individual stoats (nine males and nine females) have been radio tracked. Two kinds of transmitter collars (27 MHz) were used, a 2-battery model for males and a 1-battery model for females. Transmitter circuitry was identical to that described by Tester, Warner and Cochran (1964). The weights of the transmitters were about 10 and 6 g respectively, i.e. about 4 percent of the stoats' body weight. Battery life was generally 7-10 days and the signals were obtained within a distance of about 300 m, by use of a portable receiver. The animals were generally tracked continuously. During some periods, especially at night, the signals were obtained on a tape recorder. The timer switch recorded the signals for 20-30 s every 5 min. From the character of the signals, constant or changing in volume, periods of locomotor activity were distinguished from periods of rest.

When entering deeper parts of rodent tunnels the signals disappeared. In all cases the signals were lost for periods shorter or equal in length to periods recorded for locomotor activity when the signals were continuously obtained. Therefore the periods when the signals disappeared for less than 90 min were considered to be activities associated with hunting in deeper parts of the tunnel system.

RESULTS AND DISCUSSION

One subadult male was radio tracked in February, 1975. He exploited an area containing wet meadows populated by water voles, *Arvicola terrestris* L. (Fig. 1). No other stoats were in that area at this time as was evident from snow tracking. During the first day the male moved about 600 m and then stayed for the rest of the tracking period (seven days) within a very restricted area. During the first four days he exploited a colony of water voles and hunted for them in their underground tunnels. He then moved about 50 m to another colony. When radio tracking was finished I examined the exploited areas by digging up parts of the tunnel systems. The stoat had occupied the voles' nest and improved the nest covering with fur from captured and consumed voles. Close to the nest, where the stoat had spent the last three days, a half-eaten vole was found and two more voles had been killed and brought to the nest.

The daily activity was characterized by shorter or longer spells of activity throughout the 24 hour period (Fig. 1). No movements occurred outside the hunted areas so the stoat evidently devoted all his movement activity to hunting. On average the stoat was active about one fourth of the time recorded, i.e. about six hours per day. He was considerably more active at night than during daytime; 32 percent of time recorded at night (43.6 h) was activity compared with 18 percent during daytime (55.3 h).

Another male was radio tracked in April, 1975 (Fig. 2). He was not confined to any definite area but behaved as a transient. His movements were outside the areas occupied by two resident male stoats. A fortnight after radio tracking the male was killed by a vehicle on a road about 6 km away. The male generally changed hunting area every day moving about 600-800 m. The stay in the hunted areas generally lasted for 30-40 h. The movements from one area to another occurred during daytime.

The daily activity pattern was similar to that of the male radio tracked in February, i.e. shorter or longer spells of activity alternated with periods of rest (Fig. 2). However, in contrast to the male tracked in winter this male was mostly active during daytime. At night the activity generally occurred in short periods (mostly 10-15 min). The diurnal activity included one or two longer periods of activity connected with movements to a new hunting ground. The average time of activity per day was roughly the same as for the stoat tracked in winter; 27 percent of recorded time (107.5 h) was locomotor activity. The main part concerned activity in the hunted grounds but roughly one fourth of active time was associated with movements from one place to another.

A third male established (spring, 1977) in an area containing a breeding female. This was evidenced from trapping. During radio tracking in May the male regularly patrolled his territory, about 35 ha in size. Some restricted parts were used as hunting grounds and associated with these areas there were nest sites in regular use. Extensive movements occurred during daytime. At night the activity generally occurred in the restricted hunted grounds.

This male showed a high level of locomotor activity; more than one third of recorded time was as locomotor activity, i.e. more than eight hours per day. He was more active during daytime (38 percent of recorded 57.8 h) than during the night (29 percent of recorded 33.3 h). The difference was mainly due to extensive diurnal

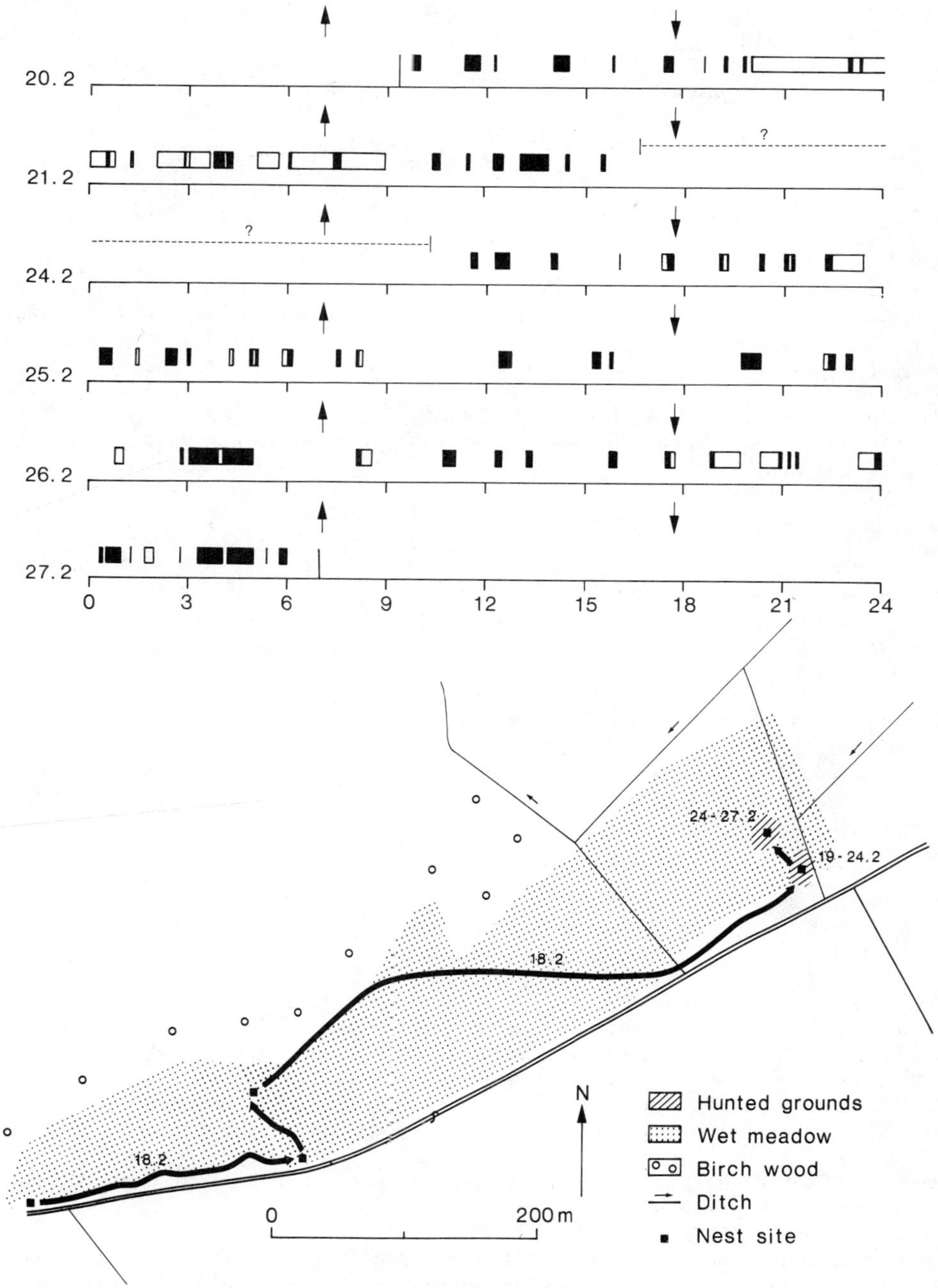

Fig. 1. Daily activity pattern and movements of a male stoat (♂ 34) radio tracked in February, 1975. Activity is denoted by filled bars. Open bars signify time when no signals were obtained. This happened when the stoat entered deep tunnels. Sunrise and sunset are denoted by vertical arrows.

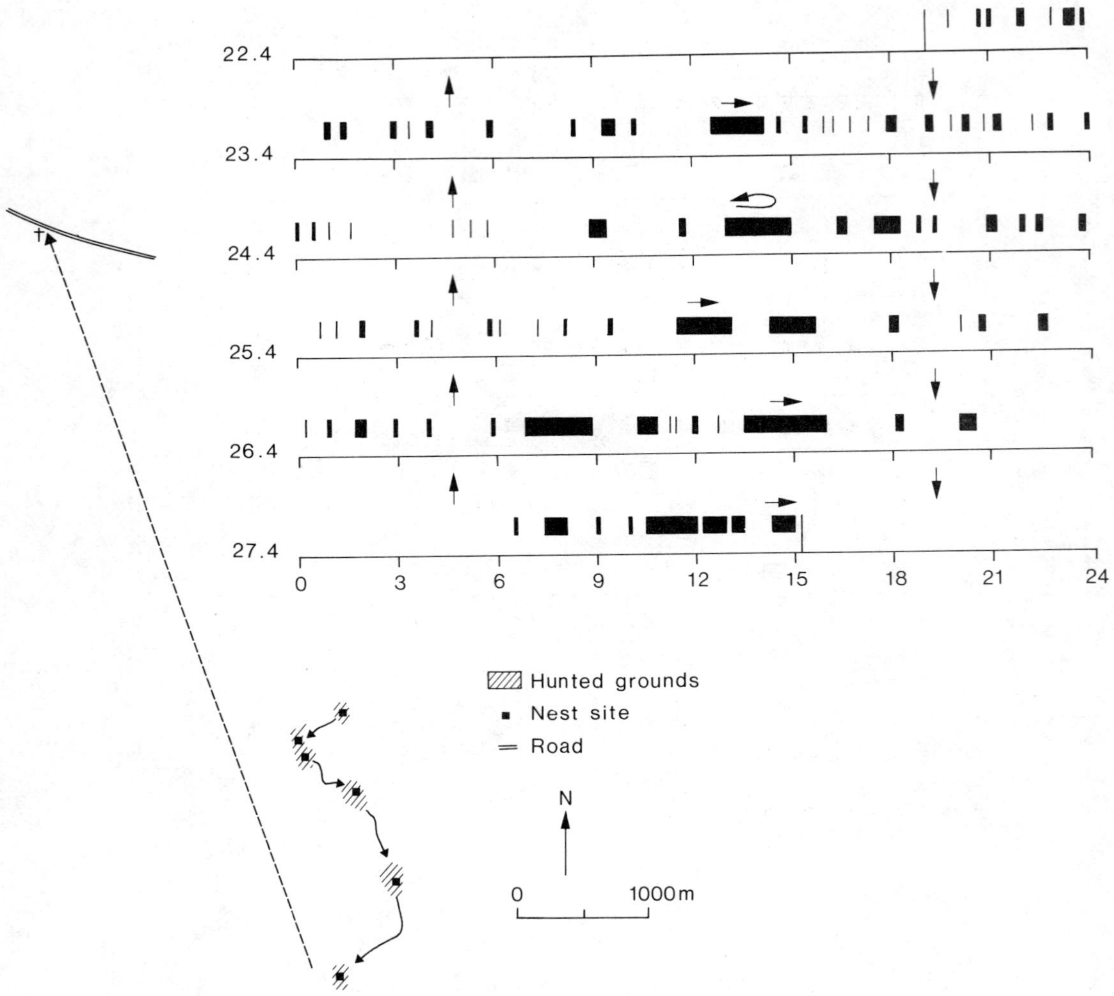

Fig. 2. Daily activity pattern and movements of a transient male stoat (♂ 22) radio tracked in April, 1975. The symbols are the same as in Fig. 1. Horizontal arrows signify movements from one site to another; rounded line denotes movements and return to the same spot.

movements. Hunting activity in the restricted areas had similar duration during the light and the dark periods.

A fourth male was radio tracked in October, 1974 (Fig. 4). He was established in an area containing smaller wet meadows intermingled with pine plantations and grazed fields. His territory included two females and adjoining to his territory two juvenile males occurred. The confinement of the male to a definite area was evidenced from trapping throughout autumn. During radio tracking he regularly patrolled his territory and covered the main part once every day. The transfers from one marsh area to another occurred during daytime but movements between different parts of the marshes also occurred at night.

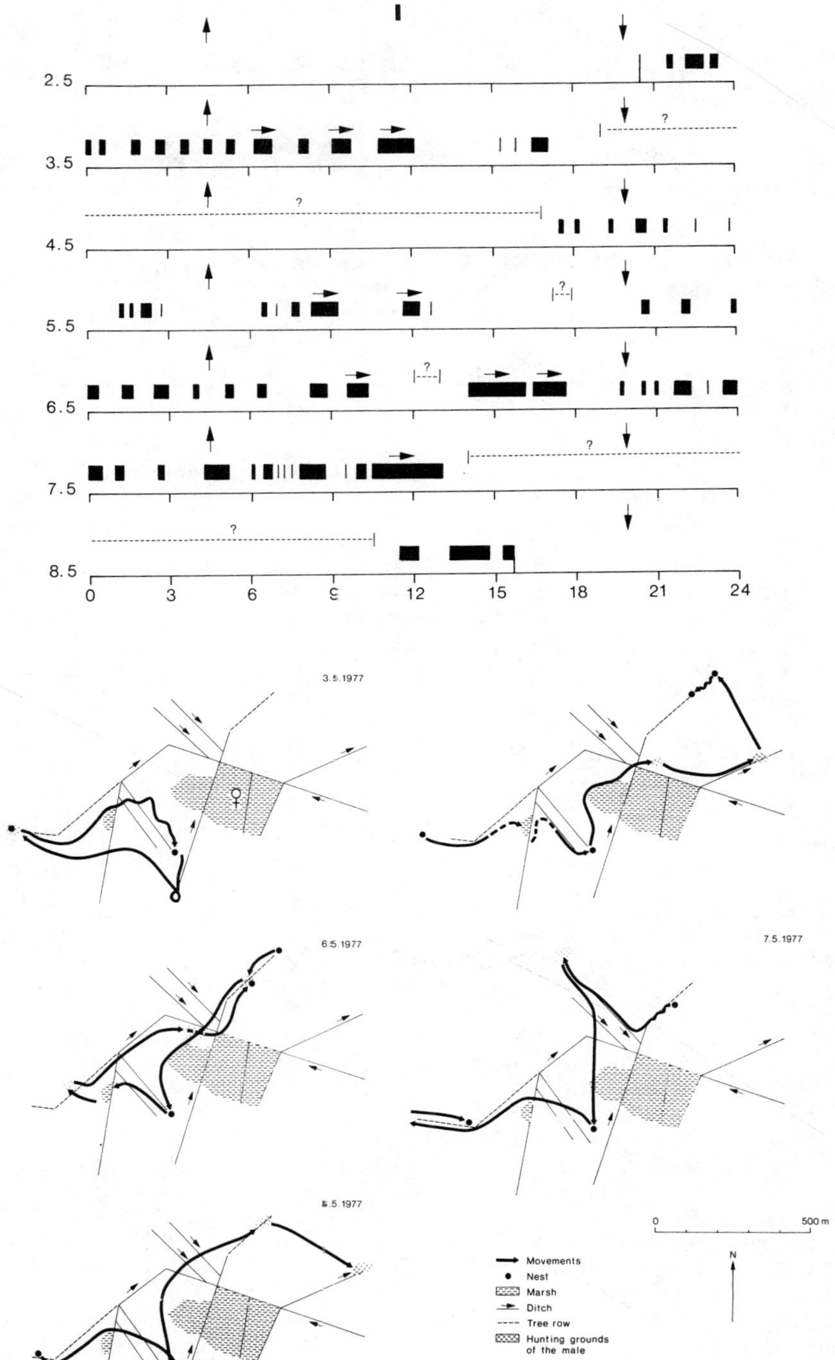

Fig. 3. Daily activity pattern and movements of a resident male (♂ 62) racio tracked in May 1977. The symbols are as in Figs. 1 and 2.

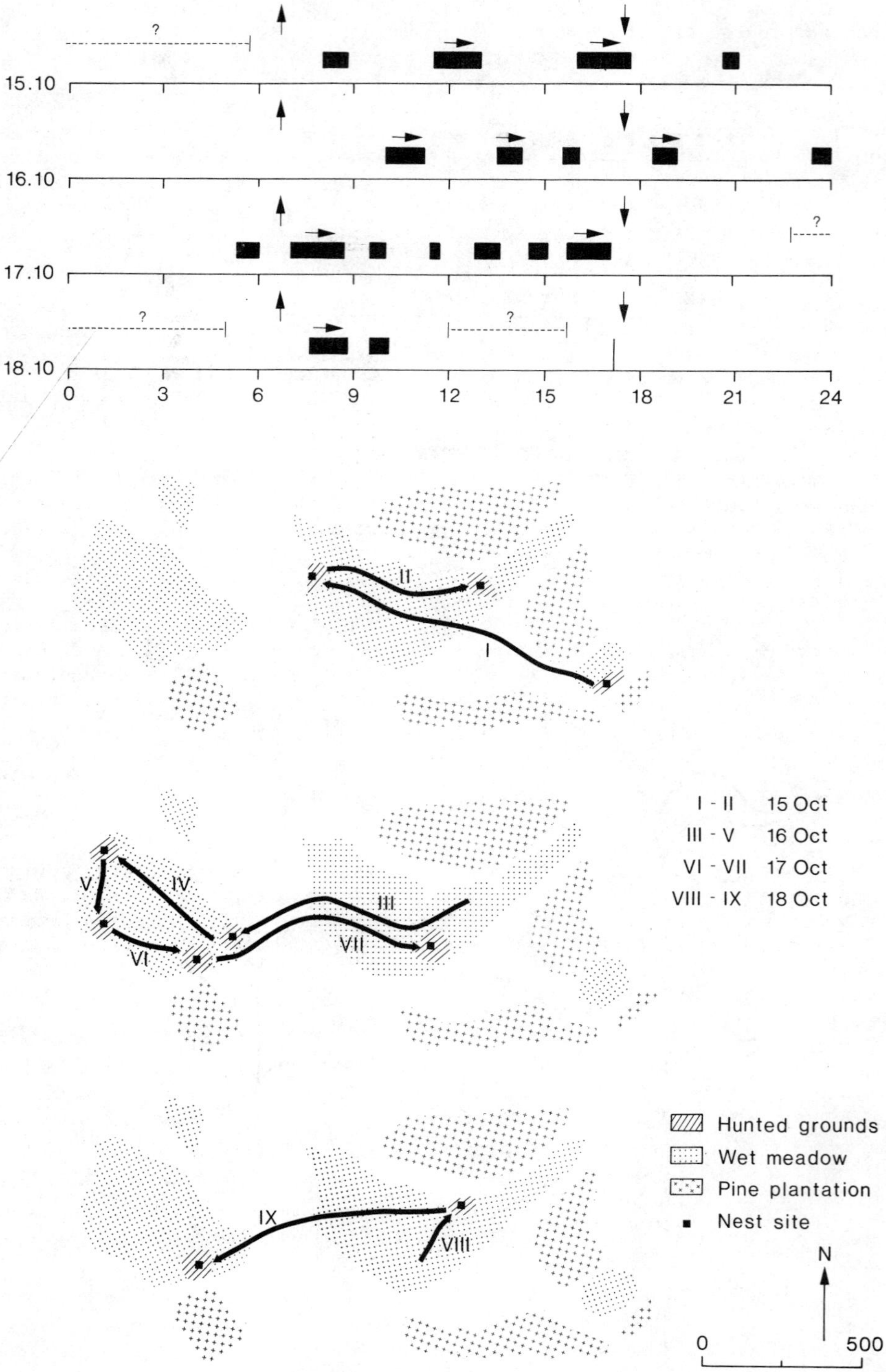

Fig. 4. Daily activity pattern and movements of a resident stoat (♂ 15) radio tracked in October, 1974. The symbols are as in Figs. 1 and 2.

This male devoted less time to locomotor activity, and including movements this was less than one fourth of recorded time (72.0 h). Most activity (80 percent) occurred in daytime.

Radio tracking data illustrate three movement patterns in male stoats: 1. The movements of a resident male regularly patrolling his territory. 2. The movements of a male confined to a definite area (home range) without displaying any patrolling activity. In such cases a stoat may return to a given part of his range at intervals of two or three weeks (Nyholm, 1959, personal observation). 3. The movements of a transient animal, staying for short periods in restricted areas for hunting and then moving to new areas.

The daily activity pattern was characterized by a short term rhythm: spells of activity and rest alternated more or less regularly in periods of 3-5 h. Locomotor activity generally lasted for 10-45 min but sometimes as long as 3-4 h. Some of the recorded locomotor activity probably concerned activity in the nest including eating and grooming. Stoats observed in an enclosure generally used 5-10 min for a meal and usually fed about every fourth hour (Erlinge, unpublished).

A short term rhythm similar to that in the stoat occurs in the weasel (Erkinaro, 1972a) and in voles (Erkinaro, 1961, 1969, 1972b; Stebbins, 1975, and Lehmann, 1976). In the small mustelids the short term activity pattern is probably an adaptation to their high food requirements (Scholander *et al.*, 1950; Iversen, 1972). The radio tracked stoats exemplify the seasonal change in daily activity pattern which occurs in stoats (radio tracking data and cage experiments, Erlinge, unpublished); a tendency to be nocturnal in winter and diurnal in summer. The changes occurred in March and November. In voles there is a similar change but they are diurnal in winter and nocturnal in summer (Erkinaro *op. cit.*). In stoats the increased daytime activity in summer was associated with increased movements. The duration of hunting activity during daytime and nighttime was less different.

As the stoats can hunt for the voles also when they are resting in their tunnels their hunting is independent of the activity rhythm of the voles.

Acknowledgements — I thank Mr. G. Göransson for help with the telemetry equipment, Mr. I. Nilsson and Dr. N. Stenseth for reading the manuscript and Mrs. S. Douwes for preparing the illustrations. Grants were obtained from the Swedish Natural Science Research Council and the National Swedish Environment Protection Board.

REFERENCES

Erkinaro E. (1961) The seasonal change of the activity of *Microtus agrestis*. *Oikos* 12, 157-163.

Erkinaro E. (1969) Der Phasenwechsel der lokomotorischen Aktivität bei *Microtus agrestis* (L.), *M. arvalis* (Pall) und *M. oeconomus* (Pall). *Aquile Ser Zool.* 8, 1-31.

Erkinaro E. (1972a) Seasonal changes in the phase position of circadian activity rhythms in some voles and their endogenous component. *Aquile Ser. Zool.* 13, 87-91.

Erkinaro E. (1972b) Fasväxling hos djur. (English summary). *Fauna och flora* 67, 215-219.

Iversen J. A. (1972) Basal energy metabolism of mustelids. *J. Comp. Physiol.* 81, 341-344.

Lehmann, U. (1976) Short-term and circadian rhythms in the behaviour of the vole, *Microtus agrestis* (L.). *Oecologia* 23, 185-199.

Nyholm E. (1959) Kärpästä ja lumikosta ja niiden talvisista elinpiireistä. (Stoats and weasels and their winter habits: in Finnish.) *Suomen Riista* 13, 106-116.

Scholander P. F., Hock R., Walters V. and Irving L. (1950) Adaptation to cold in arctic and tropical mammals and birds in relation to body temperature, insulation and basal metabolic rate. *Biol. Bull.* 99, 259-271.
Stebbins L. L. (1975) Short activity periods in relation to gapperi. *Oikos* 26, 32-38.
Tester J. R., Warner D. W. and Cochran W. W. (1964) A radio-tracking system for studying movements of deer. *J. Wildl. Mngmt.* 28, 42-45.

Summer Activity Patterns of Radio Marked Beaver, *Castor canadensis*

R. A. LANCIA,* W. E. DODGE** and J. S. LARSON*

**Department of Forestry and Wildlife Management, University of Massachusetts, Amherst, Massachusetts 01003, U.S.A.*
***Massachusetts Cooperative Wildlife Research Unit, Amherst, Massachusetts 01003, U.S.A.*

Abstract — A continuous recording, radio telemetry system was used to monitor activity patterns of 7 beavers in 2 colonies located near Amherst, Massachusetts, U.S.A. Activities were recorded from April to October, 1976 and 1977 (1 colony each year); up to 4 individuals were simultaneously monitored. Individuals followed 'typical' 24.0 h activity rhythms; onset of activity occurred at 17.2 ± 0.7 h ($\bar{x}$ ± S.D.) in one colony and at 17.0 ± 1.0 h in the other. Mean cessation of activity times was 6.1 ± 0.8 for one colony and 4.1 ± 1.3 h for the other. Onset of activity tended to be earliest in mid summer and latest in spring and fall. Sunset and length of twilight were statistically significant predictors of onset of activity for most individuals. Total activity period was partitioned into predominately active and predominately inactive portions; activity (12.8 ± 1.0 h) exceeded inactivity (11.1 ± 1.0 h) in one colony and inactivity exceeded activity (12.9 ± 1.3 and 11.2 ± 1.3 h, respectively) in the other. Animals showed biomodal peaks of activity (dusk and dawn), and spent the majority of the night's active period away from the colony site. In one colony when kits were present, time away from the colony site for the adult female was less than that for a comparable prepartum period.

INTRODUCTION

Most early investigators of animal activity rhythms studied the influence of laboratory controlled conditions on behavior patterns of captive individuals. Few studies were made of general activity patterns of free ranging animals because longterm observations in the field were difficult. The advent of radio telemetry techniques provided the first opportunity for intensive, longterm monitoring of individual activities under conditions when other techniques were useless. The objective of the present study was to determine year long activity patterns of individual, free ranging beaver (*Castor canadensis*) by radio tracking techniques.

Beaver colonies monitored in this study were located on the Prescott Peninsula, Quabbin Reservation, New Salem, Massachusetts. The peninsula is drained by 42 km of streams divided into 6 major systems. Beaver activity was first noted in 1952, and the population subsequently increased to an estimated 300 individuals in 47 active colonies by 1975.

METHODS

TRAPPING AND HANDLING

Animals, captured in Bailey and Hancock live traps, were anesthetized (Lancia, Brooks and Fleming, 1978) at the colony site, fitted with radio collars and released. Individuals were grouped into sex and age classes and marked with numbered metal and color coded, plastic ear tags for later identification (Lancia, 1979).

ACTIVITY MONITORING TELEMETRY SYSTEM

A radio telemetry system (Lancia and Dodge, 1977) was developed to continuously monitor beaver activity. Variations in received signal strength caused by movement of a radio tagged animal were used to determine presence or absence of activity (Sunquist and Montgomery, 1973). Whenever animals were in receiver range, regardless of their location, activity was continuously monitored on 2 stripchart recorders. Up to 4 channels (individuals) were recorded simultaneously.

Low frequency (30 MHz), Colpitts oscillator, transmitter collars were constructed as described by Lancia and Dodge (1977). The weight of the transmitter collar averaged 149 g or about 1 percent of mean body weight and had an *in situ* life of 4 to 6 months. As collars failed, individuals were recaptured when possible and reinstrumented.

Time corrected chart records were transcribed, by 15 min intervals, to a computer compatible format with coded values for predominant activities (e.g., activity, inactivity, out-of-range, and no data). Code assignment was periodically verified by comparing transcribed data with simultaneous field observations. In addition, times of initiation (emergence) and cessation (return) of the daily active period were determined to the nearest 5 ± 2 min.

RESULTS

Summer activities of colonies 5 and 6 were recorded from April to October, 1976 and 1977, respectively as a part of a year-round study (Lancia, 1979). In colony 5 the adult pair, a subadult female, and in colony 6 the adult pair, a 2-year old male and a yearling male were monitored for at least a portion of the colony monitoring period.

Total activity period (τ) for both colonies was nearly 24 h and was partitioned into portions of activity and inactivity (Table 1). In colony 5 activity exceeded inactivity; as a result, colony 5's active period was about 16 percent longer than colony 6's. Since τ during the summer months averaged essentially 24 h, emergence and return times followed similar patterns. However, due to slight daily differences in τ, monthly emergence and return patterns were not identical.

Return times for colony 5 from April to August tended to be about 1 h after sunrise. During September and October return times advanced to slightly before sunrise. Return times for colony 6 were before or only slightly after sunrise and appeared to coincide with sunrise until September when return occurred well before sunrise.

Sunrise, sunset, and other photoperiod and meteorological characteristics were used as independent variables to predict emergence time for 3 individuals each in colonies 5 and 6. Prediction equations accounted for 25.8 to 59.9 percent of the observed variation. In general, variation in response to these characteristics was substantial; however, for most individuals 2 photoperiod variables — length of twilight and time of sunset — were statistically significant ($P < 0.05$) predictors. For 5 of 6 beavers tested, twilight was inversely related to emergence time. Sunset was positively correlated with emergence times. All individuals followed similar 24 h activity patterns, and biomodal peaks of activity in the colony site occurred at evening emergence and dawn return. Activity peaks in the evening tended to be higher in amplitude and longer in duration than corresponding peaks at dawn. Animals spent a greater percentage of the nocturnal period 'out-of-range' of the recording system than active in the colony site, and periods of 'out-of-range' were evenly distributed throughout the night or were concentrated in late night or early morning hours.

Table 1.
Lengths of periods of activity and inactivity within the total activity period for colonies 5 and 6

Colony	Total activity period (h) ($\bar{x}$ ± S.D.)	Activity (h) ($\bar{x}$ ± S.D.)	Inactivity (h) ($\bar{x}$ ± S.D.)
Colony 5	24.0 ± 0.6	12.8 ± 1.0	11.1 ± 1.0
Colony 6	24.0 ± 0.7	11.2 ± 1.3	12.9 ± 1.3

When kits were present at colony 5, the adult female departed from the general pattern described above. Time 'out-of-range' throughout the night was substantially reduced. In addition, inactivity during the night, presumably in the lodge, was higher than for corresponding prepartum periods. The evening peak in activity for the adult male was longer than for periods before parturition, or after kits were large enough to regularly leave the lodge.

DISCUSSION

Individuals in a species population must occupy a similar 'temporal niche' to maintain conspecific integrity. To unify this 'temporal niche', individuals initiate the daily cycle at about the same time. Natural selection operates to adjust onset of activity via the following mechanisms: (1) temporal food availability, (2) predator avoidance, (3) thermoregulation, and (4) behavioral interactions. Food availability can explain initiation of activity at sunset for some insectivorous bats (DeCoursey and DeCoursey, 1964); however, onset of activity for beavers cannot be explained this way because their food (herbaceous material) does not vary in availability throughout the day.

Onset of activity for some lagormorphs also closely coincides with sunset (Mech, Heegen and Siniff, 1966); however, this synchrony is probably a mechanism to avoid predation rather than a result of food availability. Beavers have adopted an alternate strategy — they display an exceptional array of sophisticated behavioral predator avoidance mechanisms, e.g. lodge and pond construction and alarm signals. Selective pressures may have operated to evolve these behavioral activities rather than to adjust the time of onset of activity. Furthermore, it is unlikely that, in summer, initiation of activity is consistently involved in thermoregulation since these animals are homeothermic. Consequently, another explanation for onset of activity is required.

Because colony members are an integrated social group (Hodgdon, 1978), social interactions are the most plausible selective factor affecting emergence times. Mean emergence times for colonies 5 and 6 were nearly the same and followed similar monthly patterns even though colony 5 was active about 1.5 h more per night than colony 6. This suggests time of onset of activity may be a relatively fixed event and may initiate the daily activity cycle. Busher (1975) also found emergence times were 'very consistent' for the 2 colonies he observed and radio tagged.

A behavioral factor may operate to select intercolony synchrony of emergence times based on colony territoriality; but, how this interaction would operate is obscure. Beaver territories are (1) spatially distinct, (2) probably mutually exclusive, and (3) may be passively defined by scent mounds. These factors are thought to reduce intraspecific aggression by decreasing active territorial defense. Consequently, the need for concurrent onset of activity for the population is reduced. Nevertheless, intraspecific strife does occur, although it may not be frequently observed, and does require the presence of the participants at the same time and place. Thus, intercolony behavioral interactions can provide selective pressure toward synchronization of onset of activity for the population as a whole.

For both colonies 5 and 6, photoperiod cues involving twilight and sunset accounted for statistically significant portions of observed variation in emergence time for most individuals. The negative relationship between twilight and emergence suggested emergence would be earliest when twilight was longest, i.e. in mid summer. In addition, the sunset (positive) and twilight (negative) effects were seemingly contradictory. Apparently, the twilight effect was overriding in mid summer when emergence times were earliest and twilight was longest, and the sunset effect was predominant in fall when emergence times nearly coincided with sunset.

Individuals in colonies 5 and 6 partitioned τ into inactive and active portions in different ways. In colony 5 mean active period exceeded mean inactive period whereas the reverse was true for colony 6. Tevis (1950) visually observed one colony in New York state (U.S.A.) from June, to September and determined active and inactive periods to be 13.5 and 10.5 h, respectively. Busher (1975) used telemetry and determined active and rest periods for 3 colonies in California state (U.S.A.) to be 12.5 and 11.5 h. Results of these studies were relatively consistent and were similar to our findings for colony 5. In comparison, the reverse situation where inactivity exceeded activity (colony 6) seems anomalous. Additional data are required to establish mean and extreme period lengths before generalizations can be made.

A reduction in time spent away from the colony site by the adult female in colony 5 indicates a change in activity associated with maternal care of the neonates requiring her continued presence throughout the night. When kits were small, the adult male spent more time in the colony site in the evening than before parturition or after kits became more mobile. Presumably, the adult male was also involved with care of the young and perhaps made more trips to bring food to the lodge in early evening than at other times of the night. His activity through the remainder of the night was similar to prepartum patterns.

Acknowledgements — This is a contribution of the Massachusetts Cooperative Wildlife Research Unit jointly supported by the U.S. Fish and Wildlife Service, the Massachusetts Division of Fisheries and Wildlife, the University of Massachusetts, and the Wildlife Management Institute. Partial funding provided by the Massachusetts Agricultural Experiment Station (McIntyre-Stennis).

REFERENCES

Busher P. E. (1975) Movements and activities of beavers, *Castor canadensis*, on Sagehen Creek, California. M.A. Thesis. San Francisco State University. 86 p.

DeCoursey G. and DeCoursey P. (1964) Adaptive aspects of activity in bats. *Biol. Bull.* 126, 14-27.

Hodgdon H. E. (1978) Behavior of an unexploited beaver (*Castor canadensis*) population. Ph.D. Thesis. University of Massachusetts. 292 p.

Lancia R. A. and Dodge W. E. (1977) A telemetry system for continuously recording lodge use, nocturnal and subnivean activity of beaver (*Castor canadensis*). In F. M. Long (ed.), *Proc. First Internat. Conf. Wildl. Biotelemetry*, pp. 86-92.

Lancia R. A., Brooks R. P. and Fleming M. W. (1978) Ketamine hydrochloride as an immobilant and anesthetic for beaver. *J. Wildl. Mngmt.* 92, 946-948.

Lancia R. A. (1979) Year-long activity patterns of radio-marked beaver (*Castor canadensis*). Ph.D. Thesis. University of Massachusetts. 146 p.

Mech L. D., Heegen K. L. and Siniff D. B. (1966) Onset and cessation of activity in cottontail rabbits and snowshoe hares in relation to sunset and sunrise. *Anim. Behav.* 14, 410-413.

Sunquist M. E. and Montgomery G. G. (1973) Activity patterns and rates of movement of two-toed and three-toed sloths (*Choloepus hoffmanni* and *Bradypus infuscatus*). *J. Mammal.* 54, 946-954.

Tevis L., Jr. (1950) Summer behavior of a family of beavers in New York State. *J. Mammal.* 31, 40-65.

Radio Telemetric Determination of the 24-Hour Feeding Activities of Sea Otters, *Enhydra lutris*

T. R. LOUGHLIN

Office of Marine Mammals and Endangered Species, National Marine Fisheries Service, Washington, D.C. 20235, U.S.A.

Abstract — Traditionally sea otters (*Enhydra lutris*) were considered to be diurnally active but some casual night time observations indicated that some nocturnal feeding may occur. Seven free ranging sea otters were fitted with radio tracking collars to determine their 24 h feeding activities near Monterey, California. The transmitters used were FM blocking oscillators which operated in the 164 MHz band. Sea otters in California rarely leave the water. By using this characteristic, their stereotyped feeding behavior, and the tendency for VHF transmissions to be highly attenuated in seawater, three surface behavior patterns were determined by the strength and consistency of the transmitted signal. These behaviors included inactivity, active but not feeding, and feeding. The results indicated that animals fed as much by night as by day (45 percent of all feeding was at night) and that they typically returned to the same feeding location during both phases. These data help explain how sea otters obtain sufficient food to sustain their uncommonly high metabolic rate.

INTRODUCTION

Early investigations inferred that sea otters (*Enhydra lutris* L.) were diurnal and that they slept through the night after a late afternoon or crepuscular period of intensive feeding (Fisher, 1939, 1940; Kenyon, 1969), but casual observations of four night time dives reported by Miller, Hardwick and Dahlstrom (1975) and others (Bergman, 1927, in Barabash-Nikiforov, 1947; Shimek and Monk, 1977) suggested that sea otters might be active more at night than previously thought. Physiological studies have also shown that sea otters have a high metabolic rate (Morrison, Rosenmann and Estes, 1974), require large amounts of food (up to 25 percent of body weight per day), and have only a small energy store in the form of subcutaneous fat (Kenyon, 1969). Therefore, from considerations of energetics, it seemed improbable that sea otters would be inactive for a 10 to 12 h period during the night (or any other time) without an intervening feeding period.

The object of this study was to confirm whether or not sea otters foraged at night. Recent advances in radio tracking techniques on free ranging marine mammals (e.g. Siniff *et al.*, 1975) allowed specific questions to be asked concerning nocturnal foraging of sea otters. By attaching radio transmitters to the sea otter's neck, it was confirmed that they performed the same basic activities both diurnally and nocturnally and that approximately 45 percent of feeding was at night.

STUDY AREA AND METHODS

The study was conducted in the vicinity of Monterey, California, U.S.A., (36° 36' N, 121° 53' W) between Seaside and Point Joe. The subtidal topography varied from sand to mud to rock with extensive *Macrocystis pyrifera* kelp beds on the rock substrate to depths of 20 m.

The 24 h feeding activity was determined by attaching radio collars to six sea otters (three males and three females). The transmitters used were frequency modulated (FM) blocking oscillators which operated in the 164 MHz band. They typically had a pulse rate of 110 min^{-1} pulse width of 35 ms, and an average current drain of 0.42 mA h^{-1} (Fig. 1). The transmitters and lithium battery package were encapsulated in a waterproof dielectric resin (Stycast 1090, Emerson and Cuming, Inc., or Scotchcast

Fig. 1. The circuit diagram (A), transmitter (B), and encapsulated transmitter with lithium battery package (C) used in this study. The attachment collar is not shown.

XR-5235, 3M Co.) and attached to a rubber collar (1.6 × 0.3 × 58.4 cm) which was placed around the animal's neck. The total package when attached to the animal weighed no more than 95 g (in air). The attachment buckle was held to the rubber collar by No. RW-35 round wire staples which decompose in seawater allowing the collar to fall off after two to three months.

Radio tagged animals were monitored from the shore with a model 1080 Panasonic 3-band receiver and/or a portable, crystal controlled, directional receiver purchased from the Cedar Creek Electronics Laboratory, University of Minnesota. A four-element Yagi antenna was attached to the directional receiver. The animals were located by rotating the antenna towards the strongest signal. By moving the receiver and antenna along the shore in both directions, it was possible to get a reliable fix on the animal to within ± 50 m. Each radio tagged animal was monitored for at least 96 continuous hours and its approximate location determined for four days. Data were taken at least every 15 min for inactive animals and active animals were monitored continuously.

California sea otters, unlike those in Alaska, rarely leave the water and float on their backs in near shore kelp beds during rest periods. One advantage of using VHF transmission on totally aquatic mammals is the ability to deduce different behaviors depending on whether the animal is on the surface or underwater. VHF is highly attenuated in seawater but only slightly attenuated in air (Mackay, 1970; Terhune, 1967). Therefore a transmitter that emits VHF could only be monitored when above the water. It was possible to deduce three surface behavior patterns for sea otters by the strengh and consistency of the transmitted signal. The correlation between the animal's behavior and the transmitted signal (or lack of signal) was confirmed by direct observations.

Rest or inactive: This behavior was characterized by a steady, strong signal but with an occasional break of less than two seconds (due to wave action).
Active but not foraging: A strong to weak signal with intermittent breaks of one to six seconds indicated that the animal was grooming, swimming, playing, etc.
Active and foraging: A strong to weak signal with breaks of 10 s to three min indicated that the animal was diving for food.

The diurnal behavior patterns of animals with rear flipper tags (Loughlin, 1977) but without collars were monitored as controls to assess the influence of the collar on the behavior of telemetered sea otters. Radio tagged animals were allowed about 8-12 h to become accustomed to the collar before recording data.

Sea otters were captured while resting in the kelp with a SCUBA, hand held trap developed by the California Department of Fish and Game (Wild and Ames, 1974). The animals were transferred to a boat where they were weighed, tagged, measured, and fitted with radio tracking collars. They were then released at the capture site after 40 min.

RESULTS

The telemetry study indicated that sea otters had active periods throughout the day, that they foraged at any time of the day, and performed the same activities both diurnally and nocturnally (Fig. 2).

It was characteristic of the 24 h activity pattern that different behaviors occurred in long time segments. For example, no foraging period of less than one hour duration was recorded. Typically an otter was inactive for a few hours, had a short grooming bout, foraged for a few hours, had a longer grooming bout, then was inactive until the next grooming and foraging sequence. On the average, sea otters were active 46 percent of the time, of which 34 percent was spent feeding and 12 percent in other

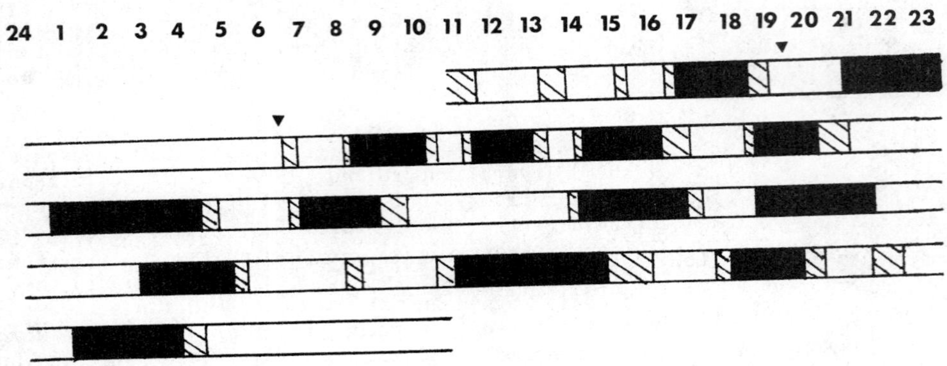

Fig. 2. The 24 h feeding activity pattern for a 19 kg, male sea otter equipped with a radio collar. Data are shown for 96 continuous hours during August 13-17, 1976. The activity pattern shows that the sea otter foraged almost as much by night as by day and that it performed the same activities nocturnally and diurnally. Solid bars = foraging; cross hatching = active but not foraging; open areas = inactivity; closed triangles = dawn and dusk.

activities (Table 1). A significant amount of feeding time (45 percent) occurred at night (Table 1, column D).

There appears to be a general trend for an otter to feed during a particular time of the day. For instance, some showed a nocturnal feeding tendency, some were diurnal, while others seemed to spend as much time foraging by night as by day. Feeding activities were examined by recording the duration of foraging dives, duration of foraging period, and the location of foraging. The durations of foraging dives were recorded to within ± 1 s by monitoring the signals transmitted. Of 2,725 recorded dives, the mean duration from all dives was 57 s. This compares favorably with reports of 66 s (Calkins, 1978), 60 s (Kenyon, 1969), and 50 s (Estes, 1974) for Alaskan otters and to the 55 s and 60-90 s reported by Hall and Schaller (1964) and Limbaugh (1961), respectively, for California sea otters. The longest dive recorded during this study was 4 min 25 s.

Table 1.
The percent time spent in various behaviors during day and night for radio tagged sea otters

Sea otter	Active (A)	Foraging (B)	Active but not foraging (A-B)	Active at night[a] (C)	Foraging at night[b] (D)	Active time foraging (B/A)
$\bar{X}$ for males	43.3	32.8	10.5	20.4	51.2	75.5
$\bar{X}$ for females	50.2	35.9	14.3	39.7	39.2	71.0
$\bar{X}$ total	46.7	34.4	12.4	30.0	45.2	73.3

[a] The percent of time active (A) in the night time, therefore A-C equals percent of time active in daytime.
[b] The percentage of (B) spent in night time; (B)-(D) equals percentage of (B) in daytime.

The mean duration for dives of males was about 9 s less at night than during the day, whereas night dives of females averaged almost 16 s more at night than those during the day. Student's *t*-test showed that the difference between the duration of night dives for males and females was significant at the 0.005 level.

Data were recorded for 69 complete feeding periods of which 37 were diurnal and 32 nocturnal. The mean duration of all feeding bouts was 2 h 29 min. There was a negligible difference between the average duration of diurnal and nocturnal feeding bouts, with the nocturnal ones being 9 min longer. Feeding bouts for males at night tended to be only 14 min longer than those for females. The otters averaged three feeding bouts per day (Loughlin, 1977). Since feeding bouts averaged 2.5 h, 31 percent of the day was spent feeding; this reasonably approximates the 34 percent value of time spent feeding presented above.

The otters were monitored during feeding periods to determine movement patterns to and within the feeding area. Individual otters left the rest area for a feeding site usually located within 0.5 to 2.0 km from the rest area, fed, then returned to the rest area (Fig. 3). They usually fed in, and adjacent to, *Macrocystis* kelp beds but some otters preferred to feed in sandy areas. Feeding typically occurred from shore to 200-300 m offshore and in water from 1 to 20 m deep. They fed in the same approximate location both night and day.

DISCUSSION

Numerous studies have dealt with sea otter feeding in one way or another, and for further discussion of prey items, feeding behavior, etc., the reader is referred to these additional references: Ebert, 1968; Estes and Palmisano, 1975; Lensink, 1962; Lowry and Pearse, 1973; Vandevere, 1969.

As stated in the Introduction, casual night time observations have indicated that nocturnal feeding of otters may be more common than previously thought. Nocturnal feeding is also expected from an analysis of the energetics and thermoregulation of sea otters. Morrison *et al.* (1974) found in laboratory experiments, that Alaskan sea otters exhibited a basal metabolic level (M_b) more than 2.5 times the standard (average basal) level for a mammal of its size. This elevated level conforms to that seen in other marine mammals (pinnipeds and cetaceans), but the reason for this has evaded explanation (although it may pertain to thermoregulatory functions).

Concomitant with this elevated M_b in sea otters is a voracious appetite and food consumption of up to 25 percent of the body weight per day (Kenyon, 1969). This high level of food consumption is probably a result of the elevated M_b with its increased demands for energy. For California sea otters, these energetic demands are met by numerous diurnal and nocturnal feeding periods. The animals fed an average of 7.5 h each day, spreading this time out into three feeding bouts of 2.5 h duration, each occurring at varying times of the day and night.

The ingestion of food increases the metabolic rate, and in man the metabolic rate is 10 to 20 percent higher after eating. This effect of food on metabolic rate is known as specific dynamic action (SDA). Protein gives the greatest effect, carbohydrate and fat much less (Vander, Sherman and Luciano, 1975). The observations of the 24 h feeding activity in sea otters, as presented in this study, may explain the query posed by Morrison *et al.* (1974) regarding sea otter's SDA. It was their belief that the observed M_b of 2.5 times the standard level could be explained if the SDA were spread throughout the day, or "... that the SDA is not expended in a metabolic 'bulge' following a meal as in other carnivores but is rationed out over the day." They continue, "We have no information as to how this might be affected. Delayed absorption from the gut would meet the requirements but is hardly compatible with observations of the passing of food, including undigested flesh, within 3 hr."

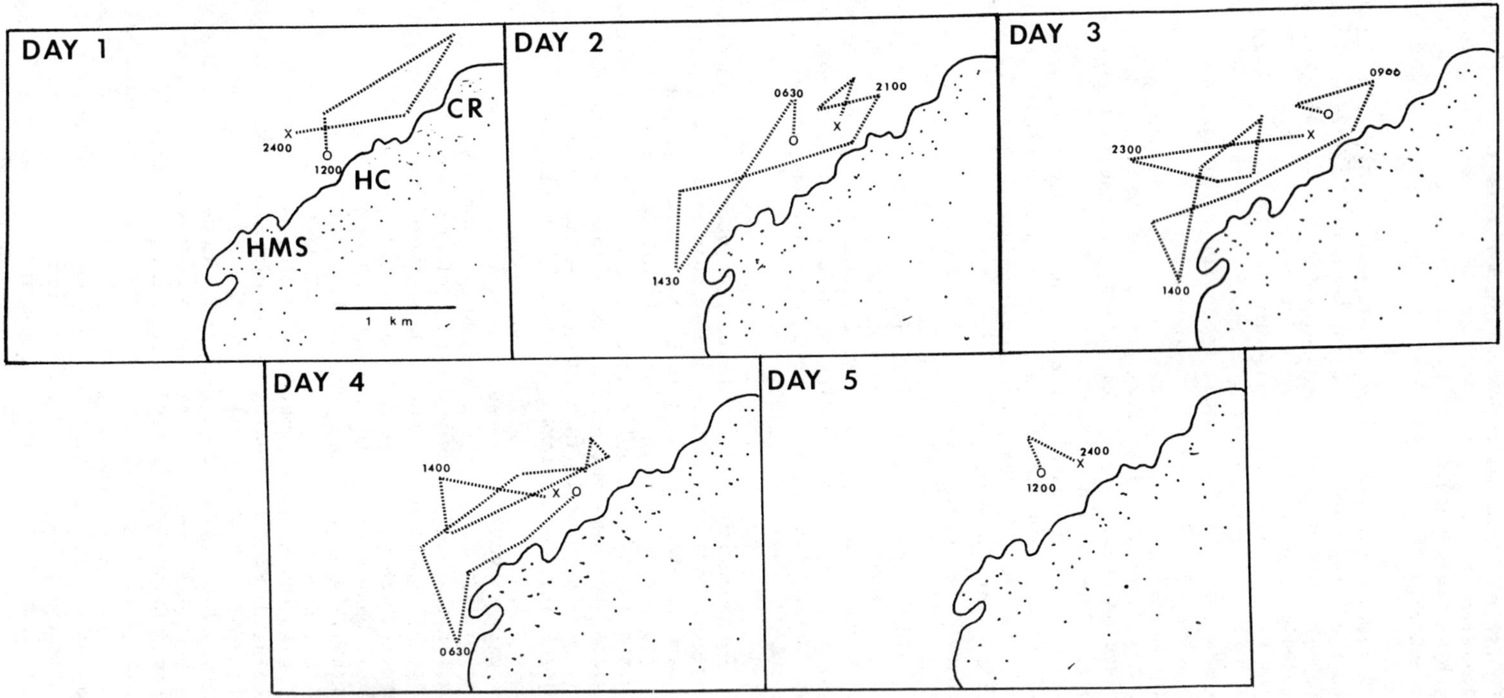

Fig. 3. A typical movement pattern for a male sea otter. The sequence begins at 1200 h of Day 1 at a rest area near Hopkins Marine Station (HMS) and ends at 1200 h of Day 5 back in the same location. The pattern shows that this animal repeatedly returned to the same area off Hopkins Marine Station to feed and then returned to the rest location. Hovden Cannery (HC) and China Row (CR) are two easily identifiable local landmarks near Monterey, California, U.S.A. O = Start; X = End of sequence.

One explanation for the requirement of the SDA to be expended throughout the day, rather than in a metabolic 'bulge', is to eat at frequent intervals throughout the day. This may be accomplished in sea otters by numerous, long, feeding periods during the day and night.

In summary, the results from this study confirm that California sea otters spend a substantial amount of time feeding at night and during the day. These findings help explain how sea otters obtain large amounts of food required to sustain their high basal metabolic level.

Acknowledgements — I am grateful for the generous field assistance of B. Shade, A. Williams, M. Flippo, E. Byrant, M. Reidman, D. Costa, J. Vandevere, J. Ames and R. Jameson. The staff at the California Department of Fish and Game, Monterey, and Hopkins Marine Station of Stanford University gave valuable logistical support. The study was financed by contract numbers MM6AC004 and MM6AC024 from the U.S. Marine Mammal Commission.

REFERENCES

Barabash-Nikiforov I. I. (1947) *The Sea Otter* (translated from the Russian by Israel Program for Scientific Translations, Jerusalem, 1962). 227 p.

Calkins D. G. (1978) Feeding behavior and major prey species of the sea otter, *Enhydra lutris*, in Montague Strait, Prince William Sound, Alaska. *Fish. Bull.* 76, 125-131.

Costa D. P. (1978) The ecological energetics, water and electrolyte balance of the California sea otter, *Enhydra lutris*. Ph.D. thesis, University of California, Santa Cruz. 75 p.

Ebert E. (1968) A food habits study of the southern sea otter, *Enhydra lutris nereis*. *Calif. Fish Game* 54, 33-42.

Estes J. A. (1974) Population numbers, feeding behavior and the ecological importance of sea otters in the western Aleutian Islands, Alaska. Ph.D. thesis, University of Arizona. 125 p.

Estes J. A. and Palmisano J. F. (1975) Sea otters: their role in structuring near-shore communities. *Science* 185, 1058-1060.

Fisher E. M. (1939) Habits of the southern sea otter. *J. Mammal.* 20, 21-33.

Fisher E. M. (1940) The sea otter in California. *Sixth Pac. Science Congress*, 231-240.

Hall K. R. L. and Schaller G. B. (1964) Tool using behavior of the California sea otter. *J. Mammal.* 45, 242-252.

Kenyon K. W. (1969) The sea otter in the eastern Pacific Ocean. U.S. Dept. of Interior, Bureau of Sport Fisheries and Wild., North Amer. Fauna No. 68. 352 p.

Lensink C. J. (1962) The history and status to sea otters in Alaska. Ph.D. thesis, Purdue University. 188 p.

Limbaugh C. (1961) Observations on the California sea otter. *J. Mammal.* 271-273.

Loughlin T. R. (1977) Activity patterns, habitat partitioning, and grooming behavior of the sea otter, *Enhydra lutris*, in California. Ph.D. thesis, University of California, Los Angeles. 110 p.

Lowry L. and Pearse J. (1973) Abalones and sea urchins in an area inhabited by sea otters. *Marine Biol.* 23, 213-219.

Mackay R. S. (1970) *Bio-medical Telemetry*. New York: John Wiley. 533 p.

Miller D. J., Hardwick J. E. and Dahlstrom W. A. (1975) Pismo clams and sea otters. Calif. Dept. Fish and Game, Mar. Resources Tech. Report No. 31. 49 p.

Morrison P., Rosenmann M. and Estes J. A. (1974) Metabolism and thermoregulation in the sea otter. *Physiol. Zool.* 47, 218-229.

Shimek S. J. and Monk A. (1977) The daily activity of the sea otter off the Monterey Peninsula, California. *J. Wildl. Mngmt.* 43, 277-283.

Siniff D. B., Reichle R., Hofman R. and Kuehn D. (1975) Movements of Weddell seals in McMurdo Sound, Antarctica, as monitored by telemetry. *Rapp. P. -v. Reun. Cons. int. Explor. Mer.* 169, 38-393.
Terhune J. M. (1967) Audibility aspects of sonic tracking marine mammals. *J. Mammal.* 571, 179-180.
Vander A. J., Sherman J. H. and Luciano D. S. (1975) *Human Physiology — The Mechanisms of Body Function.* New York: McGraw-Hill. 412 p.
Vandevere J. E. (1969) Feeding behavior of the southern sea otter. In T. Poulter (ed.), *Sixth Ann. Conf. on Biolo. Sonar and Diving Mammal*, pp. 87-90. Stanford Res. Inst., Menlo Park, California.
Wild P. W. and Ames J. A. (1974) A report on the sea otter, *Enhydra lutris* in California. Calif. Dept. Fish and Game, Mar. Resources Tech. Report No. 20. 93 p.

Home Range Concept and Activity Patterns of the European Wild Boar (*Sus scrofa* L.) as Determined by Radio Tracking

R. MAUGET

Centre d'Etudes Biologiques des Animaux Sauvages, 79360 Beauvoir-sur-Niort, France

Abstract — Data were obtained on 8 radio equipped wild boar monitored by an automatic radio tracking system in the Chizé Forest, France. Examples of data processed by computer are presented. It is possible to go beyond the conventional home range determination in terms of size and shape. A dynamic home range concept may emphasize the occupation of the biotope. Analysis of data involved the quantification of the following home range parameters — *maximum home range size* (240-425 ha for females, 700 ha or more for males) — *total dimensions of frequented areas* (203 to 342 ha) — *feeding areas* during nocturnal activity (105 to 203 ha) — *activity volumes* (related to daily activity lengths) — *occupation indices* (volume/square unit). This latter parameter, embracing both activity length and the size of activity areas, may be an index of food availability. Activity rhythm was also investigated. Boars exhibit a typically nocturnal activity. The onset of activity is correlated with sunset.

It has been stated that the home range is the area *occupied* by an animal, or a group, during a given period (S.F.E.C.A., 1969). Therefore, the home range must be studied in terms of its occupation characteristics and not only in terms of its shape and size. The fixed, continuous, radio tracking system developed in the Chizé Forest (Midwestern France) may objectively emphasize this qualitative and quantitative utilization of the natural environment. In a previous study (Mauget, 1979) we determined the spatial organization of a natural population of wild boar in the Chizé National Forest. In this Reserve, groups (mainly adult sows with their litters) are established in ranges, the areas of which vary from 100 to 400 ha. Home range of wild boar or wild hog, in open areas, cover from 100 to 2000 ha (Lewis, 1966; Kurz and Marchinton, 1972; Pine and Gerdes, 1973; Martin, 1975). It is difficult to compare the numerical parameter 'home range extent', the values of which may vary with the different methods used.

During 4 years, we have studied 8 radio equipped wild boar. The radio tracking procedures, the equipment employed and the automatic data processing are discussed (Deat *et al.*, 1980, this volume). A large amount of information was obtained (1184 days recorded, individual animals studied for periods ranging from 2 to 6 months). In this paper, we are only presenting examples of data processing illustrating the utilization of space and the activity rhythm of wild boar.

Utilization of space may be approached in several ways. Home ranges, as determined by a line connecting outermost locations, makes the spatial distribution of animals conspicuous (Fig. 1a). The ranges of the 8 tracked boars overlap. The areas within each boundary vary from 240 to 425 ha for the females, reaching 700 ha for an adult male.

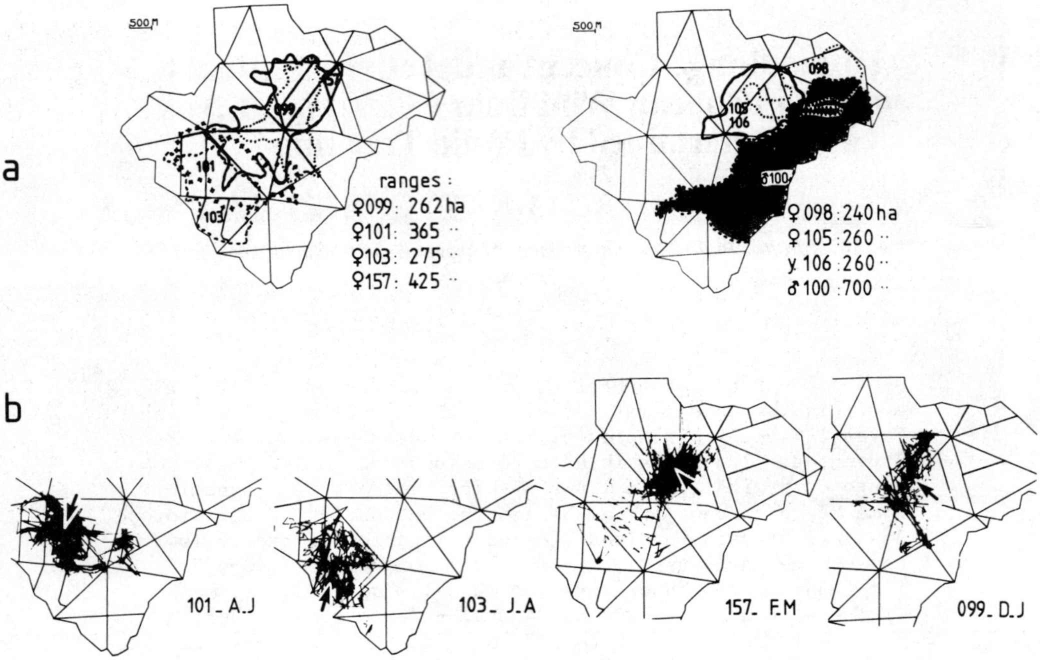

Fig. 1. Home ranges of 8 radio tracked wild boar in the forest (thin lines represent the road system). (a) Boundaries of overall home ranges; (b) superposition of daily trackograms (arrows = resting areas). A-J: April-June, J-A: July-August, F-M: February-March, D-J: December-January.

Display of home ranges determined by the superposition of daily 'trackograms' during periods of stability (2 months presented here) gives a detailed impression of the boar's spatial distribution (Fig. 1b). In every range, a heterogeneity appears in the pattern of use. Areas frequented more or less often are shown. A color discrimination of day and night locations allows the distinction of resting areas (arrows on figure).

The 3 dimensional representation of home ranges (Fig. 2) is based on a squaring of 100 × 100 m in XY coordinates and frequency of locations inside each square in the Z coordinate. Peaks corresponding to the major activity areas (feeding) and bedding areas appear. In the data gathered at night, bedding peaks are suppressed. Numerical activity parameters can be obtained from this representation: whole of frequented areas (203 to 342 ha), size of activity areas (105 to 203 ha), volume of activity (duration of nocturnal activity) and *occupation index:* volume/square unit. This last parameter is used to characterize the home range utilization. It could be related to food availability but further statistical analyses are needed.

Activity rhythm is typically biphasic: rest during daylight, movements during the night. The onset of activity was correlated with sunset (Fig. 3). There appeared to be seasonal variations. During fall and winter, the activity began around sunset while, during spring and summer, it always started before sunset. The delay sometimes

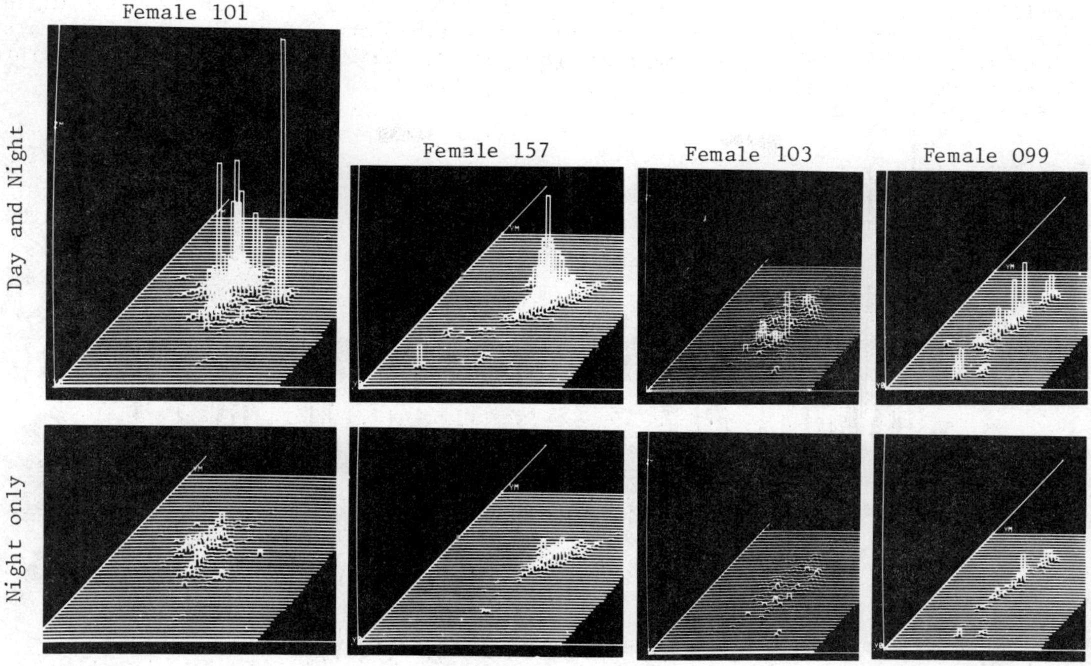

Fig. 2. 3-dimensional analysis of home range.

Numerical parameters

	Female 101	Female 157	Female 103	Female 099	
Total number of 1 ha squares	342	284	262	203	Whole of frequented areas
Total squares during night	283	218	244	148	Activity and traveling areas
No. of squares frequency = 1	80	57	64	43	Traveling areas
No. of squares frequency ≥ 2	203	161	180	105	Activity (feeding areas)
Area of daily activity	127.3	198.4	106.2	223.6	Length of daily activity
Occupation index = area/activity square unit	0.63	1.23	0.59	2.13	Food availability index

reached 100 min. This could be due to the fact that during the day, boars rest in dense cover, where the luminosity decreases far before sunset. This biphasic activity rhythm is altered during the nursing period when it becomes polyphasic.

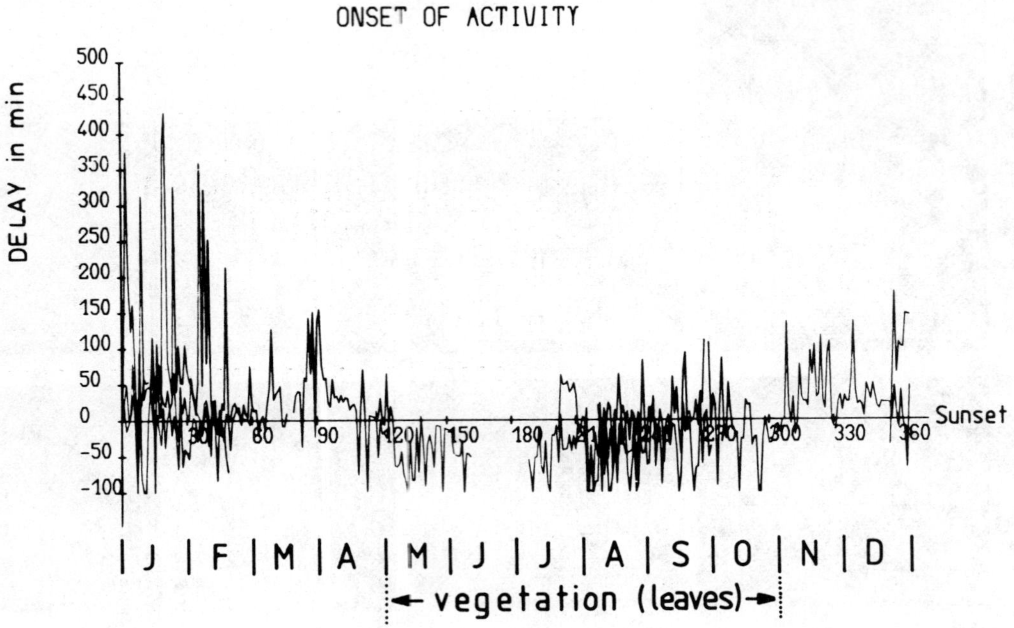

Fig. 3. Onset of activity correlated with sunset.

REFERENCES

Déat A., Mauget C., Mauget R., Maurel D. and Sempéré A. (1980) The automatic continuous and fixed radio tracking system of the Chizé Forest: theoretical and practical analysis. In C. J. Amlaner, Jr. and D. W. Macdonald (eds.), *A Handbook on Biotelemetry and Radio Tracking*, pp. 439-451. Oxford: Pergamon Press.

Kurz J. C. and Marchinton R. L. (1972) Radiotelemetry studies of feral hogs in South Carolina. *J. Wildl. Mngmt.* 36, 1240-1248.

Lewis J. C. (1966) Observations of pen-reared European hogs released for stocking. *J. Wildl. Mngmt.* 30, 832-835.

Martin J. T. (1975) Movement of feral pigs in North Canterbury, New Zealand. *J. Mammal.* 56, 914-915.

Mauget R. (1979) Mise en évidence, par captures-recaptures et radiotracking du domaine vital chez le Sanglier (*Sus scrofa* L.). *Biol. Behav.* 4, 25-41.

Pine D. S. and Gerdes G. L. (1973) Wild pigs in Monterey County, California. *Calif. Fish Game* 59, 126-137.

S.F.E.C.A. (1969) Proc. of meeting, *cit.* Le comportement territorial chez les vertébrés, In P. B. Richard (ed.), *Territoire et Domaine Vital, Entretiens de Chizé*, pp. 1-19, Vol. 1. Paris: Masson.

The Use of Radio Tracking in an Ecological Study of the Lesser Flamingo in the Eastern Rift Valley

C. H. TUITE

The Wildfowl Trust, Slimbridge, Gloucester, U.K.

Abstract — Radio tracking was used to study diurnal activity of free living lesser flamingoes in East Africa. Data were also obtained on daily range and migratory behavior. The method of packaging and attaching the transmitter, battery life duration and reception range are given and the advantage of using telemetry over other behavioral sampling methods is discussed. The results of time budget analyses are given and mention made of their use in developing energy budget models to examine the energetics of diet selection, ranging and breeding initiation.

INTRODUCTION

Many recent ecological studies of birds have been concerned with developing energy budgets for populations or individuals. Such an approach may be used to examine a number of fundamental questions, such as to what extent is energy availability related to demographic variables (breeding success, mortality, dispersion) and of what significance are populations or communities in the trophić structure of ecosystems (Kushlan, 1977; Furness, 1978). At the level of the individual, energetic considerations may provide information relevant to questions about feeding, ranging and breeding strategies and examples and further discussion of this energy optimization approach are given in Ake Norberg (1977) and references therein.

The Eastern Rift Valley of Kenya and northern Tanzania forms an important part of the range of the lesser flamingo (*Phoeniconaias minor*, Geoffroy). These birds depend on a series of closed basin alkaline saline lakes, which have formed along the Rift Valley floor and lie along an approximately north-south line stretching for about 1000 miles. Lesser flamingoes are algivorous filter feeders, the diet being made up almost entirely of the planktonic blue green alga *Spirulina platensis* and benthic diatoms (Jenkin, 1957; Tuite, 1978). At any time, birds can only exploit one or other of these two food resources as they grow in separate parts of the lake systems so that both the feeding site and behavior appropriate for grazing them are different (Tuite, 1978).

Because of the limited diet diversity and the filter feeding habit, the energy obtained from feeding and food availability are unusually simple to measure and model. I wanted to use energetic models to examine the following questions: when does the diet change from *Spirulina* to diatoms and why, what are the energetic constraints on ranging between the different lakes and is energy availability likely to be a significant factor limiting breeding? To develop energetic models relevant to these problems it is necessary to determine the metabolic cost of free existence (E_{fe}). One method of estimating E_{fe} without using direct physiological telemetry of metabolism is by measuring the amount of time individuals engage in activities with different metabolic costs (see also Schwartz and Zimmerman, 1971; Kushlan, 1977; Furness, 1978; Tuite, 1978; for further discussion and examples of this method).

RADIO TRACKING EQUIPMENT AND METHODS

The transmitters were model SB2 as supplied by AVM Instrument Co., each powered by a lithium 2.7 V battery, giving a rated life of 6 months-1 year and transmitting between 148.350-148.625 MHz. The antenna was a whip type, about 25 cm in length, made from steel guitar string. Particular care was taken to make the radio pack as waterproof as possible because of the corrosive properties of the extremely alkaline saline water of the lakes which flamingoes inhabit. All the soldered connections were covered with beeswax and epoxy resin (Rapid Araldite). The antenna was also protected by application of a thin coating of epoxy. The battery and transmitter were then wrapped in plastic insulating tape and finally packaged in plastic impregnated cloth material. The seams were sewn with nylon thread and sealed with epoxy. The total mass of the package was 70-90 g, which represents ca. 5-8 percent of the total body mass of an adult flamingo.

The transmitter was attached to birds as a back pack with the antenna protruding posteriorly. It was held in place by two pairs of straps (ca. 1 cm in width) made from the plastic impregnated cloth material. The anterior pair were passed round the breast/base of the neck and joined. The second pair from the back of the package ran behind the wing attachment and joined the anterior straps laterally (see Fig. 1). The joints were sewn with nylon thread and glued with epoxy while the bird was being held. This configuration allowed completely free movement of the wings and did not appear to impair flying ability.

The receiver was an AVM model LA12 (12 channel) and was used with a hand held Yagi directional antenna. The reception range varied with the height of the antenna above the transmitter and ranged from ca. 1 km (antenna hand held by observer on the ground) to ca. 30 km (antenna mounted on aircraft). By use of the directional antenna and the signal reception strength meter, it was possible to locate the position of a bird to within 50 m. The radio pack could then be seen with binoculars or a telescope and the bird observed.

RESULTS AND DISCUSSION

Three birds were radio tracked between January and May, 1976 at Lake Nakuru, Kenya. Daily presence or absence at the lake was recorded and diurnal activity of one of the three birds was monitored continuously on five separate observation days. Data were also obtained on the range of lake shore used by individuals.

The most important results were from the time budget analyses, the results of which are summarized in Table 1. These figures were incorporated into metabolic equations to estimate the mean rate of free existence metabolism.

Fig. 1. Diagram to show harness configuration of radio package used on lesser flamingoes. (Drawn by Debbie Tuite).

Table 1.
The mean proportion of 12-hour daytime periods during which radio tagged birds were engaged in different activities

Activity	Mean percentage of time	S.E.
Feed	60.6	± 3.8
Preen	5.1	± 1.2
Rest/stand	24.4	± 4.0
Fly	0.2	± 0.04
Other	9.7	± 4.2

REFERENCES

Ake Norberg R. (1977) An ecological theory on foraging time and energetics and choice of optimal food searching method. *J. Anim. Ecol.* 46, 511-529.

Furness R. W. (1978) Energy requirements of seabird communities: a bioenergetics model. *J. Anim. Ecol.* 47, 39-53.

Jenkin P. M. (1957) The filter feeding and food of flamingos. *Phil. Trans. R. Soc.* (*B*) 240, 401-493.

Kushlan J. A. (1977) Population energetics of the american white ibis. *Auk* 94, 114-122.

Schwartz R. L. and Zimmerman J. L. (1971) The time and energy budget of the male dicksissel. *Condor* 73, 65-76.

Tuite C. H. (1978) The lesser flamingo (*Phoeniconaias minor*, Geoffroy): aspects of its ecology and behaviour in the Eastern Rift Valley of Kenya and northern Tanzania. Ph.D. thesis, Univ. of Bristol.

Radio Tracking in Australia

S. WHITEHOUSE

Agriculture Protection Board of Western Australia, Jarrah Road, South Perth, Western Australia 6151

Abstract — A research project to study the movements and patterns of land use of dingoes (*Canis* (domestic)) in arid and semiarid areas of Western Australia was begun in 1970. The study was faced with an elusive carnivore living in rough and inaccessible terrain, a research team of only one person and a limited budget. The solution of radio tracking dingoes from an aircraft provided frequent accurate locations with minimal disturbance to the animal. After the initial capture and fitting of the animals with radio transmitters, labor costs were minimal. Aerial radio tracking also meant that animals were not lost when they moved out of range of fixed ground towers.

RADIO TRACKING DINGOES FROM AIRCRAFT

Wherever man and any of the wild canids have come into contact, there has frequently been a severe conflict of interests between the predatory habits of the carnivorous canids and man's interests in raising domestic stock. This conflict has occurred to varying degrees with all members of the genus *Canis* and the dingo — *Canis* (domestic) (Clutton-Brock, Corbet and Hills, 1976) — is no exception.

In Western Australia, dingo populations have been eradicated from the more densely settled agricultural areas and are currently restricted to the rangelands. Here, dingoes have traditionally been reputed to inflict severe economic losses on sheep flocks and cattle herds — particularly on sheep. Dingoes have been viewed by property managers as both predators of sheep and as also thought to contribute additional economic losses by causing mismothering of lambs, reduction in weight gain and reduction in wool yield.

Until recently (Coman, 1972; Corbett and Newsome, 1975; Green and Catling, 1977; Newsome *et al.*, 1973; Shields, 1972; Whitehouse, 1977a,b), there has been little factual information on dingo biology and attempts by Government and landholders to manage dingo populations and thereby reduce dingo-induced economic losses, have been largely based upon folklore. In an effort to define the true nature of the impact of dingoes upon rangeland domestic animal husbandry in Western Australia (especially sheep husbandry), the State Government initiated a research program in 1970. One of the projects undertaken was a study of dingo movements and patterns of land usage.

Dingoes are a medium sized, cursorial carnivore. They avoid the high temperatures of the day and tend to be most active during late afternoon, night and early morning. The rangeland areas, where the study was to be undertaken, are arid or semiarid (mean annual rainfall between 125 and 300 mm); extremely hot (daily summer maxima in excess of 40°C are common); sparsely populated (the population of a single small English village spread over an area larger than the United Kingdom); and the terrain is extremely rugged.

The combination of the elusive nature of the dingo and the difficult terrain in which to work, compounded by limited staff and other resources at the disposal of the project, meant that the study techniques were restricted.

At the time the study commenced, the State Government of Western Australia employed more than 40 field staff as full time dingo trappers (called 'doggers'). A few of these men were utilized in the study, to catch dingo pups alive at nest sites, ear tag them and then release them unharmed. More than 100 pups were marked in this way and ear tags were returned when dingoes were lated killed by doggers or station personnel (Whitehouse, 1977b). A technique for live trapping adult dingoes unharmed was also perfected and some twenty adults were captured, ear tagged and released.

However, the disadvantages of a mark, recapture program soon became apparent. The only information being gathered was on release and recapture points. There was no knowledge of the actual distance travelled by the animal, the preferred routes of travel or any other patterns of land use. Radio tagging individual dingoes and obtaining repeated locations was the obvious answer.

As described earlier, the terrain, resources available and the nature of the animal being observed placed severe constraints upon the type of radio tracking that could be undertaken. Tracking from vehicles was not possible as few vehicle roads gave limited access to the study area generally and particularly to hills. This would have severely reduced the usable range of the tracking system and the accuracy of the locations obtained.

The establishing of fixed receiving towers also presented problems. Apart from the capital costs involved, each tower would need to be 'manhandled' to the top of a hill, erected and then staffed with at least two people when operating. At least two towers would be needed and this would still leave the strong possibility of the animal moving out of range resulting in the need for more towers. Preliminary data from ear tagged dingoes already suggested movements in excess of 20 km may be common, and yet the ground to ground range of the transmitters was only in the vicinity of 5 to 10 km. Obviously the cost of erecting and staffing sufficient towers to cover several dingo home ranges would be prohibitive.

The techniques available at that time for radio tracking from aircraft had been developed in the United States. A Yagi antenna was either taped underneath the wing of a high wing monoplane aircraft or else held out of a window by hand (Mech, 1974). Both of these techniques are prohibited by the civil aviation authorities in Australia. These authorities also require that any large external fixtures must be fastened securely to the airframe and not merely to the skin. For a Yagi antenna, this would require extensive internal modifications to the wings and involve considerable cost.

It was decided to develop an end fire monopole array. The technical data have been fully described elsewhere (Whitehouse and Steven, 1977), and so only a brief outline will be given here. The array consists of four elements each one quarter physical wavelength long and spaced one quarter physical wavelength apart. Coaxial cable of $\frac{3}{4}$ *electrical* wavelength connects each element and cable from the outboard element is then fed into the aircraft cabin (Figs. 1 and 2). A four element array is on each wing and the arrays are more sensitive to signals to the side of the aircraft rather than fore or aft (Fig. 3).

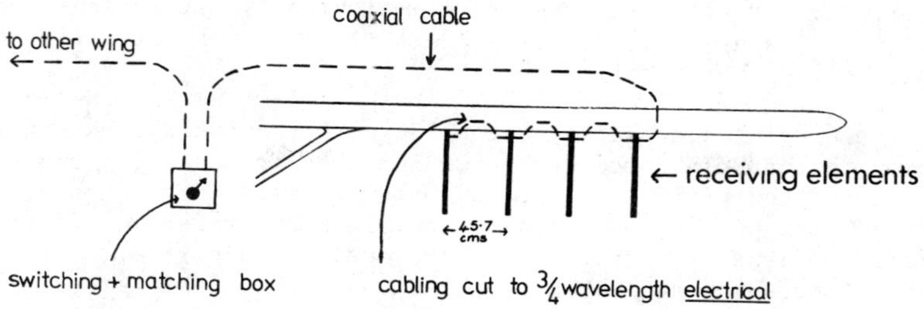

Fig. 1. Schematic diagram of end-fire monopole array showing positioning on aircraft (from Whitehouse and Steven, 1977).

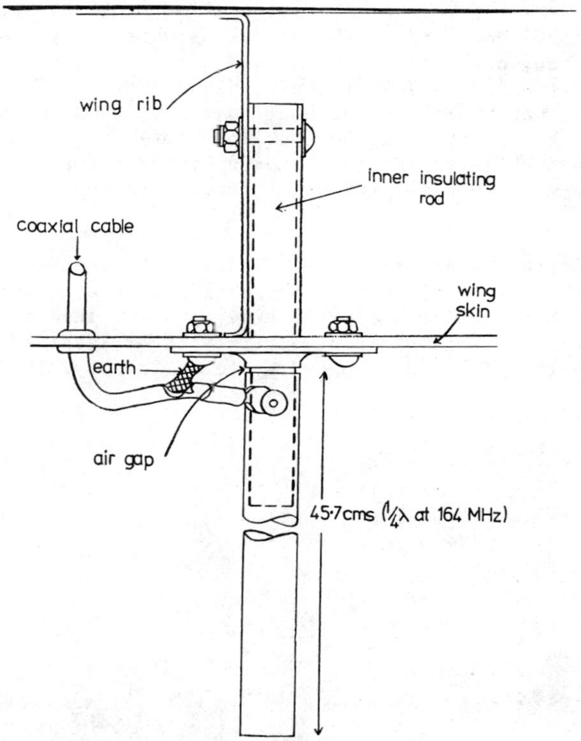

Fig. 2. Detailed representation of an individual element in end-fire monopole array showing method of fixing onto wing rib (from Whitehouse and Steven, 1977).

Fortuitously, at the chosen frequency, 164 MHz, the physical distance required between each element (45.7 cm) corresponded exactly with the spacing between wing ribs on the Cessna 172 being used. This meant that no expensive modifications to the internal wing structure were required as each element could be attached directly to a wing rib. Total costs in 1975 for installation of an array on each wing, plus wiring to the cabin, was under $200 Aus. In late 1978, the fabrication and attachment of two new arrays cost $720 Aus. The effect on aircraft performance was negligible and there was no difficulty in obtaining approval from the civil aviation authority.

The technique of using the system has also been fully described elsewhere (Whitehouse and Steven, 1977). Briefly, it consists of matching signal strength being received by the array under each wing, orienting the aircraft in the direction of the transmitter and following the signal to the source.

Night radio tracking uses a similar technique with the added constraint that a greater minimum ground clearance must be maintained for safety purposes. This, combined with the reduced visibility, means that night time accuracy may vary between ± 20 m and ± 200 m depending on weather conditions and terrain. If the animal is beside a prominent landmark and the sky is cloudless, accurate locations may be obtained, but if no such landmark is available and the night is overcast, larger errors may occur. In this study, night radio locations are used to supplement the intensive day tracking and confirm the general patterns which are emerging.

The system described above, has now been in field use in the north of Western Australia radio tracking dingoes for over three years. Most of the radio locating has been done during daylight, apart from the week spanning the full moon when night tracking is also carried out. Apart from the pilot, the only staff needed is a single radio operator. However, on the occasions when locations are obtained on a 24 h basis over a 7 day period, two pilots and two radio operators are used working in shifts. Currently a Cessna 172 aircraft, plus pilot is hired from a commercial company for $45 Aus. per hour. In one hour up to twelve animals can be located spread over the 650 km^2 study area, however, the rate of obtaining locations depends upon many variables such as the distance moved by the animal since it was last located. These costs compare more than favorably with the costs involved in a ground tracking system.

Aerial radio tracking has other advantages besides the reduced costs. The accuracy of daytime locations is commonly of the order of ± 20 m and, in practice, the resolution of the available maps of the study area has been the final determinant of overall system accuracy. This order of accuracy is at least as good, if not better, than the values reported for ground based systems (Yerbury, 1977).

Fig. 3. End-fire monopole array mounted on Cessna 172 aircraft (from Whitehouse and Steven, 1977).

The system also gives the option with medium or large animals of confirming the location by direct observation and simultaneously, additional data on behavior can frequently be recorded. With dingoes it has been possible to observe group social behavior and many kills of prey. These have been subsequently located by ground teams and additional data recorded.

The aerial radio tracking system described here is currently being expanded to track red kangaroos (*Megaleia rufa*) (Oliver, personal communication). A previous radio tracking study of red kangaroo movements using a ground based system was carried out in the summer of 1968/69. Two towers were constructed on small hills with the receiving antennas 3.7 km apart. The maximum range at which signal strength was sufficient for obtaining an accurate bearing was 24 km. However, later analysis showed that at a range of greater than 11 km, the errors involved were too large to allow the data points to be used. It was also found that the technical competence of the operators and hence their accuracy, varied greatly. For 24 h radio locating, over a three-week period, a staff of seven people were needed. The maximum number of transmitters being tracked at any one time was 11.

However, using the aerial radio tracking system developed for the dingo study, 24 kangaroos have been instrumented and are currently being tracked. We found the animals as often from the air as we had done previously with the ground system, and accurate locations could be obtained wherever the animals were located as they are always within range. Variability between the results obtained by different operators is minimal (Whitehouse and Steven, 1977). Locations can be confirmed by direct observation when necessary and additional behavioral data obtained. Staff and capital costs are also greatly reduced.

The aerial radio tracking system has not been without its problems. As mentioned above, the chosen frequency of 164 MHz had a large advantage when considering antenna construction and attachment on light aircraft in Australia. Also, transmitters in this frequency range were available as 'off-the-shelf' items from the U.S.A. (AVM Instrument Co., Champaign, Illinois). However, the frequency was also used for a radio-telephone link by a nearby iron ore mining company, and on occasions their use of the frequency caused interference. Some interference was also received on summer afternoons and evenings, caused by tropical electrical storms.

Ghost signals were occasionally received. These occur when signals from the transmitters are reflected off a solid object (such as the 'iron stone' hills) received by the antenna, and then apparent location of the transmitter (by radio tracking) is very different from the true position. However, because the technique involves following the signal to its source, ghost signals are easily distinguished and are never a serious problem.

Dingoes sometimes avoid extreme daytime temperatures by entering caves and this may cause severe attenuation of the radio signal. On some occasions, signals have abruptly appeared and disappeared and ground investigation has shown the presence of deep caves. Fortunately, complete disappearance of the signal is rare and the more frequent mild attenuation has just lengthened the searching time involved.

Navigation and plotting the locations has caused a few problems. Available maps of the study area are not accurate and aerial photographs have to be used. Also, navigation at night was difficult, especially during cloudy periods, as there are no aircraft navigation aids.

BRIEF NOTES ON SOME AUSTRALIAN WILDLIFE STUDIES WHICH HAVE USED RADIO TELEMETRY

Radio tracking was first used in wildlife studies in Australia in the mid 1960s. Fullagar (1967) has reported on the use of radio telemetry in Australian biological research during 1966. He reports some 18 projects on 17 different species ranging from elephant seals (*Mirounga leonna*) to ants (*Iridomyrex detectus*) and included in the swamp tortoise (*Pseudemydura umbina*), the echidna (*Tachyglossus aculeatus*) and the wedge-tailed eagle (*Aquila audax*). Of the 18 projects, 13 involved movement studies, seven were obtaining physiological information and two were measuring microclimate. At that time no results had been published.

Since then, radio tracking has become widely used in wildlife studies. Its most common use is still in movement studies and some of the published results have been on the echidna (Augee, Ealey and Price, 1975), the salt water crocodile (*Crocodylus porosus*) (Webb and Messel, 1978), the common wombat (*Vombatus ursinus*) (McIlroy, 1976), the iguana (*Varanus gouldii*) (Green and King, 1978) and woylies (*Bettongia penicillata*), and tammar wallabies (*Macropus eugenii*) (Christensen, 1978). Physiological data have also been collected using radio telemetry. The species studied include the echidna (Augee, Ealey and Spencer, 1970) and the iguana – *Varanus gouldii; Varanus varius* (Green and King, 1978).

Transmitters used in Australian wildlife studies have most frequently operated in the 27 MHz range using the original design of Cochran and Lord (1963) or variations of this design (Tester, Warner and Cochran, 1964). The 97 MHz band was used for a study of red kangaroo (*Megaleia rufa*) movements (Oliver, personal communication), the 151 MHz band for the woylies and tammars, the 164 MHz band in the dingo study reported here and the 1.25 GHz range is being used in the study of salt water crocodiles (*Crocodylus porosus*) in northern Australia.

The transmitters in the 97 MHz and 1.25 GHz ranges were specially developed for the studies being undertaken. The 151 and 164 MHz transmitters were available from commercial manufacturers in U.S.A.

Power sources have varied. Many of the studies undertaken using the 27 MHz equipment have been carried out by graduate students at universities with limited funds. These studies have used alkaline or mercury cells and the life of the transmitters have been variable, but comparatively short. More recent studies have had the advantage of improved technology (and often larger budgets). Lithium cells have been used in the dingo study referred to here. The cells supply 2.7 V to transmitters with an average current drain of between 0.28 mA and 0.44 mA (0.40 mA is the modal value). Each transmitter has a characteristic pulse rate between 51 and 99 pulses m^{-1}, and is on a separate frequency 100 KHz apart. Transmitter lifetimes in excess of 18 months have been achieved in the field.

The 1.25 GHz system developed for the crocodile study uses rechargeable batteries and solar panels. The batteries powering the transmitters by themselves given the constraints of transmitter size and range, could have an expected lifetime of 3-6 months. However, the combination of solar panels and rechargeable cells gives a predicted life in the range of 1.5-5 years. This system also uses pulsed transmitters on discrete frequencies, the pulse rate being one pulse lasting 0.05 s every 2 s. However on each frequency one of five tones is superimposed and so five individually identifiable animals can be tracked on each channel.

This brief look at some of the wildlife studies in Australia that have used, or are using, telemetry is by no means exhaustive. Biotelemetry has been widely used in many different areas but the budgets of the individual studies have proved a major constraint. Because of this, most studies have used 27 MHz equipment based on the Cochran and Lord (1963) circuit, they have been short range and of short duration.

Acknowledgement – This paper was presented in Oxford by Mr. D. G. Burnside, Rangeland Management Advisor with the Western Australian Department of Agriculture, and I most gratefully acknowledge his cooperation.

REFERENCES

Augee M. L., Ealey E. H. M. and Price I. P. (1975) Movements of echidnas, *Tachyglossus aculeatus*, determined by marking – recapture and radio-tracking. *Aust. Wildl. Res.* 2, 93-102.

Augee M. L., Ealey E. H. M. and Spencer H. (1970) Biotelemetric studies of temperature regulation and torpor in the echidna, *Tachyglossus aculeatus*. *J. Mammal.* 51, 561-570.

Christensen P. E. S. (1978) The biology of *Bettongia penicillata*, Gray, 1837, and *Macropus eugenii*, Desmarest, 1804, in relation to fire. Ph.D. Thesis, University of Western Autralia.

Clutton-Brock J., Corbet G. B. and Hills M. (1976) A review of the family *Canidae*, with a classification by numerical methods. *Bull. Br. Mus.* (*Nat. Hist.*) *Zoology* 29, 117-199.

Cochran W. W. and Lord R. D. (1963) A radio-tracking system for wild animals. *J. Wildl. Mngmt.* 27, 9-24.

Coman B. J. (1972) Helminth parasites of the dingo and feral dog in Victoria with some notes on the diet of the host. *Aust. Vet. J.* 48, 456-461.

Corbett L. K. and Newsome A. E. (1975) Dingo society and its maintenance: a preliminary analysis. In M. W. Fox (ed.), *The Wild Canids: Their Systematics, Behavioural Ecology and Evolution*, pp. 369-379. New York: Van Nostrand Reinhold.

Fullagar P. J. (1967) The use of radio telemetry in Australian biological research. *Proc. Ecol. Soc. Aust.* 2, 16-26.

Green B. and Catling P. (1977) The biology of the dingo. In H. Messel and S. T. Butler (eds.), *Australian Animals and Their Environment*, pp. 49-60. Sydney: Shakespeare Head Press.

Green B. and King D. (1978) Home range and activity patterns of the sand goanna, *Varanus gouldii* (Reptilia: Varanidae). *Aust. Wildl. Res.* 5, 417-424.

McIlroy J. C. (1976) Aspects of the ecology of the common wombat, *Vombatus ursinus*. I. Capture, handling, marking and radio-tracking techniques. *Aust. Wildl. Res.* 3, 105-117.

Mech L. D. (1974) Current techniques in the study of elusive wilderness carnivores. *Proc. Int. Congr. Game Biol.* 11, 315-322.

Newsome A. E., Corbett L. K., Best L. W. and Green B. (1973) The dingo. *Aust. Meat Res. Council Rev.* 14, 1-11.

Shields J. W. (1972) Acclimatization and energy metabolism of the dingo (*Canis dingo*) and the coyote (*Canis latrans*). *J. Zool. Lond.* 168, 483-501.

Tester J. R., Warner D. W. and Cochran W. W. (1964) A radio tracking system for studying the movements of deer. *J. Wildl. Mngmt.* 28, 42-45.

Webb G. J. W. and Messel H. (1978) Movements and dispersal patterns of *Crocodylus porosus* in some rivers of Arnhem land, Northern Australia. *Aust. Wildl. Res.* 5, 263-283.

Whitehouse S. J. O. (1977a) The diet of the dingo in Western Australia. *Aust. Wildl. Res.* 4, 145-150.

Whitehouse S. J. O. (1977b) Movements of dingoes in Western Australia. *J. Wildl. Mngmt.* 41, 575-576.

Whitehouse S. J. O. and Steven D. (1977) A technique for aerial radio tracking. *J. Wildl. Mngmt.* 41, 771-775.

Yerbury M. J. (1977) Telemetry and crocodiles. In H. Messel and S. T. Butler (eds.), *Australian Animals and Their Environment*, pp. 313-332. Sydney: Shakespeare Head Press.

Methods and Experience Gained from Biotelemetry Investigations of Mouflon (*Ovis ammon musimon* Schreber 1782)

H. WITT

Institut für Haustierkunde, D-23 Kiel, F.R.G.

Abstract — Since February 1977, I have been investigating the behavior of Mouflon. The study site is near Hanover, F.R.G.. It consists of a viewing enclosure of about 3 ha (7.5 acres), a large compound of about 1200 ha (3000 acres) and about 3500 ha (8750 acres) of natural forest. All areas, with the exception of a few meadows, are completely forested (mainly beech, oak and pine). The terrain is of a highland character with elevation between 90 and 400 m above sea level.

Since January 1978, radio tracking equipment has been used to obtain details concerning activity periods and the size of the living territory. 10 animals of various age and sex were marked.

EQUIPMENT USED

1. Manufacturer: Telonics, 1048 East Norwood, Mesa, Arizona 85203, U.S.A. (the technical data can be obtained from the sales catalog of this company). 10 transmitter collars, configuration 5B, with model MK-IV (rev. 1) transmitter unit, type B-2 H battery, and magnetic on-off switching; including special position sensing circuitry. Frequencies 150.100-150.550 MHz.

   2 receivers, model TTR-1 (Rev. 4 C) — 150-10 with 10 channels.

   1 digital data processor, model TDP-1.

   1 antenna, receiving, 2-element beam, model RA-2A-K, with pouch and 48" feedline.

   1 headset, noise-cancelling, type RA-2.

   1 recorder-driver, analog, dual-channel (amplitude and period) model TAP-II.

   Different adaptors, chargers etc.

2. Manufacturer: Gulton Industries Inc., East Greenwich, Rhode Island 02818, U.S.A.

   2 recorder, strip chart, pressure sensitive, dot printing model 288/F 12, 0-1 ma/ 12 VDC UNR/2/3/A.

3. Manufacturer: Dr. G. Hauser, Oberer Triftweg 33, D 338 Goslar, F.R.G.

   1 device for standardizing the level during the automatic registration of activities

from radio tagged animals, with adjustable, selective frequency filter for increasing the receiver sensitivity by a factor of 2-4.

RELIABILITY AND DURABILITY

One transmitter failed after four days. The cause is unknown, as the animal continued to wear the collar for visual marking, although it has not been seen since December, 1978. All other nine transmitters worked perfectly for one year, until February, 1979; then 8 animals were killed, and the transmitters removed. The last animal will wear the transmitter until it fails. The following was noted concerning the worn collars: on 3 collars the antenna was broken off at the base, even so these animals were still easy to locate. On both the male animals, the metal boxes in which the transmitters were fitted, were very badly dented, most probably due to knocks received from rivals. Despite this, these transmitters still worked perfectly. In January, 1979, one animal was found dead, lying with its collar under water at the bottom of a frozen stream. The transmitter was unaffected by this. The durability of the transmitter and the collar was therefore extremely good.

There were only minor failures noted on the reception side. Two direction finding antennas and two vehicle antennas were used during the investigation period, these were often bent in the field and broke during the course of this study. A defective soldered joint was noticed in one receiver, otherwise the units worked without failure. One must however note that both the receivers used, although of the same type, had very differing reception sensitivity. One receiver was found to be very insensitive compared to the other. The automatic receiver unit for recording the activities was checked daily. The following faults were noted:

1. The frequency drifted considerably, especially through temperature fluctuations, so that the receiver had to be retuned nearly every time. Also the automatic gain control reacted to temperature fluctuations, and this device must be exactly tuned to the receiver. This effect is amplified, so that often the stationary unit noted no reception even though the transmitters were within reception range.

2. The Gulton recorders do not have a deflection arrester for the pointer. Often the pointer stuck at the edge of the paper strip and several hours of activity periods were not recorded.

3. The devices can only be adjusted locally, i.e. outdoors. This is very difficult and time consuming, and in rain, fog and snow it proved impossible to prevent moisture from entering the equipment.

Summing up one can state that the durability of the devices is very good, even under unfavorable conditions, although the three faults specified above were found to be very troublesome in the automatic recorders.

Finally a brief description on the method of activity measurements: on the first recorder the field strength fluctuations are noted, which give an indication as to whether the animal has moved and the extent of movement. At the same time a second recorder notes whether the head of the animal is up or down. A mercury switch in the transmitter alters the pulse rate from about 700 ms (85.71 bpm) for the raised head to about 900 ms (66.67 bpm) for the lowered head. As a wild ruminant normally lifts its head at short intervals when feeding, to detect scent, one can in addition note periods of feeding with these recordings, and also differentiate between rest (with raised head) and sleep (with lowered head).

Satellite Radio Tracking of Polar Bears Instrumented in Alaska

A. L. KOLZ,* J. W. LENTFER** and H. G. FALLEK***

*U.S. Fish and Wildlife Service, Denver Wildlife Research Center, Denver, Colorado, U.S.A.
**Game Division, Alaska Department of Fish and Game, Juneau, Alaska, U.S.A.
***Handar Company, Santa Clara, California, U.S.A.

Abstract — The NIMBUS 6 satellite system successfully tracked three instrumented polar bears for 8, 20, and 390 days as these bears traveled distances exceeding 330, 500, and 1650 km, respectively, from their release sites on the arctic ice north of Barrow, Alaska. Electronic equipment developed under U.S. Fish and Wildlife Service funding by commercial companies was designed as a 5 kg collar-shaped instrument package of appropriate dimensions to fit 180 to 250 kg polar bears. The primary biological objective of locating a pregnant bear in a den site was not achieved because the instrumented bear traveled off the shore of the Soviet Union, into an area not accessible to U.S. scientists.

INTRODUCTION

In 1974, the National Aeronautics and Space Administration (NASA) authorized the U.S. Fish and Wildlife Service to track polar bears with the NIMBUS 6/Random Access Measurement System (RAMS). The feasibility of satellite tracking of wild animals was demonstrated by Craighead *et al.* (1972) when an elk was tracked for 28 days with the NIMBUS 3 weather satellite. Unfortunately, equipment developed for this study was too large and heavy for a polar bear and not compatible with the NIMBUS 6 system. Hence, a program was initiated to design a small satellite transmitter, develop a suitable attachment method, and evaluate the practicality of satellite tracking for arctic research. The biological goal was to track a pregnant female polar bear to a den site that was almost impossible to find by any other means.

DESCRIPTION OF THE SATELLITE SYSTEM

The NIMBUS 6/RAMS satellite was primarily designed to measure meteorological and oceanographic phenomena from mobile platform transmitters moving randomly throughout the world. A platform consists of a transmitter and its associated transducers for measuring environmental parameters. Each radio tag transmits a 1 s, radio frequency (RF) carrier pulse on 401.2 MHz at a rate of about one pulse every minute. A portion of each RF pulse (640 ms) is encoded by phase modulation to identify individual platforms and transmit 32 bits of sensor data. The unmodulated portion of the RF carrier is used to measure the Doppler frequency shift relative to the satellite and to calculate the transmitter's earth position coordinates. The system is capable of calculating a platform's position and velocity with an accuracy of ± 5 km and ± 1 m s^{-1} respectively

Simultaneous pulse transmissions from as many as eight radio tags can be received and stored by the present satellite processing equipment. The signals are separated in both time and frequency since the tags transmit asynchronous pulses which are shifted in frequency by both Doppler and equipment tolerance effects. A predicted 200 transmitters may be in view simultaneously with a probability of satellite detection of 0.95. In other words, 200 transmitters may be within reception range of the satellite at any time, and it can receive signals from 8 of these tags simultaneously. The satellite storage memory is designed to accommodate 1000 tags per orbit.

The NIMBUS satellite travels in a sun synchronous, polar orbit and circles the earth about every 108 min. Its altitude is approximately 1100 km and it is in view of a ground receiving station in Fairbanks, Alaska, at least 10 times a day. Upon command, the contents of the satellite's memory are transmitted to this station. The raw data are then transmitted over ground lines to the Goddard Space Flight Center in Maryland for processing and dissemination to approved scientific investigators (Fig. 1).

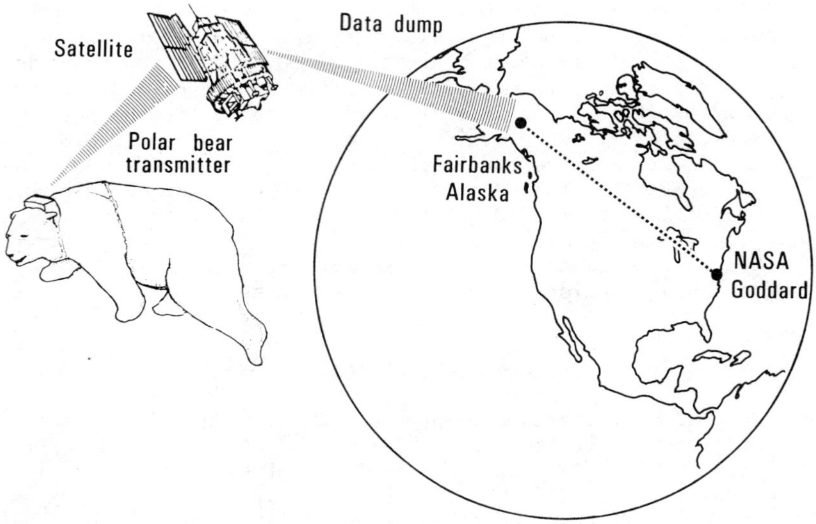

Fig. 1. Operation of NIMBUS satellite tracking system.

The NIMBUS system has been used extensively to monitor high altitude winds, barometric pressure, temperature, drifting ocean buoys, movements of the polar ice cap, and weather conditions around the world. The polar bear experiment is one of the few approved radio tracking projects using NIMBUS (see also Jennings and Gandy, 1980, this volume).

NIMBUS TRANSMITTER DESIGN CRITERIA

All NIMBUS transmitters are tested and certified to meet NASA's performance standards before field usage. This testing assures the experimenter that his data format is compatible with the satellite system and that the transmitter's frequency stability is adequate for accurate positional calculations. Table 1 summarizes these design requirements.

Table 1.
NIMBUS transmitter requirements

Frequency (nominal)	401.2 MHz ± 6 KHz
Power output	+ 33.9 dBm (2.5 W) ± 1 dB
Oscillator stability	$< 1 \times 10^{-8}$ (4.2 Hz)/15 min
RF pulse width	980 ms
Pulse rate	about 1 min^{-1}
Duty cycle	about 1/60
Modulation	± 60° ± 6° PSK
Modulation rate	100 bps (Manchester)
Antenna polarization	right hand circular

DESIGN CONSIDERATION FOR POLAR BEAR TRANSMITTERS

Package weight and volume were given prime consideration in development of component parts for the polar bear transmitters. The final design included six basic components: timing control circuits, data encoder, temperature controlled oscillator, transmitter, antenna, and battery pack (Fig. 2). One additional accessory, a VHF (164 MHz) radio tag was included within the electronics enclosure to provide day to day tracking from light aircraft.

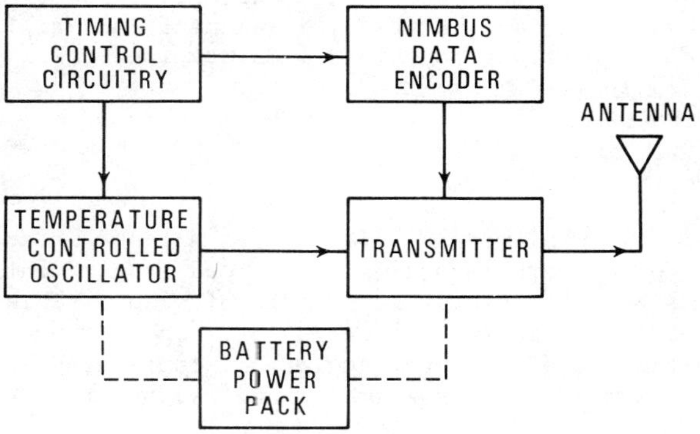

Fig. 2. Satellite — polar bear electronics.

To eliminate the need for a multiplexer and analog to digital converters we decided against measuring data from external sensors (thermocouples, pressure gauges, current monitors, etc. These sensors would have increased the circuit complexity by requiring a 25 percent increase in the number of integrated circuits. Thus, the final transmitter design was strictly for location purposes. A functional description of each transmitter component is given below.

TIMING CONTROL CIRCUITS

All clock functions necessary for transmitter operation are contained in the timing control circuits. For the transmitter to operate throughout an entire year with an acceptably sized battery pack necessitated a design permitting intermittent operation. A timing cycle, developed with CMOS circuits (similar to those in a digital watch) operates the transmitter for a continuous 8 h period during each 4 day interval. An error of a few seconds per month was allowed and the 8 h transmission period was timed to occur only during the optimal reception by the satellite on any given day. The basic clock circuit also operated the data encoder and formating circuits.

DATA ENCODER

The encoder circuits generate a synchronizing data code (which tells the satellite when to start taking data) and the required number of logic bits for the satellite to individually identify each transmitted pulse (a recognition code). The NIMBUS rejects as spurious noise any signal received without this coded modulation.

TEMPERATURE CONTROLLED OSCILLATOR (TCXO)

The precision of a location calculated for a NIMBUS platform is directly related to the frequency stability of its transmitter. The TCXO developed for this program proved to be effective. The oscillator operated from -40° to $+50^{\circ}C$ and would tolerate a $10^{\circ}C$ thermal shock with an output frequency shift of $<$ 4.2 Hz/15 min (Table 1). Although the TCXO does not require an oven stabilized crystal and thereby reduces battery requirements, the timing control circuit powers the TCXO 30 min before transmitter turn-on which allows circuit stabilization.

TRANSMITTER

The transmitter circuit employed microminiature components and a hybrid integrated power amplifier. The TCXO signal (40.12 MHz) was multiplied by a factor of 10, phase modulated, filtered with a distributed filter element, and amplified to produce a 3 W output. The transmitter was contained in a machined housing and occupied a volume of 41 cm^3 (2.5 in^3).

ANTENNA

The final antenna design employed a coplanar slot antenna measuring 12.7 × 12.7 × 1.2 cm. The radiation pattern is a single overhead lobe radiating a right hand circular polarized signal and exhibiting a gain of -6 dB. This low gain had to be compensated by a 3 dB increase in RF input power. The antenna shape permitted its incorporation into the lid of the rectangular box containing electronic components. Thus, the antenna had no protruding elements and was as small as the gain specification would permit.

BATTERY PACK

Eleven organic lithium batteries (10 A h capacity each) power the transmitter. It was necessary to derate the advertized battery capacity at $25^{\circ}C$ by a factor of 50 percent for $-40^{\circ}C$ operation. Also, a voltage regulator circuit was installed to compensate for a measured 4:1 internal resistance change. The battery pack had sufficient capacity to operate the transmitter for 1 year.

RF LOCATOR BEACON

The RF beacon was a totally self-contained circuit operating at 164 MHz and powered by two AA inorganic lithium cells with a calculated 18 month service life. The beacon antenna was a tuned loop machined into the sides of the electronics enclosure which produced an output of less than +10 dBm. Signal range from a search aircraft was generally in excess of 24 km. This beacon locator was an invaluable accessory for the polar bear field program by allowing close observation of instrumented bears whenever flying conditions permitted.

ELECTRONICS PACKAGING

The electronics enclosure consisted of a solid piece of polycarbonate plastic (Lexan) machined to house the various components. All sides were $\frac{1}{4}$ in thick and the antenna formed the lid of the assembled rectangular box with outside dimensions of 14 × 14 × 5 cm. This box was screwed onto a flat surface formed into a molded plastic collar that encapsulated the battery pack. Connections were made to the electronics enclosure through the two mating surfaces, and a watertight seal was formed with rubber gaskets and silicon rubber. The battery power pack and electronics enclosure were thus integrated into a single rigid structure with no exposed wiring for optimal packaging reliability.

The inside surface of the collar formed an elliptical cylinder with a perimeter of 80 cm. This shape gave the collar a self-righting capability on the bear's neck with the major axis remaining vertical. Batteries weighing 946 g were encapsulated into the bottom of the collar and overbalanced the weight of the electronics box on the top. Thus, the weight distribution in conjunction with the eccentric shape allowed the collar to ride in an upright and stable position on the bear. Each plastic collar (elastameric urethane Adiprene type L 167) was hand finished to remove all sharp corners and ensure smooth surfaces. A 1.27 cm thickness of nonabsorbing, closed cell Etha-Foam was glued to the inside surface of the collar as padding for the bear's neck. The integrated transmitter-collar measured 27.9 w × 40.6 h × 14 d cm and weighed about 5.6 kg (Fig. 3).

ATTACHMENT DESCRIPTION

Although collars are commonly used to attach radio tracking equipment to wildlife, the collar developed for the polar bear was unique in that it fitted freely over the animal's head and was completely loose around the neck. Four cables extending

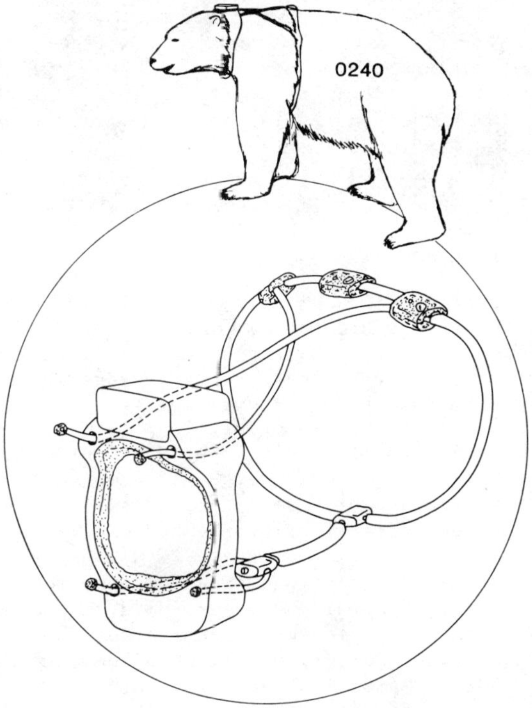

Fig. 3. Transmitter collar and harness for polar bears.

posteriorly from the collar attached to a girth cable fitted behind the bear's front legs which kept the collar in position (Fig. 3). All cables were 4 mm diameter stainless steel covered with two layers of Dacron reinforced Tygon tubing to prevent ice formation. The metal fittings were covered with plastic tubing to ensure no direct contact between the metal and the bear. The collar and harness were both white and did not adversely affect the camouflage of the bear (Fig. 4).

Fig. 4. Mr. L. Kolz with instrumented polar bear.

FIELD OPERATIONS

The Naval Arctic Research Laboratory (NARL) at Barrow, Alaska, provided the research facilities necessary for instrumenting the polar bears. Bears were located from light aircraft by following their tracks or sighting them directly on the sea ice. They were immobilized by injecting 1.6 mg kg^{-1} body weight of Sernylan (phencyclidine hydrochloride) and 0.07-0.11 mg kg^{-1} body weight of Acepromazine. The bears were transported by fixed-wing aircraft or helicopter. Cages were available at NARL for holding the animals before and after fitting the radio collar.

Tagged bears were observed as they regained consciousness to study their reactions to the harness and collar. In general, the bears paid little attention and only occasionally shook their heads, necks and forequarters. The transmitters weighed 2.5 to 3.0 percent of the bear's weight, which is not an excessive load. In fact, in wildlife telemetry studies, mammals often carry packages exceeding 5 percent of their weight for extended periods of time without deleterious effects.

A Cessna 180 aircraft was equipped with two 3-element Yagi antennas for tracking the 164 MHz locator beacon. The antennas were mounted under each wing and directed away from the fuselage and downward at about a 30° angle. In flight, the search pattern extended the maximum reception distance off either side of the aircraft. Both antennas could be monitored simultaneously by using a hybrid junction, or a switch could be manually operated to monitor the antennas individually. An experienced operator could easily listen to the signal intensity on each antenna and direct the pilot to fly toward the RF beacon. This aircraft tracking system was an effective way of routinely locating the bears for position verifications with the satellite system and to ensure that the instrument package was not restricting or hindering normal behavior of the bears.

RESULTS AND DISCUSSION

Collars were tested after final design and fabrication. Position accuracy tests were run against a standard production system whose RF carrier was stablized with an atomic frequency standard. The tests were run at Santa Clara, California, (37° N. latitude) with normal night-to-day temperature variations of 20 to 30°C. Two days of testing showed no measurable difference in the root mean squared (rms) position error of the polar bear system compared with the high stability system. In each instance, the same antennas were used with identical input powers. The rms position error varied from 1 to 3 km from actual position. Tests at Barrow (71° N. latitude) with collars on the ground and on captive bears at temperatures from -35 to 5°C gave positions from the satellite within 1 to 14 km of actual positions. It would have been desirable to have simultaneous location data on a free-roaming bear from both the tracking aircraft and satellite system, but conditions never permitted this. Nevertheless, satellite data did allow us quickly to establish aerial search boundaries.

Three mature female bears were fitted with transmitters in March and June 1977. They were judged to be in estrus and possibly pregnant because they were not accompanied by young, external genitalia were swollen, and in one case a female was accompanied by a mature male. Described below are the field data recorded for each of these bears which are identified by Nos. 1793, 1794, and 1795.

POLAR BEAR NO. 1793

We originally planned to instrument two polar bears in March 1977. However, two unexpected electronic problems occurred in Barrow. First, the reset function on the timing control circuit was intermittent, making it impossible to initiate properly the 4-day transmit cycle. In addition, the internal resistance of the batteries at temperatures below -30°C was greater than expected, and the voltage applied to the biphase modulator and TCXO were marginal. We decided to disable the 4 day timer and operate the transmitter continuously at the expense of reduced service life. Adjustments were then made on the modulator to accommodate the low battery voltage.

These changes allowed us to instrument a polar bear and determine the efficacy of the collar-harness attachment. We had no previous opportunity to try this attachment on a free-roaming bear. There are only a few weeks during the spring and fall in the arctic when ice and weather conditions permit this type of research.

Bear No. 1793 was instrumented on 24 March and placed in a wooden cage to allow transmission of radio signals to the satellite. Three days later, NASA verified that transmissions had been received. The bear was then tranpsorted 47 km NNW of Point Barrow and released on 28 March. The bear was sighted from a search plane the following day about 54 km north of the release site, accompanied by a much larger (male?) bear.

The satellite transmiter ceased operation (due to a depleted battery pack) 8 days after the bear was released. On 9 April, the bear was located via the RF beacon, immobilized, and the collar attachment examined. One small rubbed spot was found where a cable end had been left exposed. Hence, cable ends were carefully covered on subsequent harnesses. The harness and transmitter were removed and the animal was released.

Detailed location mapping (Fig. 5) indicates an average daily movement of 56 km for bear 1793, and a maximum 1 day movement of 82 km. The bear traveled for 12 days and was recaptured 335 km north of her release site.

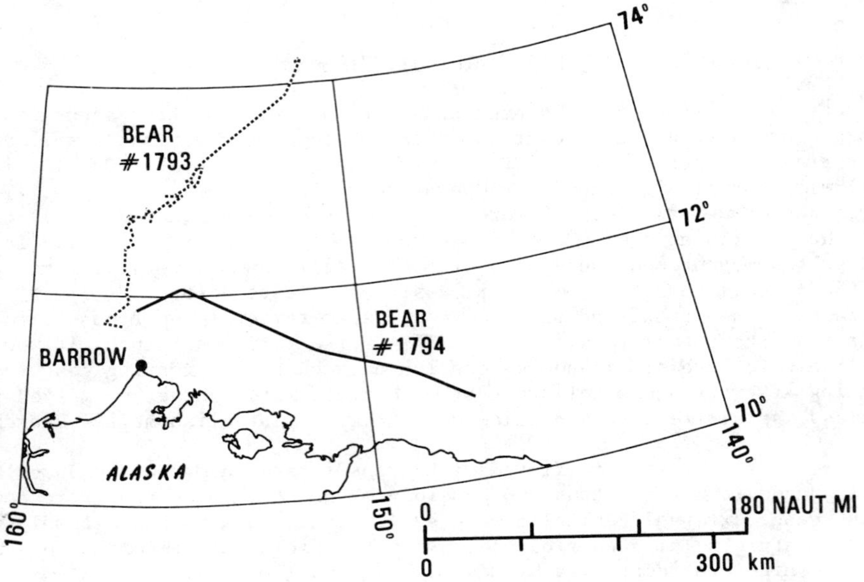

Fig. 5. Travel paths for polar bears nos. 1793 and 1794.

POLAR BEAR NO. 1794

Bear 1794 was one of two females captured in April and held at Barrow until engineering changes could be perfected for an improved transmitter-collar. Delayed release was also necessary because ice packs break up in May and tracking by helicopter was no longer effective until June. A new battery pack and plastic collar were designed while engineering refinements were made on the electronics circuitry. Two transmitter-collars were completed in June and sent to Barrow for attachment.

Polar Bear No. 1794 was released 56 km north of Point Barrow on 18 June. Weather conditions were poor and no attempt was made to track the bear after release. The satellite provided position data every fourth day through 8 July, a period of 20 days. The last transmission was received on 12 July, but signal quality was too poor to provide location data. In November an aircraft search was conducted on the assumption that the location beacon was still functioning, but the search was severely hampered by weather conditions and no signals were received. During the 20 days of satellite tracking, polar bear No. 1794 had traveled 370 km east, against the direction of ice movement (Fig. 5).

POLAR BEAR NO. 1795

Bear No. 1795 was the second bear held in captivity at Barrow; she was released

50 km north of Point Barrow on 18 June. The bear traveled generally east for 10 days, and then turned west toward the Soviet sector roughly following the 73.5° parallel of latitude until mid-November (Fig. 6). The data indicates that movement decreased markedly during the winter months. We can only speculate that the bear entered a den for the winter, and the movement we measured was that of the polar ice cap. The bear was tracked for 228+ days and a distance of 1320 km to this apparent den site about 330 km off the coast of Siberia. Unfortunately, it was not possible to organize a ground expedition in that region.

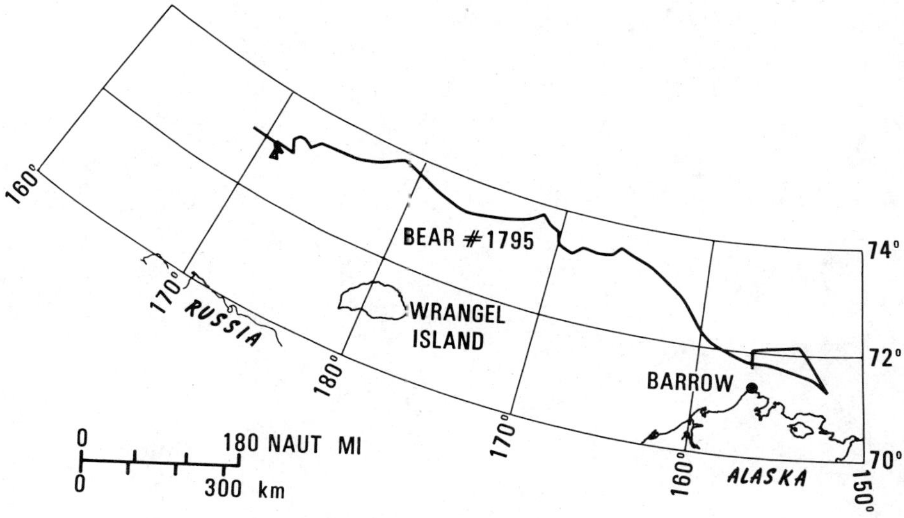

Fig. 6. Travel path for bear no. 1795.

CONCLUSIONS

Satellite tracking of wildlife can quickly provide information on animal populations anywhere in the world. Instrumenting a polar bear was an extreme challenge because of the low temperature environment, the corrosive action of seawater, the required electronics development, and the mechanical restraints for the packaging. A new satellite program, the ARGOS system, with tracking capabilities similar to those of NIMBUS, has been launched.* Scientists should give this program serious consideration in planning their future wildlife tracking studies.

Acknowledgements — We gratefully acknowledge NASA for authorization to use the NIMBUS 6 satellite and for making data available; the Naval Arctic Research Laboratory at Barrow, Alaska, and especially personnel of its Animal Research Facility; Darwin Throne of the Handar Company for his excellent engineering contributions; Fred Sorensen and Richard Johnson of the U.S. Fish and Wildlife Service for help with field and laboratory work; and George Hall, John Hall, and Richard Hensel for aid in design of methods to package transmitters and attach them to bears.

*Due to legal technicalities in the frequency assignment, those used by ARGOS (TIROS-N) are not currently available for general radio tracking. A few experimental tracking studies are all that have been authorized.

REFERENCES

Cote C. E., DuBose J. F. and Coates J. L. (1973) The NIMBUS F. random access measurement system. *The Fifth Annual Southeastern Symposium on System Theory*, March 23, 5 pp.

Jennings J. G. and Gandy W. F. (1980) Tracking pelagic dolphins by satellite. In C. J. Amlaner, Jr. and D. W. Macdonald (eds., *A Handbook on Biotelemetry and Radio Tracking*, pp. 753 -755. Oxford: Pergamon Press.

Tracking Pelagic Dolphins by Satellite

J. G. JENNINGS* and W. F. GANDY**

*National Oceanic and Atmospheric Administration, National Marine Fisheries Service, Southwest Fisheries Center, La Jolla, California 92038, U.S.A.
**National Oceanic and Atmospheric Administration, National Marine Fisheries Service, Southeast Fisheries Center, National Fisheries Engineering Laboratory, NSTL Station, Mississippi 39529, U.S.A.

Abstract — A satellite-linked tracking system is being designed for assessing numbers of dolphins involved in the yellowfin tuna fishery of the eastern tropical Pacific, U.S.A. Testing is via the NIMBUS-6 satellite. The transmitter has a 17.8 cm one-quarter wavelength stub antenna, operates at 401.2 MHz, has a power output of 0.5 W, and incorporates a seawater switch to regulate transmission. Preliminary tests in the laboratory and on captive dolphins revealed problems with the bit rate oscillator, the seawater switch, and frequency modulation. Field tests of prototype transmitters on wild Hawaiian spotted dolphins are scheduled for spring, 1979.

INTRODUCTION

Under the U.S. Marine Mammal Protection Act of 1972, the National Marine Fisheries Service (NMFS) was made responsible for conservation of dolphins involved in the yellowfin tuna fishery of the eastern tropical Pacific, U.S.A. Regulations governing the fishery, including dolphin mortality quotas, are based on determinations of the status of the stocks. Assessment of animal populations requires data on distribution, migration, and mixing of the stocks, as well as on numbers. These data are combined with age, sex, and growth information obtained from sampling the kill to determine the status of the stocks.

The area involved in the fishery is substantial — 13 million km^2. NMFS conducts aerial surveys, but the range of the relatively low flying planes limits coverage. A conventional tagging study is planned using highly visible disc tags placed on the dorsal fin. Cost of the tags is minimal, but charter cost for a tuna purse seiner (which is the only means of catching large schools of dolphins) is approximately $10,000 per day. Tagging studies generally provide data at only two locations — the site of tagging and that of recovery. Radio tracking using standard HF and VHF equipment generally involves tracking one animal for a limited time and there is always the risk of losing the animal.

PRELIMINARY TESTS

Initial development of the system concentrated on antenna selection. A 17.8 cm one-quarter wavelength stainless steel stub was chosen. Preliminary tests on a captive dolphin were conducted at Sea World, San Diego, California, in June, 1977, using a modified engineering model of a satellite-linked transmitter developed for use on buoys. These tests, as well as all other work to date, have used the Random Access Measurement System (RAMS) on the NIMBUS-6 satellite. The purpose of the test was to assess the impedance match of the antenna, transmitter, and a live dolphin. The animal was trained to surface on command for transmission every 64 s. A minimum of four successful up-links (transmission to the satellite) per satellite orbit is required to accurately determine position. Of the 14 tests conducted from the live dolphin, only 2 tests were successful with the satellite receiving 4 and 5 up-links. Of 23 verification tests conducted with the transmitter on a roof, 10 were successful with as many as 15 up-links. Failure to achieve sufficient up-links was attributed to several factors: the satellite recorders were sometimes turned off; high satellite elevation angles may have occasionally been a problem; a defective battery cell caused the voltage to drop off after 36 h when it was designed to last 72 h; and the oscillator was not adequately temperature compensated. Most importantly, it was later discovered that the RF oscillator did not fully stabilize until 15 min after power was supplied. Since we did not turn on the unit until immediately before the satellite pass, the frequency drift was too great for reception by the satellite receiver. The test did successfully demonstrate that the concept is feasible and that geographic position of the platform could be determined (Gandy, Vanselous and Jennings, 1977).

MATERIALS AND METHODS

A new prototype transmitter pack is being designed and built. The engineering model consisted of: a power supply of 18 3-VDC organic lithium cells; main timing circuit; RF oscillator and power amplifier; seawater activation switch; one-quarter wavelength stub antenna; and aluminum (6061) housing. The system is currently packaged in two cylindrical tubes each measuring 17.5 × 5 cm, which are connected by a 1.61 cm diameter neoprene tube. The pack weighs 907 g. The system operates at a frequency of 401.2 MHz with a transmission repetition rate controlled by the seawater switch. The transmission rate is one message per second if the switch circuitry is inhibited.

Laboratory tests were conducted in November, 1978 and February, 1979, using a RAMS satellite simulator, to evaluate the transmitter's operating characteristics such as frequency variation, phase modulation, proper message formating and transmission timing.

Field tests on dolphins were conducted during 8-14 February, 1979, at Sea Life Park, Hawaii, using the engineering model. The package was attached with a nylon harness to the area of the dorsal fin of a captive dolphin. A receiver was located near the habitat pool, so that the operational status of the transmitter could be monitored. Satellite data were obtained by telephone through the NIMBUS Control Center at NASA's Goddard Space Flight Center, Greenbelt, Maryland, U.S.A. A test schedule was established, based on predicted times of overpasses and elevation angles of the satellite. Angles greater than 23° were chosen in order to have maximum time during the passes. The duration of these passes was about 20 min.

RESULTS

The engineering model failed the initial laboratory test in November, 1978 due to excessive frequency drift and improper phase modulation. The unit successfully interrogated the simulator during the second laboratory test in February. However, three problem areas were identified: (1) the frequency variation was 1.5 Hz min^{-1} versus a design specification of 0.28 Hz min^{-1}; (2) the four data channels of the simulator received random messages; and (3) the seawater switch tended to lock in the 'on' position causing the transmitter to operate at the maximum repetition rate of one message per second.

Eight tests of the engineering model were conducted on a captive dolphin. Sufficient up-links to determine geographic position were achieved during five overpasses (Table 1). The average error in calculated position was 23.33 km, reflecting frequency instability in the transmitter and possibly a larger longitudinal computation error at higher satellite elevation angles. The loss of data for two overpasses was due to the failure of the seawater switch in one case and the spacecraft recorder not being turned on in the other. The third failure is unexplained.

Table 1.
Reported position of transmitter
(Actual locaticn was 21.32°N-157.66° W)

Date (calendar)	Time (GMT)	Reported position (Latitude)	Reported position (Longitude)	Computed error (km)
2-11-79	2212	21.32°N	158.07°N	44.4
2-12-79	0934	21.14°N	157.46°W	29.6
2-12-79	2125	21.35°N	157.63°W	9.3
2-14-79	0941	21.34°N	157.96°W	33.3
2-14-79	2149	21.32°N	157.66°W	0

CONCLUSION

Tests have shown that it is feasible to acquire data on geographic positions of marine animals with a satellite-linked tracking system. Four design changes were implemented in recent tests: (1) RF shielding of the seawater switch circuit; (2) an inhibit circuit to prevent the transmitter from operating continuously when the animal is at the surface; (3) temperature compensation of the bit rate oscillator; and (4) frequency stability to within 0.28 Hz min^{-1}.

Two prototype units with these modifications will be tested on wild Hawaiian spotted dolphins (*Stenella attenuata*) in offshore Hawaiian waters in the spring of 1979. Successful tracking of these dolphins is prerequisite to placing transmitters on wild dolphins of the same species in the fishery. Placement on wild dolphins in the eastern tropical Pacific could begin in late 1979. Transition to the TIROS-N satellite is being evaluated which would require some modification of the electronics.

The tracking system is being designed to be applicable to other marine animals such as large whales and sea turtles. The pack will be reduced in size, and could be designed to incorporate oceanographic and physiological sensors.

REFERENCE

Gandy W. F., Vanselous T. M. and Jennings J. G. (1977) Tracking marine animals by satellite. *Proceedings, ARGOS Meeting, Paris,* France, November 3, 1977.

Radio Tracking of Dolphins in the Eastern Tropical Pacific Using VHF and HF Equipment

R. W. BUTLER and J. G. JENNINGS

National Oceanic and Atmospheric Administration, National Marine Fisheries Service, Southwest Fisheries Center, La Jolla, California 92038, U.S.A.

Abstract — Radio tracking was used to study the dynamics of schools of dolphins involved in the yellowfin tuna fishery of the eastern tropical Pacific. Complete VHF and HF systems were assembled. The primary VHF system consisted of Model MK-IV transmitters by Telonics, Inc., a Model 320 automatic direction finder (ADFS) by Ocean Applied Research (OAR), and an OAR Model AA-380-400 Adcock Antenna. The primary HF system was based on OAR AB-245 transmitters, an OAR ADFS 210 receiver, and an OAR AA-270-400 Adcock antenna. Secondary components were assembled for both systems.

The VHF system proved most reliable in allowing three two-day tracking sequences over 90.7, 19.1 and 120.9 nautical miles. Maximum range of the VHF system was 9.2 ± 0.2 nautical miles or 14.7 ± 0.3 nautical miles depending on the receiver. The dolphin schools showed a low degree of cohesion over the period of the study.

INTRODUCTION

The primary purpose of this paper is to assist other researchers to assemble radio tracking systems by briefly describing VHF and HF systems used to track dolphins in the ocean. A tracking system was essential to studies of school dynamics involving tagging, tracking, and recapture of dolphin schools to determine the degree of school integrity. Because other investigators consistently reported problems with malfunctions of components, particularly the transmitters, two complete tracking systems were assembled using VHF and HF.

The tracking was conducted from a commercial tuna purse seiner from 12 September to 31 October, 1978 during a two month research cruise in the eastern tropical Pacific near Clipperton Island. The dolphin schools were caught by the purse seiner during routine fishing operations. The dolphins were recovered from the net and maneuvered into a comparmentalized system of net corrals and aluminum cages with chutes where they were examined, radio and disc tagged, and released. The experimental design called for placing radio tags on one or two dolphins in the school during selected sets, tracking the school for 2 days, recapturing and reexamining the school. The sequence was to be repeated and the radio packs removed at the end of the experiment.

MATERIALS AND METHODS

The VHF system was the primary one used during the study. The transmitters were Model MK-IV by Telonics, Inc. which were cubical in shape, measuring approximately 6 × 4 × 3 cm and weighing 170 g. Transmissions were 30 ms pulses at a 0.6 s pulse rate with frequencies between 148.748 and 148.850 MHz. Power output was 0.25 W and the life expectancy was 2.5 months. These transmitters did not have a salt water switch. The antennas were 48 cm stainless steel whips insulated with a plastic sheath. The transmitters passed hydrostatic testing to 300 m, although the cases did deform slightly.

The principal receiver was a Model OAR ADFS 320 automatic direction finder (ADFS) manufactured by Ocean Applied Research (OAR) which had a signal 'lock and hold' capability. This feature allows the signal to be displayed for a longer period of time on the direction finder screen thereby facilitating tracking. An Adcock antenna, OAR Model AA-380-400, was used with the OAR ADFS 320 and was mounted on top of the crow's nest, placing it about 26 m above the sea surface. A preamplifier was used to improve the signal-to-noise ratio by providing an additional gain of 10 dB. Interfaced with the ADFS 320 was a Rustrak strip chart recorder for recording surfacing frequencies of the tracked dolphins.

A secondary VHF system was set up to relocate any animals lost by the other system. It consisted of a Telonics TR-2 receiver, an omnidirectional whip antenna, a three-element quad antenna which could be hand rotated to determine bearing, and an antenna switch box. The TR-2 receiver is more sensitive than the OAR ADFS 320 but is less convenient to use for longterm tracking. The audio output of the TR-2 could be connected to a cassette tape recorder to obtain surfacing rates.

The primary HF system was based on OAR AB-245 transmitters which consisted of two tubes, each measuring about 14 × 2 cm, rigidly connected into a U-shape. A 44.5 cm polyurethane jacketed top loaded whip antenna was mounted on the front piece directly between the transmitters so that it would be in line with the dorsal fin. The transmitters with batteries weighed 237 g, had an output of 0.60 W and frequencies of 27.585 and 27.595 MHz. A salt water switch regulated transmission by turning off the transmitter when submerged.

An OAR ADFS 210 was the primary receiver with another like it serving as a back-up. An Adcock antenna, Model OAR AA-270-400, was mounted approximately 10 m above the sea surface and was connected to the ADFS 210.

Saddles of Kydex plastic were molded from a cast of the dorsal fin and upper back of a typical adult male spotted dolphin, *Stenella attenuata*. The saddles were lined with 0.64 cm closed cell, neoprene rubber. The HF transmitters were attached directly to the saddles with rivets and plastic ties. The VHF transmitters were placed on one side of the saddle and an equal weight on the other side for balance. The VHF saddles were potted with two-part rigid polyurethane foam, hydrodynamically shaped with a slightly bulbous front. Individual variation in the saddles required custom fitting to each dolphin. With the saddle in place, a 0.64 cm diameter biopsy needle was used to bore a hole through the dorsal fin used to attach the saddle. A 0.64 cm diameter stainless steel bolt was inserted simultaneously as the biopsy needle was extracted to reduce bleeding. The bolt was covered with a sleeve of biocompatible nylon which was slightly larger than the hole, thereby serving as a compress to minimize bleeding. A corrosive magnesium nut was used to assure that the pack would be released if the animal was lost.

Once the animals were released, a 24 h tracking watch was started with particular attention paid to signal strength so that the transmitter would not get out of range, yet maintaining a minimum distance of 1.5 to 2.0 nautical miles to avoid influencing the dolphin's behavior. The watch was rotated on 2 h shifts.

RESULTS

Three radio tracking sequences were successfully conducted for total distances of 90.7, 19.1, and 120.9 nm. The first two sequences were of two-day duration and the third was in two legs of two days each. During the second sequence, the saddle broke off the bolt but because it was buoyant and balanced, the pack floated with the antenna upright and was retrieved.

Successful tracking only occurred with the VHF system. The only HF transmitter used malfunctioned almost immediately whereas the five VHF transmitters functioned reliably each time.

During the first tracking sequence, an adult female and adult male dolphin were radio tagged and released with their tagged school at 16.44. By 18.55 they were headed in different directions and the female was followed. At 21.32 the last signal was received from the male. During the last radio tracking sequence, two radio tagged animals were released with their tagged school of 101 dolphins. After two days of following one of the dolphins, it was caught with a group of 150 dolphins. Only one animal had a tag. A radio pack was placed on an adult male in the new group and tracked for two additional days. It was recaptured with 54 other dolphins, only 6 of which bore tags from the previous set.

Range tests at sea on the VHF transmitters indicated a maximum of 9.2 ± 0.2 nautical miles with the OAR ADFS 320, and 14.7 ± 0.3 nautical miles with the Telonics TR-2. Tests near the coast indicated a range of about 10 miles with the TR-2. No similar range tests were conducted for the HF gear.

CONCLUSIONS

The tracking system allowed a study to be conducted on dolphin school dynamics which indicated that there was little cohesion in these dolphin schools through a short-term period (2 days). The VHF system functioned reliably and permitted successful tracking of dolphins in the open ocean. An advantage of the VHF over HF system was that there was little or no interference in that frequency which facilitated tracking.

The saddles need to be improved for longer term tracking to minimize the abrasion noted during these studies. The polyurethane potting material became water-logged as did the neoprene saddle lining. Substitutes are needed for these purposes.

Homing Behavior of Juvenile Green Turtles, *Chelonia mydas*

L. C. IRELAND

Bermuda Biological Station for Research, St. George's West, 1-15, Bermuda

Abstract — Ten juvenile green turtles were netted in the inshore waters around Bermuda, equipped with ultrasonic transmitters, and released at locations 1.5-4 km distant from their capture sites. Each turtle was tracked by investigators in a small boat fitted with an acoustic receiver and a directional hydrophone. Tracks lasted for 314-657 min. Attempts were made to relocate turtles on days following their release. Five of the turtles returned to the locations where they were netted on the same days they were released, 1 returned by the day following its release, and 1 returned within 2 days of release. Five turtles displayed site tenacity for periods of 2 weeks or longer. The study shows that acoustic telemetry is a practical technique for studying the movements of juvenile green turtles in inshore waters. Studies of the sensory cues which guide the homing behavior of juveniles may shed light on the sensory mechanisms which guide long distance, reproductive migrations of adult green turtles.

INTRODUCTION

There are many gaps in our knowledge of the life cycle of *Chelonia mydas*. Most research on this reptile has been done with nesting females and with hatchlings on their natal beach, even though the green turtle spends less than 1 percent of its life in these or any other terrestrial situations, and has been shown to be one of the world's most remarkable marine migrants (Carr, Carr and Meylan, 1978). More research needs to be done at sea, where the green turtle lives.

One of the most neglected areas of research on green turtles concerns the behavior and ecology of colonies of juveniles in certain reef and estuarine areas. At present, the best established fact about these colonies is that they exist. Programs now underway off Bermuda (Burnett-Herkes, 1974), Florida (Ehrhart and Yoder, 1978) and Hawaii (Balazs, 1976), in which juvenile turtles are netted, tagged, returned to the water, and recaptured at a later date, provide information both about the movements of juveniles and on their rate of growth. Some of the most interesting data to emerge from these tagging studies suggests that juvenile greens display homing behavior. Burnett-Herkes (1974) captured one juvenile twice and another three times. In each case, the turtle was retaken near the site of its previous capture, even though each release was at a location approximately 2 km distant from the capture site. Tagging studies, however, cannot tell us exactly how long it takes animals to return home. The shortest time between a capture and a recapture reported by Burnett-Herkes was 1 month and the longest 12 months; these data also suggest that the juveniles display site tenacity. Further, tagging experiments tell us little or nothing about an animal's actual travel path between release and recovery areas. Did the turtles travel directly back to the site of their capture or did they rediscover it by random search? The present study was designed to investigate the homing behavior of juvenile greens in greater detail, by netting them in inshore waters, displacing them to distant locations, and tracking them by means of acoustic telemetry.

MATERIALS AND METHODS

The study was conducted off Bermuda during October and November of 1978. The subjects, 10 green turtles weighing 2.9-31.8 kg, were netted at several different locations in inshore waters. Prior to release each juvenile was weighed, measured and fitted with a metal tag by Bermuda Fisheries. After that and just prior to release, each turtle was fitted with a harness of surgical rubber tubing, fastened together with rings of steel and aluminum. This arrangement sets up a dissimilar-metal corrosion cell that results in the release of the harness from the turtle after approximately 60 days. An acoustic transmitter, 1.93 cm in diameter, 7.00 cm long, and weighing 54 g in air, was attached to each turtle's harness at a point near the center of the carapace. The transmitters produced pulses of sound at a frequency of 45-55 KHz, well above the range of auditory sensitivity reported for green turtles (Ridgway *et al.*, 1969). Complete details on the transmitters appear in Ireland and Lawson (1980a, this volume). Also, an orange colored styrofoam cube, measuring approximately 8 cm on a side, was attached to the rear of each harness with about 9 m of fine cotton thread. We used light thread so that it would break easily if it became entangled with any underwater obstacle.

Each turtle was released alone at a point 1.5-4 km distant from where it was netted. We attempted to release and track each turtle as soon as possible after its capture. Adverse weather conditions, however, often delayed the start of experiments. Times between capture and release ranged from 1 to 7 days. Before release each turtle was kept in a large concrete tank supplied with running sea water. All releases and tracking were carried out during daylight or early evening hours for reasons of safety. Acoustic and visual contact with each turtle was maintained by investigators in a small boat equipped with an acoustic receiver and a directional hydrophone. Details of this equipment are given in Ireland and Lawson (1980b, this volume). While tracking, we usually remained 25 m or more behind the turtle with the engine shut off. At intervals of 15-30 min we started the engine and, guided either by the float or the signal from the transmitter, moved to within 10-15 m of the turtle in order to record its position. Position was determined by compass bearings relative to points on shore. Every few hours, we recorded wave direction and height, wind velocity and direction, and the direction and velocity of surface currents. For all but brief periods, the experiments were carried out under clear or partly overcast skies. Also, at irregular intervals when not engaged in tracking, we attempted to locate turtles released on previous days.

RESULTS AND DISCUSSION

We experienced only moderate difficulty in maintaining contact with the turtles. Individuals were tracked continuously for periods of 314-657 min. The floats provided very detailed information on the turtles' movements but were usually lost during tracking. In only 4 cases did a float remain attached for more than 1 h, and in only 1 case for an entire tracking session. In the clear, deep waters off the coasts of Bermuda the telemetry equipment proved to have a practical range of 1-1.5 km. In shallow waters cluttered with rocks and reefs, however, ranges were often reduced to

200 m or less. Nevertheless, we were never forced to stop tracking because we were unable to detect a transmitter. Tracking was terminated because of poor weather, fatigue, or the approach of darkness.

Five of the turtles traveled directly or nearly directly back to the sites of their captures, arriving within 100 m or less of these locations while being tracked. The travel path of 1 of these turtles is shown in Fig. 1. With our less than systematic

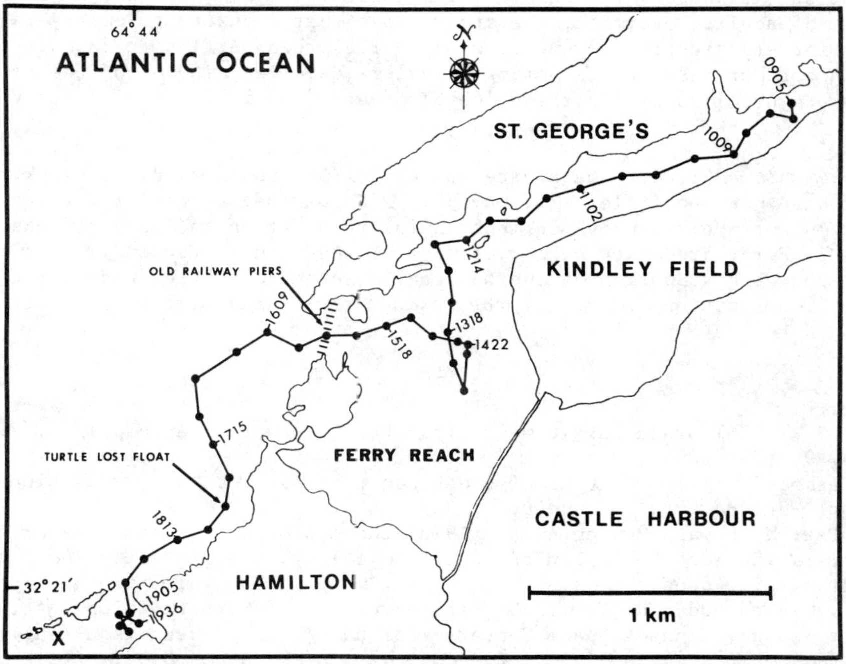

Fig. 1. Travel path of juvenile green turtle (*Chelonia mydas*) released 29 November 1978. The *X* in the lower left of the figure shows the location where the turtle was netted.

attempts to relocate turtles on days following their release, we were able to determine that 3 of these turtles remained in the vicinities of their capture sites for at least 3 weeks. The other 2 were released near the end of the project and we were unable to determine whether they would display similar site tenacity. Three other turtles moved in appropriate directions to reach their capture sites, but either moved so slowly or remained stationary for such long periods that we ended tracking operations before we learned if they would reach these areas. Of these, 1 turtle was found within 60 m of its capture site 2 days following its release and was consistently found in the same area over a period of 2 weeks, 1 remained at a point about midway between its release and capture sites for several weeks, and 1 could not be relocated. Another turtle traveled away from its capture site upon release but was found within 40 m of that location on the following day. On subsequent days, however, it could not be relocated. One turtle remained near its release site for 314 min, the total duration of the track, and could not be relocated either at its release or its capture sites on later days.

In summary, 5 turtles returned to the locations where they were netted on the same days they were released, 1 returned the day following its release, and 1 returned within 2 days of release. Five turtles displayed site tenacity for periods of 2 weeks or longer. The data indicate that there is no relationship between success in homing and the time elapsed between capture and release. Further, the directions of waves, winds, or surface currents do not appear to affect the directions of travel chosen by the turtles in any predictable way.

These results provide clear evidence of homing behavior in juvenile green turtles. Investigations of the sensory cues which enable the turtles to reach home may also shed light on the sensory mechanisms which guide the long distance, reproductive migrations of adults. Further, the study shows that acoustic telemetry is a useful technique for studying the movements of juvenile green turtles in inshore waters. Along with further research on homing behavior, detailed studies of the natural movements of juveniles within these 'developmental' habitats are certainly possible.

Acknowledgements — Without the advice and aid of Dr. and Mrs. H. C. Frick, this study would not have been possible. Mr. Campbell O'Conner helped net all of the subjects. Dr. James Burnett-Herkes provided much useful information and support. Ms. Barbara List and Mr. Peter Frederick performed most of the actual tracking. This material is based upon work supported by the National Science Foundation under Grant No. 78-18196. Contribution No. 806 of the Bermuda Biological Station for Research.

REFERENCES

Balazs G. H. (1976) Green turtle migrations in the Hawaiian Archipelago. *Biol. Conserv.* 9, 125-140.

Burnett-Herkes J. (1974) Returns of green sea turtles (*Chelonia mydas* Linnaeus) tagged at Bermuda. *Biol. Conserv.* 6, 307-308.

Carr A., Carr M. H. and Meylan A. B. (1978) The ecology and migrations of sea turtles, 7. The West Caribbean green turtle colony. *Bull. Am. Mus. Nat. Hist.* 162, article 1, 46 pp.

Ehrhart L. M. and Yoder R. G. (1978) Marine turtles of Merritt Island National Wildlife Refuge, Kennedy Space Center, Florida. In G. E. Henderson (ed.), *Proceedings of the Florida and Interregional Conference on Sea Turtles*, 24-25 July, 1976, Jensen Beach, Florida, pp. 25-30. Fla. Mar. Res. Publ. No. 33, 66 pp.

Ireland L. C. and Lawson K. D., Jr. (1980a) Assembly of a small ultrasonic transmitter. In C. J. Amlaner, Jr. and D. W. Macdonald (eds.), *A Handbook for Biotelemetry and Radio Tracking*, pp. 205-208. Oxford: Pergamon Press.

Ireland L. C. and Lawson K. D., Jr. (1980b) Construction of an acoustic receiver and hydrophone. In C. J. Amlaner, Jr. and D. W. Macdonald (eds.), *A Handbook on Biotelemetry and Radio Tracking*, pp. 225-277. Oxford: Pergamon Press.

Ridgway S. H., Wever E. G., McCormick J. G., Palin J. and Anderson J. H. (1969) Hearing in the giant sea turtle, *Chelonia mydas*. *Proc. Natn. Acad. Sci. U.S.A.* 64, 884-890.

Long Range Tracking of *Crocodylus porosus* in Arnhem Land, Northern Australia

M. J. YERBURY

Telemetry Group, Environment Physics Department, School of Physics, University of Sydney, Sydney, N.S.W. 2006, Australia

Abstract — This paper discusses progress to date with the University of Sydney's crocodile tracking program, which forms part of a detailed study of *Crocodylus porosus* in northern Australia. Following the development of a long range microwave, solar-powered tracking transmitter and a method of attachment to crocodiles, a pilot tracking program commenced towards the end of 1976. The aim of this program was to test the viability of a system operating at the chosen frequency of 1.25 GHz and involving three fixed, triangulating receiving stations located close to the Tomkinson River in Arnhem Land, Northern Territory, Australia. Some conclusions are presented and a new system, currently under development, is described in which spread spectrum and inverse hyperbolic navigation techniques are to be used.

INTRODUCTION

In 1971, the Environmental Physics Department of the School of Physics within the University of Sydney was established. An important part of the work of the Department since that time has been the longterm study of estuarine crocodile, *Crocodylus porosus*, which inhabits the northern coastal fringe of Australia. Many of this study's results have been published already (see selected list of references) and will not be reported here. Instead, this paper will describe a long range, radio tracking system under development by the Department's Telemetry Group which, in the past, has undertaken a variety of projects relating to the study of our environment.

The Group initially set about designing a radio transmitter and hand held receiver suitable for tracking *Crocodylus porosus* in its habitat. At that time very little was known about the life cycle of this animal and especially about the movements of mature individuals although some reports had been received of large crocodiles observed in the open sea hundreds of kilometers from land. Clearly, one requirement of the tracking system was a long range capability, another was a long operating lifetime, preferably in excess of one year. The original tracking system has been described by Brockelsby (1974) but since then, with increasing knowledge about the behavior of *Crocodylus porosus* and with advances in electronics technology, significant changes have been made to the original system and an entirely new approach is under development.

THE CURRENT TRACKING SYSTEM

Figure 1 shows a transmitter, now obsolescent, designed for tracking crocodiles longer than about 3 m. Figure 2 is a photograph of such a transmitter attached to the head of a male crocodile 3.7 m long and weighing 220 kg.

Fig. 1. Solar-powered tracking transmitter for crocodiles longer than about 3 m. Note the scale marked off in cm. A 3-color reflecting panel is used for visual identification also.

The operating frequency of each transmitter is set to one of 10 'channels' clustered at 24 KHz spacings around the center frequency of 1.254 GHz. One of five discrete tones is used to amplitude modulate the transmitted radio frequency carrier, making it possible to distinguish between as many as 50 transmitters operating simultaneously. A nickel-cadmium battery in the base of the transmitter stores energy supplied by three panels of photovoltaic cells arranged on the sloping sides of the outer case. Approximately two hours of direct sunlight each day are required to balance the transmitter's power consumption and, from the fully charged state, it will run for six days in complete darkness. Transmitter output consists of 10 mW, 50 ms tone burst emitted once every two seconds which is sufficient to enable the signal to be detected at a distance of up to 200 km with a receiver installed in a light aircraft flying at a height of about 3,000 m. With the receiving antenna within a meter or so of the ground, however, the range is reduced to about 2 km.

The transmitter so far described is suitable for attachment to crocodiles longer than about 3 m. Crocodiles in this size range at present are not common in northern Australia and are quite difficult to catch. Therefore, efforts were made to reduce the size and weight of the transmitter so as to enable smaller crocodiles to be tracked. The resulting transmitter can be seen in Fig. 3. Compared with the earlier type which weighs 350 g, the new transmitter weighs 140 g, has a volume of 145 ml and therefore floats in fresh water. Its power output of 20 mW is greater and its efficiency of 20 percent is a factor of 2 or 3 times that of the earlier type. Although solar panels can be used, the transmitter shown employs two, size AA, 3.4 V lithium cells capable of maintaining transmissions for approximately 500 days.

Fig. 2. A transmitter attached to the head of a 220 kg, male crocodile, 3.7 m long. The attachment method uses dental acrylic keyed to the sides of the skull platform.

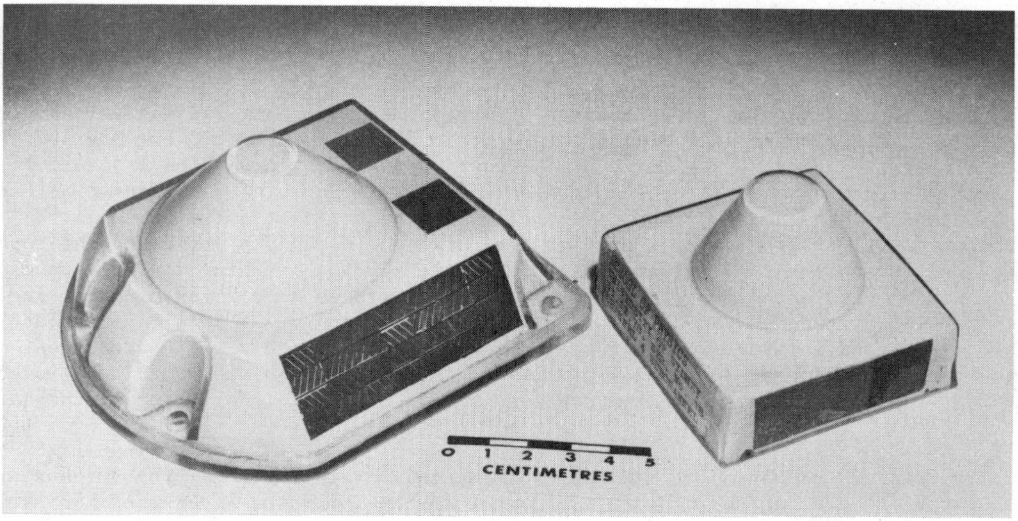

Fig. 3. For comparison with the earlier, solar-powered transmitter, a more recent design is shown on the right. Powered by lithium cells, this smaller transmitter has a greater power output and runs for about 500 days.

Figure 4 is a cross sectional diagram of this transmitter case and its antenna. The electronics compartment is hermetic with glass/metal seals where wires pass through its wall, and the entire device is enclosed in a shell of vacuum formed cellulose-acetate-butyrate (Uvex). A hybrid microcircuit (Figs. 5 and 6) designed and assembled in the Department's telemetry laboratory performs a number of functions at the transmitter. It maintains a low frequency quartz watch crystal in a state of oscillation and generates one of five discrete identification tones with which it amplitude-modulates the RF carrier for 50 ms every two seconds. Its power consumption totals 140 μW.

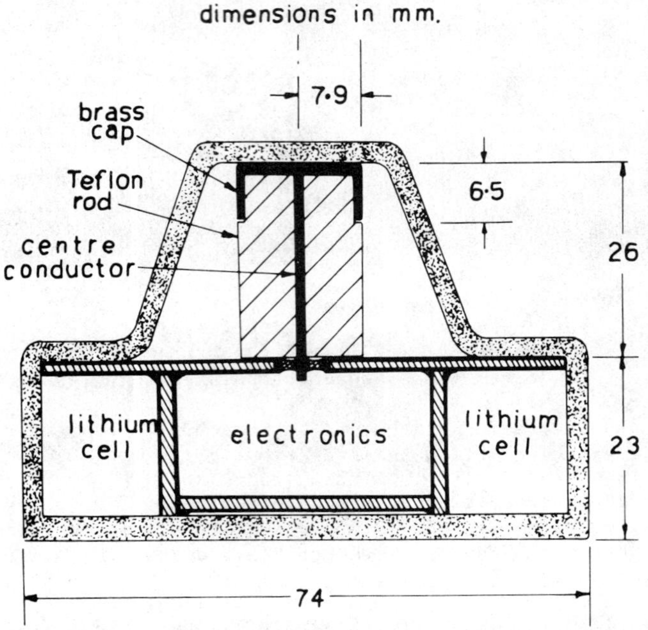

Fig. 4. Cross-section of the smaller tracking transmitter.

The transmitter is required to operate in a harsh environment. It must withstand continual immersion in saline water at depths of up to 30 m and, over a period of a few minutes the temperature may cycle between 60 and 10°C; prolonged exposure to bright sunlight, frequent mechanical shocks, abrasion and vibration are to be expected. The method of attachment to the animal also must be at least as reliable as the transmitter. For crocodiles longer than about 3 m, the bony platform on the top of the skull overhangs the earflaps sufficiently to allow a matching mould to be made and 'keyed' onto each side (Fig. 7). This process takes only a few minutes, and two sections with embedded bolts being held in position by a flat aluminum plate on which the transmitter is mounted. This procedure results in a rigid attachment requiring minimal pressure on living tissue. Also, the transmitter is positioned so that it receives the most exposure even when the animal is swimming at the surface (Fig. 8).

During October and November, 1976 an experimental, 3-station receiving antenna array was established along the eastern escarpment of the Tomkinson River near Maningrida, in Arnhem Land. Receivers were installed in towers overlooking the river at spacings of 5, 12 and 16 km and steerable paraboloids 1 m in diameter were used to obtain bearings on the maximum received signal strengths. The triangulating system was calibrated and tested during this period and it was found to be possible to locate a transmitter to within about 50 m at a distance of 5 km. Concurrently with this work,

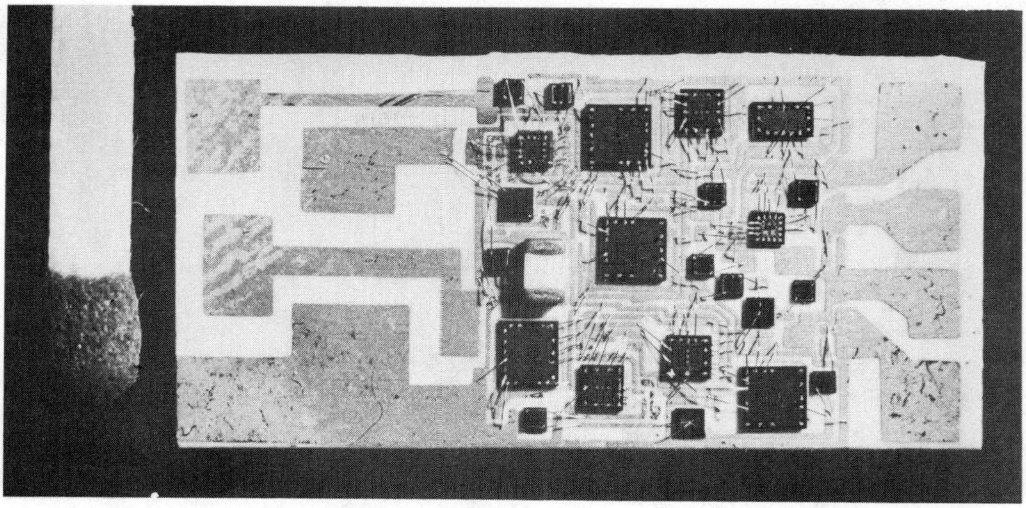

Fig. 5. Hybrid microcircuit used in the transmitter. A quartz watch-crystal (not shown) normally is mounted on the left hand section of the substrate after the rest of the circuit has been encapsulated in an epoxy resin.

an experimental and theoretical investigation was carried out over the same terrain in an attempt to determine the factors influencing the choice of an optimum radio tracking frequency. The results from this study (Horwitz, 1979) supported by experimental data from the triangulating array, suggest that a frequency closer to 500 MHz would be more suitable for tracking crocodiles. It was found that for frequencies above about 300 MHz, path loss due to scattering by foliage increases radidly and that the superposition of all the scattered waves at the receiving antenna causes fluctuations in the signal strength due to relative motion of the transmitter and foliage. Long integration times are needed therefore to establish the average signal strength received from a particular direction and this is equivalent to reducing the overall system sensitivity. To offset this preference to select lower frequencies, there is the sensitivity of the transmitting antenna to detuning by changes in its local environment which increases as λ^6 at long wavelengths. Combining these factors leads to an optimum choice for crocodile tracking which is somewhere in the range 500-800 MHz.

Although some crocodiles have been tracked using these and earlier solar powered transmitters, as yet no full scale tracking program has been started. The intention is to obtain a very reliable and preferably automatic tracking system with the capacity to collect biological, behavioral and environmental data; it would be valuable also in other wildlife studies and could find application in monitoring human and vehicle movements as well as in search and rescue operations. One important advantage of an automatic tracking system is that it dispenses with the need for maintaining staff continuously at three remote stations in an area where communications are poor and cross country travel at times is almost impossible.

THE AUTOMATIC TRACKING SYSTEM

The automatic system now being developed is designed to identify, locate and collect data from more than 100 transmitters, the number depending upon the desired tracking and data rates. It embodies the principle of hyperbolic navigation (Van Etten, 1970;

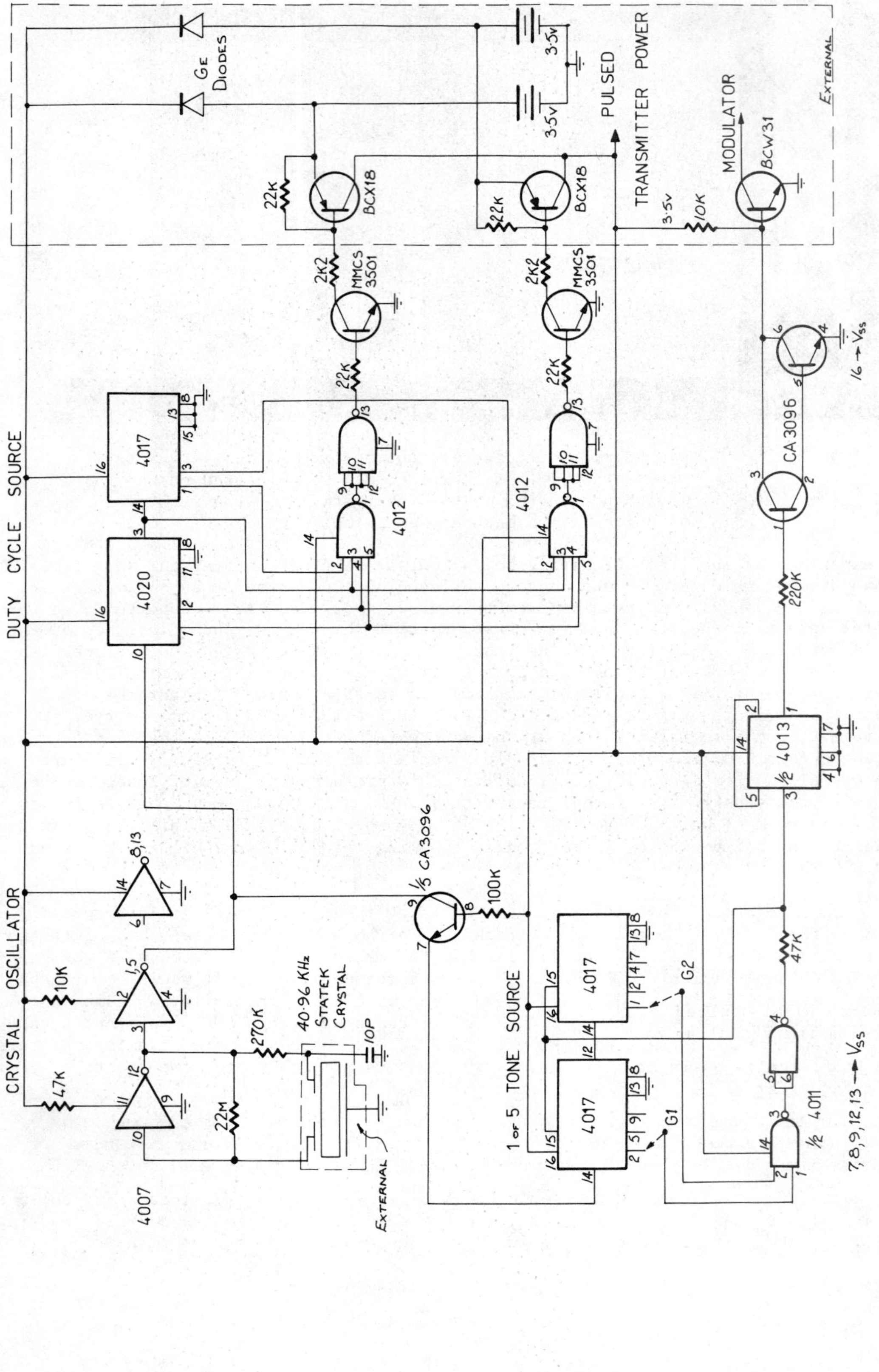

Fig. 6. Diagram of hybrid microcircuit of Fig. 5.

Fig. 7. Two moulds matching the bony protrusions above each ear take only a few minutes to form. The six bolts are used to secure an aluminum baseplate which supports the transmitter.

Beukers, 1974) applied in this case to the relatively short ranges required for tracking individual animals such as crocodiles in their habitat.

Worldwide, low frequency, hyperbolic navigation techniques have been in use since the 1940s; here a ship or aircraft equipped with a receiver, computes its own location from the differences in the times of arrival of signals from at least three widely spaced, fixed transmitting stations. The technique now being developed reverses the roles of transmitter and receiver. Instead, three, widely spaced and fixed receivers measure the differential arrival times of signals from transmitters attached to free ranging animals. Figure 9 shows the three receiving stations A, B and C with 'baselines' AB and BC of arbitrary length and orientation. Families of hyperbolas, known as lines of position, are associated with each pair of receivers.

Fig. 8. A swimming crocodile showing how the attached transmitter is well exposed.

Each hyperbola represents a constant range, or time difference, of a signal arriving at two receivers from the transmitter T. If the times of arrival of a signal from T at A, B and C are t_A, t_B and t_C, respectively, then the lines of position corresponding for example to the time differences $t_{AB} = (t_A - t_B)$ and $t_{BC} = (t_B - t_C)$ intersect at the location of the transmitter. In some cases a second, ambiguous intersection is possible but this can be resolved in a number of ways such as by reference to the previous position of the transmitter and its velocity.

One of the receiving stations is designated the 'master' and computes the transmitters' locations from the information provided by itself and the two 'slaves'. The identity, position and other data associated with each transmitter are processed and stored for later retrieval and analysis.

In order to accommodate a large number of transmitters operating simultaneously, an effective means of separating the received signals is required. Also, since receiver sensitivity is optimized by using a coherent correlation detector, then the local oscillator must be phase-locked to the incoming signal. Phase lock is difficult to achieve if the transmissions take the form of low frequency trains of narrow pulses, as each received pulse provides only one sample of the phase difference. This sample must be stored until the next arrives, the difference between the two being a measure of the phase *drift* in one pulse period. To avoid an ambiguous measurement of phase drift, the transmit frequency must not differ from the receive frequency by more than half the pulse repetition rate and if, due to noise or interference, one or more pulses are lost, phase lock will be very difficult to acquire or maintain.

If in addition to these requirements the system must be capable of handling data and

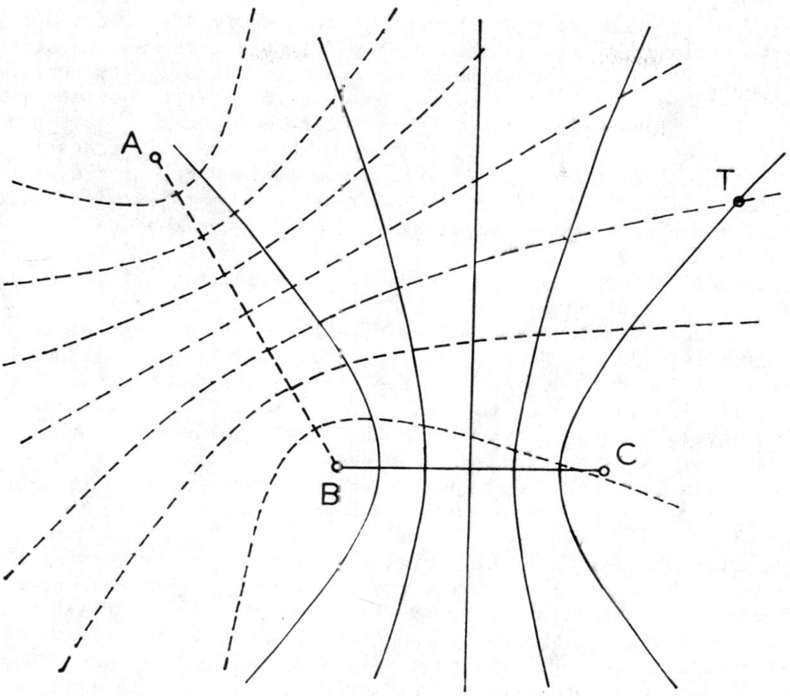

Fig. 9. Diagram showing three receivers, A, B and C and a transmitter, T. Hyperbolic lines of position associated with receiver pairs AB and BC enable the location of T to be determined at the intersection of the two appropriate hyperbolas.

have the ability to provide high range-resolution, it becomes clear that a short duration, low duty-cycle, pulse-transmission format is unsuitable.

All of the specifications can be met and the above mentioned difficulties overcome by transmitting continuously, at a low power level, while phase modulating the carrier with a very long, pseudorandom code. The resulting radio frequency transmission is known as 'spread spectrum' (Dixon, 1976) because most of the transmitted energy is dispersed into a band of frequencies about twice as wide as the code bit rate. A typical example might be a 1023-bit Gold code (Gold, 1967) in which 0 and 1 correspond to carrier phase angles of 0 and Π radians, respectively. If the code bit rate is 1 megabit s^{-1} the code repeats itself every 1.023 ms and most of the transmitted energy is contained in a 2 MHz bandwidth centered on the carrier frequency.

In order to detect such a spread-spectrum signal, the reciever generates the same code with which it phase modulates its oscillator. This is crosscorrelated with the incoming signal as the local code epoch is slowly advanced or retarded until it coincides with that of the received code. When this happens, the spectrum of the signal at the correlator output collapses to a very small bandwidth enabling a narrow band, phase-lock loop to acquire the signal.

There are many advantages to be had from this technique. In the first instance, all transmitters operate on about the same center frequency, identification being achieved simply by using a different code. Since this eliminates the need for separate frequency channels, it is not necessary to control each transmitter's frequency with great accuracy. When many transmissions are received simultaneously, the question of crosscorrelations between the locally selected code and undesired codes arises. Gold (1967) has devised a method of generating code sets containing 2^n-1 separate codes each of length 2^n-1 bits. The crosscorrelation between members of any set is $|R(\tau)| \leq 2^{(n+1)/2}+1$ bits where τ is the delay between the codes and n is a positive, non-zero integer. For example, if $n = 10$ there are 1023 codes in a set each having a peak autocorrelation of 1023 bits and a maximum crosscorrelation between any two bounded by $|R(\tau)| \leq 47$. There are many other code types with different properties having advantages and disadvantages, depending on the application.

Range resolution is a function of the code bit-rate and the ability of the receiver to track the autocorrelation peak closely. If the bit rate is 1 megabit s^{-1} for example, and the receiver tracks the peak to within ± 10 percent of one bit, the range resolution is ± 30 m. For a code length of 1023 bits, the range *ambiguity* is approximately ± 150 km. Further, if the transmitted and received codes are misaligned by as little as one bit, then the output of the correlation detector drops to a very low level. In this way, multipath signals delayed by more than one bit-interval are almost completely rejected by the receiver. For instance, with a bit rate of 10 megabits s^{-1}, those multipath signals traversing a route exceeding the direct path by 30 m or more are essentially undetected.

The spread spectrum technique is also highly effective in suppressing the effects of interference. The worst kind of interference likely to be encountered in practice is a continuous wave (CW) at a frequency lying in the receiver passband. The spectrum of this undesired signal is spread by the crosscorrelation process at the receiver so that only a small portion of its energy is accepted by the narrow postcorrelation filter; wideband interference is dispersed still further and is even less effective.

Data can be transmitted digitally by phase-shift keying (PSK), in which each *code block* is assigned a 0 or Π phase. Thus, in the previous example, one data bit would be conveyed by 1023 code bits at a maximum rate of 9775 bits s^{-1} with a code rate of 10 megabits s^{-1}.

An experimental transmitter now under construction uses an erasable, permanent read-only memory (EPROM) to store the code which can be up to 4096 bits long. The bit rate is 1 megabit s^{-1} and the radiated power is approximately 2 mW centered on a frequency close to 500 MHz. Five AA size, nickel-cadmium cells designed for high temperature operation and recharged by a solar panel provide the energy needed. There is sufficient reserve to run the transmitter for up to six days without additional charge.

At this stage of the program, there is still much to be done. The problem of acquiring the desired direct path signals at the three receiving stations, as quickly as possible and in the presence of multipath propagation, has to be tackled. Most probably this will involve an acquisition strategy developed in the software and using as much present and past information about the particular transmitter as can be made available. A minicomputer installed at the master station is required to handle this problem. Also it will solve the hyperbolic equations giving the various transmitters' positions, resolve any ambiguities which may arise, compute position errors, process and store data and make available a graphic display and hardcopy output to the operator.

The many applications of such a system justify the effort going into its development. As digital speeds increase and power consumption, size and cost of components fall, the technique presented here will become increasingly attractive.

Acknowledgements — The work of the Telemetry Group is supported by the Science Foundation for Physics and the University Research Grants Committee, both within the University of Sydney.

REFERENCES

Beukers J. M. (1974) A review and applications of VLF and LF transmissions for navigation and tracking. *Navigation* 21, 117-133.

Dixon R. C. (1976) *Spread Spectrum Systems*. New York: John Wiley.

Gold R. (1967) Optimal binary sequences for spread spectrum multiplexing. *IEEE Trans. on Inf. Theory*, IT-13. 4, 619-621.

Van Etten J. P. (1970) Low and very-low-frequency hyperbolic techniques. *Elect. Commun.* 45, 192-212.

REFERENCES RELATING TO THE CROCODILE PROGRAM IN NORTHERN AUSTRALIA

Brockelsby W. K. (1974) Microwave techniques in animal radiotelemetry. *Biotelemetry II*, 2nd Int. Symp., Davos 1974, 220-222. Basel: Karger.

Drane C. R., Webb G. J. W. and Heuer P. (1977) Patterns of heating in the body, trunk and tail of *Crocodylus porosus*. *J. Thermal. Biol.* 2, 127-130.

Grigg G. C. and Alchin J. (1976) The role of the cardiovascular system in thermoregulation of *Crocodylus johnstoni*. *Physiol. Zool.* 49, 24-36.

Grigg G. C. (1974) Tuning in to crocodiles. *Aust. Nat. Hist.* 18, 78-83.

Grigg G. C. (1977a) Ionic and osmotic regulation in the estuarine crocodile, *Crocodylus porosus*. In H. Messel and S. T. Butler (eds.), *Australian Animals and Their Environment*, pp. 335-354. Sydney: Shakespeare Head Press.

Grigg G. C. (1977b) The body temperature of crocodiles and dinosaurs. In H. Messel and S. T. Butler (eds.), *Australian Animals and Their Environment*, pp. 355-367. Sydney: Shakespeare Head Press.

Grigg G. C., Drane C. R. and Courtice G. P. (1978) Thermal time constants of heating and cooling in a dragon, *Physignathus lesueurii*, and some generalisations about heating and cooling. *J. Thermal. Biol*, in press.

Grigg G. C. (1978) Metabolic rate, Q_{10} and respiratory quotient in *Crocodylus porosus* and some generalisations about low RQ in reptiles. *Physiol. Zool.* 51, 354-360.

Horwitz C. M. (1974) Low-drain EEG amplifier. *IEEE Trans. Biomed. Eng. BME* 21, 60-61.

Horwitz C. M. (1978a) A reliable DC-DC convecter circuit. *Electron. Engng.* 50, 11.

Horwitz C. M. (1979) Optimisation of radio tracking frequencies, *IEEE Trans. Ant. Prop.*, in press.

Magnusson W. E., Grigg G. C. and Taylor J. A. (1978) An aerial survey of potential nesting areas of the saltwater crocodile, *Crocodylus porosus* Schneider, on the north coast of Arnhem Land, northern Australia. *Aust. Wildl. Res.* 5, 401-415.

Messel H. (1977) The crocodile programme in northern Australia — population surveys and numbers. In H. Messel and S. T. Butler (eds.), *Australian Animals and Their Environment*, pp. 207-236. Sydney: Shakespeare Head Press.

Messel H. and Stephens D. R. (1978) Drug immobilisation of *Crocodylus porosus* and *Crocodylus johnstoni*. *J. Wildl. Mngmt.*, in press.

Taylor J., Webb G. J. W. and Magnusson W. (1978) A method of collecting stomach contents from live crocodilians. *J. Herpet.*, in press.

Webb G. J. W., Messel H. and Magnusson W. (1977) The nesting of *Crocodylus porosus* in Arnhem Land, northern Australia. *Copeia* 2, 238-249.

Webb G. J. W. (1977) The natural history of *Crocodylus porosus*. I. Habitat and nesting. II. Growth, movement, river distributions and general comments. In H. Messel and S. T. Butler (eds.), *Australian Animals and Their Environment*, pp. 237-312. Sydney: Shakespeare Head Press.

Webb G. J. W., Yerbury M. J. and Onions V. (1978) A record of a *Crocodylus porosus* attack. *J. Herpet.* 12, 267-268.

Webb G. J. W. and Messel H. (1978a) A morphometric analysis of *Crocodylus porosus* from the north coast of Arnhem Land, northern Australia. *Aust. J. Zool.* 26, 1-27.
Webb G. J. W. and Messel H. (1977a) Abnormalities and injuries in the estuarine crocodile, *Crocodylus porosus*. *Aust. J. Wildl. Res.* 4, 311-319.
Webb G. J. W. and Messel H. (1977b) Crocodile capture techniques. *J. Wildl. Mngmt.* 41, 572-575.
Webb G. J. W., Messel H., Crawford J. and Yerbury M. J. (1978) Growth rates of *Crocodylus porosus* from Arnhem Land, northern Australia. *Aust. Wildl. Res.* 5, 385-399.
Webb G. J. W. and Messel H. (1978b) Movement and dispersal patterns of *Crocodylus porosus* in some rivers of Arnhem Land, northern Australia. *Aust. Wildl. Res.* 5, 263-283.
Webb G. J. W. and Messel H. (1978c) Wariness in *Crocodylus porosus*. *Herpetologica*, in press.
Yerbury M. J. (1977) Telemetry and crocodiles. In H. Messel and S. T. Butler, *Australian Animals and Their Environment*, pp. 215-332. Sydney: Shakespeare Head Press.

The Use of Longterm Ultrasonic Implants for the Location and Harvest of Schooling Fish

P. B. JOHNSEN

Laboratory of Limnology, University of Wisconsin, Madison, Wisconsin, U.S.A.

Abstract — Carp (*Cyprinus carpio*) tagged with implanted ultrasonic transmitters, were tracked over 3 years to determine whether this species forms aggregations during the winter. Location of large numbers of fish by this method could greatly increase fishing efficiency. Two large aggregations which form in the late autumn and persist through spring were discovered. Radio tagged carp arrived at these two areas with relative synchrony from different parts of the lake Fish introduced from another lake also aggregated at these two areas. The use of fish equipped with transmitters to locate large aggregations could add to fishing technology.

INTRODUCTION

Inland commercial fishermen are increasingly faced with rising operational costs while the price received for their catch remains fairly constant. One of the most costly aspects of fishing is the search for fish in sufficient numbers to make harvest economically feasible. In recent years, echo location (sonar) has been utilized widely, but the method requires that the fishermen search large areas to locate fish. A great reduction in search time would be achieved if the fishermen could actively be led to large concentrations of fish. To this end, a program in ultrasonic tracking was initiated.

It was commonly believed, but undocumented, that carp (*Cyprinus carpio* L.) form large aggregations during winter months. To determine whether this was true and, if so, where the fish congregate, a program was developed to tag fish with ultrasonic transmitters and follow their movements. These 'Judas fish' would, in a sense, betray their conspecifics by revealing their location. Quickly located aggregations could be exploited economically by fishermen.

MATERIALS AND METHODS

A transmitter capable of producing a signal which could be tracked at a minimum range of 2 km in water of high conductivity (450 μmhos cm^{-1}) at depths up to 26 m was required. The unit also had to function in temperatures approaching $0^{\circ}C$ for approximately 2 months. Above all, it had to be small enough to implant within the fish's body lest its visible presence caused other fish to avoid the tagged individual.

The high conductivity and depth range ruled out the use of radio transmission which suffers severe signal attenuation under these conditions. Conventional ultrasonic transmitters can operate under these conditions but required modification to meet our requirements. The use of a lithium power source which does not exhibit voltage decreases at low temperatures solved the problems of operation in cold water. To extend the life of the transmitter, the duty cycle was reduced to less than 1 percent but the pulse duration was maintained at 7 ms. This approximately equals the minimum duration necessary to produce a discernible audio tone required to discriminate the signal from background noise generated by the movement of ice. The final ultrasonic tag was a transmitter 8 × 2 cm in size with a trackable range of 3 to 9 km, depending on conditions. Operating at 70 KHz, individual transmitters were coded by pulse intervals varying from 0.75 to 1.5 s, and had an operational life of 50 to 146 days with an average of 75 days.

The transmitters were surgically implanted in the body cavity. Fish were anesthetized in 2-phenoxyethanol (0.25 mg l^{-1}) for 10 min then restrained ventral side up in an operating trough. After removing several scales anterior to the cloaca and lateral to the midline, a small scalpel incision was made and opened for 3-4 cm with scissors. The transmitter was inserted with its longitudinal axis parallel with the fish's. The incision was closed with 3 or 4 nylon sutures, and the fish were transferred to a recovery tank. Each operation took less than 5 min. All fish were kept in flow-through tanks for a 48 h recovery and observation period before release. Subsequent movements by the fish indicated that all fish survived the implant procedure.

Individual fish were identified by differences in the pulse rate of each transmitter. To determine the identity of an individual, the interval for 10 pulses was timed with a stop watch. A Smith-Root TA-50 receiver and hydrophone was used for tracking. Fish were located from a boat, airboat, or snow mobile an average of three times per week with longer intervals during inclement weather; not all fish were located in each search.

RESULTS

Tracking was conducted on an experimental basis for three winters starting in 1974 in Lake Mendota, Wisconsin, U.S.A. The results of the first two years of the program are reported by Johnsen and Hasler (1977) and can be summarized by the following observations.

Four groups of fish were tracked, one in 1974-1975 and three in 1975-1976. Fish in 1974-1975 (Group 7, 4 fish) captured off Fox Bluff, were released in University Bay. In 1975-1976, groups 2 (5 fish, Middleton Bay), 3 (4 fish, University Bay), and 4 (4 fish, James Madison Park) were captured and released at their respective sites (Fig. 1). Movement patterns of all fish released at the same site were similar, with no differences noted for different sexes or sizes. Fish did not move as a group but rather as individuals. Those released on the southern (Groups 1 and 3) and southeastern shores (Group 4) circulated counterclockwise to the northern part of the lake, while fish released on the western shore (Group 2) moved clockwise to the northern shore. This circulation was slow, with most fish taking almost 1 month in 1974 and 2 months in 1975 to arrive at the north shore.

Arrows in Fig. 1 show the generalized movement patterns. The locations of all observations were plotted, and the outside limits define the movement corridors used by the fish. Segments that are single lines do not necessarily indicate the path of an individual fish.

Tracking revealed two aggregation areas (Area A and B, Fig. 1) at the north side of the lake. During both years, large numbers of untagged fish were detected both

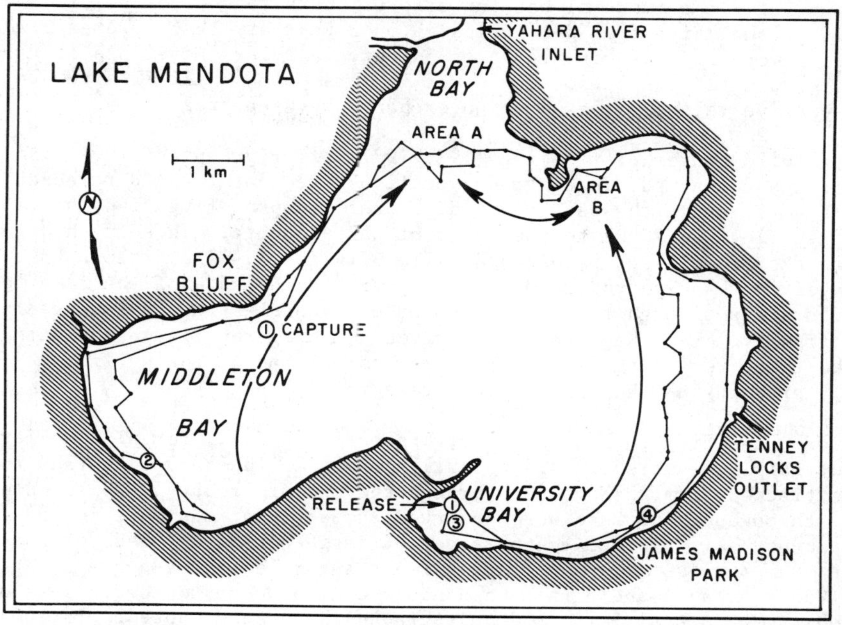

Fig. 1. Gross movement patterns of carp fitted with transmitters. All fish were found within the bounded areas. Numbers indicate capture and release sites referred to in the text. Aggregation areas are designated by A and B.

visually (activity at the water surface) and by echo location in these areas. Both areas are at depths of 5 to 7 m at the edge of extensive beds of the macrophyte, *Myriophyllum spicatum* L. A search of the littoral zone of the lake revealed no other concentrations of carp at this time.

In both years, fish tagged with ultrasonic transmitters arrived at the aggregation areas with relative synchrony, within 2 days in 1974 and 12 in 1975. On 18 November 1975, fish were still found up to 5 km from the aggregation areas. Four days later 10 of 13 instrumented fish had arrived at either Area A or B. The next observation on 30 November found all 13 tagged carp in the two areas. Fish made their most extensive movements during the interval just prior to arrival at the aggregation areas. Ten of the 17 fish tracked, crossed between the two areas at least once, indicating that fish from different parts of the lake mix freely between Areas A and B.

The occurrence of large numbers of fish in the two areas, indicated by the presence of tagged fish was demonstrated by the commercial harvest of 46,400 kg of fish from only 17,600 m of gill nets lifted. Test netting captured nothing in areas where instrumented fish were not found. The conclusions of the first two years of tracking were that carp do form aggregations, it was possible to locate them using the 'Judas fish' concept and that increased efficiency in commercial fishing efforts could be realized. Additional questions involving the length of time the fish remained in the two areas and a greater understanding of the nature of the two aggregation areas remained to be investigated by further tracking.

Persistance of the two aggregations throughout the winter was demonstrated by the tracking of 10 additional fish released during the winter of 1976. These fish, captured during the commercial netting efforts, were tagged by the same methods as the earlier fish. Five fish were released on 3 February 1976 in Area B while another five were released in Area A on 5 February 1976. Movements of these fish were monitored until the end of ice cover in late March. All fish remained in their respective release areas with no exchange between areas.

During the third winter of tracking (1976-1977), four groups of five fish each were instrumented and tracked. Three groups of fish were captured and released at Sites 2, 3, 4 (Fig. 1) as in the past. An additional group of five fish was captured in Lake Kegonsa, located 20 km to the south of Lake Mendota. These fish were transferred to Lake Mendota and released at site 3 (Fig. 1). The fifteen Lake Mendota fish behaved in the expected manner, moving around the edge of the lake to the two aggregation areas through the same movement corridors as in previous years. Of the five fish from Lake Kegonsa, two arrived at the aggregation areas with the other fish. The three remaining continued to move slowly about the perimeter of the lake eventually arriving at the aggregation areas in mid-January.

Tracking of other fish in Lake Kegonsa, Wisconsin, U.S.A., revealed no large aggregations of carp. Ten fish, ultrasonically tagged in the same manner as the Lake Mendota fish were tracked from 22 November through the end of January. While extensive movements throughout the lake were observed, no aggregations were formed. One tentative hypothesis is that the presence of large persistent beds of the aquatic plant *Myriophyllum spicatum* may play an important role in the aggregating behavior of carp. Both Areas A and B in Lake Mendota are adjacent to the largest associations of *Myriophyllum*. Aerial infra red photographs which show these two stands of plants in Lake Mendota indicate that no such accumulations of plants exist in Lake Kegonsa. Future work is required to determine if, in fact, the plants play a role in the aggregation of carp.

CONCLUSIONS

Large aggregations of fish were located by following the movements of individual ultrasonically tagged fish, with considerable savings of search effort by the fishermen. Thus the objective of using a 'Judas fish' to locate fish aggregations to increase the efficiency of commercial fishing was successfully achieved.

In addition to leading fishermen to aggregations of fish, it is possible that other environmental factors, more readily measured, which play a role in the natural history of fish may be identified. Undoubtedly biotelemetry will play a role in improving fishing technology.

REFERENCES

Johnsen P. B. and Hasler A. D. (1977) Winter aggregations of carp (*Cyprinus carpio*) as revealed by ultraonsic tracking. *Trans. Am. Fish Soc.* 106, 556-559.

The Movements of Migrating Salmonids in the Vicinity of a Heated Effluent Determined by a Temperature and Pressure Sensing Radio Telemetry System

P. B. JOHNSEN

Laboratory of Limnology, University of Wisconsin, Madison, Wisconsin, U.S.A.

Abstract — The movements of eight migrating salmonids in the vicinity of a heated effluent were monitored with a temperature and pressure sensing biotelemetry system. Fish remained at the upwind edge of the thermal plume moving in and out of the heated water. Swimming speeds were slow and turns frequent. The fish remained in the study area for 3 to 22 h with an average residence time of 13 h. After leaving the thermal plume area, swimming speeds increased and straight courses were maintained. This was typical of fish tracked before the thermal plume was present.

INTRODUCTION

Landlocked salmonids of Lake Michigan make their spawning migration along the shore to their homestreams. In recent years, nuclear-fueled electric generating stations have been constructed in these areas. Large volumes of lake water are drawn through the power stations for cooling and then discharged at 10°C above ambient temperature. The thermal discharge plumes often lie across the path of the migrating fish. To assess the possible impact of the heated effluents and to determine the salmon's response to them, a tracking program was undertaken.

MATERIALS AND METHODS

Fish-borne transmitters which transmit temperature and pressure were developed (Scidmore, Guckel and Beyer, 1975) along with the associated tracking system (Johnsen and Horrall, 1974). The transmitting units, measuring 2 × 9.5 cm, operated with a crystal controlled carrier frequency of 52.78 to 53.78 MHz with 50 KHz separating individual transmitters. A 50 ms pulse of a 400 Hz audio tone, modulated ± 200 Hz, coded the temperature information, while pulse position, or off time between pulses, coded pressure. The transmitters were sensitive to 0.1°C temperature changes and had a resolution of ± 15 cm in depth. A thermistor was used as the temperature transducer, while a silicon piezoresistive strain gauge functioned as the pressure transducer. With the standard 200 ma h battery, the transmitter life was approximately 50 h. A voltage regulator was incorporated for stability of measurements while a sample-and-hold system was used to keep battery drain to a minimum. The transmitters were encapsulated in Scotchcast 221 after priming the surface of the pressure sensor and antenna with Scotchcast XR5137 primer.

The tracking system, located on the lake shore, comprised two stations used to determine positions by triangulation. These units, six element dualbeam null-peak yagi antennas tuned to 52.78 MHz, were mounted on 6.5 m towers. Each tracking station was constructed so that the hinged supporting arms of the antennas could be lowered through a system of cables and winches to prevent damage by the frequent high winds. Operators of these stations relayed bearing information to a central station where fish positions were plotted. The central station, located midway between the two tracking stations, housed communication receivers which continuously monitored transmitted data. Received signals from a 5/8 wavelength colinear groundplane antenna determined the operational limitations of range and signal detectability of the telemetry system as a whole. With the transmitter 1 m below the surface (water conductivity = 260 μmhos cm^{-1}) the effective monitoring range was 6 km in calm weather. Surface conditions of the lake often reduced the monitoring range to 3 km or less. Audio signals from the communication receivers were recorded on magnetic tape for later analysis.

Owing to the amount and form of the data recorded, an automatic data reduction system involving analog and digital computers was required. While the system functioned well under ideal conditions, rough weather induced high signal-to-noise ratios which severely limited the effectiveness of the operation. Ultimately the costs of computer time and the problems of signal quality caused the system to be abandoned.

Tracking experiments were conducted during the autumn of 1974 at the Point Beach Nuclear Power Plant located in Manitowoc County, Wisconsin, U.S.A. During the autumn, large numbers of salmonids are attracted into the cooling water discharge structures. Fish to be tracked were captured by dipnet, radio tagged and returned to the discharge waters. Transmitters were attached to the fish by means of two stainless steel pins (Petersen Pins) passed through the base of the dorsal fin and were secured with plastic discs on the opposite side.

RESULTS

In 1974, eight fish were tracked (Table 1); three rainbow trout (*Salmo gairdneri*), two brown trout (*Salmo trutta*), two chinook salmon (*Oncorhychus tshawytscha*) and one coho salmon (*O. kisutch*). Fish generally left the discharge structure within 15 min of release while the chinook salmon remained for 60 and 99 min before entering the lake.

Table 1.
Tracking date, sex, length and residence time of fish tracked at the Point Beach Power Plant

Date tracked	Species	Sex	Length (cm)	Time spent in study area (h)
11 Oct 74	Rainbow	M	64	18.0
11 Oct 74	Chinook	M	89	17.25
25 Oct 74	Brown	M	58	22.0
25 Oct 74	Chinook	M	90	19.25
7 Nov 74	Brown	F	59	15.9
7 Nov 74	Coho	F	66	3.5
8 Nov 74	Rainbow	F	62	3.0
8 Nov 74	Rainbow	F	68	5.75

After leaving the discharge structure, fish moved out into the lake through the thermal discharge plume. All fish moved slowly along the upwind edge of the plume where the temperature gradient was the steepest. Remote sensing of surface water temperatures indicate that changes of more than 5°C may occur over less than 200 m (Scarpace and Green, 1973). Movement patterns and changes in transmitted temperatures indicated that the fish were moving in and out of the warmer water. During this time, movements were characterized by frequent turns, many of which were 180° reversals of swimming direction. Swimming speeds were slow (ca. 20 cm s^{-1}) and fish often remained at the same location for 2 to 3 position checks (measured every 15 min) before initiating movement again.

Fish remained in the study area (defined as the area within a 3 km radius of the discharge structure) after radio tagging for periods of time ranging from 3 to 22 h (see Table 1). The actual time spent by the fish in the vicinity of the plant was impossible to determine because the residence time spent before capture and tagging was unknown. Eventually all fish moved north at speeds which often exceeded 100 cm s^{-1} while maintaining straight courses. This rapid movement in a straight direction was characteristic of the movements of coho salmon tracked in the same area before plant operation (Robinson, 1973).

CONCLUSIONS

Radio tracking revealed that salmonids remained in the vicinity of the thermal discharge plume for 3 to 22 h after tagging. These fish moved in and out of the plume across the area of the steepest gradient. After leaving the area, movements were similar to those recorded in years prior to the establishment of the power plant.

The general evaluation of the tracking system was that it was too complicated for the required tasks. Additionally, the associated costs of data analysis became prohibitive and the system was abandoned in favor of simpler approaches.

Acknowledgements — This research was funded in part by the National Oceanic and Atmospheric Administration, office of Sea Grant, Department of Commerce through an institutional grant to the University of Wisconsin.

REFERENCES

Johnsen P. B., and Horrall R. M. (1974) A temperature and pressure sensing biotelemetry system. *Underwater Telemetry Newsletter* 4, 13-15.

Robinson J. P. (1973) Migratory movements of adult coho salmon (*Oncorhynchus kisutch*) in Lake Michigan as revealed by ultrasonic telemetry methods. M.S. thesis. University of Wisconsin-Madison.

Scarpace F. L. and Green T., III. (1973) Dynamic surface temperature structure of thermal plumes. *Water Resources Res.* 9, 138-153.

Scidmore A. K., Guckel H. and Beyer J. B. (1975) A pressure-temperature-locator probe for monitoring the environment of a free swimming fish. University of Wisconsin Engineering Experiment Station pub. ECE-75-1. 20 pp.

Studies on the Behavior of Fish in Relation to Power Station Operation in the U.K.

T. E. LANGFORD, J. M. FLEMING and N. P. JAMES

C.E.R.L., Marine Biological Laboratory, Fawley, Southampton, U.K.

Abstract — The behavior and movement patterns of fishes in their natural environment may be key factors determining whether an effluent or any other ecological disturbance has a significant effect on the populations. Alterations in migration, dispersal, feeding or breeding movements may, in the long term, cause adverse effects. Since 1973, ultrasonic tracking has been used to study the movements and behavior of freshwater and estuarine fishes in relation to water-use by British electricity generating stations. Three major areas of research are summarized below.

Studies in rivers heated by thermal discharges are evaluating the effects of effluent plumes on the natural homing, movement and aggregation patterns of non-diadromous (coarse) fishes. Residence times in relation to temperature suggest that the strict constraints on discharges suggested by the E.E.C. may not be ecologically necessary. Avoidance behavior has been noted in field studies. At Dinorwic in North Wales, the construction and operation of an 1800 MW pumped storage scheme may affect a lake and small river system which carry migratory salmonids. The rate of migration of these fish is being investigated in relation to patterns of flow and suspended solids. Small automatic bankside recording stations are used to monitor the passage of sonic-tagged fish. Future thermal or tidal-barrage based generation may affect at least one large estuary in Britain. To prevent ingress of migrating salmonids at water intakes and to plan locations of barrage fish passes, the routes which salmon take through an estuary are being investigated using sonic-tracking techniques. The possible 'thermal barrier' effects on migration are being studied. Problems in analysis of tracking data are discussed.

INTRODUCTION

The generation of electricity on a commercial scale requires large volumes of water, either for driving the turbines in hydrogenerating stations or for producing and cooling steam in thermal (steam) power stations. Dams and diversions on rivers cause drastic changes in their flow regimes and currents both upstream and downstream. Cooling water intakes sited on rivers, estuaries or on the coast cause localized changes in currents and sediments, while the heated cooling water discharges alter currents, sediments, temperature, and water chemistry in the vicinity of the outfalls. The extent of the physical and chemical disturbances in a habitat depends on many factors, including geography, hydrography, climatic conditions and dilution (Efford, 1975; Langord, 1974; Burnett *et al.*, 1974; Langford, in press).

The effects of the physical and chemical changes on fish populations depend to a large extent on the behavior of individuals and groups in their habitat. Their capacity to detect and avoid adverse conditions, the timing and strength of migratory stimuli, selection of specific feeding and spawning sites, and selection of preferred temperature, current velocities and chemical conditions may all be key factors. To try to understand natural behavior patterns and the responses of fishes to specific environmental disturbances, research using ultrasonic tagging techniques has been in progress since 1973 (Langford, 1974; Langford *et al.*, 1979). This paper summarizes very briefly some of the results from the three main topics, viz — (a) Responses of fish to thermal discharges. (b) The migration routes and strategies of salmonids in estuaries and rivers. (c) The rate of migration of salmonids in rivers in relation to flow alteration and regulation.

METHODS AND STUDY SITES

Both basic techniques of ultrasonic tracking fish have been used in the work so far, i.e. active monitoring of movements from a boat, *and* automatic recording from land based stations (Stasko and Pincock, 1977). Commercially available tags supplied by Smith-Root Inc. and Bayshore Systems Corporation, operating on 50 to 75 KHz, have been implanted, inserted into stomachs, or attached externally, depending on the species studied and the objectives of the project.

In the River Witham and River Thames (U.K.), the movements of resident fishes such as Bream (*Abramis brama*), and Pike (*Esox lucius*) have been monitored at daily intervals for periods of up to 50 days, in relation to the cooling water discharges (outfall) from Lincoln and Didcot Power Stations respectively. Both introduced and indigenous fish were tracked and the homing responses of the latter studied after displacement from the site of capture. In tidal waters, mainly Southampton Water, the movements and routes of salmonids have been recorded over periods of up to 37 h, by following tagged fish in a boat.

Automatic recording stations have been in use, monitoring the passage of individual ultrasonic tagged salmon and sea-trout in the River Seiont in North Wales. The river has been modified in its upper reaches to facilitate the use of a lake in a large pumped-storage hydro-electricity scheme. Six stations are spaced at intervals along the 14 km river and these record tag signals on magnetic tape and on charts (Langford *et al.*, 1977).

RESULTS

All three programs are still in progress but a large amount of information is already available, some of which has been reported (Langford, 1974; Langford *et al.*, 1977; Langford *et al.*, 1979).

RESPONSES TO THERMAL DISCHARGES

Thirty bream (*A. brama*) have been tagged and tracked in the Witham and Thames around Lincoln and Didcot outfalls usually in groups of three to six. In addition twelve pike (*E. lucius*) have been tracked in the Thames in the vicinity of the outfall from Didcot Power Station. In the Witham, indigenous bream displayed strong associations with their area of capture, though they also showed considerable individual variation in the timing and distance of daily movements. Some bream stayed near the outfall of Lincoln Power Station for up to 15 days, even though there was no discharge during this period. In the Thames, individual bream remained in the warm water plume from Didcot for 2-5 days, though they invariably left the area even though the discharge continued (Langford *et al.*, 1979). Pike have, as yet, shown little attraction to the heated water from Didcot even in winter, and displaced fish have moved both upstream and downstream past the plume. The plume at Didcot is, however, small and may not be detectable by fish moving along the opposite side of the river during high flows.

MIGRATION ROUTES IN ESTUARIES

Ultrasonic tagged salmon and sea-trout tracked in Southampton Water were found to travel often in shallow areas for long periods though some fish travelled in the deep water channel for some time. Fish have appeared to follow tidal currents and speeds very closely, but have made lateral movements, i.e. across the main current vectors, during slack water at high tide. Fish were found at times in water less than 0.5 m deep, and sometimes even in the 'surf-zone'.

MIGRATION RATES IN THE SEIONT

Records of some fifteen fish have now been analyzed from the automatic recorders, though there are, as yet, no fish recorded through the whole system. The data show that there was considerable variation in the delay period between release of tagged fish and their passage through the first recording station. Later records have shown that fish may move both upstream and downstream after release, and one fish which moved downstream waited almost one month before returning upstream. A second fish took almost ten days to traverse one 2 km reach but only 4 h to travel the next 2 km.

CONCLUSIONS

It is not possible in this short paper to present all the available data or to demonstrate logical conclusions. To generalize, however, it is obvious that ultrasonic monitoring of fish behavior can illustrate the complexity of patterns which the passive tagging and sampling methods cannot. One of the factors which will allow a species to survive habitat disturbances is the variability in the responses of individuals to chemical and physical changes in the habitat. Ultrasonic tagging techniques allow detailed studies and expression of this individuality.

Even from the very brief account given here, the variability in fish behavior is evident in the different habitats. It is hoped that from an understanding of such behavior and movement patterns, future power stations may be designed, sited and operated to produce minimal effects on important fisheries and that proposed expensive constraints on intakes and outfalls may be evaluated objectively.

Acknowledgements — We are grateful to all those colleagues in C.E.R.L. Biology Section who have helped in the routine tracking and fish catching operations. Thanks also to Drs. Peter Henderson, and Andrew Turnpenny for analysis and experimental procedures, and to all the staff of the Anglia, Southern and Welsh Water Authorities, U.K., for assistance. Finally, thanks to the Biology Group at North West Project Group, Dinorwic, for advice and facilities. This paper is published with the permission of the Central Electricity Generating Board.

REFERENCES

Burnett J. M., McMillan W., MacQueen J. F., Moore D. J. and Shepherd J. G. (1974) The cooling of power stations. Paper 2.2.5. Proc.

Efford I. E. (1975) Environmental impact assessment and hydro-electric projects. Hindsight and foresight in Canada. *J. Fish. Res. Bd. Can.* 32, 98-209.

Langford T. E. (1974) Ecology and cooling water from power stations. A review of recent biological research in Britain. Paper 2.2.4. in Proc. 9th World Energy Conference. Detroit. W.E.C. London. 32 pp.

Langford T. E., Fleming J. M. and James N. P. (1977) The tracking of salmonids in the Avon Seiont system using ultrasonic tagging techniques. C.E.R.L. Lab. Note. RD/L/N/51/77. Leatherhead, Surrey, U.K.

Langford T.E., Milner A. G. P., Foster D. J. and Fleming J. M. (1979) The movements and distribution of some Common Bream (*Abramis brama* L.) in the vicinity of power station intakes and outfalls in British rivers as observed by ultra-sonic tracking. C.E.R.L. Lab. Note RD/L/N. Leatherhead, Surrey, U.K.

Langford T. E. *The Electricity Industry and the Ecology of Natural Waters*. Liverpool University Press, in press.

Stasko A. B. and Pincock D. G. (1977) Review of underwater biotelemetry, with emphasis on ultra-sonic techniques. *J. Fish. Res. Bd. Can.* 34, 1261-1285.

The Coastal Movements of Returning Atlantic Salmon, *Salmo salar* L.

A. D. HAWKINS, G. G. URQUHART and W. M. SHEARER

Department of Agriculture and Fisheries for Scotland, Victoria Road, Aberdeen AB9 8DB, U.K.

Abstract — Six Atlantic salmon grilse were caught by bag nets at a netting station on the east coast of Scotland, tagged with ultrasonic transmitters, released and tracked by means of directional hydrophones mounted beneath two small boats. In the open sea, the fish generally moved with the tidal currents. In the vicinity of the river the behavior changed and the fish swam against the current, eventually entering the river mouth. The transition from full seawater to freshwater was rapid and without any period of adaptation. In reaching the river mouth, the salmon often passed through lines of bag and stake nets demonstrating that these nets are rather inefficient.

DISCUSSION

The return of maturing salmon to their home river from distant parts of the sea is a most interesting phase of their life, but one which is poorly understood. We have recently observed the movements of salmon on their return to coastal waters off the east of Scotland. The fish were first captured at the Rockhall station of Messrs. Jos Johnston and Sons Ltd. The station is on the coast 8 km north of Montrose, and 3 km to the north of the river North Esk. The ultrasonic tags, 1 cm in diameter and 4 cm long and transmitting 1.5 ms pulses at 75 KHz, were placed inside the stomachs of the fish where they remained for several days before being regurgitated. Each fish was left in a netting cage on the sea bed for twelve hours or so to recover and was then released and tracked by means of a directional hydrophone receiver attached to a small boat.

Six grilse (salmon which had spent a single winter in the sea) were tagged and followed, and three of these entered the river North Esk. While in the sea off the coast, the fish moved with the tidal currents and when the tide changed, the direction of movement altered. The fish did not seem to be carried along passively with the tide since they often swam faster than the current. On the other hand, they swam only at moderate speeds well within their cruising capacity. Generally the speed of the fish exceeded the predicted speed of the tide by less than 1 km h^{-1} (i.e. a swimming speed relative to the water of less than half a body length per second). Only rarely were fish observed at swimming speeds above 1.5 km h^{-1}. The fastest speed was 5.2 km h^{-1} (2.4 body length per second), when a fish swam straight from Rockhall to the mouth of the North Esk with the flood tide to enter the river.

Only one fish entered the river directly, when the river was in spate. Two fish tracked, when the river was at more typical summer levels, reached the mouth at low slack water and then swam back and forth across the river mouth before entering. Both these fish were very close to the shore on their runs past the river mouth (Fig. 1), often among breaking waves.

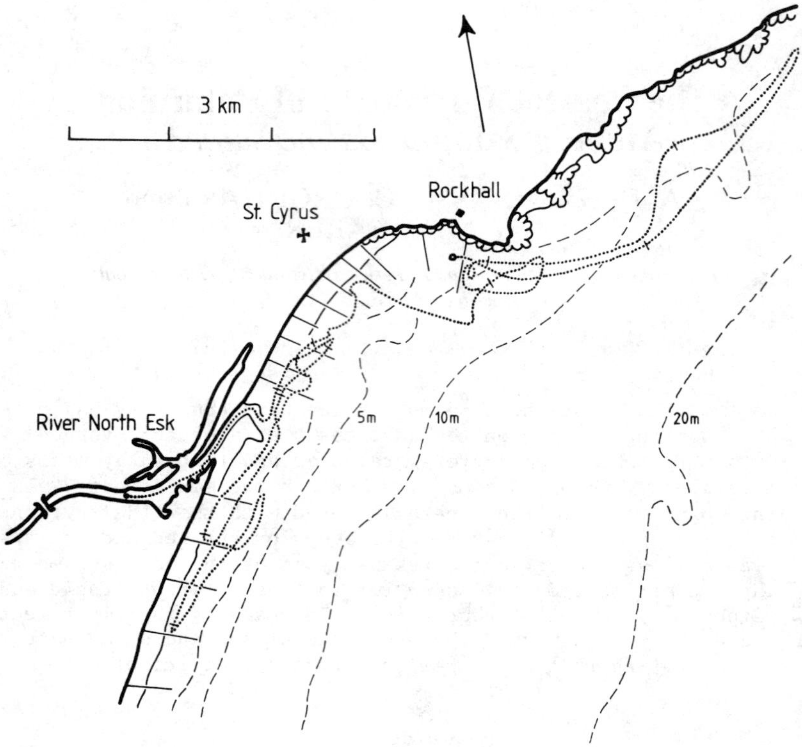

Fig. 1. Movements of a single grilse released at Rockhall, on 15 July, 1978 and followed into the river North Esk. The track of the fish is shown by a dotted line, with hourly intervals marked by a slash. Salmon nets are shown as lines running out from the shore.

Tracking within the river itself showed that all three fish reached the first major pool (the Nab pool) within twelve hours of entry, and stayed there for a while before moving on. Clearly, the water balance problem which salmon have on reaching freshwater is not a severe deterrent to the fish entering the river.

Stasko *et al.* (1973) have also reported that migrating salmon swim with the current when returning to the coast. Like us, they noted that the fish reverse this behavior in the river mouth and start to swim against the current. This change in behavior may be initiated by chemical cues from the river water. Swedish experiments in the Baltic have shown that returning salmon with their olfactory organs blocked do not home successfully (Bertmar and Toft, 1969). The chemical cue may simply be the presence of freshwater, but Solomon (1973) concluded that the return of the adults is dependent on the fish identifying particular pheromones released by fish in the home river.

Four of the salmon encountered salmon nets while moving along the coast, or into the river, and in each case successfully negotiated the hazard. Two fish, in particular, repeatedly avoided capture by both bag and stake nets as they swam to and fro along the beaches to the north and south of the North Esk.

Sometimes the fish swam to the seaward end of the line of nets and then rounded it. In many cases, however, the fish seemed to swim into the net on one side, around the end of the leader, and then out of the net on the opposite side. Other fish passed around the seaward end of a bag, then swam between the bag and the leader of the next net.

A high proportion of fish tagged by conventional means at some salmon netting stations are caught again at other stations. Of fish tagged at Rockhall, some 53 percent are subsequently recaptured. From the ease with which salmon can avoid bag and stake nets we have concluded that this high recapture rate reflects the frequency with which salmon meet salmon nets, rather than the efficiency of the individual nets *per se*.

REFERENCES

Bertmar G. and Toft R. (1969) Sensory mechanisms of homing in salmonid fish. I. Introductory experiments on the olfactory sense in grilse of Baltic salmon. *Behaviour* 35, 235-241.

Solomon D. (1973) Evidence for pheromone influenced homing by migrating Atlantic salmon. *Nature* 244, 231-232.

Stasko A. B., Horrall R. M., Hasler A. D. and Stasko D. (1973) Coastal movements of mature Fraser River pink salmon as revealed by ultrasonic tracking. *J. Fish. Res. Bd Can.* 30, 1309-1316.

Author Index

Subject Index